WindEurope Conference and Exhibition 2019

Delivering a Clean Economy for All European

Journal of Physics: Conference Series Volume 1222

Bilbao, Spain
2-4 April 2019

Editors:

Michael Muskulus
Carlo L. Bottasso

ISBN: 978-1-5108-8788-6
ISSN: 1742-6588

Printed from e-media with permission by:

Curran Associates, Inc.
57 Morehouse Lane
Red Hook, NY 12571

Some format issues inherent in the e-media version may also appear in this print version.

Printed by Curran Associates, Inc. (2019)

For permission requests, please contact the Institute of Physics
at the address below.

Institute of Physics
Dirac House, Temple Back
Bristol BS1 6BE UK

Phone: 44 1 17 929 7481
Fax: 44 1 17 920 0979

techtracking@iop.org

Additional copies of this publication are available from:

Curran Associates, Inc.
57 Morehouse Lane
Red Hook, NY 12571 USA
Phone: 845-758-0400
Fax: 845-758-2633
Email: curran@proceedings.com
Web: www.proceedings.com

Table of contents

Volume 1222

May 2019

WindEurope Conference and Exhibition 2019
2–4 April 2019, Bilbao,Spain

Accepted papers received: 11 April 2019
Published online: 21 May 2019

Preface

Papers

Assessing the wind resource and turbine performance

Digitalisation

End of life issues

Environmental impacts, social acceptance & spatial planning

Integrating wind into the energy system

Operations and Maintenance

IOP Conf. Series: Journal of Physics: Conf. Series **1222** (2019) 011001 doi:10.1088/1742-6596/1222/1/011001

Preface

On behalf of the European Academy of Wind Energy (EAWE[1]) and WindEurope (formerly the European Wind Energy Association EWEA), we hereby have the pleasure of presenting the Scientific Proceedings of the WindEurope Conference 2019.

WindEurope Conference & Exhibition 2019 took place in Bilbao on 2-4 April 2019 with the motto of "Delivering a clean economy for all European". The event is one of the largest and most important meeting places of the wind energy community. This year's conference has attracted more than 8000 participants and featured 30 thematic sessions, with contributions from more than 150 speakers and 200 poster presentations. As in previous years, the sessions mixed industrial content with scientific works, in order to increase the visibility of relevant scientific research and to facilitate an interchange of ideas between industry and academia.

As earlier, the scientific contributions of the WindEurope Conference 2019 are published in the Journal of Physics: Conference Series, an Open Access journal managed by the Institute of Physics (IOP). All papers in this volume originate from scientific contributions presented at WindEurope Conference 2019, in the form of posters or as talks, and have been peer-reviewed in order to ensure strict, international standards, coordinated by the Scientific Proceedings Editors and aided by the EAWE WindEurope Scientific Committee and members from EAWE institutions.

The possibility to provide a contribution to these proceedings was optional. This year, a total of 51 original papers were submitted, of which 46 were eventually accepted for publication.

The papers were roughly organized in ten topics, not all of which received submissions:

- Assessing the wind resource and turbine performance
- Developments in turbine technology
- Digitalisation
- Electrification and sector coupling
- End of life issues
- Environmental impacts, social acceptance & spatial planning
- Floating offshore wind
- Health and safety
- Integrating wind into the energy system
- Operations and maintenance

The manuscripts were handled by the following Section Editors:

- Stephan Barth, ForWind Center for Wind Energy, Oldenburg, Germany
- Ola Carlson, Chalmers University of Technology, Sweden
- Ndaona Chokani, ETH Zürich, Switzerland
- Emilio Gomez-Lazaro, University of Castilla-La-Mancha, Spain

[1] http://www.eawe.eu/

IOP Conf. Series: Journal of Physics: Conf. Series **1222** (2019) 011001 doi:10.1088/1742-6596/1222/1/011001

- Athanasios Kolios, Strathclyde University, UK
- Michael Muskulus, Norwegian University of Science and Technology NTNU
- Vasilis Riziotis, National Technical University of Athens NTUA, Greece

The papers in these proceedings cover the whole spectrum of current scientific interests in wind energy. The following introduction demonstrates the variety of the contributions and mentions some of the highlights.

The section on *Assessing the wind resource and turbine performance* features 9 contributions. Costantini et al. describe an interesting experimental approach that uses temperature sensitive paint in a high-pressure wind tunnel for studying the aerodynamics of wind turbine blades at realistic Reynolds numbers. Urban et al. have studied the power loss in wake conditions due to yaw with the dynamic wake meandering model. Corrections to the classical Ainslie model that take into account turbulent mixing and the effect of wake meandering have been proposed and verified on two wind farms by Gunn. A different approach has been taken by Ruisi and Bossanyi, who study an engineering wake model with corrections for atmospheric stability. Other contributions in this section are concerned with improving mesoscale models for flow over complex terrain (Flores-Maradiaga et al., Palma et al.), wind farm layout optimization with wake effects and cabling costs (Wade et al.), a novel approach to estimate aerodynamic damping under steady wind conditions (Chen et al.), and improving forecasting with sub-seasonal to seasonal climate predictions (Soret et al.).

In the section on *Developments in turbine technology* there are 22 contributions, which are too many to discuss individually. One topic of special interest was leading edge erosion, with 3 contributions alone. A highlight here is the study of leading edge erosion effects in small wind turbine blades (Latoufis et al.). A few more papers consider blade modifications such as vortex generators and trailing edge serrations. Studies on bearings in the drivetrain and the pitch system were covered by 4 contributions. Among them, a new robustness test for gearbox bearings has been investigated by Hirt et al., and a new type of rotor main bearings was studied by Schröder et al. An approach for predicting the lifetime of pitch bearings was considered by Lopez et al. A number of contributions were concerned with improving modelling, load simulations, and control algorithms of wind turbines, both fixed and floating. One of the more interesting contributions here is the study by Elorza et al. on delta control of wind turbines, for participating in grid frequency regulation. Finally, a number of contributions investigated aspects of alternative concepts, such as hydraulic wind turbines (Mulders et al.), the noise generated by vertical axis wind turbines (Möllerström and Ottermo), and high temperature superconducting generators (Kyling et al.).

On *Digitalization* only the paper by Dimitrov and Natarajan was received, which studies the use of neural networks for obtaining load-based, site-specific remaining lifetime estimates in an offshore wind farm.

On the topic of *End of life issues* 3 contributions are featured. The use of geographic information systems in the context of lifetime extension has been discussed by Piel et al., whereas Kazemi Amiri et al. consider damage in the tower, including the effect of the door opening. Topham et al. review aspects of and experience with decommissioning offshore wind farms.

The section on *Environmental impacts, social acceptance & spatial planning* features a single contribution, where Pego et al. have put forward a SWOT analysis of the offshore energy sector in Portugal.

On *Integrating wind into the energy system* we feature 3 contributions. Manero et al. provide some results on neural network modelling for forecasting. Buhagiar et al. review generation-side energy

IOP Conf. Series: Journal of Physics: Conf. Series **1222** (2019) 011001 doi:10.1088/1742-6596/1222/1/011001

storage and investigate a lithium-ion battery with simulations. An ambitious techno-economical analysis of operating wind energy plants together with energy storage systems has been presented by González-Garrido et al.

Finally, the section on *Operations and Maintenance* contains 7 papers. Gilbert et al. discuss ideas for the visualization of accessibility forecasts that might improve decision making. Applications of neural networks to SCADA data are presented by Trstanova et al. A framework for estimating remaining useful life of, e.g. wind turbine main bearings, with neural networks has been put forward by Herp et al. Similarly, Peeters et al. discuss improvements in the use of long-term vibration data for gearbox condition monitoring, whereas Koukoura et al. illustrate the use of decision tress for the same purpose. Current signature analysis for condition monitoring of generators is discussed by Artigao et al. Ideas on how to manage operations and maintenance of wind farms has been presented by Izquierdo et al.

As with previous events, the European Academy of Wind Energy is responsible for organizing the review process for scientific contributions, has contributed to the development of the sessions, and has provided chairs for sessions with scientific content. All papers were peer-reviewed by a Scientific Committee, consisting of scientists and experts from EAWE member institutes and their associates.

We thank all authors for their contributions and for complying with the deadlines in the tight publication schedule. We thank all reviewers and section editors for their engagement and entirely voluntary work, without which we could not guarantee the scientific quality and current interest of the accepted contributions. WindEurope is responsible for the organization and logistics of the conference, and we thank their highly professional staff and their associates for the excellent collaboration.

Michael Muskulus
Editor Scientific Proceedings WindEurope Conference 2019
Scientific Programme Coordinator WindEurope Conference 2019
Norwegian University of Science and Technology NTNU

Carlo L. Bottasso
Editor Scientific Proceedings WindEurope Conference 2019
President European Academy of Wind Energy EAWE
Technical University of Munich TUM

IOP Conf. Series: Journal of Physics: Conf. Series **1222** (2019) 011002 doi:10.1088/1742-6596/1222/1/011002

Peer review statement

All papers published in this volume of *Journal of Physics: Conference Series* have been peer reviewed through processes administered by the proceedings Editors. Reviews were conducted by expert referees to the professional and scientific standards expected of a proceedings journal published by IOP Publishing.

IOP Conf. Series: Journal of Physics: Conf. Series **1222** (2019) 012001 doi:10.1088/1742-6596/1222/1/012001

A Reliable Experimental Methodology for the Study of Wind-Turbine Rotor Blade Aerodynamics

M Costantini[1], C Fuchs[1], U Henne[1], C Klein[1], V Ondrus[2], M Bruse[3], M Löhr[3], M Jacobs[3]

[1]DLR, Bunsenstrasse 10, D-37073 Göttingen, Germany
[2]University of Hohenheim, Garbenstrasse 30, D-70599 Stuttgart, Germany
[3]DNW, Bunsenstrasse 10, D-37073 Göttingen, Germany

E-mail: marco.costantini@dlr.de

Abstract. The aerodynamic performance of airfoils and blades designed for modern wind-turbine rotors, which have diameters of the order of hundred meters, must be examined at chord Reynolds numbers matching those of practical applications. In general, such high Reynolds numbers cannot be achieved in conventional wind tunnels. Moreover, knowledge on the boundary-layer transition location is essential to evaluate airfoil and blade performance at these flow conditions. This work presents an experimental methodology that can be applied at flow conditions reproducing those of real wind-turbine rotor blades and simultaneously provides aerodynamic coefficients and transition locations. The experimental methodology consists of: the Temperature-Sensitive Paint (TSP) technique for global, non-intrusive and reliable transition detection; conventional pressure measurements for the determination of the aerodynamic coefficients; and the High Pressure Wind Tunnel Göttingen (DNW-HDG) to run the experiments at Reynolds numbers matching those of real applications. The obtained results can be used to verify airfoil and blade performance and to validate numerical predictions. In the present work, the experimental methodology was applied to systematically investigate the aerodynamic performance of an airfoil designed for the mid-span sections of modern wind-turbine rotor blades. The examined chord Reynolds numbers were as high as 12 million and the angle-of-attack ranged from -14° to +20°. The presented methodology was here demonstrated to be mature for productive testing.

1. Introduction

Rotor blade aerodynamics has an essential role in overall wind-turbine performance and has continuously improved in the last decades. Airfoil classes have been specifically developed for wind-turbine rotor blades [1], with airfoil size and shape optimized for the different blade sections to satisfy aerodynamic and structural demands. Especially the airfoils for mid-span and outboard sections, however, are subjected to Reynolds number and roughness effects [1-3]; both these effects are related to the occurrence of boundary-layer transition on the airfoils under practical conditions. Transition occurring more upstream than design leads to significantly higher drag [4], lower lift and thus also to a lower lift-to-drag ratio [2,5]. It is therefore of fundamental importance to design airfoils for mid-span and outboard sections that present low sensitivity to roughness [1-2,5] and experimentally verify their performance at Reynolds numbers relevant for modern wind turbines with large rotor diameters (which can be of the order of 100 m).

In the mid-span area of modern wind-turbine rotor blades, the Reynolds number Re = $U_\infty c/\nu_\infty$ (where U_∞ and ν_∞ are the freestream velocity and kinematic viscosity, respectively, and c the airfoil chord length) can reach values of $10 \cdot 10^6$, or even higher [6]. Experiments in wind tunnels allow the systematic analysis of the aerodynamic performance of airfoils / blades and the validation of numerical predictions, but the maximal Reynolds number achievable in conventional wind tunnels is typically $6 \cdot 10^6$ [2]. While maintaining the freestream Mach number below M = 0.3, the Reynolds number can be further increased by increasing the gas kinematic viscosity. This can be accomplished in two ways: by decreasing the gas temperature, as in the case of cryogenic wind tunnels such as the Cryogenic Wind Tunnel Cologne (DNW-KKK) [6-8], or by increasing the gas pressure, as done in the High Pressure Wind Tunnel Göttingen (DNW-HDG) [9-12]. In this manner, the aerodynamic performance of airfoils developed for modern wind-turbine rotor blades can be studied in the appropriate Reynolds number range, thus avoiding that the experimental data are affected by the influence of a Reynolds number different from that achieved in practical applications.

Aerodynamic performance is typically examined in wind tunnels by measuring aerodynamic loads [1-2,5-6,8,11-12]. In general, these measurements do not provide any information (if any, only indirectly) on the airfoil / blade boundary-layer state and on the laminar-turbulent transition, which is however crucial for the evaluation and further improvement of the aerodynamic performance of rotor blades. This information can be provided by one or more rows of sensors, such as unsteady pressure sensors [8] or hot films [3], but they have intrinsically low spatial resolution; moreover, they can induce significant disturbances into the thin boundary layer developing on the wind-tunnel model surface at high Reynolds numbers. Global, non-intrusive and reliable transition measurements can be carried out by means of thermographic methods, such as Infra-Red (IR) thermography [3,13-14] and the Temperature-Sensitive Paint (TSP) technique [7,10,15-19]. IR thermography is widely used for transition detection at ambient conditions, but it cannot be applied easily at the low temperatures or at the high pressures needed to match the high Reynolds numbers of real wind-turbine rotor blades [10,15]. Transition measurements under these test conditions can be successfully conducted via TSP [6-7,10,16-19].

In the present study, the aerodynamic performance of an airfoil designed for the mid-span region of modern wind-turbine rotor blades was systematically investigated at high chord Reynolds numbers (up to Re = $12 \cdot 10^6$) for a wide range of angles-of-attack (from -14° to 20°) in the DNW-HDG wind tunnel. The present work is built on and extends the investigations presented in [19]. Boundary-layer transition on the model upper surface was measured globally by means of TSP, while the coefficients of airfoil lift, drag and pitching moment were obtained via integration of the pressure distributions measured on the wind-tunnel model surface and in the model wake.

2. Experimental Setup

The investigated two-dimensional wind-tunnel model, with a chord length of c = 0.1 m and a span width of b = 0.6 m, had the DU 91-W2-250 airfoil [1,5,11] as cross section. The examined airfoil has a maximal thickness of 25 % c and had been specifically designed for the mid-span region of modern wind-turbine rotor blades. The shape of the airfoil is shown in Figure 1a; the wind-tunnel model and its instrumentation are presented in Figure 1b. The model was equipped with TSP to measure non-intrusively the surface temperature distribution on the model upper surface, and thus boundary-layer transition; moreover, TSP also allowed to examine separated flow regions at high angles-of-attack (see Sec. 3). A layer of Carbon-Nanotubes (CNT) [7,10,16-17,19] was integrated in the TSP layer composition; the CNT layer served as electrical heating to impose a temperature difference (of approx. 1-1.5 K) between model surface and flow, thus enhancing the temperature difference between the laminar and turbulent flow domains close to the model surface and enabling global and reliable transition detection via TSP. (The temperature difference imposed between flow and model surface was proven to be sufficiently low to have no appreciable influence on boundary-layer transition.) By using CNT as integrated heating, the measurement of a whole sweep of angles-of-attack (from AoA = -14° to 20° in this work) could be completed in approx. 30 minutes. TSP sensor [20], surface quality, optical setup, acquisition and elaboration of the TSP images were the same as those discussed in [19]. In that previous work [19], TSP was also demonstrated to have no appreciable influence on the measurements of the aerodynamic coefficients, provided that the quality of the model surface with and without TSP coating was essentially the same. In the present work, the TSP surface was polished to an average roughness value of $R_a = 0.04$ µm (mean roughness depth $R_z = 0.2$ µm), i.e, to very low roughness values compatible with the thin boundary layer developing on the wind-tunnel model surface at the examined high Reynolds numbers.

As shown in Figure 1, the model was also instrumented with pressure taps to measure the surface pressure, and with a Pt100 temperature sensor to monitor the temporal evolution of the surface temperature during the experiments. The 61 pressure taps were installed in the model mid-span area and embedded in the TSP. The positions of the orifices along the airfoil contour are indicated by open circles in Figure 1a. The coefficients of airfoil lift (c_L) and pitching moment (c_m) were obtained via integration of the measured surface pressure distribution. As can be seen in Figure 1b, the pressure taps located in the region at $x/c \leq 65$ % were distributed in a staggered order, so that the measurement at any one tap was not affected by the disturbances generated by the upstream orifices, in a manner similar to that described, e.g., in [18]. The angle of these pressure tap lines about the vertical axis was approximately ±20° with respect to the freestream. The line of pressure taps located at $x/c > 65$ % was essentially parallel to the freestream: in this region, the boundary layer was expected to be turbulent at all test conditions examined in the present work, so that the measurements of these pressure taps were assumed to be not affected by the disturbances caused by the upstream orifices.

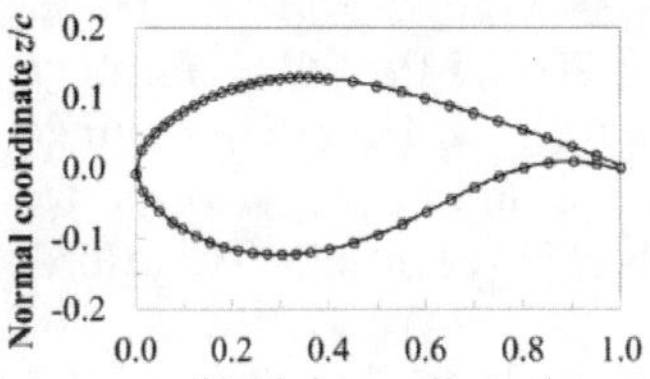

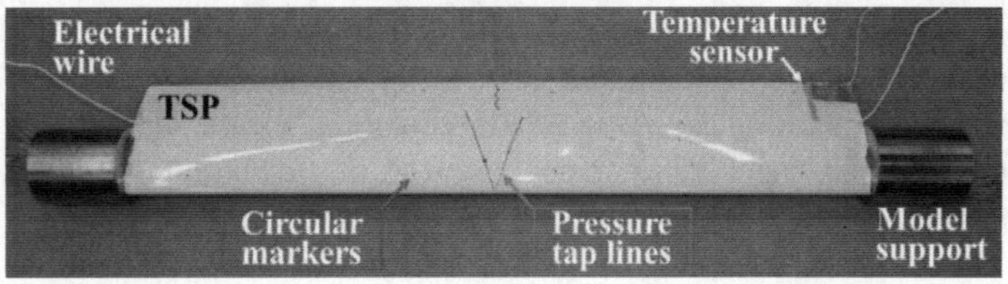

Figure 1. a: DU 91-W2-250 airfoil contour (black line) [1,5] and locations of the pressure taps (black circles); axes not equally scaled. b: top view of the instrumented model (leading edge at the bottom of the image).

The experiments were carried out in the DNW-HDG wind tunnel [9]. DNW-HDG is a closed-return wind tunnel where the air pressure can be increased up to 10 MPa. Experiments can be run at freestream velocities up to $U_\infty = 38$ m/s, thus allowing to achieve Reynolds numbers up to approx. $20 \cdot 10^6$ for a model chord length $c = 0.1$ m (as that used in the present work). A sketch of DNW-HDG is shown in Figure 2a. The test section is 1 m long and has a square cross-section with a side of 0.6 m. Experiments on two-dimensional models with $c = 0.1$ m can be conducted for a full polar of angle-of-attack, i.e., over the whole range of 360°. The flow turbulence intensity Tu in the test section is $0.2\ \% < Tu < 0.5\ \%$ for the examined Reynolds number range, where Tu was found to increase at higher Reynolds numbers [9,11-12]. Details on the DNW-HDG wind tunnel and on its instrumentation are given in [9-12,19].

A wake rake, equipped with 6 static pressure and 110 total pressure probes, was installed at a spanwise location $y/b = 0.67$ and at a distance $\Delta(x/c) = 3.4$ downstream of the model trailing edge. (y is spanwise coordinate, positive starting from the model port side and oriented towards the model starboard side.) The airfoil drag coefficient (c_D) was evaluated from the pressure distribution measured in the model wake according to the momentum theorem. Note here that the wake rake can be traversed in the spanwise direction to determine the drag coefficient at different spanwise locations and thus evaluate the impact of three-dimensional flow effects on the measured c_D of the airfoil.

An image of the experimental setup in the DNW-HDG test section is shown in Figure 2b. In this figure, the optical setup for TSP measurements (windows for camera and light-emitting diodes, i.e.,

IOP Conf. Series: Journal of Physics: Conf. Series **1222** (2019) 012001 doi:10.1088/1742-6596/1222/1/012001

LEDs), the wires used to supply the electrical power to the CNT and the Pt100 temperature sensor can also be seen. Note here that the electrical wires and the temperature sensor were partially exposed to the flow, but their location was at a sufficient distance from the TSP evaluation area (see Sec. 3), from the pressure taps region and from the wake rake. Therefore, the measurements of transition location and aerodynamic coefficients were not affected by disturbances induced by the wires and by the Pt100. Further details on the experimental setup and on its accuracy are provided in [19].

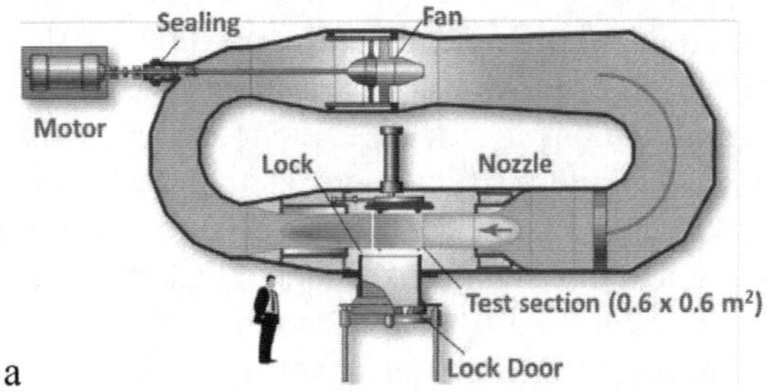

a

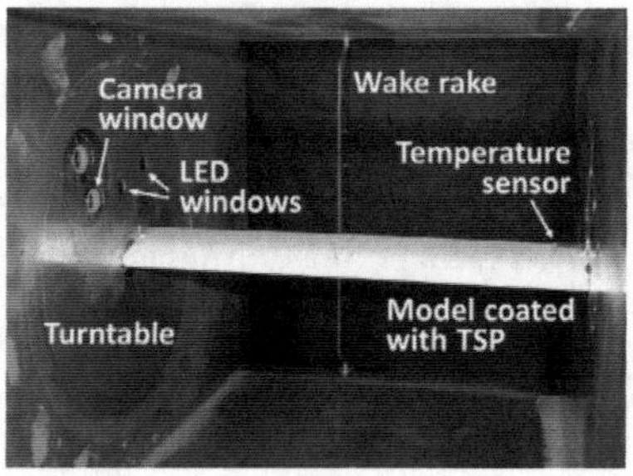

b

Figure 2. a: Principle sketch of the DNW-HDG wind tunnel. b: model mounted in the DNW-HDG test section, as seen from an upstream position.

3. Results and Discussion

The measurements were carried out at angles-of-attack from -14° to 20° for the following Reynolds numbers: Re = 3, 6, 9 and 12 · 10^6. The Mach number remained below M = 0.1 for all examined test conditions. As compared to previous work [19], experiments at Re = 6 and 9 · 10^6 were performed here also for an alternative combination of flow total pressure and freestream velocity, in order to examine the repeatability of the achieved results.

Example TSP results, obtained on the model upper surface at Re = 6 · 10^6, M = 0.06 (flow total pressure p_0 = 4.6 MPa) and various angles-of-attack, are presented in Figure 3. These TSP results show the wind-tunnel model as it would be seen from the top wall of the DNW-HDG test section; they were achieved after the TSP images, acquired by the camera mounted at the turntable of the test-section side wall (see Figure 2b), had been mapped onto a three-dimensional grid representing the model upper surface. This operation was performed using the DLR software package *ToPas* [21], in a manner similar to that described in [18,22]. The regions at $y/b \leq 19$ % and $y/b \geq 78$ % (i.e., those close to the test-section side walls) were masked white, since these regions were not completely visible in the TSP images and/or their spatial resolution was too low. (These limitations were due to the restricted optical access in the DNW-HDG test section – see [19].) In Figure 3, the flow is from the left; bright and dark areas correspond to regions of lower and higher heat flux, respectively, and therefore to regions of lower and higher wall shear stress. Bright and dark areas in Figure 3a-d (-12° ≤ AoA ≤ 7°) correspond to laminar and turbulent regions, respectively. The turbulent wedge observed in the mid-span area was caused by disturbances originating from the pressure taps in the leading-edge region (see [18] and [22] for a similar finding). The transition locations x_T/c and the corresponding uncertainties reported in the caption of Figure 3 were obtained as average and standard deviation of the transition locations detected at five different spanwise positions [19]. At each spanwise position, the transition location was detected as the location corresponding to the maximal gradient of the streamwise temperature distributions; this operation was performed automatically by means of a reliable algorithm described in [22]. An increase in the angle-of-attack (from left to right in Figure 3) led to a more pronounced adverse pressure gradient on the model upper surface, which induced stronger amplification of boundary-layer disturbances and thus a movement of the laminar-turbulent transition towards a more upstream location [19]. Nevertheless, the upstream shift of x_T/c in Figure 3a-c was relatively small, in spite of the considered large steps in AoA. The relative variation of the transition location as a function of the angle-of-attack remained small up to AoA = 5°; the upstream movement of x_T/c became then more pronounced as the angle-of-attack was increased to AoA > 5°, as can be seen by a comparison of the TSP results in Figure 3c and Figure 3d. At the four highest angles-of-attack, AoA = 11°, 12°, 13° and 15° (Figure 3e-h), the flow was essentially turbulent over the whole model upper surface. At AoA = 11°, the flow was still attached over the whole model upper surface, while turbulent flow separation occurred at AoA ≥ 12°: dark and bright areas in the TSP results correspond here to attached and separated flow regions, respectively. The TSP results in Figure 3f-h also show that the separation front was not a straight line in spanwise direction. Significant spanwise variations were especially observed at stall conditions (AoA = 12°) and in the near post stall regime (13° ≤ AoA ≤ 16°), while at higher angles-of-attack (AoA > 16°) the spanwise variation of the separation front became less marked. The use of the global, non-intrusive TSP measurement technique revealed this complex flow scenario, which was very likely due to the effect of three-dimensional flow structures in the separated flow region, such as those described as "mushroom-like" structures or "owl eyes" in [23]. The dynamics of these flow structures, however, could not be captured with the adopted experimental setup, since the TSP measurements were averaged in time [19].

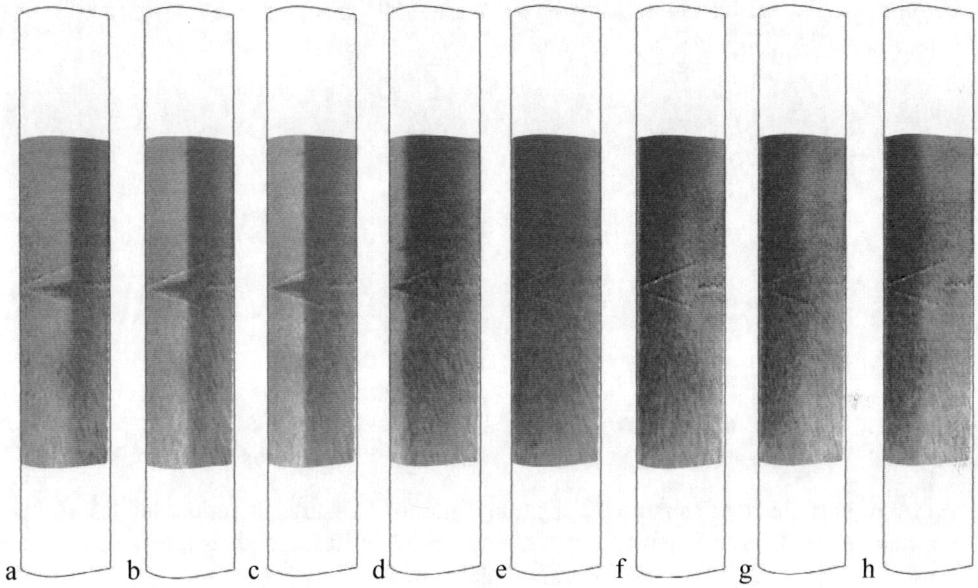

Figure 3. TSP results obtained at Re = $6 \cdot 10^6$ (M = 0.06, p_0 = 4.6 MPa) for various angles-of-attack. a: AoA = -12°, x_T/c = 54.8 ±1.0 %; b: AoA = -6°, x_T/c = 48.2 ±0.9 %; c: AoA = 0°, x_T/c = 42.6 ±0.7 %; d: AoA = 7°, x_T/c = 17.4 ±0.5 %; e: AoA = 11°; f: AoA = 12°; g: AoA = 13°; h: AoA = 15°.

The results obtained in the present study are summarized in Figure 4. The coefficients of lift, pitching moment and drag are presented as a function of the angle-of-attack in Figure 4a, Figure 4b and Figure 4c, respectively. The transition locations measured on the model upper surface by means of the TSP technique are plotted as a function of AoA in Figure 4d. In these figures, different colors show data obtained at different Reynolds numbers. Open and filled symbols correspond to the data measured at flow total pressures p_0 = 4.6 and 7 MPa, respectively. (In the legend, the corresponding data sets are indicated with (I) and (II), respectively.) The cyan rectangle in Figure 4d indicates the approximate range of AoA at which a variation in the transition location has a crucial influence on the aerodynamic performance of airfoils designed for the mid-span region of wind-turbine rotor blades [1-2,5].

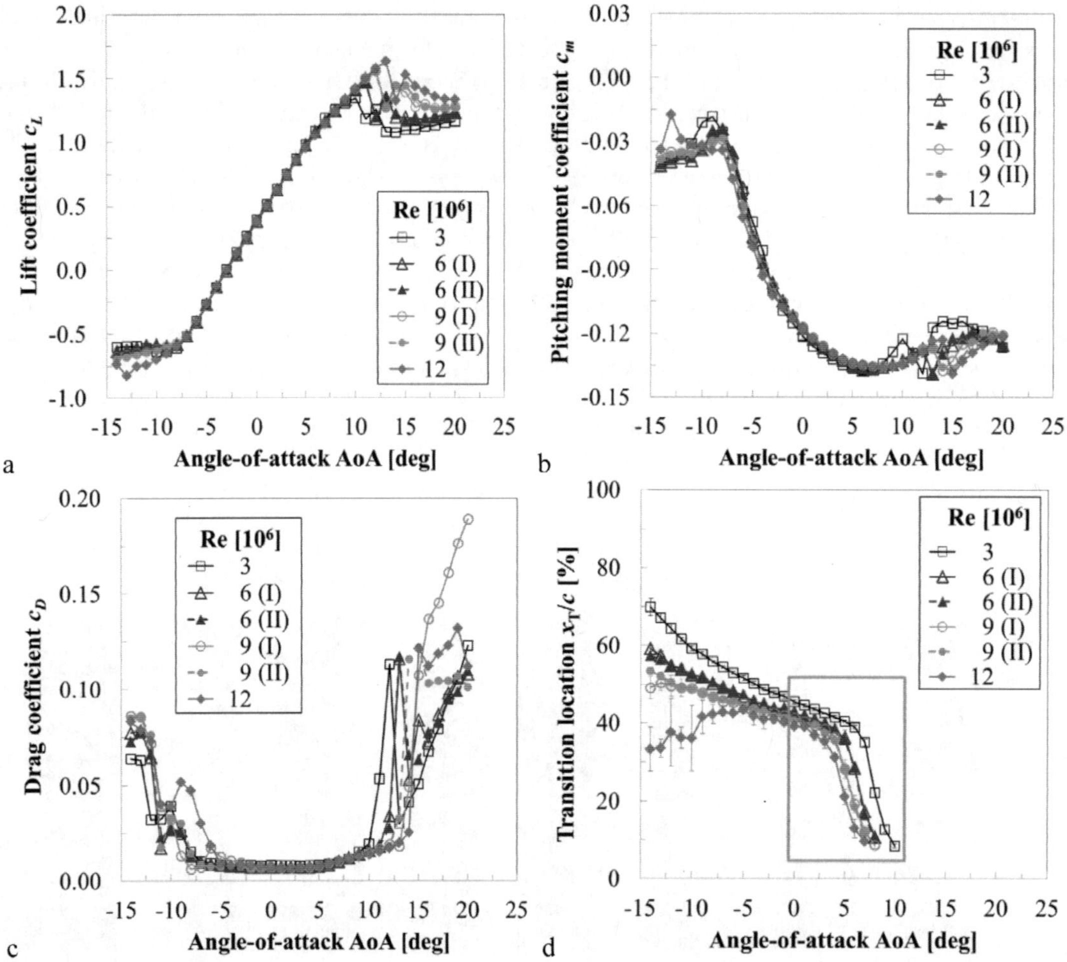

Figure 4. Aerodynamic coefficients and transition locations measured at various angles-of-attack and Reynolds numbers. a: lift coefficient; b: pitching moment coefficient; c: drag coefficient; d: transition location (upper surface).

The agreement of the lift coefficients measured at the different Reynolds numbers (Figure 4a) was excellent in the quasi-linear range of the c_L curve ($-6° \leq \text{AoA} \leq 6°$). The relative differences in the values of c_L remained below 5 % up to AoA = 10°, i.e., up to the angle-of-attack corresponding to $c_{L,\max}$ for $\text{Re} = 3 \cdot 10^6$. For each increase $\Delta\text{Re} = 3 \cdot 10^6$ in the Reynolds number, the maximal value of the lift coefficient and the value of the angle-of-attack corresponding to $c_{L,\max}$ increased consistently by approx. $\Delta c_{L,\max} = 0.1$ and by $\Delta\text{AoA} = 1°$, respectively: from $c_{L,\max} = 1.35$ (at AoA = 10°) for $\text{Re} = 3 \cdot 10^6$ to $c_{L,\max} = 1.63$ (at AoA = 13°) for $\text{Re} = 12 \cdot 10^6$. These findings are in agreement with those reported in earlier work [11]. Small differences of less than 5 % (and in most cases of less than 2 %) were observed in the pitching moment coefficients (Figure 4b) for the range of angles-of-attack $-3° \leq \text{AoA} \leq 9°$. Moreover, the evolution of c_m as a function of the angle-of-attack in the range $-8° \leq \text{AoA} \leq 10°$ was similar for all examined Reynolds numbers. In the range of angles-of-attack $-3° \leq \text{AoA} \leq 9°$, the differences in the drag coefficient (Figure 4c) were generally higher than those found for the lift and pitching moment coefficients: they were of the order of 5-10 % of the measured values of c_D. In any case, the evolution of the drag coefficient at $-4° \leq \text{AoA} \leq 10°$ was similar for all examined Reynolds numbers. At all considered angles-of-attack, transition on the model upper surface was found to occur at a more upstream location as the Reynolds number was increased (see Figure

4d). This was due to the enhanced amplification of boundary-layer disturbances at higher Reynolds numbers, which led to earlier transition. In general, the transition location was also shifted more upstream as the angle-of-attack was increased. For all examined Reynolds numbers, the relative change in transition location as a function of AoA was relatively small up to a certain angle-of-attack, corresponding to AoA = 6°, 5°, 4° and 3° for Re = 3, 6, 9 and 12 · 10^6, respectively. This was shown exemplarily for Re = 6 · 10^6 by the TSP results in Figures 3a-c. At higher angles-of-attack, the gradient $\partial(x_T/c)/\partial$AoA became more pronounced (see Figures 3c-e), until transition occurred in the leading-edge region and the boundary layer could be considered as essentially turbulent over the whole model upper surface.

Transition locations at $x_T/c < 7$ % are not reported in Figure 4d: in these cases, the evolution of boundary layer and aerodynamic coefficients was dominated by the evolution of the three-dimensional separated flow regions. As shown in Figure 3f for Re = 6 · 10^6, separation onset was found (at AoA = 12° at this Reynolds number) in the mid-span model region. However, as the angle-of-attack was increased to AoA = 13° (Figure 3g), separation was observed to occur at a more downstream location in the mid-span model area, whereas it was shifted towards a more upstream location in the neighboring regions, and in particular in the region corresponding to the wake rake location. This evolution of the separation front with increasing angle-of-attack was similar also for the two higher Reynolds numbers Re = 9 and 12 · 10^6, and provides an explanation for the different angles-of-attack corresponding to the local minima of the lift coefficient (stall conditions) and to the local maxima of the drag coefficient. In fact, as the AoA was increased up to that corresponding to stall conditions, separation occurred earlier in the pressure taps region (where c_L was evaluated) than in the wake rake region (where c_D was determined); therefore, the increase in c_D was still moderate, while c_L had significantly decreased. Similar, but opposite considerations apply for a further increase in angle-of-attack by ΔAoA = 1°: the increase in c_D was marked, while c_L had increased again. At higher angles-of-attack, the spanwise variations of the separation front became less pronounced, and the evolution of the aerodynamic coefficients was in line again. Note that the evolution of the separation front, and therefore of the aerodynamic coefficients, in the near post stall regime was slightly different at Re = 3 · 10^6, as compared to that described above for the higher Reynolds numbers. This different evolution has been discussed in [19].

The evolution of the aerodynamic coefficients at AoA < -6° was dominated by the evolution of the boundary layer on the model lower surface, and in particular by the evolution of the separated flow region on this model side for AoA < -8°. The significant increase in the drag coefficient observed at Re = 12 · 10^6 and AoA < -5° was likely due to the occurrence of one (or more) turbulent wedge(s) on the model lower surface, in a region affecting the drag measurements conducted by means of the wake rake. Unfortunately, no global information was available on the model lower surface to verify these expectations. In future investigations, TSP measurements will be carried out also on the model lower surface.

The lift and pitching moment coefficients obtained at Re = 6 and 9 · 10^6 with different combinations of flow total pressure and freestream velocity were in excellent agreement for the whole examined range of angles-of-attack, except for the data points at AoA = 15° and 16° and Re = 9 · 10^6. These differences in c_L and c_m were mainly due to small variations in the separation front and were still relatively small: within 6 % of the corresponding values of the aerodynamic coefficients. The drag coefficients were in excellent agreement at both Reynolds numbers in the range of angles-of-attack -2° ≤ AoA ≤ 11°. (At Re = 6 · 10^6, the agreement in c_D was excellent even for -10° ≤ AoA ≤ -2°.) Differences in the drag coefficient AoA ≥ 12° (i.e., at stall conditions and in the post-stall regime) were likely due to the presence, in the cases at p_0 = 4.6 MPa, of one or more turbulent wedges on the model lower surface, in the region where the wake rake was positioned. (The flow evolution on the model upper surface was essentially the same for both combinations of p_0 and U_∞.) The occurrence of one or more turbulent wedges was probably the reason also for the differences in c_D observed for Re = 9 · 10^6 at AoA < -2°. Unfortunately, the validity of this conjecture could not be verified in this work, since TSP measurements were not carried out on the model lower surface (see above). The transition locations detected on the model upper surface at Re = 6 and $9 \cdot 10^6$ with different combinations of flow total pressure and freestream velocity were in excellent agreement for the range of angles-of-attack -12° ≤ AoA ≤ 8°. At larger angles-of-attack, the boundary layer was essentially turbulent over the whole model upper surface. The reason for the differences in the transition locations seen at AoA < -12° is currently unclear. At these test conditions, the pressure distribution in the transitional region was rather flat, leading to an enhanced sensitivity of boundary-layer transition to a number of factors [22]. This topic will be the focus of future work, where experiments will be conducted at various combinations of p_0 and U_∞ for all examined Reynolds numbers. The results of these investigations may provide an explanation also for the unexpected upstream movement of the transition location observed on the model upper surface at AoA < -5° and Re = $12 \cdot 10^6$ (see Figure 4d) [19].

4. Conclusions

This contribution presents an experimental methodology to systematically analyze the aerodynamic performance of airfoils and blades designed for modern wind-turbine rotors. The methodology can be applied at flow conditions in the same range as that of real wind-turbine rotor blades and simultaneously provides aerodynamic coefficients and transition locations to verify airfoil / blade performance and to validate numerical predictions. The experimental methodology consists of conventional pressure measurements for the determination of the aerodynamic coefficients, of the TSP technique for global, non-intrusive and reliable transition detection, and of the High Pressure Wind Tunnel Göttingen to run the experiments at the same Reynolds numbers as those of real applications.

In this work, the presented methodology was demonstrated to be ready for systematic studies and productive testing. Experiments on a DU 91-W2-250 airfoil model were conducted at four different chord Reynolds numbers (from 3 to $12 \cdot 10^6$) and a wide range of angles-of-attack (from -14° to 20°), focusing on the boundary-layer evolution on the model upper surface. The data acquisition (TSP data included) for a whole sweep of angles-of-attack at a given Reynolds number was completed in approximately 30 minutes. On the model upper surface, transition was found to occur at a more upstream location as the Reynolds number was increased and, in general, also as the angle-of-attack was increased. The measurements at the two intermediate Reynolds numbers, 6 and $9 \cdot 10^6$, were performed at two different combinations of flow total pressure and aerodynamic coefficients. The aerodynamic coefficients and the transition locations, measured with the two different combinations of flow parameters, were shown to be in excellent agreement for most of the examined test conditions.

TSP was verified to be a global, non-intrusive and reliable measurement technique for transition detection at high Reynolds numbers. TSP also enabled the analysis of the evolution of the separated flow regions as a function of the angle-of-attack, thus providing an explanation for the variation of the aerodynamic coefficients observed in the near post stall regime. In fact, at these test conditions, the separation front presented significant variations in the spanwise direction, which were found to depend on angle-of-attack and Reynolds number.

Acknowledgments

The authors are grateful to: N. Falk and S. Hock (previously at DLR and TU Berlin) for the support to the TSP activities in DNW-HDG during their student projects; T. Kleindienst (DLR) and R. Lesjak (DNW) for the support during preparation of the wind-tunnel model; M. Ekert (RWTH Aachen), H. Sato and N. Tajima (Tohoku University) for the conduction and analysis of the TSP coating thickness and roughness measurements; and G. Schewe (DLR) for the helpful discussions on the boundary-layer separation behaviour of thick airfoil models at high angles-of-attack.

References

[1] van Rooij RPJOM and Timmer WA 2004 Design of Airfoils for Wind Turbine Blades *GCEP Energy Wksh.* Stanford University 26 Apr 2004.
[2] Timmer WA and Schaffarczyk AP 2004 The effect of roughness at high Reynolds numbers on

the performance of DU 97-W-300Mod *Wind Energy* **7**(4) pp 295–307.
[3] Ehrmann RS and White EB 2014 Influence of 2D Steps and Distributed Roughness on Transition on a NACA 63_3-418 *AIAA Paper* 2014-0170.
[4] Schlichting H and Gersten K 2000 Boundary–Layer Equations in Plane Flow; Plate Boundary Layer *Boundary-Layer Theory* (8th ed, Berlin: Springer–Verlag) chapter 6.
[5] van Rooij RPJOM and Timmer WA 2003 Roughness Sensitivity Considerations for Thick Rotor Blade Airfoils *AIAA Paper* 2003-0350.
[6] Rebstock R, Zhai J and Schaffarczyk AP 2013 Experimental Investigation of Reynolds Number Effect on Wind Turbine Profiles in the Cryogenic Wind Tunnel Cologne DNW-KKK *J. Energy and Power Eng.* **7** pp 1957–65.
[7] Klein C, Henne U, Sachs WE, Beifuss U, Ondrus V, Bruse M, Lesjak R., Löhr M, Becher A and Zhai J 2015 Combination of Temperature-Sensitive Paint (TSP) and Carbon Nanotubes (CNT) for Transition Detection *AIAA Paper* 2015-1558.
[8] Schaffarczyk AP, Winkler H, Freudenreich K, Kaiser K and Rebstock R 2003 Reynolds Number Effects on Thick Aerodynamic Profiles for Wind Turbines *Proc. Eur. Wind Energy Conf. and Exhib.* (Brussels: EWEA) 15.
[9] Försching H, Melzer E and Schewe G 1981 Ein neuer Windkanal für gebäudeaerodynamische und -aeroelastische Untersuchungen bei Reynoldszahlen bis 10^7 *Konstr. Ingenieurbau* **35/36** pp 127–33.
[10] Klein C, Henne U, Sachs WE, Beifuss U, Ondrus V, Bruse M, Lesjak R and Löhr M 2014 Application of Carbon Nanotubes (CNT) and Temperature-Sensitive Paint (TSP) for the Detection of Boundary Layer Transition *AIAA Paper* 2014-1482.
[11] Llorente E, Gorostidi A, Jacobs M, Timmer WA, Munduate X and Pires O 2014 Wind Tunnel Tests of Wind Turbine Airfoils at High Reynolds Numbers *J. Phys.: Conf. Ser.* **524** 012012.
[12] Pires O, Munduate X, Ceyhan O, Jacobs M and Snel H 2016 Analysis of high Reynolds numbers effects on a wind turbine airfoil using 2D wind tunnel test data *J. Phys.: Conf. Ser.* **753** 022047.
[13] Quast A 2006 Detection of Transition by Infrared Image Techniques *Tech. Soar.* **30**(1/2).
[14] Gartenberg E, Johnson WG, Johnson CB, Carraway DL and Wright RE 1990 Transition Detection Studies in the Cryogenic Environment *AIAA Paper* 1990-3024.
[15] Tropea C, Yarin AL and Foss JF (eds) 2007 Transition-Detection by Temperature-Sensitive Paint *Springer Handbook of Experimental Fluid Mechanics* (Berlin: Springer-Verlag) chapter 7.4.
[16] Goodman KZ, Lipford WE and Watkins AN 2016 Boundary-Layer Detection at Cryogenic Conditions Using Temperature Sensitive Paint Coupled with a Carbon Nanotube Heating Layer *Sens.* **16**(12) 2062.
[17] Klein C, Henne U, Yorita D, Beifuss U, Ondrus V, Hensch A-K, Longo R, Hauser M, Guntermann P and Quest J 2017 Application of Carbon Nanotubes and Temperature-Sensitive Paint for the Detection of Boundary Layer Transition under Cryogenic Conditions *AIAA Paper* 2017-0336.
[18] Costantini M, Fey U, Henne U and Klein C 2015 Nonadiabatic Surface Effects on Transition Measurements Using Temperature-Sensitive Paints *AIAA J.* **53**(5) pp 1172–87.
[19] Costantini M, Fuchs C, Henne U, Klein C, Ondrus V, Bruse M, Löhr M and Jacobs M 2019 Experimental Analysis of a Wind-Turbine Rotor Blade Airfoil by means of Temperature-Sensitive Paint *AIAA Paper* 2019-0800.
[20] Ondrus V, Meier R, Klein C, Henne U, Schäferling M and Beifuss U 2015 Europium 1,3-di(thienyl)propane-1,3-diones with Outstanding Properties for Temperature Sensing *Sens. Actuator.* A: Phys. **233** pp 434–41.
[21] Klein C, Engler RH, Henne U and Sachs WE 2005 Application of pressure-sensitive paint for determination of the pressure field and calculation of the forces and moments of models in a wind tunnel *Exp. Fluids* **39** pp 475–83.

IOP Conf. Series: Journal of Physics: Conf. Series **1222** (2019) 012001 doi:10.1088/1742-6596/1222/1/012001

[22] Costantini M 2016 Experimental analysis of geometric, pressure gradient and surface temperature effects on boundary-layer transition in compressible high Reynolds number flow *PhD Thesis* RWTH Aachen University. https://elib.dlr.de/117965/.

[23] Schewe G 2001 Reynolds-number effects in flow around more-or-less bluff bodies *J. Wind Eng. Indust. Aerodyn.* **89** pp 1267–89.

IOP Conf. Series: Journal of Physics: Conf. Series **1222** (2019) 012002 doi:10.1088/1742-6596/1222/1/012002

The effect of wake position and yaw misalignment on power loss in wind turbines

Albert M. Urbán[1], Jaime Liew[1] , Ebba Dellwik[1] and Gunner Chr. Larsen[1]

[1] Department of Wind Energy, Technical University of Denmark (DTU), Frederiksborgvej 399, 4000 Roskilde, Denmark

E-mail: amur@dtu.dk

Abstract. For a single wind turbine, the efficiency of extracting energy from the wind depends on the ability to align the wind turbine with the dominating wind direction. Considering average power production, yaw misalignment is relevant when the wind turbine operates with maximum power coefficient. On the other hand, the power production is less sensitive to yaw misalignment in high wind speeds, where the available energy in the wind field is higher than the maximum wind turbine capacity. In a wind farm, the interaction between nearby wind turbines alters the flow, and the power production is reduced. The present study investigates how yaw misalignment affects the power production in these wake situations compared to yaw misalignment effects for a wind turbine in the free-stream. Two generic cases are presented in this paper, offshore and forest, where the atmospheric conditions alter the morphology of the wake and, therefore, the power output of a yawed wind turbine operating in wake conditions. The results show that, for a conventional downstream spacing further than 3 rotor diameters, yaw misalignment results in larger power loss in wake situations than in free-stream. In wake situations, the presented results also show that the spatial distribution of the deficit influences the relative power loss when the wind turbine is operating in yawed conditions.

1. Introduction

To achieve the maximum profitability in a wind farm, the wind turbine design, farm layout, and operation strategy must be optimised in an integrated manner. An efficient wind farm design takes into account the influence of the wind turbines on each other, typically with a focus on wake effects. The detrimental effects of wakes on power output and loads in a wind farm can not be completely eliminated, and therefore it is important to consider the best way of operating a wind turbine in these situations. This paper presents a thoughtful explanation on the effect of yaw misalignment on the power output of a wind turbine when operating in a wake case. The findings can be used in the design of both the wind farm and its control system.

The yaw angle of a wind turbine relative to the incoming wind direction plays a significant role in the loads and power production. Significant fatigue reductions can be achieved with an active yaw controller as shown in [1] and [2]. These strategies, although beneficial in reducing loads by purposefully inducing a yaw misalignment, sacrifice power output in the partial load regime. A key finding, which is presented in this paper, is how the power deficit as a function of yaw angle varies in wake conditions.

In literature, power loss due to yaw misalignment is typically defined as being proportional to a factor of $\cos^{\alpha}\gamma$ [3], where γ is the yaw angle and α is constant exponent term. In uniform flow conditions, $\alpha = 3$. However, measurements and simulations show that the value of α varies for unsteady wind speeds. Different studies have found the α parameter ranging from $\alpha = 1.8$ to $\alpha = 5.14$ [4][5], depending on atmospheric and turbine operational condition. As part of the investigation in this paper, the value of α is determined for wind turbines in full wake situations.

2. Methodology

Wind turbine simulations for this investigation are performed using the HAWC2 aero-servo-elastic code [6]. HAWC2 is based on blade element momentum theory extended with sub-modules such as dynamic stall, dynamic inflow or skew angle among others [7]. The structural part consists of a multi-body Timoshenko beam formulation defined with a floating reference system.

A generic 2.3MW wind turbine model is coupled with the open-source DTU wind energy controller [8]. The simulated wind turbine is a collective pitch regulated, variable speed turbine with a rotor diameter of 90 m and a rated power of 2.3 MW at 12 m/s. The spectral tensor Mann model [9] was used to generate the background wind field. The electrical power output is obtained by means of HAWC2 as averaged time-series data.

A wake situation is generated by performing simulations with an upstream, wake-generating wind turbine, and a downstream wind turbine located at a predetermined location behind the first wind turbine. The unsteady inflow conditions experienced by the upstream wind turbine are transported to the downstream wind turbine using the Dynamic Wake Meandering Model (DWM) [10, 11]. Thus, the downstream wind turbine feels a disturbance of the original turbulent inflow due to the wake.

The simulated cases are run for an offshore case ($TI = 6\%$) [12], and a high-turbulence forest case ($TI = 22\%$) [13] [14]. The different turbulence intensities were achieved by scaling the turbulent wind fields generated with the Mann model, as explained in [6]. For both the offshore and the forest case the Mann turbulence box parameters are defined according to the IEC-61400-1 standard, $\gamma = 3.9$ and L = 29.4. The wind turbine spacing as well as the yaw angle of the downstream wind turbine are varied to characterize the effect of yawing in a wake situation. Six ten-minutes simulations with different stochastic realisations (ie. based on different random seeds) per case are used to improve the statistical significance of the results. The investigated scenarios include mean wind speeds ranging from 4 m/s to 25 m/s.

The wake modeling in HAWC2 is based on the DWM model. The wind turbine wake includes the combination of three items: the wake deficit, the wake meandering and the wake self-generated turbulence. First, the wind turbine induction is computed for the given atmospheric condition, which results in a spatial averaged thrust profile. Then, the wake deficit evolution is computed based to the steady-state velocity and turbulence intensity as function of downstream position [10]. The HAWC2 DWM model has been validated against measurements from the Egmond aan Zee Wind Farm [15], and the power production has been compared to measurements and other engineering models in [16].

3. Results

The result section is divided in four parts. The first part describes the effect of the wake on the wind field immediately before an aligned downstream wind turbine as well as its effect on the power output of this wind turbine. The second part presents the power loss due to yaw misalignment, when a downstream wind turbine experiences the wind fields as described in the first section. The third part introduces the loss in annual energy production (AEP) obtained to characterize the power loss due to yaw misalignment. Finally, the model limitations are presented.

3.1. Effect of wind turbine spacing on power production loss

This section compares the influence of wind turbine spacing on the rotor mean wind speed and power output of downstream wind turbines. The rotor mean wind speed is defined as the wind speed averaged over time and space, i.e. over both the rotor plane and the duration of the simulation. The simulated results are generated for full wake conditions and no yaw misalignment.

First, the rotor mean wind speed deficit (relative to the free-stream case) is compared in Figure 1. In both the forest and offshore cases, it is possible to observe the wake recovery as a function of wind turbine spacing. As wind turbine spacing increases, the wind speed deficit observed by the downstream wind turbine decreases due to turbulent mixing. The rotor mean wind speed deficit is larger for the offshore cases compared to the forest case due to the lower offshore turbulent intensity, leading to modest mixing of the wake. This result is aligned with the formulation used in [10].

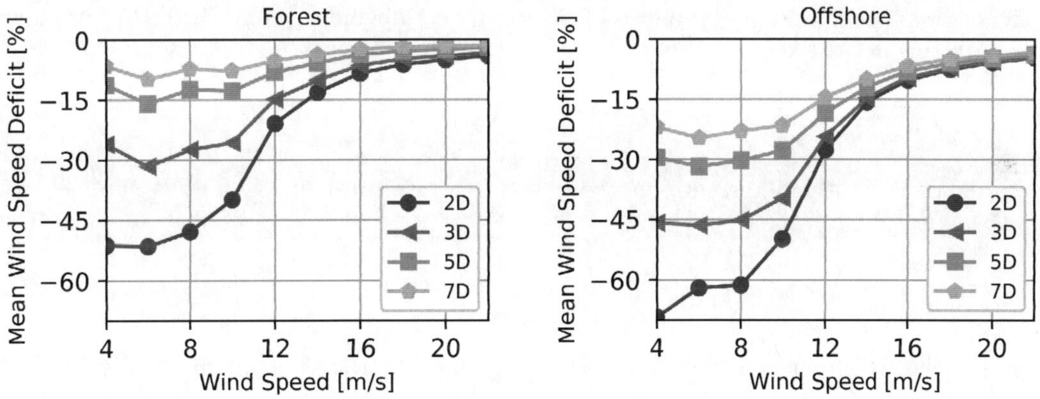

Figure 1: Influence of wake position on rotor-averaged wind speed deficit for respectively the forest and the offshore case for different downstream positions

The associated power output of the downstream wind turbine is compared in Figure 2. For both cases, the highest power production over all operating wind speeds is found in the no-wake case, whereas the lowest is obtained in the wake case with smallest the wind turbine interspacing. Close to rated power (11.2 m/s), the forest case produces less power than the offshore case. This phenomenon is a result of the higher turbulent intensity, which causes the power curve to be 'smoothed out' due to the large turbulent variations in wind speed. Conversely, at low wind speed, i.e. 8 m/s, the same effect causes the forest wind turbine to produce more than the offshore wind turbine.

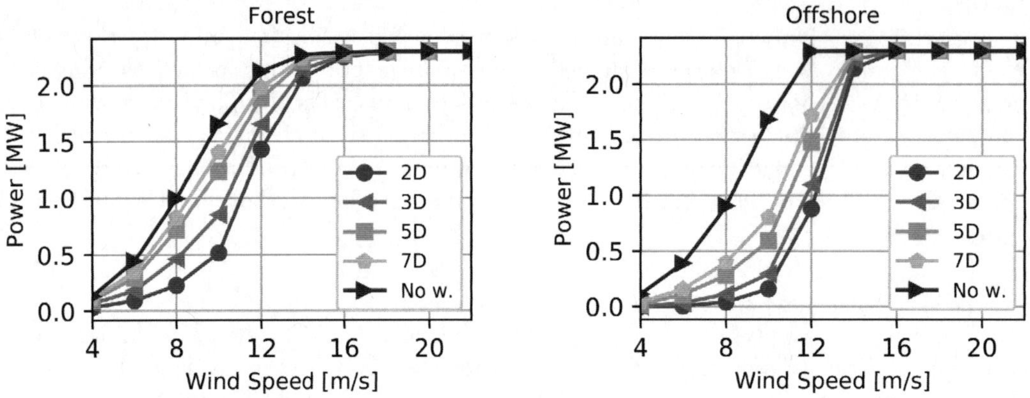

Figure 2: Influence of wake position on electrical power production for the forest and the offshore case for different downstream wind turbine positions

3.2. The effect of yaw misalignment on power production loss in wake situations

Using the cases from the previous section as a reference, the downstream wind turbine is now yawed in order to quantify the effect of yaw on power production in wake situations. As shown above, the presence or absence of wake and the atmospheric conditions during below rated operation influence the power production. In order to model the power loss due to yaw misalignment, the power output is normalised against the aligned case. For a given yaw misalignment angle, γ, and power output of the downstream wind turbine, $P(\gamma)$, the normalised power output is defined as:

$$\eta(\gamma) = \frac{P(\gamma)}{P(0)}. \tag{1}$$

In accordance with other literature on power loss in yaw situations [3], $\eta(\gamma)$ is parameterised using the approximation,

$$\eta(\gamma) \approx cos^{\alpha}\gamma, \tag{2}$$

where α is the fit exponent which describes $\eta(\gamma)$ the best. Based on momentum theory, the value of α is 3 [3]. However, in [4], it is reported that measurements involving both wind tunnel models and fullscale wind turbines tested at Vattenfall/FFA and DNW have indicated that the exponent may vary between 1.8 and 5.14 for the NREL Phase VI rotor. Such studies have also consistently shown that the BEM theory over predicts the power at yawed conditions. This is possibly related to the limitations of the linear momentum equation modified for yawed flows.

The sensitivity of the power as function of yaw angle is only relevant below rated wind speeds. For wind speeds higher than rated, the wind turbine produces rated power regardless the yaw misalignment. Thus, the analysis of α in that region lacks interest because it corresponds to a fixed value of zero. The results for forest and offshore case at 6 m/s are presented in Figure 3 to ensure the wind turbine operation is below rated.

Figure 3 shows the relative power loss due to wake misalignment for the forest and offshore cases respectively. A free inflow simulation is also displayed, which provides the reference or no wake case. It can be observed for the forest case that the relative power loss in a yawed case is

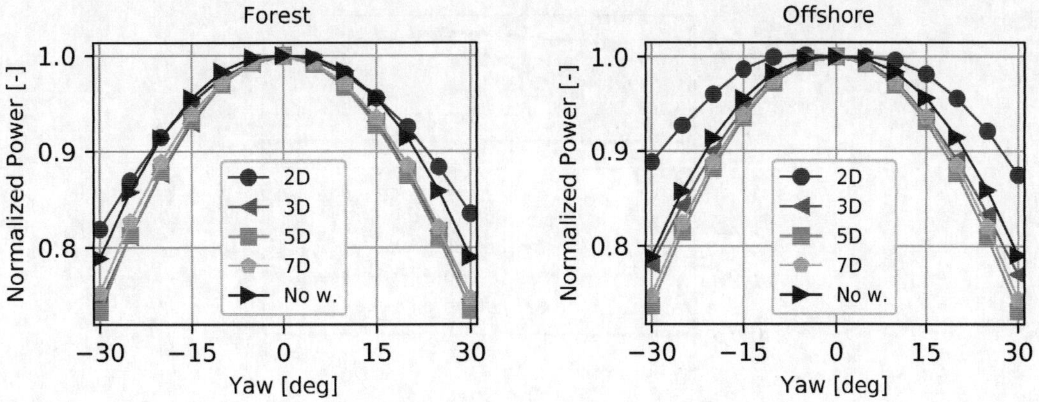

Figure 3: Power normalization for forest and offshore case at 6 m/s for different downstream distances

higher for the 3D, 5D and 7D while less power is lost if the wind turbines are placed 2D. For example, the 3D case at a yaw angle of -30° presents a power reduction of 26%, i.e. produces only 74 % of the available power, while the free-stream case for the same angle presents a reduction of 21.5%. On the other hand, the 2D case shows a reduction of 17.5%, which is lower than the free-stream case. Therefore, it appears that the relative power loss due to yaw misalignment is diminished when a turbine is in a near-wake scenario. The cause of this surprising phenomenon is a result of the spatial distribution of the wake deficit, as explained below. It should be noted, however, that it is unlikely to observe turbine spacing less than 3D in modern wind farms.

Similarly, for the offshore case, the 2D case produces 88.5 % of the available power at a yaw angle of -30°, while the freestream case produces slightly below 79%. The rest of the downstream distance, 3D, 5D and 7D, produces lower power than the free stream case. This phenomenon can be further explained by determining the relative power loses due to yaw, which can be obtained by computing the α value in Equation 2 found by minimizing the sum of square errors between the fitting and the actual data. α provides the sensitivity of power loss as function of yaw angle. The higher the value of α, the higher the power loss for a given yaw angle. The α fit for the offshore and forest case at 6 m/s is shown as function of wind turbine spacing in Figure 4.

Figure 4 presents the α fit as function of downstream distance for the free stream, offshore and forest case. The free stream value of α is used as a reference. The α value for the 2D downstream distance is lower in both the offshore and forest case compared to the free stream case, which can also be deducted from Figure 3 where the 2D case presents relatively lower power loss due to misalignment. In contrast, α is larger than the freestream case for wind turbine spacings greater than 3D. The α is expected to converge to the reference value for large wind turbine spacings, as the wake dissipates due to turbulent mixing.

In searching for an explanation of the variation of α as a function of wind turbine spacing, it was found that α is highly dependent on the radial distribution of wind speeds over the rotor plane, which corresponds to the shape of the wind field created by a wake. It is normally assumed that the wind field is homogeneous, or contains only vertical non-uniformity (such as wind shear or tower shadow). In these cases, yawing the wind turbine simply reduces the wind speed projected onto the rotor plane by a cosine factor. There is, in fact, an additional effect,

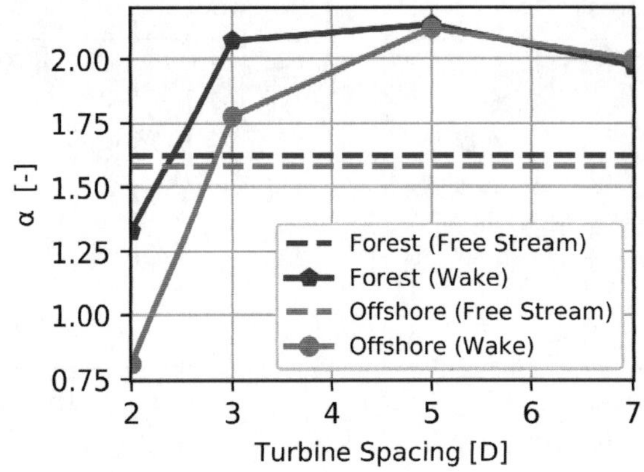

Figure 4: α fit for free stream, offshore and forest case at 6m/s as function of downstream wind turbine spacing (distance normalized with wind turbine diameter)

which is that each point on the rotor point is, on average, shifted towards the center of the rotor plane. This commonly neglected effect, combined with a radial non-uniformity in the wind field due to yaw misalignment, can have a significant impact on the power output. An example of the flow observed from a downstream wind turbine is shown in Figure 5, where 2D and 5D can be observed. For low wind turbine spacing, a ring shaped wake exists, which causes low values of α. As the ring shape dissipates, α increases.

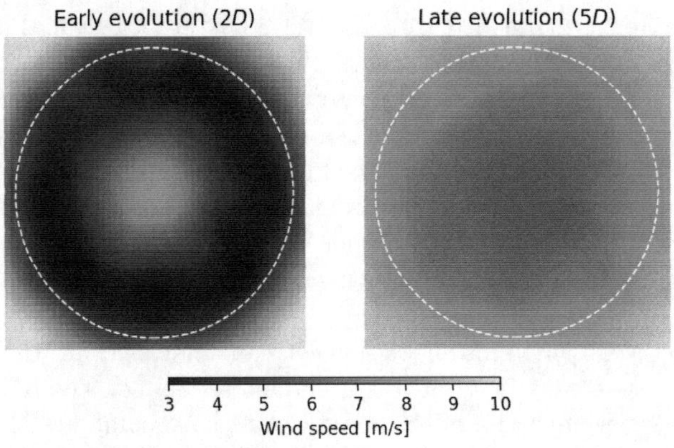

Figure 5: 10 minutes averaged wind fields observed from downstream wind turbine at 2D and 7D (forest case, $U = 10m/s$).

3.3. Annual Energy Production Analysis

From the results above, it is clear that a combination of wind turbine spacing and turbulent mixing plays an important role on the power loss that occurs, when a wind turbine is yawed in a wake condition. To better quantify this, the reduction in annual energy production (AEP) is determined for varying wind turbine spacing while fixing the yaw misalignment to 15° as an example. The results of this analysis, shown in Table 1, illustrate a key finding in this paper.

Namely, the power loss is maximum at a wind turbine spacing of approximately 3D. Surprisingly, the relative power loss in a near wake scenario (2D) is lower than the 3D to 7D cases. This is due to the early evolution of the wake shape, which has a ring shape compared to a more mixed wake experienced at 3D and beyond as observed in Figure 5.

Table 1: Loss in AEP due to 15° yaw error in full wake situation (Weibull parameters, $A = 8.46$, $k = 2$).

	Wind turbine spacing (L_{long})				
Case	No wake	2D	3D	5D	7D
Offshore	2.70%	2.97%	4.55%	4.48%	4.15%
Forest	3.24%	3.64%	4.95%	4.90%	4.50%

It should be noted that AEP calculations are performed for a fixed wind direction, which corresponds to a wake case. The analysis does not take into account the whole complexity of a wind farm, since multiple wake scenarios in such depends on the wind direction and farm layout.

3.4. Model Limitations

The results of this paper rely on The Dynamic Wake Meandering model in HAWC2 to propagate the wake in simulation. The Dynamic Wake Meandering model is a medium-fidelity engineering model capable of predicting the non-stationary wind farm flow field and thus the wake deficit and its dynamics behind a wind turbine. The limiting assumptions of the model [10] relevant to this paper relate to the near-wake approximation and radial symmetry. In the near wake regime - here assumed to extend to 3D downstream of the rotor plane - the wake is obtained by expanding the wake resulting from BEM modelling using a simple mass and momentum conserving pressure recovery approximation [11]. The model moreover assumes the wake deficit to be circular symmetric. Due to these simplifying assumptions, the complex behaviour in the near wake regime and non-rotationally symmetric wakes are not represented in this study. Additionally, the analysis in this paper assumes the wake intercepts the downstream wind turbine with no meandering in order to isolate the effect of the wake profile on power output. Meandering is not considered, but is a logical step in the future work of this investigation.

4. Conclusion and future work

The effect of power loss due to yaw misalignment in wake situations is presented in this paper using a 2.3MW wind turbine model simulated in HAWC2. Two contrasting site conditions, an offshore and forest case, are analysed to emphasise the consistency of the results despite large variations in the ambient turbulence intensity. In both atmospheric condition cases, a larger power reduction is found when there is yaw misalignment in a wake situation at least 3D downstream. On the other hand, yaw misalignment at distances below 3D present a lower reduction than the free stream case. To quantify this effect, the exponent term, α is numerically determined, representing the power sensitivity to yaw misalignment.

Variations on α are attributed to the wake deficit profile which varies due to two main factors: ambient turbulence intensity, and wind turbine spacing. Higher levels of turbulence intensity cause the wake to dissipate faster due to a higher rate of turbulent mixing, whereas larger wind turbine spacing provides a longer period for the mixing to occur. In the near wake region, the wake profile presents a bimodal diametric distribution, which is, surprisingly, found to increase the power output of a downstream wind turbine when yawed. As the wake profile dissipates beyond the near wake region, the bimodal shape breaks down into a single coherent velocity deficit. Wind turbines in the region of approximately 4D-5D experience the greatest reduction in power output due to yaw misalignment. α converges to the freestream value as wind turbine spacing increases further.

The presented results emphasize the importance of the wind turbine alignment in below rated conditions when in full wake conditions. In particular, the percentage reduction in AEP due to yaw misalignment is largest for wind turbines spaced at approximately 4D. The presented work provides valuable information for wind farm planning when wake situations are present.

5. Acknowledgements
This research was supported by the EUDP funding agency through the project Lidar detection of wakes for wind turbine optimization.

References
[1] Knud Abildgaard Kragh and Morten Hartvig Hansen. Load alleviation of wind turbines by yaw misalignment. *Wind Energy*, 17(7):971–982, 2014.
[2] Albert M. Urban, Torben J. Larsen, Gunner Chr. Larsen, Dominique P. Held, Ebba Dellwik, and David Robert Verelst. Optimal yaw strategy for optimized power and load in various wake situations: Paper. volume 1102. IOP Publishing, 2018.
[3] Knud Abildgaard Kragh and Morten Hartvig Hansen. Potential of power gain with improved yaw alignment. *Wind Energy*, 18(6):979–989, 2015.
[4] JG Schepers. EU project in german dutch wind tunnel. In *DNW', presented at IEA Expert Meeting on Aerodynamics*, 2001.
[5] H. Aagaard Madsen, Niels N. Sørensen, and S. Schreck. Yaw aerodynamics analyzed with three codes in comparison with experiment. In *AIAA Paper 2003-519*. American Institute of Aeronautics and Astronautics, 2003.
[6] Torben J. Larsen and Anders Melchior Hansen. *How 2 HAWC2, the user's manual.* Risø National Laboratory, 2007.
[7] K. Boorsma, J.G. Schepers, S. Gomez-Iradi, I. Herraez, T. Lutz, P. Weihing, L. Oggiano, G. Pirrung, H.A. Madsen, W.Z. Shen, H. Rahimi, and P. Schaffarczyk. Final Report of IEA Wind Task 29 Mexnext (Phase 3). Technical Report ECN-E-18-003, ECN, May 2018.
[8] Morten Hartvig Hansen and Lars Christian Henriksen. Basic DTU wind energy controller. 2013.
[9] J. Mann. Wind field simulation. *Probabilistic Engineering Mechanics*, 13:269–282, 1998.
[10] Gunner Chr. Larsen, Helge Madsen Aagaard, Kenneth Thomsen, and Torben J. Larsen. Wake meandering: A pragmatic approach. *Wind Energy*, 11(4):377–395, 2008.
[11] H Aa Madsen, Gunner Chr Larsen, Torben J Larsen, Niels Troldborg, and R Mikkelsen. Calibration and validation of the dynamic wake meandering model for implementation in an aeroelastic code. *Journal of Solar Energy Engineering*, 132(4):041014, 2010.
[12] Torben J. Larsen, Helge Aagaard Madsen, Gunner Chr. Larsen, and Kurt Schaldemose Hansen. Validation of the dynamic wake meander model for loads and power production in the egmond aan zee wind farm. *Wind Energy*, 16(4):605–624, 2013.
[13] Johan Arnqvist, Antonio Segalini, Ebba Dellwik, and Hans Bergström. Wind Statistics from a Forested Landscape. *Boundary-Layer Meteorology*, 156(1):53–71, jul 2015.
[14] A Chougule, Jakob Mann, Antonio Segalini, and Ebba Dellwik. Spectral tensor parameters for wind turbine load modeling from forested and agricultural landscapes. *Wind Energy*, 18(3):469–481, 2015.
[15] Torben J. Larsen, Helge Aagaard Madsen, Gunner Chr. Larsen, and Kurt Schaldemose Hansen. Validation of the dynamic wake meander model for loads and power production in the egmond aan zee wind farm. *Wind Energy*, 16(4):605–624, 2013.
[16] Rolf-Erik Keck. Validation of the standalone implementation of the dynamic wake meandering model for power production. *Wind Energy*, 18(9):1579–1591, 2015.

IOP Conf. Series: Journal of Physics: Conf. Series **1222** (2019) 012003 doi:10.1088/1742-6596/1222/1/012003

Improvements to the Eddy Viscosity Wind Turbine Wake Model

K Gunn

E.ON Climate & Renewables, Westwood Way, Coventry, UK, CV4 8LG

E-mail: kester.gunn@eon.com

Abstract. This paper presents improvements to the eddy viscosity wake model to significantly improve its fidelity without compromising its low computational cost. The eddy viscosity wind turbine wake model, proposed by Ainslie, has been a staple component of the wind yield toolchain for decades. Its low computational cost and excellent predictive power are likely to keep it as such for years to come. However, the standard implementations of the model rely heavily on several heuristics which, although well validated, may not scale well for large turbines or large windfarms.

In this paper, some non-physical heuristics are examined, and alternatives proposed which better reflect the underlying physics the model seeks to represent, which is shown to provide much better fidelity at no additional computational cost. The proposed model allows for better predictive capabilities of the performance of the next generation of offshore wind farms. The two key areas investigated are: first the impact of the ambient condition (TI, stability, ...) on wake development; and second a re-examination of the wake combination method, leading to the inclusion of three additional processes in the model. It is shown that modifications to these areas provide a more verifiable wake model than the baseline implementations, and allow identification of the remaining physical processes to be further incorporated into future versions.

1. Introduction

The Eddy Viscosity (EV) wind turbine wake model, proposed by Ainslie [1], is a common component of commercial modelling packages [2, 3]. Many implementations rely on heuristics which, although well validated, may not scale well for large turbines or large windfarms and have been shown to produce artefacts which could lead to suboptimal windfarm designs when such models are used for layout optimisation [4].

In this paper, some of the non-physical heuristics are examined, and alternatives proposed which better reflect the underlying physics the model seeks to represent, resulting in much better fidelity at no additional computational cost. The resulting model is compared to measurements from two offshore windfarm, and is shown to improve agreement.

1.1. The Baseline and New Model

In order to illustrate the effect of non-physical heuristics on the results of wake models, two implementations of the EV mode are compared in this paper:

Baseline The "baseline" implementation is an approximation to commercially available implementations [2, 3]. It is given as a comparison to the "new model" and windfarm data allowing the impact of the heuristics to be understood.

New Model The "new model" replaces the non-physical heuristics with better representations of the physical processes. It is used to show how such improvements yield results with fewer modelling artefacts and better general agreement to measured data.

Initially, the single turbine wake model is discussed:

Section 2 presents a brief introduction to the underlying EV wake model (as defined by Ainslie [1]), and the numerical scheme used to solve it (derived by Anderson [5]). This model only represents the development of a single turbine's wake but is the basis for both the baseline and new model implementations.

Section 3 discusses the "ambient mixing". It is shown that the baseline formulation, taking its length scale from the turbine rotor diameter, is inappropriate and can result in unrealistic artefacts in results. A new formulation is recommended in its place.

Moving to multiple turbines, the wake combination method and its interaction with other physical processes is then discussed, yielding three corrections:

Section 4 analyses the "maximum deficit" wake combination method used in the baseline implementation. It is shown why, despite its lack of physical meaning, this method is often able to produce good results. Three improvements are derived in the following three sections. These, in combination, are required to correctly account for the physics which the "max deficit" heuristic is able to, in some cases, approximate:

Section 5 introduces the wake combination method used to replace "max deficit" in the new model which is based on conservation of kinetic energy. This modification, in isolation, yields significant over-estimations of wake loss. As such it must be used in combination with the two remaining modifications.

Section 6 introduces a heuristic to account for turbine-wake interaction, i.e. the impact of a downstream turbine on the wake of an upstream turbine. This process reduces wake losses of rows of turbines when they are aligned into the wind.

Section 7 re-introduces a physical process, meandering, which was first introduced by Ainslie [1] but does not appear to be in the commercial models on which the baseline implementation is based.

These three corrections are developed such that they need only be applied at locations where the wind speed needs to be estimated (i.e. at turbine locations). As such, they have very low impact on the computational cost of the model. Finally:

Section 8 shows results from the models and comparisons to measurements from two offshore windfarms; and

Section 9 draws conclusions and makes recommendations for further work .

Initially, though, it is worth discussing the philosophical approach which has been taken to identify these corrections: verification[1].

1.2. Validation or Verification?

It is common to strive for a "verified and validated" model, despite this being an unachievable goal [6]. Verification is the process of ensuring a model has a correct interpretation of the relevant physical processes, while validation seeks to show that a model produces results within a required level of error.

[1] Not to be confused with "verificationisum" which, with logical positivism, has been abandoned.

Due, in part, to the commercial necessity of wake modelling (i.e. it is required to estimate the commercial value of a windfarm), much emphasis has been placed in the past on *validating* wake models (e.g. [2, 7, 8]). This has resulted in a lack of emphasis on *verifying* the models.

Within a given operational envelope (e.g. turbine spacing, farm size, control strategy, or rotor diameter), even a model with relatively poor representation of the underlying physics can be made to provide excellent results with appropriate tuning parameters. Such a model can therefore pass the required hurdles of a validation process [9]. However, without appropriate modelling of the underlying physical processes, the value of the model for extrapolating beyond the validated envelope remains unknown.

For example, the pioneering work of Jensen [10] accounts for conservation of momentum, but discards all other physical processes in favour of heuristics[2] yielding a computationally efficient model. Many derivatives of this model have "passed" validation exercises. However, when used to extrapolate (e.g. to windfarm control), unrealistic results have been observed.

In this paper model results have been compared with measured data. Conditions where the model *does not fit* have been examined and this has been used to hypothesise what physical processes are not appropriately modelled. Thus, the model derived here should be considered better *verified* than the baseline model, in that the objective is to improve the representation of the physics. This model *should not be considered validated*, and any validation process must be performed with new datasets.

1.3. A Brief Note on Nomenclature

Coordinates are given either as $[x, y, z]$ being the stream-wise, cross-stream and vertical coordinates with the origin at a turbine's hub (when using polar coordinates, r is the radial) or as global coordinates $[X, Y, Z]$, with Z origin at the ground.

To aid clarity and avoid "dimensionality errors", parameters denoted with a tilde ($\tilde{X}$) are nondimensionalised by the inflow wind velocity at hub height (U_0) and rotor diameter ($\eth$), as is common in wake modelling:

$$\tilde{U} = \frac{U}{U_0},\ \tilde{x} = \frac{x}{\eth},\ \tilde{\epsilon} = \frac{\epsilon}{U_0 \eth}$$

where ϵ is the eddy viscosity. Upper case velocities $(U, V, \ldots)$ are means, while lower case $(u, v, \ldots)$ are turbulent components.

2. The Single Turbine Model

Ainslie [1] derived the EV model for the wake of individual turbines by solving (in cylindrical coordinates) the momentum equation:

$$U\frac{\partial U}{\partial x} + V_r\frac{\partial U}{\partial r} = \frac{1}{r}\frac{\partial}{\partial r}\left(\epsilon r\frac{\partial U}{\partial r}\right) \tag{1}$$

(where the "eddy viscosity", ϵ, is used to describe the shear stress: $-\overline{uv_r} = \epsilon \partial U/\partial r$ and V_r is the radial velocity) and the continuity equation:

$$\frac{\partial U}{\partial x} + \frac{1}{r}\frac{\partial}{\partial r}(V_r r) = 0 \tag{2}$$

[2] The "top hat" wake profile is an obvious example which Jensen [10] himself shows not to hold. A less obvious one is the linear rate of wake width expansion. Examination of results from more verifiable models (e.g. Larsen [11], the EV model, see Fig. 1b and CFD [4]) or measured data (e.g. measurements by Richardson given by Taylor [12]) show wake expansion will not be linear.

Ainslie, and most commercial implementations [2, 3] solve the equations by finite difference, using the Thomas (tridiagonal matrix inversion) algorithm[3]. As shown by Anderson [5], however, if the wake profile is assumed to be self-similar at all locations downstream (which can be shown by examining the results from the Thomson algorithm implementation), and invoking conservation of momentum, then Eqn. 1 can be reduced to:

$$\frac{d\tilde{U}_c}{d\tilde{x}} = \frac{16\tilde{\epsilon}\left(\tilde{U}_c^3 - \tilde{U}_c^2 - \tilde{U}_c + 1\right)}{\tilde{U}_c C_t}$$

This can be solved using a Runge-Kutta method (here the classic 4th order variant is used) to give identical results to the Thomas algorithm implementation. The off-centre flow speed can then be calculated knowing that the wake takes a Gaussian profile:

$$\tilde{U}(\tilde{r}) = 1 - (1 - \tilde{U}_c)\exp\left(-\tilde{r}^2/\tilde{w}^2\right) \tag{3}$$

where $\tilde{w}$ is the wake (half-) width[4], which is calculated using conservation of momentum as:

$$\tilde{w}^2 = \frac{Ct}{4\left(1 - \tilde{U}_c\right)\left(1 + \tilde{U}_c\right)} \tag{4}$$

The model is initialised at $\tilde{x} = 2$ as defined by Ainslie [1], where the flow is no longer dominated by pressure gradients.

3. Ambient Mixing

The EV wake model assumes that the rate of recovery of wind turbine wakes is governed by two mixing components: a "local" component (K_l) generated from the shear of the wake, and an "ambient" component (K_a) generated by the shear in the atmospheric boundary layer:

$$\epsilon = F(K_l + K_a)$$

where F is a "filter function" accounting for the build-up of turbulence close to the turbine (and is here taken as $F = 1$ inline with WindFarmer [3][5]). Here $K_l = k_l w(U_0 - U_c)$ where $k_l = 0.015\sqrt{3.56}$ as estimated by Ainslie [1].

3.1. *The Ainslie Formulation*

The ambient mixing in terms of the "normal boundary layer parameters" is estimated as:

$$K_a = \kappa U_* Z/\phi_m(Z/L) \tag{5}$$

where $\kappa = 0.4$ is the Von Kármán constant, U_* is the friction velocity and $\phi = 1$ for neutral conditions.

[3] The Larsen [11] model solves the same equations analytically, making further simplifications at the expense of generality.

[4] Ainslie uses $\tilde{d}$ as the measure of approximately the full wake width, while here, w is used (equivalent to "2 standard deviations" of the Gaussian profile, thus approximately a half wake width) to remove the need for an arbitrary constant:

$$\tilde{d} = \sqrt{3.56}\,\tilde{w}$$

[5] It is not clear why this function can be neglected. It is possible that the blade-induced turbulence [13], which would not have been present in many of the wind tunnel permeable disk datasets available to Ainslie, counters the lack of shear-induced turbulence. An improved "near wake" model accounting for pressure effects and blade turbulence may improve accuracy.

Invoking the "Law of the Wall" to substitute for U_*, Ainslie [1] then gives the ambient mixing in neutral conditions as:

$$K_a = h\kappa^2 U_0 / \ln{(h/z_0)}$$

where z_0 is the roughness length. However, he next presents the complete (including K_l) nondimensionalised form of the eddy viscosity, wherein K_a has become:

$$\tilde{K_a} = \frac{K_a}{U_0 \eth} \approx \frac{\kappa^2}{\ln{(h/z_0)}} \tag{6}$$

where the hub height (h) has been cancelled by the rotor diameter ($\eth$). This approximation ($h \approx \eth$), although reasonable in 1988, is unnecessary and, with modern windfarms can lead to some unrealistic artefacts (see Section 3.4).

3.2. The Baseline Formulation

The WindFarmer [3] and OpenWind [2] Theory Manuals give the ambient mixing as[6]:

$$\tilde{K_a} = \kappa^2 I_0 \tag{7}$$

This formulation has the benefit that the mixing is a function of only known or measured quantities (the turbulence intensity (I_0) is normally measured while the roughness length (z_0) is harder to obtain). However, it is not clear whence this equation derives[7] and if it has any basis in theory. As with the Ainslie formulation, this formulation is incorrectly nondimensionalised by the rotor diameter (i.e. assuming $h \approx \eth$).

3.3. An Alternative Formulation

Townsend [16] gives a turbulence model in the form:

$$\overline{u^2}/U_\star^2 = B_1 - A_1 \log(z/\delta)$$

where u is the unsteady component of stream-wise flow, δ is the thickness of the shear layer, and A_1 and B_1 are non-dimensional constants. This equation is well grounded in theory (a detailed derivation can be found by Perry and Chong [17]), but in reality is only valid over a limited range of heights (see, for example, Marusic et al. [18]) and is unlikely to be satisfactory in non-neutral conditions. However, again substituting for U_* in Eqn. 5, it allows the relation between K_a and I_0 to be investigated:

$$K_a = \kappa I_0 U_0 h / \sqrt{B_1 - A_1 \log(h/\delta)} \tag{8}$$

Requiring, as it does, two poorly understood constants (for those wishing to test this formulation, $A_1 = 1.25$, $B_1 = 2.2$ are not unreasonable estimates [18]), and site measurements of δ, it is not recommended that this formulation be used. However, there is now at least a good theoretical basis on which to place the use of I_0 in estimating K_a. Instead, the following is recommended:

$$K_a = k_a I_0 U_0 l_0 \tag{9}$$

[6] There is some uncertainty here as all documents which present the mixing in this form seem to make a dimensional error, where the non-dimensional ambient mixing is added to the dimensional local mixing [2, 3, 14]. It is assumed that actual implementations are correct and this is only an error in the documentation.

[7] WindFarmer [3] cites it to Ainslie [1], who presents no such relation. It is possible it can be derived based on Eqn. 5 and some derived values from wind standards such as EN1991-1-4 [15], however it is not clear if such an approach is soundly based in theory.

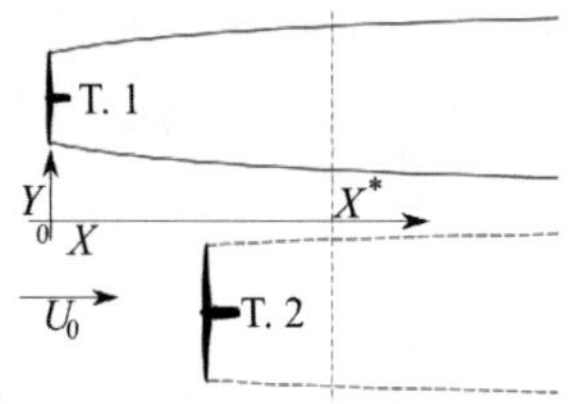

(a) Illustrative setup with two rotors.

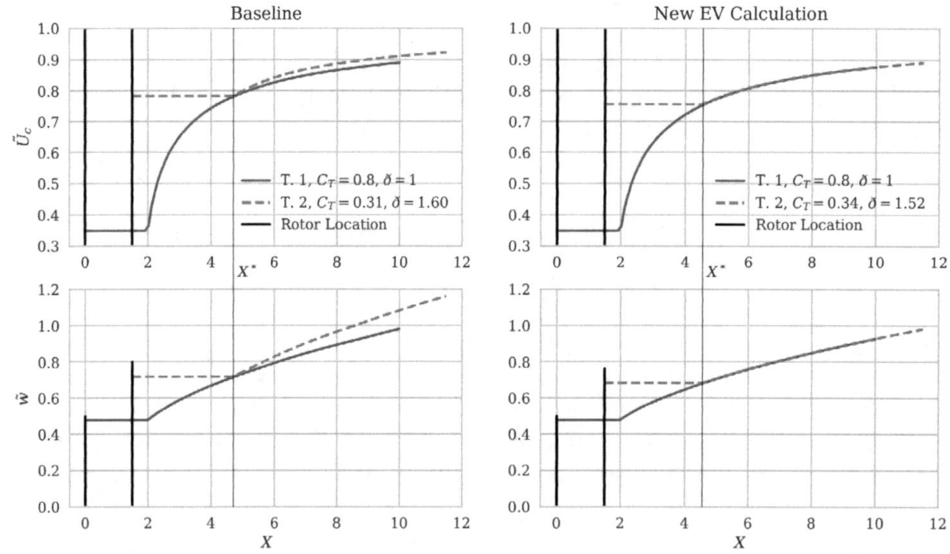

(b) Illustration of the different rates of wake expansion resulting from different rotor diameters (left), and the modified EV calculation resulting in consistent expansion (right).

Figure 1: The impact of selecting the wrong length scale for the ambient mixing.

where $k_a \approx 0.5$ is a dimensionless constant, and I_0, U_0 and l_0 (the length scale) are all derived from site measurements (e.g. deriving l_0 from frequency analysis of anemometer data). For neutral conditions, $l_0 \approx \kappa h$.

3.4. Impact of Ambient Mixing Length Scale

Both the Ainslie and baseline formulations for K_a (Equations 6 and 7 respectively) take the length scale as the rotor diameter ($\eth$) rather than any feature of the natural flow field. Thus, the rate of expansion of a wind turbine's wake, kilometres downstream of the wind turbine which formed it, is dependent upon the size of the rotor.

This can be shown to be inappropriate by considering a system of two rotors, as illustrated in Fig. 1. The first rotor ($\eth = 1, h = 0.8$ at $X = 0$) is operated at close to optimal thrust coefficient of $C_T = 0.8$. The second rotor is positioned at $X = 1.5$ but sufficiently far to the side that the turbines do not interact (illustrative map in Fig. 1a). It is shown that one can size the second rotor and operate it with a certain thrust such that the width and deficit of both wakes will, for some X (X^*), be identical. Thus both should now develop identically. However (left panels on Fig 1b), the two wakes will develop at different rates due to the different length scales the wind "remembers"[8] from the rotors. Conversely, referring to the right panel, where the length scale is

[8] A homeopathic interpretation.

IOP Conf. Series: Journal of Physics: Conf. Series **1222** (2019) 012003 doi:10.1088/1742-6596/1222/1/012003

taken as a function of the hub height (Eqn 8, neutral conditions), the two wakes now develop identically.

4. The "Maximum Deficit" Wake Combination Method

Much research has been performed on the subject of "wake combination". Many researchers have concluded that the "maximum deficit" method (as used in OpenWind [2] and WindFarmer [3]) is optimal despite its distinct[9] lack of physical meaning, in which:

$$(U_0 - U) = \max_i (U_0 - U_i)$$

where U_i is the speed modified by the wake of (only) turbine i, and U is the speed once the wakes are "combined". However, high resolution Computational Fluid Dynamics (CFD) analyses has failed to corroborate this approach [4].

Before selecting a more appropriate wake combination method, it is worth examining why the maximum deficit method is able to produce good results. Examination of the underlying physics shows several processes which are not modelled in the baseline implementation. Instead, two non-physical heuristics combine to approximate these actual processes in specific cases.

The two heuristics in the baseline implementation are:

(i) scaling the wake of a waked turbine so that it returns to U_0 at the edges;

(ii) "max deficit" wake combination, which due to (i) now approximates the deficit from both turbines.

These two heuristics are illustrated in Fig. 2. In such models, the scaled wake becomes a representation of the total wake in the system (i.e. the energy deficit integrated over the scaled wake is closer to the total deficit than the deficit generated by one turbine). Thus, the single scaled wake (which has the maximum deficit) is selected as an approximation for the total wake.

The three physical processes which these heuristics balance to approximate are:

(i) Conservation of energy should be applied when combining wakes (Section 5)

(ii) The impact of downstream turbines on wakes will have a destructive impact on the wake (Section 6)

(iii) Wake meandering, caused by large-scale horizontal turbulence, will cause a reduction in the perceived wake (Section 7)

As shown in Fig. 2f the two heuristics above provide a reasonable approximation to reality when turbines line up. When this is not the case, however, the max deficit approach provides poor results.

5. Wake Combination Conserving Kinetic Energy

Katic et al. [19] suggests using conservation of Kinetic Energy (KE) when combining wakes, and propose:

$$(U_0 - U)^2 = \sum_i (U_0 - U_i)^2$$

[9] While other wake combination methods such as "linear" or "sum of squares" may also lack a physical interpretation, the individual wakes are *combined*, whereas for "max deficit" the wakes exist independently, one of which becomes *true* at the point it is sampled[10].
[10] The Copenhagen interpretation.

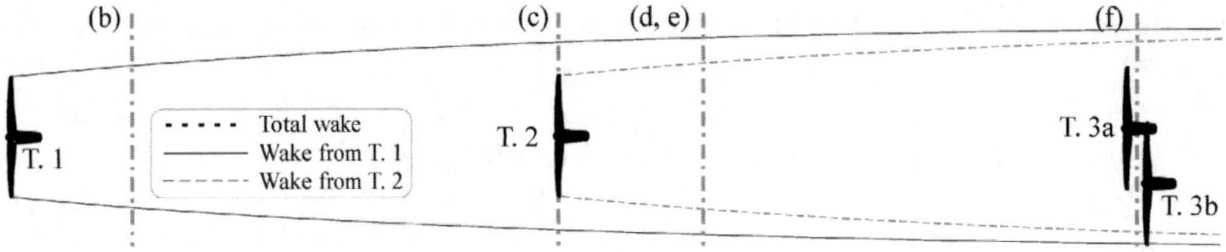

(a) An illustrative windfarm layout. Letters in brackets show the X locations of sub-figures (b) to (f).

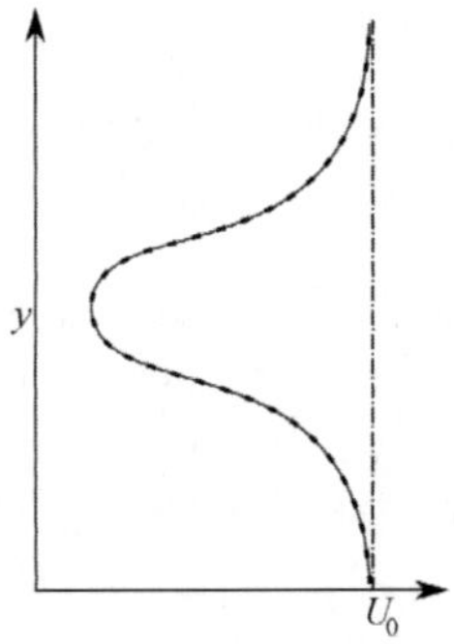

(b) The wake from the first turbine, T. 1, is initialised as prescribed by Ainslie [1]

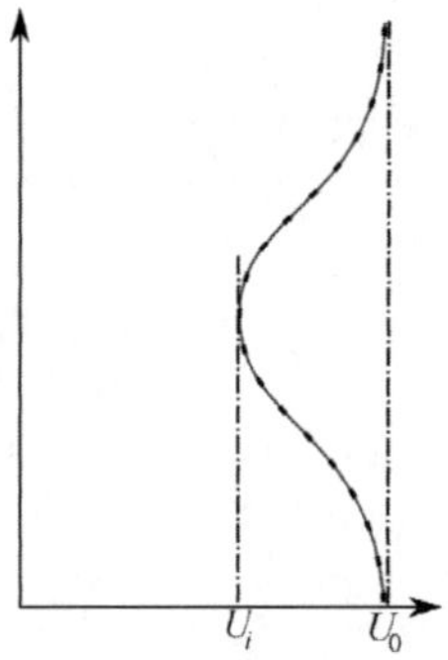

(c) The wake is propagated to the location of T. 2. The wind speed in the wake is sampled as the input speed for T. 2 (U_i)

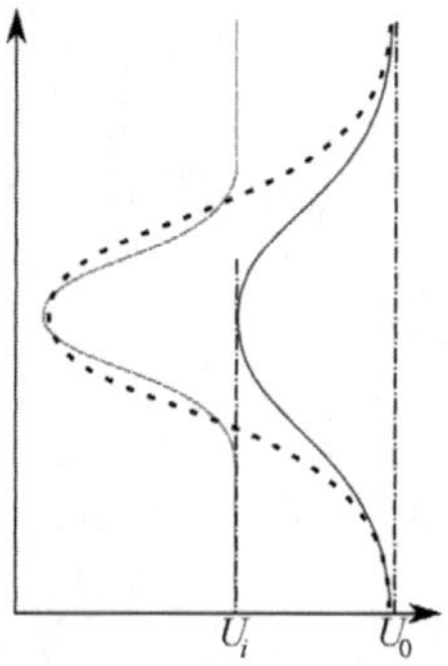

(d) T. 2's wake is initialised as prescribed by Ainslie [1], but taking $U_0 = U_i$ when calculating the profile.

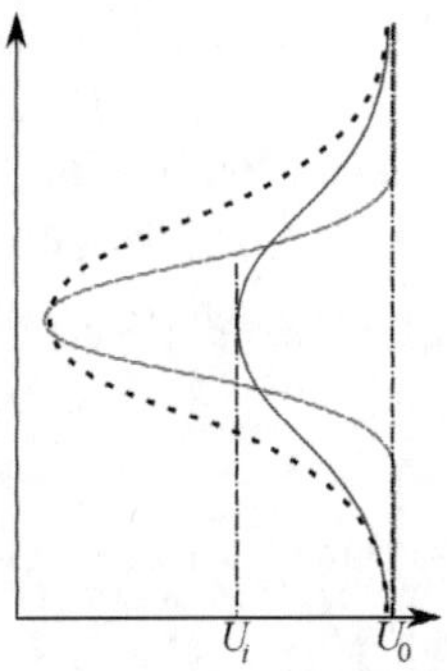

(e) As the profile in Fig. 2d does not return to U_0 at the edges the wake is "scaled" to return to U_0 rather than U_i [2, 3]. Note that the profile for T. 2 now more closely represents the *total* deficit (dashed black line) than the deficit of T. 2. Integrating the energy deficit in the re-scaled wake would yield an answer closer to the total energy deficit than that which is attributed to T. 2

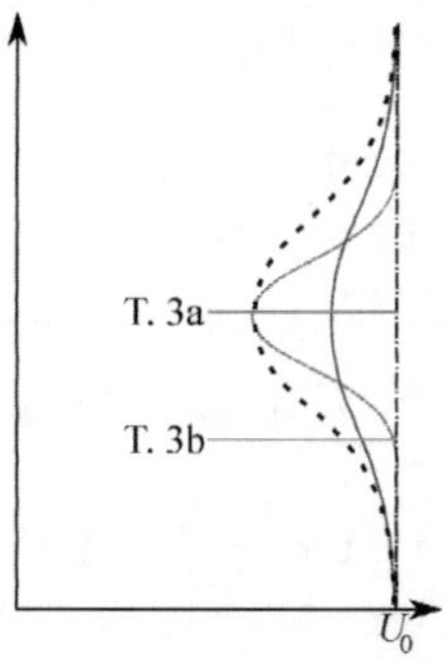

(f) At T. 3, the deficits from the two turbines are "combined". As the scaled wake from T. 2 is an approximation to the *total* wake in the system the "worst" wake is taken. Any attempt to sum the two wakes would double count the deficit from T1. This results in a good approximation when the turbines are aligned (T. 3a), but is much poorer otherwise (T. 3b).

Figure 2: Illustration of the interaction of wake "scaling" during the initialisation of a wake, and the "max deficit" wake combination method as used in the baseline implementation and by several commercial implementations [2, 3]. The total wake is illustrative, based on high resolution CFD results [4].

However, this formulation does not correctly account for KE[11]. The correct formulation, as presented by Kuo et al. [20] amongst others, is:

$$(U_0^2 - U^2) = \sum_i (U_0^2 - U_i^2)$$

When used in isolation (i.e. replacing "max deficit" and wake scaling) this approach will result in a significant over-estimation of wake losses (see Section 8.2.2). However, when combined with the turbine wake interaction (Section 6) and wake meandering (Section 7) the correct magnitude of loss is obtained (see Section 8).

6. Turbine-Wake Interaction

In the baseline wake model implementation (as in most engineering wake models), there is only a "one way coupling" between wakes and turbines. After a wake has been generated by a turbine, that wake will affect the operation of downstream turbines (wake-turbine interaction) but downstream turbines will have no impact on the development of that wake (turbine-wake interaction). As illustrated in Fig. 3, the expanding streamtube around a downstream turbine will result in the expansion of the wake from an upstream turbine. From conservation of momentum[12] it is clear that the expansion of the wind turbine wake will result in a reduction of the deficit in the wake. Thus, the approach taken here is:

(i) Where the wake from turbine T. i reaches the X location of turbine T. j, calculate the expansion of the stream-tube (e_j) around turbine T. j (Eqn. 10)

(ii) Calculate the expansion this will result in for the wake from T. i. ($e_{i,j}$, Eqn. 11)

(iii) Calculate the width of wake T. i using Eqn. 4, and add $e_{i,j}$ (i.e. increase the width of the wake from T. i)

(iv) Update the centreline deficit of the wake form T. i to comply with conservation of momentum (Eqn. 12)

From conservation of mass the expansion of the streamtube can be calculated as:

$$\frac{e_j}{\eth} = \sqrt{1 - a_j}\left(\frac{1}{\sqrt{1 - 2a_j}} - 1\right) \tag{10}$$

where a_j is the induction factor of turbine T. j (calculated as $a = 1/2\left(1 - \sqrt{1 - C_t}\right)$). It is now necessary to estimate what impact this stream tube expansion will have on the wake of T. i. When the two turbines line up exactly, the wake will be expanded significantly. From CFD results it has been observed [4] that when the turbines are just off alignment, the upstream wake is actually marginally "squeezed" rather than expanded. Thus, the direct application of this expansion is complex. Here, the squeezing of the wake is neglected, and the magnitude of the wake expansion is taken as a Gaussian distribution, with maximum expansion if the two turbine line up, moving to zero when the turbines are separated in the cross stream direction:

$$e_{i,j} = c_0 e_j \exp\left(-\frac{(Y_i - Y_j)^2}{c_1^2 \eth_j^2}\right) \tag{11}$$

[11] Despite no physical interpretation, this formation remains comman in the literature but is refereed to as "sum of squares" rather than "conservation of kinetic energy"

[12] Note that conservation of momentum does not hold within the expanding streamtube as the turbine applies a thrust to the flow. However, the momentum deficit introduced by the turbine is wholly accounted for by the introduction of the wake from that turbine. Thus, conservation of momentum does apply to the wake from the upstream turbine.

IOP Conf. Series: Journal of Physics: Conf. Series **1222** (2019) 012003 doi:10.1088/1742-6596/1222/1/012003

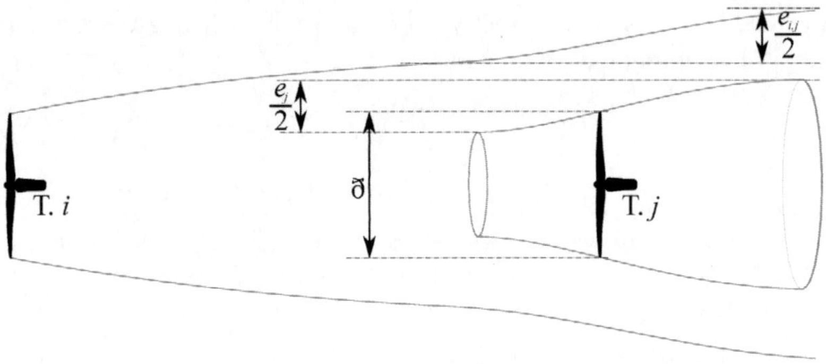

Figure 3: Stream-tube expansion and its impact on an aligned upstream wake. The streamtube around T. j expands by e_j resulting in the wake from T. i expanding by $e_{i,j}$.

where Y_i and Y_j are the cross stream positions of turbines T. i and T. j respectively. In reality the correct function is likely to be "squarer" towards the centre, and dip below zero at the edges of the turbine. Further work is required to estimate and parametrise such a function. Two parameters in Eqn. 11 (c_0 and c_1) are tuning parameters and further work is required to fix their values appropriately. Here $c_0 = 2$ and $c_1 = 1.5$ have been found to work adequately to verify the concept if not validate the model[13].

The centreline velocity for the wake from T. i is then calculated as:

$$\tilde{U}_c = \sqrt{1 - C_t/4\tilde{w}^2} \tag{12}$$

7. Meandering

Meandering impacts wakes by diverting them to either side through large scale fluctuations in the flow direction (i.e. with length scales larger than those driving wake recovery). The wake structure remains unchanged. Recent work has been done on the impact of wake meandering (e.g. Johnson et al. [21] and Larsen et al. [22]). However, the correction from Ainslie [1] is used here[14].

The motion of the wake centre-line can be interpreted in similar terms to particle diffusion by turbulent flow. Under this framework (and in a Lagrangian reference frame) the variance ($\sigma^2_{Y_c}$) of the cross-flow position of the wake (Y_c) can be estimated. General solutions under this approach (particularly in the complex non-isotropic atmospheric boundary layer) are hard to define (requiring knowledge of the turbulent Lagrangian autocorrelation, which is hard to measure), however it can be shown [12] that:

$$\sigma^2_{Y_c} = \begin{cases} \sigma_v^2 t^2 & t \ll T_v^L \\ 2\sigma_v^2 T_v^L t & t \gg T_v^L \end{cases}$$

[13] Values of $c_0 > 1/2$ may seem odd. Sadly, this is an introduction of an additional heuristic: a second physical process which is not accounted for here is the fact that the local mixing K_l will be increased when the two wakes align (as the combined wake has a larger deficit). Accounting for this explicitly has been found to be extremely computationally expensive as the flow feels must be calculated more often, and the integration step size reduced. Thus, taking this effect as being linearly proportional to the effect of the expanding stream tube is a necessary evil. It is hoped that a better relation can be derived through high resolution CFD modelling in the future.

[14] Although this correction is presented in the "original" EV model paper (cited to an earlier paper by the same author), it does not seem that it is implemented in the commercial models, or at least it does not appear in the theory manuals for them [2, 3].

where T_v^L is the (cross-stream) Lagrangian (integral) time scale, σ_v is the standard deviation of the cross-stream velocity[15] and t is time. These results effectively treat the turbulent velocity as constant ($t \ll T_v^L$) over the timescale of interest (thus the distance the centreline moves is simply vt), or as a Wiener process ($t \gg T_v^L$) where "steps" are uncorrelated (so offset grows with the square root of the time).

To identify which approximation ($t \ll T_v^L$ or $t \gg T_v^L$) is more appropriate, estimates of t and T_v^L must be made. The time for the wind to travel from the waking turbine to the waked turbine is $t = x/U_0$. Taking $x \approx 1$km as the order of distances of interest, and $U_0 = 10$m/s as a speed of interest, $t = 100$s. Appropriate values for T_v^L are hard to find, but $T_v^L \approx 200$s is not unreasonable [23].

Thus, the correlated case ($t \ll T_v^L$) is more appropriate, but this is marginal. Given this, and that closer turbine wake interactions (smaller x) are more important, $\sigma_{Y_C}^2 = \sigma_v^2 t^2$ is used here. However, with better analysis of site measurements (to better estimate the autocorrelation function), improved results are likely to be possible, especially for (spatially) large wind farms.

Taking $v \sim \mathcal{N}(0, \sigma_v^2)$, then $Y_c \sim \mathcal{N}(0, \sigma_{Y_C}^2)$. From the above, $\sigma_{Y_C}^2 = \sigma_v^2 t^2 = \sigma_v^2 x^2/U_0^2$. Having identified the Probability Density Function (PDF) of the wake centreline offset, the effective wake can now be calculated by convolving this PDF with the wake profile. To simplify this process rather than calculating this for all Y, it need only be calculated for the mean centreline (zero offset) and the remainder of the profile can be reconstructed by conservation of momentum using Equations 3 and 4

Noting that the deficit is also a Gaussian form, the convolution yields:

$$U_0 - U_c^* = \frac{(U_0 - U_c)}{m}$$
$$m = \sqrt{1 + 2\sigma_\theta^2 \frac{x^2}{w^2}} \tag{13}$$

where σ_θ is the standard deviation in wind direction[15] which is the result presented by Ainslie [1][16].

7.1. *Meandering and Histograms*

It is common practice to perform yield estimates by undertaking a simulation for each "bin" of a 2D histogram of speed and direction, and multiplying the result by the probability of that "bin". When analysing windfarm data, the inverse approach is often taken (i.e. the individual flow cases are "binned" and the mean calculated within each bin). In such cases θ will not be normally distributed, but will be the convolution of the true variability of θ ($\mathcal{N}(0, \sigma_\theta^2)$) and the distribution within the bin (for narrow bins, approximately uniform, $\mathcal{U}(\ldots)$). As such it is hard to differentiate the effects of meandering from the effects of the histogram. Results here (Section 8) are obtained using a local polynomial regression rather than "binning" and so the histogram effect should not exist.

[15] Although σ_v is not often measured, there is good evidence that $\sigma_v \approx \sigma_u \approx I_0 U$ (i.e. the turbulence is horizontally isotropic). For low levels of turbulence ($I_0 \lesssim 15\%$), the standard deviation of wind direction (in radians) is:

$$\sigma_\theta \approx \sigma_v/U \approx I_0$$

by the small angle approximation.

[16] Note that the constant 7.12 from Ainslie [1] (Eqn. 10) is given as 2 here for consistency between the different definitions of wake width, w and $\tilde{b}$

8. Results

Results of the baseline and improved models are illustrated in Figures 4a, 4b and 5, with some additional details shown in Fig. 6. The "Baseline" model is implemented with "max deficit", "wake scaling" and the ambient mixing is calculated using Eqn. 7. The "KE" model is implemented with conservation of kinetic energy wake combination, no "wake scaling", and Eqn. 9. "All Improvements" comprises "KE" with the new turbine-wake interaction heuristic, and wake meandering.

8.1. Wake Rose Figures

In Figures 4a 4b and 5, results are presented in the form of a "wake rose" wherein the radius shows the power generated by a turbine divided by the power that turbine would generate if unwaked. The angle shows the direction from which the wind blows. Green dots show the locations of other turbines in the farm, with the turbine at the centre (marked with an X) being the one for which the results are presented. Thus, for example, in Fig. 4a, at $\sim 90°$ the wind is flowing over one turbine before reaching the turbine of interest, while at $\sim 270°$ it is flowing over two. Figs. 4a and 4b are for a turbine at low and high speeds at Scroby Sands. Fig. 5 is for low speeds at Rodsand II. Fig. 6 shows an enlargement from Fig. 4a and illustrates the effects of the various modifications to the model.

Measured results are obtained from SCADA data, with the inflow estimated from the mean of the unwaked turbines (with some corrections applied). The line of best fit, 95% confidence and 95% prediction intervals are obtained using a multi-dimensional (direction and speed), robust (bisquare), local polynomial (2nd order in direction, 1st in speed) regression with Gaussian weighting (kernel widths of $\pi/40$rad and 1m/s for direction and speed respectively).

8.2. Discussion

It is clear from a review of Figs. 4a, 4b and 5 that the new model gives a better general agreement with the measured data than does the baseline (although, as already discussed, this does not comprise a *validation*). From the perspective of verification, it is worth examining specific features of the results, with the aid of Fig. 6.

8.2.1. Effect of Thrust Coefficient Figures 4a and 4b show relatively low (7m/s) and high (12m/s) wind speeds, and as such the front row of turbines will be operating at very different thrust coefficients. The new model appears to fit well to both the low and high thrust cases, as does the baseline model. Although not the focus of this paper, it is worth noting that this is because both models have an appropriate physical interpretation of the thrust coefficient (C_t) which is not universally true amongst wake models: where coefficients which should be functions of C_t have been estimated based on measured data, and thus the relationship has been lost.

8.2.2. Effect of Conservation of Energy in Wake Combination As discussed in Section 4, the "max deficit" is often found to be the best approach for wake combination. With reference to (a) in Fig. 6 it is clear why conservation of energy is not generally used, as it results (when used in isolation) in a vast increase in wake loss, giving a deficit far larger than measured. However, when used in combination with the other corrections in this paper, this over-estimate in wake loss is reduced.

8.2.3. Off-line turbines In Section 4 it was postulated that the "max deficit" wake combination fails when turbines do not line up. Scroby Sands is a particularly good windfarm to test this hypothesis as the array is fairly deep, but turbines are not aligned to a grid. With reference to (b) in Fig. 6, it is clear that when the wind flows through a large number of poorly aligned

IOP Conf. Series: Journal of Physics: Conf. Series **1222** (2019) 012003 doi:10.1088/1742-6596/1222/1/012003

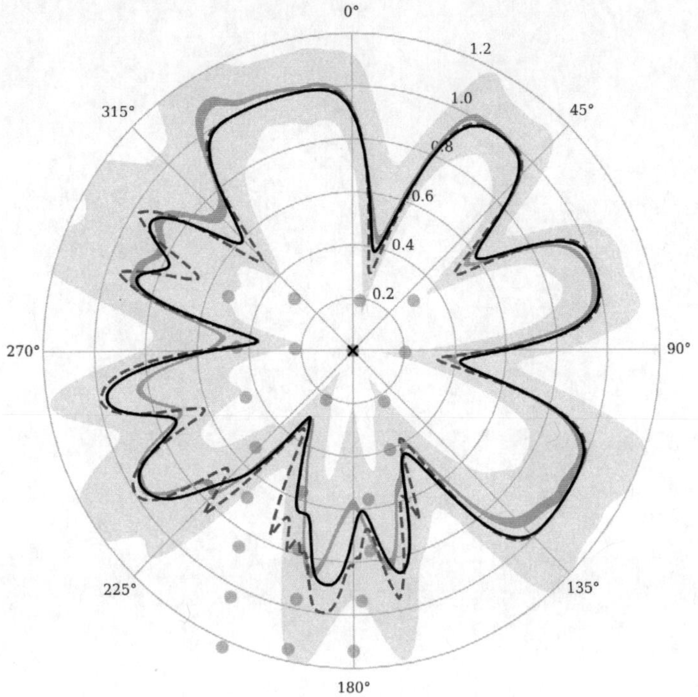

(a) $U_0 = 7\text{m/s}$

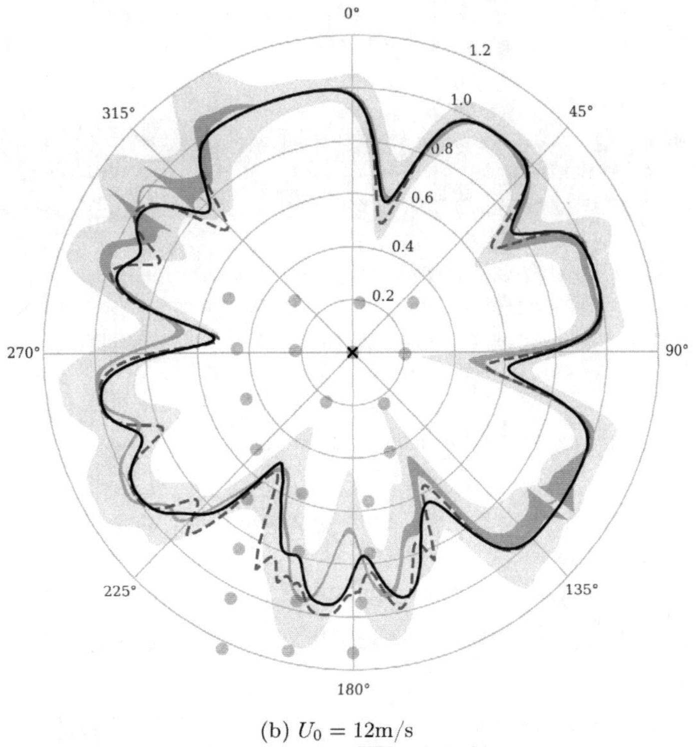

(b) $U_0 = 12\text{m/s}$

Figure 4: Illustrative results for a turbine at Scroby Sands at low and high wind speeds (see Section 8.1)

IOP Conf. Series: Journal of Physics: Conf. Series **1222** (2019) 012003 doi:10.1088/1742-6596/1222/1/012003

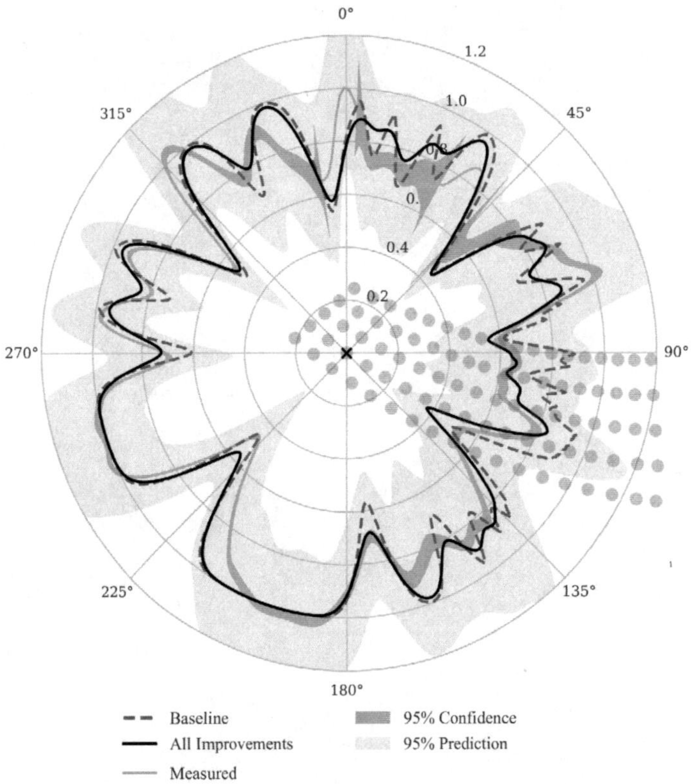

Figure 5: Illustrative results for a turbine at Rodsand II. (see Section 8.1)

turbines, "max deficit" significantly under-estimates the wake loss. This is a particular problem during array layout optimisation, as optimising using this model will result in a larger benefit from offsetting turbines. It is clear that the new model significantly reduces this problem.

8.2.4. Effect of Turbine-Wake Interaction The over-estimate in wake loss from using conservation of energy wake combination is removed, in-part, by the turbine-wake interaction model defined in Section 6. In Fig. 6, for westerly winds (c) the turbine of interest is waked by two upstream turbines. As can be seen, the over-estimation has been significantly reduced.

8.2.5. Effect of Meandering Meandering has two key effects on wakes. With reference to Fig. 6:

d.i with a similar effect to moving average in direction, peaks and troughs in the deficit are reduced in magnitude[17].

d.ii it results in a reduction in directionally-narrow effects.

Both of these changes have a significant impact in closing the gap between the model and the measured data.

[17] Note, however, that meandering cannot be applied as a moving average as a post processing step unless the wind angle is highly autocorrelated over the distances between turbines.

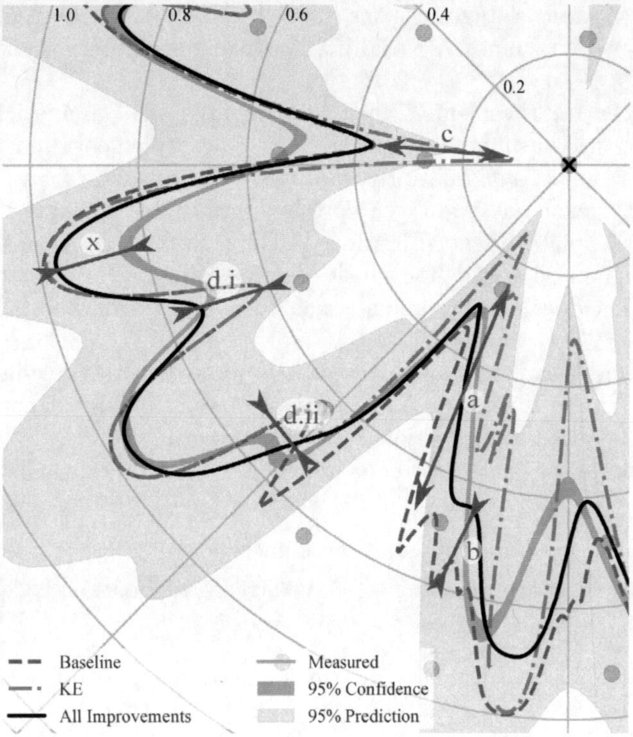

Figure 6: Enlargement from Fig. 4a showing features of the wake model. Comment for (a) is in Section 8.2.2, (b) in Section 8.2.3 , (c) in Section 8.2.4, (d) in Section 8.2.5, and (x) is discussed in the conclusions (Section 9).

8.2.6. "Large Array" effects With reference to Fig. 5, for easterly winds ($\sim 100°$) the wind flows through the majority of the windfarm before reaching the turbine of interest. In this direction, the baseline model significantly under-predicts the wake losses. This could be seen as evidence for the requirement of a "large array correction", such as WindFarmer's "Correction for Large Wind Farms" [3] or OpenWind's "Deep Array Wake Model" [24]. However, the new model does fits very closely to the measured data without any "large array correction". Further research is required to test if these corrections are well justified or are merely an additional heuristic to account for errors introduced by the use of "max deficit".

8.2.7. Increased Errors In some directions (e.g. $\sim 95°$ in Fig. 4b) the new model seems to produce a marginally lower quality fit than the baseline model. The cause for this is not yet clear and further investigation is required. Some avenues to explore include: regularity within the eddies causing meandering at a length scale of the distance between the turbines [12], the interaction of wake meandering and the filter function, the form of the turbine-wake interaction function, or simply errors in the measured data.

9. Conclusions

This paper shows that the EV wind turbine wake model can, with a few modifications, be significantly improved in fidelity. These improvements will lead to better layout designs for windfarms, and improved yield estimates.

First, a new method of estimating the ambient mixing is proposed. Although this method results in relatively modest changes to the overall results, it will result in more realistic simulations for large future turbines, especially when turbines of different sizes are present in the same simulation.

The remaining three improvements, when implemented in combination, replace the non-physical heuristics for wake combination in the baseline implementation ("wake scaling" and "max deficit"). These result in a significantly improved wake estimates. Furthermore, given their stronger basis in reality, the model's ability to predict performance outside of the validated range of conditions (i.e. extrapolate) is much improved. For example, while most wake models work well for windfarms aligned on a grid (for which they are validated), it has been shown that the baseline implementation does not extrapolate well to non-gridded wind-farms (such as Scroby Sands).

The three improvements are all implemented at points when the flow-field is estimated (i.e. the locations of downstream turbines). As such, they come at no significant computational cost (furthermore, if the hypothesis in Section 8.2.6 is found to hold, then these corrections will significantly reduce computational cost, removing the need for a "large array correction"). These three corrections are:

(i) combine wakes accounting for conservation of energy;
(ii) account for the effect of the induction of down stream turbines on wakes; and
(iii) apply a wake meandering model.

Finally, it should be emphasised that this paper does not claim to produce a *validated* model, but rather a "better *verifiable*" model. The approach, rather than testing the accuracy of the model on mean power production, has been to seek conditions in which the models do not work, and review for what physical processes the model does not correctly account. This approach has yielded significant improvements in the model, and as such, a similar approach should be applied to improving wake models going forwards. For example, the error (x) in Fig. 6 has not been significantly reduced by the corrections here. It is possible that this is the result of wake-wake interaction (where the deficit from one wake recovers more slowly due to a neighbouring wake drawing on the same source of energy [4]) which will require a new way to estimate the local mixing K_l. Thus, further work is still required on:

- wake-wake interaction;
- global blockage effects;
- autocorrelation of turbulence for wake meandering;
- tuning the turbine-wake model (Section 6); and, finally,
- validating the model again new datasets.

References

[1] Ainslie J 1988 Calculating the flowfield in the wake of wind turbines. *J. Wind Eng Ind Aerod* **27** 213–224.

[2] 2010 OpenWind theoretical basis and validation. Tech. rep. AWS Truepower. Version 1.3.

[3] 2009 *GL WindFarmer Theory Manual.* Version 4.0.

[4] Gunn K, Stock-Williams C, Burke M, Willden R, Vogel C, Hunter W, Stallard T, Robinson N and Schmidt S R 2016 Limitations to the validity of single wake superposition in wind farm yield assessment. *J. Phys. Conf. Ser* **749** 012003.

[5] Anderson M 2009 Simplified solution to the eddy-viscosity wake model. Tech. rep. RES.

[6] Popper K R 1959 *The Logic of Scientific Discovery.* Routledge.

[7] 2014 Validateion report WindFarmer. Tech. rep. DNV-GL. Version 5.3.

[8] Gaumond M, Réthoré P E, Bechmann A, Ott S, Larsen G, Pena Diaz A and Hansen K 2012 Benchmarking of wind turbine wake models in large offshore windfarms. *The science of Making Torque from Wind 2012.*

[9] Gunn K and Stock-Williams C 2013 On validating numerical hydrodynamic models of complex tidal flow. *International Journal of Marine Energy* **3-4** e82 – e97. ISSN 2214-1669. Special Issue - Selected Papers - EWTEC2013.

[10] Jensen N 1983 A note on wind generator interaction. Tech. Rep. Risø-M-2411 Risø.

[11] Larsen G December 1988 A simple wake calculation procedure. Tech. Rep. Risø-M-2760 Risø.

[12] Taylor G I 1922 Diffusion by continuous movements. *P. Lond Math Soc* **s2-20** 196–212.

[13] Nishino T and Willden R H J 2014 Low-order modelling of blade-induced turbulence for rans actuator disk computations of wind and tidal turbines. *Wind Energy - Impact of Turbulence* ed Hölling M, Peinke J and Ivanell S pp 153–158. ISBN 978-3-642-54696-9.

[14] Larsen G, Madsen Aagaard H, Larsen T and Troldborg N 2008 Wake modeling and simulation. Tech. Rep. Risø-R-1653(EN) Risø.

[15] 2005 EN 1991-1-4 (2005) (English): Eurocode 1: Actions on structures - part 1-4: General actions - wind actions. Standard. The European Union.

[16] Townsend A 1980 *The Structure of Turbulent Shear Flow.* Cambridge Monographs on Mechanics. Cambridge University Press. ISBN 9780521298193.

[17] Perry A E and Chong M S 1982 On the mechanism of wall turbulence. *J. Fluid Mech.* **119** 173–217.

[18] Marusic I, Monty J P, Hultmark M and Smits A J 2013 On the logarithmic region in wall turbulence. *J. Fluid Mech.* **716** R3.

[19] Katic I, Høystrip J and Jensen N 1986 A simple model for cluster efficiency. *European Wind Energy Association Conference and Exhibition.*

[20] Kuo J Y J, Romero D A and Amon C H 2014 A novel wake interaction model for wind farm layout optimization. *ASME 2014 International Mechanical Engineering Congress and Exposition* volume 6B.

[21] Johnson P B, Jonsson C, Achilleos S and Eames I 2014 On the spread and decay of wind turbine wakes in ambient turbulence. *J. Phys. Conf. Ser* **555** 012055.

[22] Larsen G C, Aa H, Madsen F, Bingöl J, Mann S, N Ott J, Sørensen V, Okulov N, Troldborg M, Nielsen K, J Thomsen T, Larsen R and Mikkelsen T 01 2007 Dynamic wake meandering modeling. Tech. Rep. Risø-R-1607(EN) Risø.

[23] Dosio A, Guerau de Arellano J V, Holtslag A A M and Builtjes P J H 2005 Relating Eulerian and Lagrangian statistics for the turbulent dispersion in the atmospheric convective boundary layer. *J. Atmospheric Sci.* **62** 1175–1191.

[24] 2016 *OpenWind User Manual.* Version 1.8.

IOP Conf. Series: Journal of Physics: Conf. Series **1222** (2019) 012004 doi:10.1088/1742-6596/1222/1/012004

Engineering models for turbine wake velocity deficit and wake deflection. A new proposed approach for onshore and offshore applications.

Renzo Ruisi, Ervin Bossanyi

DNV GL, One Linear Park, Avon Street, Bristol, BS2 0PS, UK
renzo.ruisi@dnvgl.com, ervin.bossanyi@dnvgl.com

Abstract. An engineering wake model based on the Ainslie model is proposed. The eddy viscosity term associated with momentum diffusivity is modified to take into account the effects of atmospheric stability. The parameters used are typically available from high-quality on-site measurement campaigns and the effects of atmospheric stability are based on empirical models for the estimation of the Monin-Obukhov length. The dependence on physical quantities only is particularly advantageous for fast wake modelling, since no further parametrical tuning is needed for each specific case. The proposed wake model is initially compared to wind tunnel data and CFD simulations to test the chosen Obukhov lengths for specific flow conditions. Wind farm production data and concurrent meteorological data at one onshore site are then used to validate the model for specific on-site flow conditions, obtaining good results. Two offshore wind farms are also used to assess the model in a large-scale wind farm scenario: results look promising although some reservations are expressed on the effect of the wake superposition model. Models for the prediction of wake centreline deflection due to yaw are compared at different yaw angles using wind tunnels data and CFD simulations. Although the EPFL model showed some advantage, especially in non-neutral conditions, both models give satisfactory results. Furthermore, it was showed that the wake deflection caused by the rotating wake for non-yawed turbines can have a large impact on the predictions of the centreline deflection.

1. Introduction

The need for fast and increasingly accurate wake models for wind turbines is promoted by the quest for higher accuracy in energy yield predictions, the increased size of wind farms, and by the need to exploit the potential for increasing wind farm energy production through active wake control methods which are currently being developed. The increased interest of the industry towards wind farm optimisation control algorithms, for example wake steering, highlights even more the importance of reliable wake models which can run fast enough for control design optimisation and testing.

Numerous efforts have resulted in various wake models available in literature: from the seminal works of Lissaman [1], Jensen [2] and Ainslie [3, 4], to the more recent works of Bastankhah & Porté-Agel [5] and Niayfair & Porté-Agel [6], Ishihara [7] and Gebraad [8, 9, 10]. These wake models have been developed and validated against wind farm production data, computational simulations and wind tunnel data. . Engineering wake models rely, to different extents, on wake measurements and subsequent fitting of empirical parameters, which exposes the models to the risk of over-fitting, potentially producing erroneous results when extrapolated to different configurations of turbine characteristics, wind farm layouts and atmospheric conditions.

More advanced models based on a Computational Fluid Dynamics (CFD) approach have been used in the recent years. From simulations based on Reynolds-Averaged-Navier-Stokes (RANS) equations and related turbulence models, to the more complex use of Large-Eddy-Simulations (LES) using Actuator-Line turbine models [11, 12, 13]. Their main common drawback is the need of large computational resources and time, which does not suit the market needs for a reliable prediction of wake losses during the planning phase of a wind farm, usually including different turbine models and configurations.

In the wind energy industry, it is common to use RANS simulations associated to the $k - \epsilon$ turbulence model and generally using an Actuator Disk Model (abbreviated as ADM-R or ADM-NR, where the suffix R indicates a rotating disk model and NR indicates a non-rotating disk model) to allow the calculation of loads and wind speed deficits due to the presence of the rotor. It is also noted that amendments to the $k - \epsilon$ model have been recently proposed for the use in wind turbine applications in [13]. A more computationally expensive approach, used for high-fidelity analysis, is Large Eddy Simulations (LES) in which only the eddies of smaller scales are modelled, whereas the larger eddies scales are numerically resolved using statistical techniques based on the Kolmogorov energy cascade analogy [14]. High fidelity simulations are usually associated with Actuator Line Models (ALM) considered to be more accurate in the calculation of body forces at the rotor [11].

Another important factor influencing turbine wakes is the atmospheric characteristics at the site and their change with time. The most commonly used wake models do not take atmospheric stability into consideration. However, some commercial wake models currently used in the offshore wind market such as FUGA [14] and FarmFlow [15], do use some simplified methods to estimate the effects of stability on the wake development. Deliverable 1.4 [16] of the CL-Windcon [17] project shows a comparison between different engineering wake models against a field test of a utility-scale wind turbine in different atmospheric conditions.

The lack of atmospheric stability corrections in common wake models used for long-term energy yield calculations can easily be explained: taking the average of a sufficiently long time of turbine operation, the non-neutral atmospheric conditions average out to give neutral atmospheric conditions, at least for sites where very stable or very unstable conditions are not predominant. What is more, the estimation of the occurrences of different stability characteristics at the site can be a difficult task, since requiring the measurement of specific wind characteristics.

1.1 Objectives and outline

The main objective of this paper is to propose a new approach for the wake modelling of predominantly stable or unstable atmospheric conditions, detailed in Section 3. The proposed model is a modification of the Ainslie model [4], taking the effects of atmospheric stability into account. This model does not introduce any additional experimental parameter hence not requiring additional calibration against experimental data, and only requires an estimate of the Monin-Obukhov length (or MOL) [14]. A validation of this model against a set of experimental and wind farm production data (detailed in Section 2) is also presented, and an overview of the methodology for atmospheric stability classification is provided. In addition to the above, another objective of this study is to increase confidence in the use of existing wake deflection models for the purpose of wind farm control and optimisation.

It is also highlighted that this paper will only focus on simplified engineering models and steady-state turbine wakes, therefore not considering dynamic affects such as wake meandering. All the tests presented in this paper have been performed using LongSim, a software created and used internally in DNV GL for advanced studies on wind farm steady-state and dynamic simulation, set-point optimisation and wake steering optimisation.

2. Experimental and numerical data used for the calibration of models

A set of different datasets have been used to perform wake model calibrations and validations. These are detailed in the following sections.

2.1 Wind tunnel measurements and setup

Wind tunnel experiments were carried out at the Politecnico di Milano between 2017 and 2018, as part of the CL-Windcon project [17]. The wind tunnel is a closed loop, boundary-layer type and different turbulence intensities and wind shear profiles can be reproduced. A turbine model with a rotor diameter of 1.1 m, named G1, was designed and built (a detailed description of the turbine model can be found in the public Deliverable no. 3.1 [18]). Experiments were carried out for an isolated turbine and for groups of turbines, but only the single-turbine results have been used in this paper. Free flow wind speed

IOP Conf. Series: Journal of Physics: Conf. Series **1222** (2019) 012004 doi:10.1088/1742-6596/1222/1/012004

at the turbine's hub height (0.825 m or 0.75 rotor diameters) is measured using a Pitot tube located 3 rotor diameters upstream of the turbine, and a full inflow-plane investigation was previously carried out to characterise the free flow (available in public Deliverable 3.4 [19]). Hot-wire probes were used to record flow measurements (with a sampling frequency of 2 kHz) which were averaged over a minute to obtain steady-state effects at 5, 7.5 and 10 rotor diameters behind the turbine location. All the tests were carried out with a mean wind speed of 5.5 m/s.

Two different flow conditions were reproduced in the wind tunnel experiments: the *offshore* condition is characterised by an averaged wind shear profile corresponding to a power law exponent of 0.08 and lower flow-wise turbulence, whereas the *onshore* wind shear condition corresponds to a power law exponent of 0.20 and higher flow-wise turbulence (a full description and characterisation of the wind tunnel experiments is presented in the public Deliverables [18, 19]). The standard deviation of the flow-wise free-stream wind speed component σ_U is shown in Figure 1, measured at the wind tunnel test section with no turbine models installed and measured at the location of the turbine rotor plane. The measured σ_U remains almost constant with height for the offshore case whereas it shows a noticeable decrease (for increasing height) for the onshore case across the rotor area. Turbulence stochastic models [20] generally assume that σ_U is constant with height. Although it is suggested to use the value of turbulence intensity averaged across the rotor, it is anticipated that this might cause some discrepancies between the wake models and the measured wake profiles.

Flow measurements were carried out both for a zero-yaw configuration and for different yaw angles, spanning from -40° to 40° with a 10° step. Detailed investigations have also been carried out on the measured time-series at different locations relative to the wind turbine, in particular to analyse the energy spectra, the estimated integral length scale and the measurement scattering. The spectral analysis (performed using a Welch approach with the help of a Hanning filter and a de-trending filter) has not shown unexpected features and the turbulent dissipation rate of the inertial subrange seems to agree well with the Kolmogorov law. These results are not shown here, but indicate that the turbulent flow development within the wake behaves as expected and shows no unexpected artefacts in the measurements provided, which are deemed of good quality.

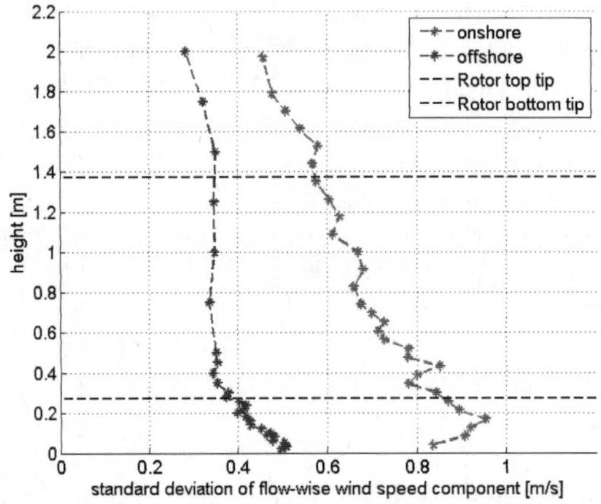

Figure 1. Variation with height of the standard deviation of the horizontal wind speed component for both offshore and onshore case, measured for the characterisation of the free-stream conditions in the wind tunnel at Politecnico di Milano.

2.2 CFD simulations

A set of time-averaged results from the open-source SOWFA computational solver [12] was provided by the National Renewable Energy Laboratory (NREL). SOWFA uses a detailed Actuator Line Model (ALM) to produce high-fidelity unsteady simulations of wakes and turbine loads. The simulation entailed a single prototype 5MW turbine with a rotor diameter of 126 m and a hub height of 90 m, with averaged free-stream wind shear profile with power law exponent of approximately 0.18, averaged inlet

IOP Conf. Series: Journal of Physics: Conf. Series **1222** (2019) 012004 doi:10.1088/1742-6596/1222/1/012004

wind speed of 8 m/s and turbulence intensity estimated to be approximately 10 % (therefore characterised by overall neutral flow conditions). The inflow conditions are generated in a precursor simulation run using a one-equation eddy-viscosity model for the definition of the subgrid scale, therefore the averaged wind characteristics can show variations in the free-flow.

2.3 Wind farms production data

For the purpose of comparing the proposed wake model to other wake models and test its applicability to a real case scenario, production data from one small onshore wind farm, Wieringermeer, and two large offshore wind farms, Horns Rev I and Nysted, were utilised.

The Wieringermeer Wind Farm site consists of 5 wind turbines installed in a test site in Netherlands. These are variable speed and pitch turbines of 80 m diameter, 2.45 MW rated power and hub height of 80 m. TNO provided the data as part of the UPWIND project and gave permission to DNV GL to use mast measurements and power turbine production data (for a total of approximately 3.2 years of available data) at a single turbine, namely T6: this is the second turbine from the west, being waked by turbine T5 for westerly winds. The atmospheric conditions at the site were measured at a mast located in the vicinity of the turbines and were made available for wind directions ranging approximately between 230° and 320°. The instruments installed on the mast include a thermocouple measuring the temperature difference between 37 m and 10 m heights, 3 sonic anemometers and 3 cup anemometers mounted at heights of 108 m, 80 m and 52 m, and two wind vanes at 79.2 m and 51.2 m. Further details about the on-site measurements are provided in [21]. As the individual pressure and temperature signals were not provided, these were retrieved from the ERA5 [25] reanalysis dataset at the node centred approximately 4 km to the south of the mast. Although the uncertainty given by such dataset are considered larger than the ones associated to direct site measurements, these are considered acceptable for the purpose of the estimation of the atmospheric stability at the site. The results presented in this paper show binned mast and turbine data using 2° wind direction bins, as well as filtering out wind speed outside the range 6-8 m/s. Also, as the purpose of this analysis is to compare the model to measured data where highly stable and unstable atmospheric conditions are present, only the summer months (June to August) were considered, and the dataset has been subdivided into daytime and night-time periods, respectively ranging between 7-18 hours and 20-5 hours.

Horns Rev I and Nysted are large offshore wind farms situated respectively 20 km off the west coast of Denmark and 15 km off the south coast of Lolland Island, Denmark. Details on the turbine model, hub height and inter-turbine spacing are detailed in Table 1. Both wind farms have an oblique rectangle layout as shown in Figure 2. The atmospheric stability conditions at the sites have been classified based on 12 months of data recorded at the Horns Rev I site masts and based on 3 years of data recorded at the Nysted site mast: these were for the UPWIND project [22] from the calculation of the Richardson number based on the temperature difference and wind speed difference at two heights. The stability classification for Horns Rev I is subdivided into seven categories (very unstable, unstable, neutral/unstable, neutral, neutral/stable, stable, very stable), whereas in Nysted six categories were used (very unstable, unstable, neutral/unstable, neutral/stable, stable, very stable): these are shown in Figure 2. The production data and the averaged wind conditions (mean wind speed, wind direction and

	Horns Rev I	**Nysted**
Turbine model	Vestas V80 2.0 MW	Bonus 2.3 MW
Diameter [m]	80	82.4
Hub height [m]	70	68.8
Inter-turbine spacing [D]	7	10.3 east-west 5.8 north-south
Available SCADA data	2 hours of data for each wind direction and wind speed bin	18 hours of data
Directional bins	255° - 285°, 5° step	263° - 293°, 5° step
Reference turbine	T07	A5

Table 1. Summary of characteristics of Horns Rev I and Nysted wind farms.

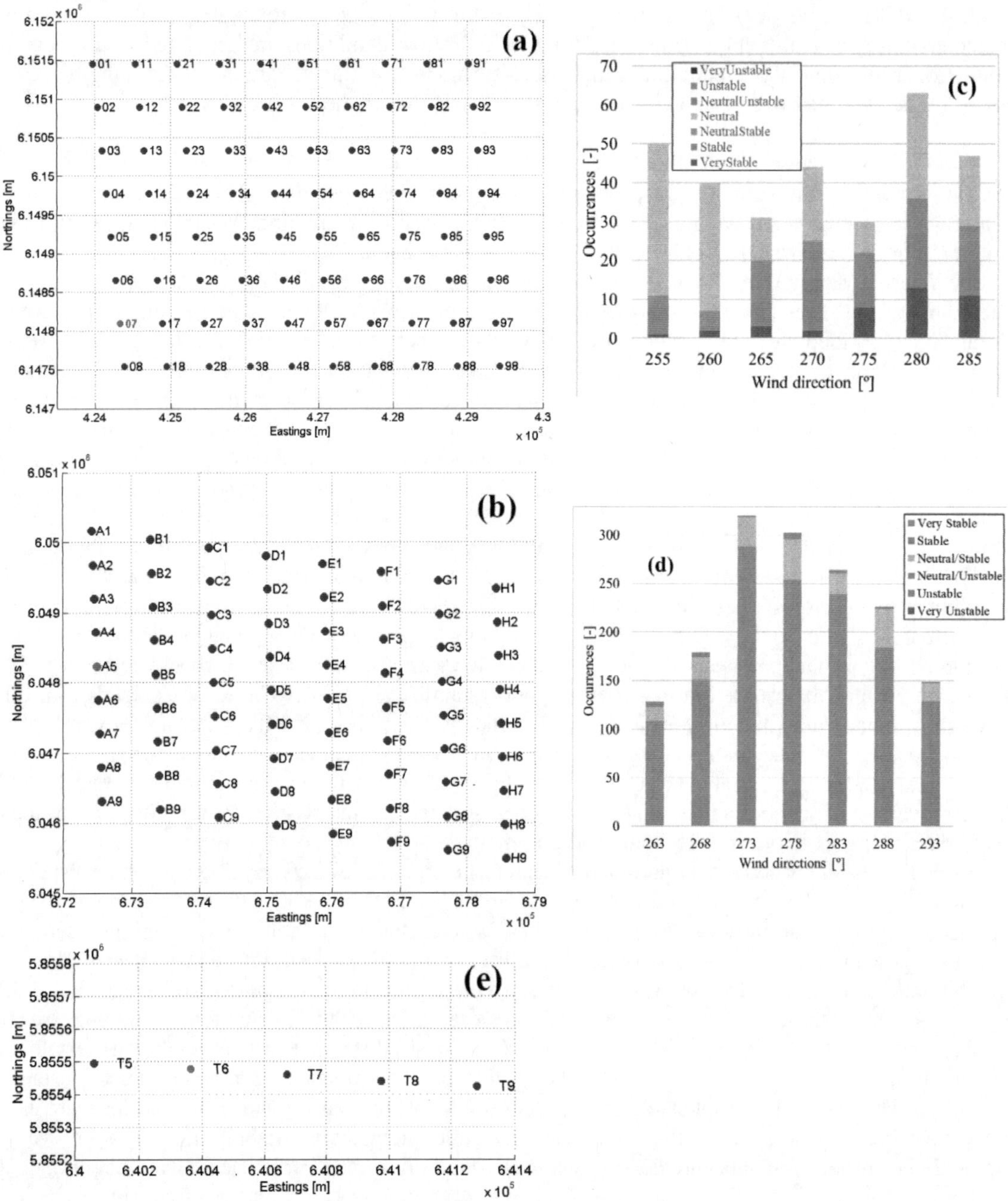

Figure 2. Layout of the Horns Rev I (a) and Nysted (b) offshore wind farms, and respective histograms of the occurrence of different stability classes (c.d) for the 10 m/s wind speed bin. Layout of the Wieringermeer wind farm (e). Reference turbines in red.

mean turbulence intensity) recorded at the sites have been provided for three wind speed bins (6±1 m/s, 8±1 m/s and 10±1 m/s) and 7 direction bins as detailed in Table 1. The power production data for each turbine was provided as normalised to the reference upwind turbine for each respective site. Details on the data used for the modelling of the wind farms' production are given in Section 4. Production and wind condition data for both Horns Rev I and Nysted wind farms have been provided as part of the UPWIND project [22, 23, 24].

3. Wake models

3.1 Wake deficit model

The Ainslie model is part of that class of turbine wake models called *Gaussian models* and it is sometimes referred to as eddy viscosity model (or EVM). It is obtained solving the Reynolds-Averaged-Navier-Stokes (RANS) equations in two dimensions and formulated in cylindrical coordinates, using the assumptions for incompressible, stationary and axisymmetric flow and neglecting gravity forces and pressure gradients. By using the axisymmetric flow and the Bousinnesq assumptions, the viscous forces can be simplified to

$$-u_x u_r = \frac{\mu}{\rho}\frac{\partial u_x}{\partial r} = \nu\frac{\partial u_x}{\partial r} \tag{1}$$

In equation (1), ρ is the air density, μ is the dynamic viscosity of the air, x and r are respectively the axial and radial coordinates, and u_x and u_r are respectively the axial and radial velocity components. Also, the kinematic viscosity (or momentum diffusivity) indicated as ν in the equation above can be expressed as $\nu = \nu_{\mathrm{amb}} + \nu_t$, where the first is the ambient kinematic viscosity (small for air in standard conditions, in the order of $1.5 \cdot 10^{-5}$ m^2/s and usually neglected), and where ν_t is the eddy viscosity term. Based on dimensional analysis, this term can be rewritten as: $\nu_t = k_1 l_w U_w + K_m$, where l_w is assumed to be of the order of the wake width (as defined by Ainslie, roughly two times the full-width-at-half-height, or FWHH, of the Gaussian self-similar wake shape), U_w is a velocity scale, K_m is the term for the ambient momentum diffusivity and k_1 is a constant which value, used throughout this study, is based on extensive internal validation from DNV GL. Using the assumption of isotropic turbulence, the ambient momentum diffusivity can be written $K_m = \frac{\kappa u_* z}{\Phi_m\left(\frac{z}{L}\right)}$ where $\kappa = 0.41$ (von Karman constant), Φ_m is the stability correction term (as a function of height, *z*, and Obukhov length, *L*) and $u_* = \sqrt{\tau_{ij}/\rho}$ is the friction velocity, where τ_{ij} indicates the shear stress. For a non-neutral flow, the friction velocity can obtained from the following equation of the non-neutral wind profile:

$$U = \frac{u_*}{\kappa}\ln\left(\frac{z}{z_0} + \Psi_m\left(\frac{z}{L}\right)\right) \tag{2}$$

with z_0 being the roughness length. The function Ψ_m is zero for neutral conditions and it can be defined as follows, according to Businger-Dyer [26] and Högström [27]:

- Stable conditions: $$\Psi_m\left(\frac{z}{L}\right) = -\frac{5z}{L} \tag{3}$$
- Unstable conditions: $$\Psi_m\left(\frac{z}{L}\right) = \ln\left[\left(\frac{1+x^2}{2}\right)\left(\frac{1+x}{2}\right)^2\right] - 2\,arctan\,x + \frac{\pi}{2} \tag{4}$$

where

$$x = \left(1 - \frac{19.3z}{L}\right)^{\frac{1}{4}} \tag{5}$$

We can rewrite the whole eddy viscosity term as follows, after normalising by $U_{\mathrm{HH}}D$ (where D is the rotor diameter):

$$\frac{\nu_t}{U_{HH}D} = k_1\left(\frac{b_w}{D}\right)\left(1 - \frac{U_C}{U_{HH}}\right) + \frac{\kappa^2\left(\frac{HH}{D}\right)}{\ln\left(\frac{HH}{z_0} + \Psi_m\left(\frac{z}{L}\right)\right)\Phi_m\left(\frac{z}{L}\right)} \tag{6}$$

In the formula above, b_w is the wake width and U_c is the wake centreline velocity. Also, note that no assumption that $HH \approx D$ is used and the stability correction term Φ_m is defined in the next section. Also, the filter function used by Ainslie for near-wake regions is set to F=1, hence this formula is valid only in the far-wake. Furthermore, using the approximation $\frac{u_*}{\kappa} \approx \sigma_u$ [23, 28], the logarithmic function above can be approximated to 1 / TI, where TI indicated the turbulence intensity. The proposed model will be referred in this paper as the *Stratified-EVM* wake model.

3.2 Experimental laws for atmospheric stability

Different experiments were carried out to describe the wind and temperature profiles in the atmospheric boundary layer (ABL), most notably the Kansas study [20], from which the Businger-Dyer experimental model was obtained [26], and some tall mast experiments [29] from which the Högström [27] and the Gryning [30] models originate. The Businger-Dyer relationship (3,4) is used in this study with the revised parameters by Högström. This defines the correction term Φ_m as follows:

- Stable conditions: $$\Phi_m = 1 + \frac{5z}{L} \tag{7}$$
- Unstable conditions: $$\Phi_m = \left(1 - \frac{19.3z}{L}\right)^{-0.25} \tag{8}$$

The set of equations above depend on the Monin-Obukhov length, L (or the Monin-Obukhov stability parameter z/L). This parameter can be difficult to estimate, as it changes throughout the day (and therefore should be seen rather as a distribution or a time-series) especially for sites with extreme weather conditions. Another common parameter used for the classification of stability in the atmosphere is the bulk Richardson number, which can be defined as:

$$Ri_B = \frac{g(\bar{\theta} - T_s)\Delta z}{T_s \Delta U^2} \tag{9}$$

where g is the gravitational acceleration constant, θ is the potential temperature, which can be defined as $\theta = T\left(\frac{P}{P_{ref}}\right)^{-\left(\frac{Rd}{Cp}\right)}$, where R_d is the gas constant for dry air and C_p is the specific heat constant for air, P is the atmospheric pressure, P_{ref} the pressure reference and T_s the temperature near the ground expressed in Kelvin.

The ranges of MOL, Obukhov stability parameter and bulk Richardson number typically used in the industry are summarised in the following table.

Stability class	Pasquille [31]	L [m] [14]	z / L [-] [32]	Ri$_B$ [33]
Very unstable	A	-100 < L < -50		Ri$_B$ ≤ -0.023
Unstable	B	-200 < L < -100	-1.4 < z/L ≤ -0.35	-0.023 ≤ Ri$_B$ < -0.011
Neutral/Unstable	C	-500 < L < -200		-0.011 ≤ Ri$_B$ < -0.0036
Neutral	D	|L| > 500	|z/L| < 0.35	-0.0036 ≤ Ri$_B$ < 0.0072
Neutral/Stable	E	200 < L < 500		0.0072 ≤ Ri$_B$ < 0.042
Stable	F	50 < L < 200	0.35 ≤ z/L < 7.0	0.042 ≤ Ri$_B$ < 0.084
Very stable	G	10 < L < 50		Ri$_B$ ≥ 0.84

Table 2. Summary of atmospheric stability classes used in literature, expressed using different metrics.

3.3 Turbine interaction

Every turbine affected by the wakes of a turbine upstream will experience inlet flow characteristics which cannot be approximated to the ones experienced in the free, undisturbed flow. Two main

IOP Conf. Series: Journal of Physics: Conf. Series **1222** (2019) 012004 doi:10.1088/1742-6596/1222/1/012004

characteristics are taken into consideration when modelling the wake at a turbine downstream: the added turbulence model and the superposition model.

The added turbulence model allows to estimate the turbulence intensity seen by the affected turbine. Two approaches are typically used: the Quarton-Ainslie model [34] and the Crespo-Hernandez model [35]. They model the turbulence intensity at the affected turbine as the combination of the free-stream turbulence intensity and an added turbulence, which is modelled as a function of turbulence intensity and thrust coefficient (in the case of the first model) or as a function of turbulence intensity and axial induced velocity (in the case of the second model). Moreover, it is noted that the Crespo-Hernandez model is considered valid only in the far-wake, assumed to start at 5 rotors diameters downstream of the turbine, whereas this limitation is not considered in the Quarton-Ainslie model.

Different superposition models are commonly used in order to obtain combined effects from superposition of turbine wakes, hence predicting combined wake velocity deficits and wake added turbulence. Examples of these models are dominant wake and sum-of-deficits models, however none of it has been shown to have a clear advantage in the literature (see [36, 37] for more details), and certainly none of these are based on proper physical laws, but rather mathematical or pragmatic laws.

The models for added turbulence intensity and turbine wake superposition are not discussed and validated in depth in this publication. However, it is worthwhile to stress how these models are fundamental for the overall correct modelling of wind-farm-wide wake effects. For instance, when additional deep-array effects are introduced, such as the large wind farm corrections used in Windfarmer [38], materially different predictions can be obtained when different superposition models are used. It is also underlined that different assumptions on added turbulence affect the wake dissipation of downstream turbines, which is important to keep in mind when comparing different models: for instance, the increased wake dissipation downstream due to the added turbulence is not considered in the model from Bastankhah & Porté-Agel [5], contrarily to the wake model used in Windfarmer.

3.4 Models for yawed rotors

The theory behind the flow characteristics for a yawed rotor has roots in the helicopter industry since used to calculate the rotor loads during forward flight. Glauert's correction for yawed rotors to the Rankine-Freude actuator disk model [39] and Coleman's study on skew deflection angle in the rotor near-wake (based on a vortex line approach) [40] are the main examples. More recently, two theories have been extensively used in literature to model the wake past yawed rotors: the Jimenez model [41] and the model proposed by Bastankhah & Porté-Agel (commonly referred to as the EPFL model, also in this paper) [5,6]. It is worth mentioning that other, more recent studies have appeared in literature, trying to study the wake deflection effect from a vorticity point of view [42, 43], but these lie outside the scope of this paper. The Jimenez model is based on a momentum balance approach and does not take into account the tangential losses caused by the three-dimensional characteristics of the wake flow, as well as any effect from atmospheric stability; it is therefore considered to overestimate the wake deflection. Moreover, it uses one empirical parameter which needs to be calibrated against measurements, even if a link between this parameter, K_d, and the turbulence intensity has been suggested in [28].

The EPFL model uses two Gaussian distributions aligned to the vertical and lateral axes (with respect to the plane normal to the flow direction) to allow for different velocity deficit distributions in the two axes. After approximating the wake growth rate within the far-wake region as linear and estimating the length of the potential core, used to define the inlet boundary conditions for the far-wake region (based on Coleman's approximations for the wake skew angle), it is possible to calculate the lateral wake deflection due to a given yaw angle, depending on both the incoming turbulence intensity and the thrust coefficient. The EPFL model needs two empirical parameters for the wake velocity deficit calculations and two more for the lateral deflection calculations: these were estimated from wind tunnel experiments and LES simulations [5]. However, more validation on real-scale turbines would be beneficial to obtain reliable sets of empirical parameters to be used for different classes of turbines. Furthermore, it is worth commenting on the importance of the definition of the velocity and thrust vectors in the different models, which might differ from one another. For example, in the Ainslie model, the wind speed is defined as perpendicular to the rotor plane, and this will need to be taken into account when calculating the thrust coefficient for a yawed rotor, as the final formula will depend on which direction the velocity and the thrust force vectors are projected onto.

Additionally, the effects of wake centreline displacement due to the combined effect of wake rotation and non-uniform inflow shear have been described in [44] and these are discussed in the test cases used for this study.

4. Results and discussion

4.1 Wind tunnel tests

4.1.1 Wake velocity deficit comparisons (zero yaw) for an isolated turbine.

Comparison between the measured, steady-state wake velocity deficit plots and the wake model predictions are performed at a horizontal plane at hub height (corresponding to 0.75 rotor diameters). The standard Ainslie model, the EPFL model and the Stratified-EVM model are compared to the measurements at the three downstream planes, as shown in Figures 3 to 5.

In Figure 3, the standard Ainslie model is shown not to fit the experimental data. The EPFL model is compared to the wind tunnel measurements in Figure 4. A considerable difference between the plots labelled (a,b) and the ones labelled (c,d) is visible. By using the EPFL parameters originally obtained by Bastankhah & Porté-Agel [5], the wake model manages to fit reasonably well the measurements with low turbulence (a), whereas it performs poorly in the other test case with higher turbulence intensity (b). A new set of parameters was optimised from the measurements using a cost-minimising function approach during the CL-Windcon project [16]. As it can be seen in plots (c, d) this new set of parameters improves the model fit noticeably. Although the optimisation of the set of parameters could constitute a good mathematical approach to calibrate the wake model for any specific case, it could be unfeasible when used for wake assessments where wake measurements are not available. The two sets of parameters are shown in Table 3.

	α	β	k_a	k_b
Proposed in [5]	2.32	0.154	0.3837	0.003678
Optimised in [16]	1.2973	0.2375	0.1177	0.0194

Table 3. Sets of parameters used in the EPFL model, used for the wind tunnel comparisons

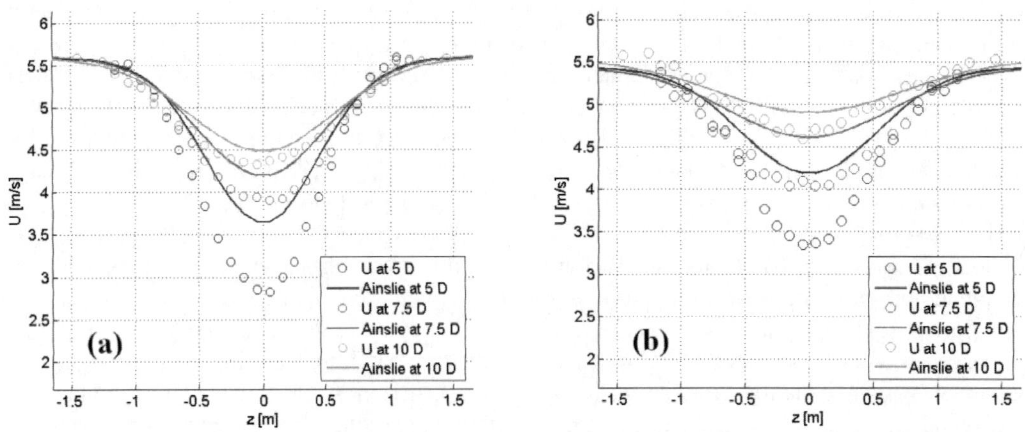

Figure 3. Standard Ainslie model applied to the wind tunnel tests, both for the offshore (a) and onshore (b) flow case.

The *Stratified-EVM wake model* is compared to the same test case, as shown in Figure 5. Based on the broad analogies between the inflow conditions and the different stability classes in the literature (see

Table 2), different Obukhov lengths were used to fit the measurements. For the offshore case, characterised by a low turbulence intensity and predominantly stable conditions, a MOL of 200 m has been found to give the best fit with the measurements at the most upstream plane. An Obukhov length characteristic of stable conditions, however, results in the overprediction of the wind speed deficits at the two most downstream measurement planes, for which neutral characteristics were imposed to obtain a reasonable fit with the data, using a MOL of 2000 m. For the onshore case, instead, using a MOL of 450 m representative of slightly stable conditions allows to obtain a reasonable fit to the experimental data at the three measurement planes, with a small wind speed deficit underprediction for the most upstream measurement plane and a larger wind speed deficit overprediction for the most downstream measurement plane. As it can be observed comparing the onshore and the offshore test cases, the wake centreline wind speeds at the two more downstream planes are similar. This might suggests that the wake in the offshore case tends to recover quickly to the same wind speeds observed in the onshore case, which might explain the need to use different Obukhov lengths for different measurement planes in order to obtain a good fit for this test case. This might also explain the not satisfactory fit found using the standard Ainslie model for both the offshore and the onshore cases.

It is noted that the MOL values mentioned above are kept as an approximated round number, and a best fit value is not calculated using optimisation techniques in order to avoid overfitting. Due to the nature of the wind tunnel tests, and hence the lack of proper atmospheric stability classification, it is not possible to obtain a precise value for the MOL. Moreover, as the atmospheric boundary layer and the turbine model are scaled down in this experimental setup, the MOL has been also scaled down by a factor of 100, under the assumption that the turbine model in question is the (approximately) 1:100 scale reproduction of a utility-scale turbine.

In conclusion, the uncertainties associated with the behaviour of the flow and the atmospheric stratification characteristics recreated in the wind tunnel do not allow to use this data to fully validate the model. Nevertheless these tests show that when an Obukhov length in line with the indicative atmospheric characteristics recreated in the wind tunnel is chosen to modify the eddy viscosity term in the Ainslie model, this can be used to obtain a satisfactory prediction of the turbine wake. In order to gain more confidence in the proposed model, this will be compared against additional test cases in this paper, including utility-scale operational wind turbines.

4.1.2 Wake centreline deflection due to rotor yaw, for an isolated turbine

The measurements for the non-zero yaw cases were analysed and compared to the Jimenez and EPFL models, as detailed in Section 2. As described above, it is known in the literature that the combination of a rotating turbine wake and a non-uniform wind shear profile deflect the wake centreline even when the turbine's yaw angle is zero. Analysing the wind tunnel measurements, no clear indication of such deflection was found and therefore it is not considered in the results presented in this section.

Different k_d parameters were tested in order to obtain a closer fit from the Jimenez model to the wind tunnel measurements. A value of 0.1 was deemed to be the optimal value. This value is lower than the other ones found in literature for utility-scale wind turbines [8, 41], which has the effect of increasing the centreline deflection. When using the EPFL model, the four optimised parameters for this wind turbine model test will be used for these comparisons. The measurements are compared to the Jimenez and the EPFL models in Figure 6 for negative yaw angles. The Jimenez model better fits the measurements for the onshore test case, whereas the largest discrepancy is obtained for the largest absolute yaw angles in the offshore test case. A possible explanation to this could be the lack of any relationship to the variation of atmospheric stability within this model: as the flow becomes more stable, the overall wake deflection will be larger due to the reduced effect of flow mixing and entrainment along the wake. This indicates that in order to obtain a better fit for specific atmospheric conditions using the Jimenez model, specific k_d parameter should be chosen for each case, which might be unpractical. The EPFL model shows smaller deflection discrepancies compared to the Jimenez model: this is particularly true for the offshore test case. Although a large discrepancy is noticeable for the onshore test case at a

IOP Conf. Series: Journal of Physics: Conf. Series **1222** (2019) 012004 doi:10.1088/1742-6596/1222/1/012004

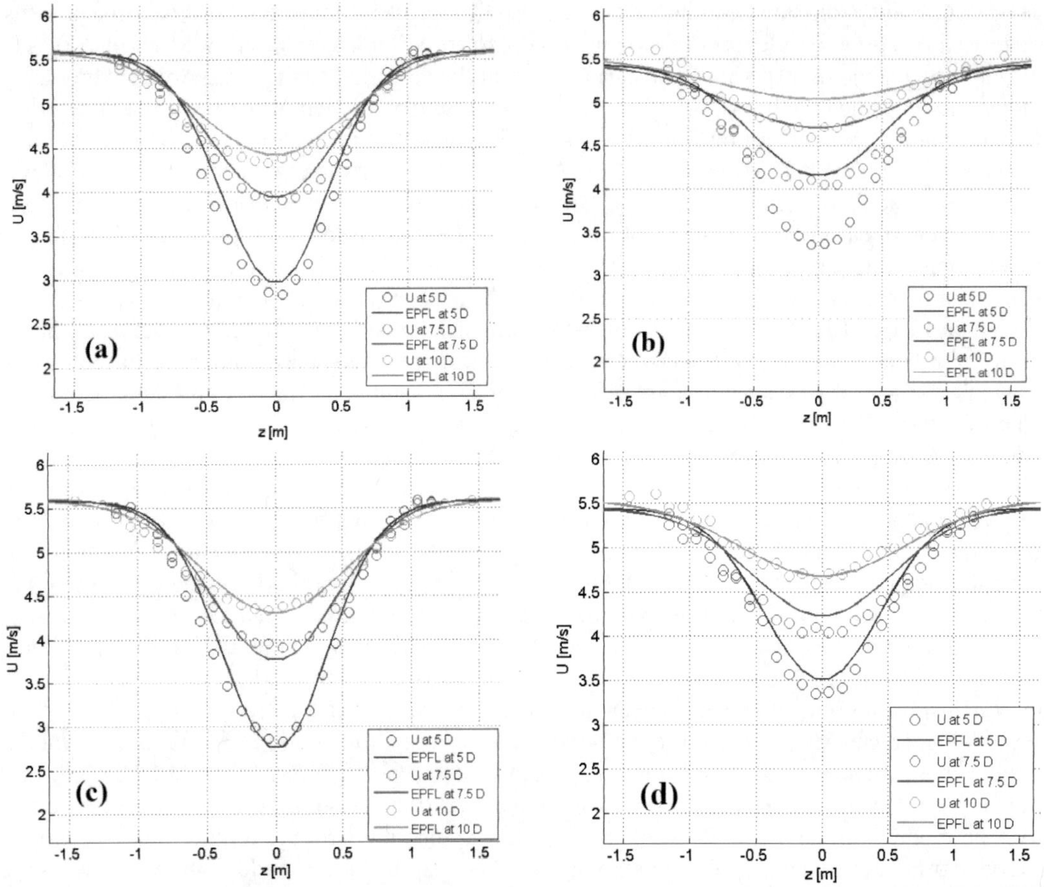

Figure 4. EPFL model applied to the wind tunnel tests, both for the offshore (left) and onshore (right) flow case. The set of parameters proposed in [5] have been used in (a-b), whereas the set of parameters proposed in [16] are used in (c-d).

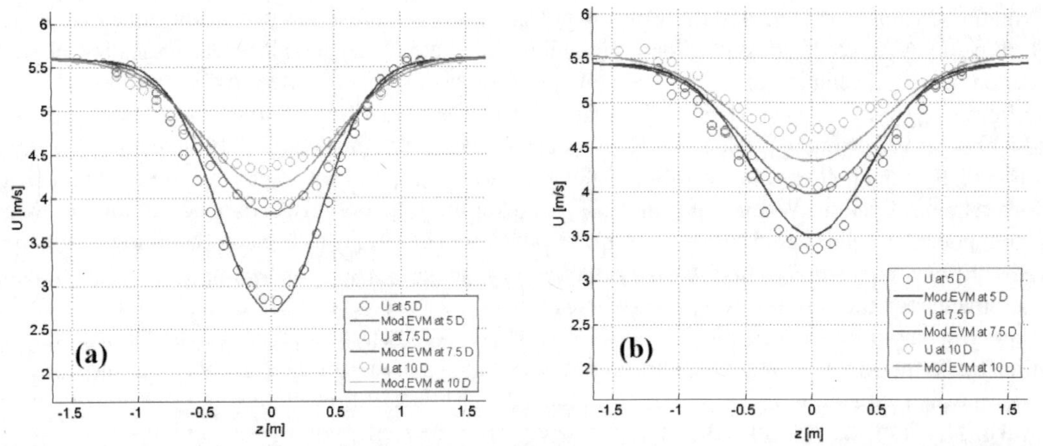

Figure 5. The *Stratified-EVM wake model* applied to the wind tunnel tests, both for the offshore (a) and onshore (b) flow case.

yaw angle of -40°, this is considered to be of lesser importance at least as far as wake steering applications are concerned, as smaller angles are usually used due to the risk of large increases in loads.

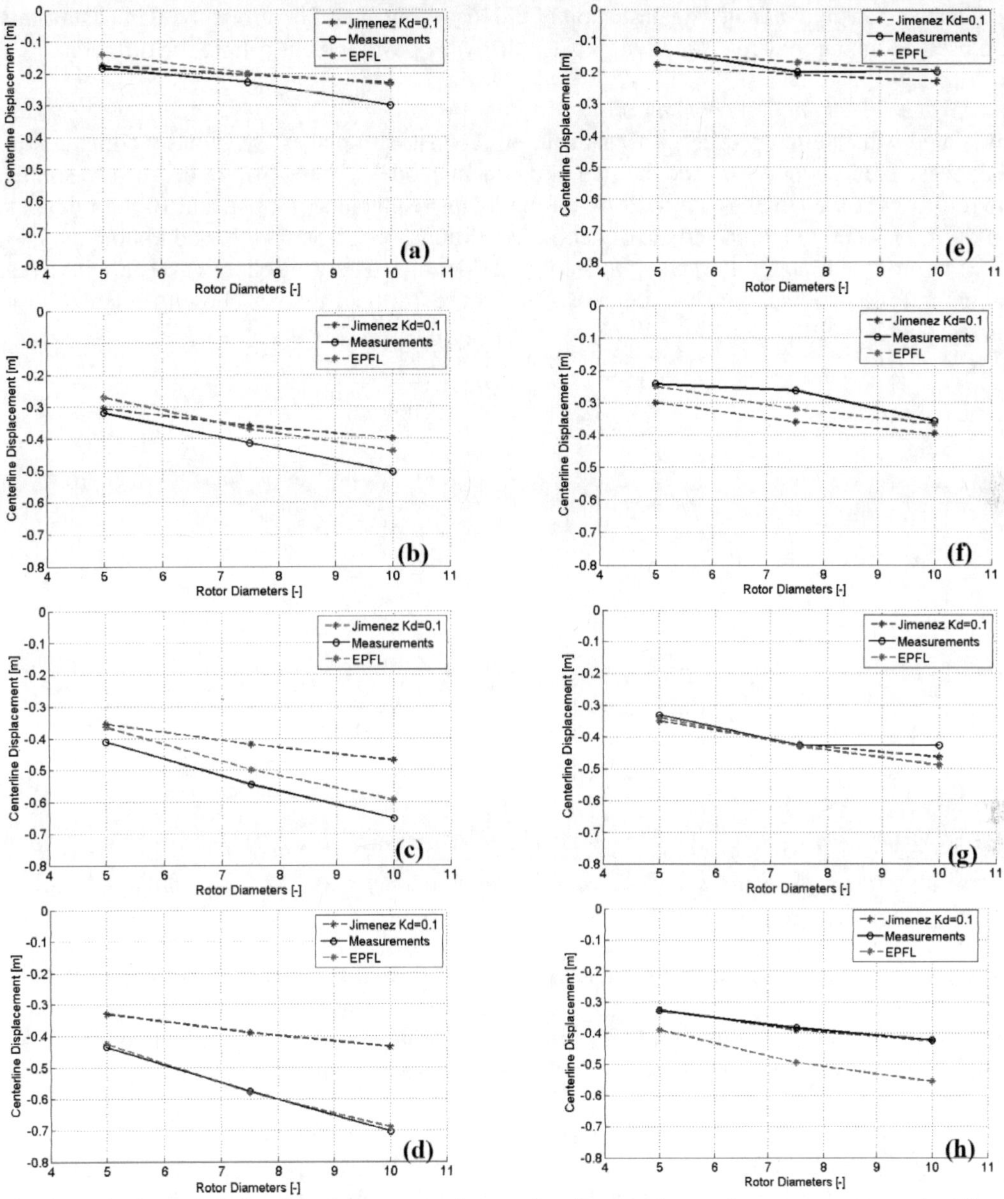

Figure 6. Wake lateral deflection due to yawed rotor for the wind tunnel test case, comparing Jimenez and EPFL models. (a,b,c,d) are for the offshore case, respectively at -10°, -20°, -30°, -40°; (e,f,g,h) are for the onshore case, respectively at -10°, -20°, -30°, -40°.

Although both models compare reasonably well to the measurements, it is highlighted that additional validations are needed to increase confidence on the parameters to be used for real-world applications, in both models.

4.2 CFD simulations

In this section, data from CFD simulations at different yaw angles are compared against the two Jimenez and EPFL wake deflection models. It is worth underlying these CFD results are not representative of any field test and are not validated against actual field measurements. Therefore, the CFD data analysed here are used assuming they realistically represent the wake development behind the turbine.

4.2.1 Wake velocity deficit comparisons

The simulated wake deficit profiles at four different downstream planes are initially compared to the proposed wake model (Figure 7a) and the standard Ainslie model (Figure 7b), for the turbine simulation with no rotor yaw. Based on the comparisons at four downstream planes (4, 6, 8 and 10 rotor diameters), using a MOL of -1000 m (characteristic of neutral conditions, see Table 2) allowed to obtain a good fit between the proposed model and the simulated wake deficits. The centreline wake deficits from the simulation and the two wake models are shown in Figure 7b: both models show to predict the wake deficits reasonably well.

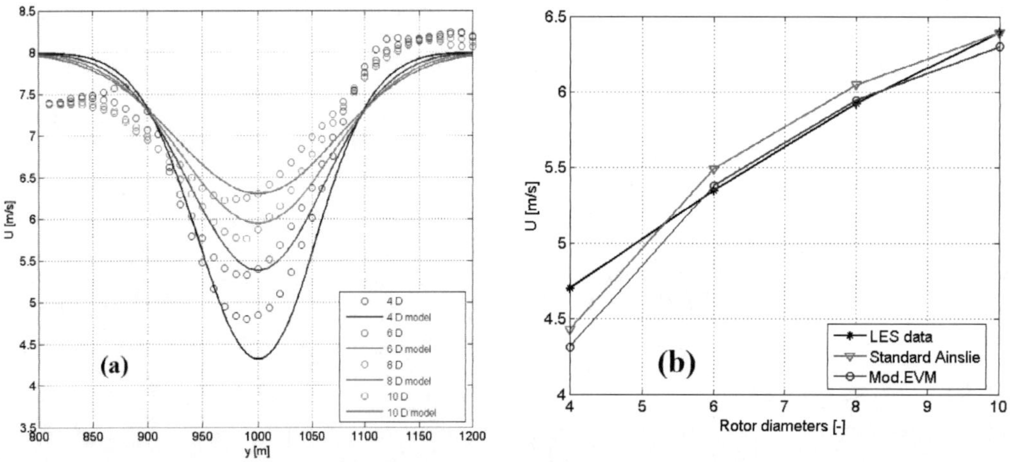

Figure 7. Wake velocity deficit comparisons for the CFD test case, shown at 4, 6, 8 and 10 rotor diameters downstream: (a) *Stratified-EVM wake model* with MOL=-1000 m, (b) Standard Ainslie model and *Stratified-EVM wake model.*

It is visible from Figure 7 how the centreline of the velocity deficits tends to shift towards the left for increasing distance downstream: this is caused by the combined effect of the rotating turbine wake and the non-uniform inflow wind shear profile. In the next section, it is shown how the knowledge of this component of wake displacement allows to obtain better wake deflection predictions.

4.2.2 Wake centreline deflection due to rotor yaw, for an isolated turbine

Both the Jimenez and the EPFL models were used to predict lateral wake displacement due to yawed rotor for this CFD test case in neutral flow. Simulations of the same turbine analysed above have been carried out at the following yaw angles: -25°, -20°, -15°, 10°, 15°, 20°, 25°.

The parameters originally proposed by Bastankhah & Porté-Agel [5] have been used in these tests for the EPFL model, whereas different values of k_d have been used in the comparisons in Figure 8 and Figure 9. In order to take into account the wake centreline deflection for non-yawed rotor visible in Figure 7a, the centreline deflections from the LES data were linearly summed to the deflection component obtained for the no-yaw case, an approach which is also used in the software FLORIS [9, 10]. The plots in Figure 8 and 9 show both the deflections obtained from the LES data (indicated simply as 'LES' in the plots) and the summed deflection components (indicated with 'LES + Linear deflection').

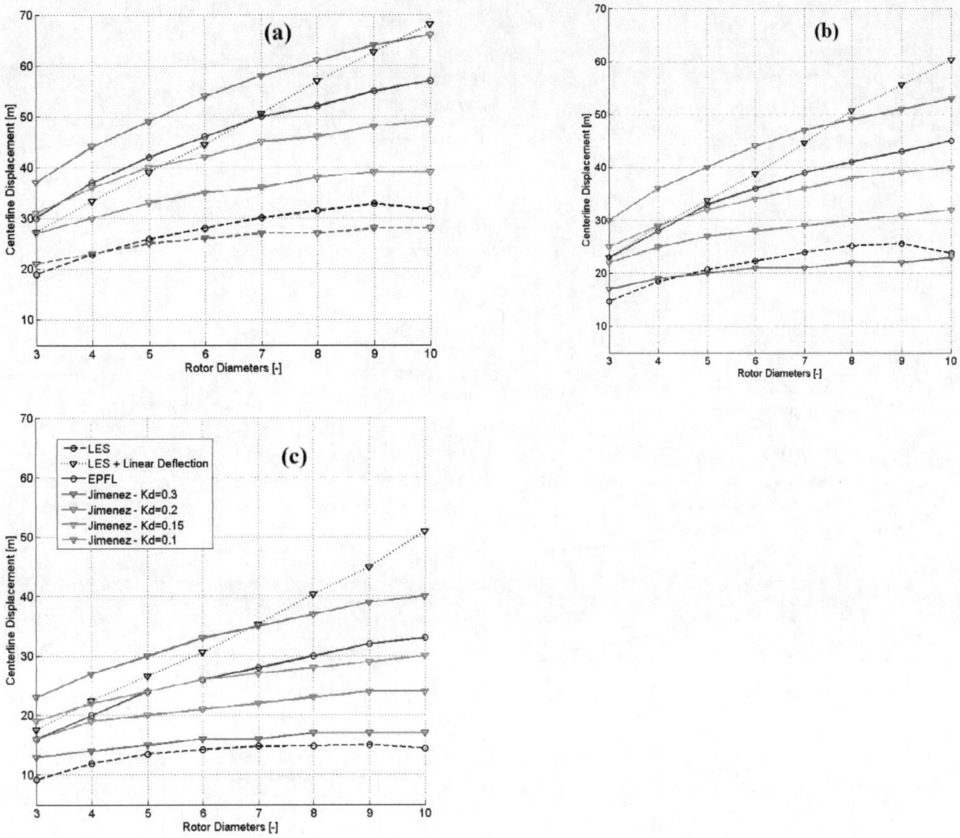

Figure 8. Wake lateral deflection for the CFD test case, shown for yaw angles -25°(a), -20° (b), -15° (c) and including the EPFL model and the Jimenez model (with different k_d parameters).

The plots clearly show that if the no-yaw component of centreline deflection is not taken into account, the EPFL model largely overpredicts (for negative yaw angles) and underpredicts (for positive yaw angles) the wake deflection. The Jimenez model is shown to be able to better approximate the wake deflection downstream of the turbine, however very different values would need to be used for the k_d parameter for positive and negative angles.

When the effect of the non-yaw deflection component is removed via linear sum, it can be seen how both the EPFL model and the Jimenez model better approximate the data from the simulations. Particularly for positive yaw angles, both the EPFL model and the Jimenez model (with a k_d parameter of 0.15) predict well the wake deflection downstream of the turbine for all the angles used in this investigations. When the yaw angles are negative, an acceptable fit of the two models with the data is visible only for the -25° and -20° yaw angle cases and up to approximately 5 rotor diameters downstream of the turbine.

These comparisons show how both the models used for this test case can predict reasonably well the wake deflection in neutral conditions, although the centreline deflection for no-yaw needs to be estimated to give reliable results. To the best knowledge of the authors, no low-fidelity or medium-fidelity models are available in the literature to help predict this deflections effect.

4.3 Wind farms production data

In order to further increase confidence in the described methodology, wind farm production data is used in this section to perform comparisons with the proposed model.

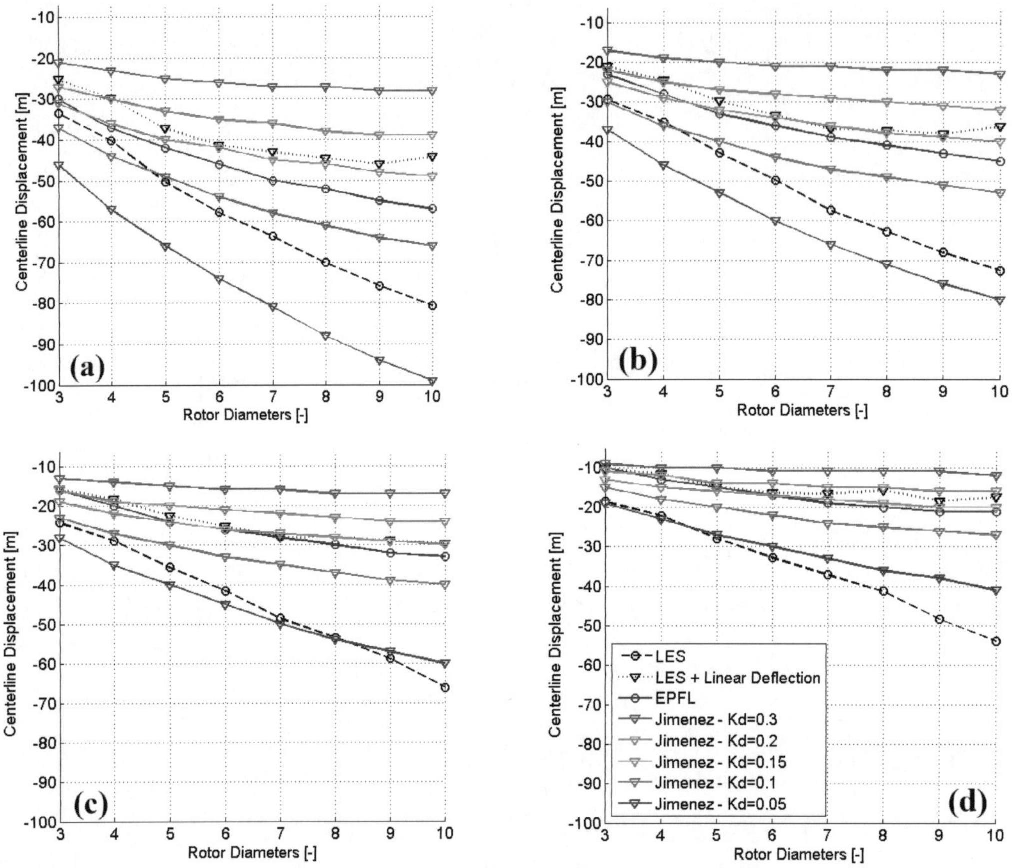

Figure 9. Wake lateral deflection for the CFD test case, shown for yaw angles 25°(a), 20° (b) 15° (c) 10° (d), including the EPFL model and the Jimenez model (with different k_d parameters).

4.3.1 Wieringermeer onshore wind farm

This dataset was used to test the *Stratified-EVM* wake model against on-site meteorological mast data and SCADA data at a site where the effects of atmospheric stability and instability are not negligible. The power production data for Turbines T06 are normalised by the production of the leading turbine for westerly winds, T05. The data is binned for the specific wind speed range 6-8 m/s and was also binned for wind directions every 2°, between 230° and 320°. As a first test, the whole production dataset was compared to the proposed wake model using a range of MOL values characteristic of neutral atmospheric conditions: as it can be seen in Figure 11(a), a good fit is obtained using an MOL ranging from 750 m to 1000 m, and the results from the Windfarmer model are also superimposed for reference. In order to compare the model to measured data where highly stable and unstable atmospheric conditions are present, only the summer months (June to August) were considered, and the dataset has been subdivided into daytime and night-time periods (respectively ranging between 7-18 hours and 20-5 hours). It is expected that the summer-daytime atmospheric characteristics exhibit signs of unstable conditions, whereas the summer-night-time atmospheric characteristics exhibit signs of stable conditions. In order to estimate the on-site atmospheric stability, the bulk Richardson number was calculated for each 10-minute averaged record from the mast signals and its distribution is shown for both daytime and nigh-time in Figure 10. Binned wind speeds, binned wind directions and binned turbulence intensities were used as model inputs for each directional bin and were compared to the measured, filtered data. The comparison shows how the daytime and night-time filtered production data at Turbine T06 exhibit different wake effects due to the predominance of unstable and stable atmospheric effects, respectively. The median bulk Richardson number for each distribution has been calculated and empirical formulas have been used to relate them to the correspondent Obukhov lengths, as described in Holtslag [29]. The results obtained from the *Stratified-EVM* wake model are shown in Figure 11(b): the model appears to be able to correctly predict the power deficit due to wake of the preceding turbine for both predominantly stable and unstable conditions.

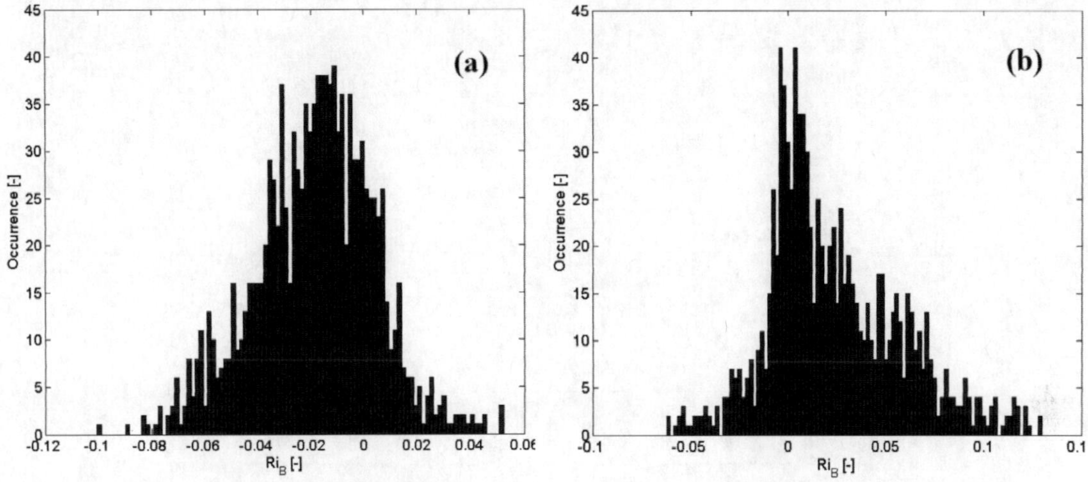

Figure 10. Distribution of bulk Richardson number for daytime(a) and night-time (b), both for the summer period, at the Wieringermeer site.

4.3.2 Horns Rev I and Nysted offshore wind farms

Turbine production data at the Horns Rev I offshore wind farm, provided as part of the UPWIND project [22], have been used to further validate the proposed *Stratified-EVM wake model* against a real-world scenario. In order to model the production for all the wind turbines of the wind farm, the choice of the added turbulence model and superposition model required particular attention. The Crespo-Hernandez model [35] was shown to perform well on previous wake model validations based on a large offshore wind farm, like in [6]. Since preliminary tests indicated this added turbulence model performed better than the Quarton-Ainslie model for the specific test setup, and since the relatively large inter-turbine spacing at the Horns Rev I wind farm falls within the indicated range [35], this model was selected for this test case. Both the dominant wake and the sum-of-deficits superposition models have been tested on this wind farm validation exercise and the latter has shown to have a noticeable improvement in the turbine row power predictions (this is also in accordance to what is shown by Niayifar & Porté-Agel [6]). Based on the available information on the stability classification at the site and the occurrence of each stability class, the *Stratified-EVM wake model* was run using three different MOL setting and the results were weighter-averaged according to the neutral, stable and unstable occurrences at the site, as shown in Figure 2. The measured power production for each turbine row is compared to the modelled power production in Figure 12(a). The comparison is shown for the specific 10±1 m/s wind speed bin. The same added turbulence model and wake superposition model used for the Horns Rev I wind farm above, were applied to the Nysted offshore wind farm too. Measured power production for each turbine row and representative of the 10±1 m/s wind speed bin is compared to the modelled power production in Figure 12(b). For this specific test case, the results are the weight-averaged of two runs of the proposed model, to better follow the provided stability classification at the site (i.e. one run using an Obukhov length representative of all the stable and neutral/stable subclasses, and another using an Obukhov length representative of all the unstable and neutral/unstable subclasses).

The power production predicted at the two sites using the DNV GL's Windfarmer wake model is also shown for comparison in Figure 12, which fits the production data better than the proposed model.

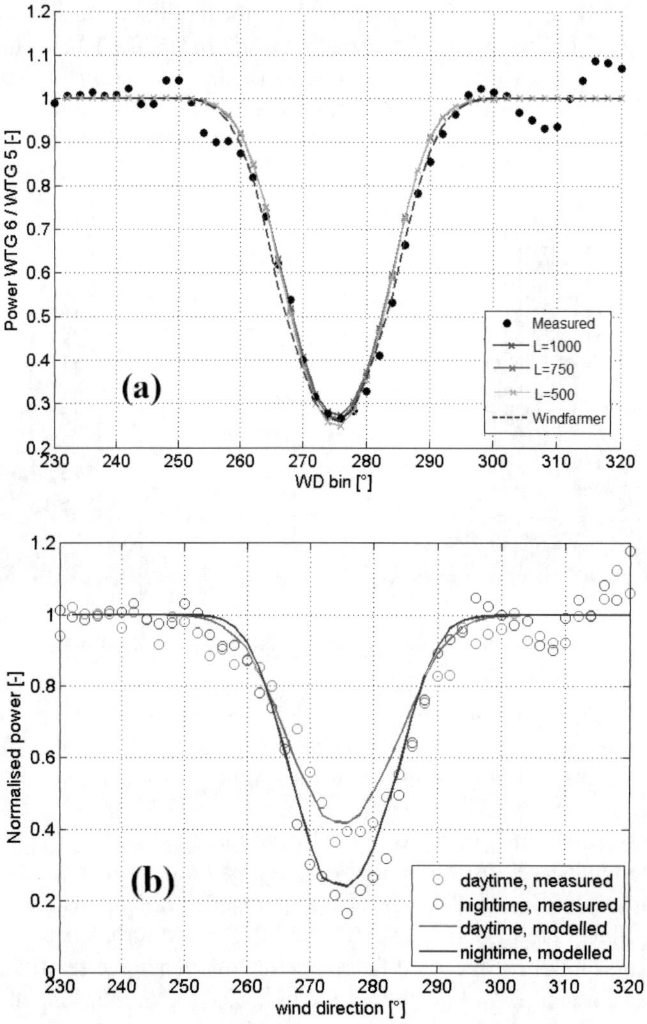

Figure 11. Wieringermeer: power production at Turbine T06, normalised by power at turbine T05, shown for filtered and binned data. (a) neutral case, (b) stable and unstable cases.

Nevertheless, the proposed model is deemed to perform reasonably well especially for the first turbine rows. The largest discrepancies between the production data and the proposed model are found in the last turbine rows in both sites. It is noted that the turbine production data and the atmospheric classification are provided as averaged values, therefore it was not possible to make a time-series-based comparison by filtering and isolating stable and unstable occurrences to be compared against the model (as it was done for the Wieringermeer test case). As the greatest discrepancies are visible in the last wind farms' rows, it is suspected the choice of a specific wake superposition model can heavily influence the results, and it is therefore considered by the authors as an area where further investigations are needed.

5. Conclusions

A wake model based on a formulation of the Ainslie model for stratified flow has been proposed in this work: the eddy viscosity parameter characterising the Ainslie model was modified to take into account the effects of atmospheric stability. This model has been initially calibrated against wind tunnel tests carried out for two different inflow conditions (characteristic of neutral onshore and stable offshore

IOP Conf. Series: Journal of Physics: Conf. Series **1222** (2019) 012004 doi:10.1088/1742-6596/1222/1/012004

scenarios), and against the results from an LES simulation of a utility-scale wind turbine in neutral atmospheric conditions. The proposed model showed to be able to predict reasonably well the wake deficits for these test cases, once an appropriate Obukhov lengths characteristic of specific atmospheric stability conditions is selected. Due to the nature of the wind tunnel tests and the LES simulations, it was not possible to obtain an estimate of the Obukhov length from the data. Nevertheless, it is shown that when a value of Obukhov length is chosen in line with the indicative atmospheric characteristics of each test case, hence modifying accordingly the eddy viscosity term in the Ainslie model, this can be used to obtain a satisfactory prediction of the turbine wake.

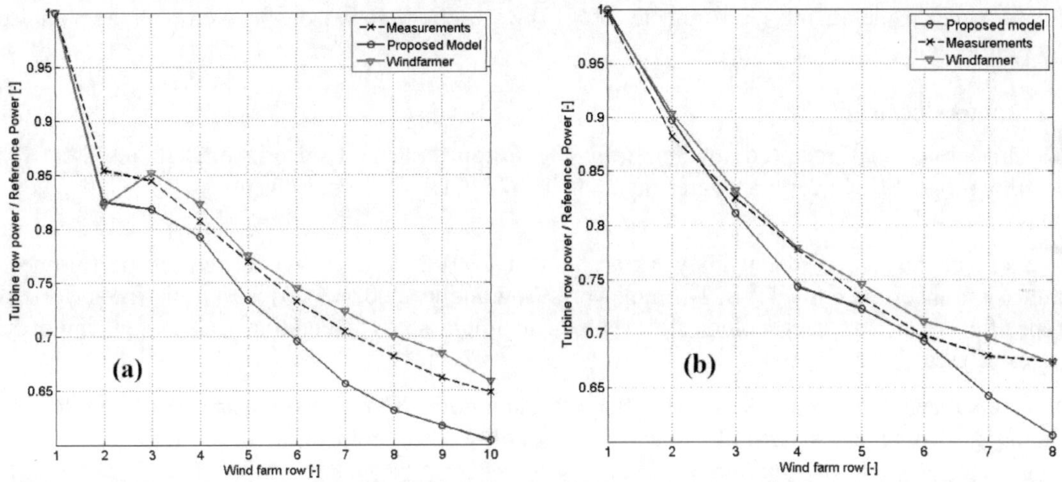

Figure 12.s Measured and modelled turbine power averaged for each turbine row, normalised by reference power. (a) Horns Rev I wind farm, (b) Nysted wind farm.

Production data from three wind farms have also been used to increase confidence in the validity of this model. The Wieringermeer wind farm was used to further validate the model for neutral and non-neutral atmospheric conditions. The proposed model was calibrated against the whole production dataset characterised by neutral atmospheric conditions, and a MOL ranging between 750 m and 1000 m allowed to predict well the production at the test turbine. Furthermore, the measured production data and the meteorological data were further filtered to isolate daytime and night-time data during the summer months. The bulk Richardson number was estimated using the filtered data and a correspondent Obukhov length was obtained for both time periods. The use of a characteristic Obukhov length on meteorological measurements for both time periods allowed to obtain a good agreement between modelled and measured power production.

In order to test the proposed model against large wind farms, production data from two large offshore wind farms (Horns Rev I and Nysted) were used. The proposed model was used alongside the Crespo-Hernandez model for added turbulence and the sum-of-deficits wake superposition model. The stability classification at these sites was available as probability of occurrence for each considered wind speed and direction bin, therefore a weighted average approach was used to combine simulations representative of neutral, stable and unstable conditions, using representative Obukhov lengths. The proposed model is deemed to fit reasonably well to production data especially for the initial turbine rows, although the wake model used in DNV GL's Windfarmer is shown to better fit the production data. As the greatest discrepancies are visible in the last wind farms' rows, it is suspected the choice of a specific wake superposition model can heavily influence the results, and it is therefore considered by the authors as an area where further investigations are needed.

Additionally, the EPFL and the Jimenez models for wake deflection have been compared using both the wind tunnel and the high-fidelity CFD simulation test cases. For the wind tunnel experiments, which were carried out for two different flow conditions, the EPFL model was found to better predict the wake deflections compared to the Jimenez model, especially for the experiments carried out with low ambient turbulence. However, it is found that both models heavily rely on numerical parameters and additional validation is required to increase the confidence in their tuning. The data from the LES simulations exhibited a clear wake centreline deflection for non-yawed rotor, an effect which is caused by the rotating wake and the impinging non-uniform wind flow. It was shown in this study that when the effect of this deflection component is not addressed, both the EPFL and the Jimenez models do not predict well the deflections downstream of the turbine. After the non-yaw deflection component is linearly added to the overall deflection obtained from the simulation, it was noticed how both the EPFL model and the Jimenez model (using a k_d parameter of 0.15) were able to fit reasonably well with the obtained deflections.

6. Acknowledgements

This project has received funding from the European Union's Horizon 2020 research and innovation programme under grant agreement no. 727477 (CL-Windcon, website: www.clwindcon.eu).

Some of the data used for this work are part of UPWIND, a project funded by the European Commission under the 6th (EC) RTD Framework Programme (2002- 2006) within the framework of the specific research and technological development programme "Integrating and strengthening the European Research Area".

The authors thank the National Renewable Energy Laboratory (NREL) and in particular Paul Fleming for kindly providing results from LES simulations of utility-scale prototype wind turbines.

References

[1] Lissaman P B S 1979 Energy Effectiveness of arbitrary arrays of wind turbines, *Journal of Energy*, **3**(6):323–8

[2] Jensen A 1983 Note on wind generator interaction, *Risø DTU*

[3] Ainslie J F 1985 Development of an eddy viscosity model for wind turbine wakes, *Proc. 7th British Wind Energy Association conference*, Oxford

[4] Ainslie J F 1988 Calculating the flow field in the wake of wind turbines, *Journal of Wind Engineering and Industrial Aerodynamics*, **27**(1–3):213–24

[5] Bastankhah M, Porté-Agel F 2016 Experimental and theoretical study of wind turbine wakes in yawed conditions, *J. Fluid Mech.*, **806**, pp 506-541

[6] Niayifar A, Porté-Agel F 2015 A new analytical model for wind farm power prediction, *J. of Physics*, 625, 012039

[7] Ishihara T, Qian G W 2018 A new Gaussian-based analytical wake model for wind turbines considering ambient turbulence intensities and thrust coefficient effects, *Journal of Wind Engineering and Industrial Aerodynamics,* **177**, 275–292

[8] Gebraad P M O, et al. 2016 Wind plant power optimization through yaw control using a parametric model for wake effects a CFD simulation study, *Wind Energy*, **19**:95–114

[9] NREL, FLORIS software, http://github.com/wisdem/floris

[10] TU Delft, FLORISSE software, https://github.com/TUDelft-DataDrivenControl/FLORISSE_M103

[11] Sanders B, et al. 2011 Review of computational fluid dynamics for wind turbine wake aerodynamics, *Wind Energy*, **14**(7):799–819

[12] Churchfield M, et al. 2014 Overview of the simulator for wind farm application (SOWFA), https://nwtc.nrel.gov/system/files/SOWFA_tutorial_05-20-2014.pdf

[13] van der Laan M, Sørensen N, et al. 2015 An improved k-ϵ model applied to a wind turbine wake in atmospheric turbulence, *Wind Energy*, **18**(5):889–907

[14] Ott S, Nielsen M 2014 Developments of the offshore wind turbine wake model Fuga, *DTU Wind Energy E*, No. 0046
[15] Ozdemir H, Versteeg M C, Brand A J 2013 Improvements in ECN wake model, *ICOWES2013 Conference*
[16] CL-Windcon project 2018 Deliverable D1.4, Classification of control-oriented models for wind farm control applications, http://www.clwindcon.eu/
[17] CL-Windcon website http://www.clwindcon.eu/
[18] CL-Windcon project 2018 Deliverable D3.1, Definition of wind tunnel testing conditions, http://www.clwindcon.eu/
[19] CL-Windcon project 2018 Deliverable D3.4, Testing in the wind tunnel of wind turbine controllers, http://www.clwindcon.eu/
[20] Svensson G, et al. 2011 Evaluation of the diurnal cycle in the atmospheric boundary layer over land as represented by a variety of single-column models: The second GABLS experiment, *Boundary-Layer Meteorol*, **140**(2):177206
[21] Schepers J G, et al 2011 Analysis of wake measurements from the ECN Wind Turbine Test Site Wieringermeer, EWTW, *Wind Energy*, **15**(4):575–91
[22] Barthelmie R J, et al. 2011 Flow and wakes in large wind farms. Final report for UpWind WP8, *Risø DTU,* 1765(EN)
[23] Frandsen S, et al. 2007 The shadow effect of large wind farms: measurements, data analysis and modelling, *Risø DTU*, 1615(EN)
[24] DNV GL 2014, Windfarmer Validation report version 5.3, http://www.dnvgl.com/
[25] Hersbach H, et al. 2018 Operational global reanalysis: progress, future directions and synergies with NWP, *ECMWF*
[26] Dyer A J 1974 A review of flux-profile relationships, *Boundary-Layer Meteorology*, **7**(3):363
[27] Högström U 1988 Non-dimensional wind and temperature profiles in the atmospheric surface layer: a re-evaluation, *Boundary-Layer Meteorology*, **42**(1–2):55–78
[28] Peña A, et al. 2016 On the application of the Jensen wake model using a turbulence-dependent wake decay coefficient: the Sexbierum case, *Wind Energy*, **19**(4):763–76
[29] Holtslag A A M 1984 Estimates of diabatic wind speed profiles from near-surface weather observations, *Boundary-Layer Meteorology*, **29**(3):225–50
[30] Gryning S E, et al. 2007 On the extension of the wind profile over homogeneous terrain beyond the surface boundary layer, *Boundary-Layer Meteorology*, **124**, p. 251-268
[31] Pasquill F 1974 Atmospheric diffusion 2nd ed., *Halsted Press*
[32] Hansen K S, Larsen G C, Ott S 2014 Dependence of offshore wind turbine fatigue loads on atmospheric stratification, *Journal of Physics: Conference Series*, **524**:012165
[33] Mohan M 1998 Analysis of various schemes for the estimation of atmospheric stability classification, *Atmospheric Environment*, **32**(21):3775–81
[34] Quarton D, Ainslie J 1990 Turbulence in wind turbine wakes, *J. Wind Eng.*, **14**(1):15-23
[35] Crespo A, Hernandez J 1996 Turbulence characteristics in wind-turbine wakes, *Journal of Wind Engineering and Industrial Aerodynamics*, **61**(1):71–85
[36] Gunn K, et al. 2016 Limitations to the validity of single wake superposition in wind farm yield assessment, *Journal of Physics: Conference Series*,**749**:012003
[37] Machefaux E, Larsen G C, Leon J P M 2015 Engineering models for merging wakes in wind farm optimization applications, *Journal of Physics: Conference Series*, **625**:012037
[38] Schlez W, Neubert A 2009 New developments in large wind farm modelling, *Garrad Hassan Deutschland GmbH*
[39] Burton T, Jenkins N, Sharpe D, Bossanyi E 2011 Wind Energy Handbook, *John Wiley & Sons*
[40] Coleman R P, et al. 1945 Evaluation of the induced velocity field of an idealised helicopter rotor, *NACA*, L5E10
[41] Jimenez A, Crespo A, Migoya E 2010 Application of a LES technique to characterize the wake deflection of a wind turbine in yaw, *Wind Energy*, **13**:559-572

[42] Martínez-Tossas L A, Annoni J, Fleming P A, Churchfield M J 2018 The aerodynamics of the curled wake: a simplified model in view of flow control, *Wind Energy Science Discussions*, **7**:1–17

[43] Fleming P, et al. 2018 A simulation study demonstrating the importance of large-scale trailing vortices in wake steering, *Wind Energy Science*, **3**(1):243–55

[44] Zahle F, Sørensen N. 2008 Overset Grid Flow Simulation on a Modern Wind Turbine, *26th AIAA Applied Aerodynamics Conference*, Honolulu, Hawaii

[45] Sakagami Y, Santos P A A, et al. 2017 Low turbulence intensity and low wind shear effects on annual energy production over the northeast coast of Brazil, *Wind Europe Conference*

IOP Conf. Series: Journal of Physics: Conf. Series **1222** (2019) 012005 doi:10.1088/1742-6596/1222/1/012005

Enhanced modelling of the stratified atmospheric boundary layer over steep terrain for wind resource assessment

A Flores-Maradiaga [a*], R Benoit [b], C Masson [b]

[a] Department of Mechanical Engineering, Federico Santa María Technical University, Av. España 1680, casilla 110-V, Valparaíso, Chile
[b] École de Technologie Supérieure, Université du Québec, Notre-Dame O 1100, H3C 1K3, Montréal, Québec, Canada

* E-mail: alex.floresm@usm.cl

Abstract. The Mesoscale Compressible Community (MC2) model [1], devoted for weather forecasting and used in the Wind Energy Simulation Toolkit (WEST) [2], performs well for simulations over flat, gentle and moderate terrain slopes but is subject to numerical instability and strong spurious flows in presence of steep topography. To remove its inherent computational mode and reduce the wind overestimation due to terrain-induced numerical noise, a new semi-implicit (N-SI) scheme [3] was implemented to discretize and linearize the non-hydrostatic Euler equations with respect the mean values of pressure and temperature instead of arbitrary reference state values, redefining as well the buoyancy to use it as the thermodynamic prognostic variable. Additionally, the climate-state classification of the statistical-dynamical downscaling (SDD) method [4] is upgraded by including the Brunt-Väisälä frequency that accounts for the atmospheric thermal stratification effect on wind flow over topography. The present study provides a real orographic flow validation of these numerical enhancements in MC2, assessing their individual and combined contribution for an improved initialization and calculation of the surface wind in presence of high-impact terrain. By statistically comparing the wind simulations with met-mast data, obtained within the Whitehorse area of the Canadian Rocky Mountains, it is confirmed that these numerical enhancements may reduce over 40 percent of the wind overestimation, thus, attaining more accurate results that ensure reliable wind resource assessments over complex terrain.

1. Introduction

Currently, the fast growing wind industry requires accurate and less uncertain wind resource assessment over diverse topographic configurations for wind farm planning and development. For this purpose, the Canadian Research Laboratory for Nordic Environment Aerodynamics of Wind Turbines (NEAT) has upgraded the mesoscale compressible community model (MC2) [1] as part of the latest improvements for the Wind Energy Simulation Toolkit (WEST), a widely used open-source software for wind resource assessment [2, 4]. Recently, the MC2 model has been upgraded with a built-in large eddy simulation (LES) methodology and advanced physical parameterization schemes [1, 5], which increase its capabilities to resolve nonlinear atmospheric flow phenomena over high impact orography.

However, it has been observed that MC2, similar to other mesoscale models that employ terrain-following coordinates, generates non-negligible spurious flows for wind simulations over complex terrain with moderate and steep slopes [4-7]. An important aspect of models that employ terrain-following σ-coordinates is that the pressure-gradient force may be miscalculated near steep slopes, mostly due to the computational grid-cell deformation [8]. As demonstrated by Klemp [9], Zängl [10] and Blocken et al. [11], this terrain-induced spurious forcing is enhanced when an instable discretization scheme is applied, causing numerical noise and the overestimation of the surface wind speed at the mountain crests and lee-sides. This aspect is quite preoccupying for multiscale wind modelling, since wind farm engineering and financial decision-making depends heavily on the resource assessment results. Pinard et al. [12-13] and Gasset et al. [4] studies suggest that terrain-induced spurious forcing along with inappropriate initialization procedures significantly degrade the model's accuracy for real orographic flow realizations in highly complex sites.

Pinard's et al. [13] uncertainty diagnosis (i.e., modelled versus observation data) of MC2's performance for strongly stratified high-shear ABL flow simulations over mountainous topography in the western Canadian Rockies, confirm how the original semi-implicit (O-SI) discretization [1] in combination with the original wind-climate classification (OC) [2, 4] yield an overestimated mean velocity, directional shifting and strong spurious noise that degrade the consistency and accuracy of real case wind resource assessments.

The latter problem is corrected precisely with the new semi-implicit (N-SI) scheme [3], presented in section 2, which discretizes the model equations about the mean values of temperature (T_0) and pressure ($q_0 = \ln p_0$) and redefines the buoyancy -prognostic thermodynamic variable for MC2-, yielding a profound restructuration of the non-linear residuals to ensure numerical stability and control the computational mode. Additionally, the Brunt-Väisälä buoyancy frequency is added to the wind-climate new classification (NC) to account effectively for the thermal stratification along with wind speed, direction and shear. The SLEVE vertical coordinate [14] is also employed to ensure smoother terrain-conforming grid levels aloft so the irregular surface signal does not propagate unnecessarily into the domain's interior. Hence, the objective of the present study is the validation with real orographic flow of the combined enhancements for MC2, obtained from the N-SI time discretization scheme and the introduction of the Brunt-Väisälä buoyancy frequency in the SDD initialization scheme.

The combined enhancement obtained from the N-SI scheme and the new SDD wind-climate classification is expected to improve mostly the surface layer modelling and reduce the initialization errors for vertical momentum transport. It should enable an efficient and more accurate integration with high-order numerical algorithms, such as the semi-implicit semi-Lagrangian (SISL) solver [15-16]. The model is also supplemented with a second-order accurate Robert-Asselin-Williams (RAW) time-filter [17] to conserve better the system's total energy. The proposed enhancements are general enough to be applicable in any other multiscale model with similar numerical schemes.

2. Modified Model Equations

In general terms, the MC2 kernel solves the governing equations by separating the material derivatives applied on the prognostic variables ($d\mathbf{\Psi}/dt$) and linear terms (**L**, treated implicitly) from the non-linear terms (**R**, treated explicitly), external forcing and source terms (**F**), expressed in matrix form as follows:

$$\frac{d\mathbf{\Psi}}{dt} + \mathbf{L} = \mathbf{R} + \mathbf{F}. \tag{1}$$

The semi-implicit semi-Lagrangian (SISL) discretization is applied on the first three terms of matrix system (1) to calculate the fluid particle's trajectory over three time-levels and, then, the external forcing and source terms are added in a fractional-step procedure [1]. Thus, the fundamental improvement introduced in system (1) is the restructuration of the linear **L** and non-linear **R** terms, in order to remove the computational mode and terrain-induced noise in the new formulation. As explained in [5], the momentum and heat turbulent fluxes are included in the **F** terms after the governing equations are filtered, as required for large-eddy simulations. These turbulent fluxes need to be modeled using a particular SGS scheme in order to close the equation system.

The core system based on [1] includes the following momentum, energy and continuity equations:

$$\begin{aligned}
&\frac{d\mathbf{v}}{dt}+[\nabla-\gamma_*\mathbf{k}]P+\mathbf{k}(b-\gamma_A P)=-f\,\mathbf{k}\times\mathbf{v}-\frac{b}{g}(\nabla-\beta_A\mathbf{k})P+\mathbf{f},\\
&\frac{d}{dt}(b-\gamma_* P)+N_*^2 w=-b\left[\beta_A w+\frac{R}{c_v}\nabla\cdot\mathbf{v}\right]+\gamma_A Q,\\
&\frac{d}{dt}\left(\frac{P}{c_*^2}\right)+\nabla\cdot\mathbf{v}-\frac{g}{c_*^2}w=\frac{Q}{c_pT},
\end{aligned}\tag{2}$$

where $d/dt=\partial/\partial t+\mathbf{v}\cdot\nabla$ represents the material derivative, $\mathbf{v}=(u,v,w)$ the velocity, $\mathbf{k}$ is the vertical direction unit vector, f the Coriolis parameter, $P=RT_*\ln(p')$ the generalized pressure, $p'=p-p_*$ the pressure perturbation, $b=g(T'/T_*)$ the buoyancy force, $T'=T-T_*$ the temperature perturbation, $\mathbf{f}=(\mathrm{f}_u,\mathrm{f}_v,\mathrm{f}_w)$ the non-conservative forces and Q heat sources. The main variables are complemented with the speed of sound $c_*^2=(c_p/c_\upsilon)(RT_*)$ and the natural oscillation frequency $N_*^2=g\gamma_*=g(\beta_A+\gamma_A)=g^2/c_pT_*$, two fundamental parameters among with the constants $\beta_A=\partial\ln T_*/\partial z$ and $\gamma_A=g/c_pT_*$.

To enhance the model's numerical stability, an eigenmode analysis was applied on this equation system (2), discretized with the original semi-implicit scheme (O-SI), which revealed that the nonlinear terms relating the buoyancy with the pressure-gradient and flow divergence contribute significantly to the origin of numerical noise [3]. Thus, to overcome this inherent deficiency and ensure the hydrostatic balance, the buoyancy was redefined as $\hat{b}=g(T'/T)=g\alpha/(\alpha+1)$ with $\alpha=T'/T_*$, yielding a restructured system with a new semi-implicit (N-SI) discretization, as follows:

$$\begin{aligned}
&\frac{d\mathbf{v}}{dt}+\overline{(\alpha+1)(\nabla P+\mathbf{k}\hat{b})}^{t}-(\gamma_*+\gamma_A)\mathbf{k}P=-f\,\mathbf{k}\times\mathbf{v}+\frac{\hat{b}}{g}\beta_A\mathbf{k}P+\mathbf{f},\\
&\frac{d}{dt}\left[(\alpha+1)\hat{b}-\gamma_*P\right]+\overline{\frac{N_*^2 w}{\alpha+1}}^{t}=-\left(\frac{\alpha}{\alpha+1}\right)N_*^2 w+\gamma_A Q,\\
&\frac{d}{dt}\left(\frac{P}{c_*^2}\right)+\overline{\nabla\cdot\mathbf{v}-\frac{g}{(\alpha+1)c_*^2}w}^{t}=\left(\frac{\alpha}{\alpha+1}\right)\frac{g}{c_*^2}w+\frac{Q}{c_pT},
\end{aligned}\tag{3}$$

where the overbar $\overline{[\ ]}^{t}$ denote the implicit time averaging operator for terms solved over three time levels (3-TL). With this N-SI scheme, system (3) contains explicitly treated terms appropriately modified to recover the linearity in the hydrostatic relationship, which ensures the model's numerical stability in the presence of steep topography.

Based on the Bousinesq hypothesis, and using Einstein's notation, the turbulent stresses and heat fluxes can be expressed in terms of eddy- and heat-mixing coefficients $(\mu_t=\rho K_M;\mathrm{Pr}_t=K_M/K_T)$, the resolved strain rate $S_{ij}=1/2\left[\partial\tilde{u}_i/\partial x_j+\partial\tilde{u}_j/\partial x_i\right]$, the sub-filter turbulent kinetic energy $\overline{\rho}k=1/2\overline{\rho u_i' u_i'}$ and the potential temperature gradient, such that:

$$-\overline{\rho u_i' u_j'}=\tau_{ij}=2\mu_t S_{ij}-2/3\left(\mu_t S_{ll}-\overline{\rho}k\right)\text{ and }-\overline{\rho u_i'\theta'}=\mu_t/\mathrm{Pr}_t\left[\partial\tilde{\theta}/\partial x_i\right].\tag{4}$$

Here the tilde represents the application of the implicit filter when the solution is projected onto the numerical grid. For this LES implementation in MC2, the classical Smagorinsky (SMAG) SGS scheme [18] is employed as a constant coefficient parametrization with stability functions based on the Richardson's number ($Ri = N^2/\overline{S}^2$) with $N^2 = (g/\tilde{\theta})(\partial\tilde{\theta}/\partial z)$ as the natural frequency of the air parcel's oscillation. The SMAG closure is formulated with the following relationships [5]:

$$
\begin{aligned}
&K_M = \lambda f_m \overline{S}, \quad K_T = \lambda f_h \overline{S}, \\
&\lambda = \left[\frac{1}{C_s\Delta} + \frac{1}{\kappa(z+z_0)^2}\right]^{-1/2}, \\
&f_m = \begin{cases} (1-16Ri)^{1/2}, & 0 < Ri \\ \left(1-\dfrac{Ri}{Ri_c}\right)^4, & 0 \le Ri \le Ri_c \end{cases} \\
&f_m = \begin{cases} \dfrac{(1-40Ri)^{1/2}}{\Pr_t}, & 0 < Ri \\ \dfrac{(1-1.2Ri)}{\Pr_t}\left(1-\dfrac{Ri}{Ri_c}\right)^4, & 0 \le Ri \le Ri_c \end{cases}
\end{aligned}
\tag{5}
$$

where λ is the characteristic length scale, f_m and f_h are the stability functions for momentum and heat transport, respectively, $\overline{S} = 2(S_{ij} - \delta_{ij}/2S_{ll})$ is the corresponding strain rate tensor modulus for compressible flow simulations [19-20], C_S is the Smagorinsky coefficient, $\Delta = (\Delta_X\Delta_Y\Delta_Z)^{1/3}$ is the filter length scale, $\kappa = 0.4$ is the von Kármán constant, z_0 is the aerodynamic roughness length, Ri_c is the critical Richardson number. A comprehensive explanation on the SGS scheme details and implementation for MC2 is provided by [5].

3. Validation of the enhanced modelling method

To demonstrate the advantageous features of the N-SI scheme to achieve enhanced numerical stability and the reduction of terrain induced noise, a canonical test with initially zero mean flow (i.e., atmosphere-at-rest simulation) is performed over flat terrain and a steep Gaussian hill with a maximum slope of 45°. An atmosphere-at-rest simulation requires an initialization with a hydrostatically balanced non-rotational atmosphere with zero horizontal velocity input, zero horizontal pressure gradient and a horizontally homogenous thermodynamic sounding. For these simulations, a free-slip condition and constant temperature are imposed on the surface. At the top of the domain a numerical sponge layer, based on Shuman's boundary condition [17], is set to restrict the spurious reflection of acoustic and gravity waves. These cases are initialized with an isothermal atmosphere, an ideal condition that simplifies the analysis of the temperature perturbation influence on the model's stability. The surface and isothermal reference temperatures are set to 250 K, which is a reasonable choice for mesoscale experiments [1].

The domain is $L_x = 10$ km long and $L_z = 5$ km high, with horizontal and vertical resolutions of $\Delta x = 100$ m and $\Delta z = 50$ m, respectively. For the time interval, preliminary testing revealed that $\Delta t =$ 10 s is a good compromise for this grid resolution. Although the simulations are performed for 72 hours, with numerical instability there is no guarantee that the model would converge to the correct solution, because the floating-point round-off or truncation errors can be magnified, instead of damped, causing the deviation from the exact solution to grow exponentially in an asymptotical fashion.

Figure 1 compares the mean flow fields obtained with the former and new model versions (Fig. 1a and 1b, respectively). Clearly, the N-SI scheme significantly reduces the terrain-induced noise and spurious flows by three orders of magnitude, which usually arise over the mountain slopes and propagate into the whole domain.

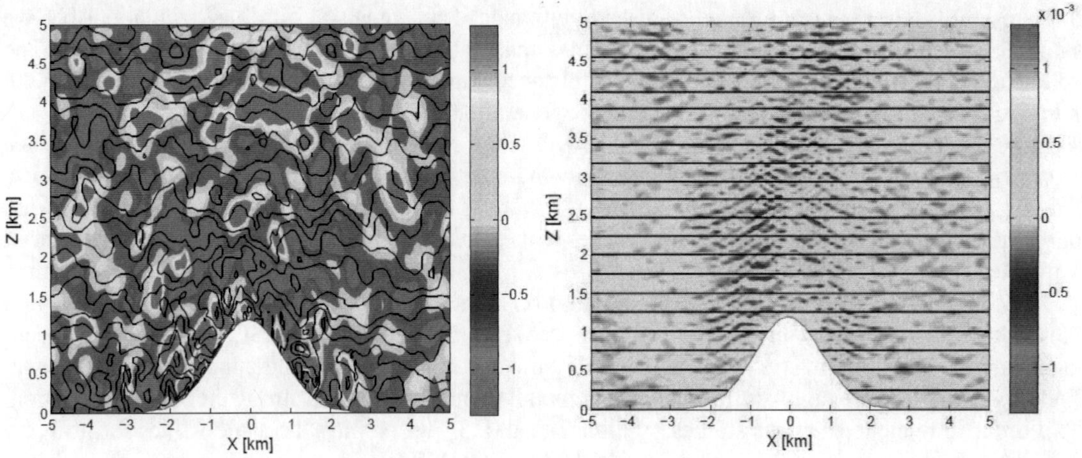

Fig. 1 – Comparison of mean wind field between the former (O-SI) and new (N-SI) model versions after 42 hours of an isothermal atmosphere-at-rest test. Solid lines denote potential temperature isentropes (interval of 2.5 K) and color shading represents the velocity (ms^{-1}). The right panel has a scale three orders of magnitude lower than the left panel since the numerical noise is significantly weaker.

Also, the maximum wind speed and noise kinetic energy spectra for both semi-implicit schemes are compared over flat and the 45° terrain slope in Figure 2, proving that the O-SI scheme is absolutely instable and highly noisy, contrary to the N-SI scheme that exhibits a controlled computational mode and low accumulation of kinetic energy for high-frequency small-scale structures. This achievement will allow the reduction of strong spurious flows behaving as fictitious turbulent wind with an energy cascade mechanism as depicted in Fig. 2b.

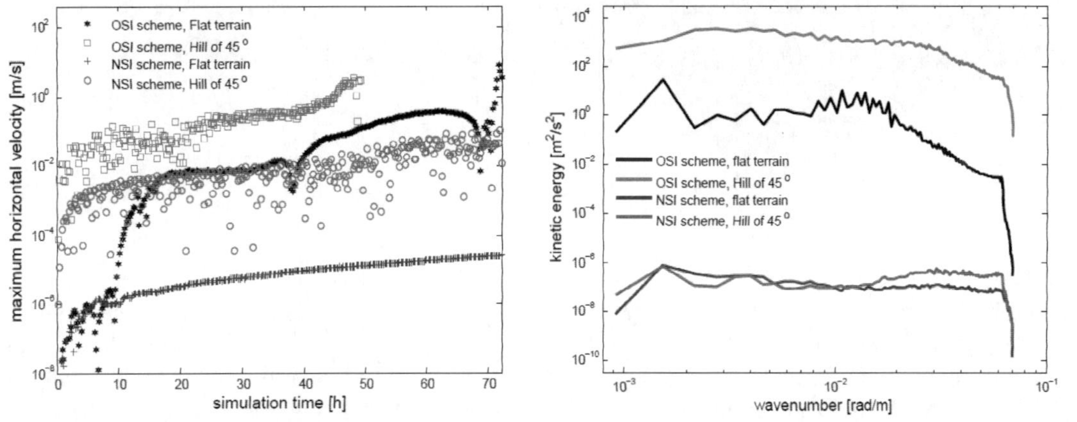

Fig. 2 – Comparison of (a) maximum wind speed and (b) noise energy spectra for flat and steep terrain the former (O-SI) and new (N-SI) model versions after 42 hours of an atmosphere-at-rest test.

The O-SI scheme generates an asymptotically growing computational mode over both flat and complex terrain, which inevitably yields an extremely noisy and unstable model. Complex terrain amplifies the high-frequency noise at least one order of magnitude and leads to an early blow-up of the unstable model. Steep hill slopes magnify spurious flows, which could erroneously be considered as mountain top speed-up or lee-side wind storms, leading to wind speed overestimation in real mesoscale applications. We distinguish an extended dissipation range ($k > 10^{-2}$ m^{-1}) in the O-SI noise spectrum, which resembles the physical turbulence dissipation range. Thus, the O-SI spurious circulations behave as numerically generated turbulence that can contaminate atmospheric boundary layer simulations. The O-SI scheme alone is notoriously unable to perform accurately and cannot yield reliable results due to its inherent instability and strong numerical noise generation.

To demonstrate that the combined N-SI and NC schemes implemented in MC2 effectively reduce the wind-speed overestimation, we simulate the wind over the cold-climate mountainous Whitehorse area (Fig. 3a) in the Yukon Territory of Western Canada. As mentioned in [12], sites of interest for wind energy development in this area have steep slopes that exceed 15°. This area is bordered and influenced by the Wrangell St. Elias Mountains, Western Costal Mountains, Mackenzie Mountains and Canadian Rocky Mountains. The landscape is covered with prominent ice-capped peaks, deep valleys, rivers and lakes, where the majority of terrain features range between 1000 and 3000 m ASL, among which Mount Logan stands out with a maximum height of 5959 m ASL. The local climate conditions and seasonal wind flow variations are highly influenced by the mountainous forcings and strong thermal stratification.

In order to relate to previous studies by Pinard et al. [13], we employ the same values for grid size (square lateral length of 100 km and height of 10 km), grid's horizontal resolution ($\Delta x = \Delta y = 1$ km, with $100 \times 100 \times 100$ grid-points), domain layering (vertically stretched with 12 layers in the first 1500 m above ground level), time interval ($\Delta t = 60$ s) and the total simulation time (16 hrs, which ensures quasi-steady convergence after 12 hrs and integration period). A column-type URANS scheme is used for turbulence parameterization. The simulation is performed with the log-law surface model, lateral boundary conditions obtained from a synoptic scale model and a Shuman [21] wave absorbing sponge at the top 10 layers to avoid gravity wave reflection that could bias the results.

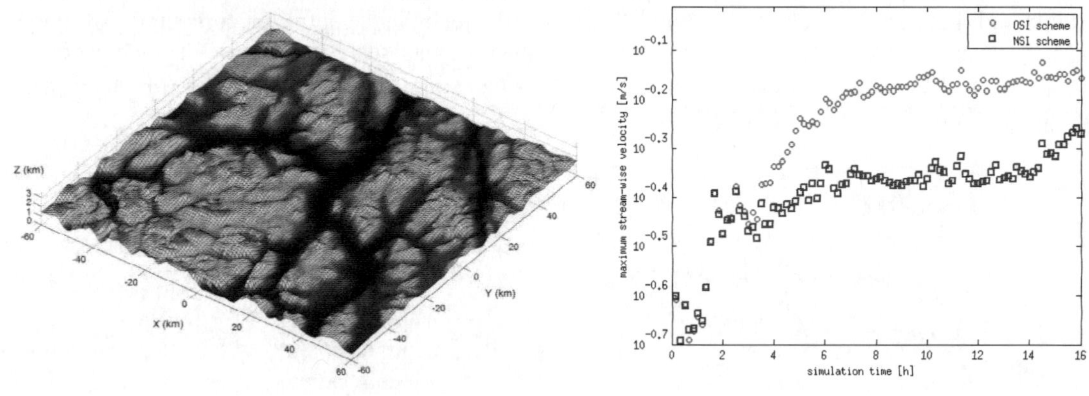

Fig. 3 – (a) Illustration of the Whitehorse valley at 1 km resolution and (b) the evolution of the numerical noise for streamwise surface wind speeds (ms^{-1}) after 16 hours of an atmosphere-at-rest simulation.

Before running the orographic flow case, a atmosphere-at-rest test is performed over this complex terrain to assess the propagating gravity oscillations under stringent conditions, with a strongly stable atmosphere initially at rest with underlying multiscale ridges. In Fig. 3b we observe that the wind speed starts with a similar magnitude for both SI schemes and grows asymptotically as the integration progresses. However, the maximum horizontal velocities obtained with the N-SI scheme display an approximate reduction of 30% with respect to the O-SI results. This noise reduction, achieved by the N-

IOP Conf. Series: Journal of Physics: Conf. Series **1222** (2019) 012005 doi:10.1088/1742-6596/1222/1/012005

SI scheme alone without any other stabilizing mechanism (i.e. decentering or filtering), certainly enhances MC2's accuracy and numerical stability by weakening its spurious computational mode.

Then, proceeding with the real orographic flow over the Whitehorse area, a full set of simulations were initialized with the SDD algorithm that automatically defines the number of situations to model, depending on the chosen wind climate-state classification (OC yields 224 cases and NC yields 896 cases). Table 1 summarizes the statistical assessment of the former or baseline (OC+O-SI) and novel (NC+N-SI) scheme combinations upon the comparison of modelled versus observed data, measured at 30 m above ground level by 16 wind-stations distributed in the Whitehorse valley. The NC+O-SI only yields a 0.045 ms^{-1} reduction of the MAE respect to the baseline combination, which represents an improvement of 4.6% obtained only by using a better SDD initialization algorithm. With the OC+N-SI combination, we achieve a reduction of the wind MAE by 0.41 ms^{-1}, which translates into a 42.4% improvement with respect to the baseline schemes since the spurious acceleration is attenuated. Finally, the two schemes together as NC+N-SI yield the best flow field readjustment, attaining a 0.44 ms^{-1} reduction of the wind MAE, i.e. 45.4 % improvement, that ensures a higher accuracy for wind resource assessment. The average and standard deviation of wind speed are reduced significantly, achieving much higher correlations between modelled and observed values and almost half the mean absolute error (MAE) with the NC+NSI combination.

Table 1 Statistics of the numerical simulations of stratified wind in the Whitehorse valley.

Schemes	**Average (ms^{-1})**	**Minimum (ms^{-1})**	**Maximum (ms^{-1})**	**σ_U (ms^{-1})**	**Correlation model-obs.**	**Mean Abs. Error (ms^{-1})**	**MAE reduction**
OC+OSI	4.3392	2.7493	6.7301	1.1107	0.7485	0.9687	-
NC+OSI	4.2709	2.8268	6.4874	1.0671	0.7669	0.9238	4.64 %
OC+NSI	3.7669	2.1601	7.0317	0.6811	0.9175	0.5584	42.35 %
NC+NSI	3.8106	2.0973	7.2712	0.6589	0.9299	0.5291	45.38 %

Replicating correctly the surface flow pattern over prominent mountain and valley features is a key challenge to achieve an effective improvement in multiscale wind resource assessment. Fortunately, the NC+N-SI scheme combination imbedded in the SDD method yields a correction of the surface flow field misrepresented with OC+O-SI. Figure 4 depicts the mean wind field realizations over the Whitehorse valley, that is expected to flow east-northward based on the observed and reanalysis data. Clearly, the OC+O-SI baseline combination (Fig. 4a) fails to obtain appropriate wind speeds and directions. In fact, it reveals that this baseline permutation significantly overestimates the wind magnitudes when the air masses interact with steep terrain slopes. Although the OC+O-SI combination is able to replicate the deep valley flow channeling, the adverse influence of the miscalculated wind aloft yields a wrong vector orientation with recirculation and flow inversion in the valleys.

IOP Conf. Series: Journal of Physics: Conf. Series **1222** (2019) 012005 doi:10.1088/1742-6596/1222/1/012005

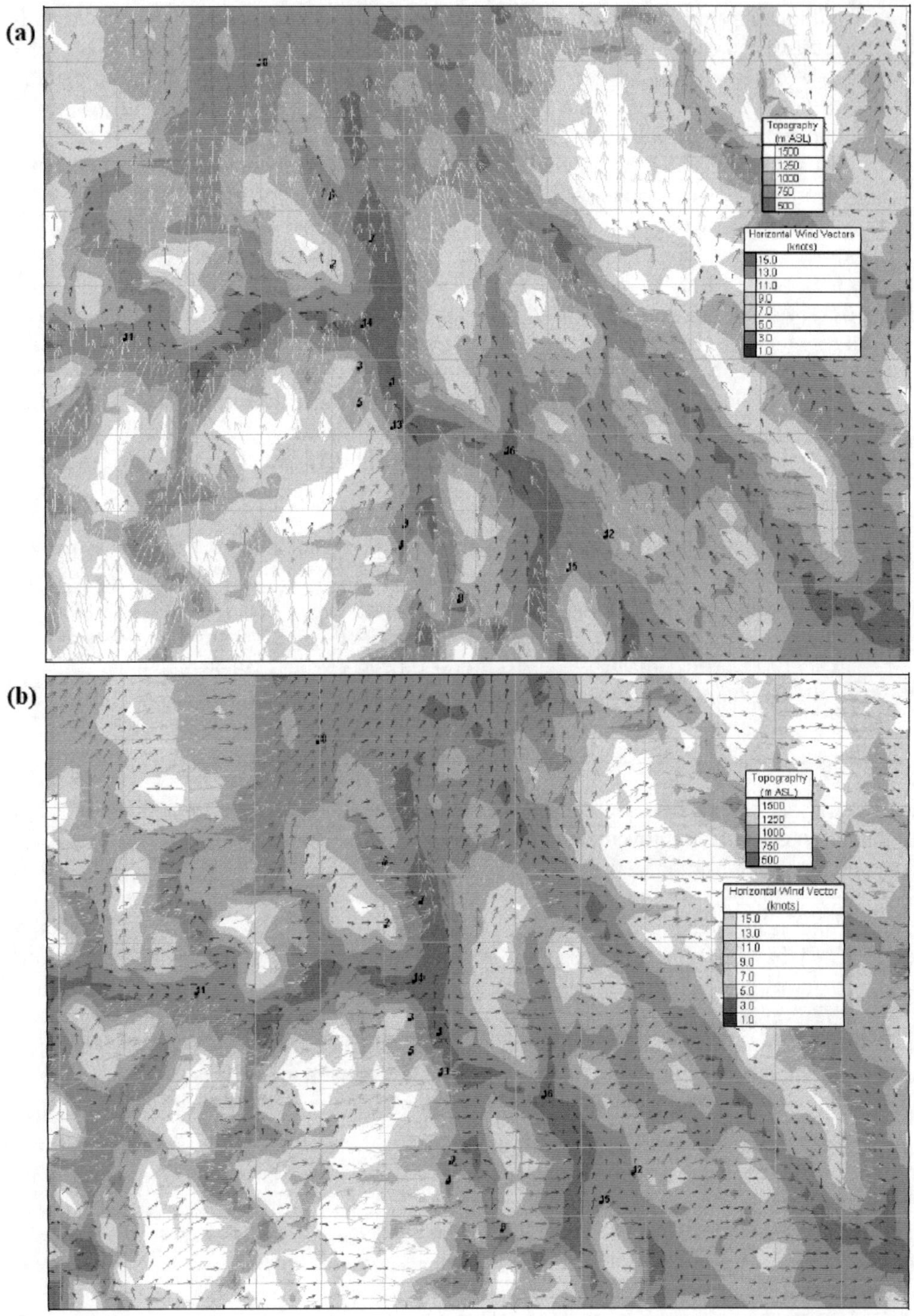

Fig. 4 – Mean wind flow patterns at 30 m AGL in the Whitehorse valley, obtained with the (a) OC+OSI and (b) NC+NSI combinations implemented in MC2.

Indeed, the cold airflow in this site should be oriented primarily east-wise, with nearly half of the magnitude, as it is evident in the NC+N-SI realization (Fig. 4b). The former scheme combination also achieves the expected mountain-valley systems, with low-level jets over the hilltops and cross channeling within the valleys. Hence, the proposed enhancements for the MC2 model and SDD algorithm will, most likely, correct the long-standing difficulties faced by WEST and similar wind resource assessment toolkits for ABL simulations over steep complex terrain.

4. Conclusions

Without any damping or filtering mechanisms, for a resting stratified atmosphere over complex terrain, the N-SI scheme yields a noise reduction of nearly 30% with respect to the O-SI results. The analysis of the ABL simulations with the proposed new semi-implicit discretization and the new wind-climate classification for SDD initialization indicates that, in presence of steep mountain slopes, MC2 effectively attains an error reduction of 40% or more as compared to the former schemes. The overall improvement, with 45.38% MAE reduction, 93% correlation and half the standard deviation, reflects that the combined NC+NSI contribution is the best for orographic flow modelling over steep complex terrain. Although modest, these improvements contribute to the uncertainty reduction for wind resource assessment over complex topography.

Acknowledgements

The authors want to thank the Canadian National Science and Engineering Research Council (NSERC) for funding this research through project WESNet 1.1-a, as well as the Chilean National Commission for Scientific and Technological Research (CONICYT) for funding this research through project FONDEF ID16I10105. We also wish to thank the Environment and Natural Resources Department of Canada and, particularly, Dr. Claude Girard and Dr. Nicolas Gasset for providing the MC2 code, technical support and counselling during this study.

References

[1] Girard C, Benoit R and Desgagné M 2005 *Finescale topography and the MC2 dynamics kernel* Monthly Weather Review **133** 1463–1477
[2] Yu W, Benoit R, Girard C, Glazer A, Lemarquis D and Pinard 2006 *Wind Energy Simulation Toolkit (WEST): A Wind Mapping System for Use by the Wind Energy Industry* Wind Engineering **30** 15-33
[3] Flores-Maradiaga A, Benoit R and Masson C 2016 *Enhanced Method for Multiscale Wind Simulations over Complex Terrain for Wind Resource Assessment* Journal of Physics: Conf. Series **753** 082030
[4] Gasset N, Laundry M and Gagnon Y 2012 *A Comparison of Wind Flow Models for Wind Resource Assessment in Wind Energy Applications* Energies **5** 4288-4322
[5] Gasset N, Benoit R and Masson C 2014 *Implementing Large-eddy Simulation Capability in a Compressible Mesoscale Model* Monthly Weather Review **142** 2733-2750
[6] Fleming P, Churchfield M, Scholbrock A, Michalakes J, Johnson K and Moriarty P 2013 *The SOWFA super-controller: a high-fidelity tool for evaluating wind plant control approaches* National Renewable Energy Laboratory NREL/CP **5000** 57175
[7] Lundquist K, Chow F K and Lundquist J 2010 *An Immersed Boundary Method for the Weather Research and Forecast Model* Monthly Weather Review **138** 796 – 817
[8] Shchepetkin A and McWilliams J 2003 *A method for computing horizontal pressure gradient force in an ocean model with a non-aligned vertical coordinate* Journal of Geophysical Research **108** 3090
[9] Klemp J 2011 *A Terrain-following Coordinate with Smoothed Coordinate Surfaces* Monthly Weather Review **139** 2163–2169
[10] Zängl G 2012 *Extending the Numerical Stability Limit of Terrain-Following Coordinate Models over Steep Slopes* Monthly Weather Review **140** 3722–3733

[11] Blocken B, Hout A, Dekker J and Weiler O 2015 *CFD Simulations of Wind Flow over Natural Complex Terrain: Case Study with Validation by Field Measurements for Ria de Ferrol, Galicia, Spain* Journal of Wind Engineering and Industrial Aerodynamics **147** 43-57
[12] Pinard J P, Benoit R and Yu W 2005 *A WEST Wind Climate Simulation of the Mountainous Yukon* Atmosphere-Ocean **43** 259-282
[13] Pinard J P, Benoit R and Wilson J D 2009 *Mesoscale Wind Climate Modelling in Steep Mountains* Atmosphere-Ocean **47** 63-78
[14] Schär C, Leuenberger D, Fuhrer O and Girard C 2002 *A New Terrain-following Vertical Coordinate Formulation for Atmospheric Prediction Models* Monthly Weather Review **130** 2459–2480
[15] Tanguay M, Robert A and Laprise R 1990 *A Semi-Implicit Semi-Lagrangian Fully Compressible Regional Forecast Model* Monthly Weather Review **118** 1970–1980.
[16] Thomas S, Girard C, Benoit R and Pellerin P 1998 *A New Adiabatic Kernel for the MC2 Model* Atmosphere–Ocean **36** 241–270
[17] Williams P D 2011 *The RAW Filter: an Improvement to the Robert-Asselin Filter in Semi-Implicit Integrations* Monthly Weather Review **139** 1996–2007
[18] Smagorinsky J 1963 *General Circulation Experiments with the Primitive Equations. Part 1: The Basic Experiment* Monthly Weather Review **91** 99–152
[19] Cuxart J, P Bougeault and J L Redelsperger 2000 *A Turbulence Scheme Allowing For Mesoscale and Large-Eddy Simulations* Quart. J. Roy. Meteor. Soc. **126** 1-30
[20] Drobinski P, Carlotti P, Redelsperger J L, Banta M, Masson V and Newsom R 2007 *Numerical and Experimental Investigation of the Neutral Atmospheric Surface Layer* J. of Atmospheric Sciences **64** 137–156
[21] Shuman F 1957 *Numerical Methods in Weather Prediction: (II) Smoothing and Filtering* Monthly Weather Review **85** 357–261

IOP Conf. Series: Journal of Physics: Conf. Series **1222** (2019) 012006 doi:10.1088/1742-6596/1222/1/012006

Unravelling the wind flow over highly complex regions through computational modeling and two-dimensional lidar scanning

J.M.L.M. Palma[1], A. Silva Lopes[1], V.M. Costa Gomes[1], C. Veiga Rodrigues[1,3], R. Menke[2], N. Vasiljević[2] and J. Mann[2]

[1] Faculty of Engineering of the University of Porto, Rua Dr. Roberto Frias s/n, 4200-465 Porto, Portugal

[2] DTU Wind Energy, Technical University of Denmark, Frederiksborgvej 399, 4000 Roskilde, Denmark

[3] Currently with Vestas Wind Systems A/S, Design Centre Porto, Rua Lionesa, 4465-671 Leça do Balio, Porto, Portugal

E-mail: jpalma@fe.up.pt, asl@fe.up.pt, vcgomes@fe.up.pt, calvr@vestas.com, rmen@dtu.dk, niva@dtu.dk, jmsq@dtu.dk

Abstract.
Computational modeling and lidar scanning data of the atmospheric flow over a double-ridge (Perdigão) were used to study a 24-hour period on 14–15 May 2017 with a maximum wind speed of more than 6 ms^{-1}. An influence of the stratification was observed throughout most of the time, in the form of a lee-wave triggered by the topography or flow separation further downstream of the ridge. The good agreement between the results obtained by lidar scanning and computational modeling increases our confidence in the use of either of these two techniques. See corresponding video at doi.org/10.11583/DTU.7863482.

1. Introduction

In this study, we show how scanning lidars and three-dimensional time-dependent computational models can be used to increase our knowledge on the flow over complex regions where separated flow, hydraulic jumps and internal gravity waves were observed. These are phenomena that may affect wind power production, but are usually not very well modeled when assessing the wind resources, if modeled at all [1, 2].

Similar phenomena were already observed during Perdigão-2015 [3] and reported in [4], in the case of southwesterly winds. Here we made use of a small set of the Perdigão-2017 field experiment [5, 6] and compared against simplified flow models, both analytical and scale laboratory models [7], with which a good qualitative agreement was found. Quantitative agreement was not sought at this stage, since it is part of an ongoing work, based on a data set larger than the one used here. Our focus is on the flow dynamics of a 24-hour period (starting on May 14 at 18:00 UTC) shown by a video [8], to be found at doi.org/10.11583/DTU.7863482, to which this paper can be considered as a guide.

IOP Conf. Series: Journal of Physics: Conf. Series **1222** (2019) 012006 doi:10.1088/1742-6596/1222/1/012006

2. The case study

Perdigão is comprised by two parallel ridges (129.1° and 132.5°, SW and NE ridges), about 1400 m apart and 215 m above the surrounding region, Figure 1. Two field campaigns took place in the region, in 2015 [3] and 2017 [5], with the participation of European and US research groups. The high density of scientific equipment in 2017 has led to a database, still under organization, which is expected to become an important tool for validation and development of computational models.

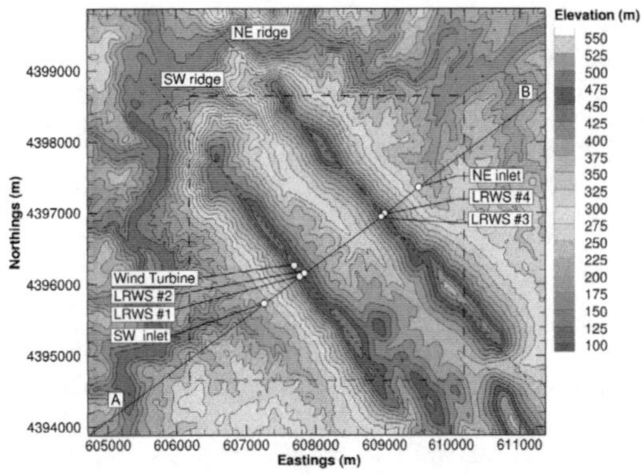

(a) Reference locations (WGS84 UTM 29N): SW and NE ridges and inlet locations, wind turbine and the four long-range wind-scanners (LRWS) along transect A–B (54.7°, relative to the E-W axis). High resolution region, 4 km × 4 km, centred at 4396649N 608198E.

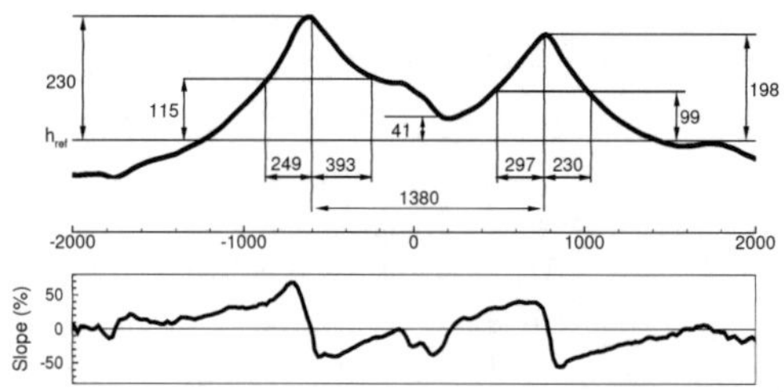

(b) Terrain profile (distance and height in meter, $h_{ref} = 250$ m a.s.l.) and slope along transect A–B.

Figure 1: Perdigão site.

3. Techniques

Computational modeling provides views of the wind flow pattern over large areas, enabling the identification of flow structures beyond the height of meteorological masts. There is no experimental technique other than lidar scanning that can provide similar information.

3.1. Computational modeling

To cover the whole range of scales, a chain of computational models was used, from synoptic with input data from the Global Forecast System [9], passing by WRF [10] mesoscale calculations on three nested domains (37.5, 7.5, and 1.5 km grids), to micro scales resolved by VENTOS®/M [11]. The Reynolds averaged and the $k - \varepsilon$ two-equation turbulence model equations were discretised in a domain of 18 km × 18 km × 10 km; a region (dashed square in Figure 1a) was horizontally resolved by a regular mesh (40 m), 4 m height close to the ground [4]. Conditions at the boundaries were updated every 5 min by the WRF results on the 1.5 km grid. The time step was equal to 1.5 s. The 24 hour (57600 individual time step calculations) animation [8] consists of 288 frames averaged to 5-minute period (200 time steps), running at 5 frames per second.

3.2. Lidar scanning

For this study we deployed measurements from four scanning-Doppler lidars [12] on top of the ridges (LWRS #1 to #4, in Figure 1a). They performed range-height-indicator (RHI) scans along transect A–B (SW–NE direction), covering a distance of up to 3 km on both sides outside

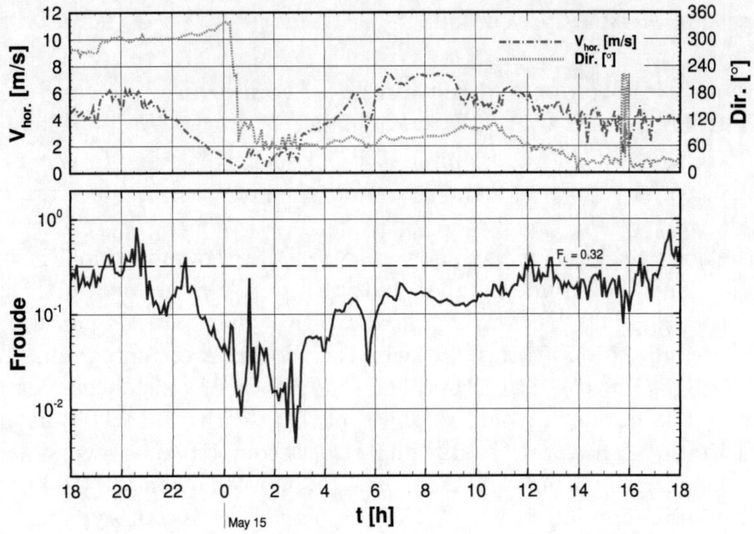

(a) At SW inlet (in Figure 1a), 715 m SW of SW ridge.

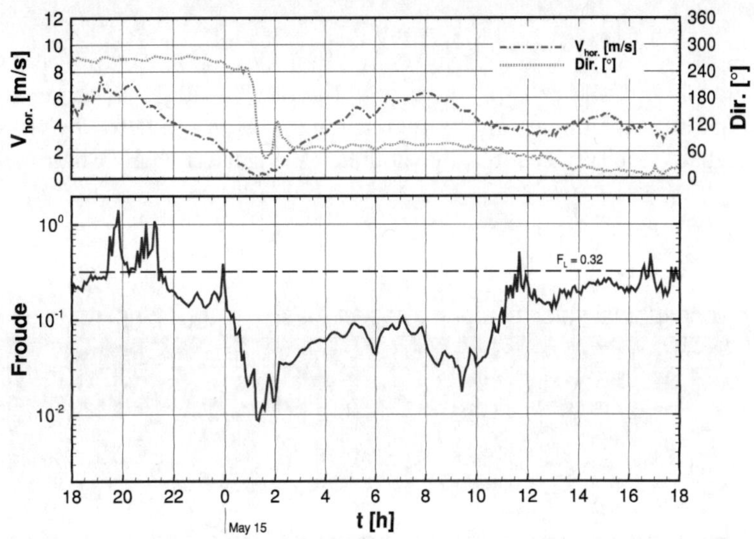

(b) At NE inlet (in Figure 1a), 715 m NE of NE ridge.

Figure 2: Froude number, wind speed and direction at $h_{agl} = 215$ m agl, from 14 May 18:00 UTC until 15 May 18:00 UTC. $F_L = 0.32$, critical Froude number.

of the valley and the region in between the ridges (Figure 1b). Measurements were taken continuously during the upward movement of the scanning beams, with an averaging elevation speed of 0.75° per second over a range of 3°. Range gates were placed every 15 m up to a distance of 3000 m and starting at a distance of 100 m from the scanners. The probe length (full-width at half maximum) of the scans was about 30 m along the laser light propagation path. Thus measurement volumes overlapped and the effective size of the measurement volumes increased with distance because of averaging in the transverse direction while scanning. A single scan took about 24 seconds to complete. We filtered the data using several consecutively applied methods to remove measurements that are influenced by hard targets and other artefacts. The 24-hour animation consists of 3168 individually acquired scans averaged to 5-minute intervals (11 scan) and runs at a speed of 5 frames per second.

4. Wind flow conditions: speed, direction and Froude number

Stratification was characterised by the Froude number, $F = U/(Nh_{agl})$, where U is the wind speed at h_{agl} (215 m agl) and N, the Brunt-Väisälä frequency, is equal to $\sqrt{g/\theta\,\partial\theta/\partial z}$, where g is the gravitational acceleration, and θ and $\partial\theta/\partial z$ are the potential temperature and its gradient at the ridge height (h_{agl}).

To monitor the flow development, we show (Figure 2) the computational results of the wind speed and direction, and the Froude number southwest of SW ridge and northeast of NE ridge, from 18:00 (14 May) to 18:00 (15 May). There are two main periods: until and after 00:00. The northwesterly wind (270°), rotates clockwise to NE (60°), remaining almost perpendicular ($\approx 70°$) to the NE ridge (129.1°) until 13:00. The wind rotation, which occurs within a relatively short period of time (30 minutes), is noticed first (one hour earlier, at 00:00) in the south side of the SW ridge (Figure 2a). The rotation, though fast, is accompanied by a wind deceleration from more than 6 ms^{-1} to nearly no wind (less than 1 ms^{-1}), between 00:30 and 03:00, with the Froude number reaching a minimum of 0.003 at 02:45. The nighttime period, as expected, displays the features of a stably stratified atmosphere, with a smoother wind speed time series compared with the spikier shape during the afternoon, typical of an unstable convective boundary layer.

5. Flow patterns

In this section, we present the flow patterns over the double-ridge, based on the longitudinal velocity along transect, streamlines, and potential temperature predicted by the computational model, and the radial velocity measured by the lidars in transect A–B, Figure 1a.

5.1. *Phase 1: single layer structure*

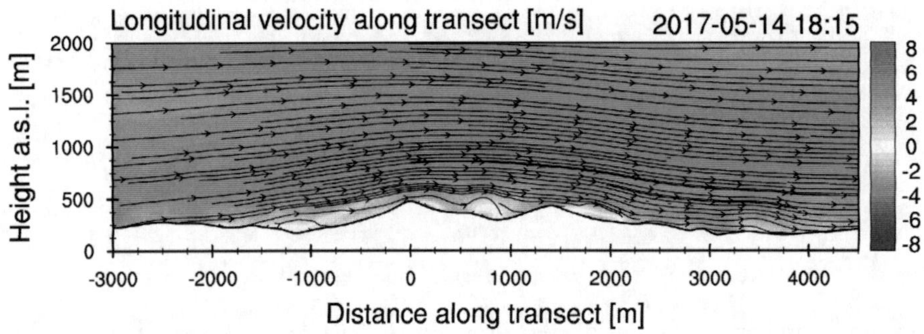

Figure 3: 14 May 2017 18:15 UTC. Streamlines and longitudinal velocity. Computational results are VENTOS®/M micro-scale results (40 m grid) coupled with WRF results (1.5 km grid), based on input data from NCEP; the terrain is plotted in white.

At 18:15 May 14, the westerly ($\approx 270°$, Figure 2a) wind shows in Figure 3 from left to right. Separated flow regions appear on lee-side of both ridges, starting at the top, due to adverse pressure gradient; i.e., the flow displays characteristics of a neutral atmosphere. Based on [7], assuming two-dimensional laminar flow, the critical Froude number $F_L = D/(L\pi)$, below which no waves are possible and flow separation is boundary-layer controlled, is equal to 0.32; here, D, the water tank depth in the laboratory experiment [7], was taken as the free atmosphere height, equal to 1500 m, also equal to L, the width of a single hill in Perdigão. Given the differences between the real flow and the laboratory experiment [7], the agreement between F_L and the

Froude number in Figures 2a and 2b can be considered as a rule of thumb that nevertheless will be used as an indication of nearly unstratified flow between 18:00 and 22:00 May 14.

5.2. Phase 2: the low-level jet from NE and the two-layer structure

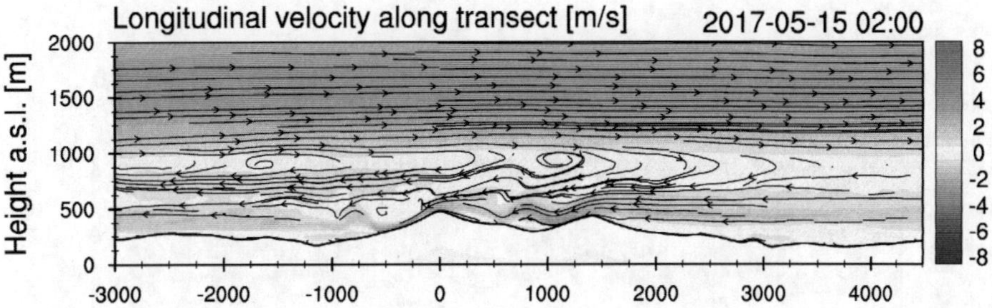

(a) Streamlines and longitudinal velocity. Computations as in Figure 3.

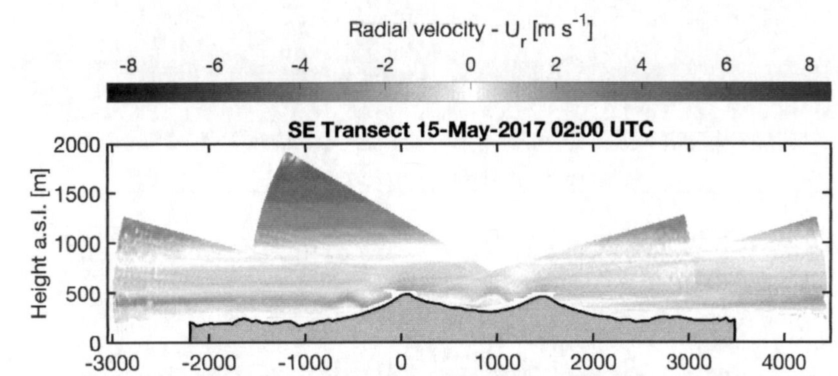

(b) Radial (along the laser beam) velocity, minus sign indicates flow towards the southwest (from right to left in the plot). Scanning lidar results are composite of four range-height-indicator scans; the terrain is plotted in grey.

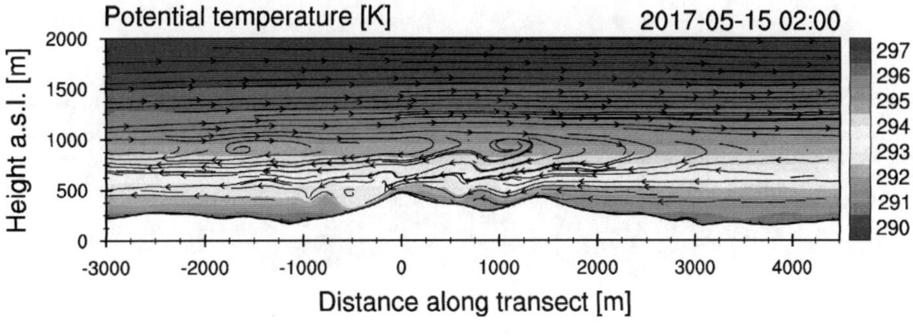

(c) Potential temperature. Computations as in Figure 3.

Figure 4: 15 May 2017 at 02:00 UTC.

During the next four hours (22:00–02:00 UTC), a two-layer structure develops (Figure 4a and 4b): an upper layer (above 1000 m a.s.l.), free atmosphere, made of southeasterly winds, decoupled from the bottom layer (below 800 m) of low wind speed flow, driven by a northeasterly jet flow of cold air at 500 m a.s.l. (Figure 4c). As the jet strength increases, the flow accelerates

IOP Conf. Series: Journal of Physics: Conf. Series **1222** (2019) 012006 doi:10.1088/1742-6596/1222/1/012006

over the two-ridge topography, and internal lee-waves emerge. Separated flow regions do not occur on the lee-sides, they are not set by pressure gradient and the Froude number is about 0.03 (Figures 2a and 2b). According to [13] the low-level jet (LLJ) originates from the plateau between Serra da Estrela and Sierra de Gata mountain ranges, N and NE of Perdigão and the turning of the wind direction is induced by slope and valley winds under weak synoptic conditions.

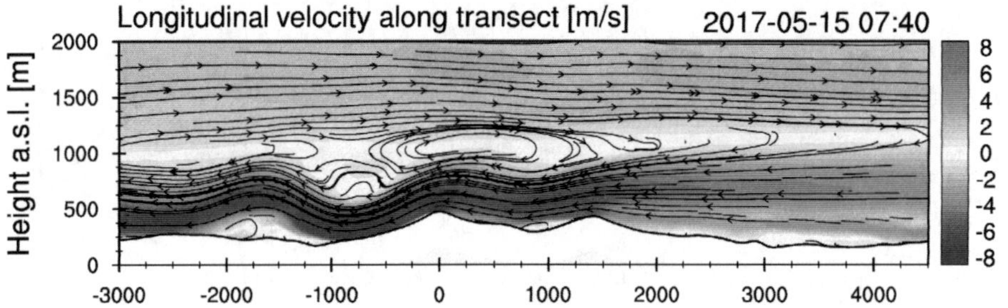

(a) Streamlines and longitudinal velocity. Computations as in Figure 3.

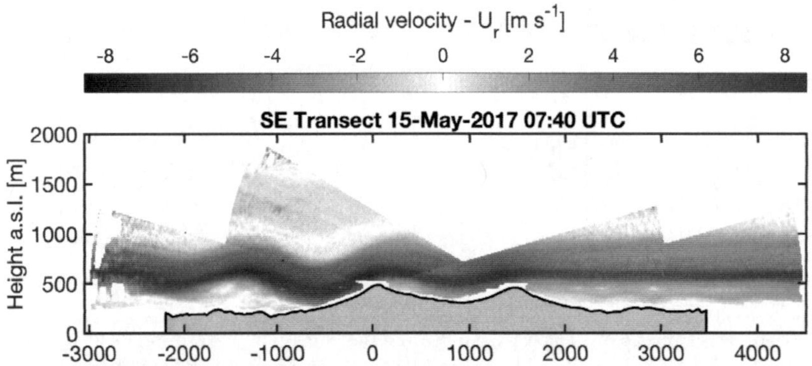

(b) Radial (along the laser beam) velocity, minus sign indicates flow towards the southwest (from right to left in the plot). Scanning lidar results are composite of four range-height-indicator scans; the terrain is plotted in grey.

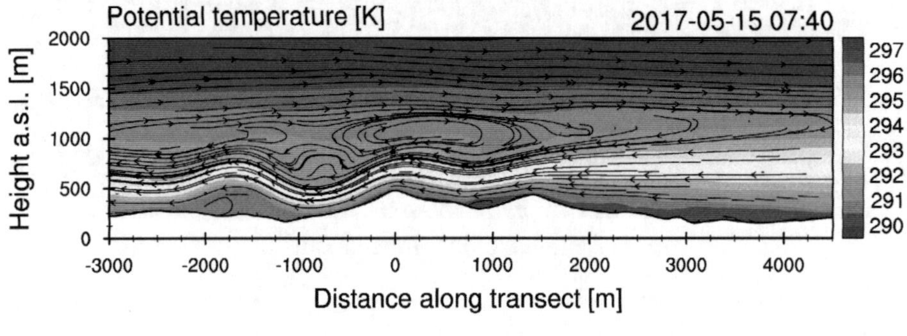

(c) Potential temperature. Computations as in Figure 3.

Figure 5: 15 May 2017 at 07:40 UTC.

The lee-wave pattern is the characteristic of most of the nighttime period, with Froude number about 0.1 (Figures 2a and 2b). Figure 5, at 07:40, was selected among many, by its quality to illustrate the atmospheric phenomena being present during this period and the

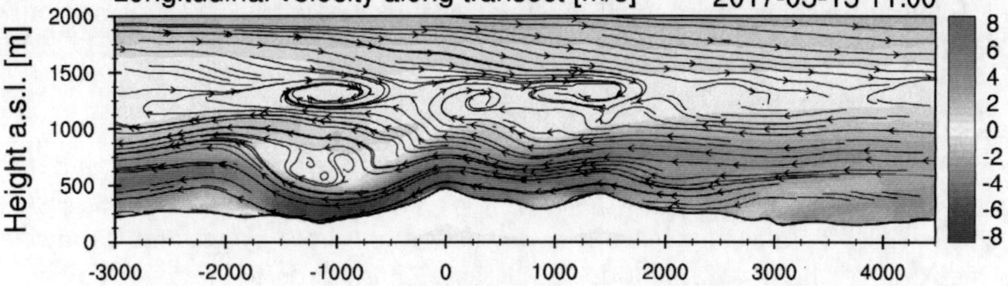

(a) Streamlines and longitudinal velocity. Computations as in Figure 3.

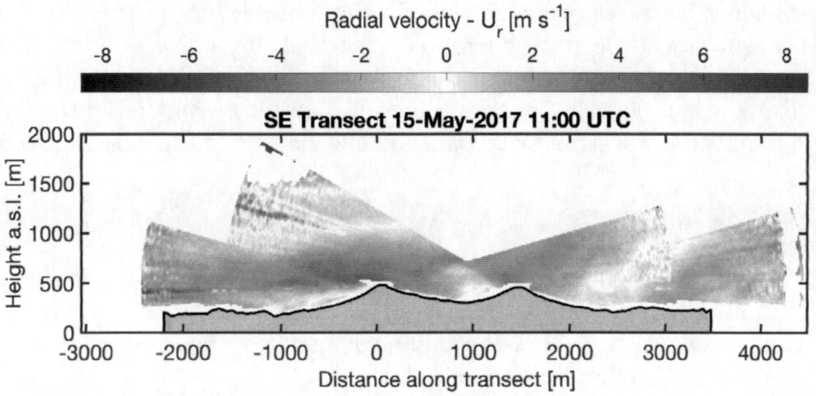

(b) Radial (along the laser beam) velocity, minus sign indicates flow towards the southwest (from right to left in the plot). Scanning lidar results are composite of four range-height-indicator scans; the terrain is plotted in grey.

Figure 6: 15 May 2017 at 11:00 UTC.

ability of both the remote sensing and the computational techniques to capture them. The lee-waves, triggered by the first (NE) ridge, have a wavelength of about the distance between ridges ($\lambda \approx 1400$ m, Figure 1b) and reach a height of 1000 m; i.e., they are trapped upwards by the top homogeneous layer of southwesterly winds, with a transition layer in between, made of large and highly stretched (length/height ratio higher than 10) vortex structures. The waves have a major impact on the flow, yielding a high wind speed (6 m/s) at the top of the ridge, at a distance of about 100 m agl, with consequences on the operation of wind turbines. The flow is almost fully attached in the valley. In the lee-side of the second hill there is no wake vortex, as would be expected under neutral conditions; on the other hand, after the hill at $x = -1500$ m and close to the ground, a lee-wave-type rotor [14] was observed and predicted. Both lidar scanning and computational model agree on the time of occurrence, length, height and location of this region, which is also a large mass of cold air, as shown by the potential temperature in Figure 5c. The agreement goes through the whole nighttime, from the beginning until the end of the lee waves, including cycles of creation and destruction of this rotor structure via hydraulic jump. This is also an indication of the sensitivity of the flow pattern to the Froude number variation experienced during the night.

5.3. Phase 3: the convective boundary layer and the end of the two-layer structure

The time of the establishment of an unstable atmosphere is typically 3 to 5 hours after sunrise (at 6:20), as can be seen on the Froude number increase after 10:00 (Figure 2). At 11:00, Figure 6, both the computer model and the lidar measurements display high activity in the transition layer, about 1000 m, reaching up to 1500 m. The wave-like structure is weakened, after which a new cycle will be reinitialised.

6. Conclusions

The observations by scanning lidars of the Perdigão site showed a large variety of flow phenomena, namely an accelerating low-level jet over two ridges, gravity waves with a length equal to the inter-ridge distance, attached flow downstream of both ridges and a stationary rotor 1500 m downstream of the downwind ridge. All these characteristics were well reproduced by the coupled models.

The good qualitative agreement between the results obtained by scanning lidar and computational modeling increases our confidence in the use of either of these two techniques. Many aspects of the physics of the atmosphere were disclosed that are relevant to wind resource analysis and wind turbine operation, which cannot be revealed by point-based measurements.

As a final note, because Figures 3–6 cannot convey a comprehensive view of the flow, with all its dynamics and complexity, we strongly recommend the viewing of the video [8] corresponding to this study.

Acknowledgments

The Perdigão-2017 field campaign was primarily funded by the US National Science Foundation, European Commission (ENER/FP7/618122/NEWA), Danish Energy Agency, German Federal Ministry of Economy and Energy, FCT-Portuguese Foundation for Science and Technology, (NEWA/1/2014), and US Army Research Laboratory.

We are grateful to the municipality of Vila Velha de Ródão, landowners who authorized installation of scientific equipment in their properties, the residents of Vale do Cobrão, Foz do Cobrão, Alvaiade, Chão das Servas and local businesses who kindly contributed to the success of the campaign. The campaign would not have been possible without the alliance of many persons and entities, too many to be listed here and to whom we are also grateful.

References

[1] G A M van Kuik and *et al.* Long-term research challenges in wind energy – a research agenda by the European Academy of Wind Energy. *Wind Energ. Sci.*, **1**:1–39, doi:10.5194/wes-1-1-2016, 2016.

[2] J M L M Palma, F A Castro, L F Ribeiro, A H Rodrigues, and A P Pinto. Linear and nonlinear models in wind resource assessment and wind turbine micro-siting in complex terrain. *J. Wind Eng. Ind. Aerod.*, **96**:2308–2326, doi.org/10.1016/j.jweia.2008.03.012, 2008.

[3] N Vasiljević, J M L M Palma, N Angelou, J C Matos, R Menke, G Lea, J Mann, M Courtney, L F Ribeiro, and V M M G C Gomes. Perdigão 2015: methodology for atmospheric multi-Doppler lidar experiments. *Atmos. Meas. Tech.*, **10**:3463–3483, doi.org/10.5194/amt-10-3463-2017, 2017.

[4] C V Rodrigues, J M L M Palma, N Vasiljević, M Courtney, and J Mann. Coupled simulations and comparison with multi-lidar measurements of the wind flow over a double-ridge. *J. Phys. Conf. Ser.*, **753**:032025, doi.org/10.1088/1742-6596/753/3/032025, 2016.

[5] H J Fernando and *et al* The Perdigão: peering into microscale details of mountain winds. *Bull. Am. Meteorol. Soc.*. (In press), doi:10.1175/BAMS-D-1-0227.1, 2019

[6] J Mann and *et al.* Complex terrain experiments in the new european wind atlas. *Philos. Trans. Royal Soc. A*, **375**:(2091):20160101, doi.org/10.1098/rsta.2016.0101, 2017.

[7] J C R Hunt and W H Snyder. Experiments on stably and neutraly stratified flow over a model three-dimensional hill. *J. Fluid Mech.*, **96**:671–704, 1980.

[8] R Menke, J M L M Palma, J Mann, N Vasiljević, A S Lopes, V M C Gomes and C V Rodrigues. Animation of lidar and simulation data of complex flow over the Perdigão measurement site, doi.org/10.11583/DTU.7863482, 2019.

[9] Global Forecast System (GFS). https://www.ncdc.noaa.gov/data-access/model-data/model-datasets/global-forcast-system-gfs.

[10] W C Skamarock, J B Klemp, J Dudhia, D O Gill, D M Barker, M G Duda, X -Y Huang, W Wang, and J G Powers. A description of the advanced research WRF version 3. NCAR/TN-475+STR. Technical report, Mesoscale and Microscale Division, NCAR, Boulder, Colorado, USA, 2008.

[11] C V Rodrigues, J M L M Palma, and Á H Rodrigues. Atmospheric flow over a mountainous region by a one-way coupled approach based on Reynolds-averaged turbulence modelling. *Boundary Layer Meteorol.*, **159**:407–437, dx.doi.org/10.1007/s10546-015-0116-7, 2016.

[12] N Vasiljević, G Lea, M Courtney, J-P Cariou, J Mann, and T Mikkelsen. Long-range windscanner system. *Remote Sens.*, **8**(11):896, 2016.

[13] J Wagner, T Gerz, N Wildmann, and K Gramitzky. Long-term simulation of the boundary layer flow over the double-ridge site during the Perdigão 2017 field campaign. *Atmos. Chem. Phys.*, **19**:1129–1146, doi.org/10.5194/acp-19-1129-2019, 2019.

[14] L Strauss, S Serafin, and V Grubišić. Atmospheric rotors and severe turbulence in a long deep valley. *J. Atmos. Sci.*, **73**:1481–1506, doi.org/10.1175/JAS-D-15-0192.1, 2016.

IOP Conf. Series: Journal of Physics: Conf. Series **1222** (2019) 012007 doi:10.1088/1742-6596/1222/1/012007

Investigation of offshore wind farm layouts regarding wake effects and cable topology

Bryce Wade[1], Ricardo Pereira[1,2] and Cameron Wade[3]

[1]Instituto Superior Técnico, University of Lisbon, Lisbon, Portugal
[2]TUDelft, Delft, Netherlands
[3]Institute for Integrated Energy Systems, University of Victoria, Victoria, BC, Canada

Abstract. Offshore wind energy is emerging as a large contributor to installed renewable energy capacity. In order to continue the momentum of its development, the offshore wind industry is looking to continually lower the levelized cost of electricity (LCOE). One area being explored in an effort to lower the LCOE of offshore wind generation is the optimization of the wind farm layout. Many of the offshore wind farm layout designs that exist today are structured in a rectilinear form where turbines are spaced evenly along columns and rows. This research explores the economic advantages of removing rectilinear constraints and optimizing the positions of the individual turbines within an offshore wind farm. At the core of achieving the research objective was the development of a model that is capable of simulating an existing offshore wind farm by converting representative wind farm data into an LCOE. The positions of the turbines within the wind farm can be modified using an optimization framework with the intent to minimize the LCOE. The model comprised of the Jensen Wake Model, a hybrid cable layout heuristic and a cost scaling model. The wind farm layout was optimized using a genetic algorithm. The cost estimation model and optimization framework were applied into two case studies to analyze the results of the wind farm layout optimization of two wind farms, Horns Rev and Borssele. In both case studies the optimized layouts provided higher AEP, shorter intra-array collection cable lengths and ultimately a lower LCOE than the baseline rectilinear layouts.

1. Introduction

1.1. Background

Developing carbon free sources of electricity has become a global focus in recent decades. As the understanding of the importance of this shift within the energy sector spreads, increased interest and investment have been directed towards renewable energy sources in an effort to expedite the transition. Wind energy is a leading technology in the development of large-scale renewable energy generation and is a significant percentage of the electricity generation of many nations. [1] [2]

Offshore wind energy in particular is an emerging technology that is becoming increasingly viable as improvements in technology are driving down the levelized cost of electricity (LCOE), making it more competitive with onshore wind technology and other forms of electricity generation.

One particular area being explored in an effort to lower the LCOE of offshore wind generation is the optimization of the wind farm layout. There are multiple parameters that are considered when designing a wind farm array including but not limited to: wake effects, wind profile, bathymetry, soil types, cable installation costs, transmission losses, environmental concerns, permitting restrictions and aesthetics.

While acknowledging these parameters, many of the resulting wind farm layout designs that exist today are structured in a rectilinear form where turbines are spaced evenly along columns and rows. This is the case for many notable large projects such as Horns Rev, London Array, Gemini Wind Farm and Dudgeon.

Wind farm designers have considered the value of minimizing wake effects, as the rows and columns of wind farms are often oriented into an optimal direction based on the local wind climate. However, the inconsistent nature of environmental factors such as the local wind climate, bathymetry, and soil types would conclude that these are not driving factors in the structured layouts.

It has been hypothesized through many academic research efforts that keeping rectilinear formats may cost a great deal when measuring the annual energy production over the lifetime of the windfarm [3].

When determining the spacing between turbines in an offshore windfarm layout, there are two key parameters that oppose with regards to profitability; wake effects and the collection cable length. As the turbine spacing increases, the wake effects decrease which results in an increase in the annual energy production. Though increased space between turbines results in longer collection cables and in turn, higher transmission losses and increased cable costs. The opposing nature of these design variables result in some layouts being more profitable than others.

Abandoning the rectilinear constraints described above and understanding how the wake effect and cable cost variables are related may be utilized in the wind farm design process to improve the economic viability of offshore wind development.

1.2. Research Objective

This work explores the advantages of optimizing the positions of the individual turbines within an offshore wind farm. The optimization seeks to minimize the objective function which is the LCOE of an offshore wind farm. At the core of this work was the development of a model that is capable of the following:

- Simulating an existing offshore wind farm by converting necessary input data into a representative LCOE.
- Providing the ability to optimize the positions of the individual turbines within the wind farm in order to lower the LCOE.
- Providing the ability to perform sensitivity analysis for different input parameters for the offshore wind farm that can provide insight to the importance of the different design variables.

The model was validated using measured data from an existing wind farm. Upon validation, the model was used to estimate the difference in LCOE's for different wind farms with the existing and optimized wind farm layouts and conclusions were drawn from the results.

2. State of the art

A wind farm consists of an array of wind turbines that convert the kinetic energy from the wind to electricity. Wind turbines can be arranged within a wind farm in a certain orientation in order to increase the energy the wind farm yields. Traditional layouts are typically in a rectilinear format, identifiable by the turbine's arrangement in equally spaced rows and columns. The majority of the largest wind farms in operation today are in a grid format, rotated in an orientation with respect to the wind direction to maximize energy yield (for the set formation). However there has been a growing field of research into truly maximizing the energy yield for a given wind farm site by abandoning the rectilinear constraints. The area of this research has been termed the Wind Farm Layout Optimization Problem (WFLOP), where interested parties are creating models and using optimization algorithms to better understand how different wind farm orientations can provide higher energy yields under different environmental and physical circumstances. [3]

2.1. Wind Farm Layout Optimization Problem (WFLOP)

The WFLOP requires a defined objective function and an optimization strategy. The objective function is calculated based on the design parameters, and is minimized through optimization of the wind farm design.

The optimization strategies use algorithms that iterate through various scenarios in search of satisfying constraints and minimizing the objective function. The optimization algorithms can be categorized into gradient methods [4], genetic algorithms [5] [6], viral algorithms [7], particle swarm algorithms [8] and greedy heuristic algorithms [9]. [10]

The WFLOP was first explored by Mosetti et al. [5] by considering a wind farm that did not have a defined desired power output, but a fixed number of potential turbine positions. The wind farm consisted of a 10x10 grid where the wind turbine could be placed in the center of each quadrant or not, in search of the most profitable design based on the objective function. Since the work of Mosetti et al. there have been numerous attempts to determine the best methods for designing a wind farm layout. The research that followed Mosetti et al. can largely be categorized into two paths: (1) a track focused on the development and testing of different optimization strategies i.e. optimization algorithms, [6] [8] [9] and (2) a track focused on improving the wind farm model accuracy [11] [12]. [3] The second path concentrates on the inclusion of more comprehensive sub-models that consist of an increasing number of parameters in an effort to make the simulations more representative and applicable.

Hebert et al. [3] provided a comprehensive review of the state of the art of wind farm design and optimization. Hebert et al. reviewed over 150 works and found that the most common objective functions being used were the maximum AEP and minimum COE. The most common wake model used within WFLOP research to date is the Katic-Jensen model. This is in large part due to its modelling simplicity which lessens the computational demand and decreases the simulation time. Hebert et al. concluded that computational technologies are typically the limiting factor for the treatment of the WFLOP problem. Beyond improvement of computational restraints, Hebert et al. concluded that the most important research trends in WFLOP are focused on improving the formulations describing the expected energy conversion of the wind farm, the wind turbine wakes and turbulence impacts, the wind turbine structural fatigue and degradation, the various types of environmental impacts, the overall electrical system losses and reliability and the uncertainty and risk management.

3. Model Development

The objective of the model optimization, is to optimize the positions of the turbines within a wind farm in an effort to minimize the LCOE. There are two variables within Equation 1 that are dependent on the positions of the turbines (all else were considered independent within this work); Initial Capital Costs (ICC) and the annual energy production (AEP). More specifically within the ICC, it is the intra-array cable costs that are dependent on the wind turbine positions.

$$LCOE = \frac{ICC * FCR}{AEP} + AOE \tag{1}$$

The modelling framework developed in this work is explained easiest by breaking down the three core components. The physical component included sub-models that provided collection cable lengths and annual energy production for a given wind farm based on its layout. The economic component utilized the cable length and AEP to generate an LCOE. The optimization component used the LCOE as its objective function and modified the turbine positions in search of an optimal layout. A schematic of the wind farm layout optimization framework can be seen in Figure 1.

3.1. Physical Component

The physical component accounts for the environmental and physical characteristics of the wind farm and provides the annual energy production and collection cable layout. The two core sub-models within the physical component pertain to the wake effects and the cable topology.

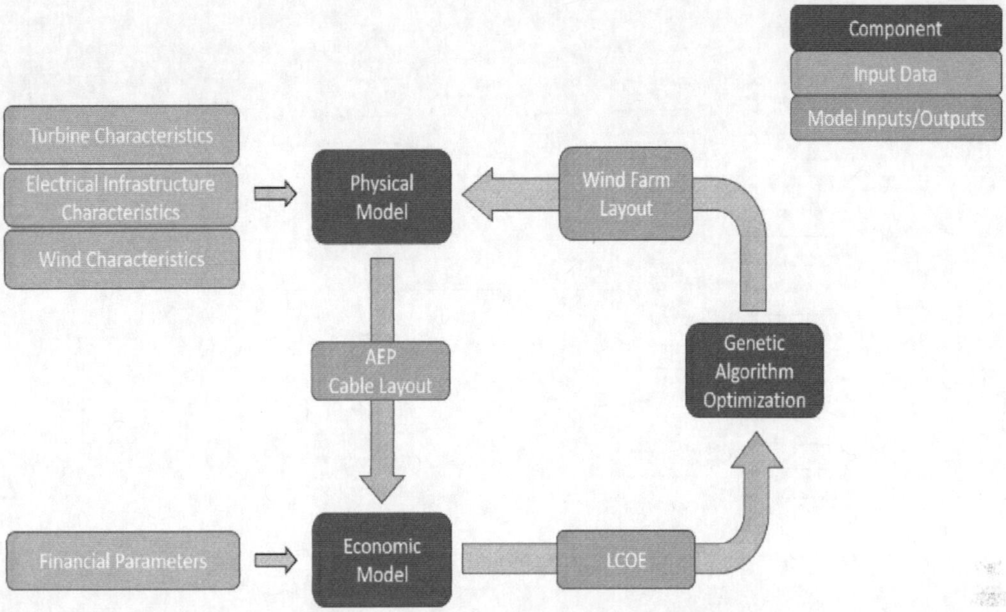

Figure 1: Optimization framework schematic.

3.1.1. *Wake effects*

As a turbine extracts energy from the wind and converts it to electrical energy, it leaves a wake behind which is characterized by reduced wind speeds and increased turbulence in the flow [13]. The wake from a turbine expands and gradually returns to the free stream condition as it continues downstream. If the wake of a turbine intersects with the swept area of another turbine downstream prior to reaching the free stream condition, the downwind turbine is considered to be shadowed by the upstream turbine [14]. If a turbine is shadowed, it experiences a lower wind velocity than the free stream, which can result in less energy conversion than if the turbine was not shadowed.

3.1.2. *Jensen Model*

The Jensen Model is one of the oldest and most widely used wake prediction models. A benefit to the use of the Jensen model is that it is a relatively simple model that is less computationally demanding than others and has still been proven to provide comparably accurate results [15] [16] [17].

A key assumption to the Jensen single wake model is that mass[1] is conserved in the direction of the turbine axis [18]. The total velocity deficit of the wind velocity at position x (see Figure 2), due to the wake effects from an upstream turbine can be given by Equation 2 [18].

$$1 - \frac{v_1}{v_0} = \frac{1 - \sqrt{1 - C_t}}{\left(1 + \frac{2\,\alpha\,x}{D_0}\right)^2} \tag{2}$$

Equation 2 assumes the complete downstream turbine is within the wake. When the downstream turbine is only partially shadowed by an upstream turbine, a proportionate amount of the wake with respect to the area shadowed is considered.

When having multiple wakes interacting with one another, it is assumed that the kinetic energy deficit of a mixed wake is equal to the sum of the energy deficits for each wake acting on a given position. This can be expressed by Equation 3 for a turbine in the wake of two others [18].

[1] Katic et al. [18] incorrectly state "momentum is conserved".

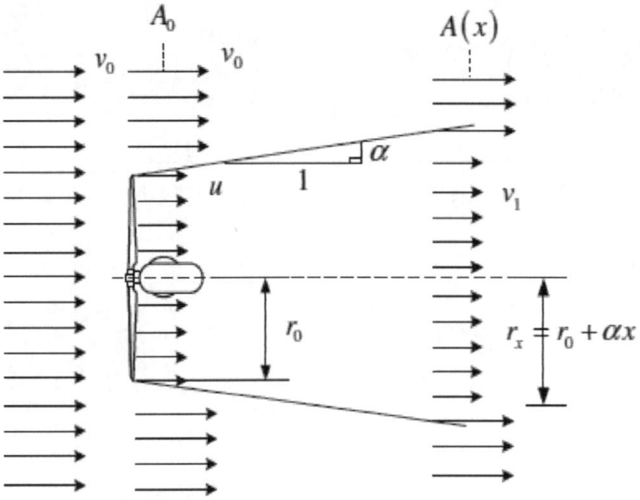

Figure 2: Jensen wake model diagram.

$$v_i = v_o \left[1 - \left(\sum_{i=1}^{N_t} \left(1 - \frac{v_i}{v_o} \right)^2 \right)^{\frac{1}{2}} \right] \tag{3}$$

This approach of using Equations 3 is valid for both scenarios seen in Figure 3.

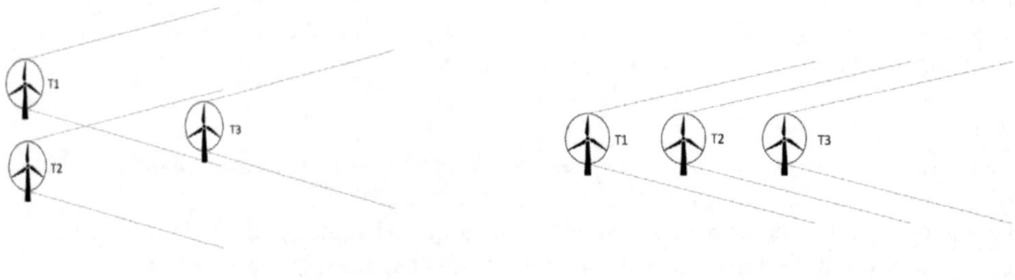

(a) (b)

Figure 3: Multiple wake effects diagram.

In Figure 3 (a), both Turbine T1 and Turbine T2 cause wakes that interact downstream and affect Turbine T3. In Figure 3 (b), both Turbine T1 and Turbine T2 cause wakes that affect Turbine T3, however in this scenario Turbine T2 doesn't experience the free stream velocity. Even in this arrangement, Equation 3 holds and the free stream velocity is used to calculate the combined wake effects for Turbine T3 [19].

3.1.3. Intra-array collection cable

The intra-array collection cable system gathers the electricity from each individual turbine and connects it to a substation. The subsea cables drop from the individual turbines to the sea bed and are typically buried 1 to 2 meters in the seabed. Current collection systems are medium voltage, and depending on the project can range from 10 to 66 kV. There are a number of design strategies for the layout of collection cables, each with their own strengths and drawbacks.

The simplest design strategy for a collection system is the radial design which connects a series of turbines on a single cable. The number of turbines per cable is determined by the maximum power output from each turbine and the maximum rated capacity of the cable. The radial design is considered to be advantageous because it uses minimal cable, but its drawback is that it does not account for any redundancy [20].

The branch design is less common for offshore wind farm designs, but is frequently used in communication networks and has desirable features for use in collection cable design. The branch design, provides opportunity for lower cost and higher reliability than the currently most common approach, the radial design. [21]

In order to adequately compare wind farm layouts, the model required a cable layout method that provided consistent radial/branched results between rectilinear layouts and irregular layouts.

3.1.4. Cable layout heuristic

A hybrid cable layout heuristic developed by Katsouris [21] that combined the planar open savings (POS) heuristic [22] and the Esau-Williams heuristic [23] was implemented. The heuristic acted as a nested optimization within the physical model that efficiently designed a near optimal cable layout using the strengths of the POS heuristic and the EW heuristic.

The POS heuristic was developed by Bauer and Lysgaard and is one of the most efficient heuristic algorithms for the radial internal connection of offshore wind farms [21]. The POS heuristic begins with an initial solution of every wind turbine being connected directly to the substation. In every step the savings heuristic considers merging two routes (cables) into one cable. The savings associated with this merge are logged, and in every step the heuristic greedily chooses the merge with the highest savings resulting in a route not exceeding capacity [22]. The heuristic continues until all savings are achieved under the given constraints and capacities and a near optimal solution is reached.

The EW heuristic is a saving procedure which, similar to that of the POS heuristic, starts from a star tree formation of the turbines and substation connections. In each iteration, the merging of two routes is determined by the option that provides the largest savings [23]. Relative to the POS heuristic, the EW provides more freedom and possibilities for merging routes since there is no limitation on the position of the turbines in their corresponding routes. The main characteristic of the EW heuristic is that it attempts to connect the turbines further from the substation into clusters first, and then upon a cable reaching its full capacity, the algorithm continues with creating a new cluster closer to the substation [21].

3.2. Economic Component

The economic component's function is to use the AEP and cable layout obtained from the physical component and to assign values to the outstanding variables required to provide the LCOE for the wind farm (see Equation 1). These outstanding variables include the annual operating expenses (AOE) (see Equation 4), bottom lease cost (BLC), operations and maintenance (O&M), levelized replacement cost (LRC), fixed charge rate (FCR) and most significantly the remaining parameters within the ICC. The ICC is the sum of the turbine system costs and the balance of station costs.

$$AOE = BLC + \frac{O\&M + LRC}{AEP} \tag{4}$$

The Wind Turbine Design Cost and Scaling Model developed by Fingersh et al. [24] was selected to be the basis of the economic model for all of the fixed costs because of the simplicity of the parameters required. The Fingersh model is intended to provide reliable cost projections for wind generated electricity based on different turbine and wind farm sizes. This was an important characteristic for implementation into a model that intends to be used for a variety of wind farm and wind turbine sizes. It is understood that at the time of publishing the Fingersh et al. model, the components directly related to offshore technology were in their infancy and the forecasts are considered extremely rough. The cost estimates are projected based on only the turbine power rating, rotor diameter, hub height, as well as a few inflation and cost of material variables, all of which remain unchanged throughout the optimization process.

Modifications were required for the electrical cost estimates within the ICC for the cost scaling model developed by Fingersh et al. The collection cable costs within the model were dependent on the rated power of the turbine, and not the cable topology which was a core feature of the model being developed. Therefore, appropriate modifications were made by using components of an additional NREL cost model developed by Green et al. [25] focused primarily on electrical collection and transmission systems for offshore wind power.

The LCOE results from the economic component are estimated values with the purpose of being used for comparison between wind farm layouts with the same economic parameters. Assumptions and limitations, as well as explanations for the derivations of the empirical cost scaling models can be found at [24] [25].

3.3. Optimization component

A genetic algorithm was selected for the optimization of the wind farm layout primarily based on the research into other WFLOP. Genetic algorithms are probabilistic search algorithms which are designed to mimic the logic of natural selection. Genetic algorithms work by generating populations of individuals (wind farm layouts) that consist of a set number of variables (x and y coordinates for each turbine). The objective function for each individual (LCOE of each windfarm layout) is calculated for the entire population. The population is evaluated based on the desired stopping criteria for the optimization and if the criteria is met the optimization process is complete. If the stopping criteria is not met, the population goes through a genetic manipulation process where the elite individuals (wind farm layouts with lower LCOE) are passed on to the following population and other individuals are selected to become parents in the following generation. Once the parent individuals have been selected, a crossover operation is used to combine the traits of the parent individuals. The final step in the genetic manipulation process is the mutation phase in which there are random alterations made to the population. These random alterations are important for incorporating new information into the population that increase the likelihood of testing more of the result surface [26]. The children individuals that were created from the parent individuals are then evaluated against the optimization stopping criteria to determine whether or not another genetic manipulation process is required. There are different parameters and strategies that can be implemented for each stage of the genetic manipulation which is described further by Pohlheim [27].

External constraints were applied to the genetic algorithm that sectioned the wind farm into tiles where only one turbine could be micro-sited within each tile as seen in Figure 4. The wind turbine could be placed anywhere within each tile so long as it also satisfied the additional Jensen model constraint which ensures no turbines were closer than 3 diameters to one another. This feature provided some structure to the wind farm which may be desirable for designers, but more importantly it limited the function result surface which reduced the computational demand of the optimization process.

IOP Conf. Series: Journal of Physics: Conf. Series **1222** (2019) 012007 doi:10.1088/1742-6596/1222/1/012007

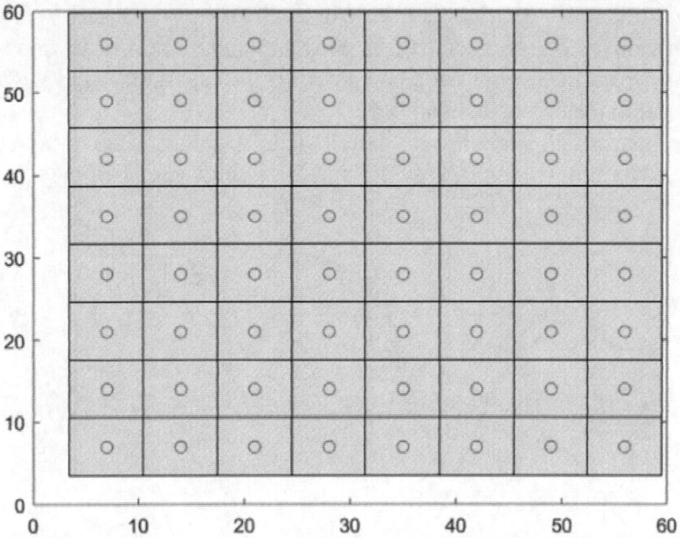

Figure 4: Wind farm layout optimization constraint.

4. Model Validation

Validation was carried out once the model was completed and model behaviour within the design space was analyzed. The wake model validation compares results from the present model with measured data from an existing wind farm.

Horns Rev wind farm (seen in Figure 5) was used for model validation due to the amount of publicly available information [28] [29] [30]. The information necessary for validation is the physical and environmental characteristics of the wind farm (turbine power and thrust curves, wind rose and Weibull distribution parameters, etc) as well as the measured annual energy production of the wind farm. Additionally, measured power data for individual machines is required to validate the wake model specifically by analyzing the wind velocity deficits at particular turbines in different locations within the wind farm.

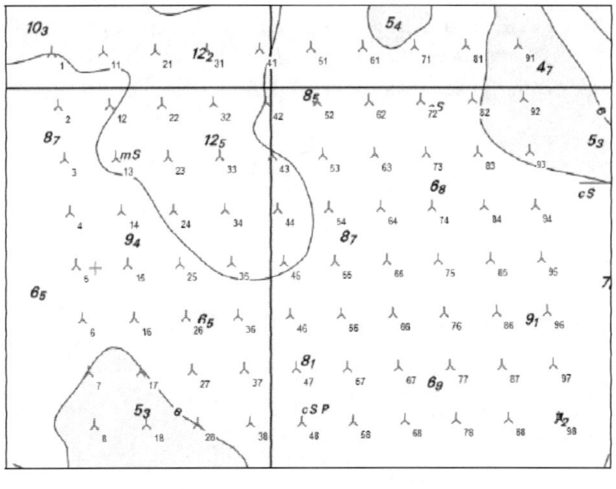

Figure 5: Horns Rev wind farm layout.

IOP Conf. Series: Journal of Physics: Conf. Series **1222** (2019) 012007 doi:10.1088/1742-6596/1222/1/012007

4.1. Wind Velocity Deficit Validation

The wind velocity deficit can be defined as the wind velocity experienced by a given wind turbine experiencing wake effects divided by a wind turbine experiencing free stream wind velocities. The wind velocity deficit results from the model were compared to the measured wind velocity deficits of two different turbines within the Horns Rev wind farm.

The results for comparing the model wake deficits with different wake decay constants of Turbine 17 with the measured data can be seen in Figure 6 and the results from Turbine 45 in Figure 7.

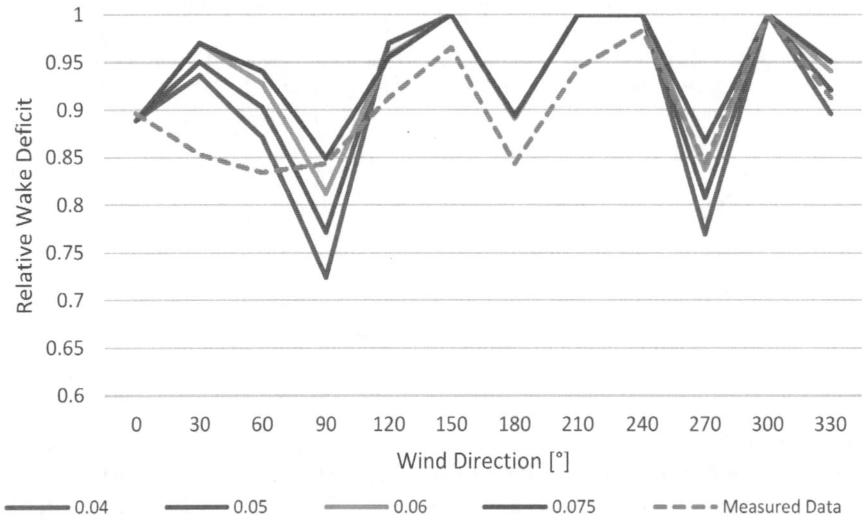

Figure 6: Wind deficit comparison (Turbine 17).

While analyzing Figure 6 and Figure 7, the model results show a trend where the highest wake effects occur when the wind approaches from 0°,90°,180° and 270°, which is expected with this rectilinear layout. In between these angles the wake deficits at the turbines tested are less, with the model results showing some orientations experiencing no wake losses.

Between 0° and 90°, the measured data does not trend with the model results as well as for other orientations. Firstly, the measured wind deficit at 0° is less than the measured wind deficit for 90°, 180° and 270°. This discrepancy is unexpected, as for example when the wind is approaching from 0°, Turbine 17 is near the back of the wind farm, compared to at 180° Turbine 17 is in the second row.

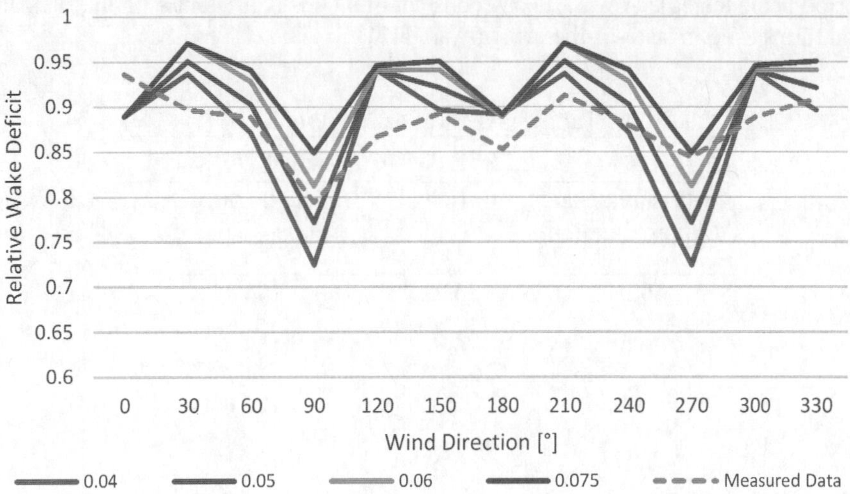

Figure 7: Wind deficit comparison (Turbine 45).

Every 90° the model results indicate a peak where the turbines experience less wake effects. The measured data matches this trend between 90° and 360°, however between 0° and 90°, the measured data inexplicably show no such peak. Additionally, Sorensen [17] indicates that for large wind farm arrays, the Jensen model over estimates the wake effects in the first rows and underestimates the wake effects in the back rows. This however does not explain the discrepancy for Turbine 45.

With regards to a comparison between wake decay constants, a wake decay constant of 0.075 always underestimates the wake effects, whereas a wake decay constant of 0.04 tends to both over estimate and underestimate depending on the direction of the wind approaching. As expected, the wake decay constants trend in the same order with larger wake decay constants providing lower wind deficit estimates and smaller wake decay constants providing higher wind deficit estimates. There are exceptions to this rule when in some orientations larger wake decay constants have an expansion angle that includes some turbines that the smaller wake decay constants do not (seen at 120°).

4.2. Limitations in Wind Deficit Validation

The model results do not match the measured data exactly for the following reasons:

- A potential influence on the difference between the measured and model results could be attributed to the Weibull distribution parameter assumption for the wind distribution between 7-10 m/s.
- Inconsistencies in the measured data that cannot be explained by relying on second-hand data.

Further model validation with other offshore wind farm data sets which were not available would provide more conclusive results.

4.3. AEP Validation

When comparing the model AEP results to the measured results for the overall wind farm output, the model was a better predictor than it was for the wind deficit measurements of individual turbines. This was expected as the aim of the Jensen model is to give an estimate of the energy content of a wind farm rather than to describe the velocity field accurately [18].

The AEP was estimated using four different wake decay constants to determine which was the best fit for the measured data. The wake decay constant that fit the measured data the best was 0.04 which gave AEP results within 1.1% of the approximate AEP that was observed at Horns Rev wind farm in

2005 (the period being tested). A wake decay constant of 0.04 is consistent with the standard value for offshore wind farms used in state-of-the-art software [17].

Table 1: AEP validation results.

k	**AEP (GWh)**	**Relative AEP**
Measured Data (Approximation)	630	100.00%
0.04	636.7	101.1%
0.05	650.8	103.3%
0.06	661.6	105.0%
0.075	669.5	106.3%

4.4. Model Behaviour Observations

4.4.1. Cable Layout Heuristic

The cable layout heuristic was proven to be more effective than the traditional radial design that collects the power from all turbines in a column and transmits it to the substation. A visual comparison of the traditional layout and an improved layout from the cable layout heuristic on a rectilinear wind farm layout can be seen in Figure 8.

The heuristic results were compared with traditional designs for different numbers of turbines, different rows and column spacing and different intra-array cable capacities. In all scenarios the heuristic outperformed the traditional design for a rectilinear wind farm layout by a reduction of 11-19% in cable length.

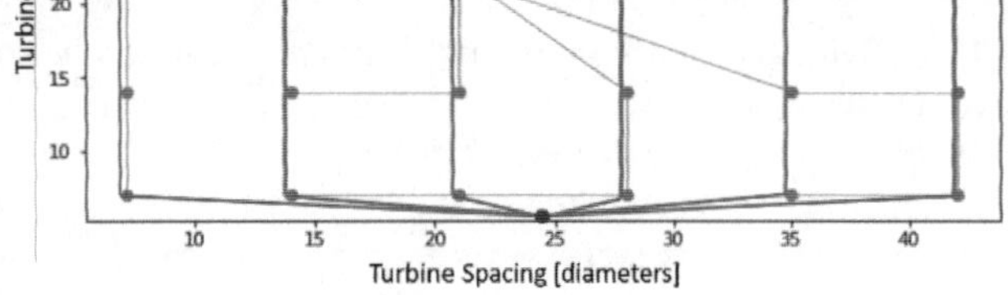

Figure 8: Cable layout comparison (rectilinear wind farm layout).

The cable layout heuristic becomes of increased value for non-rectilinear layouts where the function developed for generating the traditional layout struggles to provide reasonable results. Figure 9 shows the traditional cable layout function results along with the improved cable layout results. The improved cable layout results in only 68% of the traditional cable length required.

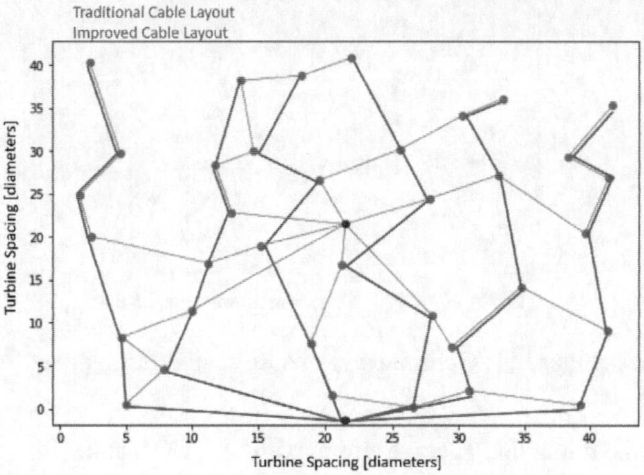

Figure 9: Cable layout comparison (irregular wind farm layout).

The cable layout heuristic function consistently provided a cable layout design with shorter overall cable length than the traditional layout function and was more robust and effective in dealing with non-rectilinear layouts.

5. Results & Analysis

5.1. Case Study I: Horns Rev Wind Farm

The optimization of the Horns Rev Wind Farm evaluated 900 individual wind farm layouts in total, over 9 generations and determined an optimal layout after the minimum LCOE from 4 consecutive generations averaged a difference of less than 0.0001. It is understood that the limited nature of this optimization simulation does not necessarily provide a true optimum, but these were the best results obtained due to time and computational restrictions. The existing Horns Rev wind farm is shown in Figure 10 with a cable layout that was designed using the hybrid cable layout heuristic. The cable layout heuristic was used instead of the installed cable layout to provide a valid comparison between the different wind farm layouts due to assumptions that were made in the cable cost sub-model. The improved wind farm layout can be seen in seen in Figure 11.

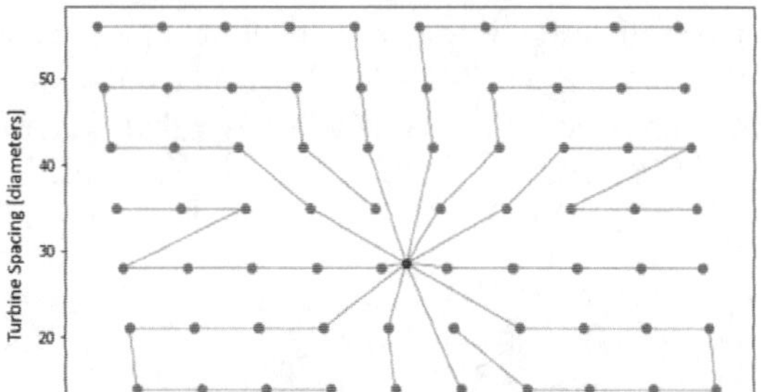

Figure 10: Existing Horns Rev layout (optimized cable design).

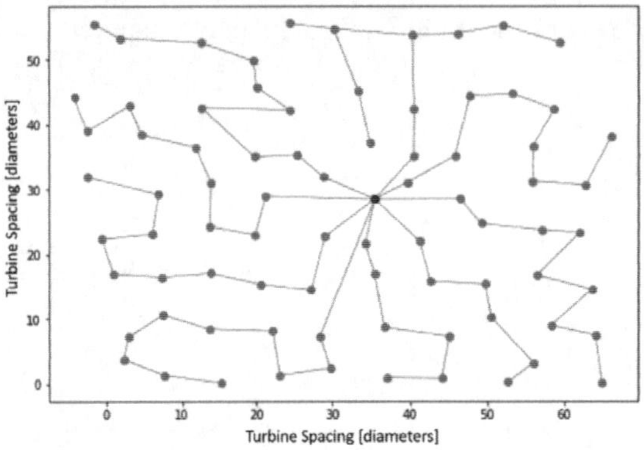

Figure 11: Optimized Horns Rev wind farm layout.

Figure 12 displays the minimum, average and maximum LCOE value for each generation. None of the individuals that were evaluated resulted in an LCOE as high as the LCOE from the existing layout (0.0904 $/kWh).

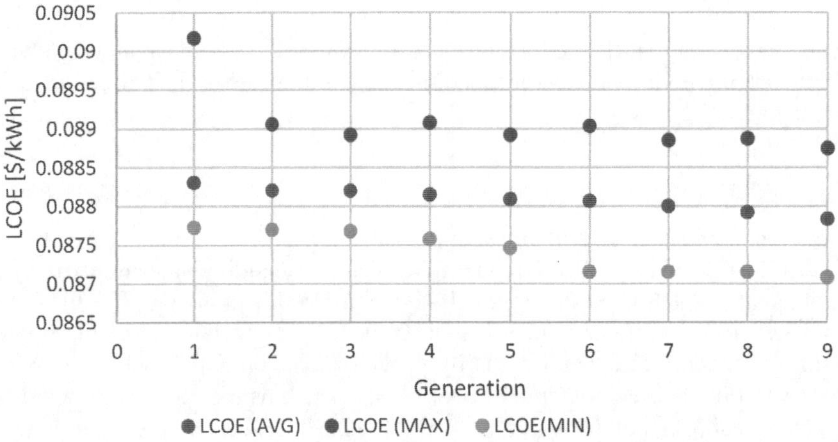

Figure 12: LCOE evolution of Horns Rev wind farm optimization.

The array efficiency for the final design is 90.5% compared to the existing design which was 86.5%. A comparison of the cable length, AEP and LCOE of the two layouts can be seen in Table 2. The total collection cable length of the improved design is 8.6% shorter than the original layout. The improved design results in an estimated 4.7% increase in AEP. As both the cable length and AEP improve, the resulting LCOE for the improved layout is 3.7% lower than the existing Horns Rev layout.

Table 2: Horns Rev wind farm layout comparison.

Parameter	**Existing Layout**	**Improved Layout**
Cable Length [m]	65611	59943 (91.4%)
Cable Cost [$10^6$$]	38.96	35.59 (91.4%)
AEP [GWh]	626.91	656.37 (104.7%)
LCOE [$/kWh]	0.0904	.0871 (96.3%)

5.2. Case Study II: Borssele Wind Farm

The second case study analyzed the Borssele Wind Farm. The Borssele wind farm zone is divided into 5 smaller wind farm sites. Borssele III and Borssele IV are considered for this paper and will be referred to as the Borssele Wind Farm. What is different about the Borssele wind farm compared to Horns Rev is that the Borssele wind farm has not yet been constructed. So, although an existing design cannot be validated like it was for Horns Rev, the data made available for the design tender [31] is used along with a proposed baseline wind farm design obtained from Perez-Moreno et al. [32] to analyze different layouts. The baseline wind farm layout (seen in Figure 13) was a rectilinear layout that comprised of 74 5-MW turbines.

Similar to the first case study, the analysis of Borssele wind farm was also limited by time constraints on the optimization. In the case for Borssele wind farm, 2200 individual wind farm layouts were evaluated over 11 generations, with the best layout seen in Figure 14.

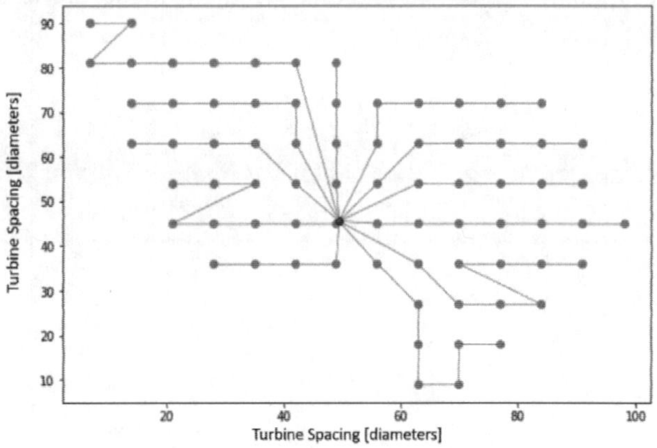

Figure 13: Baseline Borssele wind farm layout.

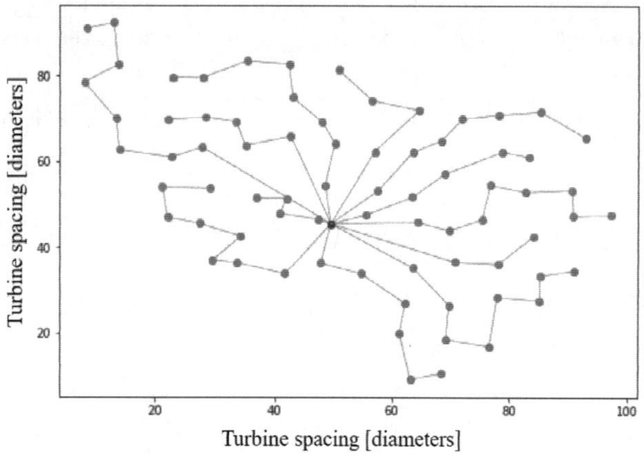

Figure 14: Optimized Borssele wind farm layout.

Figure 15 shows the minimum, average and maximum LCOE result for each generation, all of which are lower than the LCOE for the baseline layout (0.0955 $/kWh). As was the same in Case Study I, the initial individual simulated was very similar to the baseline layout which is why the maximum value in the first generation is similar to the LCOE of the baseline layout. It should be noted that the LCOE cost for the Borssele wind farm does not consider transmission costs (or losses) because the transmission was provided by another project. Also, the project proposal has a maximum bid for the cost of electricity of 0.11975 €/kWh [31], which indicates that the results from the model are within an appropriate range of expectations.

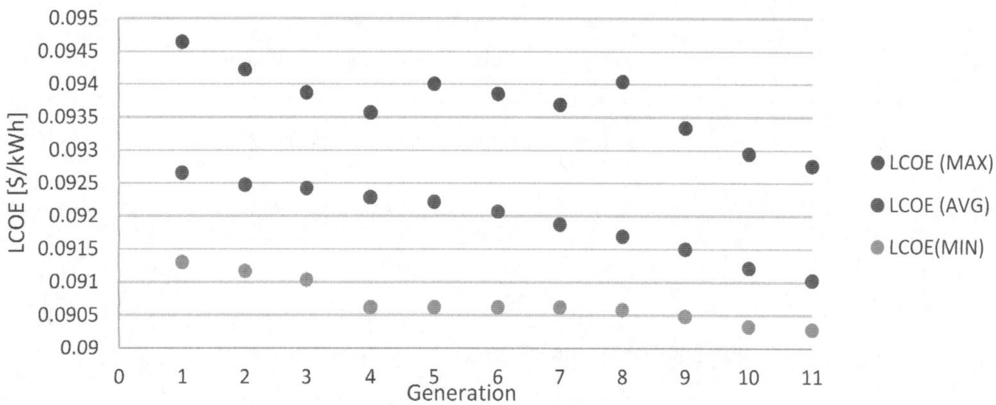

Figure 15: LCOE evolution of Borssele wind farm optimization.

The array efficiency for the final design is 93.2% compared to the existing design which was 89.0%. A comparison of the cable length, AEP and LCOE of the two layouts can be seen in Table 3. The total collection cable length of the improved design is 8.2% shorter than the baseline layout. The improved design results in an estimated 4.8% increase in AEP. As both the cable length and AEP improve, the resulting LCOE for the improved layout is 5.5% lower than the baseline Borssele rectilinear layout. The influence of each parameter on the LCOE will be explored further in the observations in the following section.

IOP Conf. Series: Journal of Physics: Conf. Series **1222** (2019) 012007 doi:10.1088/1742-6596/1222/1/012007

Table 3:Borssele wind farm layout comparison.

Parameter	Baseline Layout	Improved Layout
Cable Length (m)	107540	98705 (91.8%)
Cable Cost ($10^6$$)	89.40	82.05 (91.8%)
AEP (GWh)	1441.6	1511.2 (104.8%)
LCOE ($/kWh)	0.0955	0.09028 (94.5%)

5.3. Case Study Observations

In both case studies results indicate that optimizing the positions of the individual turbines within the wind farm provides both a higher AEP and lower cable costs. The original expectation was that these two variables were contrasting and that one may have to worsen in order for the other to improve and lower the LCOE. However, in both cases the contrasting variables both improved making it difficult to know which variable has a greater impact on the LCOE.

Upon analysis of Figure 16 and Figure 17, the average AEP results for each generation trend upwards (improves) as the optimization continues for both wind farms. The average cable cost on the other hand, has a less distinct trend for Borssele but trends upwards (worsens) for Horns Rev. This would indicate that in the case of Horns Rev, increasing the AEP was of more influence than lowering cable costs in an effort to minimize the LCOE.

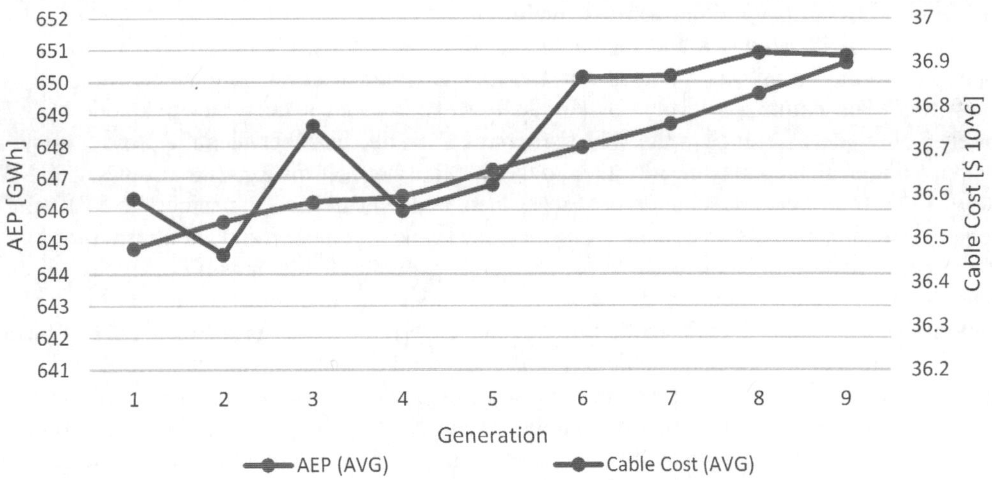

Figure 16: Horns Rev wind farm – AEP and cable cost evolution.

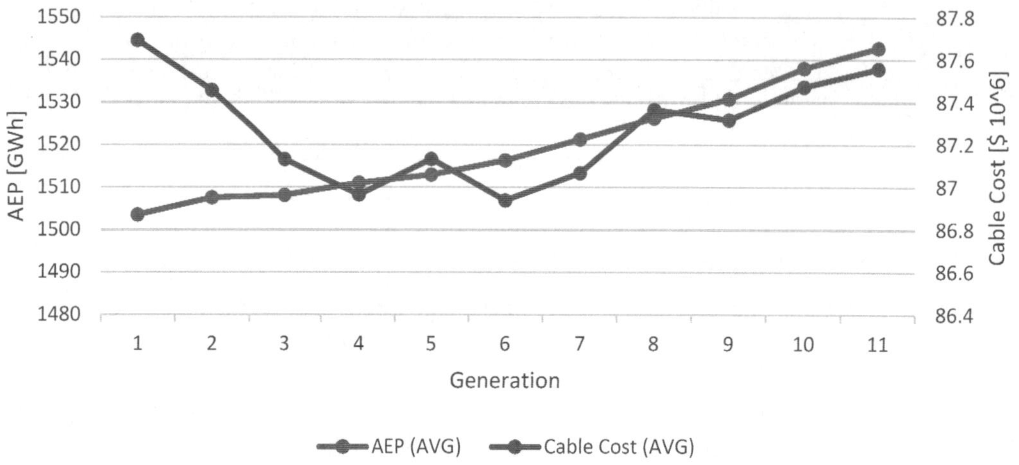

Figure 17: Borssele wind farm – AEP and cable cost evolution.

This relationship can be observed in Equation 1 which shows that the LCOE corresponds to the relationship of the ICC divided by the AEP. As noted above, the intra-array cable costs are only a fraction of the ICC which results in the LCOE being more sensitive to variations in the AEP than the collection cable costs.

The AEP for each wind farm varied no more than 7% once the optimization had abandoned the original rectilinear layout. The cable costs varied much more than the AEP, by 18% and 19% for the Horns Rev and Borssele wind farms respectively.

As both the AEP increased and the cable cost decreased for each design relative to the existing layouts, both wind farms saw a decrease in the LCOE which was the anticipated outcome from the optimization. The optimized layout for Horns Rev experienced a decrease in LCOE of 3.7% and Borssele 5.5%. The difference in the WFLO-driven LCOE decrease between the present work and the results of Pillai et al. [12] may be attributed to the fact that both Horns Rev (80 turbines, 160 MW) and Borssele (74 turbines, 370 MW) are much larger than Middelgrunden (20 turbines, 40 MW) and would experience much more wake interaction. The increased wake interaction and variability for cable design of the larger farms would be expected to provide larger potential for savings if the layout is optimized.

The results of two case studies indicate that rectilinear wind farm designs are not optimal with regards to minimizing the LCOE. Despite these findings, which agree with the prior research, rectilinear wind farm layouts are still the industry standard for offshore wind farms. This may be a result of the industry being conservative in nature and as such, hesitant to transition away from the traditional designs. Although the observations from this work indicate that the traditional designs are not optimal from a cost of energy standpoint, they may be considered optimal for other valued variables, such as planning, financing or aesthetics. This trend may change in the future as more accurate models are developed that provide more assurance and information on new design strategies to help the industry become more economically viable.

6. Conclusions

The objective of the work was to explore the advantages of optimizing the positions of the individual wind turbines within an offshore wind farm. In order to perform analysis on different wind farms, a model was developed that is capable of simulating an existing offshore wind farm by converting external input data into a representative LCOE. The model which outputs LCOE is used in an optimization framework to optimize the positions of the individual turbines within the wind farm with the objective of minimizing the LCOE. Upon consideration of accuracy, availability of data and computing restrictions, the Jensen Wake-deficit Model and a hybrid cable layout heuristic were incorporated into the physical component and an NREL cost scaling model was included into the economic component. A genetic algorithm was selected for the wind farm layout optimization component.

Horns Rev and Borssele wind farms were used as case studies to explore the advantages of optimizing the positions of the individual turbines within an offshore wind farm. The wind farm layout optimization results show that:

- Optimized wind farm layouts resulted in a reduction of collection cable length relative to the baseline rectilinear layout of 8.6% for Horns Rev and 8.2% for Borssele.
- Optimized wind farm layouts resulted in an increase in the Annual Energy Production relative to the baseline rectilinear layout of 4.7% for Horns Rev and 4.8% for Borssele.
- Optimized wind farm layouts resulted in a decrease in the Levelized Cost of Electricity relative to the baseline rectilinear layout of 3.7% for Horns Rev and 5.5% for Borssele.

The initial expectation was that the contrasting nature of the wake effects and the collection cable costs would result in the improvement of one value at the expense of the other that would ultimately lead to an improved LCOE. However, the results from the model indicate that both the wake effects and collection cable costs were improved by applying an optimized layout in both case studies. The LCOE savings were higher than the results of Pillai et al. for Middelgrunden wind farm (1-3.5%). This difference may be explained by the size and shape of the Horns Rev and Borssele wind farms relative to the Middelgrunden wind farm. The larger wind farms studied in this paper provide increased potential for savings through farm layout optimization layout because of the increased wake interaction and cable design variability.

The results observed from the model in both case studies indicate that there is potential for reducing the levelized cost of electricity for offshore wind farms by optimizing the positions of individual wind turbines within the layout. However, it is important to consider the assumptions and limitations of the implementation of the model used for analysis.

The following assumptions and limitations should be considered as they relate to the validity of the model and the qualitative results:

- The model was validated against the measured data from one wind farm due to limitations on available data. It is important that the current model is compared with measured data from more wind farms in order to compare the results and determine the model's accuracy.
- The assumption that the foundation costs for each turbine were the same regardless of the bathymetry and soil conditions may cause large discrepancies in the LCOE results. However, the rectilinear constraints are unlikely to provide a better result if the individual turbine foundation costs are considered.

The subsequent assumptions and limitations should be considered as they relate to the accuracy of the model and the quantitative results:

- The optimization framework was constrained by time which limited the effectiveness of the application of the genetic algorithm. Using a more powerful computing device and having longer simulation times would allow for improved optimization results.
- The accuracy of the LCOE results are limited by the uncertainty of the input values incorporated into the economic model which vary based on the existing market and wind farm location. This is less relevant for relative comparisons, however could provide increased value by having the model LCOE be representative of the actual wind farm LCOE.
- The application of a simplified cable loss sub-model provides limited accuracy for the total electrical losses experienced in the collection cable network. However, in both case studies the cable length was reduced relative to the existing layout, therefor it is expected that improving this sub-model would result in increased value for optimizing the turbine locations.

Despite the limitations and assumptions, the results provide valuable insight into the benefits of wind farm layout optimization as the same parameters and conditions were applied to each simulation.

References

[1] R. Perveen, N. Kishor and S. Mohanty, "Offshore wind farm development: Present status and challenges," *Renewable and Sustainable Energy Reviews (29),* pp. 780-792, 2014.

[2] S. Rodrigues, C. Restrepo, E. Kontos, R. T. Pinto and P. Bauer, "Trends of offshore wind projects," *Renewable and Sustainable Energy Reviews (49),* pp. 1114-1135, 2015.

[3] J. Herbert-Acero, O. Probst, P.-E. Rethore, G. Larsen and K. Castillo-Villar, "A Review of Methodological Approaches for the Design and Optimization of Wind Farms," *Energies (7),* pp. 6930-7016, 2014.

[4] M. A. Lackner and C. N. Elkinton, "An Analytical Framework for Offshore Wind Farm Layout Optimization," *Wind Engineering,* vol. 31, no. 1, pp. 17-31, 2007.

[5] G. Mosetti, C. Poloni and B. Diviacco, "Optimization of wind turbine positioning in large windfarms by means of a genetic algorithm," *Wind Engineering and Industrial Aerodynamics (51),* pp. 105-116, 1994.

[6] S. Grady, M. Hussaini and M. Abdullah, "Placement of wind turbines using genetic algorithms," *Renewable Energy (30),* pp. 259-270, 2005.

[7] C. M. Ituarte-Villarreal and J. F. Espiritu, "Optimization of wind turbine placement using a viral based optimization algorithm," *Procedia Computer Science,* vol. 6, pp. 469-474, 2011.

[8] C. Wan, J. Wang and Z. X. Yang G., "Optimal Micro-siting of Wind Farms by Particle Swarm Optimization," in *Advances in Swarm Intelligence ICSI 2010*, Berlin, Heidelberg, Springer, 2010, pp. 198-205.

[9] K. Chen, M. Song, Z. He and X. Zhang, "Wind turbine positioning optimization of wind farm using greedy algorithm," *Journal of Renewable and Sustainable Energy* , vol. 5, pp. 023128 1-15, 2013.

[10] A. Tesauro, P.-E. Rethore and G. Larsen, "State of the art of wind farm optimization," in *European Wind Energy Conference & Exhibition*, 2012.

[11] C. Elkinton, Offshore Wind Farm Layout Optimization, Ph.D Thesis, University of Massachusetts Amherst, 2007.

[12] A. Pillai, J. Chick, M. Khorasanchi, S. Barbouchi and L. Johanning, "Application of an offshore wind farm layout optimization methodology at Middelgrunden wind farm," *Ocean Engineering (139),* pp. 287-297, 2017.

[13] T. Burton, D. Sharpe, N. Jenkins and E. Bossanyi, Wind Energy Handbook, Chichester: John Wiley & Sons Ltd, 2001.

[14] F. Gonzalez-Longatt, P. Wall and V. Terzija, "Wake effect in wind farm performance: Steady-state and dynamic behaviour," *Renewable Energy (39),* pp. 329-338, 2012.

[15] R. Shakoor, M. Y. Hassan, A. Raheem and Y. Wu, "Wake effect modeling: A review of wind farm layout optimization using Jensen's model," *Renewable and Sustainable Energy Reviews,* pp. 58 1048-1059, 2016.

[16] J. Vermeer, J. Sorensen and A. Crespo, "Wind turbine wake aerodynamics," *Progress in Aerospace Sciences (39),* pp. 467-510, 2003.

[17] T. Sorensen, M. L. Thogersen and P. Nielsen, "Adapting and calibtration of existing wake models to meet the conditions inside offshore wind farms.," EMD International A/S, Aalborg, Denmark, 2008.

[18] I. Katic, J. Hojstrup and N. Jensen, "A Simple Model For Cluster Efficiency," in *European Wind Energy Association Conference and Exhibition*, Rome, 1986.

[19] M. Zaaijer, "Great expectations for offshore wind turbines," TU Delft (PhD Thesis), Delft, 2013.

[20] S. Lumbreras and A. Ramos, "Offshore wind farm electrical design: a review," *Wind Energy (16),* pp. 459-473, 2013.

[21] G. Katsouris, "Infield Cable Topology Optimization of Offshore Wind Farms," Delft University of Technology (Master Thesis), Delft, 2015.

[22] J. Bauer and J. Lysgaard, "The offshore wind farm array cable layout problem: a planar open vehicle routing problem," *JORS (66),* pp. 360-368, 2015.

[23] L. Esau and K. Williams, "On teleprocessing system design: part ii a method for approximating the optimal network," *IBM Systems Journal ,* vol. 5, no. 3, pp. 142-147, 1966.

[24] L. Fingersh, M. Hand and A. Laxson, "WindTurbine Design Cost and Scaling Model," National Renewable Energy Laboratory, Golden CO, 2006.

[25] J. Green, A. Bowen, L. Fingersh and Y. Wan, "Electrical Collection and Transmission Systems for Offshore Wind Power," National Renewable Energy Laboratory (NREL/CP-500-41135), Houston Texas, 2007.

[26] MathWorks, "How the Genetic Algorithm Works," [Online]. Available: https://www.mathworks.com/help/gads/how-the-genetic-algorithm-works.html. [Accessed 1 October 2018].

[27] H. Pohlheim, "GEATbx Introduction - Evolutionary Algorithms: Overview, Methods and Operators, Version 3.8," 2006. [Online]. Available: www.geatbx.com. [Accessed 25 September 2018].

[28] D. R. VanLuvanee, "Investigation of Observed and Modeled Wake Effects at Horns Rev using WindPRO," Technical University of Denmark (Master Thesis), 2006.

[29] M. Mechali, R. Barthelmie, S. Frandsen and P. Rethore, "Wake effects at Horns Rev and their influence on energy production," in *EWEC 2006 Proceedings*, Athens Greece, 2006.

[30] L. Jensen, C. Morch, P. Sorensen and K. Svensen, "Wake Measurements from the Horns Rev wind farm," in *EWEC 2004 Proceedings*, London UK, 2004.

[31] Netherlands Enterprise Agency, "Borssele Wind Farm Zone: Wind Farm Sites III & IV - Project and Site Description," Utrecht, The Netherlands, 2016.

[32] S. S. Perez-Moreno, K. Dykes, K. O. Merz and M. B. Zaaijer, "Multidisciplinary design anlysis and optimization of a reference offshore wind plant," *Journal of Physics Conf. Ser. 1037 042004,* pp. 1-15, 2018.

IOP Conf. Series: Journal of Physics: Conf. Series **1222** (2019) 012008 doi:10.1088/1742-6596/1222/1/012008

Novel Aerodynamic Damping Identification Method for Operating Wind Turbines

Chao Chen[1*], Philippe Duffour[1] and Paul Fromme[2]

[1]Department of Civil, Environmental and Geomatic Engineering, University College London, United Kingdom

[2]Department of Mechanical Engineering, University College London, United Kingdom

*Corresponding author: chao.chen.15@ucl.ac.uk

Abstract. This contribution introduces a novel method to determine the aerodynamic damping for operating wind turbines. Previous research typically estimated the modal damping ratios in the fore-aft and side-side directions as two decoupled degrees of freedom. This can result in misleading results, as the two directions are closely and unconventionally coupled through the wind-rotor interaction. This study proposes the identification of a novel type of aerodynamic damping matrix. This matrix arises from the linearization of the aerodynamic force resultant obtained from blade element momentum theory (blade modes not included). This linearized force is then applied to a beam finite element model of the tower with a lumped mass representing the rotor-nacelle assembly. This decoupled strategy efficiently describes the physics of the system including the coupling between the fore-aft and side-side motions. The identification of the damping matrix is shown to work for simulated wind time data.

1. Introduction

Damping is a key variable in wind turbine systems as it limits vibration amplitude around resonance [1] and significantly influences fatigue life [2]. Wind turbine tower design is often governed by fatigue limit

IOP Conf. Series: Journal of Physics: Conf. Series **1222** (2019) 012008 doi:10.1088/1742-6596/1222/1/012008

states due to the fluctuating nature of the wind load (and wave load offshore). For accurate fatigue analyses, a dynamic model with correctly identified damping is key for wind turbine design or assessing fatigue life and future maintenance needs. Among the different sources of damping in wind turbine systems, aerodynamic damping has the largest contribution when the wind turbine is in operation. To identify the aerodynamic damping, operational modal analysis (OMA), which uses the measured vibration responses caused by ambient excitation, is the preferred method since conventional experimental modal analysis requires controlled excitation, which is very difficult to achieve due to the large size of modern wind turbines. Aerodynamic damping identification for wind turbines have been carried out by researchers such as Hansen *et al.* [3], Devriendt *et al.* [4], Dong *et al.* [5], Ozbek and Rixen [6], Koukoura *et al.* [7] These studies estimated the modal aerodynamic damping ratios in the fore-aft (FA) and side-side (SS) directions using classic or modified identification methods in time or frequency domain. However, Tcherniak *et al.* [8] and Ozbek *et al.* [9] highlighted some difficulties in using OMA in the context of wind turbines.

Previous studies typically assumed that the FA and SS motions are decoupled as the damping identification techniques were applied on measured data only for FA or SS directions. This paper proposes a new damping identification method which falls under the umbrella of OMA in the sense that it uses ambient wind excitation as an input. However, it does not aim to estimate modal parameters but rather identify on a more realistic damping model. According to the authors' recent research [10], it was found that the rotating blades introduce significant damping coupling between the FA and SS directions for an operating wind turbine and this coupling can be expressed by an aerodynamic damping matrix. In terms of aerodynamic damping, the wind turbine dynamic behaviour is quite different from that traditionally assumed by decoupled FA and SS motions. Based on this new description of the damping, this contribution presents an operational identification methodology that directly extracts the aerodynamic damping matrix and applies it to a three-blade horizontal-axis wind turbine. This paper is organized as follows. Section 2 describes the modelling strategy used in the paper and then describes the identification procedure. Section 3 gives a preliminary assessment of the identification procedure on a turbine case study using simulated wind velocity time series. Section 4 concludes the paper.

2. Methodology

2.1. Model description

In this study, the wind-rotor interaction is modelled separately from the tower dynamics. The two systems are then coupled through the hub resultant forces as shown in Figure 1. The tower mainly vibrates in the FA direction (x) and SS direction (y), and the vertical vibration is neglected due to its small magnitude. The model includes the linear motions represented by $\dot{x}$ and $\dot{y}$ at the tower top as well as the angular motions represented by $\dot{\theta}_x$ and $\dot{\theta}_y$ around x and y axes respectively. The inflow wind V_0 is assumed to be uniform throughout the rotor and only has a component in the FA direction. The aerodynamic forces are derived theoretically using blade element momentum (BEM) theory in steady-state conditions ignoring any blade vibration. As a result, parametric excitation due to rotor dynamics are not included. The aerodynamic forces are linearized to the sum of terms corresponding to the forces for an assumed rigid tower, plus terms proportional to the tower top linear and angular velocities, which can be expressed as

$$\mathbf{F}_{Top}(t) = \mathbf{F}_{Top}^{Rigid}(t) - \mathbf{C}_{Aero}\dot{\mathbf{u}}_{Top}(t). \quad (1)$$

According to [10], $\mathbf{C}_{Aero}$ is

IOP Conf. Series: Journal of Physics: Conf. Series **1222** (2019) 012008 doi:10.1088/1742-6596/1222/1/012008

$$\mathbf{C}_{Aero} = \begin{bmatrix} c_{xx} & 0 & c_{x\theta_x} & 0 \\ 0 & c_{yy} & 0 & c_{y\theta_y} \\ c_{\theta_x x} & 0 & c_{\theta_x\theta_x} & 0 \\ 0 & c_{\theta_y y} & 0 & c_{\theta_y\theta_y} \end{bmatrix}, \tag{2}$$

where the off-diagonal terms representing the damping coupling terms are not symmetric and have analytical expressions derived from BEM. The terms in $\mathbf{C}_{Aero}$ are defined in Appendix 1. The aerodynamic damping matrix and $\mathbf{F}_{Top}^{Rigid}(t)$ are calculated in MATLAB for a particular parameter set of mean wind speed, rotor rotation speed and blade pitch angles and $\mathbf{F}_{Top}(t)$ as defined by Equation (1). This force is then applied at the top of a finite element (FE) model (also written in MATLAB) of the tower modelled as a cantilever made of 11 Euler-Bernoulli beams elements as shown schematically in Figure 1.

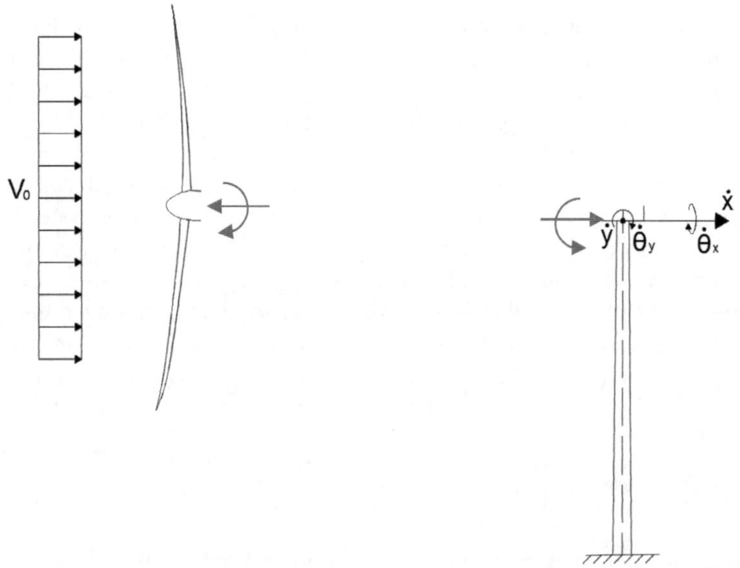

(a) Stage 1: rotor aerodynamic interaction. (b) Stage 2: tower dynamics.

Figure 1. Schematic diagrams illustrating the two-stage modelling strategy.

The equation of motion for the wind turbine system can be originally written as

$$\mathbf{M}\ddot{\mathbf{u}}(t) + \mathbf{C}_{Struc}\dot{\mathbf{u}}(t) + \mathbf{K}\mathbf{u}(t) = \mathbf{F}(t), \tag{3}$$

where $\mathbf{M}$, $\mathbf{C}_{Struc}$ and $\mathbf{K}$ are the mass, structural damping and stiffness matrices respectively. $\mathbf{C}_{Struc}$, representing standard structural damping is assumed to be a Rayleigh damping matrix. $\mathbf{M}$, $\mathbf{C}_{Struc}$ and $\mathbf{K}$ are assumed to be known either through modelling or standard OMA on a parked turbine. $\mathbf{F}(t)$ is the external aerodynamic force vector which includes $\mathbf{F}_{Top}(t)$ at the relevant degrees of freedom, and $\mathbf{u}(t)$ is the generalised displacement vector. In this study, the aerodynamic force on the tower itself is not considered, so $\mathbf{F}(t)$ is only non-zero at the tower top degrees of freedom. After implementing the two-stage modelling strategy, the equation of motion for the wind turbine system can be changed to

$$\mathbf{M}\ddot{\mathbf{u}}(t) + \mathbf{C}\dot{\mathbf{u}}(t) + \mathbf{K}\mathbf{u}(t) = \mathbf{F}^{Rigid}(t), \tag{4}$$

where $\mathbf{C}$ is the addition of the damping matrices $\mathbf{C}_{Struc}$ and $\mathbf{C}_{Aero}$ (adding the terms of $\mathbf{C}_{Aero}$ at the relevant degrees of freedom in $\mathbf{C}_{Struct}$). In Equations (3) and (4), $\mathbf{F}(t)$ or $\mathbf{F}^{Rigid}(t)$ represent the complete aerodynamic force vector applied to the flexible tower or the rigid tower, whose components corresponding to the tower top are expressed as $\mathbf{F}_{Top}(t)$ and $\mathbf{F}_{Top}^{Rigid}(t)$ in Equation (1). The model expressed by Equation (4) is referred to as the "full model", to distinguish it from the simplified two-degree of freedom (2-DOF) model which will be described in Section 2.3.

Table 1. Basic properties of the NREL 5MW reference onshore wind turbine.

Rotor Diameter, R	126m
Hub Height from MSL	87.6m
Tower Diameter, D	3.87-6.00m
Tower Thickness, t	19-27mm
Lumped Mass at Top	3.5×10^5 kg
Rated Wind Speed	12.1m/s
Natural Frequency	0.34 Hz

2.2. Model validation

This two-stage modelling strategy was validated against the fully-coupled aero-elastic package FAST provided by NREL [11] for inflow wind field with a range of constant and uniform wind speeds. Calculations were carried out using the NREL 5MW reference onshore wind turbine with main characteristics given in Table 1. Figure 2 shows time histories of the FA and SS tower top displacements following an initial displacement imposed in the FA direction. The responses generated by the full model agree well with the responses from FAST simulation.

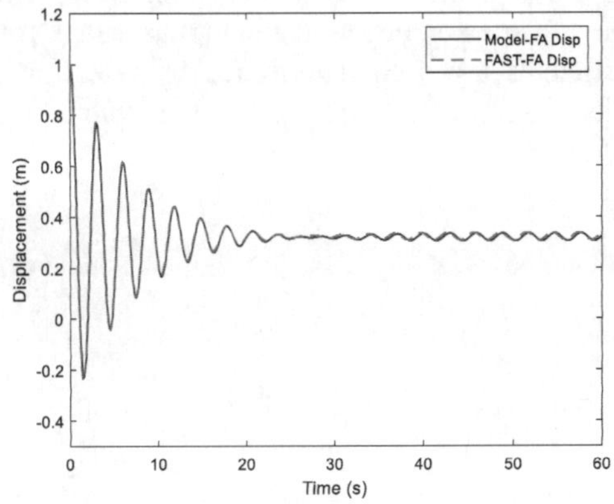

(a)

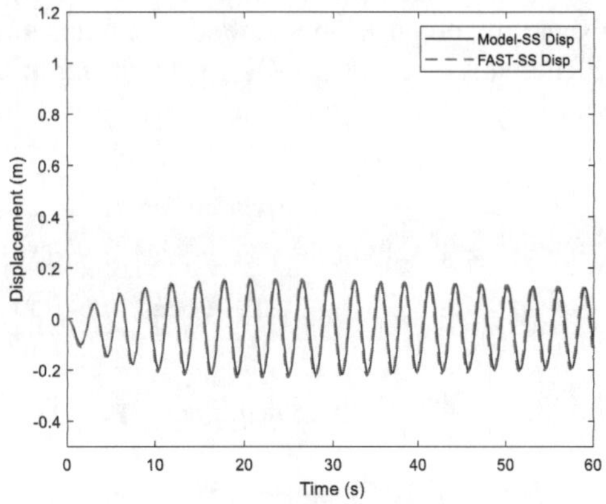

(b)

Figure 2. Comparison of the FA (a) and SS (b) vibration displacements between the full model and FAST simulation with initial displacement of the tower top in the FA direction, wind speed 10m/s.

2.3. Simplification to a 2-DOF model

In wind turbines, the FA and SS responses are dominated by the first FA and SS bending modes so the behaviour of the system can be efficiently described by only considering these two modal coordinates. Applying modal decomposition to Equations (3) and (4) using the first two bending mode shapes (FA: $\boldsymbol{\phi}_x$ and SS: $\boldsymbol{\phi}_y$), the two equations of motion for the first bending modes can be written as:

$$\begin{aligned} \bar{m}_x\ddot{\alpha}_x(t) + 2\bar{\zeta}_x\sqrt{\bar{m}_x\bar{k}_x}\dot{\alpha}_x(t) + \bar{k}_x\alpha_x(t) &= \boldsymbol{\phi}_x^T\mathbf{F}_x(t), \\ \bar{m}_y\ddot{\alpha}_y(t) + 2\bar{\zeta}_y\sqrt{\bar{m}_y\bar{k}_y}\dot{\alpha}_y(t) + \bar{k}_y\alpha_y(t) &= \boldsymbol{\phi}_y^T\mathbf{F}_y(t), \end{aligned} \tag{5}$$

where $\bar{m}_x$, $\bar{m}_y$, $\bar{k}_x$ and $\bar{k}_y$ are the modal masses and stiffnesses for the first FA/SS mode respectively, $\bar{\zeta}_x$ and $\bar{\zeta}_y$ are the structural modal damping ratios, and $\alpha_x(t)$ and $\alpha_y(t)$ are the modal coordinates for the FA and SS modes respectively. $\mathbf{F}_x(t)$ and $\mathbf{F}_y(t)$ are the linearized aerodynamic forces which also include tower top velocity terms. These terms can be combined with the structural damping ratios into a 2x2 damping matrix

$$\bar{\mathbf{C}} = \begin{bmatrix} \bar{c}_{11} & \bar{c}_{12} \\ \bar{c}_{21} & \bar{c}_{22} \end{bmatrix}. \quad (6)$$

The above modal decomposition converts the full model to a simpler 2-DOF model, the coefficients of which will be identified in the following procedure outlined next.

2.4. Damping identification

Standard OMA implicitly assumes proportional damping and therefore calculates modal damping ratios. The damping identification method introduced by [12] was used instead as it does not assume that the damping matrix is symmetrical. It operates in the frequency domain with transfer functions and is outlined in Appendix 2. In practice, wind time series and structural responses could be obtained from actual measurements but in this paper, they were simulated. Wind time series were generated by inverse Fourier transform of a Kaimal Spectrum, using customised MATLAB code or TurbSim in FAST [13]. The turbulent wind time series can be converted into aerodynamic forces on a rigid turbine using steady BEM code. The 2-DOF model can generate the responses in FA and SS directions from these forces. Given the inputs (aerodynamic forces) and the outputs (the responses), the frequency response functions (FRFs) of the 2-DOF model can be estimated from the cross spectral density (CSD) matrices of the inputs and the outputs. The method calculates an auxiliary, frequency dependent damping matrix from the FRF matrix of the 2-DOF system. Damping coefficients are then determined by selecting an appropriate frequency range and averaging the results from the auxiliary damping matrix. The identification procedure is described schematically in Figure 3.

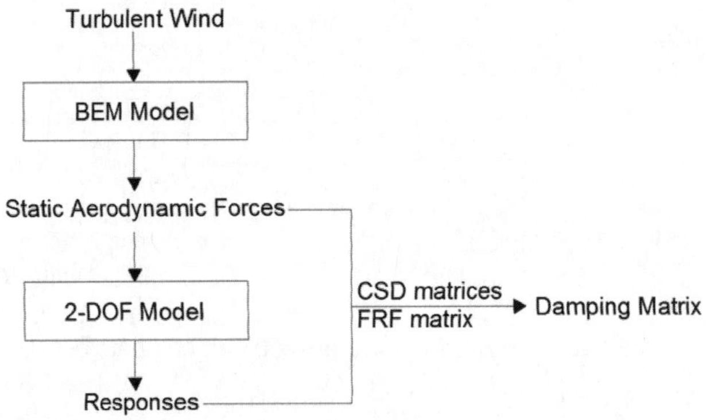

Figure 3. Schematic for the aerodynamic damping matrix identification procedure.

3. Results

The identification procedure developed was tested using the FE model established in MATLAB. The onshore 5MW NREL reference wind turbine from FAST provided the tower dynamic properties and the airfoil properties for the FE model. Given a particular mean wind speed, the terms in the 2x2 damping matrix can be calculated. These calculated values represent reference analytical values. To simulate a more realistic incoming wind flow, turbulent wind time series with same mean speed were used as input for the 2-DOF system to generate the system responses. The aerodynamic damping matrix was determined for every frequency point employing the identification procedure outlined above, using the generated responses and the estimated aerodynamic forces applied to the rigid tower. The estimation of damping coefficients is eventually made by averaging the frequency-dependent damping coefficient estimates.

As an example, a mean wind speed equal to 20m/s is chosen to test the proposed damping identification procedure. With a predefined 1% structural damping ratio [1] for the first bending modes, the structural modal damping matrix is

$$1 \times 10^4 \times \begin{bmatrix} 1.7265 & 0 \\ 0 & 1.7265 \end{bmatrix} (N \cdot s/m),$$

and the reference aerodynamic damping matrix is

$$1 \times 10^5 \times \begin{bmatrix} 1.0805 & 0.2128 \\ 0.4138 & 0.1123 \end{bmatrix} (N \cdot s/m).$$

The estimation of the FRFs was obtained by averaging 5 simulations of 1-hour duration. If the number of simulations or the simulation duration increases, the estimation of FRFs is more accurate. After obtaining the FRFs for the 2-DOF model, the damping matrix can be estimated following the method by Chen *et al.*'s method [12]. The estimated frequency-dependent damping matrix is shown in Figure 4, fluctuating around the analytical solution. Averaging the frequency-dependent matrix from 0.01Hz to 0.3 Hz, a final estimation of the aerodynamic damping matrix after subtracting the structural modal damping matrix can be obtained, which is

$$1 \times 10^5 \times \begin{bmatrix} 1.0708 & 0.2318 \\ 0.4093 & 0.1208 \end{bmatrix} (N \cdot s/m).$$

The estimated aerodynamic matrix is very close to the predefined one above and it was found that the maximum difference is for C_{12} and less than 10%. Identified values for the four damping coefficients in the 2x2 damping matrix for mean wind speeds from 8m/s to 20m/s (2m/s interval) are plotted against the analytical values, as shown in Figure 5. It can be seen that the identified values are close to the analytical values.

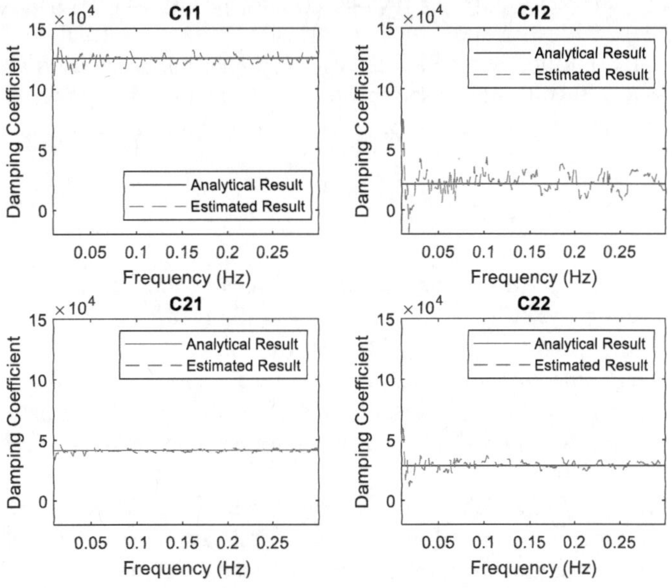

Figure 4. Comparison between analytical (blue solid) and identified results (red dashed) for the frequency dependent aerodynamic damping coefficients. (Unit: N·s/m).

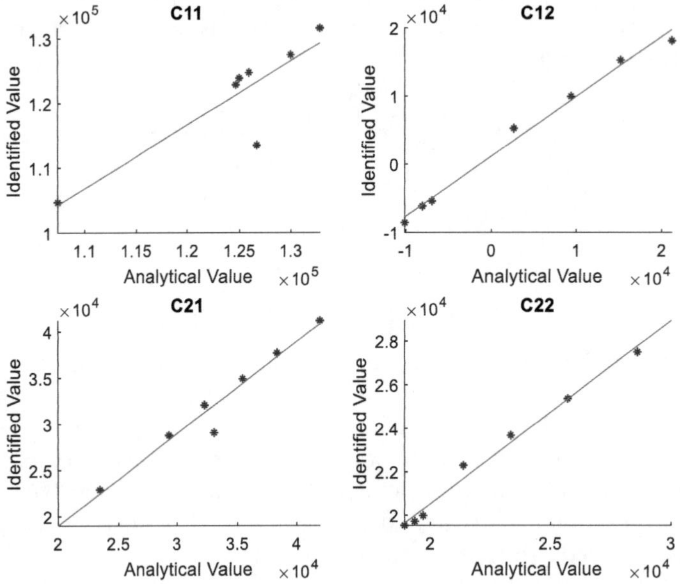

Figure 5. Analytical values against identified values after averaging for mean wind speeds from 8m/s to 20m/s in 2m/s steps (Unit: N·s/m).

4. Conclusion

This contribution presented the development of a new methodology to identify the aerodynamic damping in wind turbines in operation by using vibration responses caused by ambient excitation. The identification obtains the aerodynamic damping matrix rather than conventionally-used damping ratios. The coupling between FA and SS motions is included by the developed aerodynamic damping matrix and therefore the proposed identification method provides better information to characterise the aerodynamic damping in wind turbines, which is key for robust fatigue and reliability analyses. The methodology was devised so that it can be used in a practical situation. It requires wind time series, the dynamics response of the tower and modal properties of wind turbines in parked condition.

References

[1] C. Chen and P. Duffour, "Modelling damping sources in monopile-supported offshore wind turbines," *Wind Energy*, vol. 21, no. 11, pp. 1121–1140, 2018.

[2] R. Rezaei, P. Fromme, and P. Duffour, "Fatigue life sensitivity of monopile-supported offshore wind turbines to damping," *Renew. Energy*, vol. 123, pp. 450–459, 2018.

[3] M. H. Hansen, K. Thomsen, P. Fuglsang, and T. Knudsen, "Two methods for estimating aeroelastic damping of operational wind turbine modes from experiments," *Wind Energy*, vol. 9, no. 1–2, pp. 179–191, 2006.

[4] W. Weijtjens, Rasoul Shirzadeh, Gert De Sitter, and C. Devriendt, "Classifying resonant frequencies and damping values of an offshore wind turbine on a monopile foundation for different operational conditions," in *Proceedings of EWEA , Copenhagen*, 2014.

[5] X. Dong, J. Lian, M. Yang, and H. Wang, "Operational modal identification of offshore wind turbine structure based on modified stochastic subspace identification method considering

harmonic interference," *J. Renew. Sustain. Energy*, vol. 6, no. 3, pp. 1649–1654, 2014.

[6] M. Ozbek and D. J. Rixen, "Operational modal analysis of a 2.5 MW wind turbine using optical measurement techniques and strain gauges," *Wind Energy*, vol. 16, no. 3, pp. 367–381, 2013.

[7] C. Koukoura, A. Natarajan, and A. Vesth, "Identification of support structure damping of a full scale offshore wind turbine in normal operation," *Renew. Energy*, vol. 81, pp. 882–895, 2015.

[8] D. Tcherniak, S. Chauhan, and M. H. Hansen, "Applicability Limits of Operational Modal Analysis to Operational Wind Turbines," In *Structural Dynamics and Renewable Energy*, vol 1, pp. 317-327. Society for Experimental Mechanics. Conference Proceedings of the Society for Experimental Mechanics Series, no. 10, 2011.

[9] M. Ozbek, F. Meng, and D. J. Rixen, "Challenges in testing and monitoring the in-operation vibration characteristics of wind turbines," *Mech. Syst. Signal Process.*, vol. 41, no. 1–2, pp. 649–666, 2013.

[10] C. Chen, P. Duffour, and P. Fromme, "Modelling wind turbine tower-rotor interaction through an aerodynamic damping matrix," *J. Sound Vib.* Research Article Under Review.

[11] J. M. Jonkman and M. L. J. Buhl, "FAST user's guide," Technical Report No. NREL/TP-500-38230. *National Renewable Energy Laboratory (NREL),* Golden, CO, 2005.

[12] S. Y. Chen, M. S. Ju, and Y. G. Tsuei, "Estimation of mass, stiffness and damping matrices from frequency response functions," *J. Vib. Acoust.*, vol. 118, no. 1, pp. 78–82, 1996.

[13] B. J. Jonkman and L. Kilcher, "TurbSim user's guide version 1.50," Technical Report No. NREL/TP-500-46198. *National Renewable Energy Laboratory (NREL),* Golden, CO, 2009.

IOP Conf. Series: Journal of Physics: Conf. Series **1222** (2019) 012008 doi:10.1088/1742-6596/1222/1/012008

Appendix 1

The terms in aerodynamic damping matrix in Equation (2) are defined by the following expressions according to [10]:

$$\mathbf{C}_{Aero} = \begin{bmatrix} c_{xx} & 0 & c_{x\theta_x} & 0 \\ 0 & c_{yy} & 0 & c_{y\theta_y} \\ c_{\theta_x x} & 0 & c_{\theta_x\theta_x} & 0 \\ 0 & c_{\theta_y y} & 0 & c_{\theta_y\theta_y} \end{bmatrix}$$

$$= \begin{bmatrix} N_b \int_0^R \frac{\partial(dT)}{\partial V_0} & 0 & -N_b \int_0^R r \frac{\partial(dT)}{\partial V_r} & 0 \\ 0 & -\frac{N_b}{2} \int_0^R \frac{\partial(dS)}{\partial V_r} & 0 & -\frac{N_b}{2} \int_0^R r \frac{\partial(dS)}{\partial V_0} \\ N_b \int_0^R r \frac{\partial(dS)}{\partial V_0} & 0 & -N_b \int_0^R r^2 \frac{\partial(dS)}{\partial V_r} & 0 \\ 0 & \frac{N_b}{2} \int_0^R r \frac{\partial(dT)}{\partial V_r} & 0 & \frac{N_b}{2} \int_0^R r^2 \frac{\partial(dT)}{\partial V_0} \end{bmatrix},$$

where dT and dS are the steady-state forces in normal and tangential directions respectively applied to one blade element at distance r from the hub. V_0 is an steady inflow wind velocity in the FA direction, V_r is the tangential speed of the blade element at distance r which is caused by the rotor rotation, N_b is the number of blades, R is the radius of the blade.

IOP Conf. Series: Journal of Physics: Conf. Series **1222** (2019) 012008 doi:10.1088/1742-6596/1222/1/012008

Appendix 2

Outline of Chen *et al.*'s method [12]
Given a dynamic system with mass, stiffness and damping matrices **M**, **K** and **C** excited by an external force **f**, the equation of motion is

$$\mathbf{M}\ddot{\mathbf{x}} + \mathbf{C}\dot{\mathbf{x}} + \mathbf{K}\mathbf{x} = \mathbf{f}.$$

Rewrite this equation of motion in frequency domain:

$$(-\omega^2\mathbf{M} + i\omega\mathbf{C} + \mathbf{K})\mathbf{X}(\omega) = \mathbf{F}(\omega).$$

And the frequency response function (FRF) matrix $\mathbf{H}(\omega)$ is defined as:

$$\mathbf{H}(\omega) = (-\omega^2\mathbf{M} + i\omega\mathbf{C} + \mathbf{K})^{-1}.$$

The "normal" FRF $\mathbf{H}^{\mathbf{N}}(\omega)$ is defined with the undamped system:

$$\mathbf{H}^{\mathbf{N}}(\omega) = [\mathbf{K} - \omega^2\mathbf{M}]^{-1}$$

With the "normal" FRF, the frequency domain equation of motion can be written as

$$[\mathbf{H}^{\mathbf{N}}(\omega)]^{-1}\mathbf{X}(\omega) + i\omega\mathbf{C}\mathbf{X}(\omega) = \mathbf{F}(\omega),$$

or

$$\mathbf{X}(\omega) + i\mathbf{G}(\omega)\mathbf{X}(\omega) = \mathbf{H}^{N}(\omega)\mathbf{F}(\omega),$$

where

$$\mathbf{G}(\omega) = \omega\mathbf{H}^{N}(\omega)\mathbf{C}.$$

Therefore, the relationship between the measured FRF $\mathbf{H}(\omega)$ and the "normal" FRF $\mathbf{H}^{\mathbf{N}}(\omega)$ is

$$\mathbf{H}^{N}(\omega) = [\mathbf{I} + i\mathbf{G}(\omega)]\mathbf{H}(\omega),$$

where **I** is an identity matrix. Since $\mathbf{H}^{N}(\omega)$ and $\mathbf{G}(\omega)$ are real matrices, the imaginary part of the RHS in the above equation is zero, giving

$$\mathbf{G}(\omega) = -\mathrm{im}\big(\mathbf{H}(\omega)\big)\big[\mathrm{Re}\big(\mathbf{H}(\omega)\big)\big]^{-1}.$$

Finally, the damping matrix at any given frequency can be expressed by

$$\mathbf{C} = \frac{1}{\omega}[\mathbf{H}^{\mathbf{N}}(\omega)]^{-1}\mathbf{G}(\omega).$$

It should be noted that this damping identification method needs the prior knowledge of the stiffness and mass matrices.

IOP Conf. Series: Journal of Physics: Conf. Series **1222** (2019) 012009 doi:10.1088/1742-6596/1222/1/012009

Sub-seasonal to seasonal climate predictions for wind energy forecasting

A Soret (1), V Torralba (1), N Cortesi (1), I Christel (1), Ll Palma (1), A Manrique-Suñén (1), Ll Lledó (1), N González-Reviriego (1) and F J Doblas-Reyes (1,2)

(1) Barcelona Supercomputing Center (BSC), Barcelona, Spain; (2) Institució Catalana de Recerca i Estudis Avançats (ICREA), Barcelona, Spain

albert.soret@bsc.es

Abstract. Both renewable energy supply and electricity demand are strongly influenced by meteorological conditions and their evolution over time in terms of climate variability and climate change. However, knowledge of power output and demand forecasting beyond a few days remains poor. Current methodologies assume that long-term resource availability is constant, ignoring the fact that future wind resources could be significantly different from the past wind energy conditions. Such uncertainties create risks that affect investment in wind energy projects at the operational stage where energy yields affect cash flow and the balance of the grid. Here we assess whether sub-seasonal to seasonal climate predictions (S2S) can skilfully predict wind speed in Europe. To illustrate S2S potential applications, two periods with an unusual climate behaviour affecting the energy market will be presented. We find that wind speed forecasted using S2S exhibits predictability some weeks and months in advance in important regions for the energy sector such as the North Sea. If S2S are incorporated into planning activities for energy traders, energy producers, plant operators, plant investors, they could help improve management climate variability related risks.

1. Introduction

Understanding and quantifying climatic conditions from several weeks to months can improve the decision making of renewable energy generation and electricity demand (Figure 1). Climate predictions including both sub-seasonal (up to one month) and seasonal predictions (forthcoming months) have witnessed considerable improvements in the last decade demonstrating that probabilistic forecasting can inform better decision making for some forecast windows and regions [1]. However, despite this improvement, climate predictions come with new set of challenges for users: information is often un-tailored and knowledge of power output forecasting beyond a few days remains poor. This requires the development of robust methodologies to address decision-making needs.

The aim of this work is to assess the skill in forecasting the meteorological-driven component of wind energy production using sub-seasonal to seasonal climate predictions (S2S) and to illustrate their applications by using two different case studies. Within this context, the work done in two European projects: S2S4E (s2s4e.eu/) and NEWA (neweuropeanwindatlas.eu/) will be presented in this contribution.

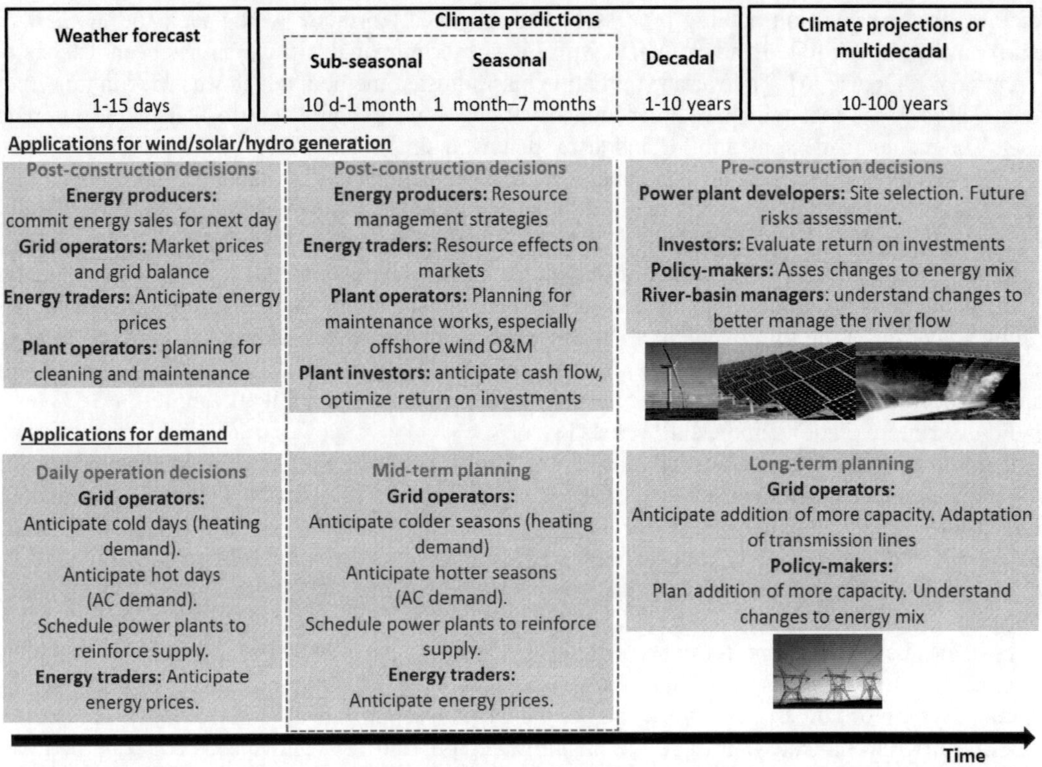

Figure 1. Forecast systems from weather forecast to climate projections including also climate predictions and their potential applications to the energy sector depending on the forecast window. Dotted line highlights the potential applications by using sub-seasonal to seasonal climate predictions.

2. Approach

In sub-seasonal to seasonal climate predictions (S2S), the atmospheric system has lost most of its memory from the initial conditions. Even if the chaotic behavior of the atmosphere does not allow predicting with accuracy the hourly changing weather beyond a few days, climate predictions are feasible because atmospheric variability on monthly/seasonal time-scales is modulated by slowly-varying boundary conditions and can retain memory from internal processes with very slow damping [2]. For a specific variable, e.g. wind speed, predictability will depend on the region, the season, lead time, forecast window and its teleconnections with the main modes of variability (e.g. North Atlantic Oscillation or El Niño-Southern Oscillation).

The forecast quality assessment was set based on two different systems, the ECMWF Extended Range [3], also known as Monthly Prediction System, and the ECMWF SEAS5 for the sub-seasonal and seasonal predictions, respectively. The Monthly Prediction System was obtained from S2S Project database (available at 1.5 degrees) and the SEAS5 from the Copernicus Data Store (available at 1 degree). The period of re-forecast data availability varies between the two systems: 1996-2015 for the Monthly Prediction System launched during the year 2016, while for SEAS5 the period 1993-2015 was used as hindcast.

The main shortcoming of climate predictions is the systematic errors that result from the inability of global circulation models to reproduce all the relevant processes responsible for climate variability and the uncertainty affecting the initial conditions [4]. Hence, climate predictions require bias adjustments in order to minimize forecast errors and produce useful information. Bias adjustment methods produce climate predictions whose statistical properties are similar to those in the underlying observational references [5], allowing the energy business to integrate bias-adjusted climate

IOP Conf. Series: Journal of Physics: Conf. Series **1222** (2019) 012009 doi:10.1088/1742-6596/1222/1/012009

predictions in their decision making processes. The predicted ten-meter wind speed and temperature forecast have been calibrated and evaluated with ERA-Interim reanalysis, which has been used as the observational reference [6]. The variance inflation bias-adjusted method, which was used in this study, produces predictions that will have inter annual variance that is equivalent to that of the reference dataset. The method is described in [4] and tested against a simple bias correction for the case of wind in [5]. The inflation of the ensemble spread ensures that predictions have reliable probabilities.

S2S forecasts have been compared with observations to assess wind predictability at different time scales in terms of the Fair Ranked Probability Skill Score (FairRPSS) [7]. FairRPSS is a measure of the predictive skill for the probabilistic forecasts for categorical events. Scores below zero indicate that the predictions are unskillful, those equal to zero are equal to the climatology forecasts, and anything above zero is an improvement upon climatology, up to 1, which indicates a 'perfect' forecast. Fair RPSS has been computed with one of many possible samples from a population, therefore the sampling uncertainty affecting this metric need to be taken into account (significant test at 95% level) to avoid over interpretations of the skill values [8].

Then, two different case studies have been analyzed to illustrate the potential application of S2S. Case studies may be understood as historical periods pointed as relevant by industrial stakeholders, periods with an unusual climate behavior affecting wind energy production. These two case studies represent different contexts and circumstances in terms of: regions, seasons and duration of the event.

3. Results

Skill estimates based on the performance of the systems in the past (20 years), may guide users about the expected performance of the future forecast. Figure 2a illustrates the skill of the Monthly Prediction System of January, two weeks before the target week in January (12-18 days). The positive values of FairRPSS indicate that there is skill in the region, and the wind speeds could be predicted two weeks in advance in many regions of interest for the wind energy sector such as the North Sea and Spain. Figure 2b characterizes the forecast skill in the Fino 1 research platform in the North Sea. Wind speed exhibits high skill (>0.3) in July and from December to March for forecast time of 5-12 days. Skill drops below 0.2 for subsequent forecast times, even though values above 0.2 are observed for forecast time of 1 month in December and of 2 months in July. Negative skill (white boxes) appears starting from forecast time of 19-25 days onward. The skill estimates shown in Figure 2 are the first insights of the skill assessment included in the NEWA wind atlas (neweuropeanwindatlas.eu/) where a comprehensive assessment will be included.

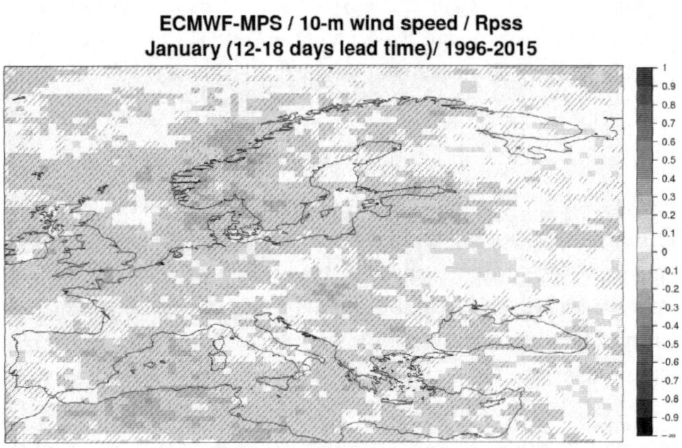

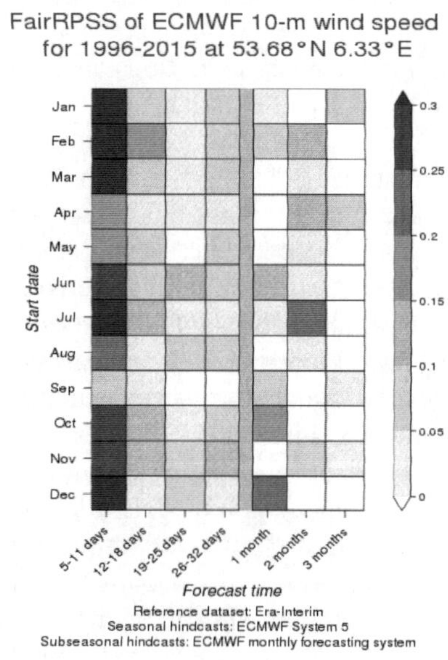

Figure 2. Left: Figure 2a, FairRPSS of 10-m wind speed for the Monthly Prediction System of January (1996-2015). Locations that passed a significant test at 95% level are hatched. Right: Figure 2b, forecast skill characterization in the Fino 1 area for different start dates (Y axis) and forecast window (X axis).

3.1. Case study 1: Cold wave over Europe and lower than usual wind power generation

During the week 17/01/2017-23/01/2017 a cold wave over Europe created a large increase in electricity demand. During this period, there were also low wind speeds over Europe and therefore lower than normal renewable energy supply. In France, under the national regulation authority request, utilities had to stop several nuclear reactors to carefully check some components. This created a situation of high risk that could have been better managed with accurate forecasts.

This case study shows that sub-seasonal predictions of temperature and wind speed (Figure 3 and Table 1) can provide added value to the current practice of using climatological forecasts. The skill increases considerably at shorter lead times and is greater for temperature than for wind speed. There is therefore confidence in anticipating episodes of high electricity demand a few weeks in advance (cold spells), although less confidence in ensuring that wind energy supply can meet the demand.

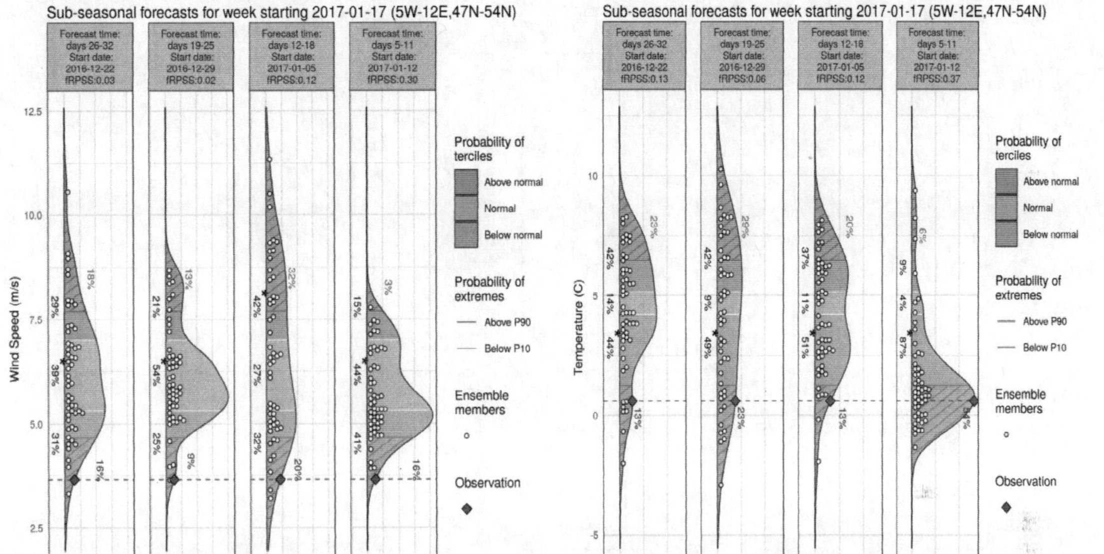

Figure 3. Sub-seasonal forecasts for wind speed (left) and temperature (right) for the week 17/01/2017-23/01/2017. Lead time ranges from 4 to 1 week in advance. Area of study: 5W-12W, 47N-54N). Yellow dots represent the members of the forecasts organized into three categories of equal size (terciles; above normal, normal and below normal) defined according to the climatology. Black percentages represent the forecast probabilities of each tercile, the most likely tercile is highlighted with a black star and the blue and red percentages represent the probabilities for p10 and p90 (hatched areas), respectively. Purple diamonds represent the observations for the target period.

Table 1. Associated probability skill (FairRPSS and Fair Brier Skill Score for percentile 10 and 90) scores are shown for wind speed and temperature for the area of study (5W-12E, 47N-54N).

Skill	Weeks before January 17-23. Wind speed				Weeks before January 17-23. Temperature			
	4	3	2	1	4	3	2	1
FairRPSS	0.03	0.02	0.12	0.30	0.13	0.06	0.12	0.37
BS P10	0.07	-0.02	0.00	0.21	-0.03	-0.04	0.26	0.59
BS P90	0.02	0.01	-0.01	0.19	-0.12	0.05	-0.15	0.29

3.2. Case study 2: Wind drought in North America

During the first quarter of 2015 the United States experienced a widespread and extended episode of low wind speeds [9]. This episode had a strong impact on wind power generation. Some wind farms did not generate enough cash for their steady payments, and the value of wind farm assets decreased.

Forecasts available for January to March 2015 did predict a lower than usual wind speeds situation over the area under study three months in advance. However, predictions were increasingly ambiguous as time passed by, reaching an almost equal probability for each tercile one month ahead of the target period (Figure 3). Despite being the weakest forecast, December's prediction showed ensemble members pointing at really low values for wind speed, not completely forecasting the actual values (purple diamond), but close to the extremely low observations of the variable. The skill associated with the forecasts was fair for all three lead times, meaning the predictions were better than plain climatology and could be taken into account (Table 2).

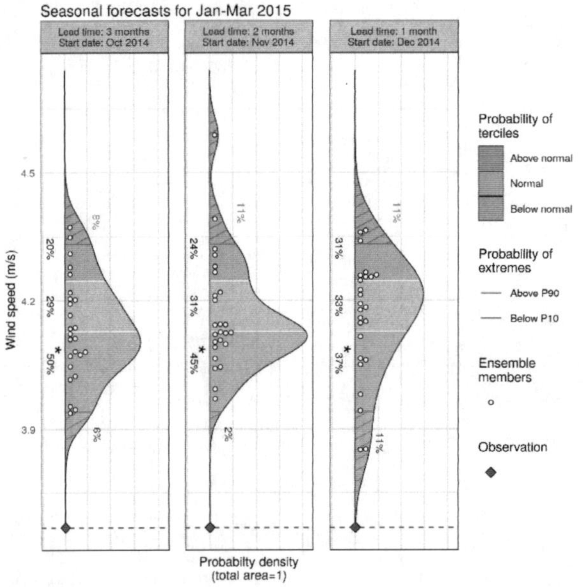

Figure 4. As Figure 3. Wind-speed seasonal forecasts for January to March 2015. Lead time ranges from 3 to 1 months in advance. Area of study: (124W-95W, 26N-44N).

Table 2. As table 1. January to March 2015. Area of study: (124W-95W, 26N-44N).

SKILL Wind	Forecast lead time		
	3-months (October)	2-months (November)	1-month (December)
FairRPSS	0.35	0.39	0.35
BS P10	-0.07	-0.27	-0.16
BS P90	0.10	0.04	0.07

4. Conclusions

Recent advances in global climate models, which simulate the physics of the whole climate system, demonstrate that probabilistic forecasting approaches can produce improved information upon retrospective climatology at certain spatial and temporal scales. Hence, energy decision makers now have a new set of tools to strengthen their decision making. Sub-seasonal to seasonal climate predictions (S2S) could be an important planning tool for energy traders, producers, plant operators, plant investors that can lead to better and timely management climate variability related risks.

The main conclusions of this work are two. First, the wind speed forecasted by S2S exhibits predictability some weeks and months in advance in important regions for the energy sector such as the North Sea. Second, current methodologies assume that future conditions will be like those of the past; an approach that makes impossible to anticipate extreme events that have never happened before.

This work aims at assessing the merits and caveats of climate prediction systems in predicting wind energy resources. To foster the uptake of S2S predictions in decision making processes, two complementary works are being performed during the preparation of this contribution. One is an economic analysis of relevant case studies (including the two presented in this contribution), understood as an evaluation of S2S predictions from the decision making point of view. The other will be an assessment of the use of S2S predictions in a real-time operational context. Both works will be public reports available through the S2S4E project website (www.s2s4e.eu).

Acknowledgements
We thank the S2S4E (GA776787), NEWA (PCIN-2014-012-C07-07), ERA4CS-INDECIS (GA690462) and ERA4CS-MEDSCOPE (GA690462) projects funding for allowing us to carry out this research. We acknowledge use of the s2dverification (http://cran.r-project.org/web/packages/s2dverification) and Specs-Verification (http://cran.r-project.org/web/packages/SpecsVerification) R-language-based software packages. We also acknowledge the ECMWF for the provision of the ECMWF SEAS5 and the Monthly Prediction Systems and the ERA-Interim reanalysis datasets.

References

[1] Alessandrini S., Sperati S., and Pinson P., 2013. A comparison between the ECMWF and COSMO Ensemble Prediction Systems applied to short-term wind power forecasting on real data. *Applied energy*, 107, 271-280.

[2] Doblas-Reyes F. J., García-Serrano J., Lienert F., Biescas A.P. and Rodrigues L. R., 2013. Seasonal climate predictability and forecasting: status and prospects *Wiley Interdis. Rev.: Clim. Change* 4 245–68

[3] Vitart F., 2004. Monthly Forecasting at ECMWF. *Monthly Weather Review*, 132(12), pp.2761–2779.

[4] Doblas-Reyes F J, Hagedorn R, and Palmer TN, 2005. The rationale behind the success of multi-model ensembles in seasonal forecasting - II. Calibration and combination. *Tellus A*, 57 (3), 234–252

[5] Torralba V., Doblas-Reyes F. J., MacLeod D, Christel I. and Davis M. 2017. Seasonal Climate Prediction: A New Source of Information for the Management of Wind Energy Resources. *Journal of Applied Meteorology and Climatology*, 56(5), 1231-1247. doi: 10.1175/JAMC-D-16-0204.1

[6] Dee DP, Uppala SM, Simmons AJ, Berrisford P, Poli P, Kobayashi S, Andrae U, Balmaseda MA, Balsamo G, Bauer P, Bechtold P, Beljaars ACM, van de Berg L, Bidlot J, Bormann N, Delsol C, Dragani R, Fuentes M, Geer AJ, Haimberger L, Healy SB, Hersbach H, Hólm EV, Isaksen L, Köllberg P, Köhler M, Matricardi M, Mcnally AP, Monge-Sanz BM, Morcrette JJ, Park BK, Peubey C, de Rosnay P, Tavolato C, Thépaut JN, Vitart F, 2011 The ERA-Interim reanalysis: configuration and performance of the data assimilation system. *Q J R Meteorol Soc* 137(656):553–597

[7] Ferro, 2014. Fair scores for ensemble forecasts. *Quarterly Journal of the Royal Meteorological Society*, 140 (683), 1917–1923.

[8] Bradley, A. A., Schwartz, S. S., & Hashino, T., 2008. Sampling uncertainty and confidence intervals for the Brier score and Brier skill score. *Weather and Forecasting*, 23(5), 992-1006.

[9] Lledó Ll., Bellprat O., Doblas-Reyes F.J. and Soret A., 2018. Investigating the effects of Pacific sea surface temperatures on the wind drought of 2015 over the United States. *Journal of Geophysical Research Atmospheres*, doi: 10.1029/2017JD028019.

IOP Conf. Series: Journal of Physics: Conf. Series **1222** (2019) 012010 doi:10.1088/1742-6596/1222/1/012010

Effects of Leading Edge Erosion on the Power Performance and Acoustic Noise Emissions of Locally Manufactured Small Wind Turbine Blades

Kostas Latoufis[1,2], Vasilis Riziotis[1], Spyros Voutsinas[1] and Nikos Hatziargyriou[2]

[1] School of Mechanical Engineering, Laboratory of Aerodynamics, National Technical University of Athens, Iroon Politechniou 9, Zografou 15772, Greece
[2] School of Electrical and Computer Engineering, Electric Power Systems Laboratory, National Technical University of Athens, Iroon Politechniou 9, Zografou 15772, Greece

E-mail: latoufis@power.ece.ntua.gr

Abstract. The local manufacturing of small wind turbines has become increasingly popular over the last decade, among international networks of renewable energy practitioners, for the implementation of rural electrification applications. Locally manufactured small wind turbines (LMSWTs), used most frequently in off-grid battery based renewable energy systems, typically utilize wooden blades and coreless axial flux permanent magnet generators due to their simple manufacturing process and local availability of materials. In this paper the effects of leading edge erosion on locally manufactured small wind turbine blades are researched experimentally, both in terms of power performance and acoustic noise emissions. From the research it is concluded that a yearly maintenance of the wooden blades of a LMSWT can lead to less acoustic noise emissions and better power performance, and with these aspect becoming more significant as the mean wind speed of the installation site increases. For practitioners, installers and manufacturers of LMSWTs, these findings can significantly guide manufacturing processes and techniques in terms of preventing leading edge erosion, while also advising preventative maintenance procedures on the field, in order to keep small wind electric systems quiet and productive.

1. Introduction

Since their introduction in 2015, the UN Sustainable Development Goals (SDG) target social and economic issues, with SDG No7 focusing on the universal access to sustainable and affordable energy. In this aspect, low cost renewable energy technologies contribute to the increase of small scale electricity production for remote areas and make such systems more accessible to rural communities. In particular the local manufacturing of such technologies significantly reduces capital costs, with the use of locally available materials, tools and manufacturing techniques, and further reduces maintenance costs [1] by providing the necessary socio-technical environment with appropriate training of the user community.

A widespread technology with such characteristics is locally manufactured small wind turbines (LMSWTs) [2]-[6] (Fig.1). Several design manuals [7] and on-line tools [8] have been a reference guide for LMSWTs, with an increasing number of NGOs, technical groups and practitioner networks implementing them on the field, in order to locally manufacture small wind turbines for the electrification of rural communities in developing countries. International practitioner networks such as the Wind Empowerment association are actively promoting the technology of LMSWTs, counting several hundreds of installations on all continents and a growing user community.

Figure 1. Local manufacturing of wooden blades using hand tools and an installed 2.4m rotor diameter locally manufactured small wind turbine [3].

LMSWTs are most frequently used in wind/solar hybrid off-grid systems, of installed generation capacity of up to 10kW. Rotor diameters span from 1.2m to 6m, respectively producing power from 200W to 5kW at a 10m/s wind speed. They are variable speed machines, consisting of a three blade wooden [9]-[10] horizontal axis rotor of constant pitch angle, of a coreless axial flux permanent magnet generator [11] with a double rotor single stator configuration and use a passive gravity furling tail system for rotor speed regulation.

In this paper, the effects of leading edge erosion on locally manufactured small wind turbine blades are researched experimentally in terms of power performance and acoustic noise emissions, with the aim of informing preventative maintenance procedures planed by practitioners, installers and manufacturers of LMSWTs. As such procedures are embedded in the socio-technical ecosystem [12] of the installations, quite operation and enduring power production can significantly contribute towards the sustainability of such systems. Additionally, this paper aims at contributing to relevant research on the effects of leading edge erosion [13] on various aspects of wind turbine technology such as rotor power performance [14] and annual energy production [15], optimal aerodynamic coefficient cp_{opt} and tip speed ratio λ_{opt} [16]-[17], and field measurements of acoustic noise emissions [18]-[19], while at the time expanding the limited literature on these themes [20]-[21] with reference to small wind turbines.

2. Experimental facilities

2.1. Field testing facility and experimental setup

A 2.4m rotor diameter LMSWT, delivering 600W of rated power at a 11m/s wind speed, was tested on a 12m guyed tower while connected directly to a 48VDC battery bank, at the NTUA small wind turbine test site in the coastal region of east Attiki. Meteorological and electrical measurements were performed with appropriate sensors, while all power performance measurements and calculations were conducted in accordance with the relevant IEC standard [22] for measuring the performance of wind turbines, and specifically Annex H which refers to small wind turbine testing.

IOP Conf. Series: Journal of Physics: Conf. Series **1222** (2019) 012010 doi:10.1088/1742-6596/1222/1/012010

(a) (b)

Figure 2. (a) Hand held sound level meter measuring position and (b) leading edge erosion of wooden locally manufactured blade.

Furthermore, all acoustic noise measurements were conducted in accordance with the relevant IEC standard [23] on acoustic noise measurements for wind turbines. The microphone and sound level meter were placed on a flat rock surface with rounded edges downwind from the prevailing wind direction (Fig.2) and at a horizontal distance of 12.5m from the tower base.

Two sets of 2.4m blade rotors were tested at the test site for power performance and acoustic noise emissions, namely a recently manufactured rotor with a smooth airfoil and a used rotor with eroded leading edges from rain and dust. The used rotor had not been maintained for a period of 18 months and was used in a highly erosive coastal environment.

2.2. Wind tunnel testing facility and experimental setup

Two scaled down versions of locally manufactured wooden rotors were tested in the wind tunnel of the laboratory of aerodynamics of the NTUA (Fig.3), namely two rotors of 1.2m in diameter, one with smooth and one with eroded leading edges. The mechanical torque on the shaft was measured using a rotary torque sensor, while the appropriate rotational speed of the rotor was provided by a variable speed DC motor drive and the wind tunnel wind speed was varied from 3m/s to 11m/s. The experimental setup was arranged so as to provide a complete range of tip speed ratio values for the rotors under test, for the complete wind speed range of the wind tunnel.

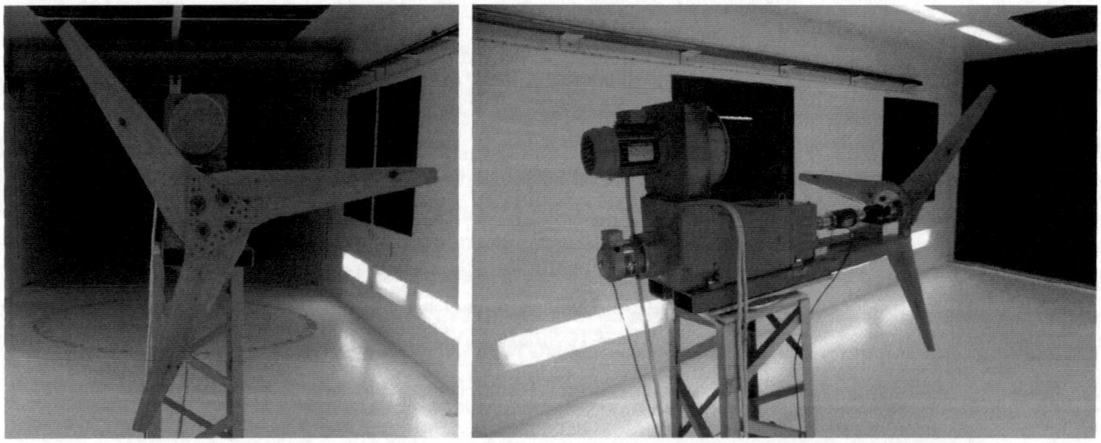

Figure 3. A wooden locally manufactured rotor in the wind tunnel of the NTUA.

3. Acoustic noise emissions

Initially, the site conditions for acoustic noise measurements were assessed. All valid data points for the sound pressure levels of wind turbine with smooth rotor and background noise emissions, for wind speeds from 4m/s to 10m/s, are shown in Fig.4. Wind turbine noise measurements display adequate separation from those of background noise, thus verifying that the test site provides a good experimental set up for acoustic noise measurement of this type.

Measurements of the sound pressure level of the smooth and eroded rotors against background nose, are displayed in Fig.5. A wide range of wind speeds, from 4m/s up to 10m/s, have produced valid measurement data points according to [23], allowing for the observation of blade erosion induced noise emissions for a range of both low and high wind speed modes of rotor operation.

Inflow-turbulence noise on a blade is caused by the interaction of the upstream atmospheric turbulence with the leading edge of the blade [24], so the eroded rotor is expected to operate with higher noise emissions. As observed in Fig.5, the acoustic noise emissions of the eroded rotor blades are increased by an average of 10% (or 4 dBA) for the whole range of wind speeds from 4m/s up to 10m/s. Individual increases in sound pressure levels for each wind speed are shown in Table 1, were a more severe influence of leading edge erosion on acoustic noise emissions can be observed for operation at higher wind speeds, in particular above 7m/s.

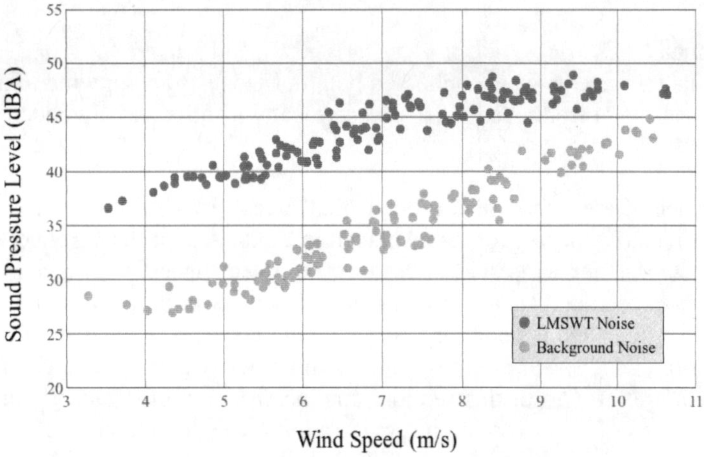

Figure 4. Sound pressure level data points of the small wind turbine with the smooth rotor against background noise for various wind speeds.

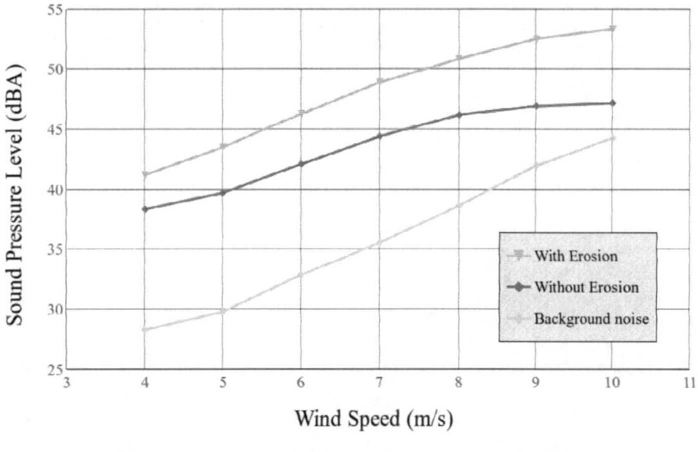

Figure 5. Processed sound pressure level data of the eroded and smooth small wind turbine rotors against background noise for various wind speeds.

Table 1. Sound pressure levels per wind speed, corrected for background noise, for the smooth and the eroded rotors.

Wind Speed (ms^{-1})	Background (dBA)	Eroded Rotor (dBA)	Smooth Rotor (dBA)	SPL Difference (dBA)
4	28.30	41.20	38.36	2.84
5	29.79	43.50	39.70	3.80
6	32.87	46.30	42.10	4.20
7	35.57	48.91	44.40	4.51
8	38.63	50.86	46.21	4.65
9	41.97	52.54	46.95	5.59
10	44.27	53.36	47.20	6.16

Furthermore, this increase of acoustic noise emissions due to leading edge erosion, shifts the acoustic noise performance of the small wind turbine further away from the acceptable noise limits for residential areas in open country, which are 44 dB for wind speeds of 8 m/s and 42 dB for wind speeds of 6 m/s, according to Danish noise level limits [25] for installation of wind turbines.

4. Power performance

Power performance measurements were conducted both in field tests and in wind tunnel tests in order to observe different aspects of leading edge erosion on locally manufactured wooden rotors.

4.1. Field tests

The power performance of the smooth and eroded small wind turbine rotors is investigated further at the test site, with the resulting power curves displayed in Fig.6. A significant reduction in rated power production can be observed for wind speeds close to the rated. Specifically a reduction of 23.7% is noted at the rated wind speed of 11m/s, while similar reductions in power production are evident for all wind speeds above 10m/s. On the contrary, an increased power performance of the eroded rotor of 7.4% is observed for the wind speed of 6m/s, with a similar behaviour for a range of low wind speeds from 5m/s to 7m/s. In the last part of this section, this increase in performance is investigated further while the wind tunnel tests of the next section provide additional experimental data.

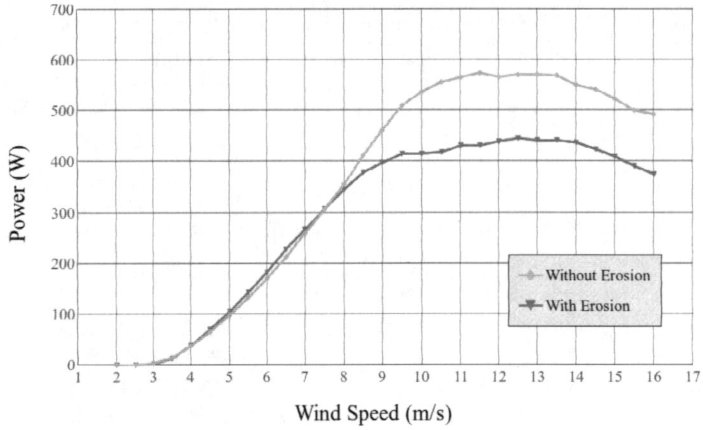

Figure 6. The power curves for the smooth and the eroded small wind turbine rotors under field testing.

The above results become more evident when considering the overall system efficiency Cp of the wind electric system, as displayed in Fig.7. For wind speeds higher than 8m/s, and especially close to the rated wind speed, the efficiency of the system is significantly reduced due to leading edge erosion, with values of up to 23.6% at the rated wind speed, specifically a reduction of efficiency from 0.154 to 0.117. Yet for lower wind speeds, i.e. from 4m/s up to 7m/s, the system efficiency of the eroded rotor displays better performance, with values of up to 8.5% at 5.5m/s, specifically in increase of efficiency from 0.287 to 0.312.

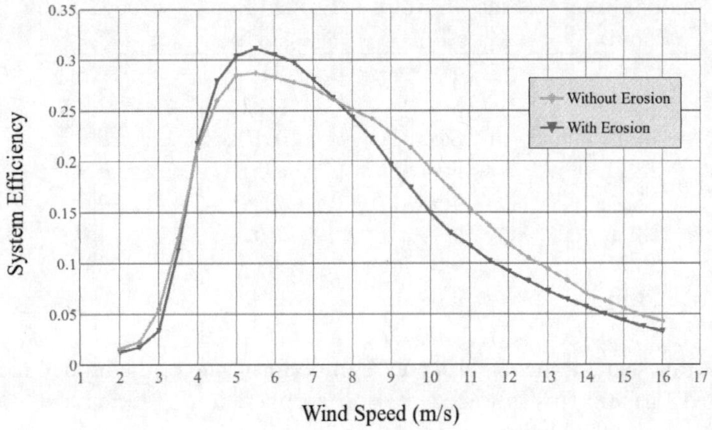

Figure 7. The system efficiency CP for the smooth and the eroded small wind turbine rotors under field testing.

The influence of leading edge erosion on the Annual Energy Production (AEP) of the wind electric system is further investigated, with the results displayed in Fig.8. The increase in system efficiency for the eroded rotor at low wind speeds which was observed in previous graphs, makes the influence of leading edge erosion of the blade rotor less pronounced in the AEP for lower mean wind speeds. As seen in Table 2, specifically for mean wind speeds from 4m/s to 6m/s the reduction in AEP gradually increases from 0.1% up to 9.6%.

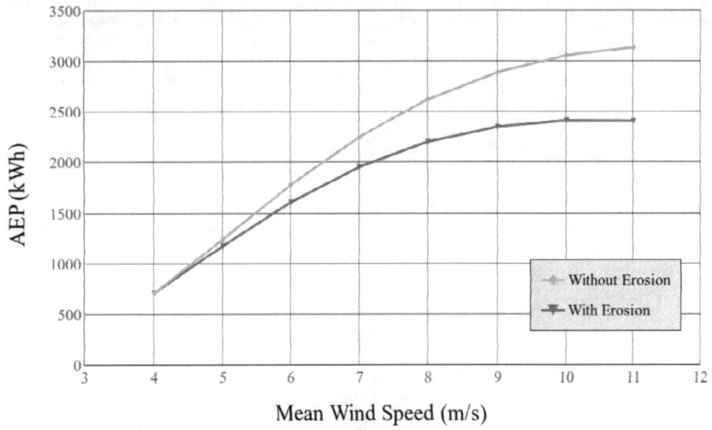

Figure 8. The Annual Energy Production (AEP) for the smooth and the eroded small wind turbine rotors under field testing.

For wind turbines of class IV or specifically for small wind turbines, which are installed closer to the ground with common hub heights ranging from 6m up to 24m, installation sites with mean wind speeds of 4m/s to 6m/s are the most frequent. From the above results it can be concluded that the reduction in AEP for LMSWTs due to blade leading edge erosion will be more significant as the mean wind speed of the site increases from 4m/s to 6m/s, which emphasizes the need for more strict preventative maintenance planning in sites with higher mean wind speeds.

Table 2. Annual Energy Production (AEP) per mean wind speed, including measurement uncertainty u_{AEP}, for the smooth and the eroded rotors.

Wind Speed (ms^{-1})	AEP Smooth Rotor (kWh)	u_{AEP} Smooth Rotor (%)	AEP Eroded Rotor (kWh)	u_{AEP} Eroded Rotor (%)	AEP Difference (%)
4	713	10.1	713	9.6	0.1
5	1247	8.1	1181	7.3	5.3
6	1782	6.6	1610	5.9	9.6
7	2252	5.6	1957	4.9	13.1
8	2624	4.9	2204	4.4	16.0
9	2891	4.4	2352	4.0	18.7
10	3057	4.0	2413	3.8	21.1
11	3133	3.8	2406	3.7	23.2

In the last part of this section, the slight increase in performance of the eroded rotor in lower wind speeds is investigated further. This is traced back to an operation of the eroded rotor closer to the optimal tip speed ratio λ_{opt} of the blades.

LMSWTs operate in low cost and low complexity wind electric systems in order to comply with the requirements of remote rural electrification applications, were maintenance and capital costs need to be low. For this reason, locally manufactured wind energy conversion systems do not perform maximum power tracking, either with varying the blade pitch angle or with the use of power electronics. Consequently, the tip speed ratio that the rotor of a LMSWT has fixed relationship with wind speed, as seen in Fig.9, and is mostly dependant on the electrical loading of the wind turbine generator.

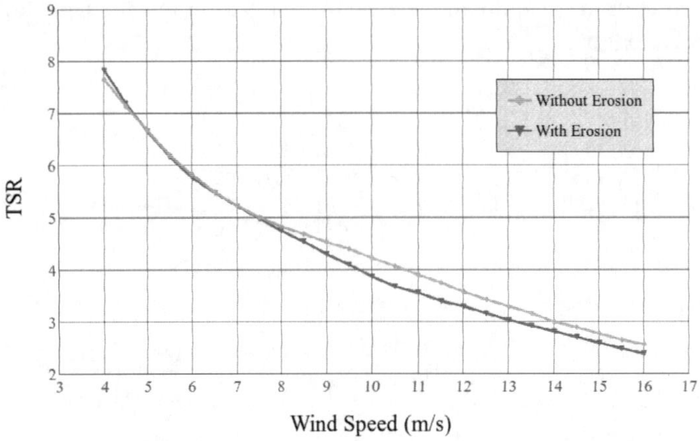

Figure 9. Differences in the tip speed ratio (TSR) of two rotors under test due to leading edge erosion.

It can be observed that for wind speeds close to 6m/s, both the smooth and the eroded rotors operate at tip speed ratios higher than the optimal, which is 5.5, as can be seen in the wind tunnel tests of the next section. The eroded rotor operates at slightly lower TSR for wind speeds around 6m/s, due to the stalling effect of higher drag forces due to leading edge erosion. Yet, this brings the operation of the eroded rotor closer to the optimal TSR, and thus operates at slightly better efficiencies than the smooth rotor. For wind speeds above 7m/s the tip speed ration of both rotors is lower that the optimal of 5.5 which brings the blades into their stall region of operation, and so the previously mentioned effect seizes to occur. Further insight into this phenomenon is provided in the next section with the assistance of wind tunnel tests performed on the two rotors.

4.2. Wind tunnel tests

The aerodynamic performance of the locally manufactured rotors is investigated in wind tunnel tests, measuring experimentally the aerodynamic coefficient over a wide range of operating points, spanning a series of tip speed ratios and wind speeds. In addition, the effects on aerodynamic performance of leading edge erosion of wooden locally manufactured blades are investigated further.

The effect of operation at low Reynolds number can be seen in the experimental results of Fig. 10, with Reynolds numbers varying from 5 x 10^4 at 3m/s up to 2 x 10^5 at 11m/s. It can be observed that the optimal value of the aerodynamic coefficient starts at low values of 0.16 at 3m/s and rises rapidly to 0.41 at 7m/s, staying stable at 9m/s and decreasing slightly at 0.39 at 11m/s. For both the smooth and the eroded rotors the optimum values of the aerodynamic coefficient cp_{opt} do not vary significantly, apart for high wind speeds at 11m/s, were the eroded rotor drops to 0.38. With respect to the optimum value of the tip speed ratio λ_{opt} it shifts from higher values of about 6 for low wind speeds to lower values of about 5 for higher wind speeds. The results for the wind tunnel measurements of cp_{opt} and λ_{opt} can be seen in detail in Table 3.

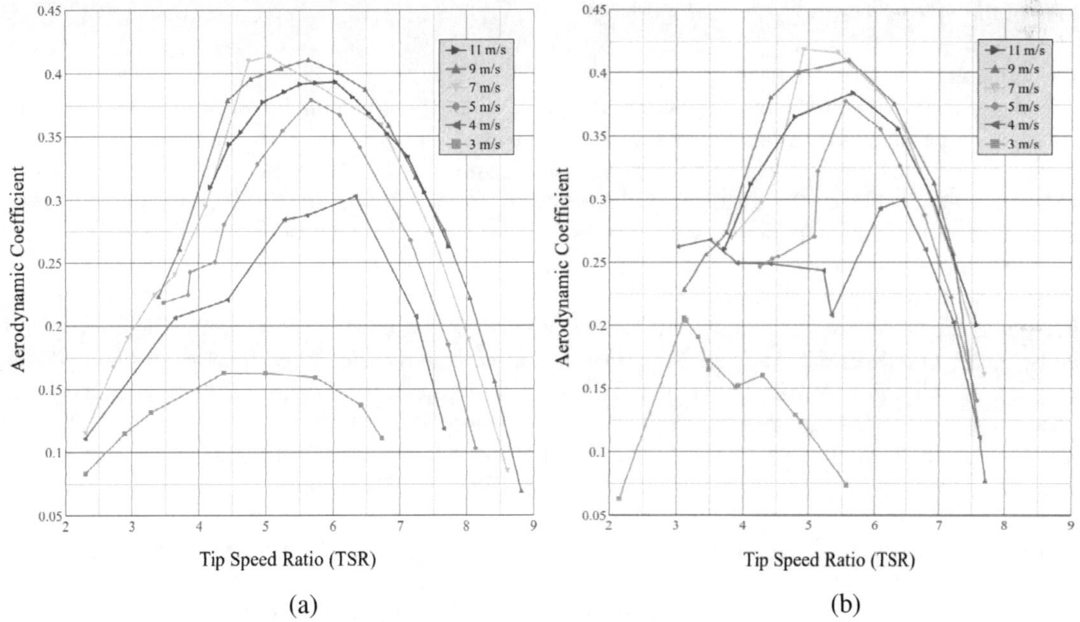

Figure 10. The aerodynamic coefficient of locally manufactured rotors, (a) without erosion and (b) with erosion, as measured in the wind tunnel.

As observed in the experimental results of Fig.10, although the maximum value of the aerodynamic coefficient is not influenced significantly due to leading edge erosion, only slightly at 11m/s, the cp vs λ curves for each wind speed are narrower and steeper, especially for wind speeds from 5m/s to 7 m/s. As a result, small changes in the tip speed ratio of the rotor, during operation in the field and within this wind speed range, can result in rapid increases or decreases of the aerodynamic coefficient.

Table 3. The optimal values of the aerodynamic coefficient cp_{opt} and of the tip speed ratio λ_{opt} as measured in the wind tunnel for the smooth and the eroded rotors.

		Smooth Rotor		Eroded Rotor	
Wind Speed (ms^{-1})	Reynolds number	Aerodynamic Coefficient cp_{opt}	Tip Speed Ratio λ_{opt}	Aerodynamic Coefficient cp_{opt}	Tip Speed Ratio λ_{opt}
3	5×10^4	0.163	4.99	0.204	3.14
4	8.5×10^4	0.303	6.33	0.299	6.43
5	9.5×10^4	0.379	5.67	0.377	5.55
7	1.2×10^5	0.414	5.05	0.418	4.92
9	1.7×10^5	0.411	5.63	0.416	5.43
11	2.1×10^5	0.393	5.74	0.384	5.66

The increased performance of the eroded rotor at low wind speeds, as analysed in the previous section, is such a case. As seen in Fig.10 and for the measurements at 7m/s, the gradients of the cp vs λ curves around tip speed ratios of 6, are significantly different between the smooth and the eroded rotors, resulting in a steeper rise of the aerodynamic coefficient of the eroded rotor for decreasing values of the tip speed ratio.

5. Conclusion

In this paper the effects of leading edge erosion on locally manufactured small wind turbine blades are researched experimentally, both in terms of power performance and acoustic noise emissions. Acoustic noise emissions of eroded rotor blades are found to increase by an average of 10% (or 4 dBA) over smooth blades for a range of wind speeds from 4m/s up to 10m/s, with the influence becoming more significant at higher wind speeds. A reduction of power production of up to 23.7% for the eroded rotor at the rated wind speed is observed, yet a slightly increased power production of 7.4% is observed at lower wind speeds from 5m/s to 7m/s. Similarly, there is a reduction of system efficiency from 0.154 to 0.117 at the rated wind speed for the eroded rotor, while for lower wind speeds, i.e. from 4m/s up to 7m/s, the system efficiency of the eroded rotor displays better performance, with values from 0.287 to 0.312 for a wind speed of 5.5m/s. This behaviour of the eroded rotor at low wind speeds is traced back to its operation closer to the optimal tip speed ratio λ_{opt} of the blades, due to stalling cased from the erosion of the leading edge. With respect to the AEP, for mean wind speeds from 4m/s to 6m/s the reduction in AEP gradually increases from 0.1% up to 9.6%. The wind tunnel tests show the behaviour of the smooth and the eroded rotors for low Reynolds numbers, concluding that the maximum value of the aerodynamic coefficient is not influenced significantly due to leading edge erosion for wind speeds below thee rated, yet the cp vs λ curves for each wind speed are narrower and steeper for the eroded rotor, especially for wind speeds from 5m/s to 7 m/s.

From the experimental results of the paper, it can be concluded that a yearly maintenance of the wooden blades of a LMSWT, which includes smoothening the airfoil and repairing the leading edge, can lead to less acoustic noise emissions and better power performance, with these aspects becoming more significant as the mean wind speed of the installation site increases. For practitioners, installers and manufacturers of LMSWTs these findings can significantly guide their manufacturing process and techniques in terms of preventing leading edge erosion, by promoting the use of durable wood species [10], adding appropriate leading edge tapes [26] or using appropriate coatings [27]. In addition, the findings of the paper can guide practitioners and installers of LMSWTs on how to plan their preventative maintenance procedures in order to keep their small wind electric systems quiet and productive, and thus further increasing the sustainability of rural electrification applications.

IOP Conf. Series: Journal of Physics: Conf. Series **1222** (2019) 012010 doi:10.1088/1742-6596/1222/1/012010

References

[1] Leary J, Howell R, While A, Chiroque J, VerKamp K and Pinedo C 2012 Post-installation analysis of locally manufactured small wind turbines: Case studies in Peru *IEEE 3rd Int. Conf. on Sustainable Energy Technologies (ICSET)* Kathmandu pp 396-401

[2] Louie H 2011 Experiences in the construction of open source low technology off-grid wind turbines *IEEE Power and Energy Society General Meeting* Detroit pp 1-7

[3] Latoufis K, Pazios T and Hatziargyriou N 2015 Locally manufactured small wind turbines: Empowering communities for sustainable rural electrification *IEEE Electrification Magazine* **3** 68-78

[4] Kostakis V, Latoufis K, Liarokapis M and Bauwens M 2018 The convergence of digital commons with local manufacturing from a degrowth perspective: Two illustrative cases *Journal of Cleaner Production* **197** 1684-93

[5] Alsop A, Eales A, Strachan S, Leary J, Persson J and Almeyda I R 2017 A global market assessment methodology for small wind in the developing world *IEEE Global Humanitarian Technology Conference (GHTC)* San Jose pp 1-6

[6] Latoufis K, Rontogiannis A, Karatasos V and Hatziargyriou N 2018 Thermal and Structural Design of Axial Flux Permanent Magnet Generators for Locally Manufactured Small Wind Turbines *XXIIIth Int. Conf. on Electrical Machines (ICEM)* Alexandroupoli pp 1001-07

[7] Piggott H 2014 *A Wind Turbine Recipe Book* (Scotland: Scoraig Wind Electric)

[8] Latoufis K, Pazios T, Chira K, Korres N and Hatziargyriou N 2018 Open design and local manufacturing of small wind turbines: Case studies in Ethiopia and Nepal *IEEE PES Power Africa* Cape Town pp 148-53

[9] Mishnaevsky L, Freere P, Sinha R, Acharya P, Shrestha R and Manandhar P 2011 Small wind turbines with timber blades for developing countries: Materials choice, development, installation and experiences *Renewable Energy* **36** 2128-38

[10] Pourrajabian A, Dehghan M, Javed A and Wood D 2019 Choosing an appropriate timber for a small wind turbine blade: A comparative study *Renewable and Sustainable Energy Reviews* **100** 1-8

[11] Rossouw F and Kamper M 2007 Use of air-cored axial flux permanent magnet generator in direct battery charging wind energy systems *7th Int. Conf. on Power Electronics and Drive Systems* Bangkok pp 1102-07

[12] Leary J, While A and Howell R 2012 Locally manufactured wind power technology for sustainable rural electrification *Energy Policy* **43** 173-83

[13] Keegan M, Nash D and Stack M 2013 On erosions issues associated with the leading edge of wind turbine blades *J. Phys. D: Appl. Phys.* **46** 383001

[14] Khalfallah M and Koliub A 2007 Effect of dust on the performance of wind turbines *Desalination* **209** 209-20

[15] Woobeom H, Jonghwa K and Bumsuk K 2018 Effects of contamination and erosion at the leading edge of blade tip airfoils on the annual energy production of wind turbines *Renewable Energy* **115** 817-23

[16] Sareen A, Sapre C and Selig M 2013 Effects of leading edge erosion on wind turbine blade performance *Wind Energy* **17** 1531-42

[17] Gaudern N 2014 A practical study of the aerodynamic impact of wind turbine blade leading edge erosion *Journal of Physics: Conference Series* **524** 012031

[18] Bertagnolio F, Madsen H and Fischer A 2017 A combined aeroelastic-aeroacoustic model for wind turbine noise: verification and analysis of field measurements *Wind Energy* **20** 1331-48

[19] Vick B and Broneske S 2013 Effect of blade flutter and electrical loading on small wind turbine noise *Renewable Energy* **50** 1044-52

[20] Monteiro J, Silvestre M, Piggott H and André J 2013 Wind tunnel testing of a horizontal axis wind turbine rotor and comparison with simulations from two blade element momentum codes *Journal of Wind Engineering and Industrial Aerodynamics* **123** 99-106

[21] Migliore P, van Dam J and Huskey A 2004 Acoustic Tests of Small Wind Turbines *Wind Energy Symposium* Nevada

[22] International Electrotechnical Commission (IEC) 2005 Power performance measurements of electricity producing wind turbines IEC 61400-12-1: Edition 1

[23] International Electrotechnical Commission (IEC) 2006 Wind turbine generator systems - Part 11: Acoustic noise measurement techniques IEC 61400-11: Edition 2.1

[24] Oerlemans S, Sijtsma P and Méndez López B 2007 Location and quantification of noise sources on a wind turbine *Journal of Sound and Vibration* **299** 869-83

[25] Miljøministeriet (Ministry of the Environment) 15th of December 2011 Bekendtgørelse om støj fra vindmøller (Statutory Order on noise from wind turbines) Nr.1284

[26] Giguère P and Selig M 1999 Aerodynamic effects of leading-edge tape on airfoils at low Reynolds numbers *Wind Energy* **2** 125-36

[27] Slot H, Gelinck E, Rentrop C and van der Heide E 2015 Leading edge erosion of coated wind turbine blades: Review of coating life models *Renewable Energy* **80** 837-48

Computational fluid dynamics methodology for characterisation of leading edge erosion in whirling arm test rigs

R Herring[1], K Dyer[1], A MacLeod[1] and C Ward[2]

[1] Offshore Renewable Energy Catapult, Offshore House, Albert Street, Blyth, NE24 1LZ, UK
[2] Department of Aerospace Engineering, Queen's Building, University of Bristol, Bristol, BS8 1TR, UK

robbie.herring@ore.catapult.org.uk

Abstract. Blade leading edge erosion has developed into a significant issue for the offshore wind industry. Protection solutions, including polymer coatings and tapes, are often applied to increase the blade lifetime. Experimental evaluation of protection systems is typically conducted in whirling arm rain erosion test rigs. Currently, there is no thoroughly validated method to relate the test results to real-world erosion performance. Furthermore, the design of rigs is not sufficiently limited to enable comparison of results between different rigs. Industry guideline, DNV-GL-RP-0171, provides a comparison method to address this issue. This paper describes the development of a droplet particle tracking Computational Fluid Dynamics methodology for rain erosion test rigs, which models the impact strike characteristics of a droplet, the number of impacts and the effect of rig aerodynamics. The methodology was applied to two rigs with different aerodynamics. Rain erosion tests were conducted in the rigs on identical coating and aluminium samples. The results were compared against predicted number of impacts from the DNV-GL guideline. Contradictory results were found, concluding that the guideline does not provide an accurate comparison between all test rigs, as it does not account for rigs where large aerodynamic effects cause droplet concentrations or droplet break-up.

1. Introduction

Blade leading edge erosion has developed into a significant issue for the offshore wind industry. Raindrops, hailstones or other particles impacting the blade leading edge cause material to be removed from the blade surface, leaving a rough profile that degrades the aerodynamic performance and impacts the structural integrity of the blade [1].

There are several erosion protection systems available to the industry that can increase the lifetime of the turbine by mitigating leading edge erosion, including coatings, tapes and soft shells. Experimental evaluation of protection systems is typically conducted in a whirling arm rain erosion test rig where a sample is fixed to the end of an arm and rotated through a rain field of a uniform droplet size [2].

Current test procedures and facilities remain limited to comparison of product performance under selected test conditions. Currently, there is no thoroughly validated method to relate test results to real-world erosion performance. Furthermore, industry standards, ASTM-G73, ISO-TS-19292-2, do not limit the design of the rigs and the choice of test parameters to allow direct comparison of test results between different test rigs. Although industry guideline, DNV-GL-RP-0171 [3], does give a comparison method it does not account for rigs where large aerodynamic effects and droplet concentrations occur. This has been observed to lead to conflicting rain erosion test results due to the elastic and viscoelastic nature of polymer coatings.

For the development of a realistic lifetime prediction model from rain erosion test results, which accounts for all protection systems and rain erosion rigs, a direct relationship needs to be established between test rig parameters and the characteristics of the erosion developed in different materials. Testing then needs to be completed in a realistic accelerated rain erosion test correlated directly to the measured offshore wind turbine droplet impact environment.

In this work, a droplet particle tracking Computational Fluid Dynamics (CFD) methodology has been developed for rain erosion tests rigs. The methodology enables modelling of the impact characteristics of a droplet, the number of impacts and the effect of test rig aerodynamics. Two whirling arm test rigs have been evaluated:

- A previous configuration of the test rig at the Energy Technology Centre (ETC).
- The test rig at Offshore Renewable Energy (ORE) Catapult. This rig is the wind industry standard design used within the DNV-GL guideline for geometrical droplet impact calculations.

2. ETC Test Rig Model

2.1. Model Description

A 3D model of the previous configuration of the whirling arm test rig at the ETC was constructed in ANSYS Fluent CFD. The ETC rig possessed two arms with either flat or aerofoil samples. This was updated to a single arm running the same aerofoil samples as the ORE Catapult rig for testing. The test room was simplified to a circular space and the motor support was approximated to an equivalent circular cone. These approximations enable the rotating rig to be modelled using a rotating reference frame within the CFD model. Since the rig is rotationally symmetrical, a periodic boundary was used along the arm centreline, halving the computational mesh size. The flow solutions have been solved in steady-state. Figure 1 presents the CFD model of the test rig with contours of velocity magnitude highlighted. The circumferential line illustrates the droplet release positions.

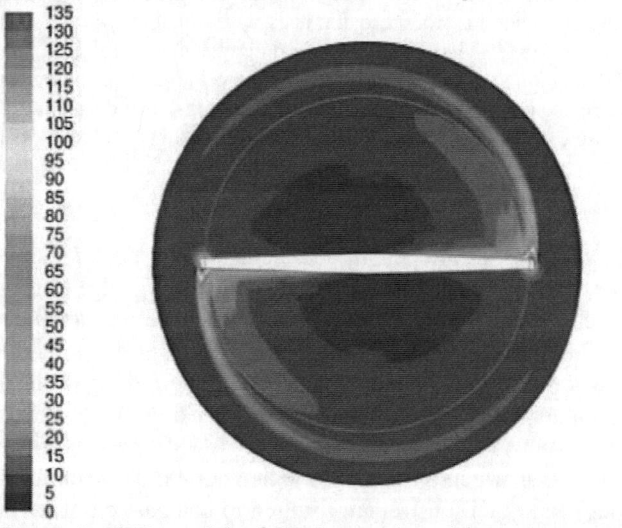

Figure 1. CFD model of ETC test rig, with contours of velocity magnitude on a horizontal plane level with the sample centreline (m/s) highlighted.

IOP Conf. Series: Journal of Physics: Conf. Series **1222** (2019) 012011 doi:10.1088/1742-6596/1222/1/012011

The local flow velocities acting on the droplet and the resulting motion of the droplet were predicted using a Matlab model. Droplets are released, and their equations of motion solved in the time domain as they pass through the varying velocity field. At each time step, the aerodynamic force on the droplet is calculated via a drag law. This drag force is added to the gravitational force and used to calculate the droplet acceleration, which is then integrated through time to calculate the trajectory. There is a risk that the aerodynamic forces experienced by the droplets as they travel through the rig may result in their breakup. To account for this, a droplet breakup criterion was implemented based on the Weber number [4]. Combining the droplet tracking with the CFD model predicts the aerodynamics effects of the test rig on falling droplets and its resulting impact characteristics.

2.2. ETC Test Rig Aerodynamics

Two test sample configurations have been repeated in the CFD model: a vertical flat plate and an aerofoil sample, both tested at 700 rpm. These cases were selected to provide a good understanding of the dominant features of the droplet behaviour and the degree to which this can be captured accurately in the modelling.

Figure 2 shows the track of streamlines over the two samples at 700 rpm. The streamlines around the flat plate sample highlight the influence of the clamping block on the flow into the centre of the sample. Furthermore, the streamlines passing over the sample indicate that the combination of the clamp and sample wake is causing rotation in the wake behind the sample. This is evident by the loop of streamlines above the sample. The aerofoil also introduces rotation into its wake due to its tip vortex. These rotational effects, along with the severe shear between the sample wake and the surroundings, influence the droplet tracks and their likelihood of breaking up along their trajectory towards the sample.

Droplets were released from needles around the full 360° of the rig at the radial location corresponding to the specimen centre. Each droplet track stops when it either strikes the sample or leaves the annular region around it. The droplet impacts on the sample were recorded and have been visualised in Figure 3.

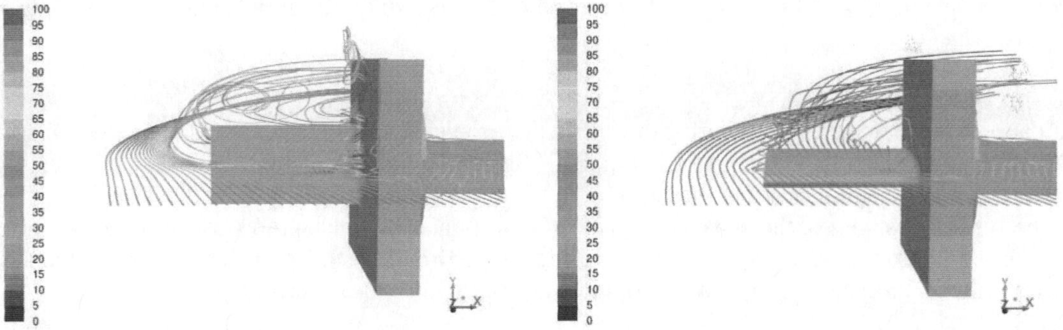

Figure 2. Flow streamlines coloured by velocity magnitude (m/s) for the flat plate (left) and the aerofoil sample (right).

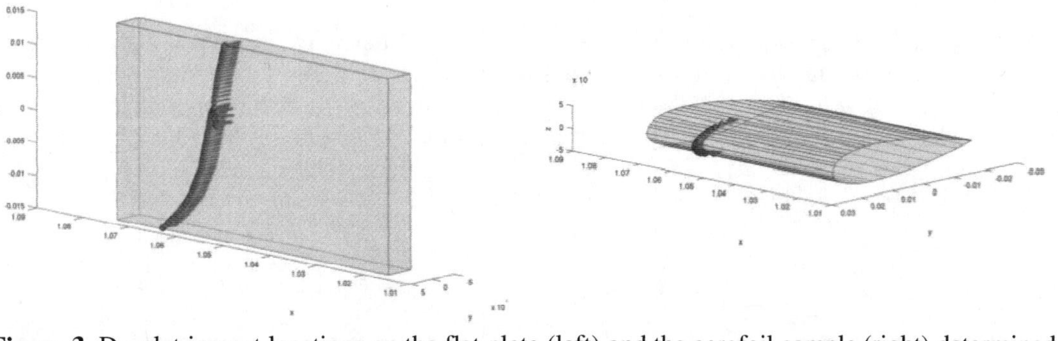

Figure 3. Droplet impact locations on the flat plate (left) and the aerofoil sample (right) determined computationally.

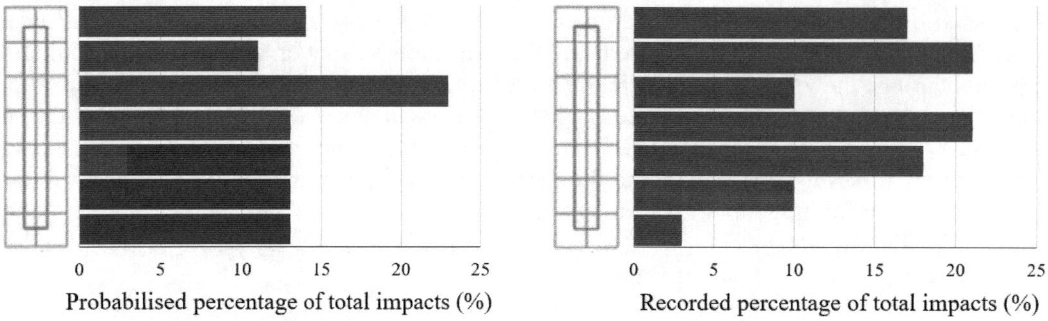

Figure 4. Histograms presenting the impact locations determined computationally (left) and experimentally (right) on the flat plate.

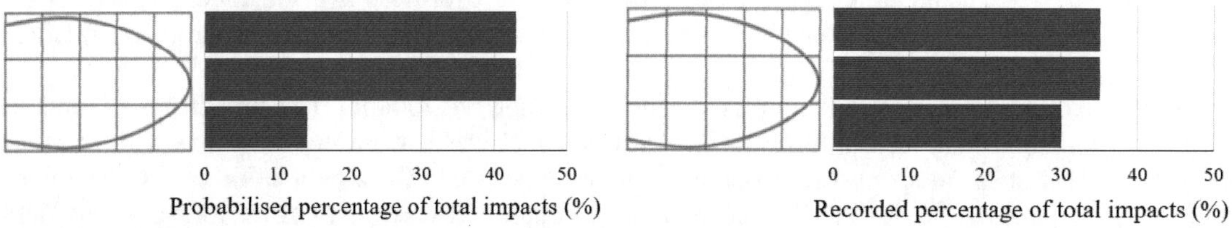

Figure 5. Histograms presenting the impact locations determined computationally (left) and experimentally (right) on the aerofoil sample.

In the flat plate results, a loop can be seen in the impact locations, introducing droplet concentrations onto the sample. The model shows that the droplets whose impacts form this loop are affected by the passage of the previous sample as they are released from the needles. This effect, combined with the uneven aerodynamic flow across the sample, cause the droplet to veer from its straight trajectory. In the aerofoil sample, the droplet impact path has moved outwards radially from the central droplet release position.

2.3. Model Validation
Validation was performed by comparing the number of droplet impacts and strike locations on a sample in the ETC test rig and the developed CFD model. High speed video cameras were used in the test rig to examine the droplet impacts and locations, both radially and longitudinally along the sample.

The vertical positions of the impact points have been collated into histograms which were compared with the results from the ETC rig (Figure 4 and Figure 5). Droplet breakup has been presented in red. The CFD model gave consistent results with the high-speed video measurements.

3. ORE Catapult Test Rig Model

3.1. Model Description
Using the validated methodology, a CFD model for the ORE Catapult test rig was developed. This rig possesses three arms with an aerofoil sample positioned along each arm. Figure 6 presents the CFD model of the test rig with contours of velocity magnitude highlighted. The circumferential lines illustrate the ten droplet release positions. The sample has a length of 0.45 m and is situated 0.8 m along each rotating arm.

IOP Conf. Series: Journal of Physics: Conf. Series **1222** (2019) 012011 doi:10.1088/1742-6596/1222/1/012011

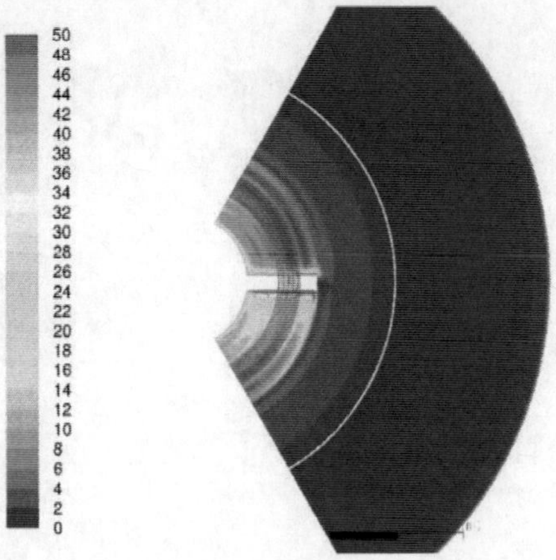

Figure 6. CFD model of ORE Catapult test rig, with contours of velocity magnitude on a horizontal plane level with the sample centreline (m/s) highlighted.

3.2. ORE Catapult Test Rig Aerodynamics

Contours of radial velocity can be seen across the plane of the sample. This demonstrates that the rotating blade causes an outward flow across the wake. Slight steps at each end of the sample result in small regions of separated flow at both ends of the sample. Figure 7 reveals that the flow over the sample and arm is streamlined and produces only a thin wake. This aerodynamic performance is significantly better than the flow results from the previous configuration of the ETC test rig. This can be attributed to the clean mounting design and aerofoil shaped sample.

The even aerodynamic flow over the sample and the thin wake is unlikely to affect the droplet trajectories. However, the outward flow and small regions of flow separation may influence the motion of the droplets. Their influence has been investigated.

Figure 8 presents the impact location of droplets along a test sample at 1000 rpm. The droplets can be seen to impact the sample at regular intervals with minimal scatter. To understand the magnitude of the scatter at different rotational velocities, the positions of the impact points have been collated into histograms (Figure 9). At all rotational velocities, the ten points of droplet release can be seen clearly. The histograms reveal that the aerodynamic forces present in the rig do cause some slight scatter. This effect becomes more pronounced at higher velocities due to a greater outward flow across the wake. However, the clear impact points demonstrate that the aerodynamic forces are minimal and do not govern the droplet flow. As a result, it can be determined that the ORE Catapult test rig does not experience aerodynamic droplet concentrations.

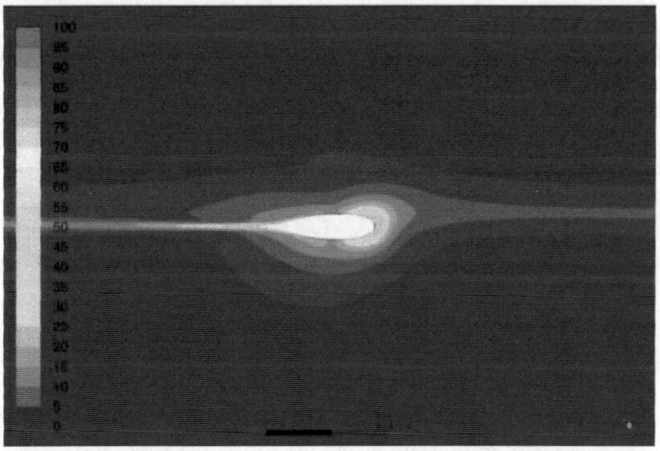

Figure 7. Contours of velocity magnitude (m/s) on the aerofoil sample used in the ORE Catapult test rig.

IOP Conf. Series: Journal of Physics: Conf. Series **1222** (2019) 012011 doi:10.1088/1742-6596/1222/1/012011

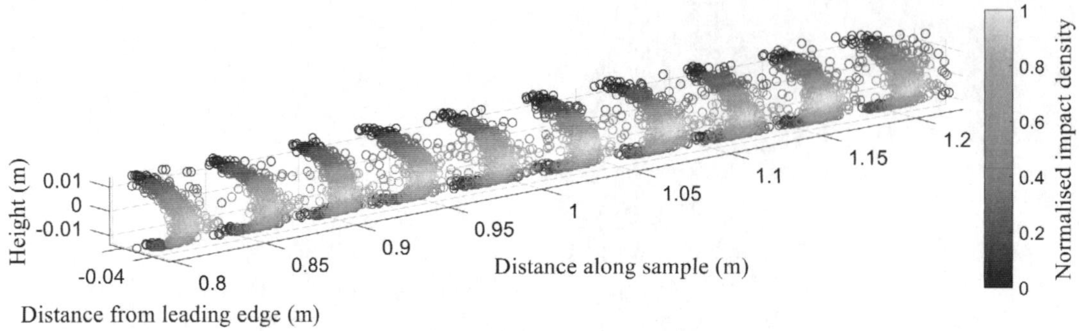

Figure 8. Droplet impact location on a sample in the ORE Catapult test rig.

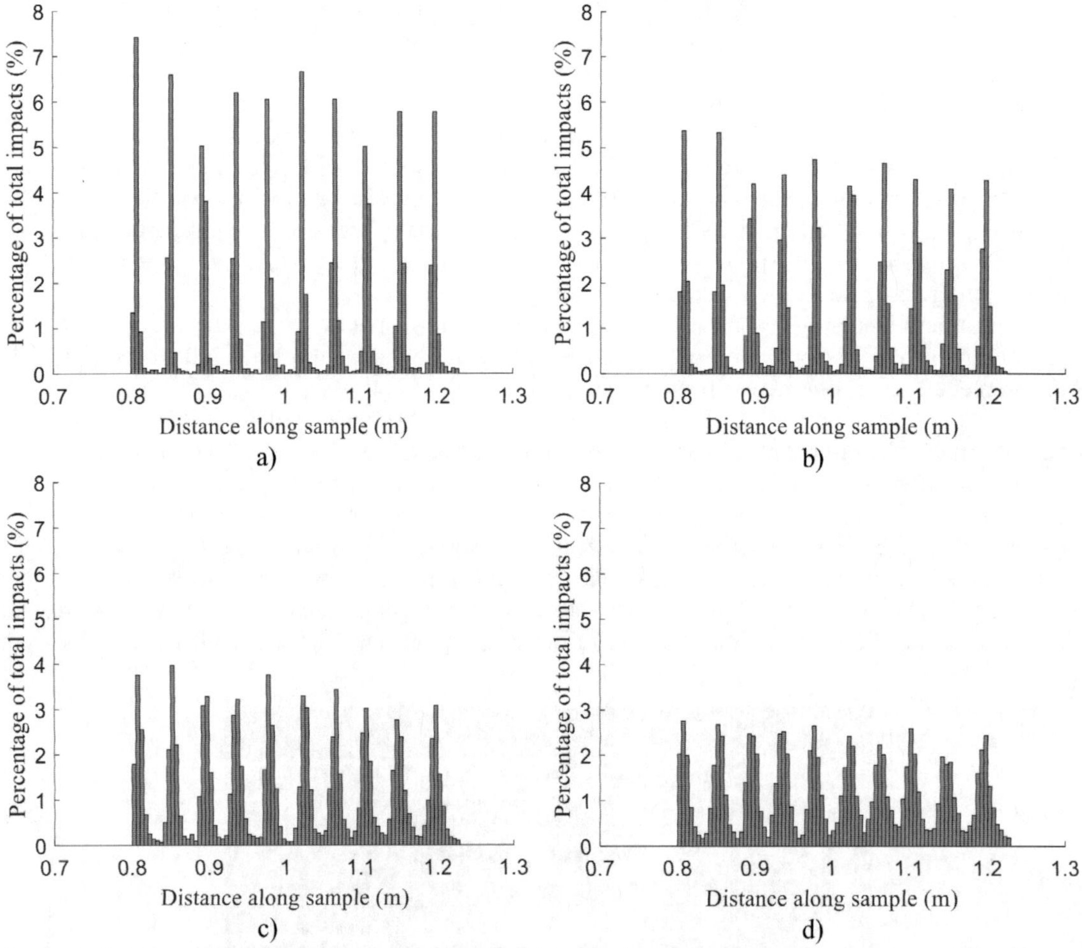

Figure 9. Histograms presenting the droplet impact locations determined computationally at rotational velocities of a) 800 rpm, b) 1000 rpm, c) 1200 rpm and d) 1386 rpm.

3.3. Model Validation

Video recording was not possible in the test rig. Instead, the CFD model was validated by evaluating the kinetic energy of each droplet impact determined computationally against the analytical solution given in DNV-GL-RP-0171. The guideline states that the droplet falling velocity is negligible compared to the sample velocity and therefore the impact velocity, v, of a droplet is given by:

$$v(r) = \omega r, \tag{1}$$

where ω is the rotational velocity of the sample and r is the radial distance of a point from the centre of rotation.

The kinetic energy, E_K, of a droplet can be expressed as:

$$E_K = \frac{1}{12}\rho\pi d^3 v^2, \tag{2}$$

where ρ is the density of the droplet and d is the droplet diameter.

Figure 10 presents the comparison of the kinetic energies of the droplets determined computationally and analytically for a range of rotational velocities. The model gave consistent results with the analytical solution.

4. Rain Erosion Testing Prediction

Rain erosion tests on aluminium samples were conducted with the growth of erosion damage characterised through mass loss and visual inspections. Aluminium exhibits repeatable erosion performance in tests and is defined as the calibration sample in the DNV-GL guidelines. During a whirling arm erosion test on aluminium, the erosion grows linearly from the tip end towards the root. Therefore, the time to damage and the pit depth is directly related to the number of droplets impacting at each point and the kinetic energy of the impact.

Figure 11 presents an aluminium sample after 10 hours at a rotational velocity of 1250 rpm in the ORE Catapult test rig. A droplet diameter of 2.37 mm was used.

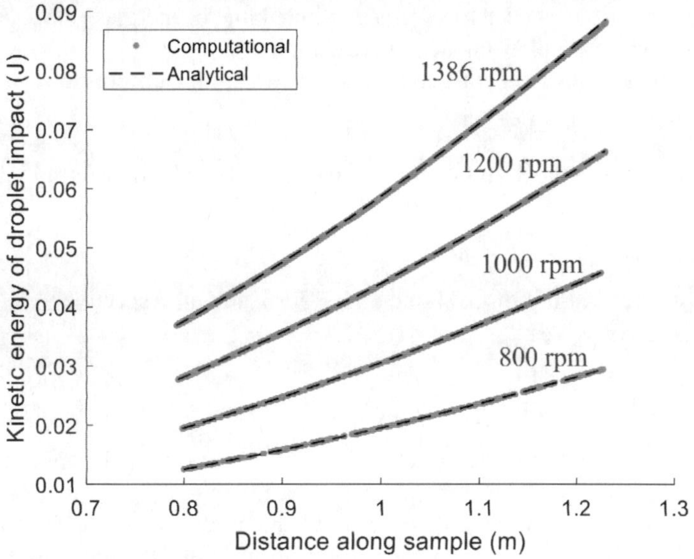

Figure 10. Kinetic energy of droplet impacts determined computationally and analytically at a range of rotational speeds

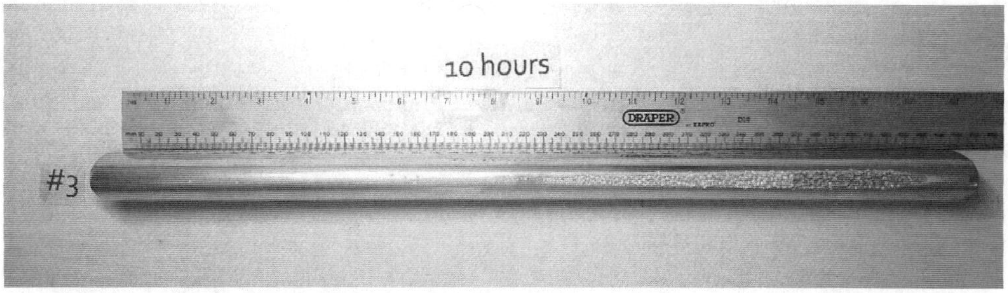

Figure 11. An aluminium sample after 10 hours in the ORE Catapult test rig at a rotational velocity of 1250 rpm.

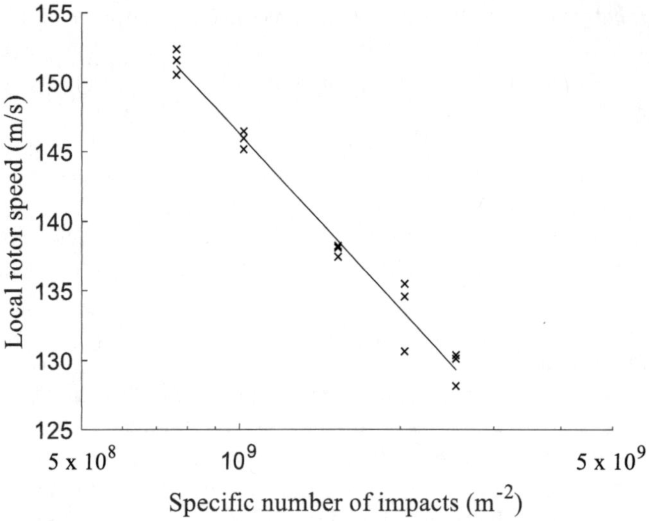

Figure 12. Incubation curve demonstrating the end of incubation at different local velocities on a sample.

Figure 12 presents the incubation curve for the aluminium samples. The recorded ends of the incubation period at different local velocities, and their corresponding kinetic energies, have been expressed in terms of the specific number of droplets impacting the sample.

The incubation curve can be developed into a generalised solution by assuming a power law:

$$N = kE_k{}^m, \tag{3}$$

where N is the specific number of impacts and k and m are constants. This can be transferred into a logarithmic form to calculate the constants k and m.

$$\log N = log\, k + m \log E_k \tag{4}$$

Fitting the test data to equation (4) gives values of -3.77 and 55660 for k and m, respectively.

Combining equation (2) and equation (3) gives:

$$N = k\left(\frac{1}{12}\rho\pi d^3 v^2\right)^m \tag{5}$$

Using equation (5), it now becomes possible to determine the incubation curves for different local velocities and droplet diameters.

The developed equation was then used to predict the damage on identical aluminium samples at a different rotational velocity, but same droplet size and then the same rotational velocity, but a different droplet diameter. A rotational velocity of 1386 rpm and a droplet size of 2.83 mm were used in the two respective rain erosion tests.

Figure 13 presents the predicted incubation curves with the results of the rain erosion tests indicated. The results show that the prediction is effective at identifying the incubation point for the aluminium samples on the ORE Catapult test rig.

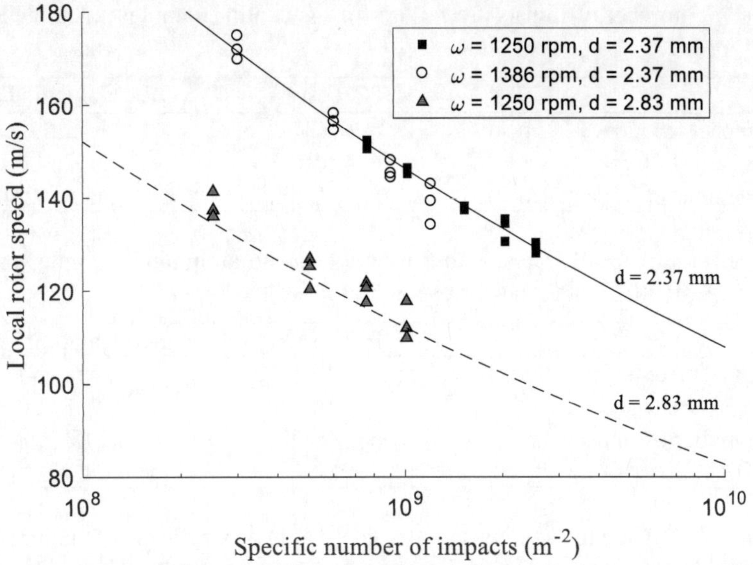

Figure 13. Incubation curves for different rotational velocities and droplet diameters with the test results highlighted.

5. Guideline Comparison

Using the methodology presented in DNV-GL-RP-0171, the two rigs were compared. Tests were performed on coating samples until failure, and on aluminium samples for 6 hours with the mass loss recorded. In both cases, the specific number of impacts was calculated using the DNV-GL guideline.

Figure 14 and Figure 15 present the rain erosion test results of the coating sample after 35 minutes in the ETC rig and 16 hours in the ORE Catapult rig, respectively. Due to the nature of the test rigs, the samples lasted considerably different periods of time. However, under the methodology presented in the DNV-GL guideline the results should be comparable. Table 1 and Table 2 present the results and the comparison between the rigs.

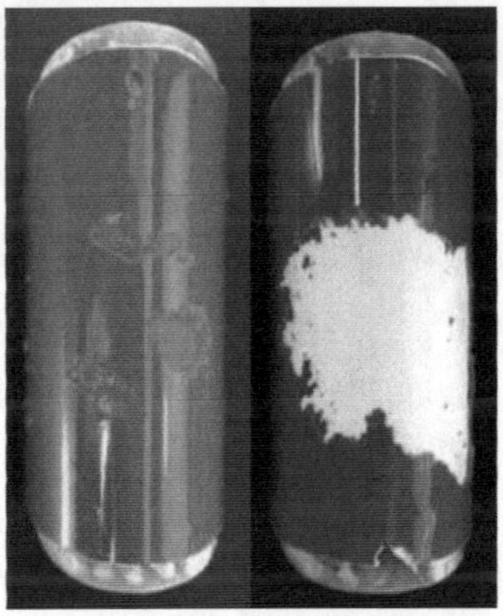

Figure 14. Coating samples after 20 minutes (left) and 35 minutes (right) in the ETC test rig.

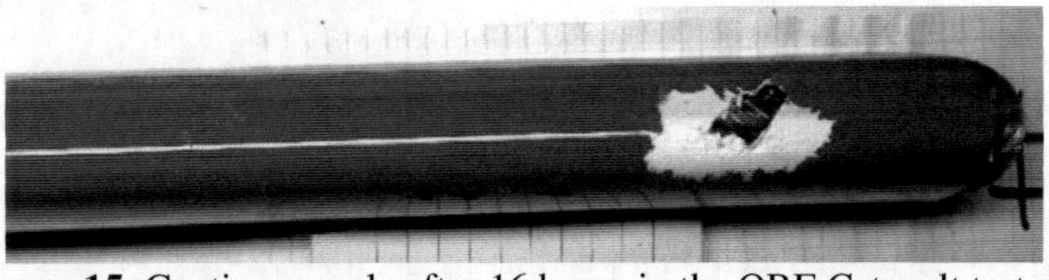

Figure 15. Coating sample after 16 hours in the ORE Catapult test rig.

Table 1. Specific number of impacts to failure for the coating samples calculated through DNV-GL-RP-0171.

	ETC Test Rig	ORE Catapult Test Rig
Specific impact frequency [s^{-1} m^{-1}]	30,877	42,699
Time to failure [s]	2,100	57,600
Specific number of impacts [m^{-1}]	64,841,700	2,459,462,400

Table 2. Specific number of impacts after 6 hours for the aluminium samples calculated through DNV-GL-RP-0171 and the corresponding mass loss.

	ETC Test Rig	ORE Catapult Test Rig
Specific impact frequency [s^{-1} m^{-1}]	30,877	57,644
Exposure time [s]	21,600	21,600
Specific number of impacts [m^{-1}]	666,943,200	1,245,110,400
Mass loss [g]	0.05	1.1

The comparison revealed that the coating samples failed after a significantly smaller specific number of impacts in the ETC test rig than in the ORE Catapult test rig. However, the results are reversed for the aluminium samples, with the ETC rig removing less mass than the ORE Catapult rig.

There is also a clear difference in the failure mode of the coating sample in the different test rigs. In the ETC rig the sample exhibits surface pitting, whilst in the ORE Catapult rig the sample experienced subsurface damage that causes large areas of coating to be removed.

The CFD models have revealed that the ORE Catapult test rig has an even aerodynamic flow over the sample, whereas the ETC test rig does not and therefore introduces a high droplet impact density in certain locations. The different characteristics of the two test rigs are thus beneficial for exploring different rain erosion materials and their elastic and viscoelastic properties.

6. Conclusions

The methodology presented enables the prediction of a droplet's impact strike characteristics, the number of impacts and the effect of rig aerodynamics. CFD models for two differing whirling arm test rigs have been developed and validated. The first model was validated against high speed video footage of the corresponding test rig, whilst the second model was validated through a comparison with the predicted kinetic energy of droplet impact against the analytical solution presented in the DNV-GL guideline.

The CFD models revealed that the previous configuration of the ETC test rig introduced aerodynamic droplet concentrations and radial outflow of the droplet path, whilst the ORE Catapult test rig did not. Repeat high speed video characterisation on the modified ETC rig showed droplet break-up still occurred indicating large aerodynamic effects. Rain erosion tests were performed on aluminium samples in the ORE Catapult test rig in line with the DNV-GL guideline. The results of one test were used to successfully predict the result of two subsequent tests performed at different conditions.

Experimental tests on identical coating and aluminium samples were conducted in the ETC rig and the ORE Catapult rig and compared to the DNV-GL guideline. Differing results were seen between the experiments, concluding that the guideline does not provide accurate comparison between test rigs as it does not account for aerodynamic induced droplet effects on the coating damage mode and material response.

The methodology in the guideline has been applied effectively in the ORE Catapult test rig to predict the results of future tests. However, the guideline has been shown to not account for large aerodynamic effects in test rigs and it has been seen that the viscoelastic properties of the rain erosion materials can be influenced by the test droplet impact conditions.

The industry goal is to develop a method for predicting the lifetime of a protection system on a wind turbine from rain erosion test results. This goal is being pursued in the DNV-GL project COBRA.

However, until aerodynamic effects, such as droplet concentration, and viscoelastic material effects are included in tests and lifetime prediction models, the lifetime prediction method is unlikely to be suitable for all test rigs and materials.

Acknowledgements
This work was supported by the Engineering and Physical Sciences Research Council through the EPSRC Centre for Doctoral Training in Composites Manufacture (Grant: EP/K50323X/1), project partners the Offshore Renewable Energy Catapult <https://ore.catapult.org.uk>, and the EPSRC Future Composites Manufacturing Hub (grant: EP/P006701/1). The authors would also like to thank the Wind Blade Research Hub for their support in delivery of this manuscript. All data necessary to reproduce the results and support the conclusions are included within this paper.

References

[1] Keegan M H, Nash D H and Stack M M 2013 On erosion issues associated with the leading edge of wind turbine blades *J. Phys. D. Appl. Phys* **46** 38 p 383001
[2] Zhang S, Dam-Johansen K, Nørkjær S, Bernad P L and Kiil S 2015 Erosion of wind turbine blade coatings - design and analysis of jet-based laboratory equipment for performance evaluation *Prog. Org. Coatings* **78** p 103-115
[3] DNV-GL 2018 Testing of rotor blade erosion protection systems
[4] Frohn A and Roth N 2000 *Dynamics of droplets* Springer-Verlag Berlin Heidelberg

IOP Conf. Series: Journal of Physics: Conf. Series **1222** (2019) 012012 doi:10.1088/1742-6596/1222/1/012012

Preliminary validation of ATOM: an aero-servo-elastic design tool for next generation wind turbines

S Scott[1], T Macquart[1], C Rodriguez[2], P Greaves[2], P McKeever[2], P Weaver[1] and A Pirrera[1]

[1] Bristol Composites Institute (ACCIS), University of Bristol, BS8 1TR, UK
[2] Offshore Renewable Energy Catapult, Blyth, NE24 1LZ, UK

E-mail: ss1870@bristol.ac.uk, terence.macquart@bristol.ac.uk

Abstract. Upscaling wind turbines has resulted in levelised cost of energy (LCoE) reductions. However, larger turbine diameters pose significant design challenges, often with conflicting requirements. For example, non-linear dynamics of aeroelastic tailored blades must be accurately predicted whilst, for the sake of efficient gradient-based design, it is also desirable to simplify the numerical definition of such blades—keeping design variables (DVs) to a minimum. This work presents and validates two features of the ATOM code (**A**eroelastic **T**urbine **O**ptimisation **M**ethods), developed at the University of Bristol, that enable accurate and efficient modelling of large-scale wind turbine blades. Both an efficient parameterisation method and high-order beam elements illustrate the capacity for increasing the speed of gradient evaluations whilst accurately predicting blade dynamics—either by reducing DVs or simulation time. As a preliminary validation, aero-servo-elastic simulations from ATOM and an industry-standard software—DNV GL Bladed—are compared against field measurements gathered from an existing 7 MW turbine.

1. Introduction

Wind energy offers an economically competitive source of renewable energy, with the average levelised cost of energy (LCoE) for onshore wind being 60 USD/MWh [1]—this being at the lower end of the spectrum for fossil fuels. Further decreases in cost of energy are desirable as they will encourage further investment in wind power generation, decreasing global reliance on fossil fuel and aiding governments in reaching climate change targets.

Upscaling wind turbines has been a reliable means to reduce the LCoE, and turbine sizes are, consequently, predicted to grow further (20 MW+) [2]. It is, therefore, imperative that industry can effectively design such machines and accurately predict their feasibility—whether from an economic, structural, or operational perspective. A key research field that is gaining much interest for wind turbine design is that of multi-disciplinary optimisation (MDO) and, in particular, gradient-based methods [2, 3, 4]. Wind turbines have a large design space with many, and often conflicting, design choices to be made. MDO techniques allow these design spaces to be algorithmically searched, minimising the cost-function (usually LCoE) with novel solutions that may not always be obvious to human designers during an iterative process.

The success of MDO in finding novel design solutions relies heavily on accurate analysis modules. Modelling aero-servo-elastic phenomenon for next-gen wind turbines becomes more challenging for a number of reasons:

- Longer, more slender blades are often more compliant, therefore, non-linear structural models are required to accurately predict large deflections, strains and other metrics required to assess structural feasibility.
- Blades with aeroelastic tailoring [5, 6] require accurate prediction of torsional dynamics as the aerodynamic loads are very sensitive to small variations in sectional angles of attack [7]. Torsional dynamics also require accurate prediction of cross-sectional properties.
- Large deflections, prebend, sweep, cone tilt and other re-orientation of blade sections require that the aerodynamic models are able to dynamically resolve wind and loading vectors in three-dimensions. Higher-order aerodynamic models such as lifting line/free-vortex wake or computational fluid dynamics offer this capability, however, are currently still too costly from a computational point of view for MDO.

MDO tools must also be computationally efficient to ensure that convergence is achievable within practical time and hardware limitations. Wind turbine standards [8] require a vast number of randomly seeded turbulent simulations to ensure statistical likelihood of capturing the full load envelope/spectrum experienced by wind turbines over their lifetime. Whilst this set of standardised load cases can be condensed for design purposes, performing a single gradient evaluation of a wind turbine design remains, to date, computationally demanding. Given these computational demands, MDOs can be made more efficient in two ways, either by reducing the time required to evaluate the fitness of a given design solution, or reducing the number of fitness evaluations. Literature offers many ways in which speed of fitness evaluations can be improved for wind turbine MDO—for example, fast and reliable BEM convergence methods [9], higher-order beam modelling to maintain accuracy of stiffness variations with fewer beam nodes [10], or reducing the number of simulations required for load evaluation [11]. Alternatively, the number of fitness evaluations can be reduced by using fewer design variables (DVs), or by using a more efficient (or better-suited) optimisation algorithm. For gradient-based algorithms, the number of fitness evaluations for one gradient evaluation through finite difference is proportional to the number of DVs. Therefore, by clever parameterisation of the turbine definition into a concise set of numerical DVs, the number of DVs can be kept to a minimum without sacrificing model accuracy or causing excessive bias in the final solution. A biased, and possibly sub-optimal, solution may be found if the parameterisation scheme favoured certain types of solutions over others that may have equal, or even better, fitness.

Given the conflicting requirements for computational efficiency and model accuracy during optimisation, this work presents a software tool developed to tackle these issues. The ATOM code (**A**eroelastic **T**urbine **O**ptimisation **M**ethods), part of ongoing development at the University of Bristol [7], is a multi-disciplinary optimisation and analysis (MDAO) tool capable of performing design studies for the next generation of wind turbines. This work presents two features of ATOM that specifically target the aforementioned issues: efficient blade parameterisation, and high-order beam modelling. The blade parameterisation highlights the potential for reducing the number of DVs whilst accurately representing blade dynamics, whereas high-order beam modelling allows more efficient structural modelling, thus, reducing simulation times. To illustrate the applicability of these methods for modern wind turbines, they are employed in a preliminary validation study of ATOM, using an existing 7 MW turbine. Simulations are run using empirical wind data taken from met-mast and anemometer recordings, field measurements from the turbine are then compared against results from ATOM and an industry standard aeroelastic software, DNV GL Bladed [12].

Section 2 begins by introducing ATOM, and the 7 MW baseline turbine used in this work.

IOP Conf. Series: Journal of Physics: Conf. Series **1222** (2019) 012012 doi:10.1088/1742-6596/1222/1/012012

Section 3 presents the efficient blade parameterisation, followed by high-order beam modelling in Section 4. Lastly, results from the preliminary validation are displayed in Section 5.

2. Modelling details

This section begins by giving details of the modelling capabilities within ATOM that are relevant to this work. Next, a brief introduction of the baseline 7 MW turbine is given.

*2.1. ATOM (**A**eroelastic **T**urbine **O**ptimisation **M**ethods)*

ATOM is an analysis and optimisation tool specifically aimed at the optimisation of aeroelastic tailored, horizontal-axis wind turbines—whereby turbine topology remains fixed during the optimisation. From an analysis point of view, a set of numerical DVs that define each aspect of a turbine can be input by the user, ATOM then performs every step from meshing of the blade and tower beams, right through to running a full set of aero-servo-elastic design load cases (DLC) and post-processing results for statistics on power, loads, cross-sectional strains and failure indices. This analysis capability is smoothly integrated within an optimisation framework with a choice of algorithms. ATOM also interfaces with the DNV GL Bladed API so as to allow equivalent Bladed models to be created from an ATOM model and then validation simulations to be performed.

A key feature of ATOM's design capabilities is the use of spline surfaces and lamination parameters [7]. These features allow completely free, un-biased design of a wind turbine rotor, with the goal of realising the full potential of composite laminates for aeroelastic tailored structures. During optimisation, DVs control a number of spline surfaces that define variations in lamination parameters and laminate thickness across the blade. This framework allows for the optimisation process to be unconstrained by conventional structural configurations and stacking sequences.

ATOM utilises a range of analysis modules, in which full detail will be given in a future publication. However, a brief summary is given here:

- The aerodynamic model utilises a modified blade element momentum (BEM) theory with corrections for dynamic wake and dynamic stall. A key feature is the 3D resolution of wind vectors, aerofoil section vectors and loading vectors to ensure that geometric deviations such as deflections, sweep, prebend, cone and tilt are all accounted for. A similar 3D formulation of BEM can be found in work by Ponta *et al.* [13].
- The structural models of the blade and tower utilise beam finite-elements (FE), with linear damping. The beam elements allow for anisotropic composite laminates with fully populated stiffness and mass matrices. A cross-sectional modeller uses 2D bar elements to generate properties from the laminate spline surfaces. ATOM allows for high-order beam modelling whereby three, four or five node beam elements can be used [10]. Application of these high-order beam elements is presented in Section 4. ATOM also allows solving for displacements of the beam models using modal decomposition, or the non-linear co-rotational framework. Lastly, beam strains can be used to extract cross-sectional laminate strains and failure indices, which can then be aggregated using Poon *et al.*'s adaptive method [14].
- The non-linear aero-servo-elastic model combines aerodynamic and structural models with a simple controller and multiple options for time-varying wind fields. The controller is embedded into the code and employs PI pitch control above rated and a simple optimal mode gain to control torque below rated. At any time step, the solver iterates through each aero-servo-elastic analysis module using a combination of explicit first- (Euler) and second-order (Heun) schemes.

Parameter	**Value**
Wind class	I_A/S_B
Rotor diameter	171.2m
Hub height	110.6m
Blade length	83.5m
Capacity	7MW
Generator	Medium (3.3kV) PMG
Convertor	Full power conversion
Drivetrain	400 rpm
Rotor speed	5.9-10.6 rpm
Wind speed	3.5-25 ms^{-1}
Rated wind speed	10.9 ms^{-1}
Design life	25 years

Table 1. Levenmouth 7 MW turbine parameters [15].

2.2. The Levenmouth 7 MW turbine

To illustrate the efficacy of this work for modelling of modern turbines i.e. large, multi-megawatt, offshore turbines with aeroelastic tailored blades, the baseline model used in this work is derived from an existing 7 MW turbine. Further, the turbine offers a starting point for future gradient-based optimisation studies that is known to be commercially-endorsed technology and thus can make the outcomes of those studies more relevant for industry applications.

The wind turbine is located off the coast of Levenmouth, Scotland, it was built by Samsung Heavy Industries and is now owned and managed by ORE Catapult for research purposes. The turbine is instrumented with a basic supervisory control and data acquisition (SCADA) system and a nearby met-mast offers meteorological data. Details of the turbine are given in Table 1 [15].

3. Structural blade parameterisation

This section presents the structural blade parameterisation method available in the ATOM code and applies it to the Levenmouth 7 MW turbine. Given this blade, the potential for minimising the number of DVs whilst accurately representing global blade dynamics is highlighted.

Large blades that have undergone detailed design phases (i.e. the Levenmouth 7 MW) have a considerable amount of information describing the geometric and structural layout. There are usually a number of spline curves that describe external geometric properties such as chord, twist, thickness, sweep and prebend, in addition to the many sets of aerofoil coordinates. Internally, the blade structure is composed of composite laminates that may include glass/carbon fibres, foam/balsa, gelcoat, lightning protection, paint and resins such as epoxy. Glass/carbon fibre layers are often used in the form of thick uni-directional (UD), biaxial or tri-axial mats. There are laminate schedules for each structural component (i.e. spar cap, web) describing the layer order, material, orientation, and spanwise start/termination locations of each ply. Then there is also definition of the component locations (i.e. distance from leading-edge (LE) to web) that often vary along the blade span. In addition to these main structural components, the detailed design may also include smaller hand lay-up patches or interfacing layers, however, these are of less importance to a MDO which, in general, is more concerned with the preliminary design phase.

Given the vast structural description generated by a detailed design phase, it is not feasible, or necessary, that all of this be accounted for during MDO. Therefore, the blade parameterisation should represent enough structural detail so as to accurately capture global blade dynamics (i.e.

stiffness/mass variations), as well as in-plane strains. Accurate global blade dynamics allow the analysis modules and optimiser to consider realistic aero-elastic effects, and in-plane strains allow for a preliminary assessment of ultimate and fatigue failure criteria.

The blade parameterisation capability in ATOM is fully detailed by Macquart *et al.* [7], however, it will be briefly described here for the sake of completeness. ATOM allows for the blade to be split into a number of regions (or components) i.e. spar cap, trailing edge reinforcement, webs etc. The location of these regions is defined by a set of key-points (KPs) that vary along the blade span and are normalised as a percentage of chord length. Each region is comprised of a number of material layers. Some layers may span a number of regions (i.e. glass TRIAX in skin) and some may be present only in a single region (i.e. UD carbon in spar). Each material layer has a set of material properties, in addition to a set of spline surfaces that define the variations in lamination parameters and laminate thickness. Additionally, there is an option to independently specify suction and pressure sides of the aerofoil, or to apply a symmetry constraint.

The 7 MW turbine blade is parameterised according to ATOM's framework. The blade structural description can be categorised into a number of conventional structural regions (see Table 2). There are a small number of laminates that don't fit exactly into a single region, such as web flanges, however, these layers can easily be absorbed into nearby regions, such as the spar cap, without significant deviation from the description. As well as region definitions, Table 2 describes the material layers present in each region and whether or not they are to be considered as DVs for optimisation. The choice for including a layer as a DV is dictated by whether the modelling fidelity allows the optimiser to effectively observe the primary purpose of that layer. For example, the optimiser will observe a relatively strong sensitivity to tip deflection constraint with respect to the thickness of a UD layer in the spar cap, thus a DV is used. However, the BIAX in the spar cap, whilst it has some contribution to global properties, is primarily used to avoid crack propagation when too many plies of the same orientation are stacked together. The structural fidelity of the model is not able to predict such phenomenon, therefore, such layers are not DVs.

After splitting of the structure into regions and material layers, thickness distributions of the layers (spline surfaces) need to be specified. An initial heuristic method was used to ensure that all material thickness distributions were matched perfectly with the minimal number of control points (CPs) along the span. This heuristic method resulted in 24 CPs. As evidence of this close match in thickness distributions, the damped modal frequencies (assumed 0.5% linear damping) of the resulting beam model match to within 2% of experimentally derived modes (see Table 3), other than the torsional mode. The blade modal testing was conducted with a hammer excitation, and strain gauge data was collected and processed using a fast Fourier transform. Therefore, the procedure was not particularly appropriate for accurately measuring a torsional mode, for which LE and TE accelerometers would be required for accurate measurement of torsional dynamics. In addition, the torsional dynamics are relatively coupled to the flapwise and edgewise dynamics of this blade, thus further complicating the successful isolation of modes. Hence, the authors deem it appropriate to discount the experimentally-derived torsional frequency. It is noted that mode shapes or damping ratios were not experimentally determined.

The potential problem with this distribution of 24 CPs is that there would be an unnecessarily large number of DVs if the design were to be used in an optimisation study. Given the 11 regions selected to be DVs, and 24 CPs, there could be up to $(24 \times 11 \times 2 =)$ 528 DVs if top and bottom skins were allowed to vary asymmetrically. As already discussed, it is desirable to reduce the number of DVs so as to reduce the computational effort of gradient evaluations. In this work, the number of DVs are reduced by removing CPs and simplifying the laminate thickness distributions. Therefore, this study aims to determine how many CPs can be removed whilst still providing a good match to the experimentally derived modal frequencies.

In order to provide a number of reduced CP distributions, a cosine spacing is used to

Table 2. Material layers present in each blade region.

Region name	Layer name	Material	DV
All regions	Skin TRIAX	Glass	**Yes**
	Gelcoat/paint	-	No
	Root UD	Glass	**Yes**
	Root BIAX	Carbon	**Yes**
	Root connection	Steel	No
LE reinforcement	LE UD	Glass	**Yes**
	LE TRIAX	Glass	No
	Adhesive	-	No
LE shell	LE core	PVC	**Yes**
Spar cap	Spar cap UD	Glass	**Yes**
	Spar cap UD	Carbon	**Yes**
	Spar cap BIAX	Carbon	No
TE shell	TE core	PVC	**Yes**
TE reinforcement	TE UD	Glass	**Yes**
	TE TRIAX	Glass	No
	Adhesive	-	No
Webs	Web BIAX	Glass	**Yes**
	Web core	PVC	**Yes**

Table 3. Percentage differences in modal frequencies and mass between experimental values and blade model with full laminate detail. Absolute values are omitted due to an NDA. 'F' = Flapwise, 'E' = Edgewise, 'T' = Torsional.

Percentage difference in metric (%)							
Mass	**F1**	**E1**	**F2**	**E2**	**F3**	**E3**	**T1**
0.11	0.53	0.41	1.86	0.33	1.66	0.63	12.39

automatically generate sets of points. The original detailed laminate thickness distributions are then interpolated over these reduced meshes. The cosine spacing results in closer spacing toward the root and tip which, at least for this particular blade, is appropriate for some of the more structurally-dominant laminates (i.e. spar cap) as they exhibit more thickness variations in these regions. It is noted that other spacings (i.e. linear spacing) may be more appropriate for other blades, and each spacing will offer its own bias on the optimised designs. However, the cosine spacing is deemed appropriate for the purpose of this example and similar results could be obtained with other similar spacings.

Results from this study are displayed in Table 4, containing percentage differences between the ATOM blade model and experimentally derived metrics, for differing numbers of CPs. The key result from this is that the number of control points, and thus the match of laminate thickness distributions, can be significantly reduced without compromising the accuracy of global blade dynamics. The mass and modal frequencies only start to deviate significantly from the experimental for 12 and fewer CPs. This result highlights that, for the purpose of wind turbine MDO, the number of DVs can be significantly reduced compared to that required for a full detailed structural representation of a blade. This reduction in information specifying the blade laminates does not diminish the global blade dynamics which are crucial for observing aero-

Table 4. Percentage differences between model and experimental modal and mass metrics.

		No of control points (CPs)										
		6	**8**	**10**	**12**	**14**	**16**	**18**	**20**	**22**	**24**	**24***
Percentage difference in comparison metric (%)	**Mass**	2.3	5.6	8.2	0.3	1.2	1.6	1.2	0.3	1.4	0.4	0.1
	F1	23.9	4.6	3.9	1.9	1.3	0.1	0.4	0.1	0.2	0.3	0.5
	E1	16.2	6.9	11.4	3.2	1.3	1.0	0.6	0.4	1.7	1.6	0.4
	F2	17.7	4.8	1.0	0.2	0.9	0.8	2.1	0.3	0.9	1.5	1.9
	E2	12.1	3.1	2.6	1.0	0.3	1.8	0.3	1.1	0.9	0.2	0.3
	F3	15.5	0.6	0.3	0.5	0.6	0.2	0.7	0.2	0.9	1.6	1.7
	E3	5.0	2.8	2.8	0.8	0.5	1.7	0.4	1.8	0.9	0.3	0.6
	T1	26.3	16.2	26.3	14.1	13.9	14.0	13.3	13.5	14.9	13.8	12.4

elastic phenomenon, blade displacements, and cross-sectional strains.

4. High-order beam modelling

This section presents the benefits of high-order beam elements, available in the ATOM code, for the efficient and accurate structural modelling of modern wind turbine blades. Again, the 7 MW turbine is used as an example to highlight the efficacy of such elements.

High-order beam elements refer to formulations of the conventional FE beam element with three, four or five nodes per element, compared to the conventional two node element [10]. Such elements allow for variations in structural properties along the element length, and thus a converged beam mesh can be achieved with fewer nodes than with conventional elements.

ATOM's beam mesh, and thus cross-section distribution, is independent of the CP mesh. Therefore, varying beam meshes can be chosen by the user and the material properties described in the spline surfaces are interpolated over the beam mesh. Given this functionality, a mesh convergence study is performed to highlight the effects of the higher-order elements. Specifically, the number of beam elements (and thus nodes) is varied, using two- and three-node elements, and certain structural metrics are compared. Two example convergence diagrams of flapwise tip displacement and the 2nd flapwise modal frequency are given in Figure 1. It can be seen that for both metrics, the three-node beam converges with far fewer nodes than the conventional two-node beam.

The comparison metrics in this convergence study are modal characteristics and deflections under a representative flapwise ultimate load case. It is noted that convergence here is not assessed against experimentally derived values, but against the asymptotically convergent values obtained numerically. The modal assurance criterion (MAC) [16] is used for comparing mode shapes, in addition to a comparison of modal frequencies. Deflection comparisons are given for both the linear (fixed linear stiffness matrix) and co-rotational non-linear (changing stiffness with geometry) beam models. The results are condensed into a table format, whereby the number of nodes for convergence is given for each metric. Here, convergence is defined as being within 1% of the final converged value. The results for convergence of modal characteristics are displayed in Table 5, where it is noted that a large number of nodes are required to model this blade due to the complexity of the internal structure. It can be seen that almost all mode shapes and mode frequencies require the same or fewer nodes for convergence when using three-node beam elements. The only exception to this is the second torsional mode shape for which the difference in number of nodes is near-negligible. Another result to note is that the mode shapes, in particular the edgewise and torsional, do not benefit as much from the higher-order elements

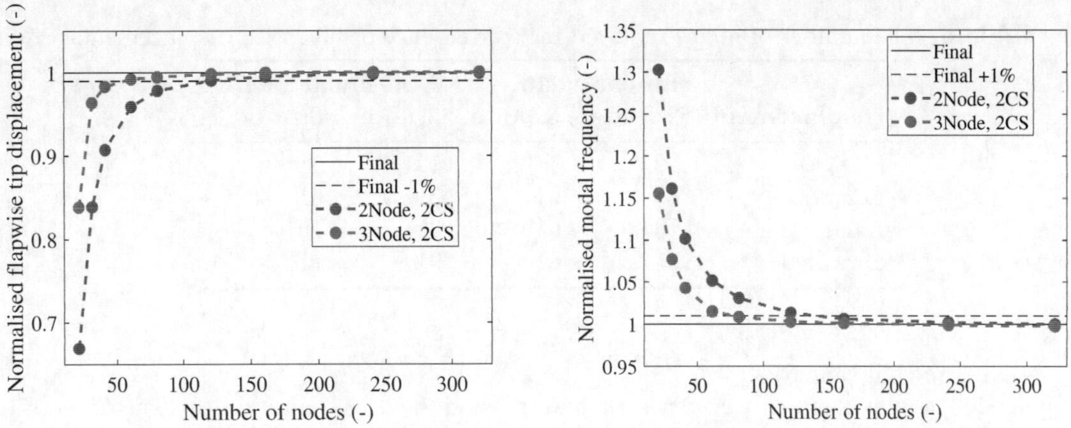

Figure 1. Convergence of blade flapwise tip deflection and 2nd flapwise modal frequency, for two and three-node elements. Values are normalised due to a NDA.

Table 5. The number of nodes required for convergence of mode shapes and modal frequencies.

	No of nodes for convergence	
Mode Type	**2-node**	**3-node**
F1	21	21
E1	21	21
F2	51	41
E2	41	31
F3	81	51
F4	111	61
E3	81	81
T1	31	31
E4	81	81
T2	41	51

Mode shapes (MAC)

	No of nodes for convergence	
Mode Type	**2-node**	**3-node**
F1	81	61
E1	71	61
F2	141	91
E2	121	71
F3	171	111
F4	181	101
E3	141	81
T1	71	61
E4	141	91
T2	221	81

Modal frequencies

as do the modal frequencies. This discrepancy further points toward high-order elements being particularly beneficial for dynamic, as opposed to steady, aero-elastic simulations.

Tip displacement results are displayed in Table 6. It can be seen that three-node elements aid convergence of displacements consistently for the non-linear beam model. However, the torsional predictions from the linear beam model do not benefit from the high-order elements—indicating that the variations in structural properties relevant to overall torsional response are represented well enough by conventional elements for this blade, and at low numbers of nodes may be misrepresented by higher-order elements.

5. Preliminary aero-servo-elastic validation

This forms the first in a series of validation exercises using the ATOM code—highlighting its applicability as a tool for optimisation and design purposes (Section 2.2). This validation exercise focuses on aero-servo-elastic simulation of the 7 MW turbine. Results from ATOM, DNV GL Bladed [12], and recorded field data are compared so as to provide code-to-code validation as

Table 6. The number of nodes required for convergence of blade tip displacements.

	Linear beam		**Non-linear beam**	
Displacement	**2 node**	**3 node**	**2 node**	**3 node**
Flapwise	121	61	121	51
Edgewise	81	61	81	61
Axial	131	61	211	101
Torsional	31	81	91	71

well as experimental validation.

Dynamic simulations are run with turbulent wind fields, generated by TurbSim [17]. Empirical met-mast wind data and temporally-matched SCADA data were recorded from the 7 MW turbine between April and September 2017 and have been made available for this validation study by ORE Catapult. Thousands of data samples were recorded, however, 84 samples have been used for this work as they show minimal yaw misalignment ($< 1°$) and minimal error between wind measurements from the nacelle and met mast ($< 5\%$). The blade model in ATOM has material spline surfaces specified at 14 control points along the span, and the beam model uses 50 3-node elements (101 nodes). Whilst the bladed model uses 80 element as a compromise between accurate beam modelling and computational efficiency, being that high-order beam elements are unavailable. Both modal and non-linear (ATOM: co-rotational framework, Bladed: multi-part beam) structural analysis modules are employed in both codes for comparison. Bladed uses a compiled dynamic link library (dll) controller designed for the turbine, however, there is a discrepancy in that the real turbine exhibits a speed exclusion zone below rated whilst the Bladed controller does not. Being that ATOM is unable to interface with Bladed-style dll controllers, a simplified version of the real controller is implemented whereby the gains are an exact match, however, the switching strategy around rated is a simplification. Both mechanical and electrical loss tables are included in ATOM and Bladed.

Validation results for key aero-servo-elastic parameters are displayed in Figure 2. It can be seen that both codes slightly over-predict power and generator torque in below-rated operation, and in a similar vein, pitch angles are also over-predicted by the codes when above-rated. This points toward the modelled rotors having a higher power coefficient than the real version in both above and below rated regimes. A possible source of this uncertainty could come from the idealised aerodynamic polars used by the BEM module, as opposed to reality where the blades may suffer from geometric defects due to manufacturing or degradation, and the aerodynamic polars may vary significantly due to 3D effects not captured by the low fidelity models. Another source may be uncertainty in the wind inputs, as only limited information was available from the met mast for generating wind field inputs.

One operational feature that is unfortunately omitted in the aero-servo-elastic simulations is a speed exclusion zone below-rated. This can be seen in the flat section of rotor speeds for the SCADA data, where a range of rotor speeds are not operated in, likely for vibrational reasons. This is another possible source of the more aerodynamically efficient operation modelled by the codes in the below-rated region.

As a preliminary validation of load predictions, flapwise and edgewise bending moments are displayed in Figure 3. Due to uncertainties in the load signals coming from the turbine, only code-to-code validation is currently available, however, reasonably good agreement can be seen.

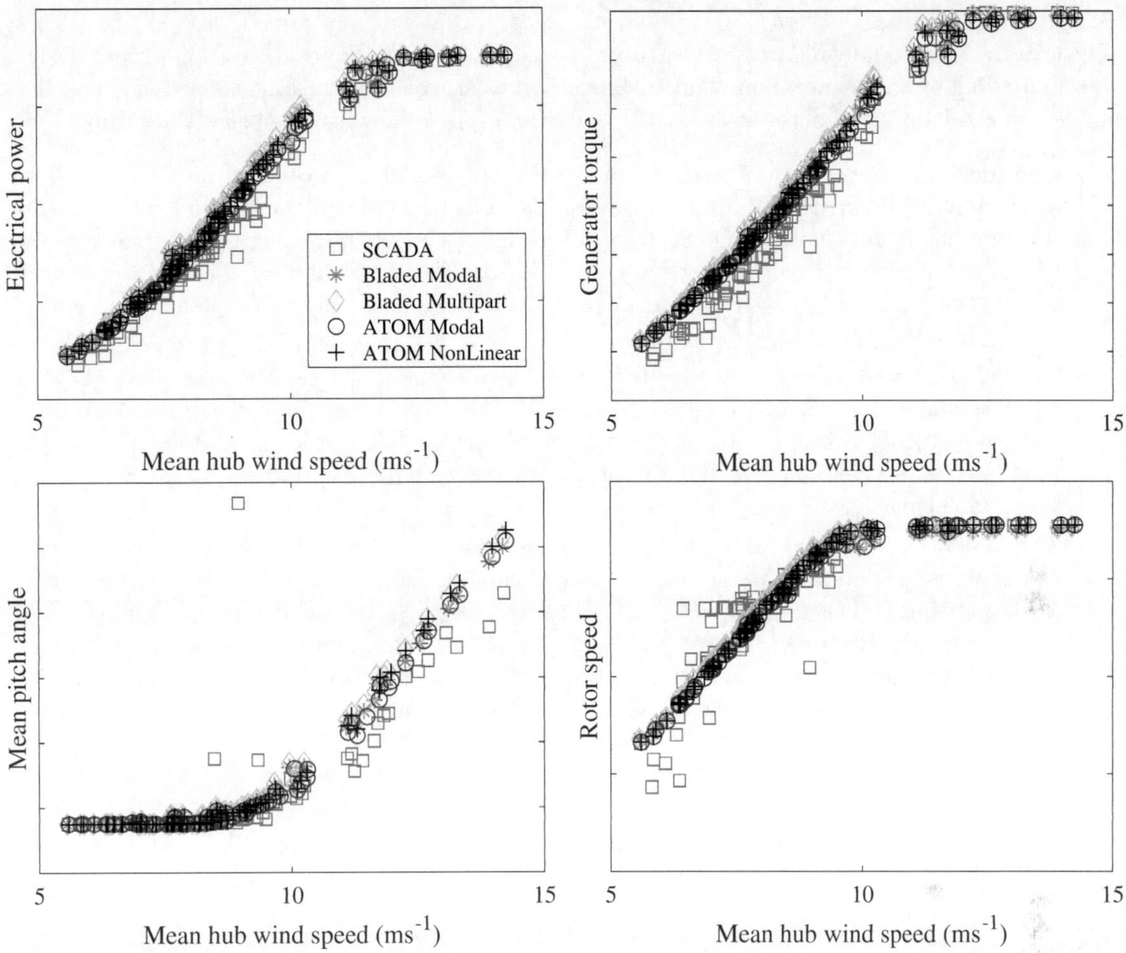

Figure 2. 10-minute averages of electrical power, generator torque, mean pitch angle and rotor speed.

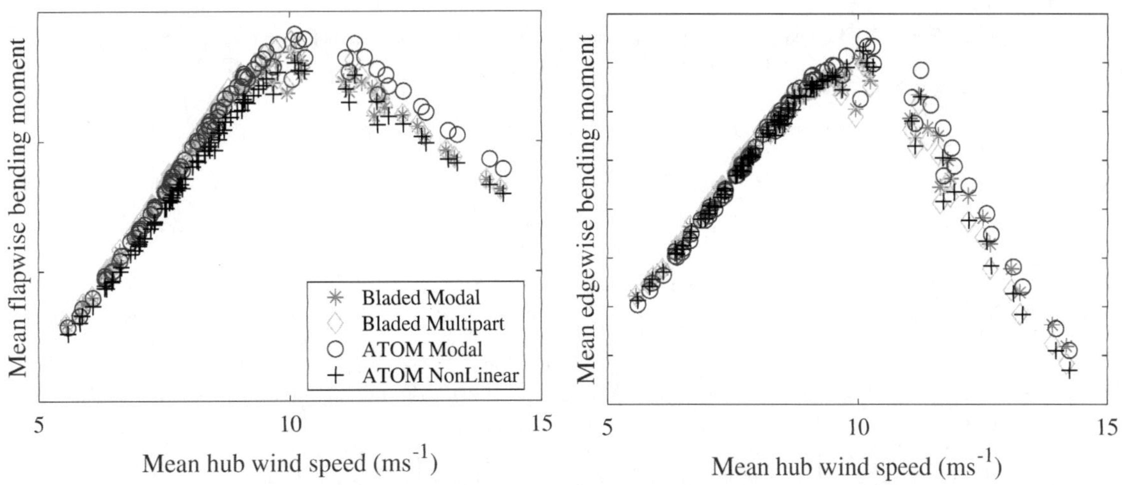

Figure 3. 10-minute averages of flapwise and edgewise blade root bending moments.

IOP Conf. Series: Journal of Physics: Conf. Series **1222** (2019) 012012 doi:10.1088/1742-6596/1222/1/012012

6. Conclusions

This work presents and validates two features of the ATOM code for accurate and efficient design and modelling of next-generation wind turbines. An existing 7 MW turbine is used as a baseline model for modelling and aero-servo-elastic validation. The following conclusions are drawn:

- Considering the vast structural detail in a real turbine blade, it would require many DVs to accurately represent the laminate thickness distributions. Given the desire for minimising the number of DVs for MDO, Section 3 highlights the potential for approximating true thickness distributions with fewer DVs. It is shown that the number of DVs can be reduced without any loss in accurate representation of blade dynamics—modal frequencies are within 2% of those experimentally derived.
- The ATOM code allows the use of high-order beam elements [10], with three-, four- or five-node elements. Section 4 highlights the efficacy of these elements for structural modelling of the 7 MW turbine. A mesh convergence displays that the high-order beam model converges in modal frequencies, shapes, and tip displacements with far fewer nodes than conventional two-node elements.
- The ATOM code, using aforementioned blade model with reduced DVs and high-order elements, is validated for its aero-servo-elastic capabilities. A set of 10-minute simulations are run using ATOM, and DNV GL Bladed, and results are compared against field measurements. Both codes show good agreement in power, generator torque, pitch angle and rotor speed predictions, whilst both slightly over predict power and torque below-rated, and over predict pitch above-rated. Root bending moments from ATOM and Bladed agree well, however, experimental load measurements are unavailable.

Acknowledgments

This project is funded by the Wind Blades Research Hub (WBRH). A joint collaboration between the University of Bristol and ORE Catapult.

References

[1] IRENA 2018 Renewable Power Generation Costs in 2017 Tech. rep. International Renewable Energy Agency Abu Dhabi

[2] Sartori L 2018 *System design of lightweight wind turbine rotors* Ph.D. thesis POLITECNICO MILANO

[3] Pavese C, Tibaldi C, Zahle F and Kim T 2017 *Wind Energy* 1–13 ISSN 10954244

[4] Zahle F, Tibaldi C, Pavese C, McWilliam M K, Blasques J P and Hansen M H 2016 *Journal of Physics: Conference Series* **753** ISSN 17426596

[5] Capuzzi M, Pirrera A and Weaver P M 2014 *Energy* **73** 15–24 ISSN 03605442

[6] Scott S, Capuzzi M, Langston D, Bossanyi E, McCann G, Weaver P M and Pirrera A 2017 *Renewable Energy* **114** 887–903 ISSN 18790682

[7] Macquart T, Maes V, Langston D, Pirrera A and Weaver P M 2017 *12th World Congress of Structural and Multidisciplinary Optimization* Feb (ISSMO) pp 1–12

[8] International Electrotechnical Commission 2005 *IEC 61400-1: Wind turbines - Part 1: Design requirements* 3rd ed (IEC)

[9] Ning S A, Hayman G, Damiani R and Jonkman J 2015 *AIAA Science and Technology Forum*

[10] Macquart T, Pirrera A and Weaver P 2017 *25th AIAA/AHS Adaptive Structures Conference* 1–16

[11] Pavese C, Tibaldi C, Larsen T J, Kim T and Thomsen K 2016 *Journal of Physics: Conference Series* **753** ISSN 17426596

[12] DNV GL 2016 Bladed Theory Manual V4.8

[13] Ponta F L, Otero A D, Lago L I and Rajan A 2016 *Renewable Energy* **92** 157–170 ISSN 18790682

[14] Poon N M K and Martins J R R A 2007 *Structural and Multidisciplinary Optimization* **34** 61–73 ISSN 1615-147X

[15] Levenmouth 7MW demonstration offshore wind turbine - Specification sheet Tech. rep. ORE Catapult

[16] Pastor M, Binda M and Harcarik T 2012 *Procedia Engineering* **48** 543–548

[17] Kelley N D and Jonkman B 2007 1–13

Wind tunnel setup for experimental validation of wind turbine control concepts under tailor-made reproducible wind conditions

V Petrović*, F Berger, L Neuhaus, M Hölling and M Kühn

ForWind – Center for Wind Energy Research, Institute of Physics, University of Oldenburg, Küpkersweg 70, Oldenburg, Germany

* Presenting author

E-mail: `vlaho.petrovic@uol.de`

Abstract. To speed up validation and prototyping of novel control and design concepts in wind energy, wind tunnel experiments should be used in synergy with numerical simulations and field experiments. The paper presents a wind tunnel setup consisting of a scaled wind turbine model MoWiTO and an active grid for generation of tailor-made and reproducible inflow conditions. To demonstrate that the presented experimental setup is suitable for fast validation of advanced control concepts, several control algorithms have been implemented and validated in various inflow conditions, such as turbulent and sheared inflows, and wind gusts. The tested control algorithms include the baseline collective pitch and torque control, individual pitch control for load reduction, and a control algorithms for compensation of static aerodynamic imbalances. All control algorithms resulted in an expected behaviour of the model wind turbine, thus emphasizing that the presented experimental setup can be used for fast and inexpensive testing of control concepts.

1. Introduction

Ever increasing dimensions and rated power of modern wind turbines and wind farms have facilitated a significant reduction in the levelized cost of wind energy, but also lead to an increase of structural loads. This further emphasizes the need for advanced control methods for load reduction. Although the scientific literature reports various control and design concepts improving different aspects of wind turbine operation, their thorough validation in realistic turbulent conditions, as well as an optimization of their parameters, is needed before such controllers can find its industrial application. This is, however, not a straightforward task, especially for field experiments, which represent a costly and time consuming solution with non-negligible chances of hazardous situations. Furthermore, non-reproducible wind conditions in the field make it more difficult to obtain conclusive results. Therefore, all concepts should be well tested before starting validation experiments in the field. To this aim, numerical simulations, such as aeroelastic simulations, are commonly used, but since they always come with certain simplifications, they might not represent the real world sufficiently well. To improve this process, we propose to use wind tunnel experiments with controllable scaled models, which can be employed both for improving numerical models and testing dynamic response of control systems in presence of repeatable and realistic turbulent conditions, thus complementing field experiments and numerical simulations. Exploiting the synergy among different validation methods can result in faster development and prototyping of new concepts and technologies in wind energy. Possible applications include testing of control algorithms for load reduction and power maximization, wake characterisation, improvement of induction dynamics models, etc.

Wind tunnel experiments have already been used for testing different wind turbine and wind farm control concepts, e.g. using a scaled blade segment [1], a scaled wind turbine model [2, 3], or using several scaled turbine models [4, 5]. We particularly emphasize the importance of active grids for generation of tailor-made and reproducible wind conditions, which is essential for a through validation and optimization of control concepts. Examples of such inflow generation can be found in [1], where inflow conditions for a blade segment were generated with the purpose of finding a control law for reducing variations in the lift coefficient, and in [3], where inflow conditions for a two-bladed wind turbine model were generated with the goal of optimizing and validating the control concept for reduction of periodic wind turbine loads. In this paper, we present a comprehensive wind tunnel testing platform for fast experimental validation of advanced control concepts and their interaction with aerodynamic effects under realistic, tailor-made and reproducible turbulent conditions. The setup is used to validate the baseline and individual pitch control algorithms in turbulent and sheared wind conditions.

The paper is organised as follows. The experimental setup is described in Section 2, and the implemented control algorithms are explained in Section 3. The obtained experimental results are summarized in Section 4, and the conclusions are given in Section 5.

2. Experimental setup

The threefold experimental setup consists of a Göttingen type wind tunnel at the University of Oldenburg, an active grid for generating the desired turbulent inflow conditions, and the aerodynamically scaled NREL 5 MW wind turbine MoWiTO (Model Wind Turbine Oldenburg).

2.1. Wind tunnel and the active grid

The dimensions of the wind tunnel are 3 x 3 x 30 m^3, with maximum wind speeds in the open test section configuration of 32 m/s. In Fig. 1, the typical turbine setup in the tunnel is sketched. The turbine is mounted on a height adjustable support to bring the turbine hub to the centre of the wind tunnel. The distance between the nozzle and the rotor plane is 4.8 m, corresponding to 2.7 D. The inflow wind velocity is measured by a hot wire positioned 2.7 m (1.5 D) in front of the rotor centre. The tunnel is operated in an open test section configuration, so that no blockage effects are to be expected [6].

To the nozzle, an active grid can be attached (see Fig. 1(b)), with which specific turbulent and transient flow patterns can be repeatedly impressed on the flow [7]. These patterns, for example, can be wind characteristics as measured in the free field by lidar systems, scaled in size and speeded up in time to match the scaling of the model turbine. Additionally, it is possible to generate a sheared flow and wind gusts, thus enabling testing of the control concept in various conditions.

2.2. Model wind turbine MoWiTO

The model wind turbine, which can be seen in Fig. 2, is a fully controllable scaled NREL 5 MW wind turbine [8], with the rotor diameter of 1.8 m [9]. The scaling is performed so that the matching of the load distribution and angle of attack variation along the blades during dynamic changes of the inflow is achieved. When preparing the inflow and control algorithms, it is important to consider the scaling factors of MoWiTO – length is reduced 70 times, and time is speeded up 46 times.

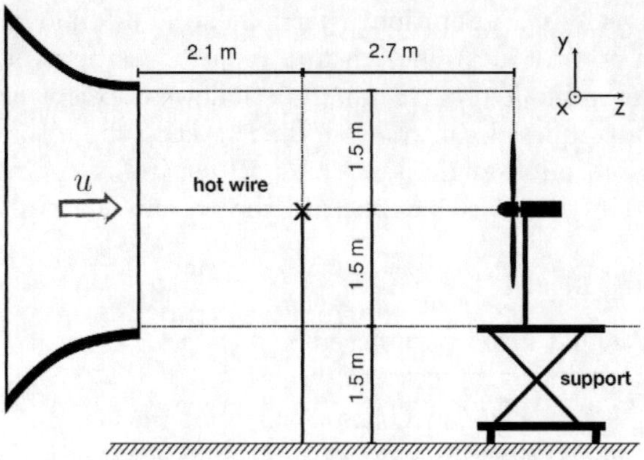

(a) Scaled turbine setup without the active grid.

(b) Turbine and active grid.

Figure 1. Experimental wind tunnel setup.

The turbine is highly sensorised, allowing for measurements of blade, shaft and tower loads, as well as pitch angles, electrical torque and rotor speed. Besides the regular blades, a set of blades equipped with rigid leading edge slats is also available [10].

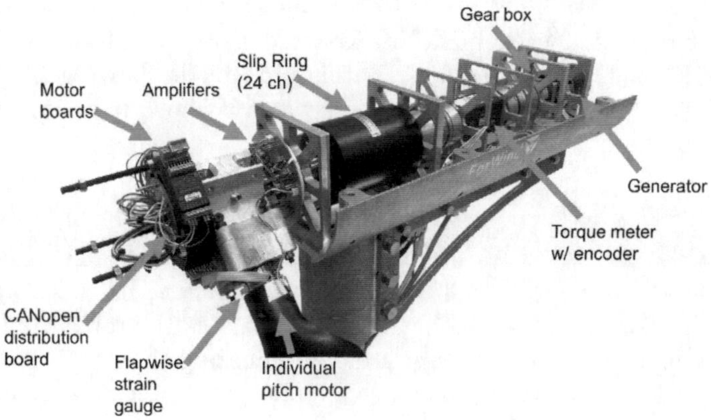

(a) Nacelle of the model turbine.

(b) Entire model.

Figure 2. Model turbine MoWiTO.

3. MoWiTO control algorithms

MoWiTO's actuators enable torque and individual pitch control, comparable to modern multi-megawatt wind turbines. Three pitch motors are controlled using their control boards, to which the reference position is sent via CANopen protocol. Control of the electrical torque, along with the rest of the control algorithms and data acquisition, is implemented on a real time system cRIO from Natiaional Instruments. The measurements are taken with the sampling time of 0.2 ms, whereas the control algorithms run at 10 ms. Note that the controller sampling time corresponds to 460 ms at the full-scale machine, which is higher than the typically used values. This value is, however, suitable considering the wind turbine aerodynamic and structural time constants, especially since there are lower demands on the protection system compared to the full-scale machines and there is no need for fast control of frequency converters.

The baseline controller has two distinct regions of operation – partial load and full load – and constraints imposed on the collective pitch and torque values ensure timely transitions between the regions. In the partial load region, the optimal torque strategy is followed, except close to the cut-in or the rated wind speed, when a torque PI controller is used to keep the rotor speed in the allowed region. In the full load region, pitch angle is set by a PI controller to keep the rotor speed at its rated value [11]. The rated parameters defined in the baseline controller are shown in table 1.

Table 1. MoWiTO's rated parameters.

Parameter	Rated value
Power	252 W
Rotor speed	560 rpm
Rotor torque	4.3 Nm

Besides the baseline controller, two additional control algorithms are implemented. Both exploit MoWiTO's capability to individually pitch each blade. Reduction of the once per revolution blade loads is described in Subsection 3.1, and the compensation of static rotor imbalances is explained in Subsection 3.2.

3.1. Individual pitch control

MoWiTO's capability to individually pitch each blade is exploited here to reduce the first harmonic (once per revolution, 1P) of the blade loads. To this aim, blade flapwise bending moments M_i are transformed with the d-q transformation, also known as Coleman transformation or multi-blade coordinate transformation:

$$\begin{bmatrix} M_\mathrm{d} \\ M_\mathrm{q} \end{bmatrix} = \frac{2}{3} \begin{bmatrix} \cos\vartheta & \cos\left(\vartheta + \frac{2\pi}{3}\right) & \cos\left(\vartheta + \frac{4\pi}{3}\right) \\ \sin\vartheta & \sin\left(\vartheta + \frac{2\pi}{3}\right) & \sin\left(\vartheta + \frac{4\pi}{3}\right) \end{bmatrix} \begin{bmatrix} M_1 \\ M_2 \\ M_3 \end{bmatrix}, \tag{1}$$

where M_d and M_q are the d-q transformations of the blade loads, and ϑ is the rotor azimuth, with $\vartheta = 0°$ corresponding to the first blade pointing upwards. Eliminating the first harmonic of the blade flapwise loads is equivalent to ensuring that mean values of M_d and M_q are equal to zero, which can be achieved with a pair of PI controllers in the d-q coordinate system [12]. The control actions determined by the two PI controllers, β_d and β_q, are then transformed into blade coordinate system:

$$\begin{bmatrix} \beta_1 \\ \beta_2 \\ \beta_3 \end{bmatrix} = \beta_\mathrm{c} + \begin{bmatrix} \cos\left(\vartheta + \varphi\right) & \sin\left(\vartheta + \varphi\right) \\ \cos\left(\vartheta + \frac{2\pi}{3} + \varphi\right) & \sin\left(\vartheta + \frac{2\pi}{3} + \varphi\right) \\ \cos\left(\vartheta + \frac{4\pi}{3} + \varphi\right) & \sin\left(\vartheta + \frac{4\pi}{3} + \varphi\right) \end{bmatrix} \begin{bmatrix} \beta_\mathrm{d} \\ \beta_\mathrm{q} \end{bmatrix}, \tag{2}$$

where β_c is the collective blade pitch angle, defined by the baseline controller, and β_i is the pitch angle of the ith blade. The parameter φ is introduced to compensate the pitch actuator dynamics, i.e. time delays caused by the blade actuation, and thus to achieve static decoupling in the d-q coordinate system [13].

3.2. Compensation of pitch offsets

A pitch offset can be caused by a faulty actuation or a faulty measurement of the blade pitch angle. Regardless of the reason, a pitch offset leads to an unbalanced rotor, and thus to increased loads and fatigue damage. Such increased loads cannot be compensated by a standard individual pitch control. Here, we used a method proposed in [13] for compensation of aerodynamic rotor imbalances, simplified to work only for static (i.e. low frequency) imbalances, such as pitch offsets. To this aim, the following load transformation is applied:

$$\begin{bmatrix} M_{\mathrm{d}0} \\ M_{\mathrm{q}0} \end{bmatrix} = \frac{2}{3} \begin{bmatrix} 1 & \cos\frac{2\pi}{3} & \cos\frac{4\pi}{3} \\ 0 & \sin\frac{2\pi}{3} & \sin\frac{4\pi}{3} \end{bmatrix} \begin{bmatrix} M_1 \\ M_2 \\ M_3 \end{bmatrix}. \tag{3}$$

The mean values of the transformed loads $M_{\mathrm{d}0}$ and $M_{\mathrm{q}0}$ correspond to the amplitude of the static rotor imbalance (c.f. [13] for more details), and it can, therefore, be compensated with two PI controllers reducing the transformed loads to zero, similar as with individual pitch control. The so determined control actions, $\beta_{\mathrm{d}0}$ and $\beta_{\mathrm{q}0}$, are transformed back to the blade coordinate system:

$$\begin{bmatrix} \beta_1 \\ \beta_2 \\ \beta_3 \end{bmatrix} = \beta_{\mathrm{c}} + \begin{bmatrix} 1 & 0 \\ \cos\frac{2\pi}{3} & \sin\frac{2\pi}{3} \\ \cos\frac{4\pi}{3} & \sin\frac{4\pi}{3} \end{bmatrix} \begin{bmatrix} \beta_{\mathrm{d}0} \\ \beta_{\mathrm{q}0} \end{bmatrix}, \tag{4}$$

Since the compensation of the static rotor imbalances acts mainly around low frequencies (i.e. close to 0 Hz), there is no coupling between the two axes in the transformed coordinate system.

Note that the inverse transformation for static rotor imbalances (4) does not change the collective value of blade pitch angles, similar as the standard d-q transformation, i.e. $\frac{1}{3}\sum_i \beta_i = \beta_{\mathrm{c}}$ for any $\beta_{\mathrm{d}0}$ and $\beta_{\mathrm{q}0}$.

4. Results

The presented experimental setup is used for thorough validation of control algorithms in various inflow conditions, including Mexican hat profiles, sudden wind gusts, and turbulent inflows with different levels of turbulence intensity and wind shear. The main focus is put on the compensation of oscillatory blade loads and pitch offsets, reaction to wind gusts, and on the operation in turbulent inflow conditions. The obtained results are summarized in the remainder of the section.

4.1. Baseline controller

The baseline controller has been tested in various inflow conditions, including different turbulent inflow and wind gusts. Figure 3 shows the probability distribution of the applied turbulent inflow velocities, and the obtained measurements of the rotor speed, torque and collective blade pitch angle. Blue dots represent individual measurements as a function of the measured wind velocity, while the red lines represent averaged static characteristics. It can be observed that the static characteristics match with the parameters from table 1, indicating that the baseline controller is working as expected. One can also observe a high variation of the rotor speed in the partial load region, where rotor speed is supposed to be adjusting to the wind conditions in order to track the maximal power production point. In the full load region, there are considerably lower variations in the rotor speed and torque, since the controller is adjusting pitch angles in order to keep the rotor speed and the power production at the rated values, which is also an expected wind turbine behaviour.

Besides in turbulent inflow, MoWiTO was also tested in the presence of wind gusts. Figure 4 shows the turbine response to two consecutive Mexican hat-like gusts. The baseline controller uses only rotor speed measurements to determine needed control actions, and it was not tuned specifically for wind gusts. This leads to high variations in the collective pitch angle and the torque, as well as in the rotor speed. Analysing a series of 10 wind gusts, pitch excursions of 11° peak to peak are observed, with the higher peak reaching on average 8.2° above and the lower

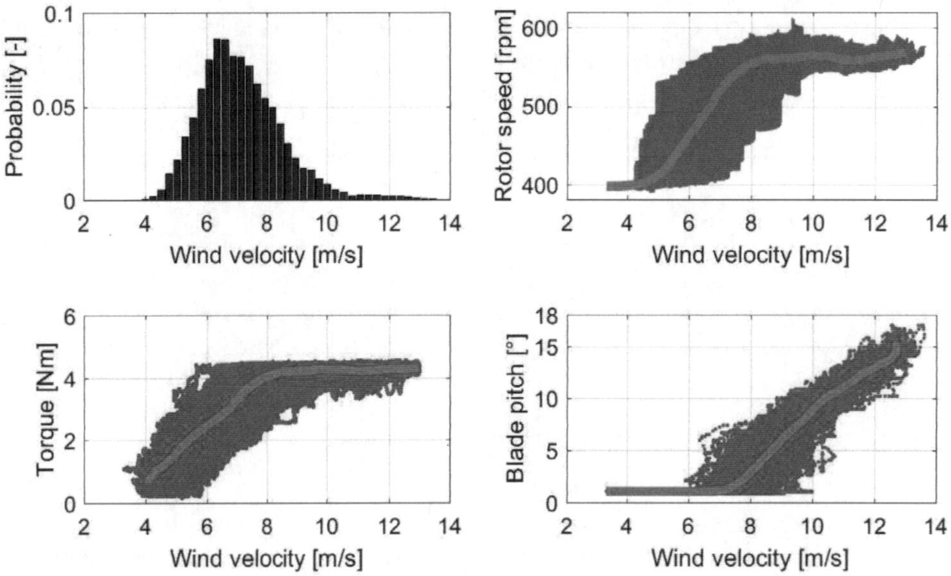

Figure 3. Probability of the different wind velocities, and the obtained wind turbine response. Blue dots represent individual measurements, while the red curves represent the averaged static characteristics of the rotor speed, torque, and blade pitch angle.

peak 2.8° below the steady state value. The overspeed was on average reaching 35 rpm, which is considered acceptable.

Testing wind turbine response to wind gusts, as shown in Fig. 4, would be extremely difficult in field experiments, which additionally emphasizes the usefulness of the presented experimental setup. However, it should be noted that, considering the time scaling of MoWiTO, the duration of the applied wind gusts is longer than the duration of the extreme operating gust defined in the IEC norm [14], which is typically used in numerical simulations.

4.2. Individual pitch control

To validate individual pitch control (IPC), the experiments were performed with MoWiTO derated to the rotor speed of 480 rpm, and power of 135 W. With this rotor speed, once-per-revolution loads are occurring at the frequency of 8 Hz, and pitch actuation at exactly this frequency is needed to reduce them. To excite the once-per-revolution blade loads, a sheared inflow was generated with the active grid.

First, a series of open loop experiments was performed, in which β_d and β_q were set manually, and the load response was measured to determine the phase offset between the blade pitch and blade bending moment signals. At the frequency of the IPC actuation, this offset was identified as $\varphi = 110°$. Using this value in the transformation (2) compensates static coupling between the axes in the d-q coordinate system [13].

Figure 5(a) shows time response of the (collective) pitch angle without IPC and the the pitch angle of the first blade with IPC. As expected, the blade is actively pitching at frequency 8 Hz (corresponding to the once-per-revolution loads) when IPC is activated, leading to higher pitch actuation. However, such pitch actuation also leads to the significant load reduction, which can be seen in Fig. 5(b), showing the spectra of out-of-plane blade bending moments with and without individual pitch control. The peak at frequency of 8 Hz is clearly reduced, leading to the 15 % reduction of the blade loads standard deviation. The results indicate that the individual

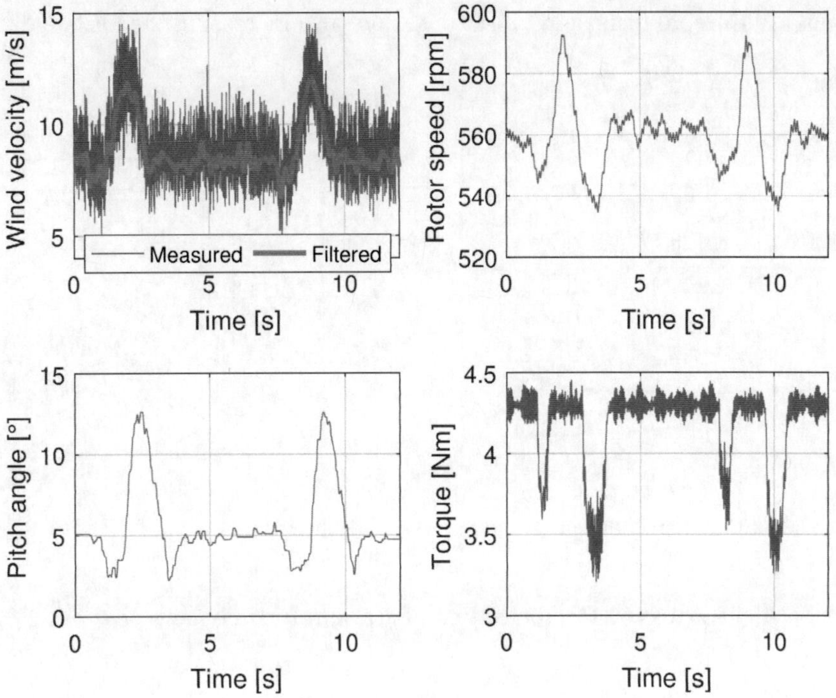

Figure 4. Wind velocity of two consecutive wind gusts, and the response of MoWiTO's rotor speed, collective pitch angle, and torque.

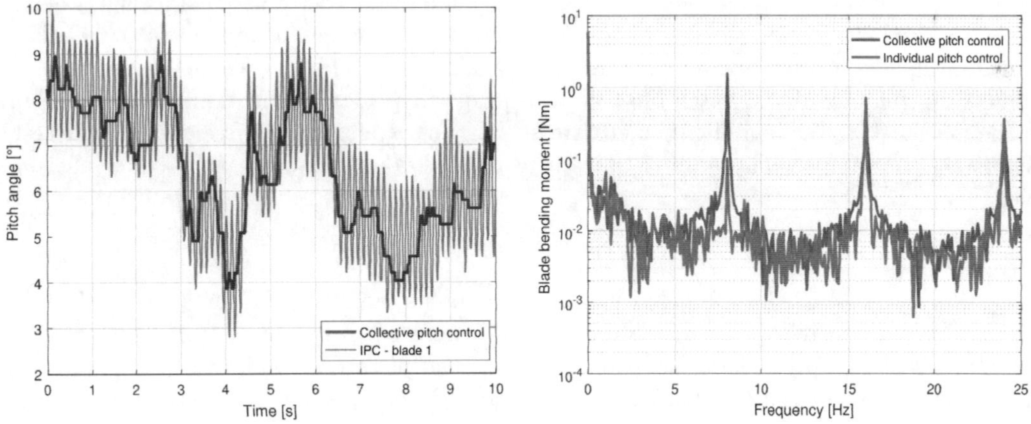

(a) Blade pitch time response.

(b) Blade bending moment spectra.

Figure 5. Time response of the blade pitch and the spectra of the blade out-of-plane bending moment with and without individual pitch control.

pitch control has been successfully implemented and tested on MoWiTO.

4.3. Compensation of static rotor imbalances

To validate the control algorithm for compensation of static rotor imbalances, pitch offsets were introduced. When the same pitch angles are measured for each blade, in fact, the second blade has a higher, and the third blade has a lower pitch angle compared to the first blade. Such an imbalance causes additional loads and vibrations, thus also increasing the fatigue damage of the turbine.

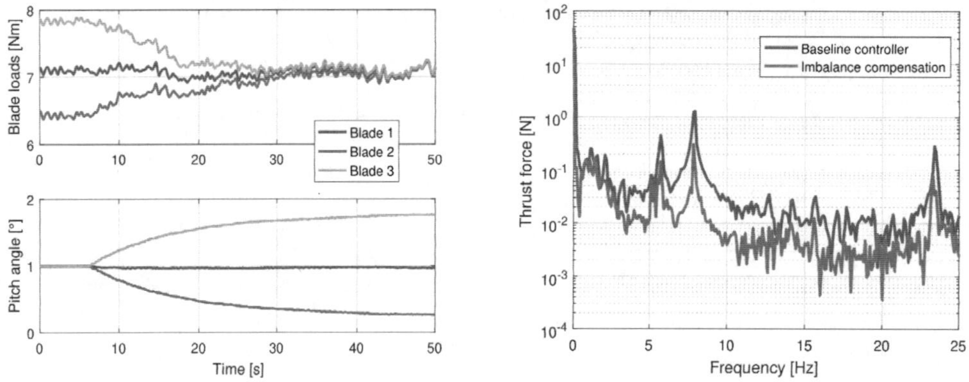

(a) Time series of the filtered blade loads and of the pitch angles.

(b) Spectrum of thrust force with and without imbalance compensation.

Figure 6. Compensation of static rotor imbalances.

Figure 6(a) shows a time response of the blade pitch angles and filtered blade out-of-plane bending moments when the control algorithm for compensation of rotor imbalances is turned on. The blade bending moments are low pass filtered to eliminate harmonic components, thus making it easier to compare the mean values of loads. At first, a significant discrepancy among blade loads can be observed, although all pitch angles are measured to be equal. This is because of the introduced pitch offsets. After the control algorithm for compensation of rotor imbalances is turned on, blades are pitched in a way to eliminate discrepancies among the blade loads, thus compensating the introduced pitch offsets. In Fig. 6(b), spectra of the thurst force computed from the measured tower bending moment are shown. As expected, an imbalanced rotor results in significant vibrations, which are significantly reduced when the imbalance compensation is applied.

5. Conclusions

The paper presented our wind tunnel setup consisting of a scaled wind turbine model (MoWiTO) and an active grid for generation of reproducible and tailored turbulent or transient inflows. As such, the presented experimental setup provides a powerful tool for improvement of numerical models, optimization of control algorithms, validation of new concepts, etc., which could lead to faster development and prototyping of new concepts and technologies in wind energy when used in synergy with numerical simulations and field experiments.

Using the presented experimental setup, several control algorithms have been tuned and validated in different inflow conditions. This includes the baseline controller tested both in turbulent conditions and in presence of wind gusts, the individual pitch controller tested in turbulent and sheared inflow, and the control algorithm for compensation of static aerodynamic imbalances. The ability to generate repeatable and tailor-made inflow conditions made comparison of different control algorithms possible and significantly less time consuming compared to field experiments. Each testing scenario showed an expected behaviour compared to the results from numerical simulations available in the literature, further emphasizing that the experimental setup with MoWiTO and the active grid is a suitable tool for validation of new control concepts.

Our future work will include control algorithms utilizing wind preview measurements and optimal control theory, such as model predictive control, as well as smart blades. The analysis of the interaction between control algorithms and blade deformations will be given a special emphasis.

Acknowledgments
This work is partly funded by the Federal Ministry for Economic Affairs and Energy according to a resolution by the German Federal Parliament (project SmartBlades 2) and by the Ministry for Science and Culture of Lower Saxony through the funding initiative »Niedersächsisches Vorab« (project »ventus efficiens«).

References

[1] Neuhaus L, Singh P, Homeyer T, Huxdorf O, Riemenschneider J, Wild J, Peinke J and Hölling M 2018 *Journal of Physics: Conference Series* **1037** 022032
[2] Bottasso C L, Campagnolo F and Petrović V 2014 *Journal of Wind Engineering and Industrial Aerodynamics* **127** 11–28
[3] Frederik J, Kröger L, Gülker G and van Wingerden J W 2018 *Control Engineering Practice* **80** 105 – 115
[4] Campagnolo F, Petrović V, Schreiber J, Nanos E M, Croce A and Bottasso C L 2016 *Journal of Physics: Conference Series* **753** 7
[5] Petrović V, Schottler J, Neunaber I, Hölling M and Kühn M 2018 *Journal of Physics: Conference Series* **1037** 032020
[6] Ryi J, Rhee W, Hwang U C and Choi J S 2015 *Renewable Energy* **79** 227 – 235 ISSN 0960-1481 selected Papers on Renewable Energy: AFORE 2013
[7] Kröger L, Frederik J, van Wingerden J W, Peinke J and Hölling M 2018 *Journal of Physics: Conference Series* **1037**
[8] Jonkman J, Butterfield S, Musial W and Scott G 2009 Definition of a 5-MW reference wind turbine for offshore system development Tech. Rep. NREL/TP-500-38060 NREL, National Renewable Energy Laboratory
[9] Berger F, Kröger L, Onnen D, Petrović V and Kühn M 2018 *Journal of Physics: Conference Series* **1104** 012026
[10] Manso Jaume A, Wild J, Hölling M and Peinke J 2015 Design and wind tunnel testing of a leading edge slat for a wind turbine airfoil *Deutsche Windenergie-Konferenz (DEWEK) 2015*
[11] Burton T, Sharpe D, Jenkins N and Bossanyi E 2001 *Wind Energy Handbook* (Wiley) ISBN 9780471489979
[12] Bossanyi E A 2003 *Wind Energy* **6** 119–128
[13] Petrović V, Jelavić M and Baotić M 2015 *Renewable Energy* **76** 418–431
[14] International Electrotecnical Commission 2005 *Wind Turbines – Part 1: Desing Requirements, Ed. 3.0* International Standard IEC 61400-1

IOP Conf. Series: Journal of Physics: Conf. Series **1222** (2019) 012014 doi:10.1088/1742-6596/1222/1/012014

A Blade Load Feedback Control For Floating Offshore Wind Turbines

Joannes Olondriz[1], Josu Jugo[2], Iker Elorza[1], Santiago Alonso-Quesada[2], and Aron Pujana-Arrese[1]

[1] IK4-IKERLAN, Paseo J.Mª Arizmendiarrieta 2, 20500 Arrasate-Mondragón (Gipuzkoa), Spain

[2] University of the Basque Country UPV/EHU, Bo. Sarriena s/n, 48940 Leioa (Bizkaia), Spain

E-mail: jolondriz@ikerlan.es

Abstract. The platform-pitch motions of Floating Offshore Wind Turbines are highly influenced by the blade-pitch controller. However, in rough sea state conditions, the platform-pitch motions are dominated by the incident wave dynamics. Then, the properties of the designed controller to reduce the platform-pitch motions can be changed due to the sea state conditions. In this study, the drawbacks of a previously designed Aerodynamic Platform Stabiliser controller performance in rough sea state conditions are analysed. Furthermore, an additional blade load feedback control loop has been designed to improve the performance of the controller in rough sea state conditions. The preliminary results presented here show the potential of the blade load feedback control loop effectiveness to improve the Aerodynamic Platform Stabiliser control loop performance in rough sea state conditions. The time-domain simulations were carried out with the NREL 5-MW wind turbine mounted on the ITI Energy's barge in the fully coupled non-linear aero-hydro-elastic simulation tool FAST.

1. Introduction

Wind energy is becoming the real green energy alternative to the conventional fossil fuel sources, as the increment of the new wind farm projects worldwide reveals. The offshore wind presents many advantages compared to its onshore counterpart [1]. The current offshore wind farms are based in the bottom-fixed wind turbine technology; however, this technology is limited by the water depth. The bottom-fixed wind turbines can only be installed in water depths less than 60 m, so can they be only installed in shallow water coastal areas. Therefore, a solution to overcome this limitation is being developing in the last decades, i.e. the Floating Offshore Wind Turbines (FOWT). Several academic research were published and many FOWT prototypes were designed and tested in the last years, e.g. BlueH, Hywind, WindFloat, SWAY, VolturnUS, Kabashima, Mitsui, Shinpuu, Hamakaze and Floatgen [2]. After the worldwide first floating offshore pilot park operating since 2017, Hywind Scotland [3], other floating farms are currently being built, such as Windfloat Atlantic [4] or the second phase of the Kincardine project [5]. The current interest in this kind of energy production confirms the tendency of increasing FOWT projects and the consolidation of the wind energy as the main green energy source of the near future.

The FOWT technology is actually in development stage before it gets mature, so that there are many different platform models designed in recent years and different platform floater

IOP Conf. Series: Journal of Physics: Conf. Series **1222** (2019) 012014 doi:10.1088/1742-6596/1222/1/012014

dispositions [6], each one holding both pros and cons. The different floating technologies considerably differ in terms of building processes as well as in performance characteristics. In general, platforms with larger hydrostatic restoring stiffness and hydrodynamic damping cost more to build, mount and deploy, while less stable platforms can affect the FOWT's performance [7]. The semi-submersible platforms provide several advantages for the platform construction, wind turbine installation and towing process. However, the coupling between the blade-pitch regulation and the platform-pitch motion, known as the platform negative damping [8][9], highly affects to the FOWT performance. Therefore, a number of different control solutions have been designed to improve the FOWT performance, such as the Aerodynamic Platform Stabliser (APS) control loop [10] designed to complement the detuned PI control loop [8] for the NREL 5-MW wind turbine mounted on the ITI Energy's barge, among others.

The detuned PI control loop designed in [8] for FOWT systems is based in the conventional onshore wind turbine generator speed control via the blade-pitch regulation. The conventional onshore generator speed control loop PI gains can be obtained by analysing the wind turbine drivetrain equation of motion, as show in [8]. However, this equation does not contemplate the platform-pitch dynamics of the FOWT system. Thus, an additional modification is required to avoid the negative platform damping effect. This modification relies on detuning the PI gains to reduce the controller-response natural frequency below the platform-pitch natural frequency, as mentioned in [11]. Although the negative effect is avoided, and a considerable generator speed regulation improvement and platform-pitch motion reduction are achieved, further performance improvements are required to reduce the tower and blade loads due to the platform motions.

Therefore, the APS control loop was designed in [10] to reduce the FOWT tower loads by reducing the platform-pitch motions while not increasing the blade loads and improving the generator speed regulation at the same time. This control loop is based on the regulation of the blade-pitch angles via the measurement of the nacelle-pitch velocity. Remarkable improvements are achieved with the NREL 5-MW wind turbine model mounted on the ITI Energy's barge, as well as with the DTU 10-MW wind turbine model mounted on the TripleSpar platform, presented in [12] and [15], respectively. The platform-pitch motions and the tower-base bending moments are reduced, not increasing the blade loads. The rotor speed regulation and, hence, the generated electric power are improved as well. The improvements, however, are registered under specific environmental conditions, i.e. in still or smooth sea state conditions. When the sea state turns rough, the APS control loop reduces the platform-pitch motions more than the natural motion induced by the waves, increasing the loads in blades. This is due to the APS control loop is not able to distinguish the platform-pitch motions produced by the wind from those induced by the waves. Thus, an additional control loop is required to compensate the negative effect of the APS control loop in rough sea state conditions and reduce the wind turbine blade loads.

2. Objectives

An additional control loop is required to counteract the negative effect of a previously designed APS control loop in rough sea state conditions. This additional control loop must alleviate the excessive APS blade-pitch control allowing the free platform-pitch motion induced by the waves. Since the FOWT tower is designed to mechanically resist the extreme wave heights of 50-years recurrence [16], the tower loads can be slightly increased during the wind turbine normal performance. This allows a blade load reduction at the cost of increasing the platform-pitch motions via the reduction of the rotor thrust in rough sea states. Furthermore, the new control loop must not be detrimental for the APS control loop performance in still and smooth sea conditions and, it must be designed to work in the above rated wind speed region.

3. Control design

The blade load feedback control loop presented in this article, named Wave Rejection (WR), contributes to the baseline Detuned PI and APS control loop blade-pitch regulation via the measurement of the blade-root-flapwise bending moments, as shown in Figure 1. Each control loop has been independently tuned: first, the APS control loop is tuned to improve the performance of the baseline Detuned PI control loop [12]; and second, the WR control loop is designed to improve the other two control loops performance in rough sea state conditions. The blade-root-flapwise bending moments of the blades are measured via an internal measurement unit, e.g. strain gauges, located in each wind turbine blade. The blade-root-flapwise input signal for the WR controller is obtained by averaging the measurement of the three blade-root-flapwise bending moments. In this way, the effect of the tower shadow in the input signal of the WR controller is minimised, improving the control performance results. This measurement provides an early detection of the thrust in the wind turbine rotor enabling a vigorous blade-pitch regulation to reduce the large blade-root-flapwise bending moments.

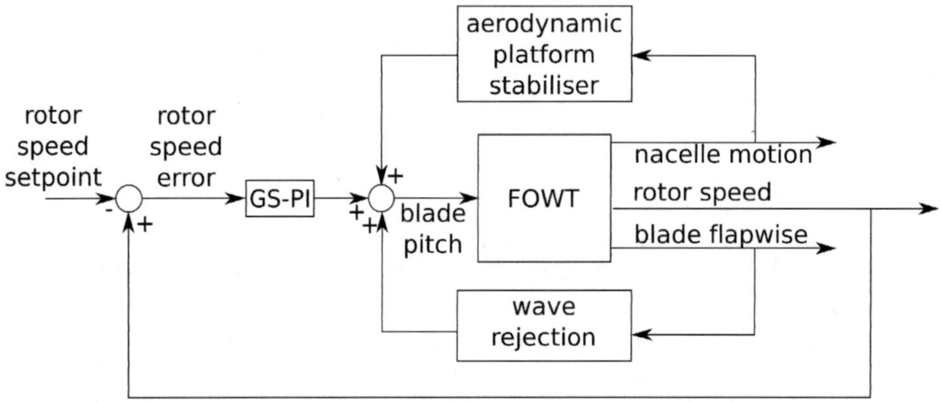

Figure 1. Control scheme block diagram.

The WR control loop has been tuned shaping the FOWT transfer function from the blade-pitch angle to the blade-root-flapwise bending moment. This controller has been designed to capture the wave dynamics through the blade-root-flapwise bending moment measurement. The DNV-GL offshore standard [16] prescribes the frequency band applicable to floating offshore structures located in the wave active zone, corresponding to the frequency range of the incoming waves, typically with periods between 4 and 25 s (0.04 - 0.25 Hz). Thus, the transfer function has been shaped in that frequency range to capture the wave dynamics. The controller is based on a band-pass filter tuned to increase the blade-pitch angle proportionally to the blade-root-flapwise bending moment throughout the mentioned frequency range, as shown in the Bode diagram of Figure 2. Then, the blade-pitch action is considerably increased in the range where the blade-root-flapwise bending moments are affected by the wave dynamics, reducing the thrust in the rotor of the wind turbine and, hence, increasing the wind turbine pitch motion.

To check the FOWT frequency-domain performance, in relation to the wind, Figure 3 shows the Bode diagrams of the closed-loop transfer functions from wind to the platform-pitch angle (top) and to the generator-speed (bottom). One can see how the platform-pitch regulation is improved with the implementation of the previously designed APS control loop, as explained in [12]. Furthermore, the WR control loop does not negatively affect the APS control improvement, while it further reduces the magnitude at the platform-pitch resonance frequency around 0.084 Hz and also the magnitude between the frequencies 0.1 to 0.4 Hz. The same improvement can

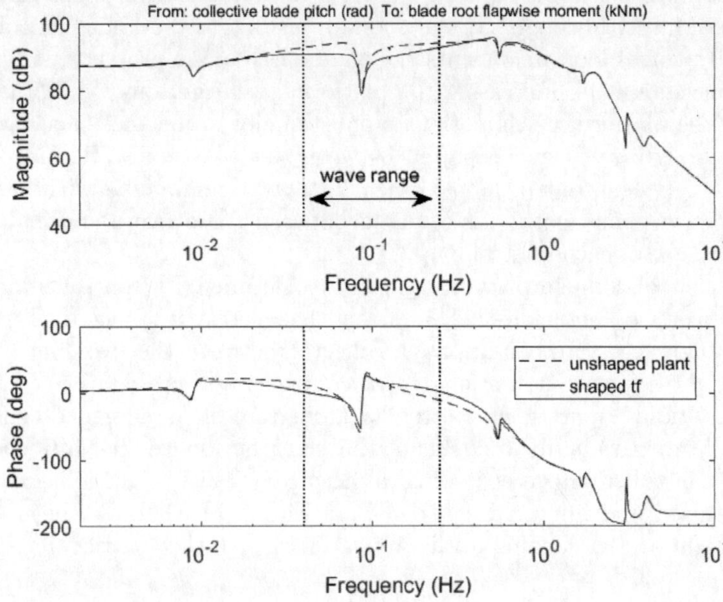

Figure 2. Wave Rejection control loop Bode diagrams.

be also seen in the wind to generator speed Bode diagram. Note that these Bode diagrams are obtained with the linearised transfer functions with a constant wind speed and still water conditions as explained in [13] and [14]. Therefore, time-domain simulations with stochastic wind speeds and irregular wave heights are run to fully check the designed controllers performance.

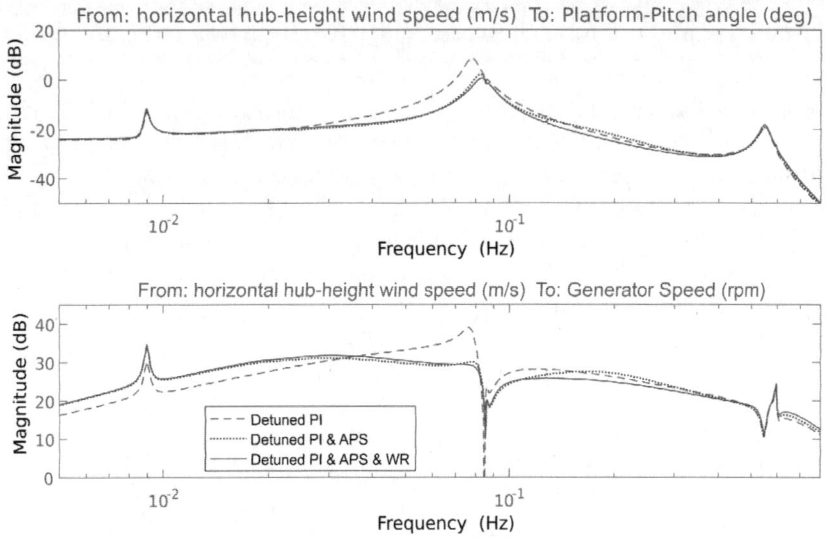

Figure 3. Bode diagrams of the closed-loop transfer functions from wind to platform-pitch angle (top) and to generator speed (bottom).

IOP Conf. Series: Journal of Physics: Conf. Series **1222** (2019) 012014 doi:10.1088/1742-6596/1222/1/012014

Some interferences are observed between the APS and WR control loops during the time-domain simulations. Mainly in rough sea state conditions due to the large control action of the APS loop. The WR control loop increments the blade-pitch angle proportionally to the blade-root-flapwise bending moments, increasing the platform-pitch motions, while the APS control loop tries to reduce the platform-pitch motions proportionally to the nacelle nod velocity. Then, opposing control actions have been registered between the APS and WR control loops. This commonly happens between control loops which do not communicate with each other [17]. Therefore, a control switching method is required between these two additional controllers to regulate the contribution of each control loop.

Since the APS control loop improves the baseline Detuned PI controller performance in still and smooth sea state conditions [12] and, on the contrary, it negatively affects to the WR controller in rough sea state conditions, a collector between the two controllers has been implemented for weighting the contribution of each controller and preserve the controller quality in any sea state condition. Figure 4 shows the block diagram of the designed control switching collector. This collector gradually increments the contribution of the WR controller while weakening the APS controller output according to the current wave height. The platform vertical acceleration is measured and filtered via a Low-Pass Filter (LPF) to obtain an estimation of the current wave height and, according to the weight factor (a), the contribution of each control loop is assigned.

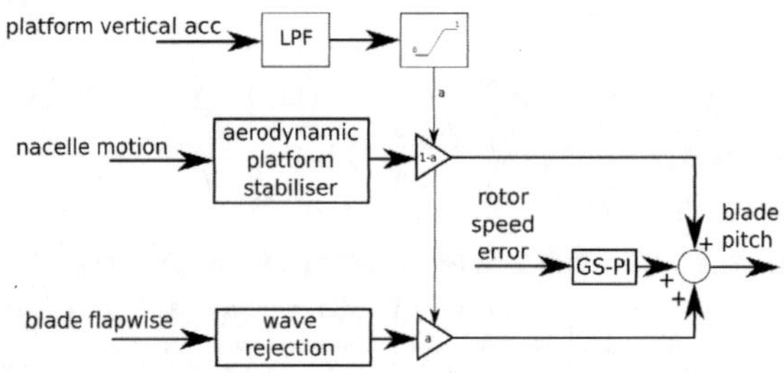

Figure 4. Aerodynamic Platform Stabiliser and Wave Rejection collector block diagram.

Furthermore, as explained for the APS control loop implementation in [12], since the WR control loop is proportional to the blade-flapwise bending moments it will be working in all wind turbine operating regions (below and above rated wind speed) unless its use is limited. Thus, the use of the WR control loop has been limited to work in the above rated wind speed region as the Detuned PI and the APS control loops actually do. For such a purpose, the conventional blade-pitch control loop is taken as a reference to enable the WR one. The final blade-pitch angle from the sum of the three control signals is:

$$\theta = \theta_{PI} + \theta_{APS} + \theta_{WR} \tag{1}$$

with:

$$0^\circ \leq \theta_{PI} \leq 90^\circ \tag{2}$$

$$-\theta_{PI} - \theta_{WR} \leq \theta_{APS} \leq 90^\circ - \theta_{PI} - \theta_{WR} \tag{3}$$

$$-\theta_{PI} - \theta_{APS} \leq \theta_{WR} \leq 90^\circ - \theta_{PI} - \theta_{APS} \tag{4}$$

where θ_{PI}, θ_{APS} and θ_{WR} are the blade-pitch angles of Detuned PI, APS and WR control loops, respectively, and θ, is the final blade-pitch angle.

IOP Conf. Series: Journal of Physics: Conf. Series **1222** (2019) 012014 doi:10.1088/1742-6596/1222/1/012014

In the block diagrams of Figures 1 and 4, one can see the gain-scheduling (GS) included in order to adapt the gains of the Detuned PI controller. The GS of the baseline Detuned PI control loop as well as the generator torque control are here adopted unchanged from [8].

4. Simulations

The simulations have been run with FAST8, from 13 to 25 m/s stochastic wind speed, and irregular sea state, from 0 to 6 m mean wave height. The time-domain simulations are 1-h (3600 s) long. All the appropriate and available FOWT Degrees Of Freedom (DOF) are enabled, 22 in total: three platform translations (surge, sway, and heave) and three rotations (roll, pitch, and yaw)(6DOF); first and second fore-aft and side-to-side tower modes (4DOF); nacelle yaw (1DOF); variable generator speed and drivetrain rotational-flexibility (2DOF); and, first and second blade-root-flapwise modes and first blade-root-edgewise mode for each blade (9DOF). The wind profiles are generated with the TurbSim module, with the highest turbulence wind type (turbine class - A) according to the IEC61400-1 standard [18]. The sea state is derived from the JONSWAP wave spectrum according to the FAST8 HydroDyn module. Figure 5 shows the Standard Deviation (STD) values of the most representative FOWT parameters for the performance evaluation in time-domain simulations with a mean wave height of 6 m. The simulations with mean wave height higher than 6 m have not been included due to the platform-pitch maximum angles exceed the 10° of inclination, taken as a threshold for the maximum inclination, as mentioned in [7].

An increment in the platform-pitch inclination (θ_{yP}) with the implementation of the WR is registered, which is the main objective of the WR control loop design as mentioned above: the increment of the wind turbine pitch motions at rough sea states and, hence, the reduction of the blade-root-flapwise bending moments (M_{yB}). One can see how the platform-pitch angle has been increased in comparison to the APS control loop results. However, in comparison to the Detuned PI, the results are higher except at high wind speeds. The blade-root-flapwise bending moment shows the best STD results with the WR control loop implementation at any wind speed conditions. A gradual improvement is shown by the blade-root-edgewise bending moment (M_{xB}), but always showing results lower than those of the other two controllers. The increment in the platform-roll angle (θ_{xP}) represents the increase in the wind turbine roll motions, however, the values are clearly smaller than those of the platform-pitch angle. The wind turbine rotor speed regulation (Ω_r) has been also improved with the WR controller at any wind speed, and hence, the generator power (P_g) production too. The generator electric torque (M_g) also shows a smoother regulation as a consequence of the improvement in the Ω_r regulation. Finally, the blade-pitch angle (θ) also shows a smoother regulation as depicted by the lowest STD results.

Figure 6 shows the extreme load values of the tower-base-pitch, blade-root-flapwise and -edgewise bending moments of the overall time-domain simulation results. The implementation of the WR control loop shows a slight increment in the maximum of the tower-base-pitch bending moment, as expected, due to the objective of increasing the platform-pitch motions. Note that the maximum value is not higher than 2.65% according to that obtained with the baseline Detuned PI controller, while the minimum it remains below, around 6.78% less. The blade-root-flapwise and -edgewise bending moments have been successfully reduced thanks to the WR control loop, not only compared to the APS control loop, but also to the baseline Detuned PI. The maximum and minimum blade-root-flapwise bending moment reductions are around 11 and 5.1%, respectively, compared with those obtained with the Detuned PI baseline controller. In addition, the maximum and minimum blade-root-edgewise bending moments have been reduced around 13.9 and 5.6%, respectively.

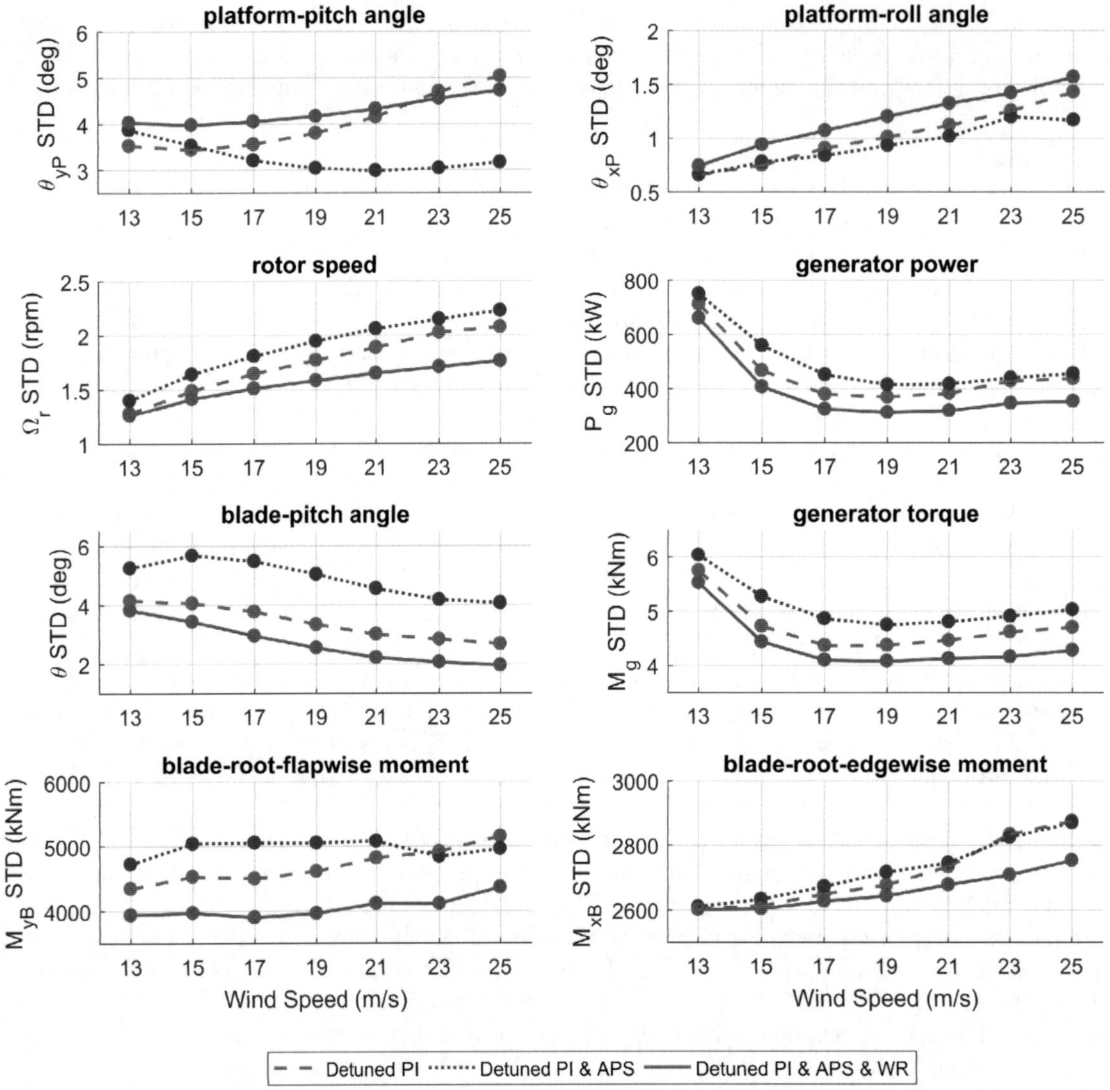

Figure 5. Standard deviation results of the time-domain simulation series at 6 m mean wave height.

5. Load analysis

The load analysis is based on the short-term fatigue calculations of the obtained time-series results. The Damage Equivalent Loads (DEL) are calculated under the same conditions for all the different FOWT control performance results, to obtain comparable values. Figure 7 shows the DEL results of the APS controller performance, normalised with respect to those of the Detuned PI baseline controller. One can see the drawback of the APS controller in the blade-root-flapwise DEL deterioration according to the wave height (H_s) increment. In still and smooth sea state conditions the APS controller improves the Detuned PI baseline DEL results; however, the loads are increased in rough sea state conditions. The worst blade-root-flapwise DEL results are registered with high waves and low wind speeds. The same tendency is shown in the blade-root-edgewise DEL results, although the deterioration is smaller. The tower-base-pitch

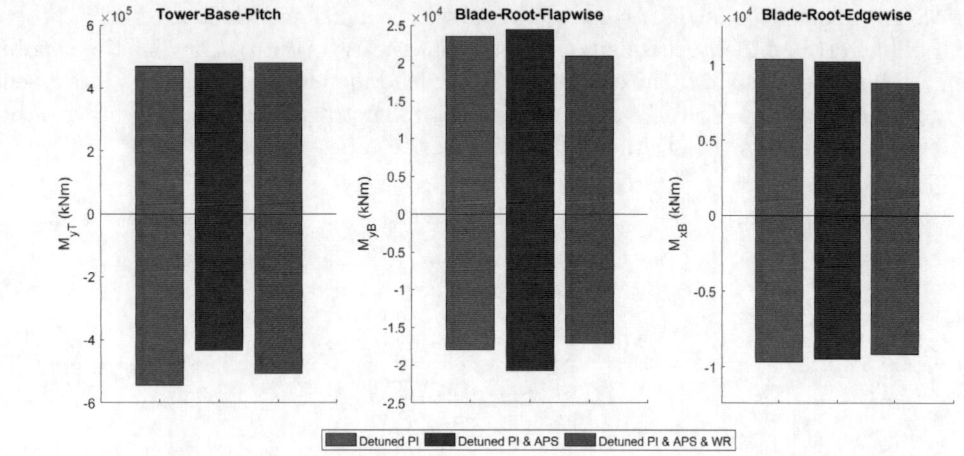

Figure 6. Extreme load values of the time-series simulations. In each subplot, Detuned PI (left, dark-green), Detuned PI & APS (centre, blue), and Detuned PI & APS & WR (right, dark-red) control results are shown.

DEL show a great reduction throughout the overall wind speed and wave height. This is due to the excessive APS control action explained above, which reduces the platform-pitch motions more than the natural motion induced by the waves. This is consistent with the platform-pitch angle STD results shown in Figure 5.

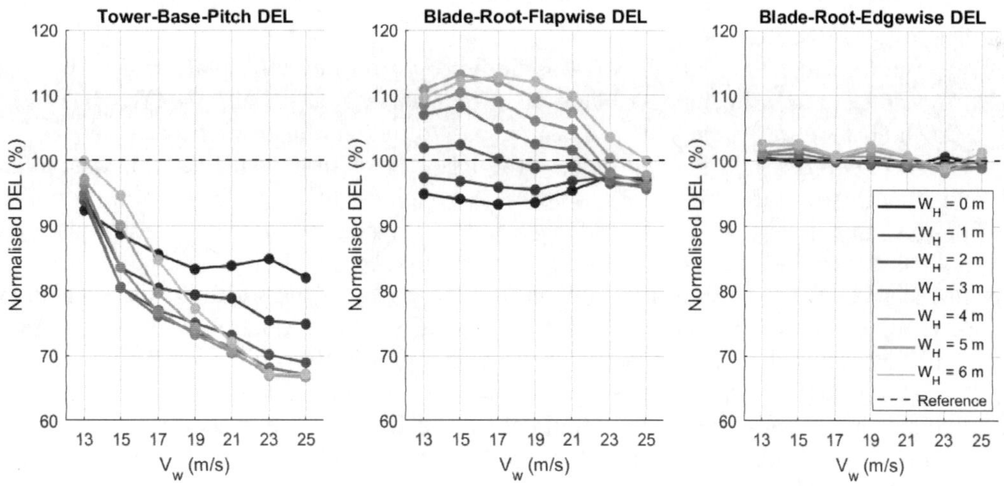

Figure 7. Normalised DEL analysis of the Aerodynamic Platform Stabiliser control loop results.

Figure 8 shows the DEL results with the implementation of the WR control loop to the APS controller. One can see how the blade-root-flapwise has been improved reducing all DEL results below the reference, corresponding to the Detuned PI controller. The blade-root-edgewise DELs have been also improved achieving similar results or even better than the Detuned PI ones. The cost of this improvement is the increment of the tower-base-pitch DELs, as expected due to the increment of the platform-pitch motions depicted by the STD shown in Figure 5. This deterioration only affects to some particular wind and wave conditions, which are hardly probable according to [8] and [19]. Finally, it is demonstrated that the WR control loop has the capability of reducing the blade loads, but at the cost of the tower load increment. Therefore, it is essential to obtain an agreement between the APS and WR controllers to achieve the reduction of both, tower and blades loads below the Detuned PI controller results.

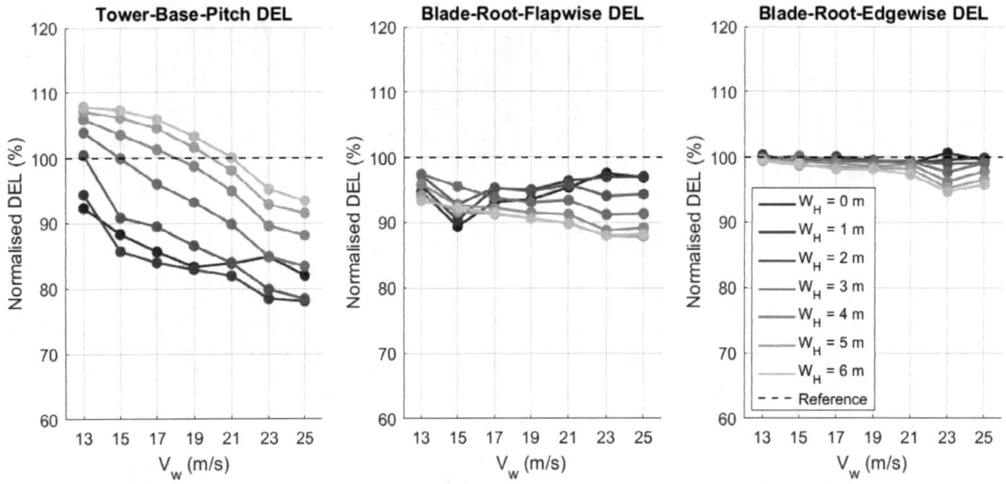

Figure 8. Normalised DEL analysis of the Wave Rejection control loop results.

6. Conclusions and Outlook

In previous studies the APS control loop was designed to improve the platform-pitch motion produced by the Detuned PI controller blade-pitch regulation in the NREL 5-MW wind turbine mounted on the ITI Energy barge [10]. The tower-base-pitch and blade-root loads were reduced whereas the rotor speed regulation and power production were improved [12]. This was possible in still and smooth sea state conditions; however, in rough seas the platform-pitch motion was reduced more than the natural motion induced by the waves increasing the loads in the blade-roots. This was due to the APS control loop, which was not able to distinguish the platform-pitch motions produced by the wind from those induced by the waves. Thus, an additional control loop was required to compensate the negative effect of the APS control loop in rough sea conditions and restrain the wind turbine blade loads. Then, in this paper, the additional WR control loop has been presented with such a purpose. Several time-domain simulations have been carried out throughout the overall above rated wind speed and irregular wave conditions. The time-domain blade extreme loads have been reduced not only below the APS results, but also below the reference Detuned PI ones. The rotor speed regulation and electric power production have been improved in the overall wind speed conditions as shown by the STD results. The blade-root-flapwise and -edgewise DEL have been reduced below the reference Detuned PI controller results. However, all these improvements have been possible at the cost of the tower-base-pitch load increment as shown by the extreme load and DELs analysis. The increased tower loads will increase the fabrication cost of the FOWT, since the substructure is expected to drive the cost as compared to the blades. Therefore, the optimisation of the controllers performance to improve the tower-base as well as the blade-root loads below the reference Detuned PI results is currently dealt with in IK4-IKERLAN.

References

[1] Bilgili M., Yasar A., Simsek E. Offshore wind power development in Europe and its comparison with onshore counterpart, *Renewable and Sustainable Energy Reviews* **2011**

[2] Rhodri J., Costa-Ros M. Floating Offshore Wind: Market and Technology Review, *Carbon Trust Report Prepared for the Scottish Government* **2015**

[3] Hanson T.D., Skaare B., Yttervik R, Nielsen F.G., Havmøller O. Comparison of measured and simulated responses at the first full scale floating wind turbine Hywind, *Proceedings of the European Wind Energy Association Annual Event* **2011**

[4] Atcheson M., Cruz J. Floating Offshore Wind Energy: The Next Generation of Wind Energy, *Springer International Publishing* **2016**

[5] Castro-Santos L., Diaz-Casas V. Floating Offshore Wind Farms, *Springer International Publishing* **2016**

[6] James R., Ros M.C. Floating Offshore Wind: Market and Technology Review, *The Carbon Trust* **2015**

[7] Wayman E.N., Sclavounos P.D., Butterfield S., Jonkman J., Musial W. Coupled dynamic modeling of floating wind turbine systems, *NREL/CP-500-39481* **2006**

[8] Jonkman J.M. Dynamics modeling and loads analysis of an offshore floating wind turbine, *NREL/TP-500-41958* **2007**

[9] Jonkman J.M. Influence of control on the pitch damping of a floating wind turbine, *NREL/CP-500-42589* **2008**

[10] Olondriz J., Elorza I., Trojaola I., Pujana A., Landaluze J. On the effects of basic platform design characteristics on floating offshore wind turbine control and their mitigation, *Journal of Physics: The Science of Making Torque from Wind Conference Series* **2016**

[11] Larsen T.J., Hanson T.D. A Method to Avoid Negative Damped Low Frequent Tower Vibrations for a Floating, Pitch Controller Wind Turbine, *Journal of Physics: The Science of Making Torque From Wind Conference Series* **2007**

[12] Olondriz J., Elorza I., Jugo J., Alonso-Quesada S., Pujana-Arrese A. An Advanced Control Technique for Floating Offshore Wind Turbines Based on More Compact Barge Platforms, *Energies* **2018**.

[13] Jonkman J.M., Marshall L., Buhl Jr. FAST User's Guide, *NREL/EL-500-38230* **2005**

[14] Olondriz J., Jugo J., Elorza I., Alonso-Quesada S., Pujana-Arrese A. Alternative linearisation methodology for aero-elastic Floating Offshore Wind Turbine non-linear models, *Journal of Physics: The Science of Making Torque from Wind Conference Series* **2018**

[15] Olondriz J., Yu W., Jugo J., Lemmer F., Elorza I., Alonso-Quesada S., Pujana-Arrese A. Using Multiple Fidelity Numerical Models for Floating Offshore Wind Turbine Advanced Control Design, *Energies* **2018**

[16] Offshore Standard DNVGL-ST-0119, Floating wind turbine structures, *Det Norske Veritas Garrad Hassan AS* **2018**

[17] Yu W., Lemmer F., Schlipf D., Chen P.W., Visser B., Links H., Gupta N., Dankemann S., Counago B., Serna J. Evaluation of control methods for floating offshore wind turbines, *Journal of Physics: Conference Series, 15th Deep Sea Offshore Wind R&D Conference, EERA DeepWind* **2018**

[18] IEC 61400-1 Ed. 3, Wind Turbines - Part 1: Design requirements, *tech. rep., International Electrotechnical Commission (IEC)* **2007**

[19] Krieger A., Ramachandran G.K.V., Vita L., Gomez A.P., Berque J., Aguirre-Suso G. LIFE50+D7.2 Design basis, *Technical Report DNV-GL* **2016**

IOP Conf. Series: Journal of Physics: Conf. Series **1222** (2019) 012015 doi:10.1088/1742-6596/1222/1/012015

Extremum Seeking Control for optimization of a feed-forward Pelton turbine speed controller in a fixed-displacement hydraulic wind turbine concept

SP Mulders[1], NFB Diepeveen[2], JW van Wingerden[1]

1. Delft Center for Systems and Control, Faculty of Mechanical Engineering, Delft University of Technology, Mekelweg 2, 2628 CD Delft, The Netherlands
2. DOT B.V., Raam 180, 2611 WP Delft, The Netherlands

E-mail: `s.p.mulders@tudelft.nl`[1], `niels@dotpower.nl`[2]

Abstract. With the sustained drive towards higher power ratings for offshore wind turbines, the size of the turbine rotor and drivetrain components scale accordingly. Compact hydraulic transmissions are widely applied in high-load systems and form a business case for application in multi-megawatt offshore turbines. The Delft Offshore Turbine (DOT) is a hydraulic wind turbine concept replacing conventional drivetrain components with a single seawater pump. In the DOT concept, pressurized seawater is directed to a Pelton turbine-generator combination, located at a central electricity generation platform. An in-field test campaign is performed using a prototype DOT turbine with a retrofitted 500 kW hydraulic drivetrain, consisting of fixed-displacement components. As a result of this configuration, a feed-forward Pelton speed controller is derived and implemented for operating the Pelton turbine at maximum efficiency. However, the controller tuning is based on estimations of physical system properties, of which the resulting optimality is unknown. For verification of the implementation, the model-free, gradient-based and data-driven Extremum Seeking Control (ESC) optimization scheme is employed. Results show fast convergence of the algorithm and an average maximum power increase of 3 %. The algorithm is well suited for application to real-world systems, due to its simplicity and ease of tuning.

1. Introduction

While the conventional drivetrain of a horizontal-axis wind turbine (HAWT) is a proven concept, the weight contribution of all components in the nacelle results in a rather high total mass. The complete turbine support structure is designed to carry this weight for the entire expected lifetime, which in turn leads to extra material and thus total cost of the wind turbine [1]. In an effort to reduce turbine maintenance requirements, complexity, weight and thus the Levelized Cost of Energy (LCOE) for offshore wind, a novel and patented [2] hydraulic concept is proposed by Delft Offshore Turbine (DOT). The eponymous turbine only requires a single water pump directly connected to the turbine rotor and replaces high-maintenance components in the nacelle, which is in effect favorable for all aforementioned aspects. In contrast to conventional wind farms, DOTs are designed to operate in a hydraulic network of multiple turbines, collectively pressurizing seawater to a central multi-megawatt generator platform. At this central location, the hydrodynamic energy is converted to electrical energy. A study on the design of the hydraulic

DOT drivetrain is performed in [3], whereas the feasibility of multiple turbines in a hydraulic network has been investigated in [4].

In 2016, extensive prototype tests have been performed with an in-field, full-scale and retrofitted wind turbine, of which the conventional drivetrain is replaced by a 500 kW hydraulic configuration [5]. The DOT500 prototype, shown schematically and physically in Figure 1, includes an intermediate oil loop providing a *hydraulic gearbox* between the low-speed oil pump directly connected to the rotor, and the high-speed water pump at the tower bottom. The pressurized water flow is converted to a high-speed water jet by means of a spear valve. The jet is then directed to a Pelton turbine wheel mechanically coupled to an asynchronous generator.

Figure 2 shows a visualization of the employed Pelton-spear valve set-up. Because all drivetrain components have a fixed volumetric displacement (apart from volumetric losses), the system pressure varies according to the turbine operating conditions. Consequently, the speed of the Pelton wheel needs to be controlled by the generator to maximize its efficiency at each operating point. As the feed-forward speed controller is derived from physical relations and its tuning is based on an estimation of system properties, the implementation might not be optimal in practice.

Extremum Seeking Control (ESC) is employed for optimization of the feed-forward controller. Results show the effectiveness of the ESC implementation for optimization of the controller on the real-world test set-up. The feed-forward gain is reduced by 13.5 % leading to a power increase of 3 %. The explanation for the rather large discrepancy between the theoretical and optimized gain later appeared to be caused by an erroneous pressure sensor reading, however, the application of ESC led to the discovery of the error.

This paper is organized as follows. Section 2 presents the principles of optimal Pelton turbine operation using a feed-forward control strategy, and subsequently elaborates on the ESC implementation for controller optimization. Section 3 presents the optimization and power extraction results obtained from the in-field prototype DOT500 wind turbine. A combined section with conclusions and a discussion on learning objectives is given by Section 4.

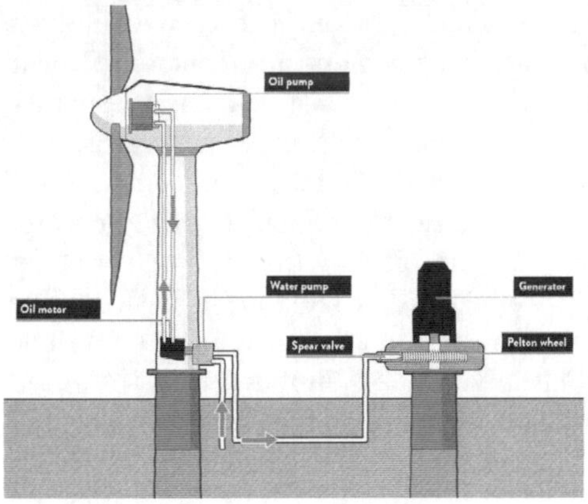

(a)

(b)

Figure 1: A schematic figure (a) of the prototype DOT500 set-up with intermediate oil loop acting as a *hydraulic gearbox* between the oil pump and water pump. The photograph (b) shows the in-field test set-up where the combined generator-Pelton turbine configuration is located at ground level.

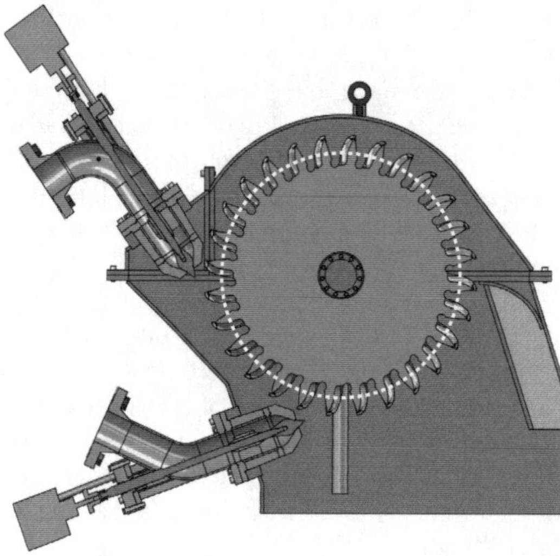

Figure 2: A Pelton wheel driven by the high-velocity water jets from two spear valves with controllable nozzle area for pressure control. The dashed circle indicates the pitch circle diameter.

2. Approach

This section is divided in two subsections. In Section 2.1, the general principles of optimal Pelton turbine operation are outlined using physical first-principles hydraulic relations. The result is used for derivation of a feed-forward speed control strategy. In Section 2.2, the ESC algorithm is described and the implementation for online optimization of the feed-forward gain is outlined.

2.1. Pelton turbine efficiency analysis and derivation of a feed-forward speed controller

Most Pelton turbines are operated under constant head pressure conditions, allowing for connection to a synchronous electrical generator. Recent works on the design of hydraulic wind turbines focus on a constant pressure discharge water flow. This is accomplished by combined controller implementations on the pump variable volumetric displacement and control of spear valves with variable effective orifice areas [4, 15, 16]. This allows the Pelton turbine to operate at its optimal fixed-speed efficiency point, and the application of a synchronous generator.

As a consequence of the DOT500 drivetrain design with fixed-displacement components, the pump discharge pressure needs to be controlled for providing for region 2 torque control substitute [5]. Resulting from this varying line pressure, the outgoing hydrodynamic water jet mass flow and velocity varies. To ensure optimal Pelton efficiency, a fixed ratio between the Pelton tangential speed and the incoming water jet speed needs to be maintained [17]. The asynchronous generator – mechanically coupled to the Pelton wheel – is supplied with the speed set point calculated by a feed-forward speed controller. The remainder of this section is devoted to the derivation of this controller.

The high-pressure water flow is converted to a high-velocity water jet. Both quantities are related by Bernoulli equation [18] for incompressible flows

$$v_{\mathrm{j}} = \sqrt{v_{\mathrm{w}}^2 + \frac{2(p_{\mathrm{w}} - p_{\mathrm{j}})}{\rho_{\mathrm{w}}}} \approx \sqrt{\frac{2p_{\mathrm{w}}}{\rho_{\mathrm{w}}}}, \tag{1}$$

in which v_{j} and p_{j} are respectively the speed and pressure of the water jet exiting the spear valve, v_{w} and p_{w} the speed and pressure of the in-line water flow, and the density of water ρ_{w} is taken constant. The approximation in Equation (1) is based on the assumption that the jet has

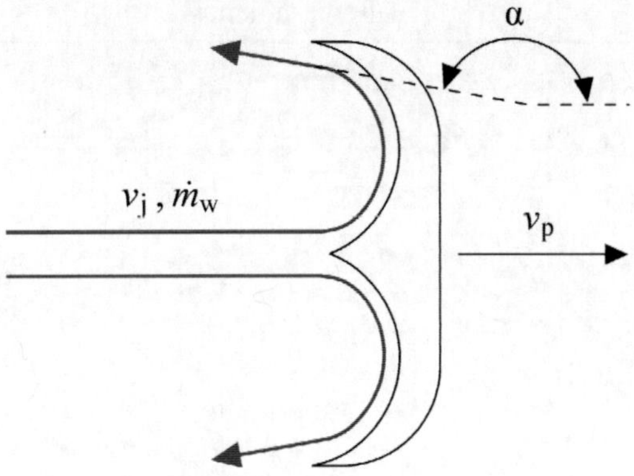

Figure 3: Schematic showing the relation between the incoming water jet speed v_j, the tangential bucket speed v_p, and their relative speed v_r.

an atmospheric pressure $p_\mathrm{j} = p_0$, and the squared speed of the in-line water flow is negligible with respect to the second term in the square root. The tangential bucket speed of the Pelton turbine v_p is related to the rotational free-run speed ω_p of the turbine by

$$v_\mathrm{p} = \omega_\mathrm{p}\frac{D_\mathrm{p}}{2}, \tag{2}$$

where D_p is the turbine pitch circle diameter (PCD), indicated by the dashed circle in Figure 2. From this, the relative velocity between the water and bucket speed is given by

$$v_\mathrm{r} = v_\mathrm{j} - v_\mathrm{p}, \tag{3}$$

where it is assumed that the water speed is maintained in the bucket [17]. The momentum of the high-speed water jet exerts a force on the moving Pelton buckets

$$F_\mathrm{p} = \rho_\mathrm{w} Q_\mathrm{w} v_\mathrm{r}(1 - \cos\alpha) = \dot{m}_\mathrm{w} v_\mathrm{r}(1 - \cos\alpha), \tag{4}$$

in which α is the angle between bucket incoming and outgoing water flow shown in Figure 3, Q_w is the water volumetric flow rate, and $\dot{m}_\mathrm{w}$ the water mass flow rate. The efficiency of the Pelton turbine is expressed as the ratio between the mechanical power available at the Pelton shaft and the kinetic power contained in the water jet: $\eta_\mathrm{p} = P_\mathrm{p}/P_\mathrm{w}$. The kinetic power of the water jet is given by

$$P_\mathrm{w} = \frac{1}{2}\dot{m}_\mathrm{w} v_\mathrm{j}^2, \tag{5}$$

and the mechanical Pelton power is defined as

$$P_\mathrm{p} = F_\mathrm{p} v_\mathrm{p} = \dot{m}_\mathrm{w} v_\mathrm{r}(1 - \cos\alpha) v_\mathrm{p}. \tag{6}$$

Now by rewriting the Pelton efficiency equation, and defining the ratio between the Pelton bucket and water jet speed as $\kappa_\mathrm{v} = v_\mathrm{p}/v_\mathrm{j}$, the following expression is obtained

$$\eta_\mathrm{p} = 2\kappa_\mathrm{v}(1 - \kappa_\mathrm{v})(1 - \cos\alpha). \tag{7}$$

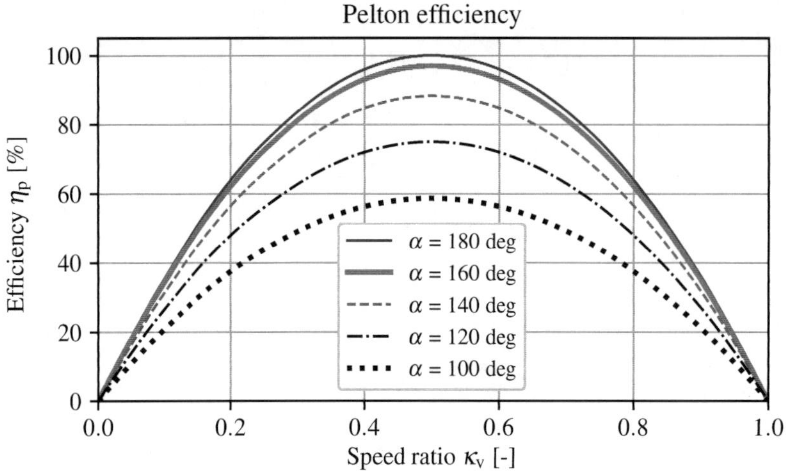

Figure 4: Pelton efficiency η_p as function of the ratio between the bucket speed and water jet. The efficiency is evaluated for multiple deflection angles α.

Solving the efficiency gradient for $\partial\eta_p/\partial\kappa_v = 0$, gives the ratio $\kappa_v = 1/2$ for maximum Pelton efficiency. An evaluation of this relation for different deflection angles α is visualized in Figure 4, and it is observed that the optimal ratio κ_v is independent of α.

As it is now known that for maximum efficiency the water jet velocity should be twice the tangential bucket speed ($v_j = 2v_p$), a feed-forward generator speed controller is derived. By this information, and substitution of Equations (1) and (2), the feed-forward generator speed controller is defined as

$$\omega_g = \sqrt{\frac{2p_w}{\rho_w D_p^2}} = K_{FF}\sqrt{p_w}. \tag{8}$$

All non-varying parameters are captured in a single constant K_{FF}, and the water pressure p_w is a measured quantity.

Remark 1 *After the water flow exits the spear valve, the aim to operate the Pelton turbine-generator combination at maximum efficiency is a decoupled control objective from the rest of the wind turbine drivetrain. This paper only considers the optimization of the Pelton-generator system. As the control of this system does not influence the wind turbine operating behavior, the relation to the extracted wind turbine energy/power is omitted. For an extensive description of the wind turbine drivetrain and its controller design, the reader is referred to [5].*

2.2. Theory on Extremum Seeking Control

The online, model-free and gradient-based optimization scheme Extremum Seeking Control (ESC) [6, 7] is applicable to parameters in a (control) system, of which the optimization objective function possesses a local optimum. The first notion of ESC is found in [8], whereas the algorithm was for the first time described in detail in the 1950's [9]. ESC has seen a revival at the beginning of the current century with a proof of stability [10]. At present, the algorithm is of great interest [7], with the trend of increasingly complex systems and the scheme's ease of implementation. ESC has shown being effective in optimizing the region 2 torque control gain of a conventional wind turbine [11, 12], but also for control optimization of multiple turbines in a wind farm setting [13, 14].

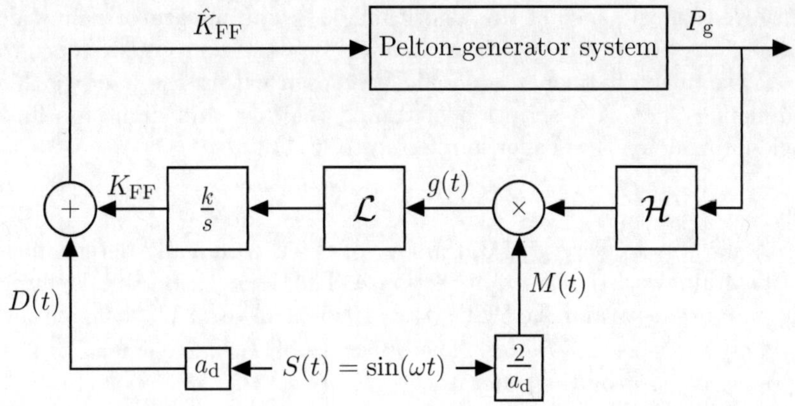

Figure 5: Block diagram of ESC implemented on the Pelton-generator system. The feed-forward gain K_{FF} is excited by the dither signal $D(t)$, and $M(t)$ is used for demodulation of the high-pass filtered generator power output signal P_g. After demodulation, the gradient information of $\partial P_g/\partial K_{FF}$ is contained as a DC-contribution in $g(t)$, which is subsequently low-pass filtered and integrated to arrive at the optimal value for K_{FF}.

Substitution of the physical system properties included in K_{FF} (Equation (8)) leads to the definition of a feed-forward Pelton turbine control implementation. By tracking the calculated speed set point ω_g results in Pelton operation at maximum efficiency for each operating pressure. However, due to inevitable discrepancies between theory and practice, the non-varying constant K_{FF} might not be optimal on the real-world set-up. For this reason, the model-free and data-driven ESC optimization algorithm is implemented.

Figure 5 presents a schematic of the continuous-time ESC implementation. The scheme aims for maximization of the generator power P_g, provided by an instantaneous sensor reading. The ESC algorithm is based on a dither-demodulation scheme by periodic excitation of K_{FF}, taken as the input of the plant

$$\hat{K}_{FF} = K_{FF} + D(t) = K_{FF} + a_d \sin(\omega t). \tag{9}$$

The dither and demodulation signals $D(t)$ and $M(t)$, respectively, are amplitude-scaled versions of the periodic excitation signal $S(t)$. By exciting the system input with the dither signal, and multiplying the generator power output P_g with the demodulation signal, gradient information is present as a DC contribution in the resulting signal $g(t)$. This can be seen by rewriting the system output signal as a first-order Taylor series expansion

$$P_g(\hat{K}_{FF}) = P_g(K_{FF} + a\sin(\omega t)) \approx P_g(K_{FF}) + a\sin(\omega t)\frac{\partial P_g}{\partial \hat{K}_{FF}} + \mathcal{O}^2. \tag{10}$$

The high-pass filter $\mathcal{H}$ removes the steady-state component, and by subsequent multiplication with $M(t)$ and by rewriting using trigonometric identities, one obtains

$$2\sin^2(\omega t)\frac{\partial P_g}{\partial \hat{K}_{FF}} = \frac{\partial P_g}{\partial \hat{K}_{FF}} - \cos(2\omega t)\frac{\partial P_g}{\partial \hat{K}_{FF}}. \tag{11}$$

The algorithm converges to a *steady-state* value of K_{FF} such that output P_g is optimized and the gradient $\partial P_g/\partial \hat{K}_{FF}$ converges to zero. As a result of phase losses in the system, the phase of the demodulation signal can be compensated to match the phase of the output excitation signal,

IOP Conf. Series: Journal of Physics: Conf. Series **1222** (2019) 012015 doi:10.1088/1742-6596/1222/1/012015

resulting in improved convergence of the algorithm [11]. The integrator gain k determines the convergence rate of the algorithm, and is a trade-off between convergence speed and stability. Furthermore, the excitation frequency ω should be chosen within the bandwidth of the system input-output dynamics. For a description of tuning the low- and high-pass filter cut-off and cut-in frequencies ω_l and ω_h, the reader is referred to [6, 11].

3. Main body

The goal of this section is to verify the optimality of the theoretically determined feed-forward gain using the ESC framework described in Section 2. The theoretical values defining the gain are respectively the water density and the PCD, $\rho_\mathrm{w} = 1025$ kg m^{-3} and $D_\mathrm{p} = 0.85$ m, and evaluation leads to $K_\mathrm{FF} = 5.2 \cdot 10^{-2}$ (rad s^{-1}) Pa$^{-1/2}$. The objective of the ESC scheme is maximization of the electrical power supplied by the generator

$$K^*_\mathrm{FF} = \arg\max_{K_\mathrm{FF}} P_\mathrm{g}(\hat{K}_\mathrm{FF}). \tag{12}$$

The frequency of the excitation signal $S(t)$ is chosen as $\omega = 0.04$ rad s^{-1}, which is well within the system bandwidth from generator speed to measured power. The low- and high-pass filters are configured to have their cut-off and cut-in frequencies at $\omega_\mathrm{l} = 0.0230$ rad s^{-1} and $\omega_\mathrm{h} = 0.0205$ rad s^{-1}. The dither amplitude is set $a_\mathrm{d} = 3 \cdot 10^{-3}$ to allow for a peak-to-peak generator speed variation of approximately 10 %. The integrator gain is set to $k = 1 \cdot 10^{-3}$, and is found empirically to provide a satisfactory trade-off between algorithm convergence speed and stability.

The performed ESC optimization experiment, of which the results are shown in Figure 6, has a duration of approximately 1 hour. The initial condition of the integrator is set to the theoretical optimal feed-forward gain of $K_\mathrm{FF,0} = 5.2 \cdot 10^{-2}$ (rad s^{-1}) Pa$^{-1/2}$. Convergence of the algorithm is observed in the first half of the experiment, after which the algorithm converges to an optimal gain of $K^*_\mathrm{FF} = 4.5 \cdot 10^{-2}$ (rad s^{-1}) Pa$^{-1/2}$. The obtained result is a decrease of 13.5 %, and shows a significant deviation from the theoretically derived result.

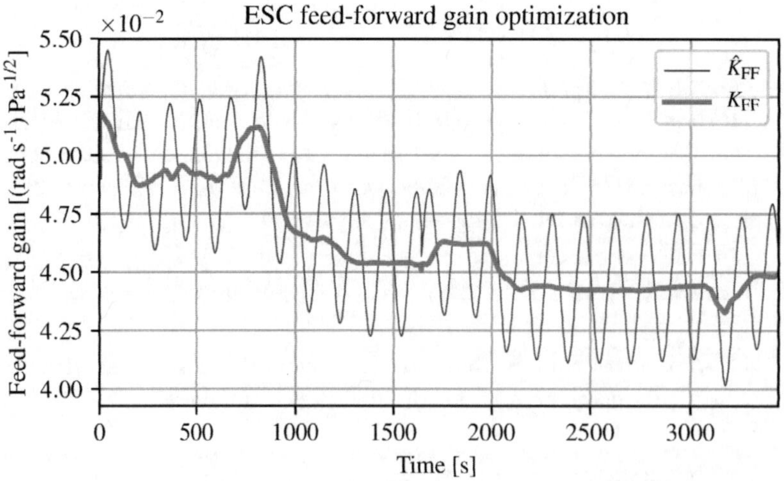

Figure 6: ESC optimization progress of the feed-forward generator speed controller gain K_FF. Convergence to a lower value is observed during the first 2000 s, while this value remains near constant in the second half of the process.

To verify whether the optimized controller gain indeed results in a higher efficiency of the Pelton turbine-generator system, two 10 minute verification data sets are performed: one with the theoretical, and the other with the optimized gain. The results are presented in Figure 7. Figure 7a illustrates the generator speed difference for the two cases, and shows a significant drop of the operational generator speed. Figure 7b evaluates the generated power for the same experiment. Because limited verification data is available, no profound statement can be made on the overall performance. However, it is possible to evaluate the average maximum power, which is defined as all the data points that are above a boundary value of 90 kW. It is observed that the average maximum power is higher for the case using the optimized controller gain. An average maximum power of 100 kW and 103 kW is obtained for the respective cases.

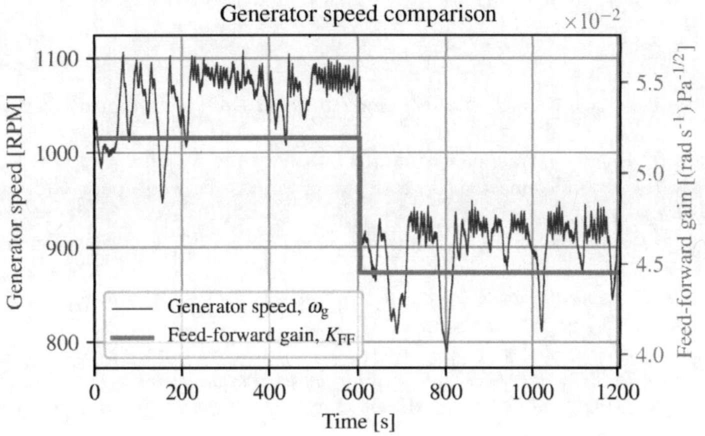

(a)

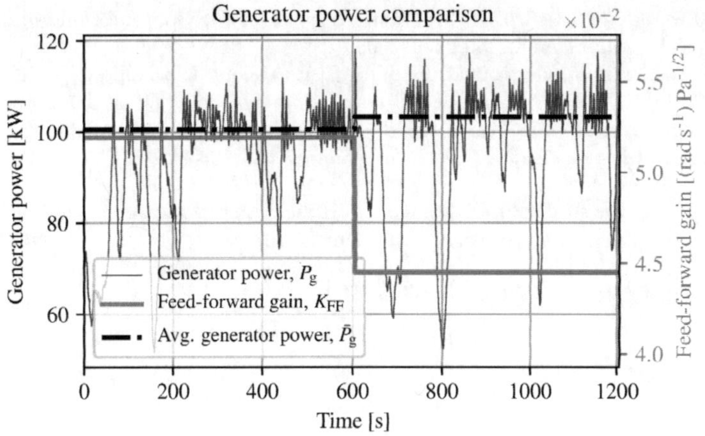

(b)

Figure 7: The effect of implementing the the theoretical and optimized gain on generator speed (a) and generated power (b). In the latter plot, an average maximum power of 100 kW and 103 kW is found for the respective cases.

IOP Conf. Series: Journal of Physics: Conf. Series **1222** (2019) 012015 doi:10.1088/1742-6596/1222/1/012015

4. Conclusions and learning objectives

This paper shows the implementation of a feed-forward speed controller for operating a Pelton turbine at its maximum efficiency under varying conditions. A theoretical estimation of the feed-forward gain can be derived from physical relations, however, the optimality on a practical real-world system is unknown. For verification, the model-free, gradient-based and data-driven optimization scheme Extremum Seeking Control is employed on the feed-forward gain on the practical test set-up. The algorithm shows fast convergence and a decrease of 13.5 % of the feed-forward gain, resulting in an average maximum power increase of 3 %. The explanation for the rather large discrepancy between the theoretical and optimized gain later appeared to be related to an error in the measured pressure value, however, the application of ESC led to the discovery of this error. It is concluded that ESC is a valuable algorithm for parameter validation and optimization. The algorithm is well suited for application to real-world systems, due to its simplicity and ease of tuning.

References

[1] Morthorst P E and Awerbuch S 2009 *The economics of wind energy* (European Wind Energy Association (EWEA))
[2] van der Tempel J 2009 Energy extraction system, has water pump attached to rotor, windmill for pumping water from sea, water system connected to water pump, for passing water pumped from sea, and generator connected to water system
[3] Diepeveen N 2013 *On the application of fluid power transmission in offshore wind turbines* Ph.D. thesis Delft University of Technology
[4] Jarquin Laguna A 2017 *Centralized electricity generation in offshore wind farms using hydraulic networks* Ph.D. thesis Delft University of Technology
[5] Mulders S P, Diepeveen N F B and van Wingerden J W 2018 *Wind Energy Science* **3** 615–638
[6] Ariyur K B and Krstic M 2003 *Real-time optimization by extremum-seeking control* (John Wiley & Sons)
[7] Tan Y, Moase W H, Manzie C, Nešić D and Mareels I M Y 2010 *29th Chinese Control Conference (CCC)*
[8] Leblanc M 1922 *Revue Generale de lElectricite*
[9] Draper C S and Li Y T 1951 *Principles of optimalizing control systems and an application to the internal combustion engine* (American Society of Mechanical Engineers)
[10] Krstić M and Wang H H 2000 *Automatica* **36** 595–601
[11] Xiao Y, Li Y and Rotea M A 2018 *IEEE Transactions on Control Systems Technology*
[12] Rotea M A 2017 *IFAC-PapersOnLine* **50** 4504–4509
[13] Boersma S, Doekemeijer B M, Gebraad P, Fleming P A, Annoni J, Scholbrock A K, Frederik J and van Wingerden J W 2017 *Proc. of the American Control Conference (ACC)* ed Sun J and Jiang Z P (United States: IEEE) pp 1–18 ISBN 978-1-5090-4583-9
[14] Ciri U, Rotea M A and Leonardi S 2017 *Renewable Energy* **113** 1033–1045
[15] Skaare B, Hörnsten B and Nielsen F G 2013 *Wind Energy* **16** 1259–1276
[16] Buhagiar D, Sant T and Bugeja M 2016 *Journal of Dynamic Systems, Measurement, and Control* **138** 081007
[17] Zhang Z 2007 *Proceedings of the Institution of Mechanical Engineers, Part A: Journal of Power and Energy* **221** 343–355
[18] White F M 2011 *Fluid Mechanics* 7th ed (McGraw-Hill)

IOP Conf. Series: Journal of Physics: Conf. Series **1222** (2019) 012016 doi:10.1088/1742-6596/1222/1/012016

Understanding of Leading-Edge Protection Performance Using Nano-Silicates for Modification

I Ouachan[1*], M Kuball[2], D Liu[2], K Dyer[3], C Ward[1] and I Hamerton[1]

[1]Bristol Composites Institute (ACCIS), University of Bristol, Bristol, BS8 1TR, UK.
[2] H. H. Wills Physics Laboratory, University of Bristol, Bristol BS8 1TL, UK;
[3] Offshore Renewable Energy Catapult (ORE), Glasgow, G1 1RD, UK;
* Corresponding author (Imad.Ouachan@bristol.ac.uk)

Abstract. Leading edge erosion caused by raindrop impact is a key problem that needs to be overcome in the wind energy sector. Solutions up to date have not proved suitable, failing prematurely in their lifecycle at a cost to the wind industry. Failure mechanisms for rain erosion are not well understood, particularly the elastic and viscous polymer properties at the resulting high strain rates (10^6 - 10^9 Hz) of raindrop impacts. The effect of the inclusion of glycidyl polyhedral oligomeric silsesquioxane nanoparticles into a commercial polyurethane coating system was studied using nanoindentation and dynamic mechanical thermal analysis (DMTA). Results show that the inclusion of POSS improves damping, providing an alternative mechanism for energy dissipation without variation of T_g and minimal loss of stiffness. This presents a way of modifying current coating systems through the incorporation of POSS. Nanoindentation obtained previously unreported properties of the coating system (hardness, modulus and short-term recovery) and highlighted a correlation between loading rate and a reduction of short-term recovery. Nanoindentation was difficult for the modified leading-edge protection (LEP) samples as two phases were formed resulting in large standard deviations. DMTA results show modification of the LEP increases damping at lower temperature ranges introducing an additional mechanism of dissipating energy. Additionally, sweep data show an increase in elasticity at higher frequencies on the modified samples.

1. Introduction

Offshore wind turbine blades are expected to remain in operation with minimal maintenance for a service life of 25 years. However, it is estimated that up to £1.3 million is spent on each turbine during its lifetime due to leading edge erosion (LEE) from the impact of rain droplets with current coating systems [1]. Erosion can be reduced by limiting blade tip speeds (currently ~100 m/s), but with reduced noise restrictions offshore, the current trend is for increasing the blade length and tip speeds to maximise the potential of wind turbines in an already competitive energy market. LEE can be caused by rain, hail, sea spray, and other particulate debris such as sand, dependent on the operating environment. For offshore wind turbines liquid droplet impingement results in significant blade erosion which is detrimental to aerodynamic efficiency and, as a result, decreased energy capture. A recent campaign by Offshore Renewable Energy Catapult aimed to provide more representative results by measuring the effect of erosion on the annual energy production of operating offshore wind turbines [1]. It was found that an uplift in annual energy production between 1.5 % and 2 % can be achieved following the repair of moderate erosion. This is seen as a significant issue as it not only decreases energy capture over time but also requires regular inspections and sometimes substantial repair.

General current offshore solutions that are available include [2, 3]:

- gelcoats *e.g.* epoxy or polyester and form a chemical bond between the two during single cure;
- flexible coatings which are applied post moulding using rollers or spray equipment.
- rain erosion shields which are premade in controlled conditions and bonded to the blade.

Cortes *et al.* reported that lower modulus flexible coatings are better against rain erosion in comparison to stiffer gelcoat types [3]. This is thought to be due to the reduction of stress at the impact surface and damping the peak value of the initial stress wave, ensuring rapid material recovery and optimised energy dissipation. Elastomeric and ductile polymer coatings generally undergo ductile erosion behaviour, dependent on the rain impact conditions, with elastomers showing a much lower weight loss [4]. This has been attributed to how the materials removal occurs – tearing and fatigue for rubbers, cutting and chip formation for ductile polymers, and brittle fracture for brittle polymers.

Significant LEE has been reported in several wind farm projects despite the use of commercially available leading-edge protection (LEP) solutions, highlighting the fact that it is a current and key area of research for the future. Repair of wind turbine blades is reported to be costly in two ways. First, there is a degradation in the aerodynamic performance, estimated to cost the European offshore wind industry between €56 m - €75 m annually [5]. Additionally there is the need for inspection, preventative maintenance, repair, and loss of production due to down time estimated to cost the European offshore wind industry €56m annually [6]. Turbines blades, especially offshore types, are inherently difficult to access due to their size, the cost of vessel hire and environmental conditions such as weather and tidal variations [7]. As a result, removal of blades is costly and carries great risk and consequently it is used only as a last resort making *in-situ* inspection and repair the most economically viable route. In-situ repairs are limited by the need for rapid curing repair materials that can be correctly cured and applied in the environmental conditions.

A raindrop impact can be broken down into three waves: the initial longitudinal compressional stress wave, the preceding transverse shear wave, and a third Rayleigh wave due to droplet deformation from pressure rebounding from the coating (containing two thirds of the total impact energy) [3]. Failure mechanisms for rain erosion are not well understood, particularly how they are affected by elastic and viscous polymer properties at the resulting high strain rates (10^6 - 10^9 Hz) of raindrop impact [8]. These high strain rates make conventional mechanical tests and analysis unsatisfactory for studying materials erosion properties as they operate at much lower strain rates (10^1 - 10^4 Hz).

Previous studies have shown that the acoustic impedance of a material is related to its ability to resist rain erosion which can be measured using ultrasonic techniques [3]. This is a parameter that can be used in the design of coatings as a potential method for in-line monitoring during manufacture or as a key parameter that can used in a predictive model to predict erosion performance of multi-layer systems. This could allow for a method of measuring individual layers of a system after curing using a high strain rate method to predict rain erosion performance and provide data for models to determine how energy will dissipate and flow through the coating system. The theory is that how ultrasonic waves propagate through the multi-layer system is relatable to rain droplet impact waves. Currently, rain erosion performance is measured by accelerating coated samples through a simulated rain droplet environment. However, this costly method prohibits detailed evaluation of changing polymer elastic and viscous properties during testing. Kuznetsova has linked nano-particle inclusion with the acoustic impedance of a sample and this work aims to tailor individual layers (composite, filler, LEP and interfaces) to evaluate improvements in rain erosion properties [9]. Coated (Polyurethane based) glass fibre reinforced polymer (GFRP) composite samples, representative of wind turbine blades, were produced and tested.

For this study a *polyhedral oligomeric silsesquioxane* (POSS) variant was selected as the nanoparticle of choice due to its liquid state, solubility, and functionality that allows for the formation of chemical bonds between the components as opposed to just a mixture of the components.

IOP Conf. Series: Journal of Physics: Conf. Series **1222** (2019) 012016 doi:10.1088/1742-6596/1222/1/012016

As a prerequisite of developing an alternative ultrasonic technique, conventional tests were used to characterise commercial materials and understand the role of viscoelasticity on multi-layered systems.

2. Methodology

2.1. Materials

Unidirectional E-Glass (06-UE-591-1200) and E-Glass biaxial +/-45° (20-EX600-127) supplied by Saertex were used in combination with the RS-M135 low temperature cure epoxy resin system and hardeners (RS-MH134 and RSMH-137) supplied by PRF Composites. The commercial coating system used is comprised of a Polyurethane (PU) based filler and LEP coating. The LEP system was modified by the addition of 10 wt. % glycidyl polyhedral oligomeric silsesquioxane (POSS) cage mixture supplied by Hybrid. This POSS variant is a hybrid molecule with an inorganic silsesquioxane at the core and organic glycidyl groups attached at the corners of the cage.

2.2. Sample Manufacture

2.2.1. Flat Panels. Two 300 x 400 mm^2 ([+/-45, 0, +/-45, 0, +/-45]) composite panels were manufactured using resin infusion. The supplier's suggested mix ratio was used and infused at room temperature. The plates were simultaneously cured using the specified cure cycle (16 hours at 60 °C) with no post curing steps. The cured panels were trimmed using a diamond saw and sanded. A filler system was applied using a smoothing tool in combination with cork guides to maintain a constant thickness. Samples were then sanded followed by the application of the LEP system using a short nap roller. Two coats were applied with a wet film thickness of between 100-125 μm with a flash off time of 45 minutes.

2.2.2. Dynamic Mechanical Analysis Samples. Stand-alone filler, LEP and modified LEP (10 wt. % POSS) were cast into 100 x 150 mm^2 mould made from cork on a tool plate and allowed to cure at room temperature. Three samples of each type were machined to 35 x 10 mm^2, with approximate thicknesses of 1 mm for the filler samples and 2.9 mm for both LEP samples.

2.3. Test Methods

2.3.1. Nanoindentation. The load-hold-unload indentation cycles were conducted using a Hysitron TI Premier equipped with Berkovich shaped diamond indenters calibrated with a fused silica reference sample. Cross sections of 1 cm^2 samples were used to allow for measurement of individual layers minimising any influence from other layers detailed in Figure 1. Both load (P) and displacement (h) were continuously recorded during the experiment and a complete P- h response was obtained. As multiple indentations were made, mean values with standard deviations were reported.

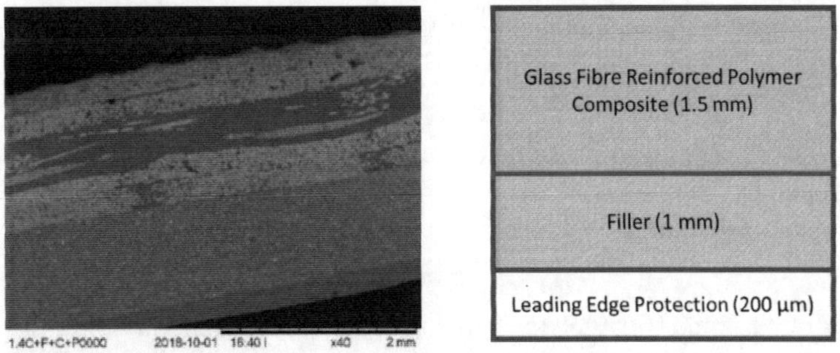

Figure 1. Cross section of test coupon. Left: SEM of sample cross section. Right: Graphic representation of cross section with approximate thicknesses.

Test 1 investigated mechanical properties of each sample layer. Testing was conducted on both the cross section and on the LEP top film surface for comparison. Loading, hold and unloading times were kept constant at 5 seconds for all indents. A series of 50 indents were made 25 nm apart in each experiment of increasing depth from 50 nm to 300 nm on each layer of the sample.

Test 2 investigated the variation of loading rate. Testing was conducted on each cross-section layer at depths of 500 nm, 700 nm, and 900 nm. A series of five indents were made 25 nm apart on each layer of the sample at loading/unloading rates of 50 nm/s, 150nm/s, and 250 nm/s with a 5 second hold time. Mechanical properties were determined from the load displacement graphs generated and summarised in Figure 2.

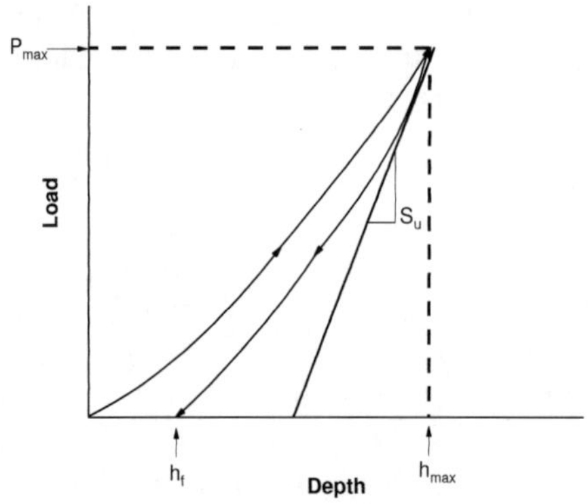

Figure 2. Example representation of indentation-load displacement data showing key parameters. P_{max}, peak indentation load, h_{max}, indenter displacement at peak load, h_f, final depth of the contact impression after unloading and S_u, contact stiffness of unloading curve calculated form the unloading gradient. Adapted from literature [10].

In this work, Oliver-Pharr analysis was used to determine the hardness and reduced elastic modulus from the unloading curve [11].

Hardness (H) is a ratio calculated by,

$$H = \frac{P_{max}}{A_c}, \tag{1}$$

where P_{max} is the maximum load and A_c is the projected area of indentation.

Unloading contact stiffness (S_u) is determined from the initial unloading slope at P_{max} by,

$$S_u = \frac{dP}{dh}, \tag{2}$$

where P and h denote the applied load and penetration depth respectively.

The reduced elastic modulus is calculated with the contact stiffness as follows,

$$E_r = \frac{S_u\sqrt{\pi}}{2\beta\sqrt{A_c}}, \tag{3}$$

where β is a factor related to the specific geometry of the Berkovich indenter.

Modulus E is determined by,

$$E = {1 - v_s^2} \Big/ {\frac{1}{E_r} - \left(\frac{1-v_i^2}{E_i}\right)}, \tag{4}$$

where E_i and V_i (1140 GPa and 0.07 respectively) represent the elastic modus and Poisson's ratio of the indenter whereas E_s and v_s (assumed as 0.5 for PU coatings and from literature [12]) correspond to specimen values.

Short Term Recovery (STR) is a measure of elastic recovery and is thought to be a key variable in raindrop impact. This was calculated by,

$$STR = \frac{(h_{max} - h_f)}{(h_{max})}, \tag{5}$$

where h_{max} and h_f denotes maximum indenter depth and final indenter depth respectively.

2.3.2. Dynamic Mechanical Analysis. 35 x 10 mm^2 samples (filler, LEP and modified LEP) were analysed in duplicate using a TA DMA Q800 dynamic mechanical thermal analyser (DMTA) equipped with a single cantilever testing attachment. Temperature scans were conducted using a 15 μm amplitude, 1 Hz frequency and while under a dynamic scan from room temperature to 150 °C at 5 °C/min. Frequency sweeps were performed from 0.1 to 200 Hz at an amplitude of 15 μm under ambient conditions.

3. Results

3.1. Nanoindentation

Determination of mechanical properties (Test 1) was performed on both the surface and cross section of samples. Results for the unmodified LEP layer are shown in Figure 3. Samples highlighted in red were omitted from statistical analysis as they were considered as an outlier. These were defined as points which are more than 1.5 times the interquartile range (either above the third quartile or below the first quartile). This also included points indented in the first 100 nm of testing due to unreliable readings at low depths, as determination of the initial contact point of low stiffness of materials is difficult to detect [13]. Similar values for both hardness and reduced modulus were obtained but the surface indentation gave results with a much lower variation in results. This is observed in the coefficient of variation of the results with an average coefficient of variation (COV) of 31 % and 11 % for cross-section and surface indentation respectively.

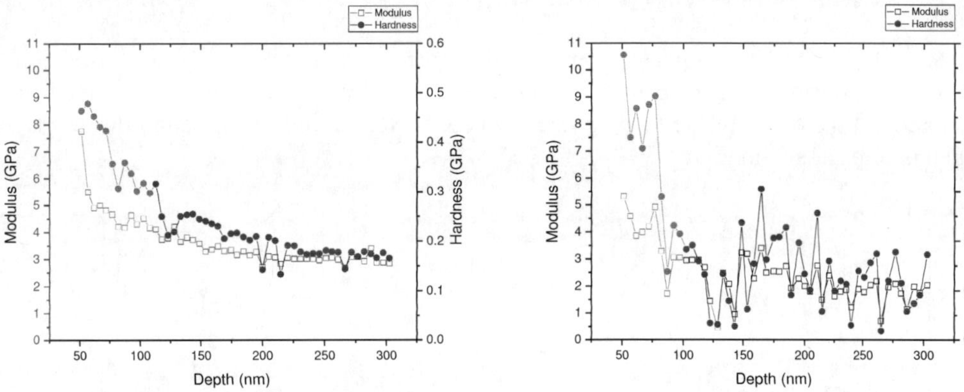

Figure 3. Nanoindentation of unmodified LEP and filler coated GFRP sample to investigate potential anisotropy. Modulus (E) and hardness (H) were determined using Oliver-Pharr analysis. Results highlighted in red are omitted from further statistical analysis. Left: Surface indentation through the LEP into the sample. Right: Indentation of cross-section of LEP film.

Contact stiffnesses and moduli for each layer cross section are summarised in Figure 4.

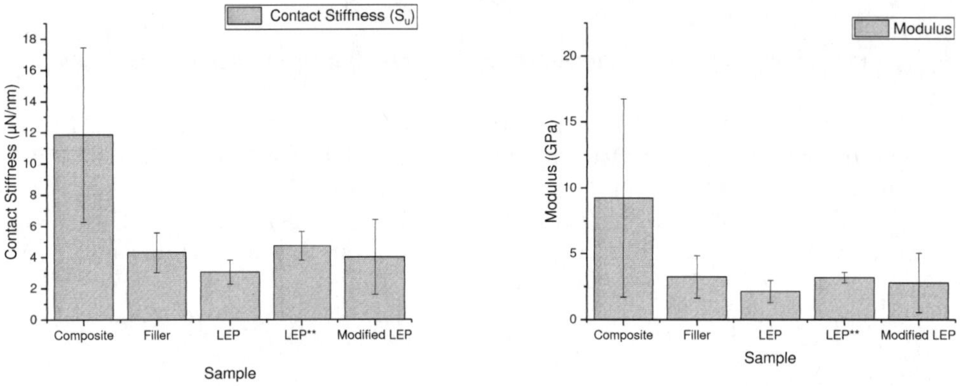

Figure 4. Nanoindentation of unmodified LEP and filler coated GFRP sample cross sections to determine mechanical properties. Contact stiffness (S_u) and modulus (E) were determined using Oliver-Pharr analysis. Contact stiffness is taken as the gradient of the unloading curve whereas modulus corrects for the compliance of the instrument in addition to assuming a Poisson's ratio of 0.5. All tests were conducted on the cross section of the sample unless denoted by ** where a surface measurement was taken. Left: Comparisons of reduced modulus. Right: Comparison of (S_u) values.

As expected, composite samples showed the highest standard deviations as they are composed of a combination of two significantly different materials (reinforcement and matrix). The modified LEP also showed high standard deviations making comparisons unviable. Statistically significant differences were seen between cross-section and surface measurements.

STR is thought to be a key parameter in rain erosion performance of materials, its correlation with loading rate could potentially be indicative of strain rates caused by varying rain impact velocities. Test 2, determination of the effect of the loading rate, was conducted on the cross section of each layer of samples and is summarised in Figure 5.

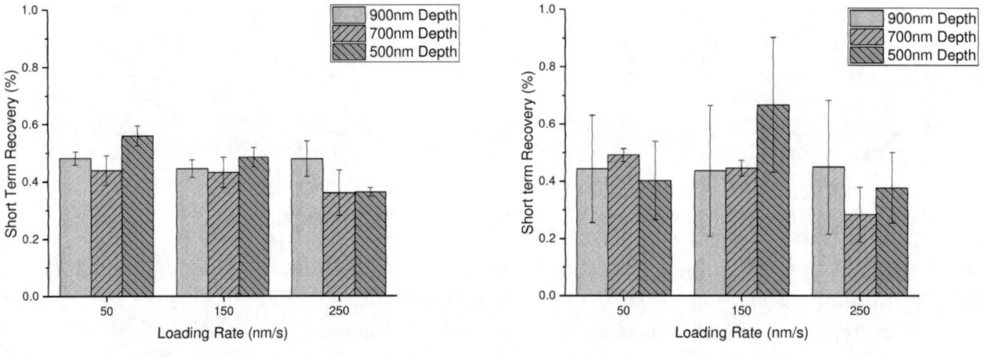

Figure 5. Nanoindentation of unmodified LEP and filler coated GFRP sample to investigate the effect of loading rate. Short term recovery was calculated from Oliver-Pharr analysis [19]. Left: Indentation of the LEP cross-section. Right: Indentation of Modified LEP (10 wt.% POSS) cross-section.

For the unmodified LEP STR at 500 nm, the increase in strain rate shows a trend of decreasing STR. The reduction in the ability of the material recover over a short period of time is thought to lead to worse rain erosion performance. This corresponds to observed increasing trends to brittle failure of coatings at increased erosion test rotational speeds. At the other indentation depths, no trend is easily observable, potentially due to insensitivity to strain rate at higher depths. As with Test 1 modified LEP results were difficult to compare, due to large standard deviations.

3.2. Dynamic Mechanical Analysis

DMTA is a technique in which a small deformation is applied to a sample in a cyclic manner allowing for the materials response to temperature, stress and frequency to be measured. A key property is the glass transition temperature (T_g), the temperature at which a polymer changes from a harder state to a more compliant or "rubbery" state. This is an important value as it could affect how materials respond to impacts at different temperatures. It is also possible to quantify the way in which a material can absorb and disperse energy using $\tan\delta$ values. It is the ratio of the loss modulus (E″), a measure of the viscous response of the material to the storage modulus (E′), a measure of the elastic behaviour of a sample. As a result, increasing the ability of a material to dissipate energy could be advantageous in minimising damage erosion caused by rain droplet impacts.
Glass transition temperature T_g values were obtained ($\tan\delta$ maximum) and onsets calculated from the storage modulus plot in Figure 6. The unmodified filler showed a lower T_g (54 °C) and lower storage modulus between 40 - 80 °C in comparison to the LEP samples. The unmodified filler also indicated that the storage modulus begins to decrease at temperatures lower than 30 °C. No significant difference between T_g values (85 °C) of LEP samples was observed however modified LEP showed a higher $\tan\delta$ value between 30 – 80 °C. This shows an increase the ability of a material to dissipate energy which could be advantageous in minimising damage erosion caused by rain droplet impacts.

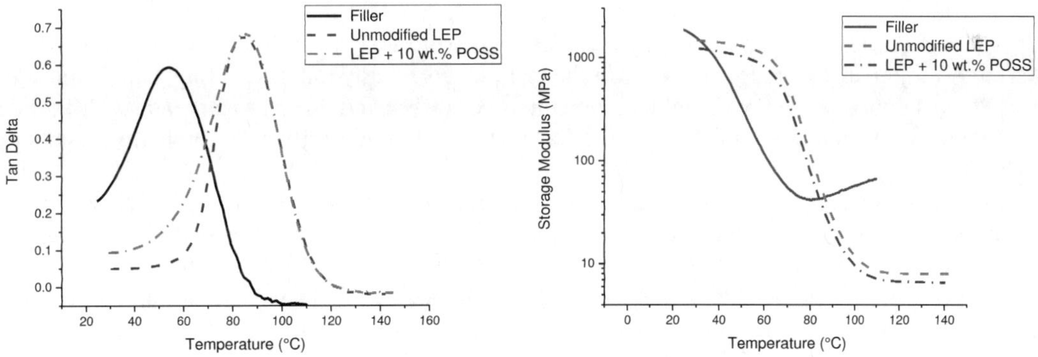

Figure 6. Comparison of filler, unmodified LEP and modified LEP (10 wt. % POSS) DMTA single cantilever temperature scans conducted using a 15 μm amplitude, 1 Hz frequency and while under a dynamic scan from room temperature to 150 °C at 5 °C/min. Left: $\tan\delta$ comparison. Right: Storage modulus comparison.

Frequency sweep data displayed that the unmodified filler $\tan\delta$ value varies greatly in comparison to the two LEP systems showing a much greater decrease over the frequency range (Figure 7). Both LEP samples showed similar trends in $\tan\delta$ values with little variation from low to high frequencies, but modified LEP consistently had a higher value at all frequencies measured, indicating improved energy dissipation. Storage modulus of both LEP samples varied accordingly (loss to storage ratio relationship) and showed similar trends of following a slight increase in storage modulus until stabilising. However, the unmodified LEP showed higher storage modulus values of approximately 300 MPa at all frequencies. The filler storage modulus also increased above both LEP samples.

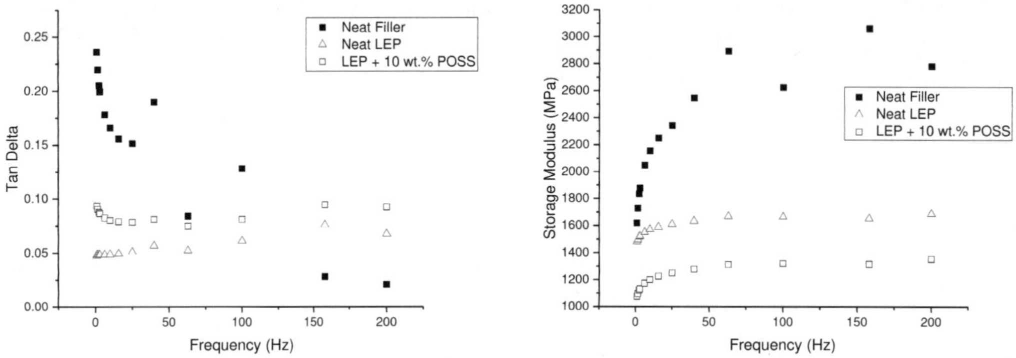

Figure 7. Comparison of filler, unmodified LEP and modified LEP (10 wt.% POSS) DMTA single cantilever frequency sweeps conducted under 0.1 to 200 Hz at an amplitude of 15 µm under ambient conditions. Left: tan δ comparison. Right: Storage modulus comparison.

4. Discussions

4.1. Nanoindentation

The use of nanoindentation offers a relatively fast, repeatable and non-destructive technique for evaluating mechanical properties of solid materials [10, 14]. Although initially used for metallic materials, its use in polymer characterisation has become more common. For the measurements to be representative of the bulk properties, homogeneity of the micro structure is required [14].

The difference between surface and cross section results could indicate either the general rule of omitting readings below 10% the total layer thickness to avoid substrate influence does not apply to this material. Alternatively, there is anisotropy of the material potentially induced during manufacturing. As the LEP cures upon reaction with ambient moisture, the surface and inner properties may differ due to a gradient of different degrees of cure. This effect could be minimised by only comparing results from the same depth.

Traditionally the dispersion of nanoparticles in polymeric materials has been problematic and commonly results in phase separation or agglomeration. The addition of nanoparticles can cause the system to become heterogenous and can explain the increase in variation seen in modified LEP samples. An example of this can be seen in studies using nanoindentation with concrete as a result of its heterogeneous nature. Studies require a large series of indentation tests (100+) to reach convergence of mechanical properties with an error of less than 5%.

Moduli were reported through the study but the Poisson's ratio of the LEP was not obtained during this study and was assumed from literature. As a result, there was no difference in value used between both LEP samples which potentially could be incorrect as the addition of crosslinks will alter the material properties, especially at high POSS concentrations. Previous work has also highlighted that the Poisson's ratio of PU varies depending on the strength of impact varying between 0.39 for weak impacts and 0.5 for increased impact strength which was not accounted for in this study [12]. This parameter could be of importance to tune rain erosion properties.

Varying loading rate showed limited trends due to large error bars, especially in the modified LEP samples. However, a correlation between increased loading rate and a decrease in STR was observed. Although not directly comparable, loading rate could be compared with strain rate seen from increasing the rotational speed of a specimen. According to literature higher tip speeds are a key factor in defining the severity of rain droplet impacts and as a result LEE damage [15]. This can be explained by the increase in kinetic energy E_K of the droplet:

IOP Conf. Series: Journal of Physics: Conf. Series **1222** (2019) 012016 doi:10.1088/1742-6596/1222/1/012016

$$E_K = \frac{1}{2}mv^2$$

where m is the mass of the droplet and v is the velocity of the particle relative to the blade surface (considered equivalent to blade speed). As a result of the square relationship it can be seen that velocity is the more important factor.

During testing it was observed from the load-displacement curves that stress relaxation was occurring during the hold phase of the test *i.e.* a decreased stress response to the same amount of strain generated in the structure. This can be linked to the viscous response of polymer as they behave in a non-linear and non-Hookean manner. Addition of reinforcements to a system generally slows the relaxation effect. This suggests that any increase in crosslinking, which hinders molecular flow of the polymer near the interface, could be used to tailor this property. However, the effect of this on rain erosion properties need to be further tested. It could be assumed that increasing the stress relaxation of the coating would improve energy dissipation and potentially minimise the effects from the combination of subsequent impacts preventing initiation of damage.

4.2. Dynamic Mechanical Thermal Analysis

The addition of nanoparticles is expected to diminish the intensity of the tan δ response. This is due to the hinderance of molecular motion of the polymer chains, resulting in an increase in the elastic response of the material and a reduction in the viscous response. There is a reduction in the tan δ intensity below 80 °C indicating an increase in elasticity. This could be further explored by the variation of POSS loadings and using localised spectral analysis to quantify dispersion.

The filler displayed a lower T_g value than that of the two LEP systems, with a reduction of the storage modulus observed at the lowest measured temperature of 30 °C. At these temperatures the filler which acts as an interfacial layer between the substrate and LEP coating can begin to lose elasticity near rain erosion test temperatures. The tan δ response shows that modification of the LEP increases damping at lower temperature ranges introducing an additional mechanism of dissipating energy, which could potentially be of benefit to rain erosion performance. This is potentially of interest as it raises questions about test methods that utilise closed vessels. This is supported by thermocouple measurements in a rain erosion test rig which show tests conducted at room temperature can see a 4 °C temperature rise over 20 minutes of testing. This is due to air friction and localised heat generated from the high strain rain impacts, which could reduce the filler elasticity as testing progresses. However, in turbine blades the temperatures are expected to be much lower due to steady state temperature of the environment which necessitates further analysis at lower temperatures.

Frequency sweeps were utilised to investigate higher strains rates occurring during both increased rotational speed (higher tip speeds) and in increasing droplet size due to the increased penetration depth imparted onto the surface upon impact. The data showed that at increasing strain rates, the filler storage modulus increases significantly resulting in an increase in its elasticity and decrease in the viscous response. The Unmodified LEP showed an increase in tan δ as frequency was increased. Modified LEP samples displayed lower tan δ values at all frequencies, with an initial decrease followed by increase in tan δ upon increasing frequency. This shows that the elasticity of the coating system increases with frequency of deformation which would result in less energy being dissipated after a rain impact.

5. Conclusions

This work obtained previously unreported mechanical properties for this commercial PU based coating system (hardness, modulus, and short-term recovery). Results show that the inclusion of POSS improves damping, providing an alternative mechanism for energy dissipation without variation of T_g and minimal loss of stiffness. This presents a way of modifying current coating systems through the incorporation of POSS.

The investigation into varying strain rate at the lowest measured depth showed a correlation between increased loading rate and a reduction in STR, which could be linked to increased rotational speeds in erosion testing. Generally, nanoindentation was difficult for modified samples as it resulted in large standard deviations. A potential reason discussed was the poor dispersion of POSS throughout the samples. The difference between results measured from the sample cross section and surface suggested potential anisotropy, a phenomenon seen already in PU foams [16]. DMTA was used to characterise the individual components of the coating system (both unmodified and modified with POSS) and identified differences between components in terms of frequency and temperature dependency. It is not clear if matching or mismatching these properties aids rain erosion performance. Testing also highlighted the low T_g of the filler which begins to lose its elasticity at low temperatures in comparison to the LEP. This could be a potential issue in closed test environments where operating temperatures can rise due to testing, especially as it is the interface between the outer coating layer and the composite substrate. Generally, results have been supported by preliminary rain erosion test results to be published in future work.

6. References

[1] Kay A. *Blade Leading Edge Erosion Programme (BLEEP): Reducing the Cost Impacts of Wind Turbine Blade Erosion*. Glasgow, 2016.
[2] Chen J, Wang J, Ni A. A review on rain erosion protection of wind turbine blades. *J Coatings Technol Res* 2018; 1–10.
[3] Cortés E, Sánchez F, O'Carroll A, Madramany B, Hardiman M, Young TM. On the Material Characterisation of Wind Turbine Blade Coatings: The Effect of Interphase Coating-Laminate Adhesion on Rain Erosion Performance. *Mater (Basel, Switzerland)* 2017; 10: 1146–1168.
[4] Dalili N, Edrisy A, Carriveau R. A review of surface engineering issues critical to wind turbine performance. *Renew Sustain Energy Rev* 2009; 13: 428–438.
[5] Wiser RH, Jenni K, Seel J, Baker E, Hand MM, Lantz E, Smith A. *Forecasting Wind Energy Costs and Cost Drivers: The Views of the World's Leading Experts*. 2016. Epub ahead of print 2016. DOI: 10.1016/S0379-6779(99)00311-2.
[6] Herring R. Integration of Thermoplastic/Metallic Erosion Shields into Wind Turbine Blades to Combat Leading Edge Erosion. In: *Fraunhofer Offshore Wind R&D 2018*. Bremerhaven, Germany, 2018.
[7] Keegan MH. *Wind Turbine Blade Leading Edge Erosion: An investigation of rain droplet and hailstone impact induced damage mechanisms*. 2014.
[8] Dyer K. Developing Tools to Improve Leading Edge Erosion Coatings. In: *Wind Blade Materials Development Forum*. Hamburg, Germany: TBM, 2018.
[9] Kuznetsova IEE, Zaĭtsev BD, Shikhabudinov AMM, Zaĭtsev BD, Shikhabudinov AMM. Effect of the nanoparticle material density on acoustic parameters of polymer-based nanocomposites. *Tech Phys Lett* 2010; 36: 759–761.
[10] Hainsworth S V., Chandler HW, Page TF. Analysis of nanoindentation load-displacement loading curves. *J Mater Res* 1996; 11: 1987–1995.
[11] Pharr GM. An improved technique for determining hardness and elastic modulus using load and displacement sensing indentation experiments. *J Mater Res* 1992; 7: 1564–1583.
[12] Tsukinovsky D, Zaretsky E, Rutkevich I. Material behavior in plane polyurethane-polyurethane impact with velocities from 10 to 400 m/sec. *J Phys IV JP* 1997; 7: 335–339.
[13] Guillonneau G, Kermouche G, Bec S, Loubet J-L. Determination of mechanical properties by nanoindentation independently of indentation depth measurement. *J Mater Res* 2012; 27: 2551–2560.
[14] Voyiadjis GZ, Samadi-Dooki A, Malekmotiei L. Nanoindentation of high performance semicrystalline polymers: A case study on PEEK. *Polym Test* 2017; 61: 57–64.
[15] MacDonald H, Infield D, Nash DH, Stack MM. Mapping hail meteorological observations for prediction of erosion in wind turbines. *Wind Energy* 2016; 19: 777–784.

[16] Huber AT, Gibson LJ. Anisotropy of foams. *J Mater Sci* 1988; 23: 3031–3040.

Acknowledgments
This work was supported by the Engineering and Physical Sciences Research Council through the EPSRC Centre for Doctoral Training at the Advance Composites Centre for Innovation and Science (ACCIS, Grant number EP/L016028/1), and The Future Composites Manufacturing Hub (Grant number EP/P006701/1). This project was also funded by the Wind Blades Research Hub (WBRH), a joint collaboration between the University of Bristol and ORE Catapult. The author would also like to acknowledge Cainan Long and Cameron Mackie for help with testing.

IOP Conf. Series: Journal of Physics: Conf. Series **1222** (2019) 012017 doi:10.1088/1742-6596/1222/1/012017

Pitch bearing lifetime prediction considering the effect of pitch control strategy

A Lopez[1], A Zurutuza[1], M Olave[2], I Portugal[2], M Muñiz-Calvente[3] and A Fernandez-Canteli[3]

[1] Laulagun Bearings, Harizti Industrialdea 201-E 20212, Olaberria, Spain
[2] IK4-Ikerlan Technology Research Centre, Mechanics Area. Pº. J. M.ª Arizmendiarrieta 2, 20500 Arrasate-Mondragon, Spain
[3] Universidad de Oviedo, Escuela Politécnica de Ingeniería de Gijón, Dep. of Construction and Manufacturing Engineering, C/ Luis Ortiz Berrocal s/n 33203 Gijón, Spain

E-mail: alopez@laulagun.com

Abstract. Ambitious worldwide renewable energy targets are pushing for wind energy to become a mainstream power source. Pitch bearings are some of the most vulnerable components of wind turbines. Inside bearings, damage in the inner raceways is significant due to the rolling contact fatigue (RCF) caused by balls. It is well known that methodologies for RCF recommended by international standards are not suitable for fatigue cases where fatigue mechanism encompass multiaxiality, as well as the loading being non-proportional, leading to unnecessary design over-costs. The objective of this work is to develop a new accurate methodology to predict the lifetime of the raceway for a pitch bearing under RCF failure mode and analyse the effect of the pitch control strategy. This methodology will help both, the manufacturer of the bearing and the manufacturer of the turbine, to design a concept focused on the overall performance of the machine.

1. Introduction

Nowadays climate change and global warming are very worrying issues for humanity. To tackle these problems, there is a rapid growth in utilizing the wind power as an electricity generation [1]. Pitch bearings are some of the most vulnerable components of wind turbines. Theses bearings are supporting high axial and radial loads while transmitting movement between adjacent elements. The Rolling Contact Fatigue (RCF) is a failure mode (Figure 1), which occurs mainly at near-end-of-expected lifetime stages, that produces the initiation of a crack under the contact surface after an alternating process of stresses and deformations that may lead to failure. Since the critical location is not known firstly, the failure time prediction becomes complex [2-4]. Most part of the literature and the standards are focused on the study of shear stresses or strains as the main cause of this failure mode, although the effect of the normal stresses and strains should not be ignored [5-6]. In the case of bearings, the international standard currently uses probabilistic methods for life prediction under RCF, such as ISO281 [7]. All hypotheses of this method are based on the rolling element rotating with a constant speed; if the rolling element oscillates; new hypotheses should be included [8]. The real stress state under the raceway is multiaxial, and as real components are subject to magnitude and direction varying load histories.

Figure 1. Example of the RCF damage in a pitch bearing [9].

In this work, a novel methodology is developed to predict the lifetime of the raceway for a pitch bearing under rolling contact fatigue failure mode. The methodology will consider the loads on the blade root obtained from simulation softwares (e.g. Bladed, Fast), the movement of the pitch angle control strategy, the multiaxial stress-state at different depths of the raceway and the flexibility of the surrounded structures. Considering the new methodology, the importance of the pitch control strategy in the rolling contact fatigue is analysed.

2. Individual pitch control vs Collective pitch control

With increase of wind turbine size the mechanical loading of its structure increases rapidly resulting in extreme values of structural loads and pronounced fatigue. Collective pitch control (CPC) defines the movement of blades simultaneously, however, Individual Pitch Control (IPC) defines the movement of each blade individually. The most important objectives of using the IPC control is to significantly reduce the axisymmetric loads observed in the rotor caused by: wind gradients, the effect of the blades´ shadow, turbulence and other factors such as misalignment effects. In general, applying this control strategy, the observed loads in the wind turbine components are reduced. But for the pitch bearing, the continuous oscillating movement that is applied increases significantly the number of cycles for RCF failure mode, reducing considerably the lifetime [10-12]. In the Figure 2 can be seen the CPC strategy against the IPC strategy.

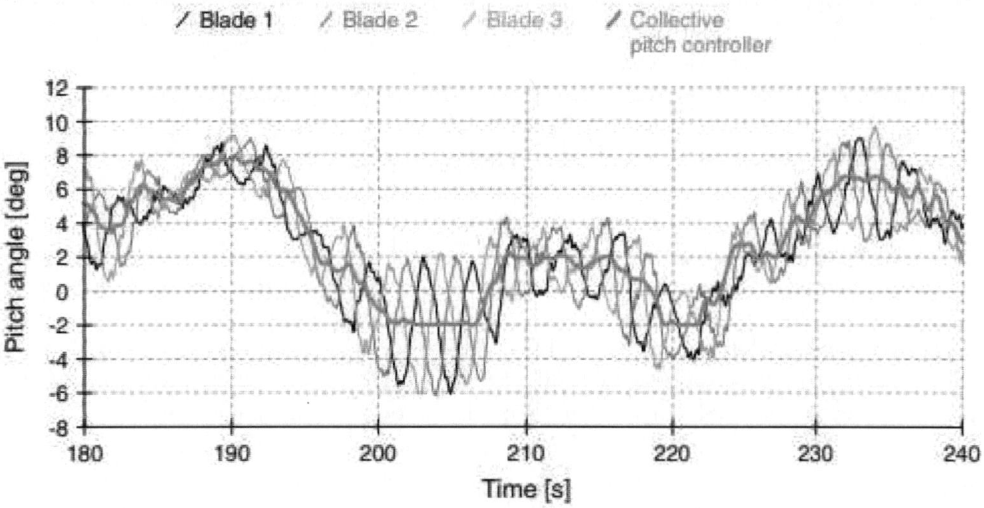

Figure 2. Comparison between CPC and IPC [11].

IOP Conf. Series: Journal of Physics: Conf. Series **1222** (2019) 012017 doi:10.1088/1742-6596/1222/1/012017

3. Methology

This section introduces the methodology developed in this work that describes in detail the fatigue life prediction and the total damage calculation procedure [13]. Furthermore, the different obstacles overcome are explained in terms of the basis of the problem and all the measures that have been taken to fix it. Figure 3 outlines the general procedure followed, from gathering loads to obtaining a damage value.

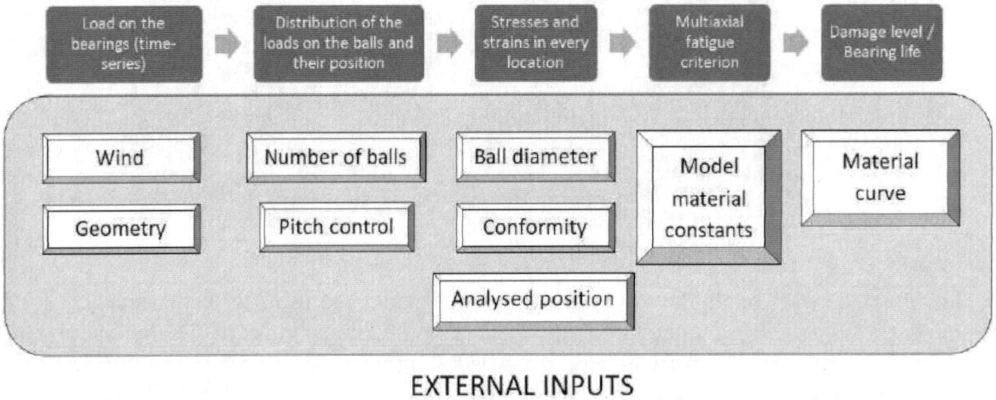

Figure 3. General procedure to obtain the damage value of the pitch bearing.

The methodology is divided in three main steps: the first is concerned with loads, the second describes the process to convert those loads into stresses and strains and the last reveals how to assess bearing damage from the former. These three steps are also divided into 10 sub-steps or blocks, shown at Figure 4 and detailed described below.

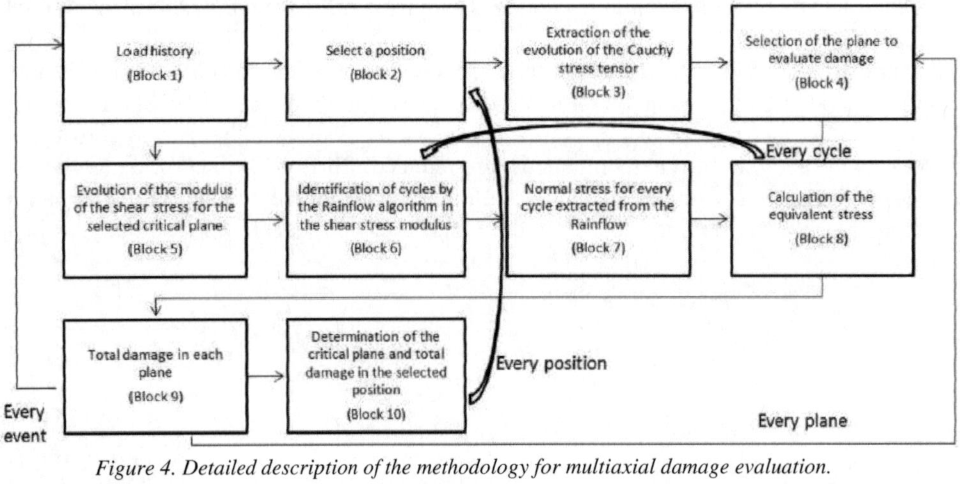

Figure 4. Detailed description of the methodology for multiaxial damage evaluation.

3.1. Load distribution along the time

Before proceeding with the fatigue analysis of the component itself, data information concerning the magnitude and direction of the different forces acting over time is collected. The input data included moments (M_x and M_y), radial forces (F_x and F_y), axial forces (F_z), time step and pitch global position. This kind of data is obtained using software like Bladed (R) (Figure 5), which is a program that provides a sophisticated numerical model of wind turbines and their operational environment.

IOP Conf. Series: Journal of Physics: Conf. Series **1222** (2019) 012017 doi:10.1088/1742-6596/1222/1/012017

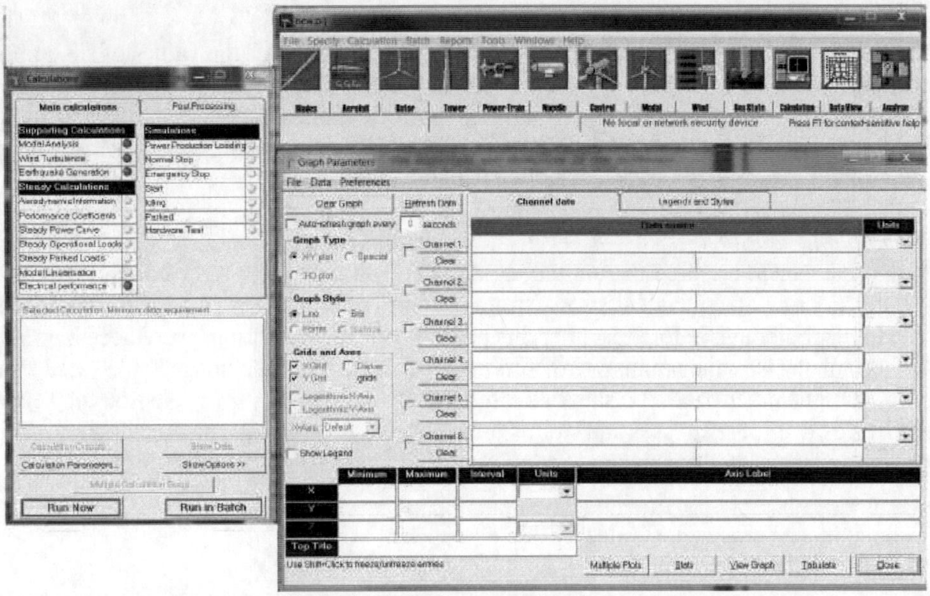

Figure 5. Bladed® program interface.

In order to simplify the calculation of the stresses in the following steps, a linearization of three main forces is performed with the purpose of converting them into a unidirectional varying vector as a function of time. Since the worst load case scenario implies, in principle, loads acting in the three principal directions, the main procedure to achieve this step goal must include superposition of those loads.

The final result of this step is a Q load matrix (Figure 6), which contains data on the loading distribution on every ball, this being a function of the rolling element of the bearing and time. Such data is obtained from complex FE model considering the bearing geometry and stiffness of the surrounding components [13]. These data are essential to calculate the stress or strain values at the subsurface of the bearing.

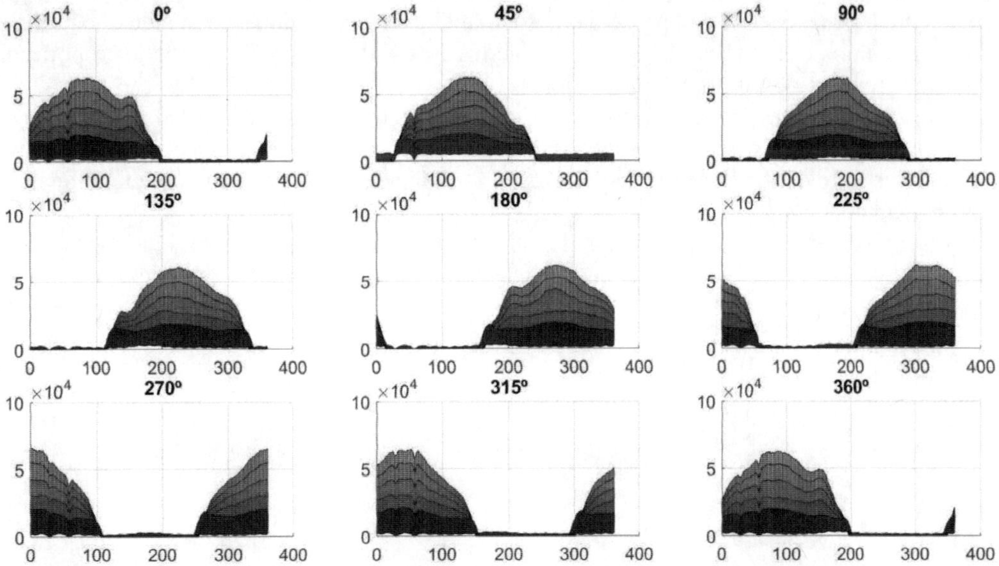

Figure 6. Example of the Q ball load for the first row and the first diagonal for each external load applied in the corresponding angle (x axis: ball position in degrees and y axis: Load [kN]).

3.2. Stresses and strains at the subsurface

An elastic simulation based on the software program ANSYS, provides the values of six components of the stress tensor. The model has to be modified to represent the required ball diameter, conformity and raceway radius. The finite element model requires the following four input parameters: the loads, the depth within the layer, the relative distance between the location of the applied load and any measured position considered and the arc angle of the raceway.

3.3. Critical plane approach

The damage assessment required the multiaxial stress and strain system to be converted into a uniaxial system by considering an equivalent stress, strain, or particular parameter, specifically selected for the fatigue model itself, though before reaching this point, some simplifications were necessary. On the one hand, the normal stresses and strains were combined to create a unique normal stress and strain vector of reference. On the other hand, the shear stresses and strains could be projected on the selected plane and simplified as two shear stresses and strains orthogonal to each other. The Papadopoulos' model [14] is used to define the normal and shear stresses for each plane rotation position (Figure 7).

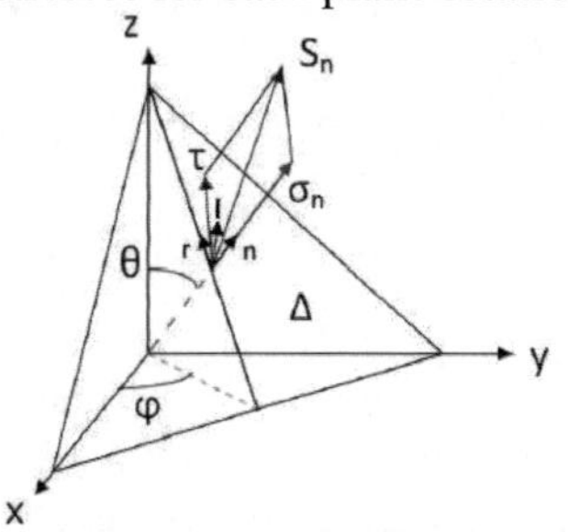

Figure 7. Stress vectors in a rotated Δ plane

Once the calculation is completed, different matrices are obtained as a function of time and each rotation plane, namely normal stresses and strains and shear stresses and strains.

3.4. Cycle definition

The first obstacle to apply the methodology proposed arose from the cycle counting method. Figure 8 a) illustrates how the peaks and valleys of the normal and shear vector do not coincide due to the non-proportional condition of the stress. The same problem arises with the strains due to the proportionality implied by the Young modulus in the elastic field.

The complex randomness of the direction and scale of the loads, however, added complexity in the derivation of stresses and strains according to the multiaxial fatigue models because it resulted in different numbers of cycles for the normal and shear stress and strain due to the unilateral analysis of both vectors (Figure 8 b)).

The resulting vectors of the shear stress and strain (τ, γ) are used to define the initiation and duration of the cycles [13], since the shear stress and strains are the main factors influencing crack initiation in RCF-type loading [14].

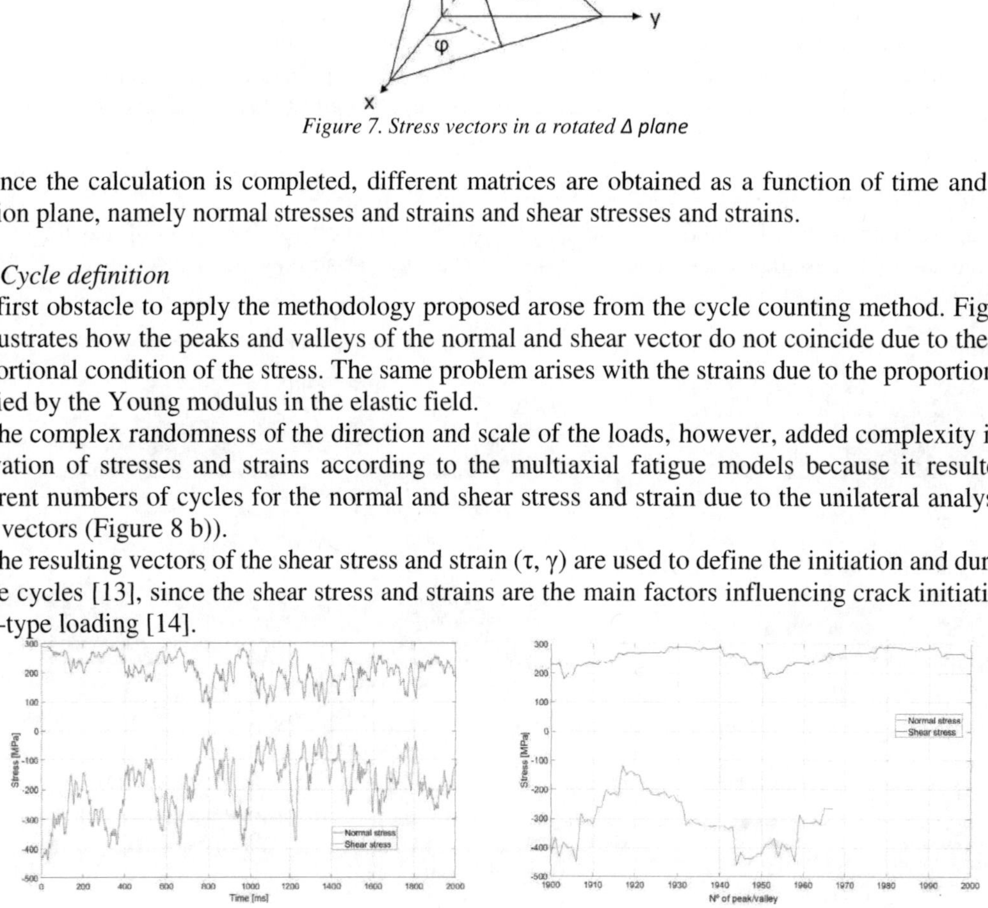

a) b)

Figure 8. a) Normal and shear stress vectors and b) time history of normal and shear stress vectors suppressing intermediate values

3.5. Fatigue models
A number of different multiaxial fatigue models are proposed as a reference parameter to evaluate the damage induced on a certain location and at a certain orientation, as a result of the particular loading acting on the component. Following a brief description of each fatigue model [15]:

3.5.1. ISO
The ISO standard defines the procedure to calculate the basic dynamic load rating for different sizes of the bearings. The standard defines the fatigue life of a bearing using a Wöhler curve and it includes a life factor to adjust the life to the desired probability of survive.

3.5.2. NREL
The design guideline 03 of NREL has modified the ISO standard for bearings with oscillating movement, which corrects the dynamic load rating $C_{a,osc}$ depending on the oscillating angle and the number of rolling elements for each row.

3.5.3. Maximum orthogonal shear stress
Lundberg and Palmgren found that when pure rolling happens on the surface of a component, the critical parameter is the maximum orthogonal shear stress.

3.5.4. McDiarmid
This criterion is included as a critical plane criteria, which are called this way since they consider the plane with the orientation that has the maximum output parameter value, in this case, identified mainly with the variation of the shear stress within a load cycle.

3.5.5. Findley
In Findley's criterion the equivalent shear stress amplitude is calculated considering the maximum value of the normal stress on the plane where the maximum equivalent shear stress is produced.

3.5.6. Fatemi-Socie
This criterion takes into account the shear strain's range instead of the stresses. Besides, it introduces as a parameter the maximum normal stress perpendicular to the plane where the maximum shear strain propagation happens.

3.5.7. Smith-Watson-Topper
The SWT parameter is defined as the maximum value of the product of the strain amplitude and the maximum normal stress between all possible orientations.

3.5.8. Brown-Miller
After observing the fast microcrack propagation, Brown and Miller proposed considering normal and shear strains in the maximum shear stress. Two strain parameters are required to describe the fatigue process since both normal and shear strains reduce the component's life.

4. Case study
The case study of 5MW-NREL wind turbine is selected for the application of the methodology developed in this work. This case study is a reference turbine defined by the National Renewable Energy Laboratory (NREL) of the Department of Energy of the United States (DOE) for the development of conceptual studies to evaluate the wind technology. For this analysis, a simplified 11 operational wind events (each one related to its concurrence value for the entire lifetime of 20 years) are selected for defining the RCF damage on the bearing. Two control types are defined: CPC and IPC.

As stated above, the IPC strategy increases the number of cycles for RCF, this difference can be seen in the Figure 9. The number of revolutions for the same tilting moment increases considerably for the IPC strategy.

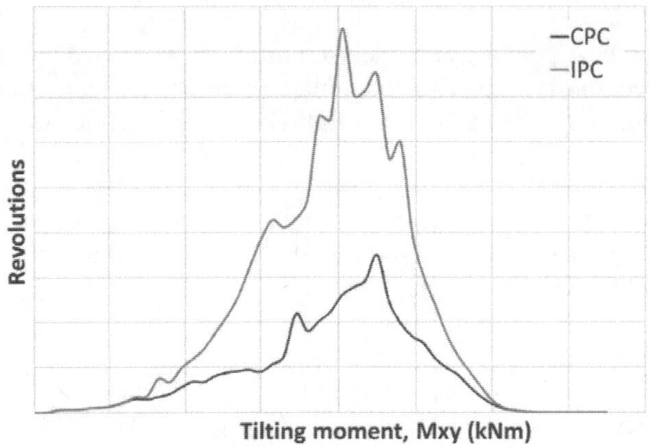

Figure 9. Revolutions against tilting moment for CPC and IPC strategies.

Figure 10, highlights the most critical values for each model, which allows us to make a direct analysis by comparing them. In this comparison, it is important to recall that the orthogonal shear model, which is the model used by ISO Standard, is the less conservative one, thus contradicting the idea of ISO as being a conservative standard. This contradiction could be assigned to the consideration of a safety factor that creates the highest damage value. Among the different models, the McDiarmid model is the most conservative. Note that because the McDiarmid and Findley models are similar though based on different parameters, the ratio between them remains constant. NREL is the most conservative methodology and there exist a wide scatter between the damage values.

Considering the results for the CPC and IPC strategy, as expected, with the IPC strategy the RCF lifetime reduces drastically. The RCF damage applying the IPC strategy is more or less 10 times the damage obtained without this control strategy.

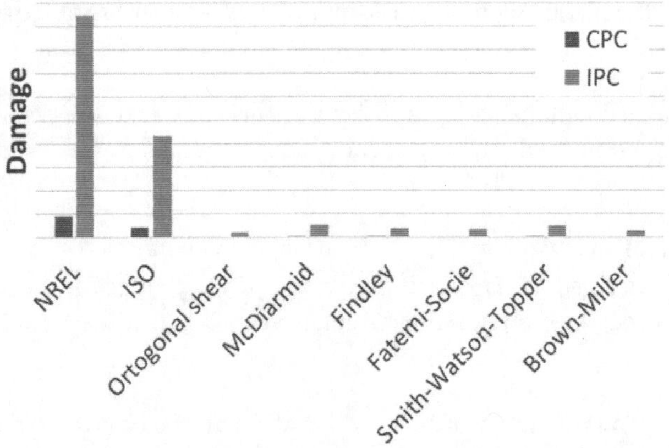

Figure 10. Maximum damage value for each fatigue model considering the pitch strategy.

5. Conclusions

The current standards for calculating the RCF failure mode do not take into account the actual conditions of the bearings during their service life. This methodology reduces the calculation uncertainties that contribute to the use of equivalent loads, the effect of the flexibility of the structures and the simplifications that are made to include the effect of the oscillatory pitch angle control. The new rolling fatigue calculation methodology allows the bearing manufacturer to iterate with the wind turbine manufacturer in:

- The optimization of the bearing to minimize the obtained damage.
- The optimization of the adjacent structures (hub, bolted connections, flexibility changes…), defining the zones to increase the rigidity.

- Defining maintenance strategies: defining the most likely starting points for failure.
- Determining control strategies for a global optimum operation of the wind turbine, detecting the worst wind event, in other words the wind event which generate the higher damage.

Regarding to the comparison between the IPC and CPC strategies, as expected the IPC strategy reduces drastically RCF lifetime. Although this control strategy can improve the lifetime of other wind turbine components, for the special case of RCF of the pitch bearing opposite situation is happened.

References

[1] Lio WH 2018 *Blade-pitch control for wind turbine load reductions* (Springer Theses)
[2] Donzella G, Mazzù A and Petrogalli C 2013 *Eng. Fract. Mech.* **103** 26–38
[3] Ciavarella M and Monno F 2010 *Tribol. Int.* **43 (11)** 2139–2144
[4] Harris TA and Kotzalas MN 2016 *Essential Concepts of Bearing Technology* (CRC Taylor & Francis Group).
[5] Otsuka A, Tohgo K, Ishida H, Yoo S, Shibata M 1992 *J. Soc. Mat. Sci.* **41 (466)** 1136–1141
[6] Papuga J 2015 *Mapping of fatigue damages, program shell of FE-calculation* (Ph.D. thesis)
[7] Halme J and Andersson P 2010 *J. Eng. Tribol.* **224 (4)** 377–393
[8] Schwack F, Stammler M, Poll G and Reuter A 2010 *J. Phys.* **753** 1–10
[9] Budny R 2016 *Improving pitch bearing reliability* (RBB Engineering)
[10] Diaz de Corcuera A, Pujana-Arrese A, Ezquerra JM, Milo A and Landaluze J 2012 *Energ* **5 (4)** 938-967
[11] Bossanyi E 2003 *Wind Energy* **6** 119-128
[12] Prinsen RCP 2011 *Investigation into reduction of the Cost-of-Energy of the Upwind 5.0 MW Wind Turbine using Higher-Harmonic Individual Pitch Control* (Ph.D. thesis)
[13] Portugal I, Olave M, Zurutuza A, López A, Muñiz-calvente M. and Fernández-Canteli A 2017 *Eng. Fract. Mech.* **185** 114–123
[14] Papadopoulos I 1998 *Fatigue Fract. Eng. Mater. Struct.* **21 (3)** 269–285
[15] Portugal I, Olave M, Urresti I, Zurutuza A, Lopez A, Muñiz-Calvente M and Fernandez-Canteli A 2019 *Eng. Struct.* **182** 112-122

IOP Conf. Series: Journal of Physics: Conf. Series **1222** (2019) 012018 doi:10.1088/1742-6596/1222/1/012018

Measuring MNm torques as part of a prototype testing campaign of a high-temperature superconducting generator for wind turbine application in the scope of the Ecoswing project

Hans Kyling, Norbert Eich and Paul Feja

Fraunhofer IWES, Am Luneort 100, 27572 Bremerhaven, Germany

hans.kyling@iwes.fraunhofer.de
norbert.eich@iwes.fraunhofer.de
paul.feja@iwes.fraunhofer.de

Abstract. Modern wind turbines have a driving torque in the range of up to 10 MNm. The measurement of such high torques poses challenges, as there is no standard measurement equipment for this torque level available today. During the EU funded EcoSwing project, the world's first superconducting low-cost and lightweight multi-megawatt wind turbine generator has been designed and tested on the DyNaLab nacelle test rig of Fraunhofer IWES. For this test campaign, a specifically designed torque measurement system has been developed to measure the torque directly at the flange of the device under test in order to evaluate the efficiency of the power train.

1. Introduction

The aim of the EcoSwing project has been to build, test and operate a 3.6 MW generator with superconducting rotor coils. Prior to the field test on a wind turbine (see Figure 1), the prototype generator has been tested in various scenarios in the Fraunhofer IWES DyNaLab test laboratory.

Figure 1: Field installation of the high-temperature superconducting generator (HTS) (Photo: Envision)

One of the key characteristics of a wind turbine is its efficiency. The higher the efficiency of the drive train, the higher is the energy yield. Due to the stochastic nature of the wind, the turbine operates primarily in different operating points. Its efficiency is not a constant value but varies with speed and torque of the drive train. The determination of an accurate efficiency map during field measurements is not possible as the wind field defining the input power can only be estimated. The availability of nacelle test benches as described in [1], however, facilitates a determination of the efficiency map at an even earlier stage within the development process. To identify the efficiency of the drive train, the mechanical input power as well as the electrical output power are needed. However, accurate mechanical torque measurement in the range of MNm is very challenging and not state of the art. The largest torque measurement calibration system in Germany gauges up to 1.1 MNm [2]. Furthermore torque measurement systems in this class are expensive. In the EU-funded Ecoswing project, a novel high-temperature superconducting generator has been developed and tested on the nacelle test bench of Fraunhofer IWES. In order to determine the efficiency map of the generator, a cost efficient project specific torque measurement system has been developed as a demonstrator and integrated into the mandatory hub adaption between the test bench and the generator rotor. The unique torque measurement system is based on a force-lever system in a symmetrical arrangement, such that the driving torque coming from the outer flanges can be measured with three calibrated load cells on its way to the driven generator rotor (see Figure 2).

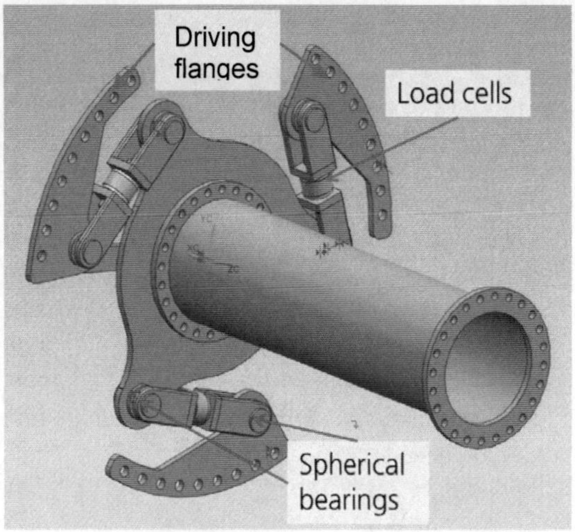

Figure 2: CAD model of the hub adaption with integrated load cells.

2. Kinematics and simulation

The adaption with the integrated measurement system has been designed in order to enable a testing scenario as close to real on-site conditions as possible. This requires the generator rotor to be loaded primarily with torque and shear forces during testing on the DyNaLab test rig. Thrust and bending moments shall not be transferred to the specimen.

2.1. Kinematics of the chosen approach

The design aim has been to create a measurement system that accounts for deformations and avoids significant bending moments in the generator shaft. At the same time, a direct and accurate measurement of the transmitted mechanical torque can be achieved via the load cells placed in the assembly's rods. The rods are spherically mounted at each end in order to create a pendulum support (see Figure 2). Therefore only pure axial loads can be transmitted by the rods and captured by the load cells, while shear forces or bending moments acting on the load cells are avoided. In this way, the torque can directly be measured as the axial force in the rods via the proposed load cells, considering the lever arm of the system as depicted on the left hand side in Figure 3 (blue line). Because of the symmetry of the design the loads in the rods can be assumed to be approximately equally distributed over the three rods.
In the real wind turbine configuration, bending loads are not transmitted to the generator due to the rotor kingpin support and a torque transmission with a flexible carbon fiber main shaft. Shear forces however might occur. To simulate this scenario, the hub adaption with the integrated measurement system was designed to enable two rotational degrees of freedom, making it impossible to transmit bending moments from the test bench load application system (LAS) to the device under test (DUT). There might however be shear forces due to axial misalignment of test rig and generator flange. To evaluate the effects of such loadings due to misalignment of the shaft, a multi-body system (MBS) simulation model has been created for further analysis. Furthermore, the developed design provides an additional degree of freedom in axial direction of the shaft that can be constrained at an adjustable position. This allows for axial positioning of the adaption and shaft and consequently of the generator rotor with respect to the stator. Using this approach possible misalignments of the flange of the test bench load application system and/or the generator mounting adaption can be compensated. The degrees of freedom are shown in Figure 3 on the right hand side.

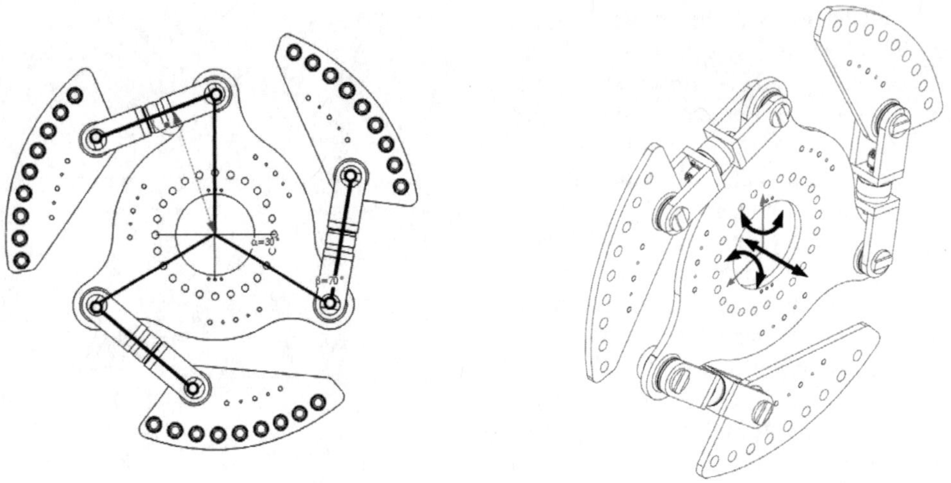

Figure 3: Schematic view of the hub adaption for torque calculation (left). Degrees of freedom of the assembly (right).

2.2. Modelling approach

The proposed design has been verified during the design process by means of a multi-body system simulation (MBS). The following topics have been addressed:

- Verification of the overall kinematics of the hub adaption system (i.e. two rotational degrees of freedom (DOF) as well as one axial DOF)
- Analysis of parasitic and/or unexpected forces on the load cells that do not act in axial rod direction, which is supposed to be the main loading direction for the load cells
- Estimation of shear forces and bending moments in the load cells due to centrifugal forces
- Assessment of the influence of gravitational loadings in a dynamic scenario.

The MBS model has been created in Samcef Field. For a purely kinematic analysis, all bodies are assumed to be rigid. As the solver requires a flexible system instead of a purely kinematic assembly, some artificial stiffness is added to the system: the six spherical bearings at the ends of the force levers are modelled as flexible bushings with both translational and rotational stiffness and damping. The load cells are modelled as flexible bushings as well, neglecting their mass and inertial forces as these are very small compared to the overall system loading. Inertial and gyroscopic forces are still acting on the load cells due to the mass of the rods. The pretension of the load cells is not considered in the analysis as only the resulting operational forces and moments are evaluated.

2.3. Simulation results

To examine the issues listed above, a simple run-up scenario was considered in the simulation as shown in Figure 4. The rotor speed increases from 0 to 12 rpm, which is roughly the rated rotor speed of the given generator, during a time span of 10 s. Simultaneously the torque increases from 0 to 3.6 MNm. After reaching these set points the operating point stays constant for another 10 s. The loading scenario of the model is achieved by imposing a rotational speed to the outer flange, while applying the torque to the inner flange of the adaption. Besides gravity no other forces or misalignments are considered in the first loading scenario.

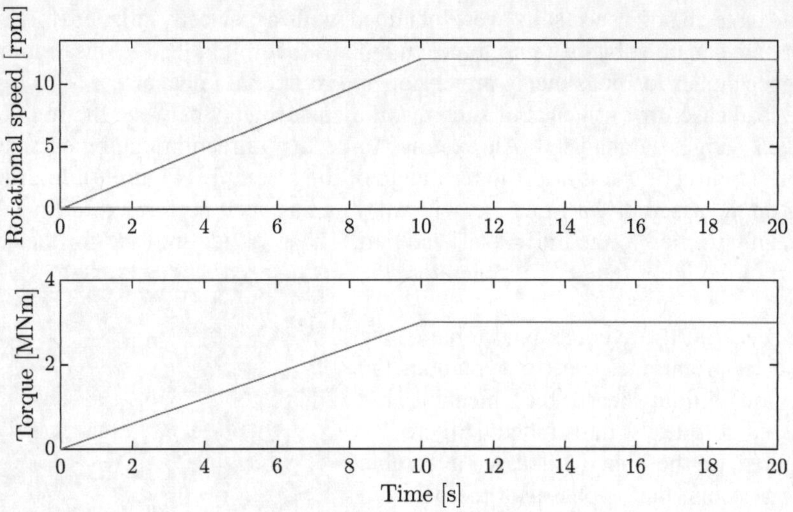

Figure 4: Loading scenario for the simulations.

The simulation results in Figure 5 show that the main axial force in each load cell F_x is almost constant over the rotational period and does only show negligible variations (see detail). The out-of-plane force component F_y is almost zero and the radial force F_z stays at about 250 N for the maximum rotational speed of 12 rpm. This component, however increases slightly, if the load cell mass is included, but it will not affect the results in a significant way. It is also visible that the bending moments experienced by the load cells are small and will not damage the load cells or compromise the measurements, as expected due to the pendulum support design.

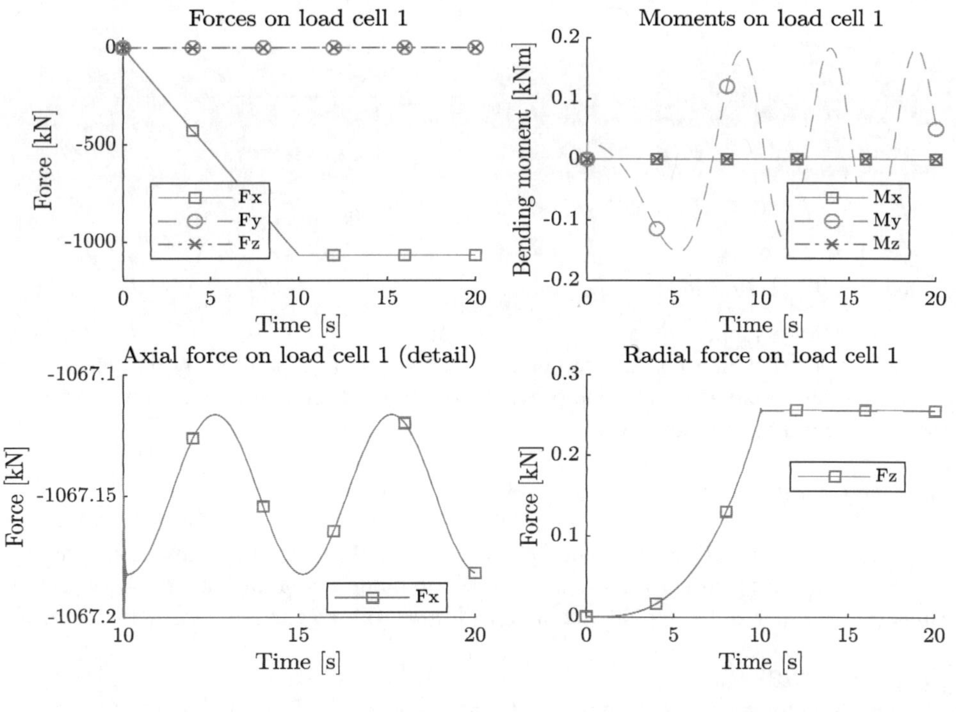

Figure 5: Results of the first load case "perfect axial alignment". Forces and bending moments on load cell 1.

It is important to note that these results were obtained with a perfectly aligned force-lever-system assembly. If there are considerable bending forces/ moments that might lead to misalignments, the load cells can experience higher loadings due to pretension and rotational imbalances.
In a consecutive load case, the influence of such radial displacements between the rotation axes of the driving and driven flanges is analysed. An external force in y-direction combined with an external bending moment in z-direction acts at the inner flange of the assembly (Figure 6). In order to evaluate realistic loading on the assembly, a finite element analysis has been performed to find the equivalent stiffness of the shaft attached to the inner rotational part. These values are used for the stiffness of the rotational support of the inner flange, i.e. generator side of the assembly.
For this loading scenario, the average axial force of the load cells at constant rotational speed stays about the same as for the case without shear forces, meaning that the same averaged torque is transmitted (Figure 7). However, compared to the first load case, a notable sinusoidal loading occurs, that is phase-shifted between the three load cells by 120° and oscillates with the frequency of drive train rotation (1P frequency). Again the non-axial loading on the load cells is negligible due to the pendulum support of the rods.

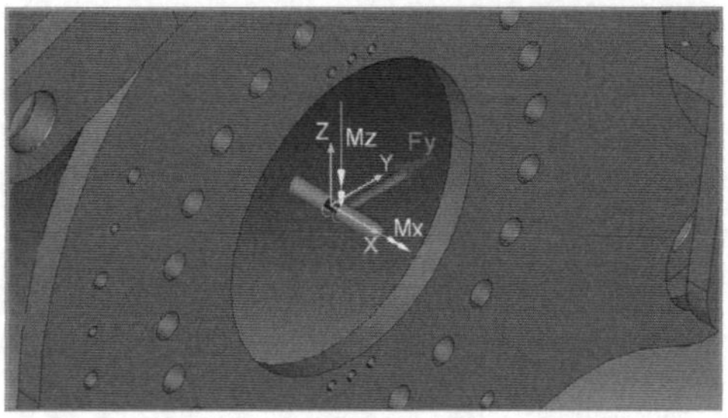

Figure 6: Loading scenario for the second load case. Driving torque, bending moment Mz (yellow) and shear force Fy (brown).

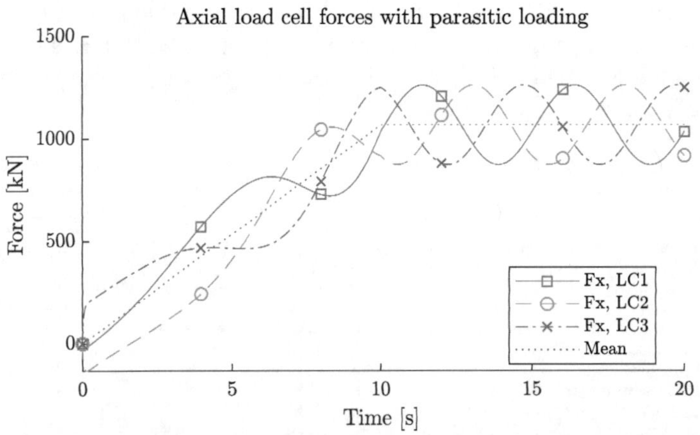

Figure 7: Results of the second load case "misalignment due to radial forces". Axial forces F_x on all load cells.

The reaction of the bearing support shows that the external shear force F_y is mainly supported via the force-lever assembly. The applied bending moment M_z, however, is supported by the generator-side bearing support, as the assembly is designed to provide a rotational degree of freedom around the z-axis and consequently cannot transmit this load.

2.4. *Kinematic analysis and mode shapes of the system*
By means of a modal analysis of the MBS assembly, the kinematic degrees of freedom of the system have been verified. As intended by the design of the hub adaption, the simulation shows three main DOF modes: two rotational modes (around y and z axes) and one mode that allows axial movement of the inner flange against the other, allowing for an axial positioning of the generator as described above.

2.5. *Conclusions of the MBS simulations*
The simulation results show that the desired kinematic behavior was fulfilled with the designed system. The loading of the load cells in axial direction can be expected to comply with the static results of the analytic calculation even for the dynamic rotational loading scenario. It was further shown that only

minor non-axial loadings act on the load cells, i.e. unexpected damaging of the load cells due to parasitic loading should not occur. The harmonic loads from gravity are negligible compared to the target loads. Even with significant shear forces, e.g. caused by a misalignment of the rotational axis, the load cells do not experience unforeseen or critical loading conditions.

3. Preparatory tests

As all rod forces are needed to calculate the resulting torque, a force transducer is installed in each rod. The load cells have been calibrated in-house in a calibrated compression-testing machine taking the whole measuring chain into account (Figure 8 left).

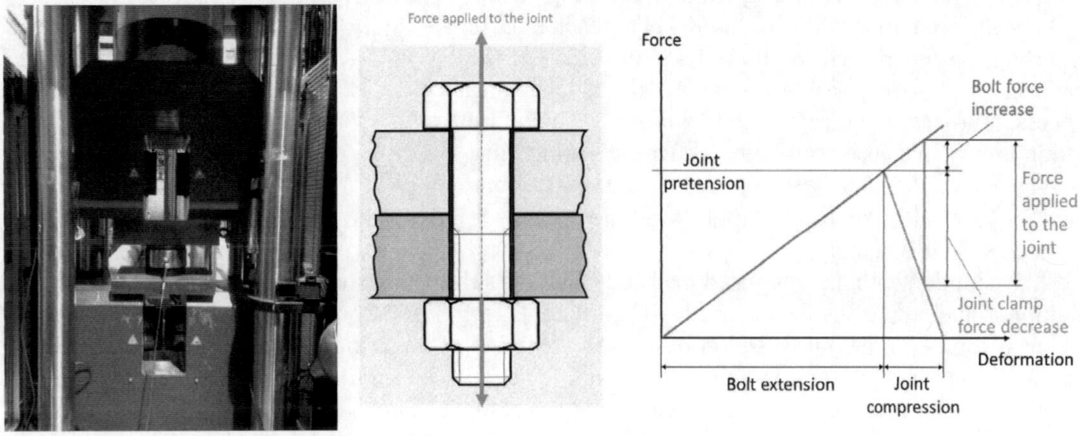

Figure 8: Compression testing machine with load cell (left). Bolted joint diagram (right)

The deployed force transducer can only measure compressive forces, but the rods needed to be designed to transmit tension forces. To enable the rods to measure operational tension forces, the load cells are adequately pretensioned with a centric bolted connection. Knowing the load factor Φ and the pretension of the bolt connection F_V, the operational load F_B on the rod can be determined by the change of the clamping force F_{Kl} on the load cell. The correlations can be expressed through the bolted joint diagram (see Figure 8 right). The operational load F_B can be calculated using the equation:

$$F_B = \frac{F_V - F_{Kl}}{1 - \Phi}$$

The pretension F_V is the reading of the load cell after assembly without operational load. During operation the readings from the load cells correspond to the joint clamp force F_{Kl}. The load factor Φ is dependent on the elastic behavior of the material in the joint. Two approaches have been used to determine the load factor. On the one hand an analytical calculation was done as described in [3] and on the other hand experiments as described below were executed in accordance to the force calibration standard [4]. At first a deviation of 17% between both approaches was found. A finite element based calculation revealed an improper use of the load introduction factor *n* in the analytical method. In this way, the deviation to the experimental results have been drastically reduced. This topic is currently further investigated. For the operational torque measurement the more confident load factor results from the experiments have been chosen as they have been acquired using calibrated measurement equipment.

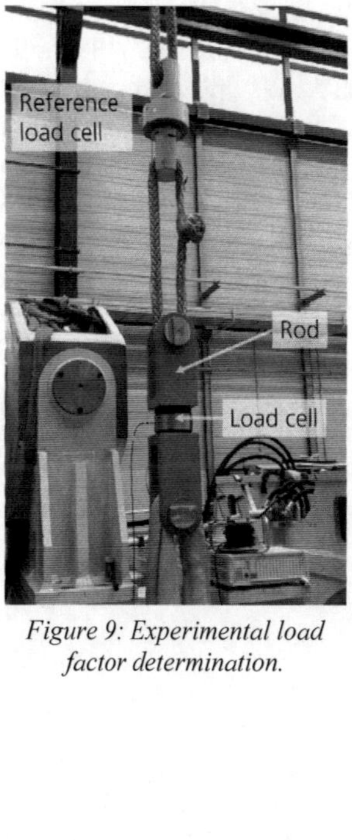

Figure 9: Experimental load factor determination.

The experiment consists of a test setup of the assembled rod in line with a calibrated reference load cell as shown in Figure 9. This setup allowed only a calibration in a partial load range (up to 27% of the total load cell measuring range) due to the limited load application capacity. The reference load cell enables the evaluation of the load factor Φ, because in contrast to the later operational use, now the operational force F_B is also measured. In accordance to [4] load steps have been applied to each of the rods and the data has been read out with two repetitions. The load levels have been held constant until steady-state has been reached as the ropes are prone to creeping. Readings of the pretension have been taken during the assembly of the rods on the ground. According to the readings the pretension was set to 1400 kN for each rod. As Superbolt tensioners have been used a fine setting of the pretension has been possible using a balanced tightening process. The screw connection tends to creep, therefore setting times have been considered. After the setting time an averaged reading has been taken as the clamping force input for the load factor Φ calculation. After the calibration process the rods have been directly mounted to the drive train assembly.

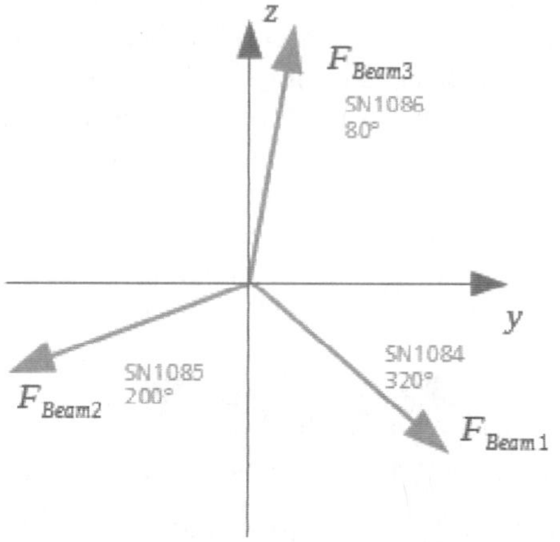

Figure 10: Rod / beam operational forces

The hub adaption with the integrated load cells facilitates the determination of the acting shear force in addition to the torque. As the connecting shaft with its length of 2.8 m is a relatively long force lever, the shear forces on the hub adaption have been identified as the critical loads for the integrity of the rotor shaft support. The rods as well as their axial forces have an angular distance of 120° to each other (Figure 10). The shear force can be calculated as the resultant force vector of the different rod forces. There are two passive effects to result in a shear force: eccentricity of the rotational axes and gravitational force, both can be quantified with the given instrumentation.

4. Assembly

For the hub adaption, a very thorough tolerance planning as well as quality assurance is needed as both aspects influence directly the quality of the measurement system. For the same reason, the assembly procedure has been planned in a step-by-step procedure. At certain steps a laser tracker system has been used to match the real positioning of the components to the designed one (see Figure 11).

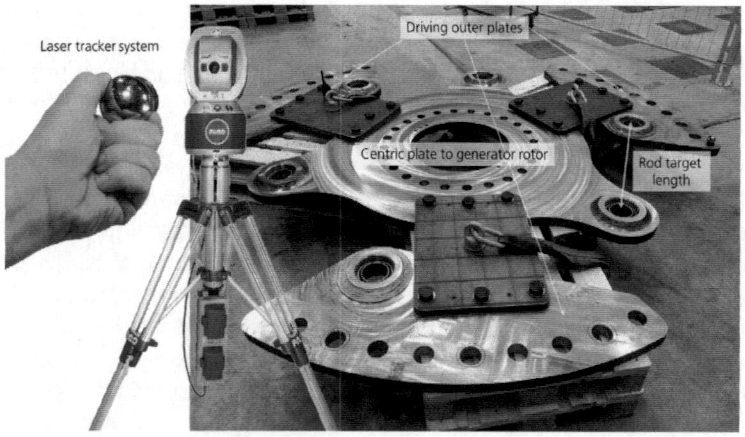

Figure 11: Assembly process of the hub adaption with help of a laser tracker

The three rods connecting the driving flange of the test bench with the rotor of the generator have been preassembled and their lengths have been controlled with a laser tracker (length x_rod in Figure 12). By means of shims the lengths have been adjusted to the distance measured between the connection points on the outer driving plate and the inner centric plate (x_dist in Figure 12), which is connected to the rotor of the generator.

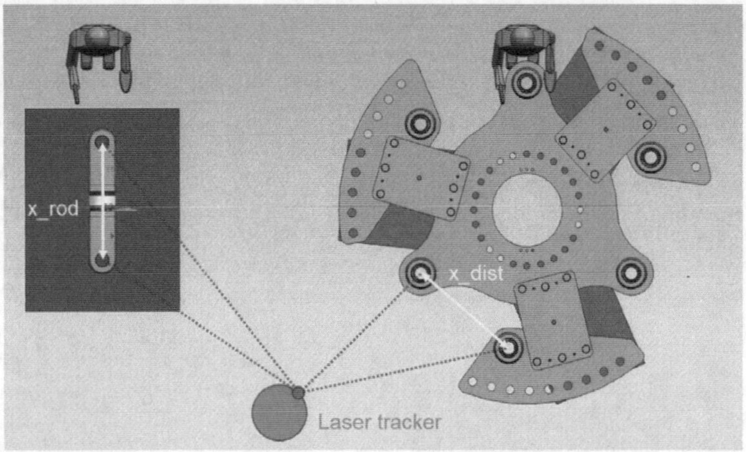

Figure 12: Procedure to determine necessary shim adaption for each rod.

By these aforementioned means, the hub adaption can be further assembled with minimum pre-stress (see Figure 13), so that the impact on the measurement quality is minimized.

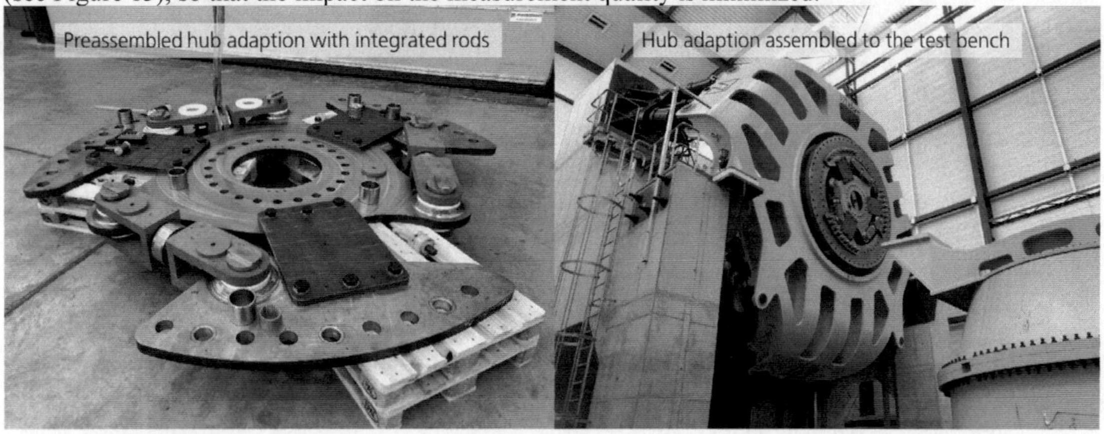

Figure 13: Hub adaption assembled prior to lift (left). Hub adaption mounted to the test bench (right).

5. Test campaign conduction and results

The assembly process of the specimen to the test bench has been done without transmission of any bending moments as strain gauges on the connecting shaft have indicated (see Figure 14). A non-critical shear force has been measured.

During conduction of the test campaign, no critical shear forces and thus bending moments have been monitored. A safe and almost free of parasitic loading operation has been possible. Figure 15 depicts the nacelle test bench with a dummy specimen adapted. This test bench in general can apply torque and parasitic loads onto the specimen, even though in the Ecoswing project the operation has been limited to the

Figure 14: Final assembly of the hub adaption.

IOP Conf. Series: Journal of Physics: Conf. Series **1222** (2019) 012018 doi:10.1088/1742-6596/1222/1/012018

torsional degree of freedom. During the Ecoswing project, a set of individual tests was performed. Unfortunately the specimen arrived later than expected and still needed to be finally assembled on site. This circumstance resulted in a much reduced testing scope compared to the original planning, such that the efficiency measurement has been limited to a couple of hours and below rated conditions. A more extensive testing of the hub adaption load measurement system has not been possible.

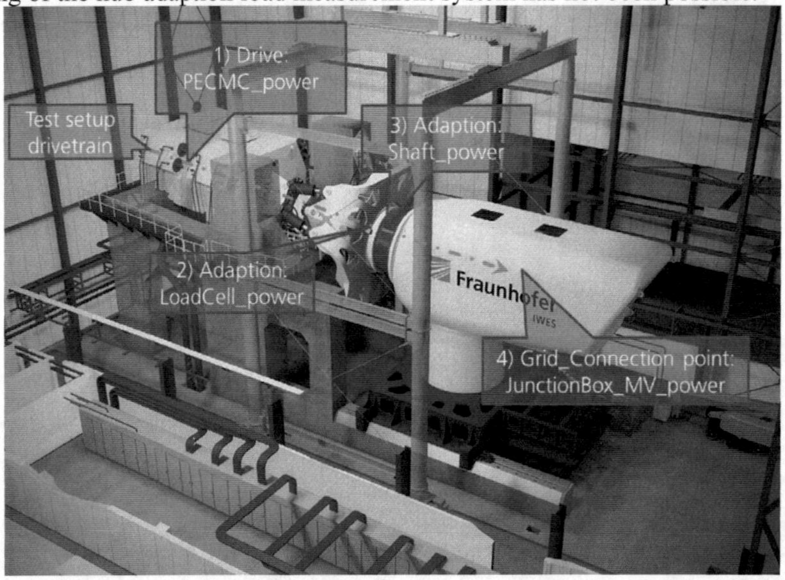

Figure 15: General overview of the whole test setup and indication of power measurement positions.

As mentioned earlier, the major aim of the integrated torque measurement is to determine the efficiency map of the high-temperature superconducting (HTS) generator. Figure 16 depicts the measured powers during the power test after low-pass filtering at the different positions along the drivetrain:

1. Test bench motor (signal PECMC_power_tp, dark blue line). The power was determined from built-in sensors from the converter. The measurement is afflicted with a high uncertainty, which is estimated as approx. 10 %.
2. Hub adaption (signal LoadCell_power_tp, red). The power was determined using three load cells as described in this paper. The uncertainty calculated according to [5] is 2.4 %.
3. Shaft adaption (signal Shaft_power_tp, light blue). The power was determined using strain gauges glued onto the shaft adaption. Again in this case, the estimated uncertainty is as high as 10 %, which must be taken into consideration in the assessment of the results.
4. Grid connection point (signal JunctionBox_MV_power_tp, pink). The power was determined by measurement of voltage and current at the interface between transformer and IWES grid simulator. The measurement in this point is considered the most accurate with an estimated uncertainty of 0.5 %.

One would expect the highest torque on the test bench motor (black) and the lowest one at the grid side of the specimen (pink), as all losses occur between these points. In contrast to this, the highest powers have been measured on the rotor adaption (red). This is not plausible, but the deviation of +9.3% remains explicable with the before mentioned large uncertainty of the test bench motor power. The measured power on the shaft adaption (light blue line) is 0.5 % below the reference power of the drive. The signature is plausible, even though one would expect a larger difference resulting from the losses in the test bench drive. The measured power on the grid connection point (pink) is, as expected, the lowest of all power values. It is 4.4 % below the reference input power on the driving side. Also in the comparison of these two signals, one would expect a larger difference as in this measurement all losses from the test bench and the specimen (generator, converter and transformer, excl. generator cooling system) are included. Again, this deviation from the expectation can be attributed to the low accuracy of the power measurement at the test-bench drive.

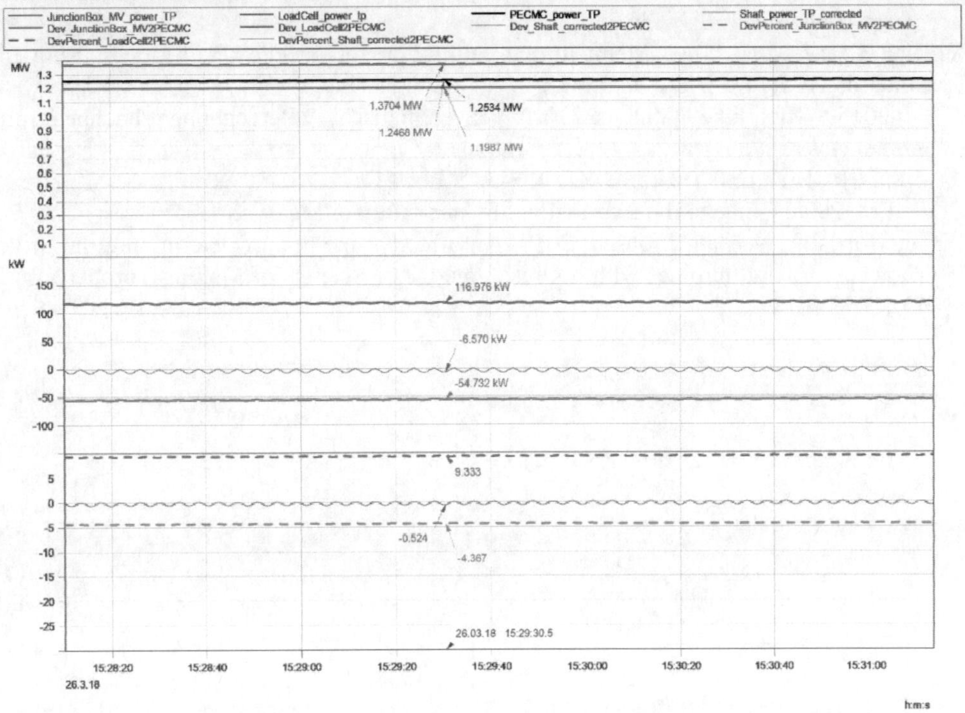

Figure 16: Measured powers for a stable operation point at the different positions along the drivetrain.

In view of the large uncertainties of the power values obtained in positions 1 and 3 (i.e. the test bench motor and the shaft adaption), it is not meaningful to determine an efficiency based on these. Taking the load cell power measurement as reference instead, a specimen system (generator, converter, transformer) efficiency of 87.5 % (1.199 MW / 1.37 MW) with an approx. uncertainty of 2.5 % is determined.

6. Conclusion

In this paper, a comparatively inexpensive way to measure torque in the MNm range is presented. Its technical feasibility could be proven during the realization of the test campaign of a HTS generator on the DyNaLab nacelle test rig of Fraunhofer IWES as part of the project Ecoswing. Based on pre-test and uncertainty calculations according to [5], the uncertainty of the proposed torque-measurement method was determined to remain below 2.4%. This is an improvement compared to the use of strain-gauge measurements on the shaft adapter to the specimen. Nevertheless the achieved accuracy is not sufficient to determine the efficiency of a specimen with a reasonable certainty. Some potential for improvements has been identified, which will be further investigated in future nacelle test bench projects. An alternative approach for the challenging efficiency determination of machines with MNm torques and low rotational speeds was developed at IWES and described in [6]. While it requires additional testing scenarios, it has the advantage of a high accuracy. This provides a good basis for further evaluating and improving the torque measurement method described in the present paper.

Horizon 2020
European Union Funding
for Research & Innovation

EcoSwing has received funding from the European Union's Horizon 2020 research and innovation programme under grant agreement No. 656024.

IOP Conf. Series: Journal of Physics: Conf. Series **1222** (2019) 012018 doi:10.1088/1742-6596/1222/1/012018

References

[1] Pilas M and Riezu M, 2017, Functional prototype testing of an 8 MW wind turbine, *German Wind Energy Conference DEWEK*
[2] Schlegel C and K H, 2016, MNm torque calibration for nacelle test benches using transfer standards, *ACTA IMEKO Vol. 5 No. 4*
[3] VDI, 2015, Richtlinie VDI 2230 - Systematische Berechnung hochbeanspruchter Schraubenverbindungen
[4] DAkkS, 2010, Richtlinie DAkkS-DKD-R 3-3 - Ablauf D
[5] ISO/BIPM, 2008, Guide to the Expression of Uncertainty in Measurement
[6] Zhang, Hongkun; Neshati, Mohsen, 2018, An effective method of determining the drive-train efficiency of wind turbines with high accuracy, The Science of Making Torque from Wind, TORQUE 2018

IOP Conf. Series: Journal of Physics: Conf. Series **1222** (2019) 012019 doi:10.1088/1742-6596/1222/1/012019

Serrations effect on the aerodynamic performance of wind turbine airfoils

Elena Llorente * **and Daniele Ragni** †

* Nordex Energy Spain S.A.U., Avenida Ciudad de la Innovacion, 3, 31621 Sarriguren, Navarra, Spain

† Delft University of Technology, Kluyverweg 1, 2629 HS Delft, The Netherlands

E-mail: ellorente@nordex-online.com

Abstract. A numerical and experimental analysis of the trailing edge serrations effect on the aerodynamic performance of the NACA64$_3$418 airfoil is presented. 3D Reynolds-averaged Navier-Stokes simulations (RANS) of the flow over the airfoil with and without trailing edge serrations have been carried out with the Transitions SST turbulence model. From these calculations an empirical law for the prediction of the aerodynamic behaviour of the airfoil with serrations is derived. The law is validated through wind tunnel experiments. The NACA64$_3$418 airfoil has been tested with trailing edge serrations installed with different flap angles to analyze the effect of the parameter on the performance of the airfoil. Results show that the trailing edge serrations cause a significant increment in lift coefficient with low penalties in drag.

1. Introduction

Reducing the noise impact of wind turbines is a main topic in the wind industry. Serrated trailing-edges are very popular in industrial applications due to their simple manufacturing, installation and maintenance compared to other solutions like brushes or porous materials [1, 2, 3]. An increasing number of wind turbines manufacturers are offering, in their blade technology, the serration devices as a solution for obtaining high power production with low noise impact, including its effect in the blade design [4]. New business opportunities have been opened due to the possibility of using these devices for retrofitting existing machines that are already operational with "ad hoc" tuning of the operational regimes for noise regulations limits. During the last years a significant number of studies have focused on understanding of the noise reduction mechanism of the trailing edge serrations [5, 6, 7] but a clear gap in knowledge about their impact on the aerodynamic performance of the machine exists. This study focuses on the analysis of the effect of the trailing edge serrations on the aerodynamic performance of a wind turbine airfoil, the NACA 64$_3$418. 3D RANS simulations have been performed using the Transition SST model [8] to quantify the changes in the aerodynamic performance of the airfoil due to trailing edge serrations. As a result of this analysis an empirical law for predicting the lift coefficient of the airfoil with serrations from the original data has been derived. This law includes the effect of serrations flap angle (the angle between the chord of the airfoil and the trailing edge extensions). This predicted law has been validated using wind tunnel experiments carried out at TU-Delft. The NACA 64$_3$418 airfoil has been tested with and without trailing edge serrations installed with different flap angles. The influence of these devices and its assembly angle on the aerodynamic behaviour of the original airfoil has been analyzed.

IOP Conf. Series: Journal of Physics: Conf. Series **1222** (2019) 012019 doi:10.1088/1742-6596/1222/1/012019

2. Computational test-case

Three dimensional simulations of the airfoil NACA$64_3$418 with and without trailing edge serrations have been carried out using the Transition SST model from ANSYS [8]. C-types structured and non-structured meshes have been generated for the discretization of the 3D domains using ICEM CFD. The airfoil has been simulated with a chord of 1 m and its shape has been modified from sharp trailing edge to a thickness of 0.2% of the chord. Serrated trailing-edges have been modeled by periodic inserts of length of 2h = 0.187 mm and amplitude of $\lambda = 0.140$ mm, as shown in Figure 1. A ratio between streamwise and spanwise length of the serrations of $\frac{\lambda}{2h} = 0.75$ has been selected for best performance in noise reduction in the mid-frequency range with small noise penalties in the high one [9].

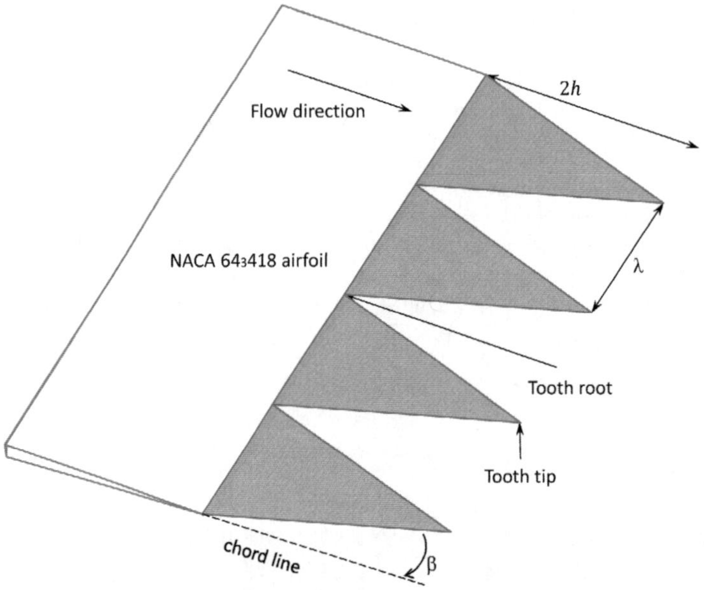

Figure 1: Scheme of the serrations with length of 2h and amplitude of λ.

3. Numerical results

In this section the aerodynamic coefficients for the original NACA$64_3$418 airfoil with and without trailing edge serrations are presented. It must be noted that for all the simulated cases, the aerodynamic coefficients are obtained using the chord of the original airfoil without including the length of the trailing edge serrations. In Figure 2 the lift and drag coefficients for different angles of attack for the airfoil with and without trailing edge serrations are compared. These results show that for all the studied angles of attack the serrated extensions do not significantly change the lift coefficient of the airfoil but cause an increment in the drag coefficient.

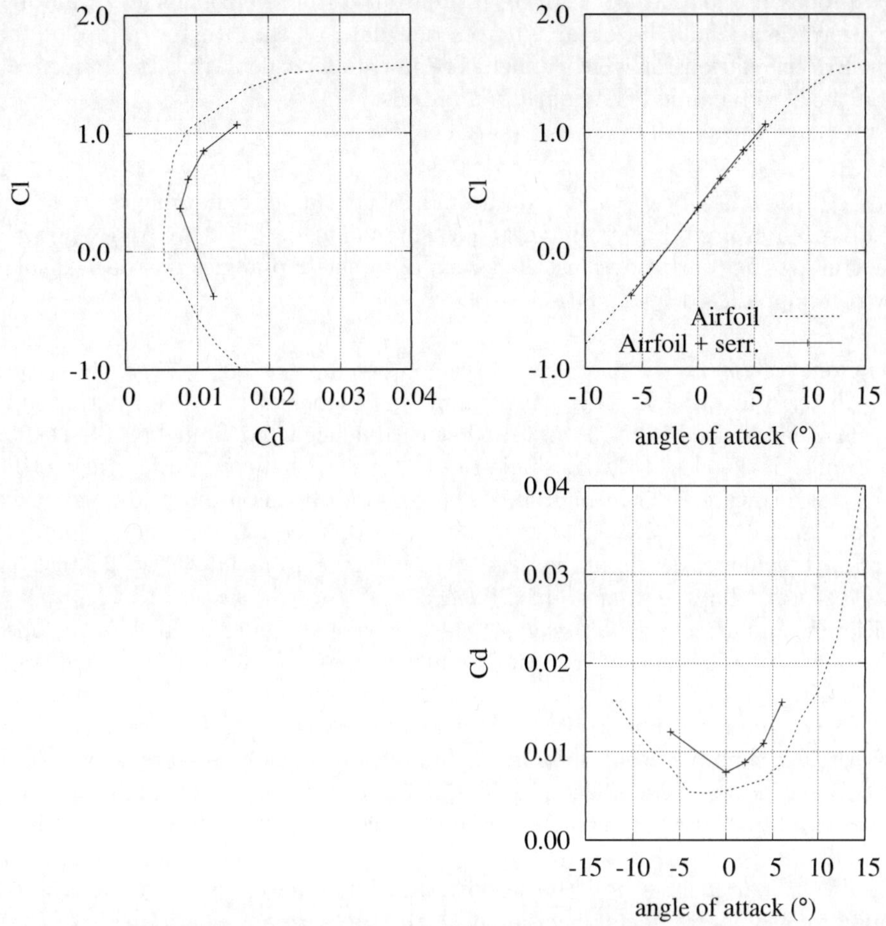

Figure 2: Polar data of the NACA64$_3$418 airfoil with and without trailing edge serrations.

The simulated results show that the trailing edge serrations affect the aerodynamic behaviour of the airfoil in which they are installed. To predict this impact trying to avoid the costly CFD simulations the linear thin airfoil theory [10] has been used to derived a law that can calculate the lift coefficient of the serrated airfoil using the data of the original profile. With this theory, first the lift coefficient of the airfoil without any extensions is derived in Equation 1:

$$Cl_a = 2\pi\alpha + Cl_0 \tag{1}$$

Applying the same methodology the lift coefficient of the airfoil with trailing edge serrations can be obtained using the Equation 2.

$$Cl_{as} = 2\pi\alpha(1 + l_s/c) + 2\pi\beta l_s/c + Cl_0 \tag{2}$$

where the lift coefficient is obtained as a function of the angle of attack (α), the flap angle (β), the length of the serrations (l_s), the chord of the original airfoil (c) and the lift at zero angle of attack of the original airfoil (Cl_0).

IOP Conf. Series: Journal of Physics: Conf. Series **1222** (2019) 012019 doi:10.1088/1742-6596/1222/1/012019

With this proposed equation an accurate estimation of the lift coefficient of an airfoil with trailing edge serrations could be obtained from the data of the original airfoil reducing the necessity of CFD calculations or wind tunnel experiments for calculating the influence of these extensions on the aerodynamic behaviour of an airfoil.

4. Validation

The previous equation has been validated through wind tunnel experiments. In this section the results obtained from the experimental measurements of the NACA$64_3$418 airfoil with and without trailing edge serrations installed with different flap angles has been displayed and compared with the predicted by Equation 2.

4.1. Experimental set-up

In order to validate the previous law, wind tunnel experiments have been performed. These tests have been accomplished in the low speed low turbulence wind tunnel of TU-Delft. It is an atmospheric tunnel of the closed-throat single-return type, with a contraction ratio of 17.8. The test section is 1.80 m wide, 1.25 m high and 2.60 m long. It has a 2.9 m diameter six-bladed fan driven by a 525 kW DC motor, giving a maximum test section velocity of about 120 m/s. The free-stream turbulence level in the test section varies from 0.02% at 25 m/s to 0.07% at 75 m/s. Electrically actuated turntables flushed with the test-section top and bottom wall provide positioning and attachment of several two-dimensional models, which can be operated at several angles of attack. The centre of rotation for all the models is set at half chord distance. The free-stream velocity is monitored by evaluation of the free-stream dynamic pressure q_0 through a calibration curve from the tunnel contraction control pressure Δp_b, measured as the difference between the settling chamber total-pressure and the wall (static) pressure at a station about half-way in the contraction. The static pressure at the contraction is evaluated as the average of 4 static-pressure taps distributed at the circumference of the tunnel. In this facility, lift and drag are typically measured with a six-degrees of freedom multi-component balance connected to the model. In this particular study, the aerodynamic forces and pressure distribution could be easily obtained from pressure ports distributed on the pressure and suction side of the models. The drag of the airfoil is instead obtained from integration of the measurements from a wake rake of Pitot tubes installed at a distance of half meter downstream the airfoil. The free-stream total-pressure is calculated as an average of all the wake-rake total-pressures outside the wake of the airfoil. Finally the free-stream static-pressure is calculated as the difference between the total pressure from the wake rake and the dynamic pressure from the tunnel controller. The maximum Reynolds number achieved for the tests was $Re_c = 1 \cdot 10^6$. The NACA$64_3$418 model was built into an aluminum wing with 250 mm of chord and a span of 1225 mm. In this model 50 pressure tabs were installed for instrumentation, 25 in each side for measuring the forces over the model, the pressure tabs features a 13-degrees offset in span-wise direction to avoid mutual pressure orifices interference. The length of the serrations was 2h = 75 mm with $\frac{\lambda}{2h} = 0.75$. Three different flap angles using the same shape have been tested, 0°, 5° and 10°. The trailing edge serrations have been manufactured in aluminum and have been installed at the pressure side of the model using screws. An aluminum tape for ensuring a smooth transition between the model and the attachment of the trailing edge extensions has been used. Due to the trailing edge curvature of the pressure side of the metallic model during the installation of the devices a correction on the flap angle has been done. In Figure 3 a picture of the model with trailing edge serrations placed in the wind tunnel test section is presented.

Figure 3: NACA64$_3$418 model tested in the wind tunnel.

4.2. Experimental results

In Figure 4 the aerodynamic coefficients for the NACA64$_3$418 airfoil with and without trailing edge serrations have been displayed. It is important to remark that to the measured data the standard wind tunnel wall corrections for lift-interference and model solid and wake blockage as given by Allen and Vincenti [11] were applied. Corrections have also been made for the effect of solid blockage of the wake rake on the test section velocity and the effect of the wake rake self-blockage on the values of the static pressures (and consequently the dynamic pressure) measured by the wake rake [12]. It is clear that the impact of the trailing edge serrations cause a significant increment of the lift coefficient mostly at angles of attack from 5° to 15°. Analyzing the drag curves, smaller differences between the serrated and the original airfoil were measured. Looking into the flap angle effect it is shown that the increase of this angle has a positive effect on lift and a slight negative effect on drag causing a gain in the efficiency of the airfoil. This effect changes for large flap angles generating efficiency looses when the flap angle is highly increased, for this airfoil the most effective flap angle is 5°.

It is remarkable the hysteresis loop at positive and negative stall angles of attack. The hysteresis means the difference in lift and drag coefficients at a given angle of attack when this angle is approached from higher or lower values. It is a sign of a large, laminar separation which in turn yields high bubble drag [13]. As it is shown in Figure 4 this hysteresis cycle is not affected by the presence of serrated extensions due to the sensitivity of the bubble to the peak pressure gradient in proximity of the nose.

IOP Conf. Series: Journal of Physics: Conf. Series **1222** (2019) 012019 doi:10.1088/1742-6596/1222/1/012019

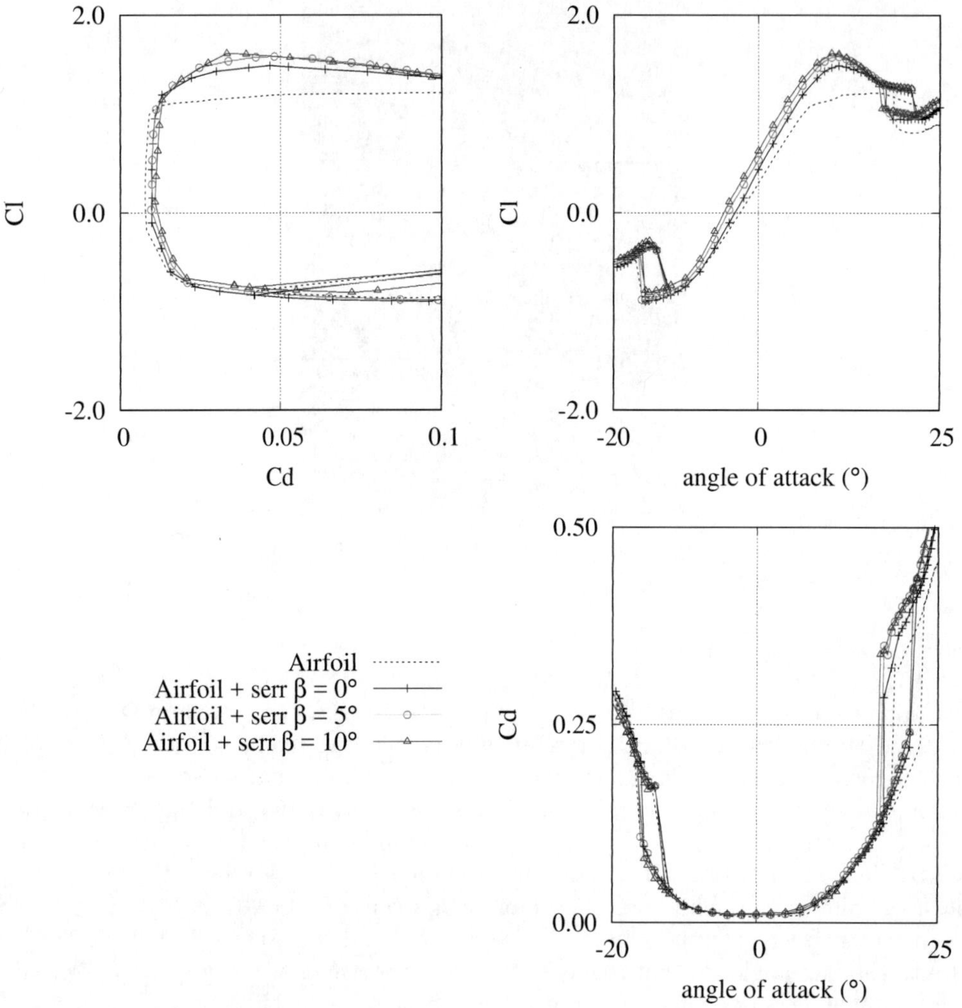

Figure 4: Aerodynamic coefficients for the NACA64_3418 with and without trailing edge serrations varying the flap angle.

In Figure 5 a comparison between the experimental results of the wind tunnel tests and the predicted by Equation 2 for different flap angles are displayed. In this plot it can be observed that for the different flap angles tested in the wind tunnel the proposed law predicts correctly the lift coefficient of the airfoil with trailing edge serrations from the data of the original profile. With this predicted law the influence of the trailing edge serrations on the whole blade can be obtained and the impact of these extensions on the power production or loads of the wind turbine can be calculated.

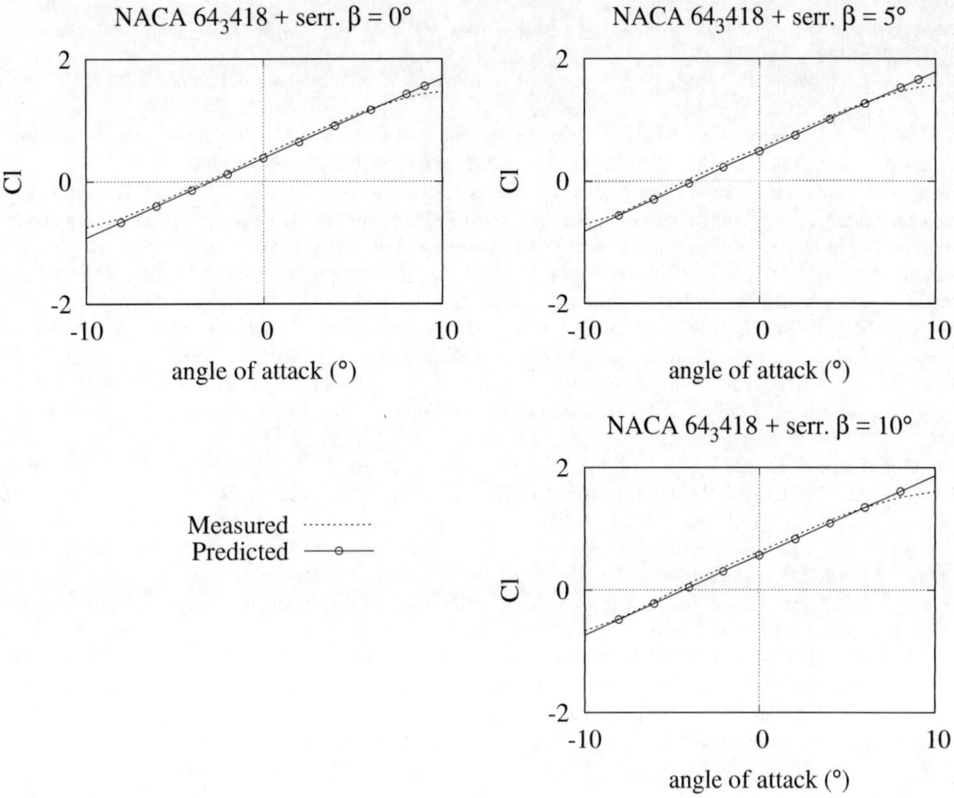

Figure 5: Comparison between the predicted and measured lift coefficient for the NACA64_3418 airfoil with trailing edge serrations varying the flap angle.

5. Conclusions

A numerical and experimental study of the aerodynamic impact of the trailing edge serrations on the NACA64_3418 airfoil has been presented. CFD simulations of the airfoil with a without trailing edge serrations show that the trailing edges serrations have an impact on the aerodynamic coefficients of the airfoil. Using these results a prediction law has been derived. With this equation the lift coefficient of the airfoil with serrations can be obtained from the data of the original profile. This equation has been validated using wind tunnel tests performed in the TU-Delft facilities and it shows a good agreement with the experimental data. The tests show a significant increment in lift coefficient due to the presence of the trailing edge extensions, the drag coefficient is also increase although at a reduced rate. In these experiments the flap angle effect is also analyzed, its increment caused a gain in lift and the penalty in drag is really slight. For this case the most effective flap angle is the intermediate one ($\beta = 5°$), increasing too much this angle cause bigger drag penalties and slights lift gains so the efficiency of the airfoil is reduced.

It can be concluded that the efficiency of the airfoil is increased by the presence of trailing edge extensions and its effect can be predicted with a simple equation that permits obtaining the lift coefficient of different airfoils with trailing edge extensions from the data of the original ones. With this information the power curve and the loads of a wind turbine with trailing edge extensions can be calculated without using costly CFD simulations or expensive wind tunnel tests, and its effect can be included in the design loop of the low noise new blades.

IOP Conf. Series: Journal of Physics: Conf. Series **1222** (2019) 012019 doi:10.1088/1742-6596/1222/1/012019

References

[1] Jaworski J W and Peake N 2013 *Journal of Fluid Mechanics* **723** 456–479 URL https://doi.org/10.1017%2Fjfm.2013.139
[2] Chong T P, Vathylakis A, Joseph P F and Gruber M 2013 *AIAA Journal* **51** 2665–2677 ISSN 0001-1452 URL http://dx.doi.org/10.2514/1.J052344
[3] Herr M 2007 Design criteria for low-noise trailing-edges *Aeroacoustics Conferences* (American Institute of Aeronautics and Astronautics) URL http://dx.doi.org/10.2514/6.2007-3470
[4] Siemens Gamesa Renewable Energy, S.A. 2018 *Siemens Gamesa 3.4/132* URL https://www.siemensgamesa.com/-/media/siemensgamesa/downloads/en/products-and-services/,onshore/brochures/siemens-gamesa-onshore-wind-turbine-sg-3-4-132-en.pdf
[5] Oerlemans S, Fisher M, Maeder T and Kgler K 2009 *AIAA Journal* **47** 1470–1481 ISSN 0001-1452 URL http://arc.aiaa.org/doi/abs/10.2514/1.38888
[6] DJ Moreau MR Tetlow L A B and Doolan C J 2010 Acoustic analysis of flat plate trailing edge noise *Proceedings of 20th International Congress on Acoustics* ed I C A (ICA)
[7] Mathieu Gruber M A and Joseph P F 2010 Airfoil trailing edge noise reduction by the introduction of sawtooth and slitted trailing edge geometries *Proceedings of 20th International Congress on Acoustics* (ICA)
[8] Menter F R, Langtry R B, Likki S R, Suzen Y B, Huang P G and Vlker S 2004 *Journal of Turbomachinery* **128** 413–422 ISSN 0889-504X URL http://dx.doi.org/10.1115/1.2184352
[9] Gruber M, Joseph P and Chong T 2011 On the mechanisms of serrated airfoil trailing edge noise reduction *Aeroacoustics Conferences* (American Institute of Aeronautics and Astronautics) URL http://dx.doi.org/10.2514/6.2011-2781
[10] Anderson J 2010 *Fundamentals of aerodynamics* (McGraw-Hill Education) ISBN 9780073398105 URL https://books.google.es/books?id=xwY8PgAACAAJ
[11] Allen H J and Vincenti W G 1944 Wall interference in a two-dimensional-flow wind tunnel, with consideration of the effect of compressibility Naca technical report National Advisory Committee for Aeronautics USA URL https://ntrs.nasa.gov/search.jsp?R=19930091861
[12] Timmer W and Van Rooij R 2003 *Journal of solar energy engineering* **125** 488–496
[13] Selig, Donovan and Fraser 1989 *Airfoils at Low Speeds* (Virginia (USA): H. A. Stokely)

IOP Conf. Series: Journal of Physics: Conf. Series **1222** (2019) 012020 doi:10.1088/1742-6596/1222/1/012020

Optimised de-rated wind turbine response and loading through extended controller gain-scheduling

Christos Galinos, Albert M. Urbán and Wai Hou Lio

Technical University of Denmark, Department of Wind Energy, Frederiksborgvej 399, 4000 Roskilde, Denmark

E-mail: cgal@dtu.dk

Abstract. An increasing number of wind turbines are expected to be able to participate in the ancillary services that would normally be provided by conventional power plants, as the penetration of the wind energy in the electricity grid is increasing. Specifically, the turbines are needed to be able to control their active power based on the operator command. Constraining the possible output power, i.e. decreasing the power to less than the maximum or rated power, is therefore important for the wind turbine control system. When a turbine is operating under a de-rating strategy, the operating points of the turbine, i.e. the steady-state blade pitch and the tip speed ratio, change compared to the normal case and this needs to be considered within the controller design. Thus, this paper is to study the influence of the de-rating on the turbine controller gain-scheduling and its effect to power output and fatigue damage of the key turbine components. The results show that the wind turbine response is refined through tuning under the different down-regulation operation points, that translate into less fatigue damage of the key turbine components. For a typical multi-megawatt turbine, it is found that the main component lifetime damage can be reduced substantially. In numerical terms, there are up to 6.5% for the tower BM in fore-aft direction and 2.7% for the blade root flapwise direction, which can lead in a prolonged operational lifetime of the wind turbine.

1. Introduction

Under normal Wind Turbine (WT) operation, the pitch controller, typically the proportional-integral (PI) controller, is responsible for governing the collective pitch of the blades in order to maintain the rated state, i.e. rated power, when the wind speed is above rated. A simple but commonly-used technique to tune the PI pitch controller is to exploit the aerodynamic gradients of the WT and the drive-train dynamics becomes a second-order closed-loop system [4]. Aerodynamic gradients with respect to the blades pitch and rotor speed are called aerodynamic gain and damping respectively. These aerodynamic gradients of the turbine depend on the operating points of the turbine. These operating point variables are the pitch angle, rotor speed and the wind speed. Consequently, the tuning of the PI pitch controller needs to be adjusted accordingly. This procedure is well-known in the wind turbine control community as the Gain-Scheduling (GS).

Traditionally, the WT gain-scheduling is only considered based on turbine's nominal operation, where the WT power output reaches the design rated power at designed wind speed. In recent years, turbines are required to de-rate or down-regulate for the purposes of providing ancillary services [1] or improving wind farm efficiency [3, 8]. However, when the WT is down-regulated, the tuning of the WT pitch controller and the GS parameters should be adjusted since the operating points are different than the designed. In this paper, the main contribution is to study how de-rating WT operation states affects the tuning of the blade pitch controller and the WT response.

The layout of the reminder of the paper is as follows. In Section 2, the theoretical model of the extended PI pitch controller GS is introduced, while in Section 3 the developed method is applied on a reference turbine. Finally, in Section 4 conclusions of the study are summarized.

2. Modelling of the GS PI Pitch Controller

The turbine GS is necessary since the dynamics of the wind turbine change with wind speed. One of the most significant parameters is the aerodynamic gain, which determines the blade pitch angle - aerodynamic power relation. Essentially, the PI controller GS is needed even in nominal operations because the dynamics of the system change substantially the performance of the closed loop from the original design. In the worst case, without GS, the speed regulation loop would affect other turbines dynamics. As the operating point changes, the PI controller gains need to be adjusted/scheduled accordingly and this can be due to changes in the incoming wind speed and the rotor speed. Changes to the controller gains alter the turbine operating point and affect the speed regulation closed-loop response.

Under normal WT operation, the blade pitch and the rotor speed are following the changes of the wind speed. Therefore, the GS should be based on the wind speed since the wind speed is the exogenous input that affects the operating point of the WT. However, due to practical reasons, the GS is done based on the pitch angle of the blades. In order to compute the values for GS, we can look into the equation of motion of a single degree of freedom WT, i.e. a stiff turbine model with only a rotational degree of freedom (rotating shaft): φ, $I\ddot{\varphi}+D\dot{\varphi}+K\varphi=0$ for more details see §4.5 [4]. Then, if the loop is closed with a PI pitch controller, the proportional and integral gains, K_p and K_i respectively, can be calculated as:

$$K_p = -\frac{2\zeta_\Omega\omega_\Omega I - \frac{\partial Q_g}{\partial\Omega} + \frac{\partial Q}{\partial\Omega}}{\frac{\partial Q}{\partial\theta}} \tag{1a}$$

$$K_i = -\frac{\omega_\Omega^2 I}{\frac{\partial Q}{\partial\theta}} \tag{1b}$$

where $\omega_\Omega = \sqrt{\frac{K}{I}}$ and $\zeta_\Omega = \frac{D}{2\sqrt{KI}}$ are the desired frequency and damping ratio of the mode associated with the rotor speed regulation, I is the rotor and drive-train inertia, K and D the 2nd order system damping and stiffness $\frac{\partial Q}{\partial\theta}$ is the derivative of the aerodynamic torque with respect to (w.r.t) the blade pitch angle or aerodynamic gain, $\frac{\partial Q}{\partial\Omega}$ is the derivative of the aerodynamic torque with respect to the rotational speed, here assumed to be negligible, and $\frac{\partial Q_g}{\partial\Omega}$ is the derivative of the generator torque with respect to the rotational speed that depends whether the turbine is regulated for constant torque ($\frac{\partial Q_g}{\partial\Omega}=0$) or constant power ($\frac{\partial Q_g}{\partial\Omega}=-\frac{P_r}{\Omega^2}$), where P_r is the rated power.

In Figure 1, the gains K_p and K_i for the pitch controller as a function of blade pitch angle for nominal WT operation for the DTU 10MW reference wind turbine (rwt) [2] are shown.

The equations (1) depict how K_p and K_i depend on the derivative of the aerodynamic torque w.r.t. the blade pitch. Since the $\frac{\partial Q}{\partial\theta}$ changes as a function of the blade pitch i.e. $\frac{\partial Q}{\partial\theta}(\theta)$, K_p and K_i become functions of the pitch angle θ. Notice that such a function $\frac{\partial Q}{\partial\theta}(\theta)$ is typically non-linear. Nonetheless, traditionally, $\frac{\partial Q}{\partial\theta}$ is fitted with a 2nd order polynomial with constants K_1 and K_2 for a WT under nominal operation, i.e. WT power output is the designed rated at

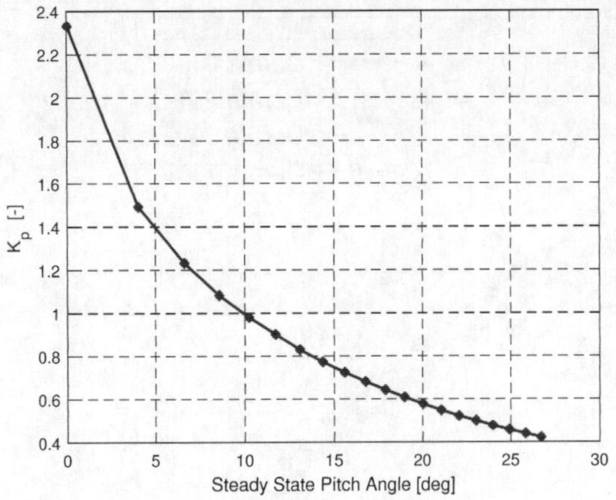

Figure 1. PI pitch controller gain K_p as a function of pitch angle under nominal WT operation.

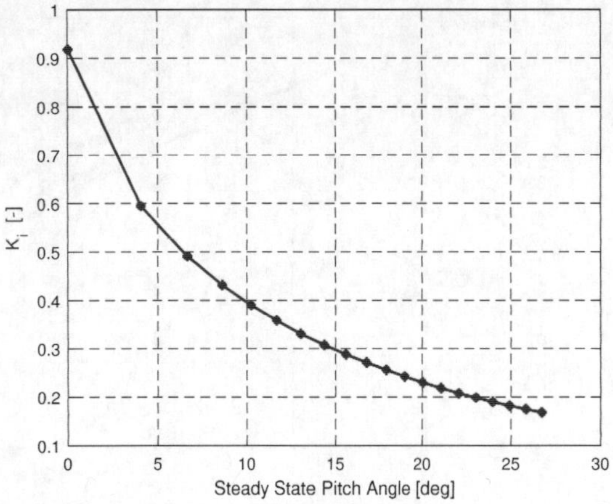

Figure 2. PI pitch controller gain K_i as a function of pitch angle under nominal WT operation

high wind speeds [10]:

$$\frac{\partial Q}{\partial \theta} = \frac{\partial Q}{\partial \theta}|_{\theta=0} \left(1 + \frac{\theta}{K_1} + \frac{\theta^2}{K_2}\right), \tag{2}$$

where $\frac{\partial Q}{\partial \theta}|_{\theta=0}$ is the gain at zero pitch. However, due to the demand of derating operation $\frac{\partial Q}{\partial \theta}$ or aerodynamic gain becomes a function of the desired power output P_r as well. Based on the de-rating percentage, the operating points of the WT can be different and therefore the aerodynamic gain which determines the value of the K_p and K_i for a specific wind speed varies.

In Figure 3, it is shown the aerodynamic gain values of the DTU 10MW rwt aero-servo-elastic model for nominal operation (power output is equal to P_r) and when the WT is forced to produce 25%, 50% and 75% of P_r. It can be observed that different constrains on power output lead to different aerodynamic gains. As a result the PI pitch controller should take these dependencies into account. The de-rating strategy followed here is the constant-Ω, for more details see [9]. The results have been obtained by using HAWCStab2 which transforms the non-linear finite beam WT aero-servo-elastic model to an analytical linearized one [5]. In the next paragraph, the developed extended GS approach is applied in the DTU reference WT aero-elastic model and its response is compared with the basic one (GS tuned only for nominal operation).

3. Application on the DTU 10MW RWT

The WT controller is the key instrument for stable and optimum WT operation under various conditions. The extended GS approach is evaluated below using the DTU 10MW rwt aero-elastic model and the open source DTU Wind Energy controller [10]. Its response and loading of main components under various simulation scenarios are compared using the basic GS approach versus the extended one. Simulations are conducted in time domain using HAWC2 aero-elastic code [7] and applying 50% down regulation.

HAWC2 performs time domain simulations of the system, where the dynamics are described by differential equations and solved using the Newmark numerical integration scheme. It is divided into modules which can be combined depending on what the user is studying. The main parts are:

- the structure where main WT components such as the blades, the tower, the rotor shaft and the substructure can be modelled by Timoschenko beam elements with Rayleigh damping

IOP Conf. Series: Journal of Physics: Conf. Series **1222** (2019) 012020 doi:10.1088/1742-6596/1222/1/012020

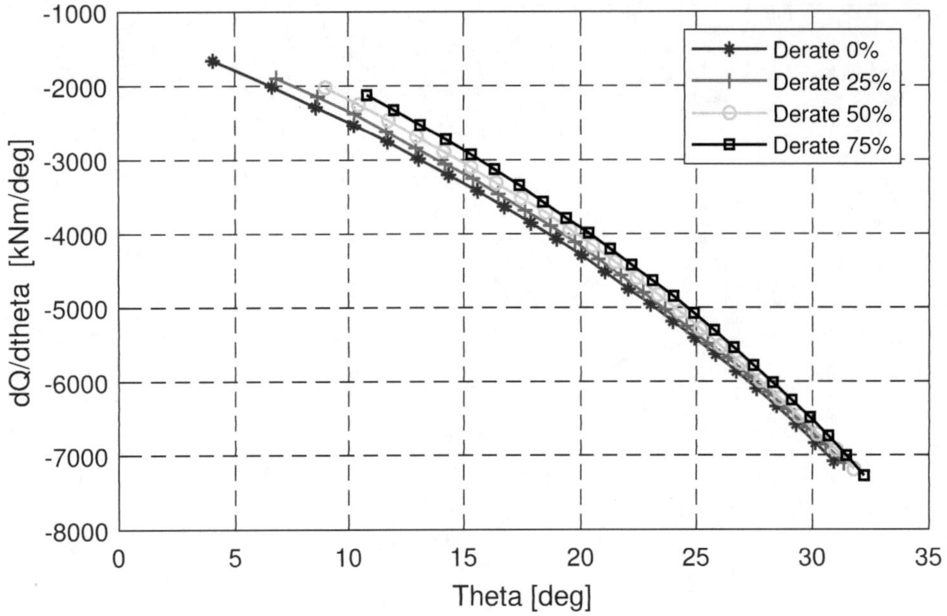

Figure 3. Aerodynamic gain as a function of blade pitch for different de-rating operation conditions.

or with super element models

- the aerodynamics which are based on the Blade Element Momentum and Actuator Cylinder models and the blade aerodynamic coefficients for computing the aerodynamic loads
- the wind field which can be deterministic, stochastic using the Veers and Mann models or custom
- the hydrodynamics which include regular, irregular waves with JONSWAP and Pierson-Moskowitz spectra, and the Morison approach for capturing the hydrodynamic loads on the components
- the soil module, where the soil-structure interaction is modelled using p-y curves
- the controller which is plugged in as a DLL file including the generator and the blade pitch system dynamics
- Finally, other special futures such as tower guy wires and nacelle lidar sensors can be implemented through custom external DLLs and link them with the code

3.1. Step response

The response of the WT is firstly evaluated by applying wind speed steps above rated wind speed where the PI pitch controller is active. The wind field is deterministic with a power law shear profile and exponent equal to 0.2 (see Figure 4). In Figures 5 and 6, the generator LSS torque, the rotor speed and the blade pitch response are plotted respectively. The results show that there is an improved response with the extended GS as the pitch angle changes more aggressively and the overshoot of the generator torque and rotor speed is smaller for each wind speed step. In Figure 7 the blade root flapwise and the tower base fore-aft Bending Moments (BM) are presented focused on a narrower time span for clarity reasons. It can be seen that due to the improved PI pitch controller response of the extended GS approach the blade load variation is smaller after each wind step while the tower fore-aft BM remains almost the same.

IOP Conf. Series: Journal of Physics: Conf. Series **1222** (2019) 012020 doi:10.1088/1742-6596/1222/1/012020

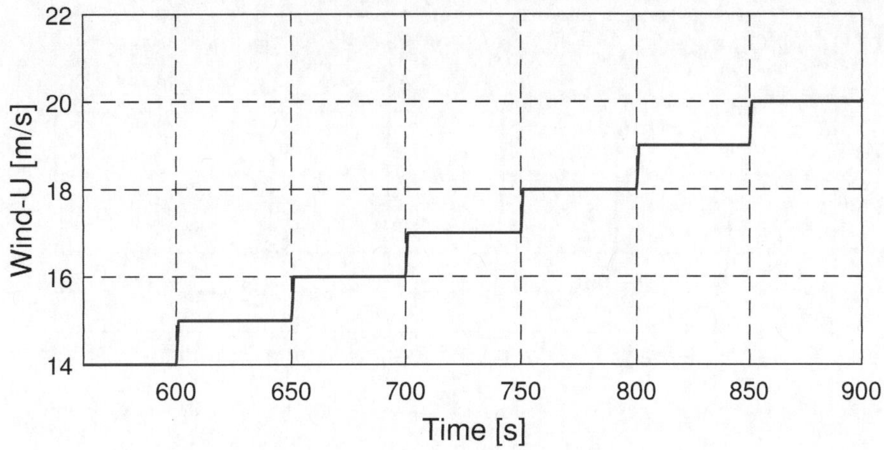

Figure 4. Applied wind steps.

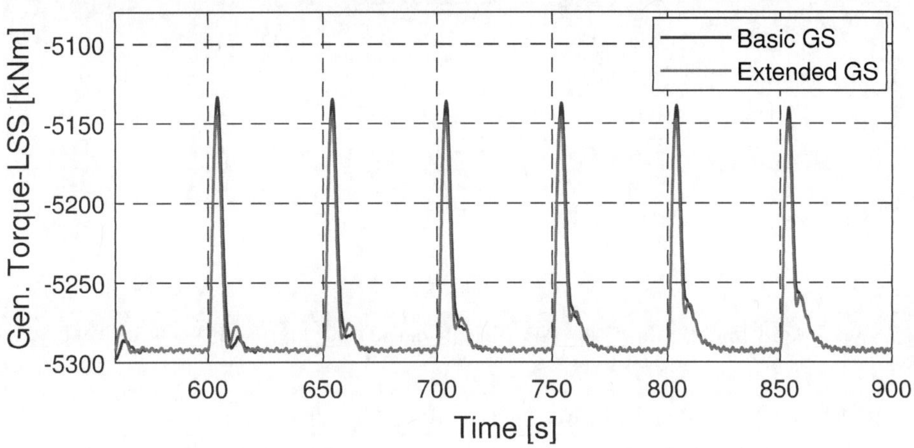

Figure 5. Generator torque expressed at LSS.

3.2. Turbulent Wind

Finally, a load comparison based on high fidelity aero-servo-elastic simulations is performed focusing on the lifetime fatigue damage of the WT main components such as the blade and the tower between the basic and the extended gain scheduling. The analysis is done following the IEC61400-1 ed.3 standard [6] and focusing only above rated where the PI pitch controller is mainly active. IEC turbulence and site conditions of Ib are assumed and eighteen 10 minute HAWC2 simulations are run per wind speed in order to capture the variations of the turbulence field. At table 1 the lifetime fatigue loads are compared after normalization with the basic GS simulation results. A Wöhler exponent of 4 and 12 are assumed for the tower and the blade respectively and a 20 year of design lifetime for the WT with 10 million cycles. The tower base fore-aft fatigue damage is reduced by 6.45% which is quite a lot and follows the blade root fatigue damage with a reduction of around 2%. The blade root edgewise damage is almost unchanged since the gravity field is the main contributor in this load direction. Finally, the tower base side-side damage is reduced slightly (around 1%). As a result, it can be concluded that even

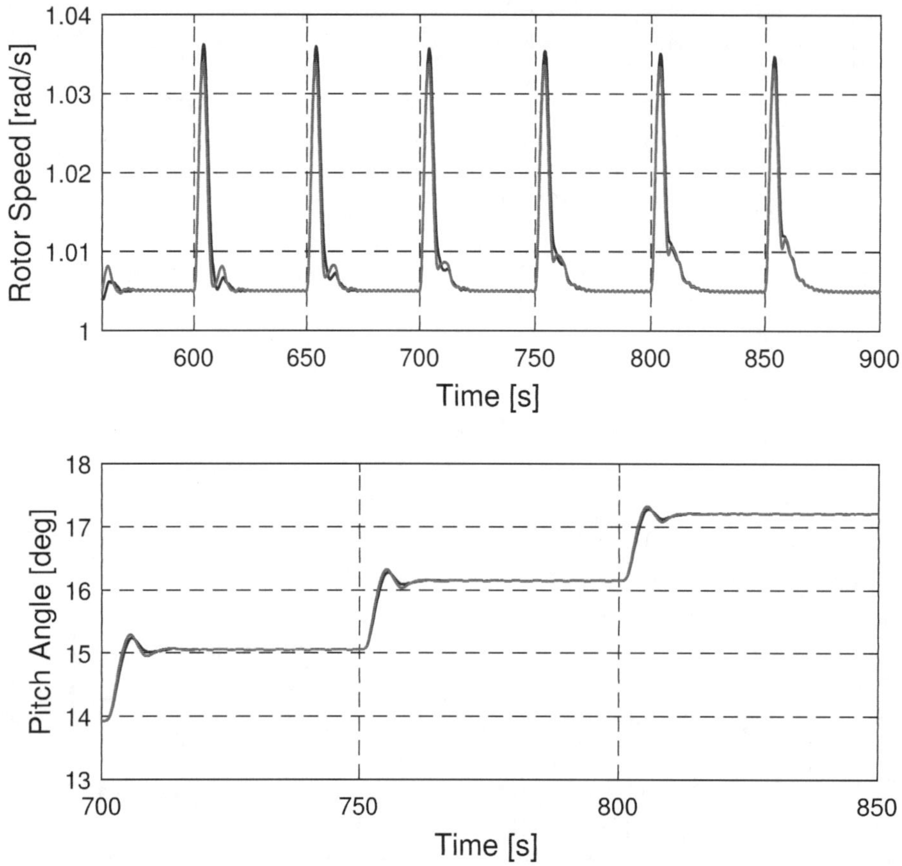

Figure 6. Rotor speed and blade pitch angle responses, upper and lower plots respectively.

Table 1. Normalized lifetime fatigue equivalent loads comparison.

Channel name	Baseline Operation [%]	Extended GS Operation [%]
Tower Base Fore-Aft BM	100	93.55
Tower Base Side-Side BM	100	98.78
Blade Root Flapwise BM	100	98.28
Blade Root Edgewise BM	100	99.99

though above rated both the basic and the extended GS approach WT simulations return the requested rated power output, the lifetime damage can be reduced substantially by extended tuning of the controller actions. This is especially important in large modern wind farms where turbines will be requested to operate in a lower rated power output more often.

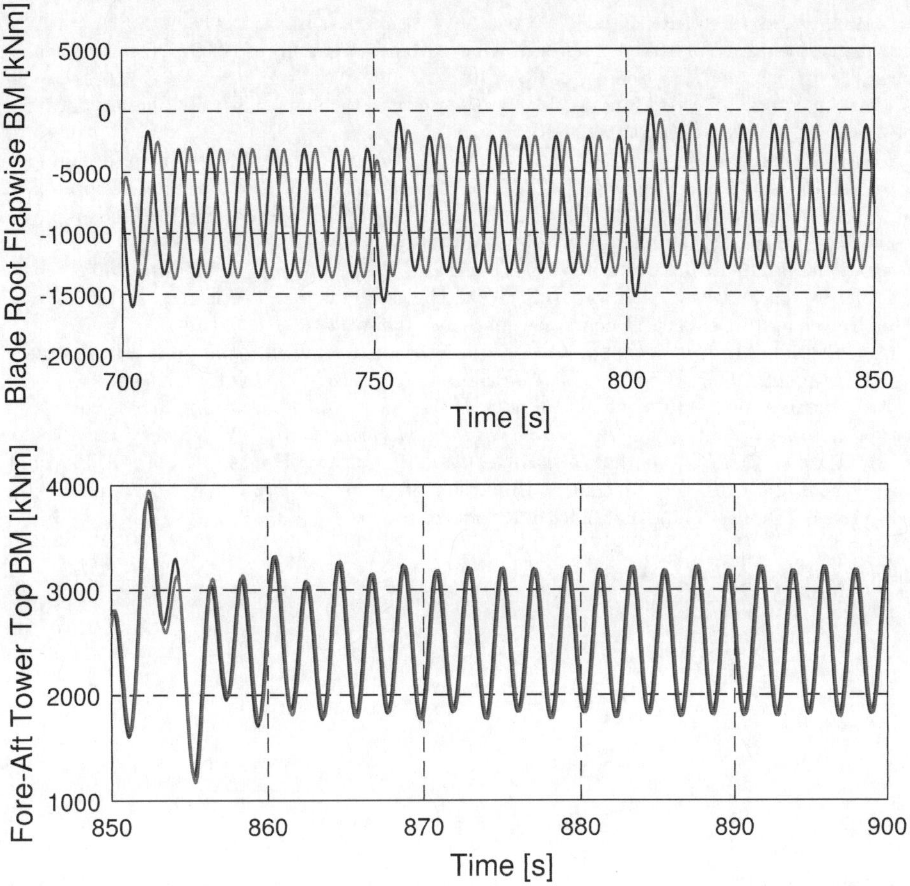

Figure 7. Blade root flapwise BM (upper plot), tower base fore-aft BM (lower plot).

4. Conclusions

In this paper is shown that de-rating WT operation affects the aerodynamic gradients of the turbine and in order to take these effects into account the PI pitch controller gain scheduling needs to be extended. In addition, is shown that the WT response is refined through tuning under the different down-regulation operation points. Finally, for a typical multi-megawatt turbine it was found that the main component lifetime damage can be reduced substantially. Up to 6.5% for the tower BM in fore-aft direction and 2.7% for the blade root flapwise direction, which can lead in a prolonged operational lifetime of the WT.

Acknowledgements

This work has been funded by the Danish Energy Technology Development and Demonstration Programme (EUDP) under project PowerKey - Enhanced WT control for Optimized WPP Operation, Grant No. 64017-0045. The support is greatly appreciated.

References

[1] Aho, J., Buckspan, A., Pao, L., and Fleming, P. An Active Power Control System for Wind Turbines Capable of Primary and Secondary Frequency Control for Supporting Grid Reliability. *AIAA/ASME Wind Symposium* (2013), 1–13.

[2] Bak, C., Zahle, F., Bitsche, R., Kim, T., Yde, A., Henriksen, L. C., Natarajan, A., and Hansen, M. Description of the DTU 10MW reference wind turbine. Tech. Rep. I-0092, Technical University of Denmark, Department of Wind Energy, 2013.

[3] Galinos, C., Larsen, T. J., and Mirzaei, M. Impact on wind turbine loads from different down regulation control strategies. In *DeepWind2018* (2018).

[4] Hansen, M. H., Larsen, T. J., Øye, S., Sørensen, P., and Fuglsang, P. Control design for a pitch-regulated, variable speed wind turbine. Tech. Rep. Risø-R-1500(EN), Risø National Laboratory, 2005.

[5] Henriksen, L. C., Tibaldi, C., and Bergami, L. HAWCStab2 User Manual. Tech. rep., Technical University of Denmark, Department of Wind Energy, 2015.

[6] IEC. International Standard IEC61400-1: Wind Turbines - Part 1: Design Guidelines, 2005.

[7] Larsen, T. J., and Hansen, A. M. How to HAWC2, the user's manual. Tech. Rep. R-1597(ver.4-7)(EN), Technical University of Denmark, Department of Wind Energy, 2018.

[8] Lio, W. H., Mirzaei, M., and Larsen, G. C. On wind turbine down-regulation control strategies and rotor speed set-point. *Journal of Physics: Conference Series 1037* (jun 2018), 032040.

[9] Mirzaei, M., Gocmen, T., Giebel, G., Sorensen, P. E., and Poulsen, N. K. Turbine control strategies for wind farm power optimization. *Proceedings of the American Control Conference* (2015), 1709–1714.

[10] Mirzaei, M., Hansen, M. H., Tibaldi, C., Henriksen, L. C., and Hanis, T. Basic DTU wind energy controller version 2. Tech. rep., Technical University of Denmark, Department of Wind Energy, 2015. `https://github.com/DTUWindEnergy/BasicDTUController`.

Load analysis of special operation conditions and implementation of derived load states in Bearing Robustness Test (BRT)

A. Hirt[1*], G. Jacobs[1], S. Neumann[1], S. Kock[1], D. Bosse[1]

[1]Chair for Wind Power Drives, RWTH Aachen, Campus-Boulevard 61, 52074 Aachen, Germany

*E-mail: andreas.hirt@cwd.rwth-aachen.de

Abstract. Premature bearing outages at the planetary and high speed shaft stage of gearboxes are responsible for long down time periods of wind turbines [1, 2]. In order to prevent premature bearing outages against certain failure mechanisms, a new robustness test procedure for original-sized WT gearbox bearings is developed within the project WT-Bearing Center.NRW. The BRT serves to qualify bearings for a damage-free operation over the entire service life [3]. It combines a fatigue load spectrum and also additional special load events [4]. This paper describes the analysis of wind turbine loads during different operation conditions and the definition of special load events, based on this analysis. The aim of including special loads to the BRT is to consider the complex real bearing loads within the test procedure as precisely as possible.

1. Introduction

Long downtime periods burden the economic efficiency of wind turbines. Most of the downtime is caused by gearbox failures. Figure 1 shows that 59 % of all downtime of a wind turbine is caused by gearbox failures. The causes for most of these gearbox failures (67 %) are bearing outages [1, 2]. Ten percent of these bearing outages occur at planetary and 70 % at high speed shaft stage [5]. That implies that presently planetary and high speed shaft bearings are the main causes for long downtime periods of wind turbines.

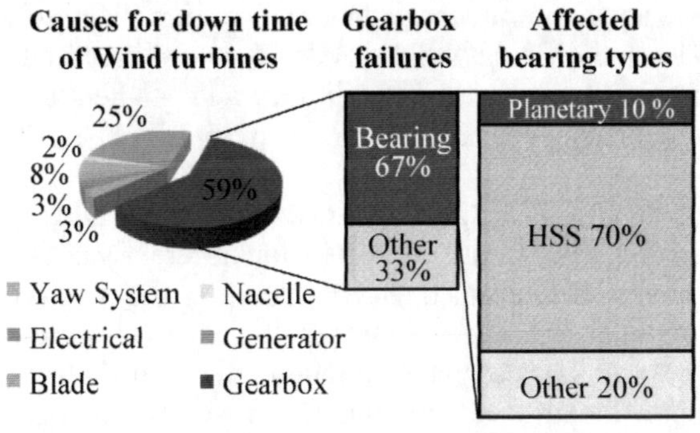

Figure 1. Causes for down time periods of wind turbines [1-3]

To increase the reliability of the bearings in wind turbines, different testing methods during the product development process are used. Actual test methods investigate isolated failure mechanism like standard fatigue tests. They do not consider complex load situations of bearings or interactions in the occurrence of damage, like preliminary damages. Consequently it is not possible to replicate realistic bearing failures with the contemporary testing methods at original sized bearings of wind turbines. In order to ensure that original sized bearings of wind turbines can be tested concerning their robustness against certain failure mechanisms, a new robustness test procedure, the Bearing Robustness Test (BRT), is developed within the project WT-Bearing Center.NRW [4]. The aim is to develop new testing standards for original-sized planetary and high speed shaft bearings of wind turbines [3].

2. Approach of load analysis

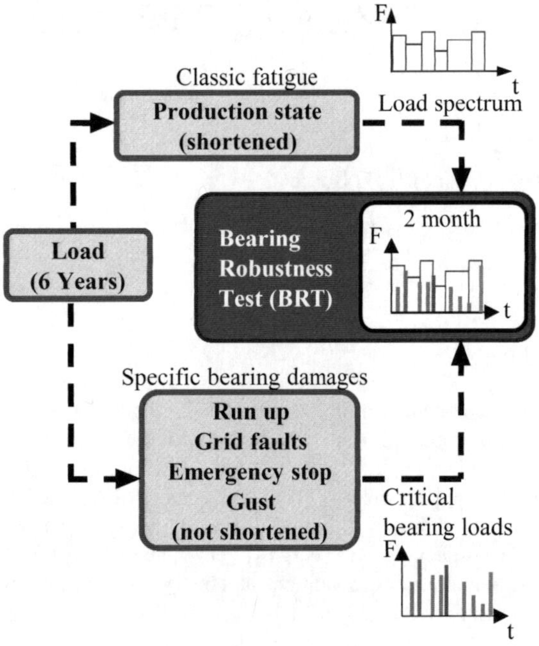

Figure 2. Schematic approach to create BRT

The BRT includes a fatigue load spectrum based on field data and also additional special load events based on simulation data. The fatigue load spectrum accelerates the loads of six years to a load spectrum of about two months and covers the realistic bearing pre-damage [6] (upper path in Figure 2). The special load events comprise loads of specific operation states to include the interaction of different load conditions. The number of considered special load events inside the BRT is restricted to the number of the events actually occurring during six years of field operation (lower path in Figure 2).

The simulative analysis of specific operation states is based on simulations with a validated multi-body system (MBS) model. The MBS model is a model of the 4 MW wind turbine test facility with a state-of-the-art 2.7 MW wind turbine. It was development and validated at the CWD [7]. Within the model aerodynamic rotor input loads are calculated based on the given curves of wind speed. Electric loads affecting the generator are also included within the model. All calculation models correspond to the Hardware-in-the-Loop (HIL) system which is also used in the real testing stand [7, 8]. A sketch of the described interaction between MBS and HIL model is visualized in Figure 3. The simulation with this model enables detailed load analyses of drive train and bearing loads during system operation. Figure 4 shows exemplary simulated input loads at hub flange (drive-, tilting- and yaw torque) by a chosen mean wind speed of 17.5 m/s. The turbulence intensity was set to 12 %. It can be shown, that extreme gradients of input loads occurring during special operation states such as gusts, emergency stops, run ups and more. Duration and absolute values of gradients vary for different operation states and different wind speeds. By analysing resulting loads of different operation states concerning absolute values, load amplitudes and gradients, it is possible to characterize the operation states. The load analyses carried out for this purpose are described in the following.

So far, production states (Design Load Case (DLC) 1.2, 1.3), gusts (DLC 2.3), run ups (DLC 3.1) and emergency stops (DLC 5.1), based on IEC-load cases [9], are simulated for different values for mean wind velocity and turbulence intensity to cover all relevant transient operation states of wind turbines. Further DLCs will be considered in a second step. After simulation, resulting drive torque loads at hub flange and resulting load curves of radial bearing loads at planetary and high speed shaft stage are characterized regarding occurring maximal/minimal loads, load amplitudes and gradients as well as rotational speed values, amplitudes and gradients. The combination of load and rotational speed characterizes the complex load situation of different operation states. These characteristics of DLCs are compared to criteria of selected damage hypothesis. The comparison shows, that some of the occurring bearing loads favor certain failure mechanism more than others and in a further step, simplified loads are derived from the analysed loads of DLCs. The derived loads are then selected as special load events for the BRT. The described approach is schematically shown in Figure 5.

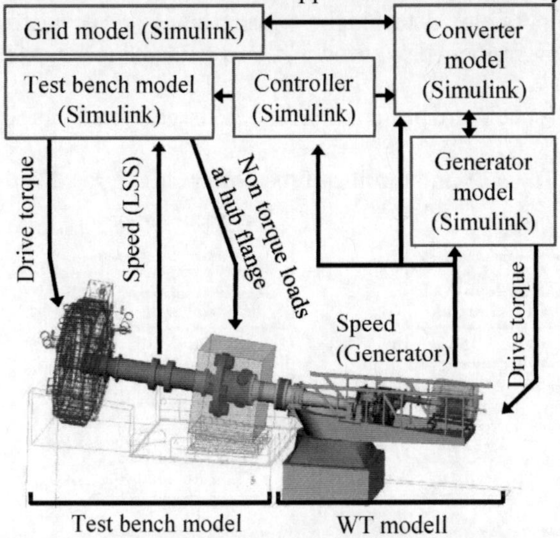

Figure 3. Interaction between multi body system model of test bench and HIL model [7]

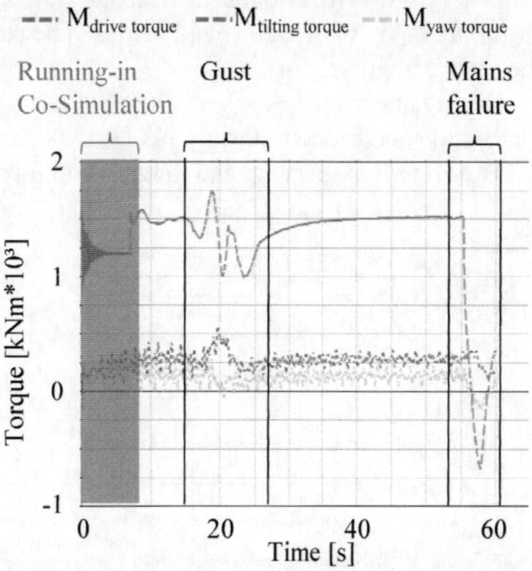

Figure 4. Exemplary load simulation with multi body system model for a mean wind velocity of 17.5 m/s.

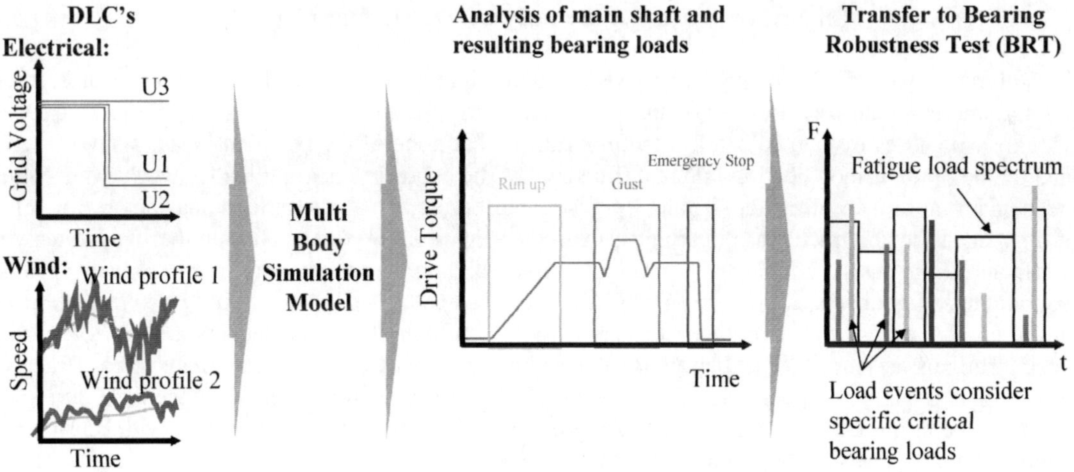

Figure 5. General approach for the characterization of loads and selection of special load events for Bearing Robustness Test

3. Load analysis of selected DLCs

The different DLCs are simulated for varying mean wind velocities. In a next step the resulting loads at hub flange as well as radial bearing loads at planetary and high speed shaft stage are analysed concerning occurring min/max values, amplitudes and gradients. Also the speed values and gradients for different stages are considered during the analysis. Figure 6 shows a summary of the load and speed ranges that occurred for the various DLCs (Run up, shut down, gust and emergency stop) at the bearing stages. The summary shows that very low loads are occurring simultaneously with low speed values during run ups and shut downs. The occurring gradients for speed and load are also very small for this operation state.

During gusts overloads at bearing stages are possible as the highest load values are occurring simultaneously with high speed values. The resulting gradients for speed and load are small to large for the regarded gusts.

During emergency stops medium loads and high speeds are present. Further, the largest gradients for loads and speeds occur during this DLC.

The shown load and speed ranges in Figure 6 refer to a wind turbine of the power class 2.75 MW.

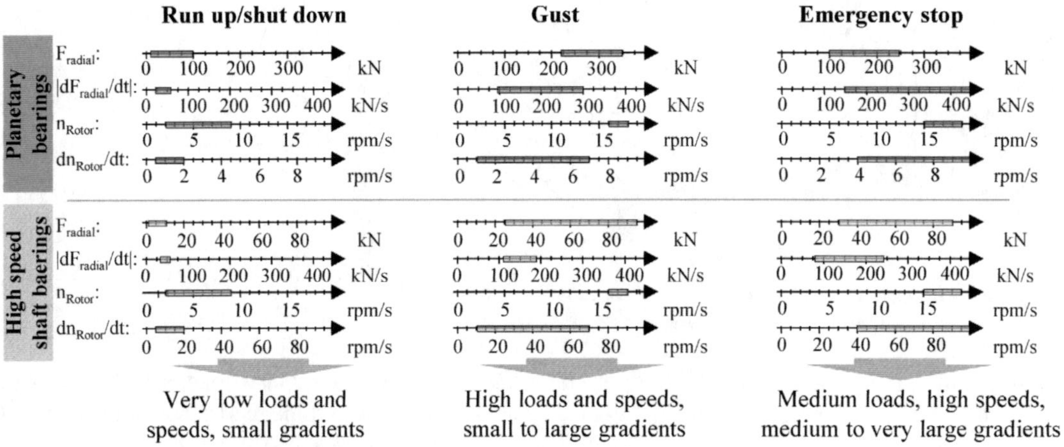

Figure 6. Occurring load and speed ranges at planetary and high speed shaft stage during different operation conditions for the analysed wind turbine (2.75 MW)

Within the project WT-Bearing Center.NRW, the focus is set to the following bearing failures: micro pitting, smearing and fractures at bearing rips. Comparing the load characteristics of analysed DLCs with known causes for named bearing failures shows, that none of the occurring loads would directly lead to one of the named bearing failures. But some of the occurring bearing loads favor certain failure mechanism more than others do. Table 1 gives a short summary about influencing factors for micro pitting, smearing and fractures at bearing rips based on literature research. The summary indicates, that overloads during gusts combined with occurring axial oscillations can encourage micro pitting. The high load and speed gradients during emergency stops can lead to high slippage values of rolling elements and in combination with an unfavourable viscosity lead to complex load states that favour smearing. During run ups and shut downs the rotational speed is very low. Combined with axial forces, fractures at bearing rips can be encouraged. Of course, none of the analysed loads that have occurred during the different operation states are critical for bearing life time directly, but in combination with bearing pre-damages and unfavourable conditions, various operating conditions can favour certain failure mechanisms. That's why it is important to consider the complex load situation of planetary and high speed shaft bearings within a test procedure. Therefore, simplified loads of the different operation states will be included to the BRT. This approach is described in the following.

Table 1. Summary about influencing factors for bearing failures [10-13]

Failure	Failure location	Characteristics of failure favouring state
Micro pitting	In the running treat of inner ring	• Small slippage states of rolling element (mainly -2 % slippage of rolling element) • Overloads • Mixed friction • Critical temperature limit T ≥ 90 °C • Axial Oscillation

Smearing	• At the contact surface of rolling element front surface and board, as well as in the rolling tread of inner ring and the rolling element • Usually in the inlet of the load zone (compiles the acceleration zone)	• ~ 40 % to 80 % slippage of rolling element • Falling below minimal load by high speed • Huge influence of viscosity (high temperatures are critical due to low viscosity)
Fractures at bearing rip		• Axial forces with low rotational speed at the same time • Tilting between inner and outer ring • Shaft deflection

Figure 7 shows exemplary loads at planetary and high speed shaft stage during the following operation conditions: Run up, gust and emergency stop.

The mean wind velocity during shown run up is 5.5 m/s. The rotor blades start to cut in the beginning and when the rotational speed reaches a specific level, the generator is switched on. The resulting loads on the bearings of the planetary and the high speed shaft stage follow the load curve on the hub flange. The wave-like load curve at the planetary stage results from the 360 degree rotation of the planet carrier, as the tilting of the planet depends on its current position.

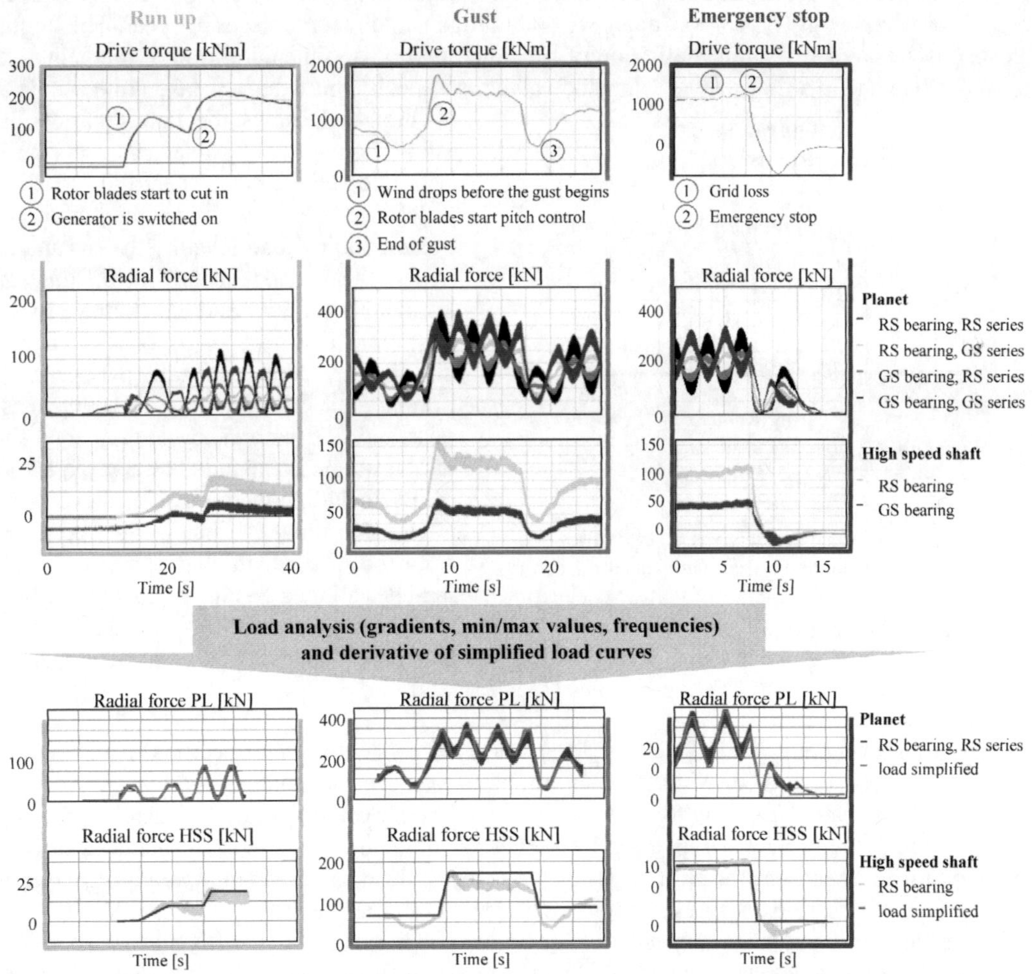

Figure 7. Load analysis of specific operation states and derivation of simplified load curves for Bearing Robustness Test

IOP Conf. Series: Journal of Physics: Conf. Series **1222** (2019) 012021 doi:10.1088/1742-6596/1222/1/012021

The shown gust occurs during a mean wind velocity of 12.5 m/s and the wind speed reaches a peak of 17 m/s. When the gust starts, the wind turbine starts pitch controlling to prevent possible resulting overloads. As the pitch control is slow, the input torque drops significant at the end of the gust again.

During the emergency stop the mean wind velocity is also set to 12.5 m/s. The stop is introduced by a mains failure and 3.2 s later the wind turbine starts with an emergency stop by disconnecting the generator load. Now the rotational speed is very high and the pitch control goes into a brake mode to reduce the speed again. During the emergency stop, very high load and speed gradients are occurring.

For the load analysis of different operation states, occurring min/max values, amplitudes and gradients are stored. Also simultaneously acting speed values at main shaft, planetary and high speed shaft stage are considered. Based on the stored values, simplified load curves are derived for the BRT (lower part of Figure 7). The simplified load curves for the planetary bearings are shown by the red lines and the simplified loads for the high speed shaft bearings are shown by the blue lines. In the next step, these simplified loads for run ups, gusts, emergency stops, etc. are included into the classic load spectrum of the BRT as special load events. Each special load event is considered in number equal to its real occurrence frequency, determined on the basis of real measurement data form comparable systems. The evaluation of wind turbines of the same performance class in Denmark has shown that approx. 2000 start-up processes take place in one year. Further approx. 200 emergency stops occur per year and correspondingly it comes to approx. 1800 normal shutdown processes per year. The frequency of gusts depends strongly on the definition of the time duration and the definition of the height of the wind speed rise and in a first step an estimated value for a specific gust shape is taken into account.

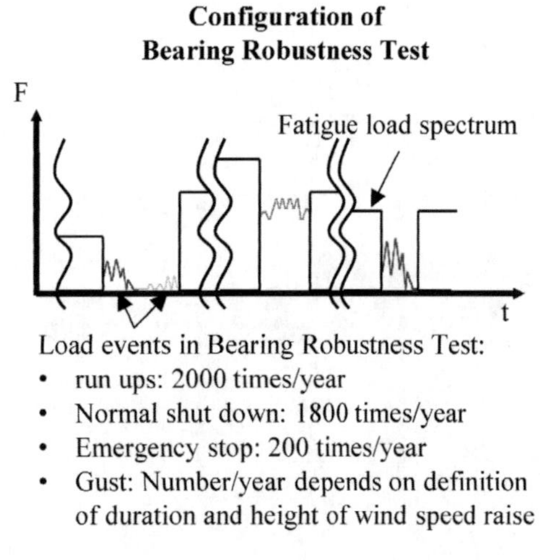

Figure 8. Configuration of BRT

Figure 8 shows the configuration of the BRT. The special load events are combined with the fatigue load spectrum, so that the accelerated loads are interrupted at an almost corresponding load level by the dynamic load events. The load events for run ups, gusts, normal shut downs and emergency stops take in total approx. 10 seconds each. By considering 2000 run ups per year, 1980 normal shut downs per year, 1000 gusts per year and 200 emergency stops per year, the total test time is expanded by approx. 80 hours.

By including the special load events in the BRT according to their real frequency of occurrence and load characteristics, the BRT includes all load conditions to which a bearing of the planetary and high speed shaft stage is exposed in the first 6 years in field operation.

4. Conclusion

Premature bearing outages at the planetary and high speed shaft stage of gearboxes are responsible for long down time periods of wind turbines. In order to prevent premature bearing outages against certain failure mechanisms, a new robustness test procedure for original-sized WT gearbox bearings is developed within the project WT-Bearing Center.NRW. In contrast to actual test methods of original sized wind turbine bearings, the BRT includes the complex load states of real operation conditions to reproduce realistic bearing damages at original sized wind turbine bearings.

To consider the complex load states of real operation conditions within the BRT, special load events are defined based on a detailed load analysis of different operation states. The chosen operation states are extracted from the defined DLCs in IEC 61400-1. A comparison of analysed loads with known causes of certain bearing failures shows, that some DLCs can favor specific bearing failures more than others. Based on the characteristics of these loads, special load events for the BRT are derived to ensure

IOP Conf. Series: Journal of Physics: Conf. Series **1222** (2019) 012021 doi:10.1088/1742-6596/1222/1/012021

that the test loads correlate with the realistic loads as good as possible. The derived special load events are scattered into the BRT according to their real frequency of occurrence, so that the BRT contains all relevant loads of the first 6 years and finally serves a new standard for bearing testing.

References

[1] Sheng S: Report on Wind Turbine Subsystem Reliability - A Survey of Various Databases, NREL/PR-5000-59111, June 2013. URL: https://www.nrel.gov/docs/fy13osti/59111.pdf
[2] Scott K G, Infield D, Barltrop N, Coultate J, Shahaj A: Effects of Extreme and Transient Loads on Wind Turbine Drive Trains, 50th AIAA Aerospace Sciences Meeting, Nashville 2012.
[3] Neumann S, Jacobs G: Reduzierung von Lagerschäden im Antriebsstrang von Windenergieanlagen - WEA-Lagerzentrum.NRW, Ingenieur-Spiegel, 2017. DOI: 10.18154/RWTH-2017-09861
[4] Kock S, Jacobs G, Hirt A, Oberdörfer S, Neumann S, Bosse D: Robustness Test for Wind Turbine Gearbox Bearings, Journal of Physics: Conference Series, Vol. 1037, Nr. 5, 2018. DOI: 10.1088/1742-6596/1037/5/052012
[5] Sheng S: Wind Turbine Gearbox Reliability Database, Condition Monitoring, and Operation and Maintenance Research, GRC Annual Meeting, Golden 2016.
[6] Oberdörfer S, Jacobs G, Kock S, Neumann S, Bosse D: Determination of a Load Spectrum for a Robustness Test of WT Gearbox-Bearings, Conference for Wind Power Drives, 2019.
[7] Matzke D, Jacobs G, Schelenz R: Full scale system simulation of a 2.7 MW wind turbine on a system test bench, Conference for Wind Power Drives 2017, Aachen 2017.
[8] Averous N R, Berthold A, Schneider A, Schwimmbeck F, Monti A, De Doncker R W, Performance tests of a power-electronics converter for multi-megawatt wind turbines using a grid emulator, J. Phys.: Conf. Ser. 753 (2016) 72015. DOI: 10.1088/1742-6596/753/7/072015
[9] Wind turbines – Part 1: Design requirements (IEC 61400-1:2005 + A1:2010); German version EN 61400-1:2005 + A1 2010.
[10] Van Lier H: Neuhärtungsgefährdung von Radial-Zylinderrollenlagern durch Lastaufschaltungen in Betriebspunkten mit Käfigschlupf, Dissertation 2015, Aachen 2015.
[11] Bajer P: Einflussgrößen auf das Schlupfverhalten von Wälzlagern in Generatorgetrieben, Dissertation 2016, Kaiserslautern 2016.
[12] Morales-Espejel G E, Brizmer V: Micropitting Modelling in Rolling–Sliding Contacts: Application to Rolling Bearings, Tribology Transactions 54 (2011) 625–643. DOI: 10.1080/10402004.2011.587633
[13] Bongardt C: Wälzlagergraufleckigkeit, Dissertation 2016, Aachen 2016.

IOP Conf. Series: Journal of Physics: Conf. Series **1222** (2019) 012022 doi:10.1088/1742-6596/1222/1/012022

Nonlinear Estimation of Important Variables using Neural Network in Wind Farms

Sung-ho Hur

School of Electronics Engineering, Kyungpook National University, South Korea

E-mail: hur.s.h@ieee.org

Abstract. Utilising additional variables, which are normally not measured, could bring significant improvements to condition monitoring and control of a wind turbine and farm. However, incorporating additional sensors to measure such variables could increase the cost significantly. As a solution, instead of equipping every turbine in the wind farm with an expensive sensor to measure such a variable, it is proposed that only one turbine be equipped with a sensor and the neighbouring turbines with an estimator that essentially replaces the sensor; that is, each estimator would subsequently estimate what the sensor would measure. Each estimator is constructed based on Neural Network, and as a result, the cost could be significantly reduced. Note that the only turbine equipped with a sensor is used to train the NN. This work presents the results of a preliminary study to examine the feasibility of the proposed approach.

1. Introduction

Control and monitoring strategies are important for wind-farm operators because of the long-lasting impact it can have on the profitability and efficiency of a wind site's operations. Having access to various wind turbine variables that are normally unavailable could improve the performance of control and monitoring strategies, but the associated costs, i.e. the costs of sensors, should be considered carefully.

For control and monitoring purposes, a wind farm could be split into a number of clusters, each of which may contain several wind turbines. In this preliminary study, it is assumed that there are 5 turbines in each cluster. Only one turbine is equipped with a (costly) sensor that "measures" tower bending moment (TBM), as an example, and each of the remaining turbines with an estimator, in place of the sensor, to "estimate" the TBM. Since these estimators would be highly nonlinear, linear estimators, such as observers or Kalman filters [1, 2], cannot be employed. Therefore, Neural Network (NN) [3, 4] is considered here for designing nonlinear estimators. In more detail, a NN-based estimator is constructed/trained based on the measurement of TBM from the only turbine with a sensor in addition to the measurements of other commonly available variables, e.g., (a various combinations of) generator torque, pitch angle, rotor speed and fore-aft acceleration (FAA), obtained from the same turbine. The trained NN-based estimator is subsequently applied to each of the 4 remaining turbines in the cluster.

It is important to point out that the NN-based estimator could be used to estimate other important variables, such a blade root bending moment (BRBM), but that usage is not discussed in this preliminary study.

IOP Conf. Series: Journal of Physics: Conf. Series **1222** (2019) 012022 doi:10.1088/1742-6596/1222/1/012022

In this work, the exemplar 5MW wind turbine model of Supergen Wind Hub in Matlab/Simulink® is employed to simulate the wind turbines and farm. The model has been used for various UK and European projects over the last decade [5, 6, 7], especially within the Supergen Wind Hub/Consortium. The latest version of the model is reported in [8].

Related work is reported in [9]. The work introduced therein focuses on load monitoring. Moreover, the work presented in [9] uses offline data and empirical models while the work presented in this paper uses the Supergen model, which is a first-principles model, and real-time data from the model.

In Section 2, how the NN works and the wind speed model required to simulate the turbine models are briefly discussed. In Section 3, a nonlinear estimator is developed based on the NN, and the NN-based estimator is subsequently tested in a Matlab/Simulink environment. Conclusions are drawn in Section 4. It should be clarified that the main contribution of the work is not in the NN. Instead, to achieve a novel objective, i.e. nonlinear estimation of useful variables using the NN (e.g. TBM and BRBM), which are often not measured, the NN is adapted.

2. Wind Speed Modelling and Neural Network

In this section, the concept of NN is briefly revised and the wind speed model used for simulation throughout this paper is reported.

2.1. Neural Network

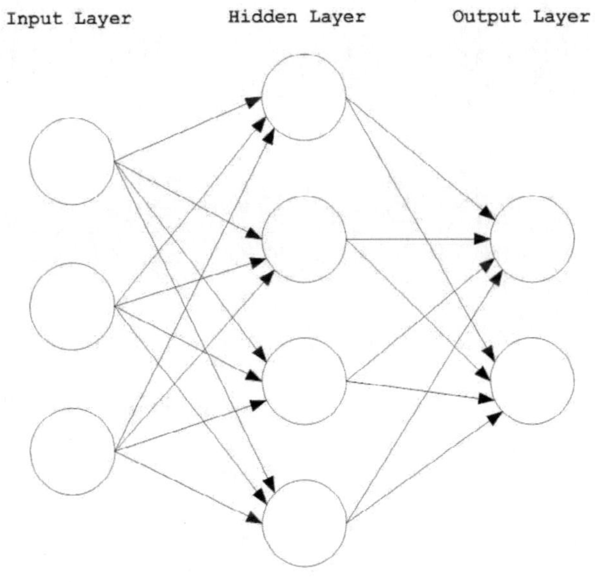

Figure 1: Neural Network.

Neural Network (NN) is an information processing paradigm inspired by the way the brain processes information and is typically organised in layers as shown in Figure 1. They are composed of a large number of highly interconnected processing elements, called 'neurones', working together to solve specific problems. A NN contains some form of 'learning rule' which modifies the weights of the connections according to the input patterns that they are provided with. Just like brains, NN learns by example by adjusting to the synaptic connections between the neurones. In order to train a NN, the weights need to be trained such that the error between the desired output and the actual output is reduced [10]. This process requires that the NN compute the Error derivative of the Weights (EW); that is, it needs to calculate how the error changes as each weight is increased or decreased slightly. The back-propagation algorithm [11] is the most common method for calculating the EW and is used for this study.

Neural Network Toolbox™Toolbox in Matlab/SIMULINK® is used here to train the NN-based estimators. In order to prevent over-fitting the early stopping technique is used. Furthermore, the following parameters are used - 3 hidden layers of size 10 (neurons) each; training time: 1000s; algorithm: 'random'; training method: 'Levenberg-Marquardt'; performance: 'Mean Square Error', etc.

2.2. Wind Speed Model

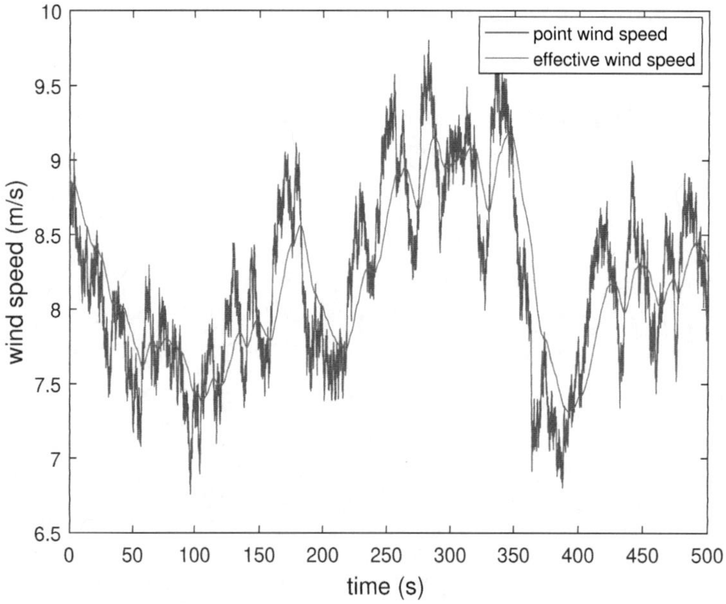

Figure 2: Point wind speed vs effective wind speed at a mean wind speed of 8 m/s.

The wind stochastically varies with time and continuously interacts with the rotor [12]. The effective wind speed is defined as wind speed averaged over the rotor plane such that the frequency response (or power spectrum) of hub torque remains uniform. It is derived here by filtering the point wind speed [13] through the filter introduced in [12]. The point wind speeds that account for the correlation of the cluster layout is obtained in DNV-GL Bladed. The effective wind speeds are required to simulate the Matlab/Simulink models (of Supergen Wind) throughout the paper. Point wind speed is illustrated in comparison to effective wind speed in Figure 2. Turbulence intensity of 10% is employed throughout this paper. The resulting effective wind speeds at a mean wind speed of 8 m/s for 5 turbines in a cluster are depicted in Figure 3. Similar results are obtained for different mean wind speeds, hence not included here.

3. Estimation using the Neural Network

An estimator is designed based on the NN introduced in Section 2.1. It is trained using the only turbine with a sensor in the cluster, and the trained NN-based estimator is subsequently applied to the remaining 4 turbines in the cluster. The full procedure for designing a NN-based estimator is summarised as follows:

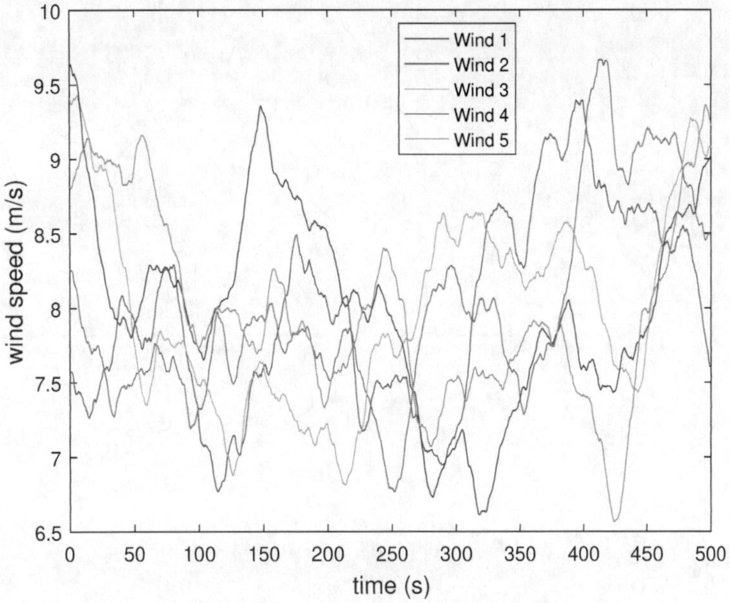

Figure 3: Correlated wind speeds for turbines 1 to 5 at a mean wind speed of 8 m/s.

(i) At regular intervals, i.e. every 1000 s in this paper, the inputs (different several combinations of torque, rotor speed and FAA as shown below) and the output (TBM) are measured and collected from (the model of) Turbine 1.

(ii) Based on the measurements from (i), a NN-based estimator is designed and trained.

(iii) Turbine 1 is equipped with a sensor and each of the rest with the nonlinear NN-based estimator from the steps above.

(iv) The above steps are repeated every 1000 s.

As previously mentioned the Matlab/Simulink model of the Supergen Wind 5MW exemplar wind turbine – which has been used by various researchers over the last decade, especially within the Supergen Wind Consortium – is employed to simulate the wind turbines. Each cluster is developed by duplicating the model, but each model would experience a different wind speed. In this paper, only one cluster is considered. The wind speed model for generating the wind speeds are described in Section 2.2.

As mentioned in Step (i) above, three different combinations of torque, pitch angle (present only in above rated wind speeds), rotor speed and FAA constitute the inputs, and the effect of a fault is also taken into account. Three different scenarios considered are as follows:

- Scenario 1: torque and rotor speed
- Scenario 2: torque, rotor speed and FAA
- Scenario 3: Scenario 2 is repeated, but it is assumed that a fault occurs, and the FAA measurement thus become unavailable. Although the FAA measurement is available when training the NN-based estimator, it become unavailable when the NN-based estimators are in operation. Therefore, a constant input of 0 for FAA, which is considered a fault, is applied, instead.

Figures 4, 5 and 6 respectively represent the results from Scenarios 1, 2 and 3. Each sub-figure in the figures depicts the measurement of TBM from each wind turbine in black in comparison

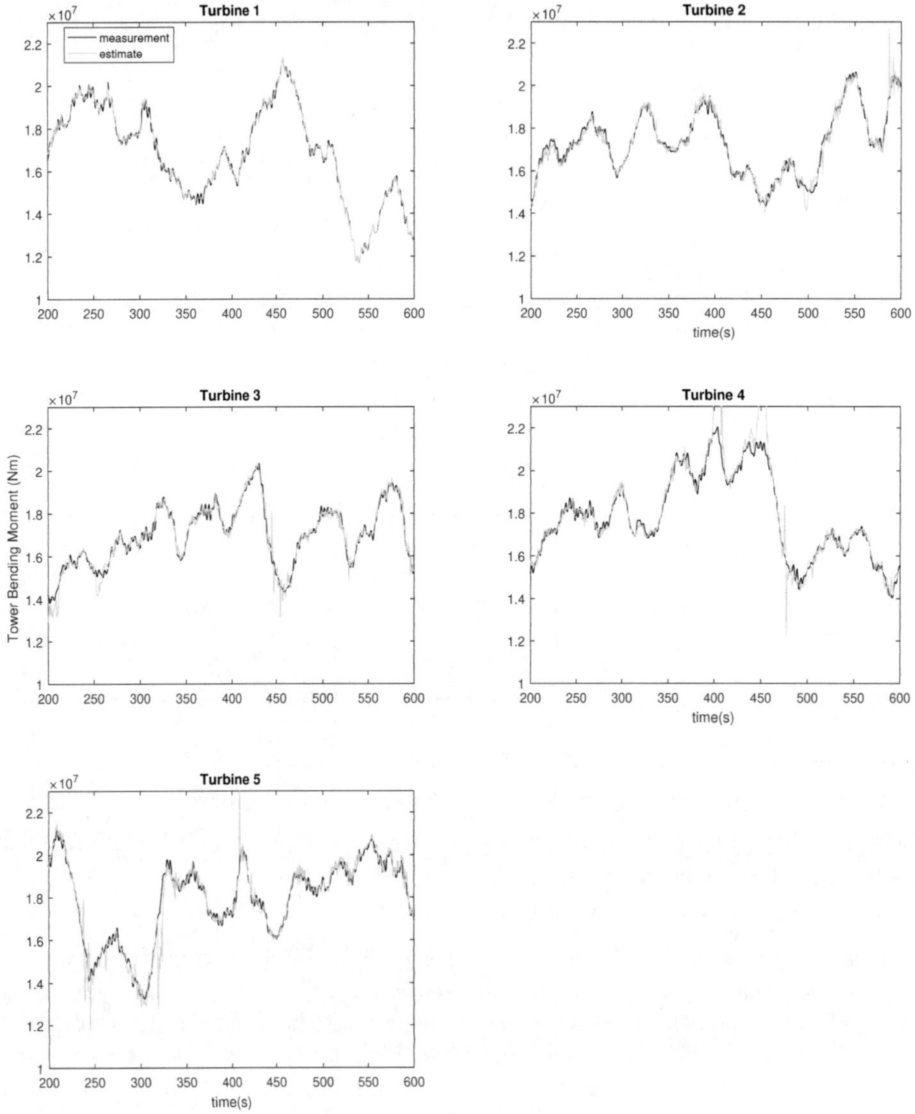

Figure 4: Scenario 1: Measurements vs NN estimates; note that the measurements for turbines 2 to 5 would not be available and are shown here for comparison purposes only.

to the estimate of TBM by the NN-based estimator in green. It is important to note that in real life, the TBM measurement would only be available from Turbine 1 since it is the only one equipped with a sensor. Moreover, Turbine 1 would not be equipped with a NN-based estimator; that is, only for comparison purposes, the measurements and estimates from each turbine are shown in the figures.

Figure 4 illustrates that in Scenario 1 the estimate from each NN-based estimator does not

IOP Conf. Series: Journal of Physics: Conf. Series **1222** (2019) 012022 doi:10.1088/1742-6596/1222/1/012022

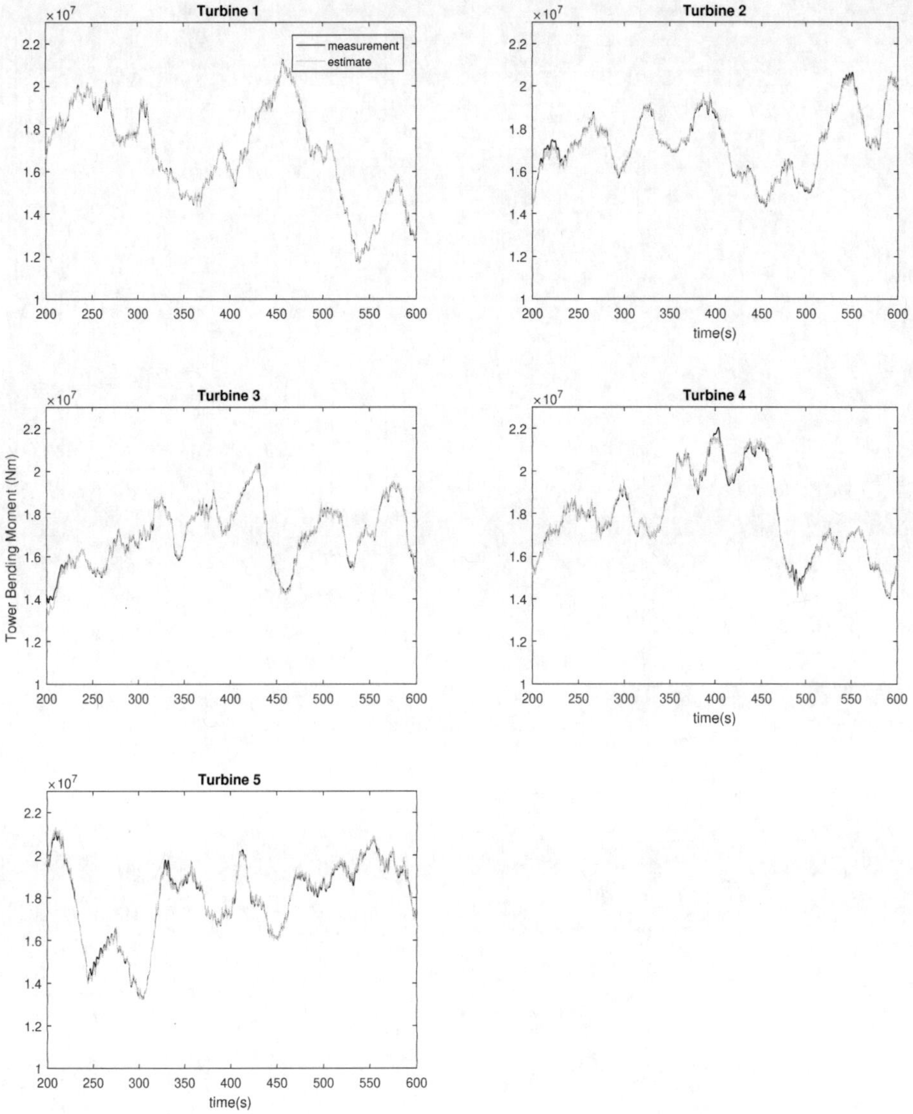

Figure 5: Scenario 2: Measurements vs NN estimates; note that the measurements for turbines 2 to 5 would not be available and are shown here for comparison purposes only.

track the measurement closely enough with Turbine 4 demonstrating the poorest performance. Figure 5 shows that in Scenario 2 the tracking is significantly improved by including an additional variable, FAA, in the training process, especially with Turbine 4 exhibiting the most significant improvement.

Even with a fault present in Scenario 3, Figure 6 demonstrates that the estimates follow the measurements more closely than Figure 4. For more detailed comparison, the power spectra of

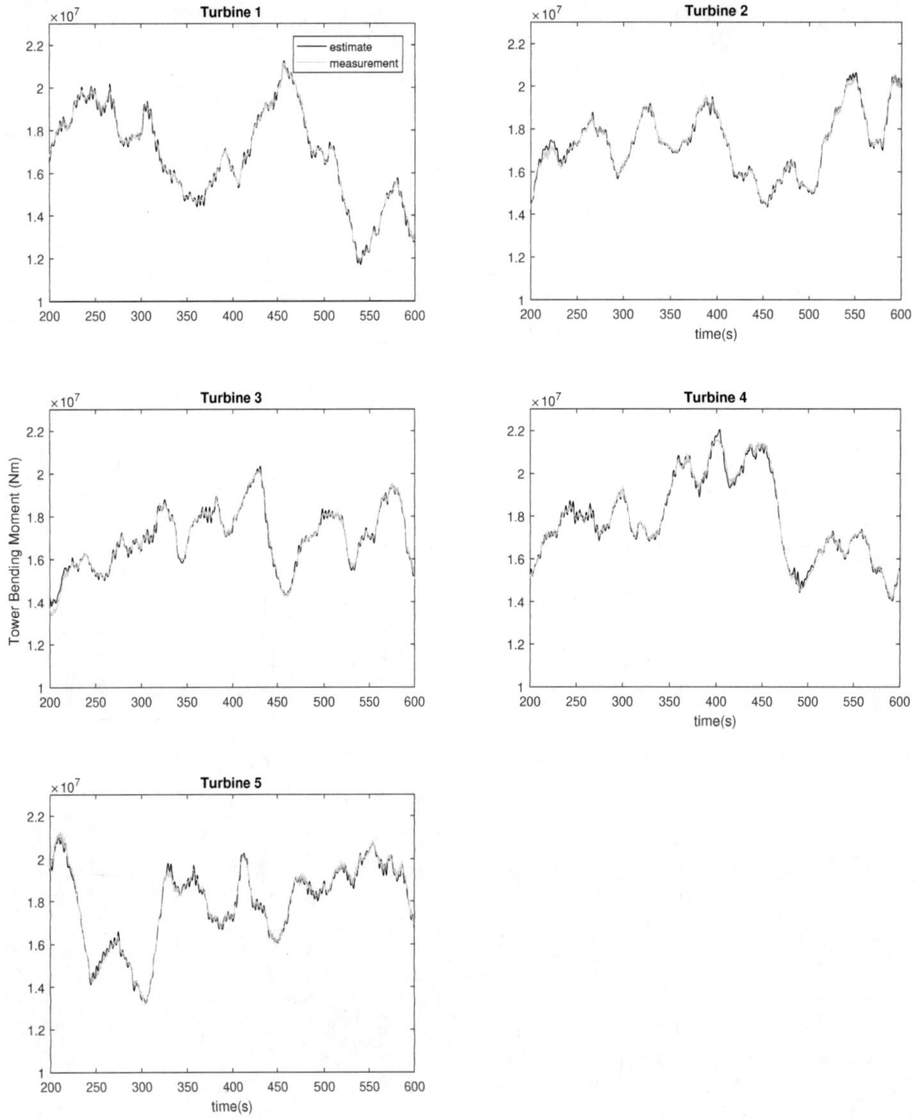

Figure 6: Scenario 3: Measurements vs NN estimates; note that the measurements for turbines 2 to 5 would not be available and are shown here for comparison purposes only.

the measurements and estimates are needed. The power spectra in Figure 7 compare Scenarios 2 and 3 in the frequency domain. Note that the power spectra are shown for Turbine 2 only here since the spectra for the remaining turbines demonstrate similar characteristics.

A more detailed examination of the time plots in Figures 5 and 6 reveals that some high frequency components are removed in Scenario 3 in comparison to Scenario 2. Now returning to the power spectra in Figure 7, the black spectrum has a peak at around 1.6 rad/s while the blue

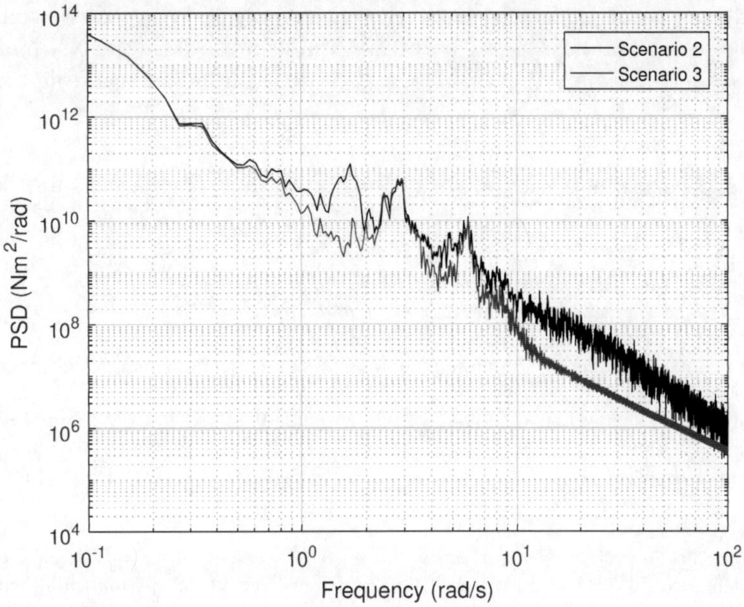

Figure 7: Power spectrum of the estimate of tower bending moment; black: Turbine 2 from Scenario 2, blue: Turbine 2 from Scenario 3.

spectrum does not. It can be deduced that the high frequency components that are removed in Figure 6 correspond to this particular peak, which is associated with the tower mode [14]. This result was expected in advance because the tower mode significantly impacts the FAA of the nacelle sitting on top of the tower. Note that the peaks at 3 and 6 rad/s shown on both the red and blue spectra correspond to 3P and 6P [15], respectively.

In summary, when the NN-based estimator is trained with torque, rotor speed and FAA, the performance of the NN-based estimator improves significantly in comparison to the NN-based estimator trained with torque and rotor speed only. Furthermore, the NN design is even robust to a fault to some degree although there are some differences in the frequency response when the fault occurs. The results of this feasibility study are significant because the proposed NN-based estimator can be applied to estimate not only TBM but also other variables, such as BBMs, which would also be expensive to measure. These estimates can be used to improve the performance of condition monitoring or control of a wind turbine and farm at a low cost.

Furthermore, by incorporating additional available variables to the array of inputs will improve the resolution of the NN-based estimators, thereby improving the estimation accuracy, i.e., other available variables besides torque, rotor speed and FAA would improve the estimators.

4. Conclusions

Having access to various different variables in wind turbines and farms that are normally unavailable could improve the performance of condition monitoring and control. This paper introduces a NN-based estimator to allow access to such variables by including the minimal number of additional sensors, which would otherwise be costly. Only one turbine is equipped with a sensor and the remaining turbines with a NN-based estimator that essentially replaces the sensor, hence only one sensor is required within a cluster.

The simulation results of this feasibility study demonstrate that the NN-based estimators estimate the TMB of each turbine within a cluster successfully. The simulations results also demonstrate that the resolution, hence the accuracy, of the NN-based estimator could be improved by introducing additional variables to the input array and that the NN-based estimator is robust to some degree to a fault. These results are significant because the NN-based estimators could be used to estimate not only the TBM but also other important variables, such as BBM.

Acknowledgments

This research was supported by Kyungpook National University Research Fund, 2017.

References

[1] Simon D 2006 *Optimal State Estimation: Kalman, H Infinity, and Nonlinear Approaches* (Wiley-Interscience)
[2] Salychev O 2004 *Applied Inertial Navigation: Problems and Solutions* (BMSTU Press)
[3] Barto A G 1995 *Neural Networks for Control* (MIT Press)
[4] Zeng Z and Wang J 2010 *Advances in Neural Network Research and Applications* (Springer-Verlag Berlin Heidelberg)
[5] Lei T, Barnes M, Smith S, ho Hur S, Stock A and Leithead W E 2015 *IEEE Transactions on Energy Conversion* **30** 1034–1051
[6] Thompson D 2018 *Mitigating size related limitations in wind turbine control* Ph.D. thesis University of Strathclyde
[7] Hur S and Leithead W E 2015 *Wind Energy*
[8] Hur S 2018 *Energy* **156** 360–370
[9] Obdam T S, Rademakers L W M M and Braam H 2009 Flight leader concept for wind farm load counting Tech. rep. We@Sea
[10] Aleksander I and Morton H 1995 *An introduction to neural computing* 2nd ed (London : International Thomson Computer Press)
[11] Rumelhart D E, Hinton G E and Williams R 1986 *Nature* **323** 533–536
[12] Leithead W 1992 *Proceedings of 14th British Wind Energy Association (BWEA) Conference, Nottingham*
[13] Munteanu I, Bratcu A, Cutululis N A and Ceanga E 2007 *Optimal Control of Wind Energy Systems: towards a Global Approach* (Springer)
[14] Leithead W and Connor B 2000 *International Journal of Control* **73** 1189 – 1212
[15] Burton T, Sharpe D, Jenkins N and Bossanyi E 2001 *Wind Energy Handbook* (John Wiley & Sons, Ltd)

IOP Conf. Series: Journal of Physics: Conf. Series **1222** (2019) 012023 doi:10.1088/1742-6596/1222/1/012023

Optimising yaw control at wind farm level

Ervin Bossanyi

DNV GL, One Linear Park, Avon Street, Bristol, BS2 0PS, UK

ervin.bossanyi@dnvgl.com

Abstract. The wind turbines in a wind farm must turn to face the wind to achieve maximum power production. Until now, each turbine achieves this individually: the nacelle-mounted wind vane measures the yaw misalignment, from which the turbine controller decides when to make a yaw correction by using a combination of heavy filtering and counters or dead-band hysteresis. In a wind farm context, each turbine could make use of information from its neighbours, and therefore, in principle, it should be possible to achieve better yaw control by using a centralised farm-level yaw control algorithm which uses input measurements from all the turbines, and then tells each turbine how to yaw. "Better" yaw control will always be a compromise between maximising energy production and minimising yaw actuator duty. Using a realistic dynamic simulation model of a wind farm which uses actual measured wind data as input, different possible yaw strategies are tested, and compared in terms of energy production and yaw actuator duty. The results indicate that centralised yaw control is likely to achieve a better compromise. There also is much current interest in the use of wake steering, where yaw setpoints are optimised to steer the wakes of some turbines away from others. Centralised yaw control may offer a better way to implement these setpoints, rather than sending yaw offsets to be acted on by the normal turbine yaw controllers with their own filtering, dead-band logic, etc.

1. Introduction

Conventionally, a wind vane on top of the nacelle measures yaw misalignment at a point which is poorly representative of the rotor-average wind direction and is also situated in highly disturbed flow, affected by flow around the nacelle and by the blade roots passing just upstream as the rotor rotates. Therefore, the wind vane signal is passed through a heavy low-pass filter, both to smooth out the short-term variations and to make the resulting signal more representative of rotor-average variations. To prevent frequent yaw manoeuvres which would cause unnecessary yaw system wear, a hysteresis dead-band is introduced. All this means that the yaw response is considerably delayed. Forward-pointing LiDAR can be used both to obtain a signal more representative of the rotor as a whole, and can also provide a small amount of preview, but it represents a considerable extra cost.

Yaw control is relatively slow and simple, and has rarely been studied in detail. Control algorithms typically use a low-pass filter and a hysteresis dead-band or a combination of counters and timers, generating discrete yaw manoeuvres from time to time. The parameters determining the yaw control are usually obtained by a combination of trial and error and field experience. The resulting yaw action depends on relatively low-frequency changes in wind direction, although they are measured at high frequency by the small wind vane. The wind plant simulator LongSim has been developed specifically to investigate control aspects of this sort, by allowing simulations with a short time-step (suitable for representing the wind vane measurements) which can run for long enough for the effects of low-frequency wind field variations to be experienced. The potential of this model to optimise single-turbine yaw control algorithms has been described previously [1]. Since then, the model has been extended to allow it to simulate a whole wind farm, accounting also for the correlations in wind field variations [2] between the turbines across the farm, and also for wake effects [3], [4], [5], including wake meandering [6] and wake deflection caused by any yaw misalignments [7]. This extended model is ideal for testing wind farm controllers, and its use to test active wake control algorithms such as induction (power set-point) control and wake steering has been presented in [8]. A more detailed description of the model can be found in [9].

In this paper, the model is used to test different wind farm yaw control algorithms and compare them against individual turbine yaw control. The results demonstrate a potential for significant reductions in yaw actuator duty and/or increased energy production for the whole wind farm.

If wake steering is to be used, part of the benefit is lost if the deliberate yaw misalignments, calculated by the wind farm controller, are applied simply as offsets and subjected to the yaw errors inherent in the individual turbine yaw control algorithms. Central yaw control as presented in this paper should be a better basis for adding wake steering control, to gain further energy capture benefits.

2. Methodology

Using the LongSim model, a time-domain simulation based on the Horns Rev 1 offshore wind farm (off the west coast of Denmark) has been set up. The turbine layout is shown in Figure 1. The 80 turbines are 80m in diameter, with a rated power of 2MW. The turbine power curve is known from public-domain information, but information about the turbine yaw behaviour is not readily available, and so realistic assumptions have been made, which may not match the actual reality for this site but are reasonable for this conceptual study. This includes modification of the power curve and rotor thrust for different yaw angles, and the actual yaw control strategy of the turbines. These assumptions are detailed in Section 2.1.

The wind conditions for the simulation are taken from publicly-available 10-minute data measured at the FINO-1 meteorological mast in the German North Sea. Although this is not the same as the Horns Rev site, it is probably reasonably representative and is certainly suitable for this conceptual study. Using a selected part of this wind record, a correlated wind field covering the entire farm has been generated, using synthetic turbulence for the higher frequencies. A half-hour simulation was used initially, and repeated with different yaw control philosophies, which are described in Section 2.2. A simulation time-step of 1 second was used. Results are presented in Section 3.

2.1. Assumptions about turbine yaw

The simple assumption that power decreases as $\cos^3$(yaw) seems not to be valid, and a number of cases have demonstrated a factor of $\cos^p$(yaw) with the exponent p being much less than 3, typically more like 1.4 – 2.0 (e.g. p=1.43 in [10]). Here the value p=1.4 has been used as a basis; however, this cannot simply be used to scale the entire power curve, as it would result in less than rated power in high winds where, in reality, pitch control would compensate for yaw to bring the power back to rated. Therefore, a factor $\cos^w$(yaw) has instead been applied to the wind speed used for the power curve look-up. For example, if $\cos^w$(yaw) = 0.95 the power at 10 m/s would be obtained from the power curve value for 9.5 m/s; and above rated, say at 20 m/s, the power would be calculated for 19 m/s, which is still rated power. For this turbine, w = 0.5 has been used as it gives a very similar result to p=1.4 in below-rated wind speeds while allowing the power curve still to reach rated, albeit at a higher wind speed than at zero yaw. The resulting power curves are illustrated in Figure 2 and Figure 3. For the turbine thrust look-up, the same factor w is assumed to apply to the effective wind speed.

For the yaw control logic, a 30-second first-order low-pass filter is applied to the wind direction, obtained as the sum of the nacelle position and the wind vane signal, and the nacelle position is subtracted from this to get the filtered yaw error (this avoids the problem of nacelle motion interfering with the filtering if it's applied directly to the wind vane signal). Then when the yaw error exceeds $\pm 8^{o}$, the turbine yaws at $\pm 0.3^{o}$/s until the filtered yaw error crosses zero.

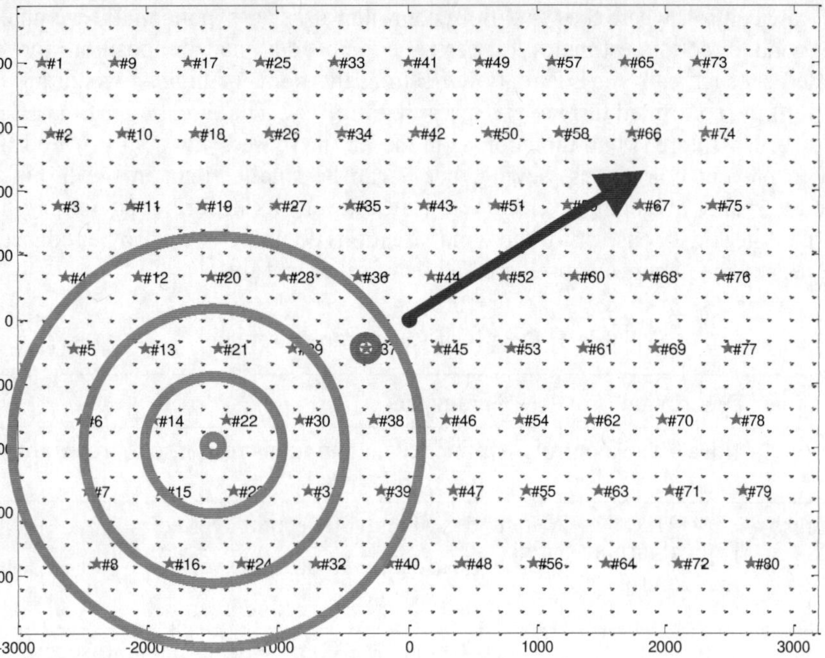

Figure 1: Turbine layout at Horns Rev 1 (axes in metres), with an illustration of central yaw control strategy. Purple arrow: approximately estimated wind direction, Red circle: controlled turbine, Green circles: equal-weighting contours, showing an upwind preview distance. The demanded nacelle direction of the controlled turbine is a weighted sum of directions at other turbines, weighted inversely with distance from the centre of the green circles.

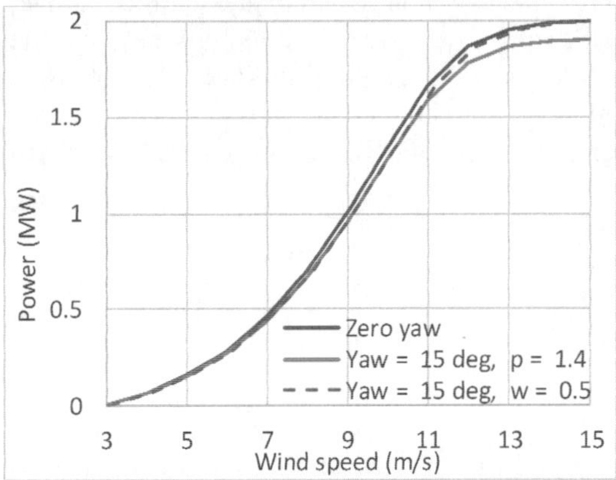

Figure 2: Assumed power curve at 15° yaw

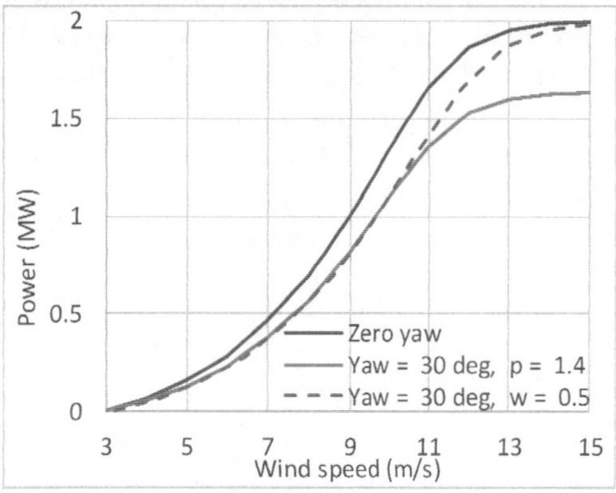

Figure 3: Assumed power curve at 30° yaw

2.2. Yaw control philosophies

The following yaw control philosophies have been simulated:

- 'Ideal': the turbine yaws continuously to maintain zero yaw misalignment. This is infeasible in practice but gives an upper limit on the possible energy capture.

- Turbine-based yaw control based on typical standard yaw control logic, using a 30-second first-order filter on the wind vane signal and various hysteresis dead-bands.
- Central yaw control, where the wind farm controller uses an exponentially-weighted average of turbine wind direction estimates to derive a demanded nacelle position for each turbine (illustrated schematically in Figure 1). Offsetting the point of highest weighting to a position upstream of the controlled turbine gives a preview effect. The direction estimates consist of the same 30-second filtered wind direction as in the turbine-based yaw control. A dead-band is still applied to prevent continuous yawing, but it can be smaller than in the turbine-based case, because the spatial averaging over adjacent turbines results in further smoothing. For the same reason, the filter time constant for the wind directions could perhaps also be reduced.

The strategies and the parameters used are detailed further in Table 1.

Table 1: Details of yaw control strategies

Strategy name	Description	Parameters
Ideal	"Ideal" yaw control	(immediate, continuous response to true rotor-average wind speed)
T1 (base case)	Typical turbine-based yaw control	8° dead-band, 30-second average
T2		5° dead-band, 30-second average
T3		2° dead-band, 30-second average
C1	Central yaw control with weighting factor exponential decay distance D and preview distance P	D=1000m, P=700m, 8° dead-band, 30-second average
C2		D=1000m, P=700m, 2° dead-band, 30-second average
C3		D=500m, P=500m, 5° dead-band, 30-second average
C4		D=300m, P=300m, 2° dead-band, 30-second average
C5		D=300m, P=300m, 8° dead-band, 30-second average

3. Results

The same half-hour simulation was repeated with each of the yaw control strategies defined above, and the total wind farm power was recorded along with the total yaw travel of all the turbines, and the total number of yawing events (in the 'Ideal' case the latter is effectively infinite because yawing is continuous). The wind speed and direction for the simulations are shown in Figure 4; this is the data from the met mast, after smoothing, which was used as a basis for the generation of the wind farm wind field.

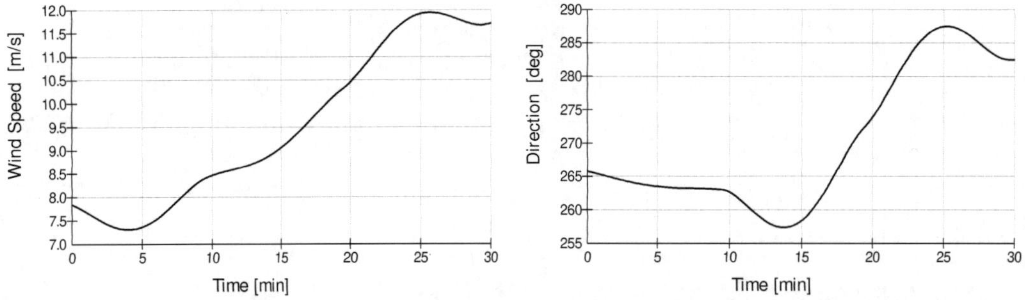

Figure 4: Wind conditions for the simulation (met mast data, smoothed)

Typical yaw time histories are shown in Figure 5, for an upstream turbine and a downstream one. The base case (T1) is compared against one of the central control cases (C3). The upstream turbine cannot benefit from any preview, because there are no turbines further upstream, but the benefit of preview is clear for the downstream one.

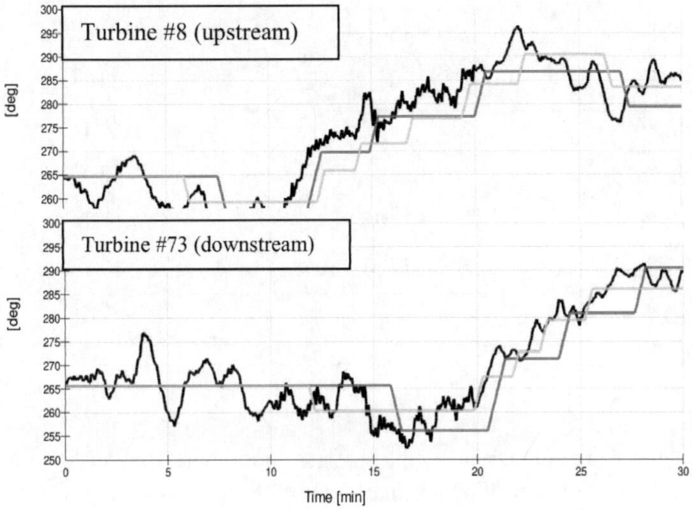

Figure 5: example turbine yaw responses at two turbines.

Black: "Ideal" yaw (actual rotor-average wind direction)

Red: Conventional turbine yaw

Green: Central yaw control; note the benefit of preview which is apparent at the downstream turbine (#73)

For all the cases, the trade-off between energy capture and yaw duty is shown in Figure 6. It is clear that the central yaw control strategies can both increase the energy capture and reduce the yaw duty, especially when measured in terms of yaw travel. The results are tabulated in Table 2.

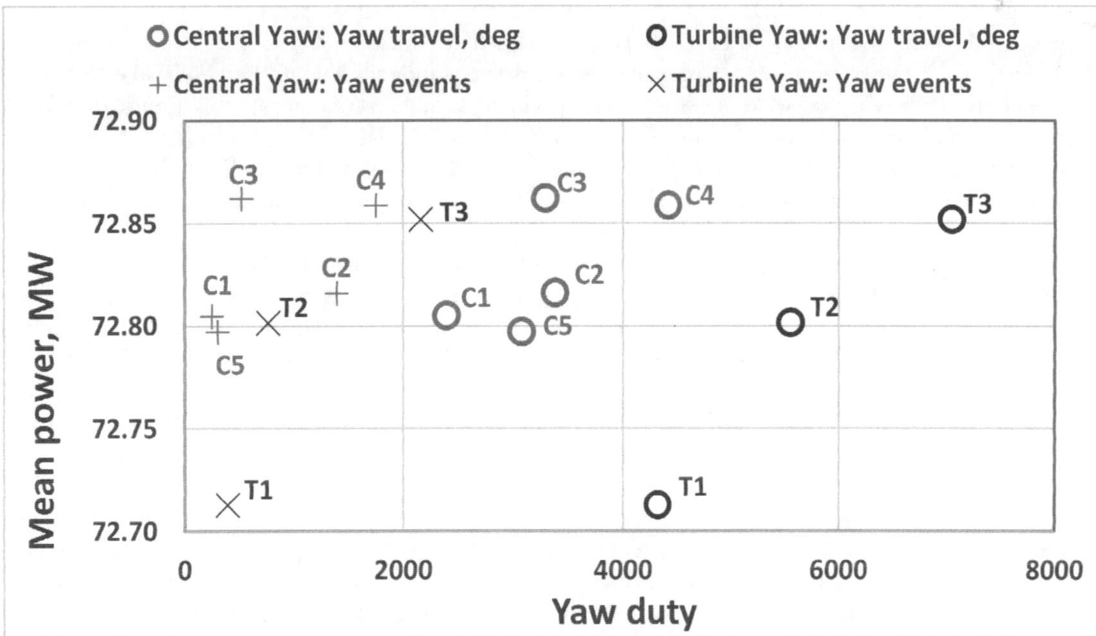

Figure 6: Results: trade-off of energy capture against yaw duty

Table 2: Summary of results

Case	Mean power, MW	Yaw travel, degrees	No. of yaw events
Ideal	77.72	36897.0	∞
T1 (base case)	72.71	4321.8	393
T2	72.80	5566.5	762
T3	72.85	7050.0	2163
C1	72.80	2397.3	250
C2	72.81	3381.9	1393
C3	72.86	3296.4	525
C4	72.86	4424.7	1752
C5	72.80	3077.7	302

4. Discussion

By combining wind direction estimates from other nearby turbines, a smoother, spatially-averaged measure of wind direction is available to each turbine, which may be used to improve the yaw control compared the conventional case where each turbine uses only its own wind vane. Furthermore, most of the turbines can benefit from preview information from any turbines which are situated further upstream. This leads to a better trade-off between energy output and yaw actuator duty, simultaneously increasing energy output while reducing total yaw travel significantly, and also reducing the number of yaw events.

The technique does rely on communication between the turbines, but this can be readily achieved through the wind farm SCADA system. It would clearly be advantageous for the yaw calculations to be carried out in a central wind farm controller, rather than at each turbine, and this is consistent with the increasing interest in using active wake control, such as induction control and/or wake steering [8]. Especially in the case of wake steering, the individual turbine yaw control can significantly detract from the implementation of the yaw setpoints which are calculated to optimise the total wind farm performance. Wake steering studies underway in the CL-Windcon project [11] are already demonstrating an improvement in wake steering performance when implemented using central yaw control. Results will be published in the near future.

Central yaw control also relies on the accurate measurement of nacelle position (also a requirement for wake steering). Currently, many turbines pay little attention to this; since the yaw control relies only on the wind vane, very crude nacelle position sensors are sufficient, and are often not even calibrated, as this is important, even for cable unwinding. Better sensors are readily available, however, such as GPS-based sensors.

The central direction estimates may have a secondary use as a way to detect yaw sensor errors on individual turbines, by detecting any large discrepancies between these direction estimates and the turbine's own measurements.

The simulations in this paper are considered reasonably realistic, and are able to use site-specific wind data to ensure that results are representative for a particular project. The simulations run reasonably fast on a single processor, which is useful for evaluating different strategies and tuning parameters, as this paper demonstrates, but it also means that longer simulations could then be run to test out a selected strategy over a longer run of data, e.g. several days or more. As ever, there are some aspects to the modelling which are not proven. For example, while the simulations are based on real met mast data, and this is used to generate representative wind fields over the whole wind farm, the met mast still only gives point data, so some assumptions are necessary. Standard models of spatial

IOP Conf. Series: Journal of Physics: Conf. Series **1222** (2019) 012023 doi:10.1088/1742-6596/1222/1/012023

coherence are assumed to apply, and the way in which low-frequency changes in conditions propagate through the farm also have to be assumed. In this case, the propagation of changes in direction through the wind farm is linked to the current wind speed and direction, which may not always be valid, and even this simple assumption needs further assumptions because those quantities are, by definition, not uniform across the farm in these cases. The assumptions are probably reasonable most of the time, but unusual events like thunderstorm fronts may be quite different in character.

5. Conclusions

In a wind farm context (even before considering any wake steering), centralised control of turbine yaw can lead to increased energy production and reduced yaw actuator duty when compared to conventional individual turbine yaw control. The example in this paper showed a very modest 0.2% increase in power production, compared to a more dramatic 24% decrease in yaw travel – in financial terms it is difficult to say which effect is the more significant, but it is a win-win situation. The technique benefits from spatial averaging, and a significant preview effect is also beneficial. These benefits are available for no additional cost, apart from the need for more accurate and properly calibrated nacelle position sensors. The paper presents a realistic dynamic simulation model which is appropriate for tuning and evaluating the performance of different yaw controllers, using site-specific wind conditions is available.

References

[1] "Long-term simulations for optimising yaw control and start-stop strategies", E A Bossanyi, T Delouvrié and S Lindahl, Proc. European Wind Energy Conference, Vienna, EWEA 2013.
[2] "Spectral structure of mesoscale winds over the water", X G Larsen et al, Q. J. R. Meteorol. Soc. 139: 685–700, April 2013.
[3] "Development of an eddy viscosity model for wind turbine wakes", J F Ainslie, Proc. 7th British Wind Energy Association conference, Oxford, 1985.
[4] "Calculating the flow field in the wake of wind turbines", J F Ainslie, Journal of Wind Engineering and Industrial Aerodynamics 27, pp. 213-224, 1988.
[5] "WindFarmer Theory Manual v5.3", DNV GL (2014).
[6] "Wake meandering – a pragmatic approach", G Larsen et al, Wind Energy 11 pp377-395, 2008.
[7] A Jiménez, A Crespo, E Migoya, "Application of a LES technique to characterize the wake deflection of a wind turbine in yaw", Wind Energy, 2010, 13:559-572.
[8] "Combining induction control and wake steering for wind farm energy and fatigue loads optimisation", E. Bossanyi, Proc. Torque from Wind conference, Milan 2018, (IOP proceedings).
[9] http://www.clwindcon.eu/wp-content/uploads/2017/03/CL-Windcon-D1.2-Wind-farm-models.pdf
[10] "Field test of wake steering at an offshore wind farm", P Fleming *et al*, Wind Energ. Sci., 2, 229–239, 2017.
[11] CL-Windcon project: http://www.clwindcon.eu.

IOP Conf. Series: Journal of Physics: Conf. Series **1222** (2019) 012024 doi:10.1088/1742-6596/1222/1/012024

On wind turbine control for participation in grid frequency regulation

Iker Elorza, Carlos Calleja and Aron Pujana-Arrese

P. J. M. Arizmendiarrieta, 2, 20500 Arrasate-Mondragón, Gipuzkoa, Spain

E-mail: ielorza@ikerlan.es

Abstract. An energy mix with a large contribution from wind power requires wind farm participation in frequency regulation. To do this effectively, it is necessary for wind turbines to curtail their power production by a certain amount, or 'delta', which serves as a reserve. Although it is relatively straightforward to achieve this in certain conditions, doing so consistently normally requires wind speed observers to estimate available power. A delta control method is proposed here, which does not rely on said observers and naturally extends popular torque and pitch control algorithms.

1. Introduction

The concept of frequency support with wind turbines is not new, and proportional de-loading for that purpose is relatively simple to achieve [1]. The idea is simple - operate turbines at a constant fraction, e.g. 75%, of the maximum power available, for any given wind conditions. Implementation is also simple, at least in certain wind conditions - modify the generator speed-torque curve or the minimum pitch, or both, for operation with a power coefficient that is a fraction, e.g. 75%, of the optimum power coefficient. In practice, ensuring proportional de-rating at all wind speeds is slightly more subtle, and some implementation choices are to be made - e.g. Loukarakis et al. [2] use two look-up tables, one 1D for pitch and another 2D for power, whereas two 1D tables are enough and no considerations are made about constant speed operating regions.

It is also arguable that delta control has a role to play in wind farm participation in grid frequency regulation [3] [4] [5] [6] [7] [8]. The idea is, again, simple - operate a turbine at a constant power delta, e.g. 1MW, below the maximum power available, for any given wind conditions. Implementation is not as simple, however - different wind speeds require operation at different power coefficient values and, as far as we have been able to review the literature, all reported constant delta methods require wind speed estimation (see, for example, [9]), with one exception (discussed below).

Kristoffersen described a Danish offshore farm's operational experience with 'Delta Control', which keeps a constant spinning reserve power [10]. However, no description is given of how said control was implemented.

Wang et al. [11] mention that proportional curtailment strategies have been demonstrated in a workshop paper by the same authors. However, they simply suppose it is technically possible to do constant curtailment 'as it is part of the regulation capacities required by the Danish [authorities for some] wind farm'.

Baccino et al. [12] use local wind estimations, without further discussion, as do Aho et al. [13].

Hansen et al. [14] also use a wind estimator, and they estimate available power based on the difference between the optimal speed-power curve and current power. This overestimates available power, because increasing generator torque to increase power output results in the rotor slowing down, wich in turn results in generator power decreasing along the optimal speed-power curve.

A delta control scheme not based on wind speed estimation is briefly described in [15], where they use a 2D look-up table, giving the generator power setpoint from generator speed and power delta. This allows overspeed-based delta control. The problem with this strategy is that generator speed quickly increases with power delta, and it does therefore not work at wind speeds high enough to result in operating points near or at the maximum generator speed, since no further over-speed is possible then. We will show that this can be done with a 1D look-up table, as well as extend it to all wind speeds.

In this paper, we take Kristoffersen's classification [10] of active power control functions as a basis for our discussion. Said classification is as follows (quotes taken from [10]):

- **Absolute power limitation**: 'the wind farm is able to operate at reduced levels with all turbines running'.
- **Balance control**: 'the wind farm is able to participate in the regional secondary control'.
- **Power rate limitation**: 'when the wind speed increases, the wind farm is able to impose a positive rate of change (dP/dt) limitation'.
- **Delta control**: 'the wind farm is controlled so the power production is an adjustable number of MWs below the possible power'. Here, we make the following further distinction:
  - **Proportional delta control**: the power production is proportional to available power, and their ratio is adjusted externally, e.g. to 85%.
  - **Constant delta control**: the difference between available power and power production is adjusted externally, e.g. to 10 MW.

We assume that all active power control functions above, although they are formally defined only at farm level, are also performed identically at turbine level, while we make the provision that setpoints may vary from one turbine to another and are decided by the farm controller. As already discussed, although the literature presents several approaches to turbine-level control functions of this sort, it is not trivial to extract from it a readily industrializable set of algorithms. Said extraction, and the addition of our own contributions, is the purpose of this paper.

2. Torque and pitch control

Following [16], we consider a base turbine controller with two generator speed regulators, e.g. PI controllers, one of which modifies the blades' collective pitch angle, β, while the other one modifies the generator torque Q_e.

The generator speed setpoint for the pitch controller, ω_β, is always the maximum operating speed, ω_{max}. The lower pitch angle saturation limit is as follows:

$$\beta_{min} = \beta_\delta (Q_e) \tag{1}$$

where $\beta_\delta (Q_e)$ is a function of generator torque Q_e which is chosen to maximize the power coefficient – we use subscript δ, rather than the more usual 'opt' because we will use this torque for delta control in section 6.

On the contrary, the generator speed setpoint for the torque controller, ω_q, is switched between ω_{max} and the minimum operating speed, ω_{min}. The choice between the two is made

IOP Conf. Series: Journal of Physics: Conf. Series **1222** (2019) 012024 doi:10.1088/1742-6596/1222/1/012024

based on proximity to the actual generator speed, ω, and the generator torque saturation limits are chosen so that:

$$Q_{\max} = \begin{cases} P_{\mathrm{n}}/\omega_{\max}, & \text{if } \omega_{\mathrm{q}} = \omega_{\max} \\ Q_{\delta}(\omega), & \text{if } \omega_{\mathrm{q}} = \omega_{\min} \end{cases} \tag{2}$$

$$Q_{\min} = \begin{cases} Q_{\delta}(\omega), & \text{if } \omega_{\mathrm{q}} = \omega_{\max} \\ 0, & \text{if } \omega_{\mathrm{q}} = \omega_{\min} \end{cases} \tag{3}$$

where P_{n} is the rated power, while $Q_{\delta}(\omega)$ is a function of ω which is chosen to make the tip-speed ratio converge to its optimal value – we use subscript δ, rather than the more usual 'opt' because we will use this torque for delta control in section 6. Figure 1a shows the resulting speed-torque curve.

3. Operation at optimal tip-speed ratio

The method for the choice of $Q_{\delta}(\omega)$ is well established, e.g. in [17]. It is based on the following simple turbine model:

$$J\frac{\dot{\omega}}{b} = Q_{\mathrm{a}} - b\left(Q_{\mathrm{e}} + Q_{\mathrm{l}}\right) \tag{4}$$

where J is the rotor inertia, b is the gearbox ratio, Q_{a} is the aerodynamic torque and Q_{l} is the torque due to mechanical losses.

The available aerodynamic power is:

$$P_{\mathrm{av}} = \frac{1}{2}\rho\pi R^2 U^3 C_{\mathrm{p}}\left(\lambda_{\mathrm{opt}}, \beta_{\mathrm{opt}}\right) \tag{5}$$

where ρ, R and U are the air density, rotor radius and wind speed, respectively, while λ_{opt} and β_{opt} are the tip-speed ratio and blade pitch angle, respectively, for which the power coefficient C_{p} is the greatest. Operating at $\beta = \beta_{\mathrm{opt}}$ is trivial - one need only choose $\beta_{\delta} = \beta_{\mathrm{opt}}$. However, λ cannot be directly manipulated, or indeed measured, since it is defined as follows:

$$\lambda = \frac{R\omega}{bU} \tag{6}$$

When operating at $\beta = \beta_{\mathrm{opt}}$ but, in general, $\lambda \neq \lambda_{\mathrm{opt}}$, the aerodynamic torque is:

$$Q_{\mathrm{a}} = \frac{1}{2}\rho b\pi R^2 U^3 \frac{C_{\mathrm{p}}\left(\lambda, \beta_{\mathrm{opt}}\right)}{\omega} \tag{7}$$

Using (6), we may rewrite (7) thus:

$$Q_{\mathrm{a}} = \frac{1}{2}\rho\pi R^5 \frac{C_{\mathrm{p}}\left(\lambda, \beta_{\mathrm{opt}}\right)}{\lambda^3}\frac{\omega^2}{b^2} \tag{8}$$

Note from (4) and (8) that $\lambda = \lambda_{\mathrm{opt}}$ becomes a fixed point if we choose:

$$Q_{\mathrm{e}} = \frac{1}{2}\rho\pi R^5 \frac{C_{\mathrm{p}}\left(\lambda_{\mathrm{opt}}, \beta_{\mathrm{opt}}\right)}{\lambda_{\mathrm{opt}}^3}\frac{\omega^2}{b^3} - Q_{\mathrm{l}} \tag{9}$$

It is also interesting to know whether (4), (8) and (9) lead to $\lambda = \lambda_{\mathrm{opt}}$ being asymptotically stable. To find out, we use (6) to rewrite (4), (8) and (9) thus:

$$\dot{\lambda} = \frac{1}{2}\rho\pi R^4 U J^{-1}\lambda^2 \left[\frac{C_\mathrm{p}(\lambda, \beta_\mathrm{opt})}{\lambda^3} - \frac{C_\mathrm{p}(\lambda_\mathrm{opt}, \beta_\mathrm{opt})}{\lambda_\mathrm{opt}^3}\right] \tag{10}$$

We then choose the following Lyapunov function candidate:

$$L = (\lambda - \lambda_\mathrm{opt})^2 \tag{11}$$

Derivation of (11) w.r.t. time and substitution of (10) yield:

$$\dot{L} = \rho\pi R^4 U J^{-1}\lambda^2 (\lambda - \lambda_\mathrm{opt}) \left[\frac{C_\mathrm{p}(\lambda, \beta_\mathrm{opt})}{\lambda^3} - \frac{C_\mathrm{p}(\lambda_\mathrm{opt}, \beta_\mathrm{opt})}{\lambda_\mathrm{opt}^3}\right] \tag{12}$$

From (12), the fixed point $\lambda = \lambda_\mathrm{opt}$ is asymptotically stable if:

$$\begin{cases} \frac{C_\mathrm{p}(\lambda,\beta_\mathrm{opt})}{\lambda^3} > \frac{C_\mathrm{p}(\lambda_\mathrm{opt},\beta_\mathrm{opt})}{\lambda_\mathrm{opt}^3}, & \text{for } \lambda < \lambda_\mathrm{opt} \\ \frac{C_\mathrm{p}(\lambda,\beta_\mathrm{opt})}{\lambda^3} < \frac{C_\mathrm{p}(\lambda_\mathrm{opt},\beta_\mathrm{opt})}{\lambda_\mathrm{opt}^3}, & \text{for } \lambda > \lambda_\mathrm{opt} \end{cases} \tag{13}$$

The fulfillment of these conditions may easily be verified from a turbine's power coefficient curve.

The choice of $Q_\delta(\omega)$ is therefore based on (9), i.e.:

$$Q_\delta = \frac{1}{2}\rho\pi R^5 \frac{C_\mathrm{p}(\lambda_\mathrm{opt}, \beta_\mathrm{opt})}{\lambda_\mathrm{opt}^3}\frac{\omega^2}{b^3} - Q_\mathrm{l} \tag{14}$$

Note from (2), (3) and (14) that (9) is fulfilled as long as $\lambda = \lambda_\mathrm{opt}$ lies within $\omega \in (\omega_\mathrm{min}, \omega_\mathrm{max})$, because Q_e saturates at Q_δ.

4. Operation at constant rotor speed

Once ω_min or ω_max is reached, it is no longer possible to operate the turbine at $\lambda = \lambda_\mathrm{opt}$. The tip-speed ratio at which we must now operate the turbine, λ_n, is determined by wind speed, U, thus:

$$\lambda_\mathrm{n} = \frac{R\omega_\mathrm{n}}{bU} \tag{15}$$

where ω_n is either ω_min or ω_max.

It is still possible, however, to choose the pitch angle, β, which also influences efficiency. We are therefore interested in finding $\beta_{\mathrm{opt}|\lambda_\mathrm{n}}$, the optimal pitch angle for the tip-speed ratio we are forced to operate at.

When operating at $\lambda = \lambda_\mathrm{n}$, the aerodynamic torque is:

$$Q_\mathrm{a} = \frac{1}{2}\rho b\pi R^2 U^3 \frac{C_\mathrm{p}(\lambda_\mathrm{n}, \beta)}{\omega_\mathrm{n}} \tag{16}$$

In steady state, (4) and (16) become

$$Q_\mathrm{e} = \frac{1}{2}\rho\pi R^2 U^3 \frac{C_\mathrm{p}(\lambda_\mathrm{n}, \beta)}{\omega_\mathrm{n}} - Q_\mathrm{l} \tag{17}$$

From (17), it is possible to numerically calculate a look-up table giving $\beta_{\mathrm{opt}|\lambda_\mathrm{n}}$ as a function of Q_e, to be applied when $Q_\mathrm{e} \notin (Q_\delta(\omega_\mathrm{min}), Q_\delta(\omega_\mathrm{max}))$. The choice of $\beta_\delta(Q_\mathrm{e})$ is, therefore, made as follows:

$$\beta_\delta(Q_\mathrm{e}) = \begin{cases} \beta_{\mathrm{opt}|\lambda_\mathrm{n}}(Q_\mathrm{e}), & \text{if } Q_\mathrm{e} \notin (Q_\delta(\omega_\mathrm{min}), Q_\delta(\omega_\mathrm{max})) \\ \beta_\mathrm{opt}, & \text{otherwise} \end{cases} \tag{18}$$

5. Absolute power limitation, balance control and power rate limitation

Absolute power limitation at rated power is often part of the torque and pitch control described in section 2. It is simply implemented by modifying (2) thus:

$$Q_{\text{max}} = \begin{cases} P_{\text{n}}/\omega, & \text{if } \omega_{\text{q}} = \omega_{\text{max}} \\ Q_{\delta}(\omega), & \text{if } \omega_{\text{q}} = \omega_{\text{min}} \end{cases} \tag{19}$$

We may generalize (19) to any power setting P_{r} thus:

$$Q_{\text{max}} = \begin{cases} P_{\text{r}}/\omega, & \text{if } \omega_{\text{q}} = \omega_{\text{max}} \\ \min\left(Q_{\delta}(\omega), P_{\text{r}}/\omega\right), & \text{if } \omega_{\text{q}} = \omega_{\text{min}} \end{cases} \tag{20}$$

This of course requires the upper torque limit to supersede the lower one, so that situations in which $P_{\text{r}}/\omega < Q_{\delta}(\omega)$ are appropriately handled. We therefore modify (3) thus:

$$Q_{\text{min}} = \begin{cases} \min\left(Q_{\delta}(\omega), P_{\text{r}}/\omega\right), & \text{if } \omega_{\text{q}} = \omega_{\text{max}} \\ 0, & \text{if } \omega_{\text{q}} = \omega_{\text{min}} \end{cases} \tag{21}$$

Note that dynamics are the same as in section 2 for $Q_{\delta}(\omega) < P_{\text{r}}/\omega$, i.e. at wind speeds insufficient to reach the power setting. However, at higher wind speeds we have, instead of (9):

$$Q_{\text{e}} = P_{\text{r}}/\omega \tag{22}$$

From (4), (6), (7) and (22):

$$\frac{J}{b^2}\dot{\omega} = \frac{1}{2}\rho\pi R^2 U^3 \frac{C_{\text{p}}(\lambda, \beta_{\text{opt}})}{\omega} - P_{\text{r}}/\omega - Q_{\text{l}} \tag{23}$$

Equation (23) has a fixed point at $\omega = \omega_0$ and $\lambda = \lambda_0 = R\omega_0/bU$, which satisfies

$$\frac{1}{2}\rho\pi R^2 U^3 C_{\text{p}}(\lambda_0, \beta_{\text{opt}}) - \omega_0 Q_{\text{l}} = P_{\text{r}} \tag{24}$$

We are, again, interested in the stability of the fixed point at $\omega = \omega_0$, so we choose the following Lyapunov function candidate:

$$L = (\omega - \omega_0)^2 \tag{25}$$

Derivation of (25) w.r.t. time and substitution of (23) yield:

$$\dot{L} = 2\frac{\omega - \omega_0}{\omega}\frac{b^2}{J}\left[\frac{1}{2}\rho\pi R^2 U^3 C_{\text{p}}(\lambda, \beta_{\text{opt}}) - P_{\text{r}} - Q_{\text{l}}\omega\right] \tag{26}$$

From (26), the fixed point $\omega = \omega_0$ is asymptotically stable if:

$$\begin{cases} C_{\text{p}}(\lambda, \beta_{\text{opt}}) > C_{\text{p}}(\lambda_0, \beta_{\text{opt}}), & \text{for } \lambda < \lambda_0 \\ C_{\text{p}}(\lambda, \beta_{\text{opt}}) < C_{\text{p}}(\lambda_0, \beta_{\text{opt}}), & \text{for } \lambda > \lambda_0 \end{cases} \tag{27}$$

and

$$\begin{cases} Q_{\text{l}}(\omega) < Q_{\text{l}}(\omega_0) & \text{for } \omega < \omega_0 \\ Q_{\text{l}}(\omega) > Q_{\text{l}}(\omega_0) & \text{for } \omega > \omega_0 \end{cases} \tag{28}$$

The fulfillment of these conditions may easily be verified from a turbine's power coefficient curve and mechanical loss model.

Balance control is essentially the same as absolute power limitation, but with a Pr that changes, while power rate limitation may be achieved by maintaining a secondary power limit which is perpetually a small increment above current power, said increment being the maximum power rate times the time step.

6. Proportional delta control

Proportional delta control at low wind speeds may easily be accomplished by modifying (14) thus:

$$Q_\delta = \frac{1}{2}\rho\pi R^5 k_\delta \frac{\omega^2}{b^3} - Q_l \tag{29}$$

where

$$k_\delta = \frac{C_\mathrm{p}(\lambda_\delta, \beta_\delta)}{\lambda_\delta^3} \tag{30}$$

Here λ_δ and β_δ are such that

$$\frac{C_\mathrm{p}(\lambda_\delta, \beta_\delta)}{C_\mathrm{p}(\lambda_\mathrm{opt}, \beta_\mathrm{opt})} = 1 - \delta \tag{31}$$

δ being the proportion of the available power to be kept as a reserve. This is implemented via two look-up tables, giving k_δ and β_δ, respectively, as a function of δ.

Note that solutions of (31) are generally non-unique, which leaves one degree of freedom for the choice of λ_δ and β_δ on the basis of criteria other than those discussed here (see, for example, [18]).

It is possible to prove that (29) leads to operation at $\lambda = \lambda_\delta$ the same way that it has been proven that (14) leads to operation at $\lambda = \lambda_\mathrm{opt}$.

When, at higher or lower wind speeds, we are constrained to operation at $\lambda = \lambda_\mathrm{n}$, the available aerodynamic power is

$$P_\mathrm{av} = \frac{1}{2}\rho\pi R^2 U^3 C_\mathrm{p}\left(\lambda_\mathrm{n}, \beta_{\mathrm{opt}|\lambda_\mathrm{n}}\right) \tag{32}$$

We may choose β_δ such that

$$\frac{C_\mathrm{p}(\lambda_\mathrm{n}, \beta_\delta)}{C_\mathrm{p}\left(\lambda_\mathrm{n}, \beta_{\mathrm{opt}|\lambda_\mathrm{n}}\right)} = 1 - \delta \tag{33}$$

It is again possible to use (17) and (33) to produce a two-dimensional look-up table giving β_δ as a function of Q_e and δ.

Lastly, for wind speeds above rated, it is necessary to limit P_r to $P_\mathrm{n}(1-\delta)$.

From the definition of δ, it is easy to estimate the reserve power as $\delta P_\mathrm{c}/(1-\delta)$, with P_c the current power.

7. Constant delta control

If we want to keep a constant power, P_δ, as a reserve, the choice of λ_δ and β_δ may appear less obvious than in section 6, because δ depends on P_av, which is not directly known:

$$\delta = P_\delta / P_\mathrm{av} \tag{34}$$

However, we need simply substitute (5) and (34) into (31) to get

$$C_\mathrm{p}(\lambda_\mathrm{opt}, \beta_\mathrm{opt}) - C_\mathrm{p}(\lambda_\delta, \beta_\delta) = \frac{2P_\delta}{\rho\pi R^2 U^3} \tag{35}$$

Because (35) makes explicit reference to U, it is not directly usable without a measurement or estimation of the wind speed. We therefore rewrite (35) thus:

$$\frac{C_\mathrm{p}(\lambda_\mathrm{opt}, \beta_\mathrm{opt}) - C_\mathrm{p}(\lambda_\delta, \beta_\delta)}{\lambda^3} = \frac{2b^3 P_\delta}{\rho\pi R^5 \omega^3} \tag{36}$$

IOP Conf. Series: Journal of Physics: Conf. Series **1222** (2019) 012024 doi:10.1088/1742-6596/1222/1/012024

From (36), we pick our strategy. We shall choose λ_δ and β_δ so that

$$\frac{C_{\mathrm{p}}\left(\lambda_{\mathrm{opt}},\beta_{\mathrm{opt}}\right)-C_{\mathrm{p}}\left(\lambda_\delta,\beta_\delta\right)}{\lambda_\delta^3}=\frac{2b^3P_\delta}{\rho\pi R^5\omega^3} \tag{37}$$

It is easy to implement (37) as two look-up tables, giving k_δ and β_δ, respectively, as functions of $2b^3P_\delta/\rho\pi R^5\omega^3$. We may then use β_δ in (1) and k_δ in (29). However, we must have $Q_\delta \geq 0$ so, from (29), there is a limit to our choice of k_δ at:

$$\frac{1}{2}\rho\pi R^5 k_\delta \frac{\omega^2}{b^3}-Q_1=0 \tag{38}$$

This, in turn, means that the actual power reserve is $\hat{P}_\delta \leq P_\delta$ when (38) is reached. It can be calculated thus:

$$\frac{C_{\mathrm{p}}\left(\lambda_{\mathrm{opt}},\beta_{\mathrm{opt}}\right)-C_{\mathrm{p}}\left(\lambda_\delta,\beta_\delta\right)}{\lambda_\delta^3}=\frac{2b^3\hat{P}_\delta}{\rho\pi R^5\omega^3} \tag{39}$$

The limit (38) can be implemented as two look-up tables giving limit values for k_δ and β_δ as a function of ω. A third look-up table giving the value of λ_δ corresponding to (38) allows us to use (39) to report actual power reserve $\hat{P}_\delta$.

We are, of course, interested in the dynamic characteristics of (4), (29) and (37), from which we now get, instead of (10):

$$\dot{\lambda}=\frac{1}{2}\rho\pi R^4 U J^{-1}\lambda^2\left[\frac{C_{\mathrm{p}}\left(\lambda,\beta_\delta\right)}{\lambda^3}-\frac{C_{\mathrm{p}}\left(\lambda_\delta,\beta_\delta\right)}{\lambda_\delta^3}\right] \tag{40}$$

There always exist $\lambda_0 \geq \lambda_{\mathrm{opt}}$ and $\beta_0 \geq \beta_{\mathrm{opt}}$ such that either (37) or (38) be satisfied for $\lambda_\delta=\lambda_0$ and $\beta_\delta=\beta_0$, so there is a fixed point of (40) at $\lambda=\lambda_\delta=\lambda_0$ and $\beta=\beta_\delta=\beta_0$. We therefore choose the following Lyapunov function candidate:

$$L=\left(\lambda-\lambda_0\right)^2 \tag{41}$$

Derivation w.r.t. and substitution of (40) yield:

$$\dot{L}=\rho\pi R^4 U J^{-1}\lambda^2\left(\lambda-\lambda_0\right)\left[\frac{C_{\mathrm{p}}\left(\lambda,\beta_\delta\right)}{\lambda^3}-\frac{C_{\mathrm{p}}\left(\lambda_\delta,\beta_\delta\right)}{\lambda_\delta^3}\right] \tag{42}$$

From (42), the fixed point is asymptotically stable if:

$$\begin{cases}\frac{C_{\mathrm{p}}(\lambda,\beta_\delta)}{\lambda^3}>\frac{C_{\mathrm{p}}(\lambda_\delta,\beta_\delta)}{\lambda_\delta^3}, & \text{for } \lambda<\lambda_0\\ \frac{C_{\mathrm{p}}(\lambda,\beta_\delta)}{\lambda^3}<\frac{C_{\mathrm{p}}(\lambda_\delta,\beta_\delta)}{\lambda_\delta^3}, & \text{for } \lambda>\lambda_0\end{cases} \tag{43}$$

Whether or not these conditions are satisfied is determined by the nature of C_{p}.

When, at higher or lower wind speeds, we are constrained to operation at $\lambda=\lambda_{\mathrm{n}}$, choose β_δ so that, instead of (37):

$$\frac{C_{\mathrm{p}}\left(\lambda_{\mathrm{n}},\beta_{\mathrm{opt}|\lambda_{\mathrm{n}}}\right)-C_{\mathrm{p}}\left(\lambda_{\mathrm{n}},\beta_\delta\right)}{\lambda_{\mathrm{n}}^3}=\frac{2b^3P_\delta}{\rho\pi R^5\omega_{\mathrm{n}}^3} \tag{44}$$

It is again possible to use (17) and (44) to produce a two-dimensional look-up table giving β_δ as a function of Q_{e} and P_δ.

Lastly, for wind speeds above rated, it is necessary to limit P_{r} to $P_{\mathrm{n}}-P_\delta$.

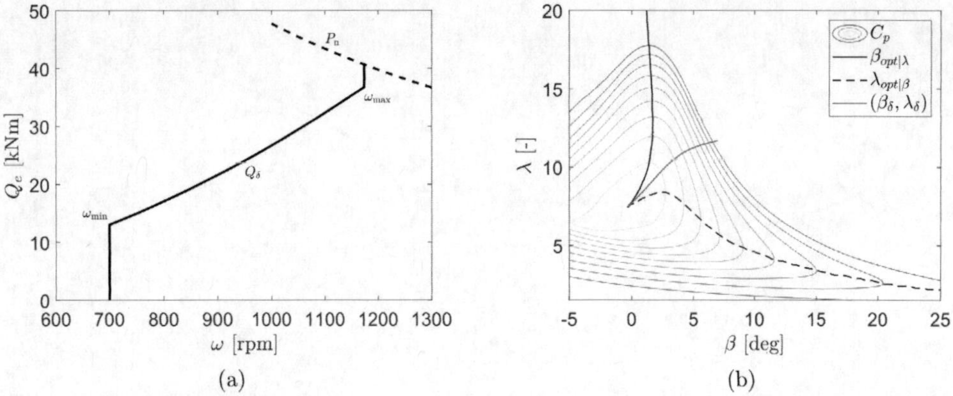

Figure 1: Torque-speed curve (a) and choice of $\{\beta_\delta, \lambda_\delta\}$ trajectory (b)

8. Example: NREL 5MW

Figure 1b shows the contour plot of the power coefficient of NREL's 5MW reference turbine model [19], as a function of tip-speed ratio and pitch angle. The lines of optimum pitch angle, given the tip-speed ratio, and optimum tip-speed ratio, given the pitch angle, are also shown, together with an arbitrary $\{\beta_\delta, \lambda_\delta\}$ trajectory, along which the operating point moves, via modification of β_δ and k_δ, up and down the C_p 'hill'.

For proportional delta control within the variable speed range, only k_δ and β_δ need be modified, in order to impose a fixed point, along the chosen $\{\beta_\delta, \lambda_\delta\}$ trajectory, which corresponds to the correct power coefficient, as dictated by (31). The look-up tables for this are shown by figure 2.

At the limits of the variable speed range, i.e. for constant-speed operation, β_δ need also change with generator torque, as dictated by (17) and (33). The two-dimensional look-up table for this is shown by figure 3a. Note that the values of β_δ on the flat sections of the lines on figure 3a correspond to the values given by figure 2b, and that said flat sections quickly shrink as δ grows. Figure 3b shows the resulting operating points.

For constant delta control within the variable speed range, again, only k_δ and β_δ need be modified, in order to impose a fixed point, along the chosen $\{\beta_\delta, \lambda_\delta\}$ trajectory, which corresponds to the correct power coefficient, as dictated by (35). The look-up tables for this are calculated from (37), and are the same as those shown by figure 2, only they are entered with $2b^3 P_\delta / \rho \pi R^5 \omega^3$, rather than δ.

At the limits of the variable speed range, i.e. for constant-speed operation, β_δ need also change with generator torque, as dictated by (17) and (44). The two-dimensional look-up table for this is shown by figure 4a. Note that the flat sections seen on the lines on figure 3a do not appear on figure 4a, since the power coefficient is now variable even within the variable speed range. Figure 4b shows the resulting operating points.

9. Conclusions

Wind farm participation in grid frequency regulation is necessary for deep wind energy penetration. Delta control will play a role there, and a seemingly viable method for that is discussed here. The distinguishing feature of our method is that it does not use wind speed measurements or estimators, and it extends a control scheme that is popular in industrial practice. Further study will evidently follow regarding the choice of the $\{\beta_\delta, \lambda_\delta\}$ trajectory and its effects on the pitch and torque look-up tables, as well as on turbine dynamics and loads.

IOP Conf. Series: Journal of Physics: Conf. Series **1222** (2019) 012024 doi:10.1088/1742-6596/1222/1/012024

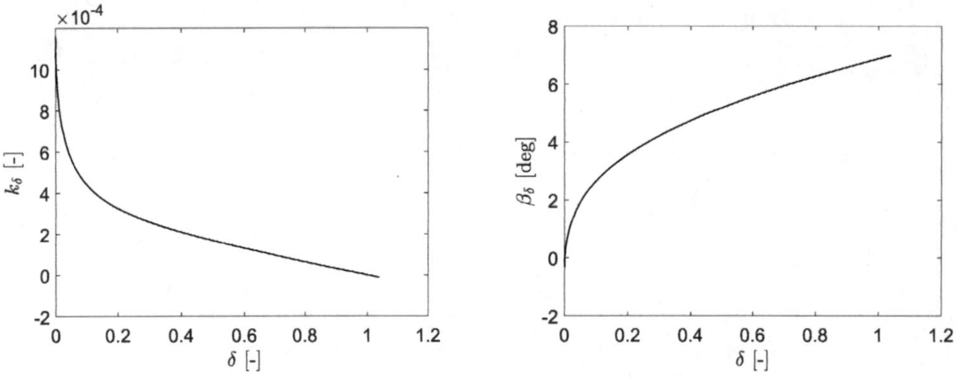

(a) (b)

Figure 2: Torque gain (a) and pitch angle (b) variation with derating ratio

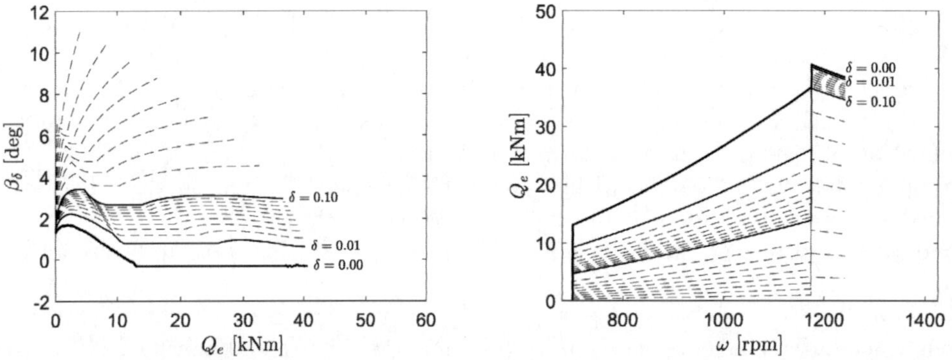

(a) (b)

Figure 3: Minimum pitch variation with torque and derating ratio (a) and operating torque at different derating ratios (b)

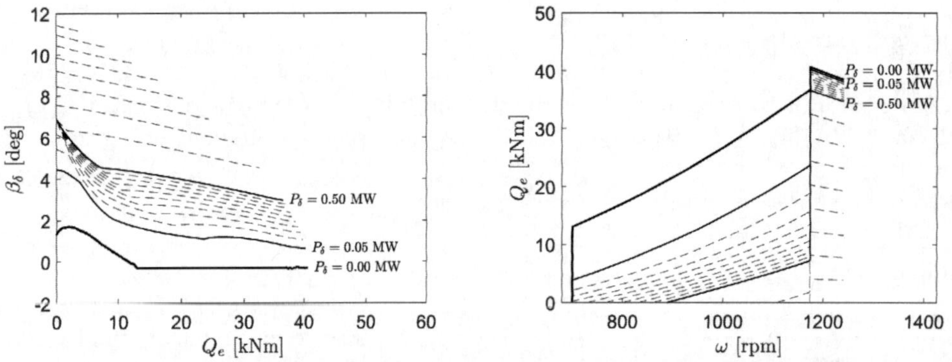

(a) (b)

Figure 4: Minimum pitch variation with torque and reserve power (a) and operating torque at different reserve power levels (b)

IOP Conf. Series: Journal of Physics: Conf. Series **1222** (2019) 012024 doi:10.1088/1742-6596/1222/1/012024

Further research is also necessary regarding the fact that, at farm level, there is in fact double accounting when wakes are ignored [20]. This is due to a turbine's power reserve being here referred to power available from incident wind, which may be in turn affected by the wakes of upwind turbines. If said upwind turbines also have power reserves, their wakes will reduce available power for the downwind turbine when the reserves are used to increase the farm's power output, i.e. the farm's reserve is less than the aggregate reserves of its constituent turbines. Some authors are already trying to fix this with farm control [21, 22].

Acknowledgments

This project has received funding from the European Union's Horizon 2020 research and innovation programme under grant agreement No 727477

References

[1] Ramtharan G, Ekanayake J B and Jenkins N 2007 *IET Renewable Power Generation* **1** 3–9 ISSN 1752-1416
[2] Loukarakis E, Margaris I and Moutis P 2009 Frequency control support and participation methods provided by wind generation *2009 IEEE Electrical Power Energy Conference (EPEC)* pp 1–6
[3] Golden R and Paulos B 2015 *The Electricity Journal* **28** 36–50
[4] Tsili M and Papathanassiou S 2009 *IET Renewable Power Generation* 1–25
[5] Martínez de Alegría I, Andreu J, Martín J, Ibañez P, Villate J and Camblong H 2007 *Renewable and Sustainable Energy Reviews* **11** 1858–1872
[6] Danish Energy Authority 2004 Wind turbines connected to grids with voltages above 100 kv - technical regulation for the properties and the regulation of wind turbines `http://www.wt-certification.dk/Common/WindTurbinesConnectedtoGridswithVoltageabove100kV.pdf`
[7] Siemens 2011 Active power control in wind parks `https://www.bpa.gov/Doing\%20Business/TechnologyInnovation/ConferencesVoltageControlTechnical/ActivePowerControlinWindParks_Siemens.pdf`
[8] Díaz-González F, Hau M, Sumper A and Gomis-Bellmunt O 2014 *Renewable and Sustainable Energy Reviews* **34** 551 – 564 ISSN 1364-0321 URL `http://www.sciencedirect.com/science/article/pii/S1364032114002019`
[9] Attya A, Dominguez-Garcia J and Anaya-Lara O 2018 *Renewable and Sustainable Energy Reviews* **81** 2071 – 2087 ISSN 1364-0321 URL `http://www.sciencedirect.com/science/article/pii/S1364032117309553`
[10] Kristoffersen J R 2005 The horns rev wind farm and the operational experience with the wind farm main controller *Copenhagen Offshore Wind*
[11] Wang Y, Bayem H, Giralt-Devant M, Silva V, Francois B and Guillaud X 2014 *IEEE Transactions on Sustainable Energy* **6**
[12] Baccino F, Conte F, Grillo S, Massucco S and Silvestro F 2015 *Sustainable Energy, IEEE Transactions on* **6** 993–1003
[13] Aho J, Pao L and Fleming P 2013 *51st AIAA Aerospace Sciences Meeting including the New Horizons Forum and Aerospace Exposition 2013*
[14] Hansen A D, Sörensen P, Iov F and Blaabjerg F 2006 *Renewable Energy* **31** 935 – 951 ISSN 0960-1481 URL `http://www.sciencedirect.com/science/article/pii/S0960148105001205`
[15] Janssens N A, Lambin G and Bragard N 2007 Active power control strategies of dfig wind turbines *2007 IEEE Lausanne Power Tech* pp 516–521
[16] Jenkins N, Burton A, Sharpe D and Bossanyi E 2001 *Wind Energy Handbook* (Wiley) ISBN 0-4714-8997-2
[17] E Johnson K 2004 *NREL*
[18] van der Hoek D, Kanev S and Engels W 2018 Comparison of down-regulation strategies for wind farm control and their effects on fatigue loads *2018 Annual American Control Conference (ACC)* pp 3116–3121
[19] JM J, Butterfield S, Musial W and Scott G 2009 *National Renewable Energy Laboratory (NREL)*
[20] Fleming P, Aho J, Gebraad P, Pao L and Zhang Y 2016 Computational fluid dynamics simulation study of active power control in wind plants *2016 American Control Conference (ACC)* pp 1413–1420 ISSN 2378-5861
[21] Shapiro C R, Bauweraerts P, Meyers J, Meneveau C and Gayme D F 2017 *Wind Energy* **20** 1261–1275 URL `https://onlinelibrary.wiley.com/doi/abs/10.1002/we.2093`
[22] Shapiro C R, Meyers J, Meneveau C and Gayme D F 2018 Coordinated pitch and torque control of wind farms for power tracking *2018 Annual American Control Conference (ACC)* pp 688–694 ISSN 2378-5861

IOP Conf. Series: Journal of Physics: Conf. Series **1222** (2019) 012025 doi:10.1088/1742-6596/1222/1/012025

Drift stability of HyStOH semi-submersible supported by airfoil shaped structures

A Manjock[1] and S Netzband[2]
[1]DNV GL Hamburg Renewables Certification, Brooktorkai 18, 20457 Hamburg, Germany
E-mail: andreas.manjock@dnvgl.com
[2]Hamburg University of Technology (TUHH), Fluid Dynamics and Ship Theory,
Am Schwarzenberg 4, 21073 Hamburg, Germany
E-mail: stefan.netzband@tuhh.de

Abstract. This study presents the results of the German research project HyStOH funded by German Federal Ministry of Economic Affairs and Energy (BMWi). The project consortium of German universities, wind turbine designers, wind farm developers and certification bodies design a novel semi-submersible steel structure with a single point mooring and self-aligning capabilities. The tower, which carries a 6 MW two-bladed downwind turbine, has an airfoil shaped cover, which supports the self-alignment of the full structure towards the main wind direction. The downwind operating wind turbine in combination with the lift force generating tower enables a passive yaw system acting against hydrodynamical impacts from waves and currents. Simulations in time domain applying the fully coupled aero-hydro-servo-elastic code Bladed_4.8 demonstrate the drift sensitivity and the self-aligning capabilities of the HyStOH design. The hydrodynamic coefficients for the simulations have been adjusted by calculations with an 1:1 model of the panel code panMARE, developed by Technical University of Hamburg-Harburg. The motivation of this analysis is to capture the complex coupled motions of an innovative FOWT by numerical simulation tools [1]. The analysis presents dynamic simulations of the HyStOH design operating in turbulent wind and irregular sea state with special focus on the yaw drift behaviour of the FOWT. The sensitivity study of yaw drift is based on numerous variations of wind-wave-current misalignments during normal operation. The results of the simulations demonstrate the weathervane capabilities of the airfoil shaped structures.

1. Introduction

The currently installed floating wind projects indicate the transition from demonstration projects towards first commercial application. The technology readiness level (TRL) for floating offshore wind turbines (FOWTs) is increasing consistently. Worldwide, especially in countries such as France, Japan, Norway and the US, new floating wind projects are being planned. Many of the projects make use of the availability of large multi mega Watt offshore wind turbines.

Different floater concepts of FOWT are currently on the market and until now and there is no distinct design concept which has been established in the industry as the most favourable concept so far. The top four floater concepts realised in the last years have been the semi-submersible, spar buoy, tension leg platform (TLP) and barge types [2].

In the German research project HyStOH funded by the BMWi, the project partners design an innovative variant of the semi-submersible floater concept. The consortium comprises German universities, wind turbine designers, wind farm developers and certification bodies. The novelty of this concept is a self-aligning floater, combined with an aerodynamical shaped tower and a downwind 6 MW rotor-nacelle assembly (RNA). A single point mooring turret below the floater structure carries five catenary mooring lines and provides the yaw degree of freedom similar to Floating Production Storage and Offloading (FPSO) station keeping systems. Thus, the RNA of this concept has no active yaw system and is directly flanged to the tower. Figure 1 illustrates the design concept of the HyStOH research project.

Figure 1. The HyStOH SelfAligner designed by the project consortium

Due to the trapezoidal floater design, the single point mooring axis has a distance of about 70 m to the tower axis which provides a large excentricity between aerodynamic forces and station keeping system. In this configuration, the airfoil shaped tower is acting as a weathervane and aero forces generate yaw moments to align the full structure into the main wind direction [3].
This study presents fully coupled aero-hydro-servo-elastic load simulations in time domain with Bladed_4.8, describing the drift and self-aligning effects during normal operational conditions of the HyStOH concept and its dynamic response to wind-wave-current misalignment.

2. Method

In a first step, the HyStOH site conditions are defined under consideration of wind-wave misalignment effects. Unidirectional approach of waves and currents is considered as most critical for the drift behaviour [4]. Therefore, a constant current of u = 0.5 m/s with a linear shear is approaching in all simulations from the same direction as the waves. The HyStOH FOWT is modelled with the fully coupled hydro-aero-servo-elastic simulation code Bladed_4.8. For adjusting the hydrodynamic system matrices reference is made to the first-order panel method of panMARE, see section 2.2. The basic environmental conditions applied in the simulations are given in Table 1.

Table 1. Environmental parameters of HyStoH Design Basis. Turbulent wind is based on von Karman spectrum with turbulence intensity according to class B [5]. Wave spectrum is based on JONSWAP spectrum.

No.	Wind speed at hub hight (10 min. mean) v_{Hub} (m/s)	Significant wave height H_S(m)	Wave peak period T_P (s)	Near surface current u (m/s)
1	1.3	0.70	6.20	0.5
2	2.6	0.80	6.30	0.5
3	3.9	0.96	6.47	0.5
4	5.2	1.20	6.72	0.5
5	6.5	1.51	7.03	0.5
6	7.9	1.88	7.41	0.5
7	9.2	2.33	7.86	0.5
8	10.5	2.85	8.38	0.5
9	11.8	3.44	8.96	0.5
10	13.1	4.10	9.60	0.5
11	14.4	4.83	10.30	0.5
12	15.7	5.63	11.06	0.5
13	17.0	6.50	11.86	0.5
14	18.3	7.44	12.72	0.5
15	19.6	8.45	13.62	0.5

In all simulations the initial position of the floater is aligned along the global x-axis so that rotor axis and the wind direction are co-linear. Thus, the rotor is approached by the wind with a yaw misalignment angle of zero and is producing optimal power output. In 9 steps the wave (and current) direction relative to the wind direction is then increased by steps of 30 degree in a range of ±120 degree.

Because of the single point mooring degree of freedom (free yaw DoF) and the motions of the floater in irregular sea state, little yaw motions and hence wind-rotor misalignments occur even under unidirectional wind-wave impact (wave direction zero degree). If the angle between wind direction and waves is stepwise increased/decreased the hydrodynamic wave forces tend to push the entire FOWT out of the wind direction. The HyStOH floater with its 8,000 tons (including ballast tanks) can reach a yaw drift velocity of about 0.2 degree/s as shown in Figure 2 (first 300 s simulation example).

IOP Conf. Series: Journal of Physics: Conf. Series **1222** (2019) 012025 doi:10.1088/1742-6596/1222/1/012025

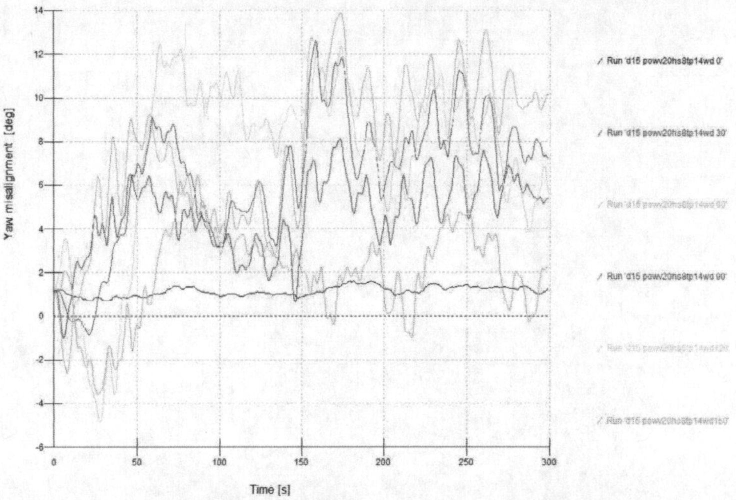

Figure 2. Initial orientation of HyStOH floater for wave directions 0 -120 degree

2.1. Aero-elastic simulation code Bladed 4.8

The computer code Bladed (developed by Garrad Hassan) performs analyses in the time domain. Time domain simulations are necessary for non-linear, dynamic and transient processes, which are important for the load analysis of FOWT. Complex hydrodynamic calculations typically consider non-linear mooring systems [6] and the effects of radiation and diffraction forces. Bladed is developed for dynamic load calculations of multibody structures with coupled aerodynamic and hydrodynamic loads. It calculates the structure dynamics with a multibody-dynamic-approach. The components such as blades and tower are modelled from flexible elements whose deformation is determined by modal analyses. This is done by a linear combination of the calculated eigenmodes resulting from a finite element method (FEM) calculation. The aerodynamic forces are calculated with the blade element momentum theory (BEMT) [7]. In addition, Bladed is able to simulate a wake behind the rotor and dynamic stall. Bladed_4.8 can simulate regular and irregular waves as well as sea currents. Regular waves are used mainly for determination of extreme loads and will be described as Airy-waves or with the stream function. For fatigue loads, a series of irregular waves is created whose amplitude and frequency is defined with a spectral density function as the JONSWAP or Pierson-Moskowitz spectrum. The hydrodynamic forces are calculated with the Morison equation. The mooring system is represented by a force-displacement ratio, linear or non-linear, and are calculated separately by the user and implemented via a stiffness matrix at the fairlead position. A detailed description of this quasi-static approach is gin in [8].

IOP Conf. Series: Journal of Physics: Conf. Series **1222** (2019) 012025 doi:10.1088/1742-6596/1222/1/012025

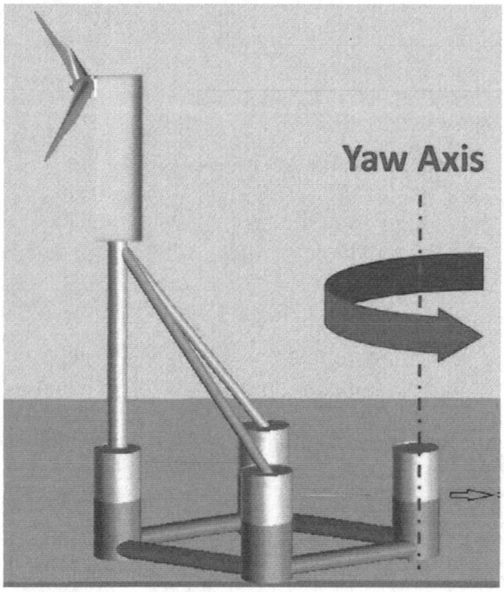

Figure 3. Bladed model including extended aerodynamical tower properties

The prediction of onshore and offshore loads from turbulent wind has been validated and proven by numerous projects and measurements over the last decades. Recent extensions of Bladed comprise modelling of elastic properties of large blades, super-elements and the introduction of tidal and floating turbine model capabilities.

2.2. Panel code panMARE

PanMARE (panel Code for Maritime Applications and REsearch) is a validated code of Technical University of Hamburg-Harburg (TUHH), institute of fluid dynamics and ship theory which has been validated in a number of commercial oil&gas and maritime projects [9]. To simulate the highly unsteady motions of the HyStOH turbine and platform, the code has been extended. Considering the hydrodynamics, panMARE accounts for the geometry of the submerged platform in detail as well as for an inviscid model of the three-dimensional hydrodynamic flow field [10]. In contrast to frequency-domain based methods, unsteady and aperiodic motions do not impair the quality of the obtained hydrodynamic results. The hydrodynamic loads are calculated by instantaneously integrating the static and dynamic pressure on the submerged platform hull. The principal ability of panMARE to directly model radiation and diffraction effects using a discretised water surface is one reason why a potential method has been chosen here [11].

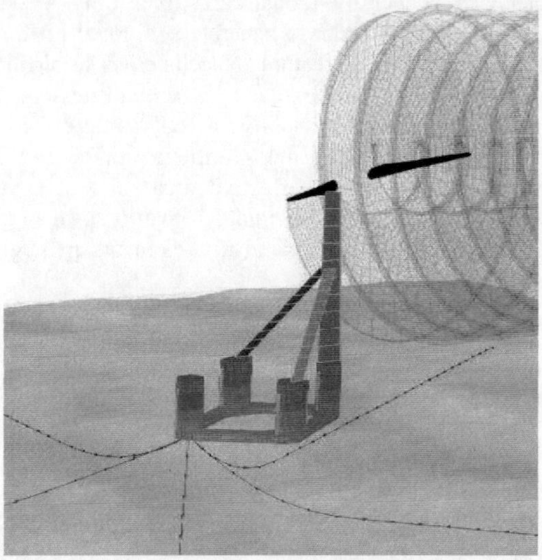

Figure 4. Model setup with platform, tower, rotor blades, rotor wake and mooring

Considering the aerodynamics, panMARE is able to account for blade-wake interaction by modelling the structure of the vortex system in the wake of the rotor using a stabilised free wake deformation method. Using the same simulation tool for the aerodynamic and hydrodynamic domains is a unique feature of panMARE and leads to a significant simplification of the coupling process. Both flow fields are solved with the same explicit fourth-order Runge-Kutta time integration scheme simultaneously [12]. Every stage of the integration scheme is performed in both flow solvers, which allows for an update of motions in every intermediate step [13]. Figure 3 is showing the model setup of the HyStOH concept in panMARE.

2.3. The HyStoH design concept

The special feature of the HyStOH design concept are the aerodynamically shaped tower and support struts. The airfoil shaped elements are horizontally symmetrical and the cord line of is aligned with the rotor axis. Wind direction misalignment with respect to the rotor axis (and thus tower cross section chord line) results in horizontal lift forces which generates a self-aligning yaw moment about the single point mooring / turret buoy. During operation, the self-alignment effect is additionally supported by the downwind rotor. During idling or standstill only the aerodynamic forces are responsible for aligning the FOWT towards the wind direction.

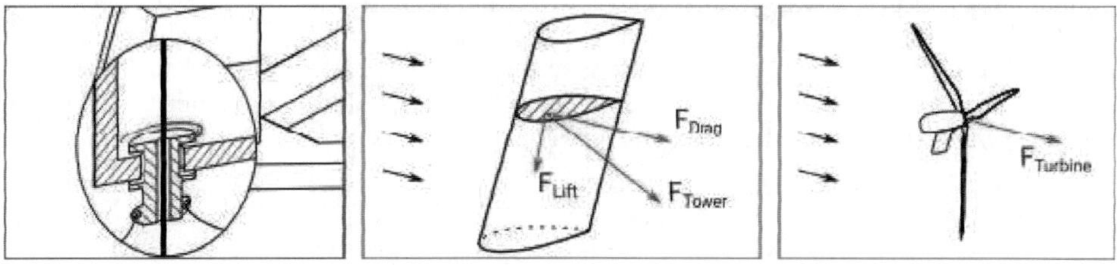

Figure 5. Single point mooring and self-aligning forces

IOP Conf. Series: Journal of Physics: Conf. Series **1222** (2019) 012025 doi:10.1088/1742-6596/1222/1/012025

As described above, the Bladed software calculate the structural loads of beam elements with the FEM approach. The beam elements are generally tubular and have only aerodynamic drag properties defined by an element drag coefficient. In order to generate additional lift forces which depending on the wind speed and angle of attack the Bladed feature "Nacelle cover as airfoil" is applied. The nacelle cover, in the HyStOH concept fixed to the tower, has been virtually extended downwards the tower. This extended nacelle surface area has a representative aerodynamic effective area of about 300 m^2 and is assigned to aerodynamic properties of the symmetric airfoil NACA0025 [14] [15]. It is conservatively assumed that the aerodynamic effective area of the HyStOH concept is located basically in the range between tower top and tower middle (position of the strut connection). Near sea level turbulence and wake effects due to the floater and strut elements are neglected.

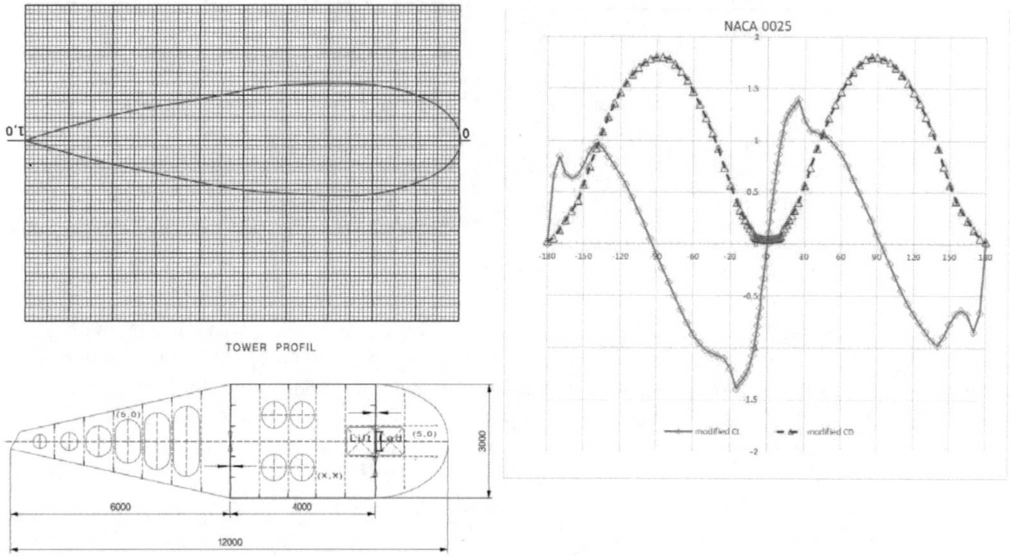

Figure 6. Approximation of tower cross section aerodynamical properties

3. Results

In order to demonstrate the stabilisation of the HyStOH concept against yaw drift, simulations with and without aerodynamic tower support are analysed. It could be shown that the in operational conditions the downwind rotor forces alone, without the aerodynamic support of the profiled tower, provide not sufficient restoring yaw moment. As expected, the yaw drift forces increase with increasing wind speeds and corresponding wave heights.

Typically, in wind turbine operation misalignment angles of 5 to 8 degrees between rotor axis and mean wind direction are accepted with respect to energy yield losses [16]. For rough sea states with significant wave heights $H_S > 7$ m and wind-wave misalignments > 30 degree the yaw drift and wind-rotor misalignment exceeded 20 degrees. Such high misalignment angles in combination with high wind speed operation will cause a loss in energy yield.

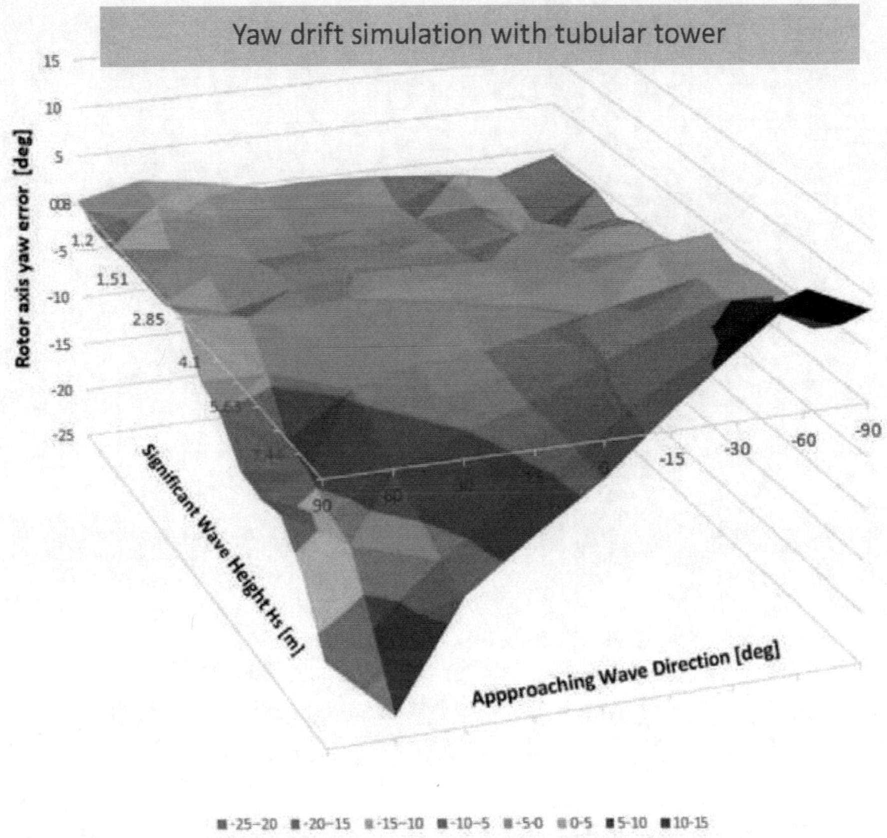

Figure 7. Yaw drift depending on wave height and wind-wave misalignment without wind-vane structures

The vertical plot axis "Rotor axis yaw error [deg]" indicates the misalignment angle between wind direction and rotor axis which has been established under the impact of a specific wave formation (600 seconds simulation time for each combination).

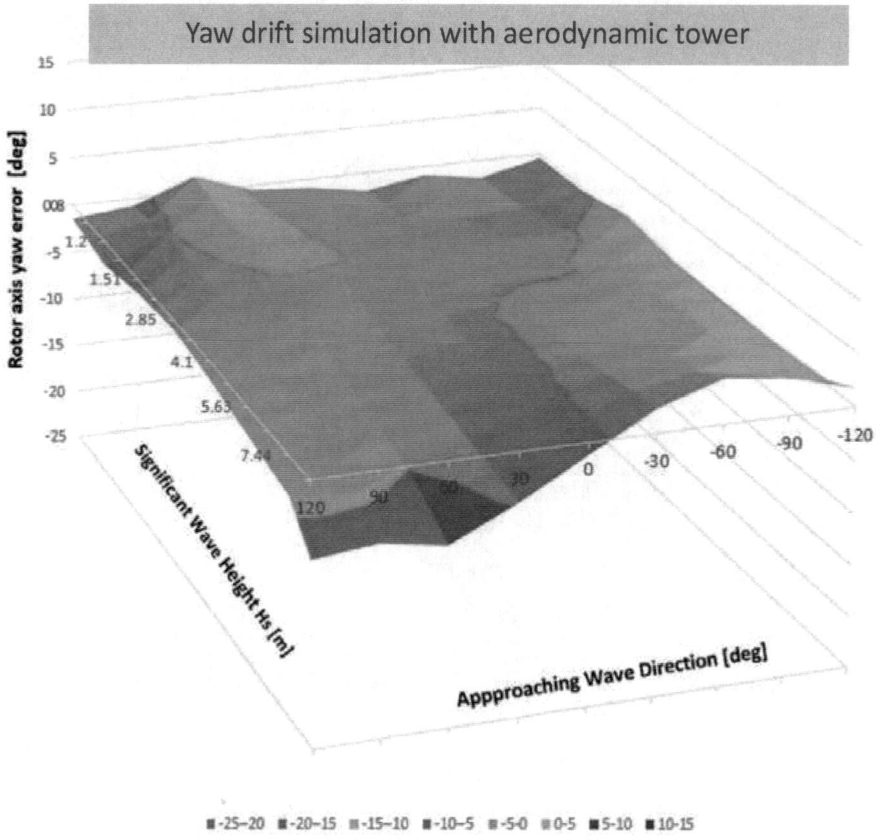

Figure 8. Yaw drift depending on wave height and wind-wave misalignment including wind-vane structures

For the trapezoidal HyStOH floater design especially wave and current directions from ±60 degree with respect to rotor axis result in large yaw drift. The deviation between the drag tower and aerodynamical tower approach is clearly visible for operating wind speed vHub = 20 m/s, corresponding to a wave height of Hs = 8.45 m.

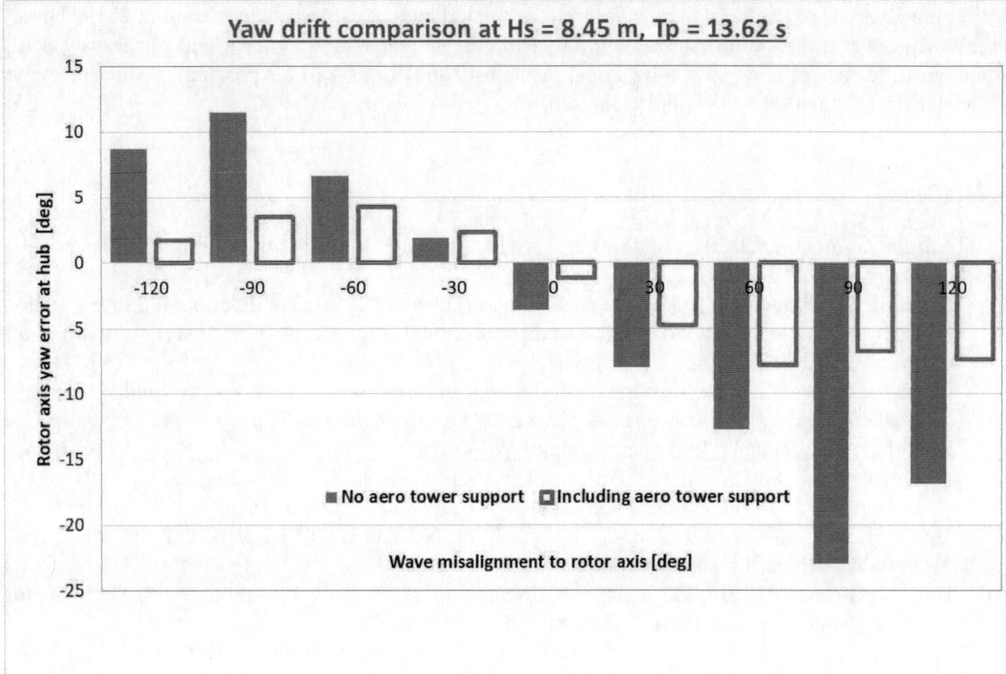

Figure 9. Comparison of yaw drift sensitivity during operation and large waves for various wind-wave misalignment

4. Conclusion

The dynamic transverse and longitudinal stability as well as the yaw behaviour of the HyStOH concept, a self-aligning FOWT, are analysed with time domain simulation tool Bladed and the first-order panel method panMARE. The simulations of the motion behaviour of the FOWT are carried out for various environmental conditions with regard to wind, natural waves and current. The mooring forces are calculated using a dynamic mooring model. The results of the investigation show that a strong coupling between aerodynamics and hydrodynamic loads as well as mooring system induced forces in the simulations is essential for the evaluation of the self-aligning behaviour of the FOWT.
As it is expected, the misalignment increases at lower wind speeds due to low aerodynamic forces. A closer look at the contributions of the different components of FOWT to the aligning moment shows that the main component is delivered by the profiled tower. Especially at higher wind speeds, the tower influence becomes more important and contribute significantly to align the platform.

5. Outlook

The present study considered normal operation load cases between cut-in and cut-out wind speed only (design load case group DLC 1.2 [5]). Furthermore, simplifications in application of wave and current directions have been made which do not correspond ideally to the site conditions of the HyStOH Design Basis. In this study wave and current have been directly linked and current speed was assumed to be constant. Therefore, the individual portion of current forces on the global yaw drift could not be identified. To pursue this study, further design load cases should be analysed, such as idling conditions and storm events. The modelling of sea current could be improved further (e.g. including current profiles and tides) and the portion on the overall yaw drift could be analysed in a separate sensitivity study.
Also, the impact of misaligned operating FOWT on power production and structural fatigue loading is an aspect which is worth to analyse in future studies.

Within the HyStOH project it is planned to perform scaled model tests in a wave basin in 2019. These tests will provide further information of the dynamic behaviour of the concept and of the yaw drift behaviour in particular. A comparison campaign against simulations will further improve the accuracy of the HyStOH simulation model applied herein.

6. References

[1] Cordle A, Jonkman J 2011 State of the Art in Floating Wind Turbine Design Tools *Renew. Energy 8:367–374*

[2] Coulling AJ, Goupee AJ, Robertson A, Jonkman J, Dagher HJ 2013 Validation of a FAST semi-submersible floating wind turbine numerical model with DeepCwind test data *Journal of Renewable and Sustainable Energy. Vol. 5. p. 23116*

[3] Netzband S et al. 2018 Assessment of floating and yaw stability of a self-aligning floating offshore wind turbine *www.cruse-offshore.de/downloads.html, 06/2018*

[4] Robertson A, Wendt F, Jonkman J, Popko W, Vorpahl F, Stansberg C T, Bouy L 2015 2nd Order Wave Forces (Wave Drift): OC5 Project Phase I: Validation of Hydrodynamic Loading on a Fixed Cylinder *Proc- 25th Int. Ocean and Polar Eng. Conf*

[5] DNV GL load standard DNVGL-ST-0437 Edition November 2016 Loads and site conditions for wind turbines *rules.dnvgl.com/ServiceDocuments*

[6] Hall M, Goupee A 2015 Validation of a lumped-mass mooring line model with DeepCwind semisubmersible model test data *Ocean Eng. 104:590–603*

[7] DNV GL, Garrad Hassan & Partners Ltd. 2016 Bladed Theory Manual Version 4.8 *www.dnvgl.com/software*

[8] Masciola M 2015 *MAP ++ Documentation*

[9] Schoop-Zipfel J, Abdel-Maksoud M 2014 Maneuvering in waves based on potential theory. *33rd Int. Conf. Ocean Offshore Arct Eng. Vol. 8B. San Francisco (CA)*

[10] Bauer M, Abdel-Maksoud M. 2012. A 3-D Potential Based Boundary Element Method for Modelling and Simulation of Marine Propeller Flows. In: 7th Vienna Int. Conf. Math Model. Vol. 7. Vienna; p. 1179–1184.

[11] Netzband S, Schulz CW, Göttsche U, Ferreira Gonzales D, Abdel-Maksoud M 2018 A Panel Method for Floating Offshore Wind Turbine Simulations with Fully-Integrated Aero- and Hydrodynamic Modelling in Time Domain *Sh Technol Res. 0:1–14*

[12] Netzband S 2017 A fully coupled Simulation Method for Floating Offshore Wind Turbines Dynamics Using a Boundary Element Method in Time Domain, TUHH, *10th IWSMH Keeling Taiwan*

[13] Katz J, Plotkin A 2001 Low-Speed Aerodynamics Cambridge (UK): *Cambridge UP (Aerospace Series)*

[14] Sarraf C, Djeridi H, Prothin S, Billard JY 2010. Thickness effect of NACA foils on hydrodynamic global parameters, boundary layer states and stall establishment *J Fluids Struct. 26:559–578*

[15] Sheidahl RE 1981 Aerodynamic Characteristics of Seven Symetrical Airfoil Sections Through 180-Degree Angle of Attack *Sandia National Laboratories, Albuquerque USA*

[16] Laino DJ, Hansen AC, Minnema JE 2002 Validation of the AeroDyn subroutines using NREL unsteady aerodynamics experiment data *Wind Energy 5:227–244*

Acknowledgments

The authors kindly thank the Federal Ministry for Economic Affairs and Energy of Germany (BMWi) for financially supporting the HySToH project [03SX409A-F].

Gefördert durch:

Bundesministerium für Wirtschaft und Energie

aufgrund eines Beschlusses des Deutschen Bundestages

IOP Conf. Series: Journal of Physics: Conf. Series **1222** (2019) 012026 doi:10.1088/1742-6596/1222/1/012026

"FlexPad" - Innovative conical sliding bearing for the main shaft of wind turbines

Tim Schröder*[1], Georg Jacobs[1]; Amadeus Rolink[1]; Dennis Bosse[1]
[1] Chair for Wind Power Drives, RWTH Aachen University
* Corresponding author: tim.schroeder@cwd.rwth-aachen.de

Abstract. This paper shows an innovative approach concerning a new type of rotor main bearings for wind turbines. The conical sliding bearing with its flexible support structure is designed for the specific loads and operating conditions of wind turbines. It is robust, compact and offers superior serviceability since individually replaceable bearing pads obviate the disassembling of the whole drivetrain during repair/maintenance.
After a short description of the bearing design the results of functional testing under real conditions in full-scale on a 1 MW system test bench are presented. The achieved load-bearing properties, the mixed friction performance and the start-up behaviour will be discussed.

1. Introduction

Wind turbines (WT) count as one of the most important techniques of renewable energy production and have a constantly increasing market share within the electric power sector. Besides the continuously increasing size and performance of wind turbines, the improvement of reliability and reduction of costs belong to the biggest current technical challenges. Especially roller bearings face their technical limits in the drivetrain of wind turbines, with a conspicuous high failure rate. Therefore, reliable, cost efficient and easy to maintain alternatives are of interest.
For the main shaft bearing this can be achieved by a sliding bearing which is, due to a conical shape of the sliding surface, capable to handle axial and radial loads in a very compact structure. By means of extensive Elasto-Hydrodynamic (EHD) simulations a new main bearing concept for wind turbines was found. Since the shaft of wind turbines is loaded with heavy tilting moments special design measures were required to avoid edge wear.
The contribution will describe the design process of this main bearing and experimental results of a full-scale demonstrator test.

2. Approach

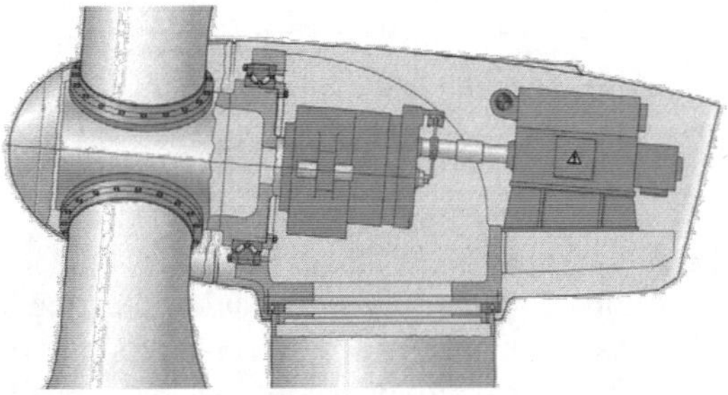

Figure 1. WT with double row tapered roller main bearing [www.skf.com]

WT drivetrains with double row tapered roller main bearings, like shown in Figure 1, have major advantages compared to a conventional three-point support as for example the drivetrain can be designed very short, which leads to lightweight components (esp. shaft) and a short load path from the rotor to the tower. On the other hand the big diameter of these bearings in combination with the high number of rollers leads to fatigue issues due to the high number of overrollings. Also accounted by the big diameter, it is a big challenge to handle tolerances in the manufacturing process and system deformations during operation. Both have a big influence on the pressure distribution and consequently also on the bearing lifetime [1, 2].The application of sliding bearings instead of roller bearings offers the chance to dismantle these barriers, which motivated the investigations on a conical (v-shaped) sliding main bearing.
EHD simulations of a pre-dimensioned rigid conical sliding bearing [3] showed distinctive issues concerning edge wear and high contact pressures already in the early design process.

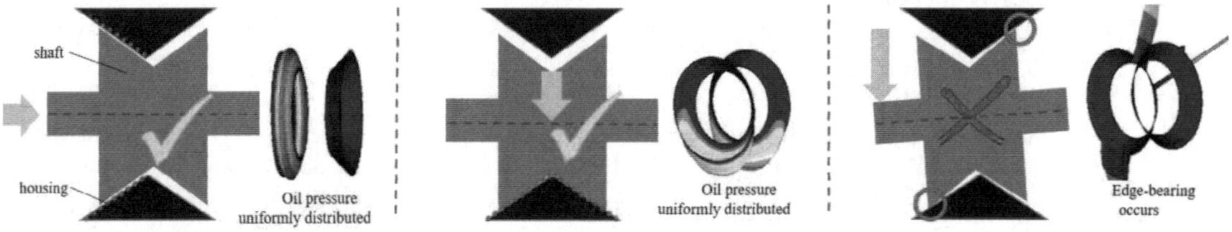

Figure 2. Load taking capability with uniaxial loads

In Figure 2 it is clearly visible by moderate hydrodynamic pressures that a conical sliding bearing is capable to handle axial loads and radial loads applied in the middle of the shaft, as in this case the bearing gap "closes" parallel. However, the bearing is not able to handle radial forces under lever or any bending moments (Figure 2, right) as the shaft tilts slightly within the bearing clearance. This effect is amplified by the short supporting width of the two cones and leads to high contact pressures resulting in edge wear and an early malfunction [5, 6]. Obviously it is essential for a WT main bearing to take all (6-DOF) loads occurring from the wind, so that specific measures are needed to solve this issue.
A flexible designed support structure of the bearing pad allows the pads to follow the movement of the shaft. By a slight deformation of the support, a smooth pressure distribution without edge wear and a long durability is ensured (Figure 3).

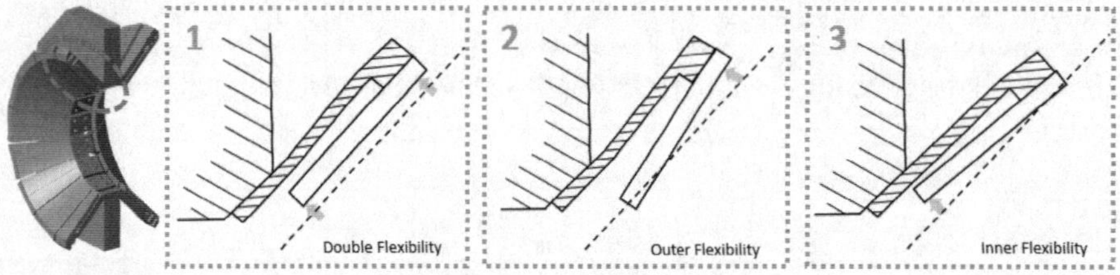

Figure 3. Flexible Support structure of the bearing pads

Depending on the actual load distribution on each bearing segment, the support structure will yield with the outer flexibility (2) and the inner flexibility (3) to a combined double flexibility (1). By this a "parallel" and well lubricated gap is ensured.
Once the functional principle of this, "FlexPad" named, sliding bearing has been proved by EHD calculations, a 1MW wind turbine has been chosen as a demonstrator platform to design, simulate and test this bearing in full-scale. Therefore, the hub-loads are calculated by a Co-Simulation of Simpack and Simulink, considering the WT controller and the wind conditions according to IEC 61400 [3,4]. Concerning this loads extensive EHD calculations have been performed in order to investigate the design

IOP Conf. Series: Journal of Physics: Conf. Series **1222** (2019) 012026 doi:10.1088/1742-6596/1222/1/012026

parameters of the bearing and especially the ideal stiffness combination of the “inner” and “outer” flexibility [5, 6].

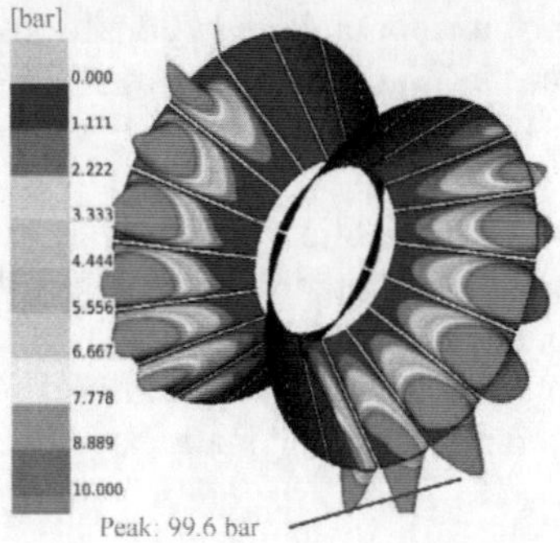

Figure 4. Pressure distribution during dynamic simulation of IEC design load case (DLC) 1.1

Figure 4 shows exemplarily the pressure distribution during a dynamic EHD simulation of IEC design load case 1.1 (NTM - Normal Turbulence Model) at the nominal wind speed of 12 m/s. It is visible, that the hydrodynamic pressure is distributed evenly over the loaded pads without any contact pressures and highest hydrodynamic pressures under 100 bar.
After concluding the design process, the bearing and the related conical shaft is manufactured and ready to be tested on a 1 MW WT system test bench. The left picture in Figure 5 shows the assembled bearing without the shaft and picture in the middle gives a detailed view on the bisected bearing demonstrating the flexible support structure of the bearing pads.

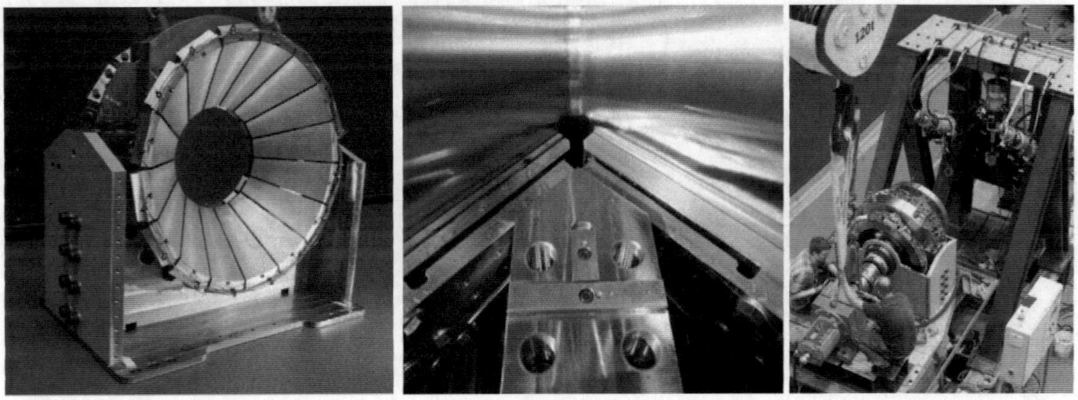

Figure 5. Pictures of manufactured bearing; bearing on a 1 MW WT machine carrier on System test bench

IOP Conf. Series: Journal of Physics: Conf. Series **1222** (2019) 012026 doi:10.1088/1742-6596/1222/1/012026

3. Concept-Validation Measurements

In order to investigate the new bearing concept under real conditions, an extensive measuring campaign has been carried out. The right picture in Figure 5 shows the "FlexPad" bearing during the mounting process on the machine carrier of the 1 MW system test bench, where it replaces the original roller main bearing.
This system test bench is equipped with a hydraulic load application unit. This unit is capable to emulate the forces and bending moments resulting from the wind exposed WT rotor and load values can be set according to the results of the hub-load analysis [3, 4].
To validate the new bearing concept, as well as the related EHD-Simulation models, extensive tests from synthetic uniaxial loads (static) to real dynamic loads have been carried out.

3.1. *Test Setup*

As the "FlexPad"-bearing is designed as a momentum-bearing (similar to SKF Nautilus, Figure 1), it takes all the loads directly without a second bearing point. That is why it is possible to simplify the test setup (see Figure 5). The Gearbox and Generator is dismounted from the machine carrier and the shaft of the main bearing is driven by a servomotor mounted at the gearbox position.

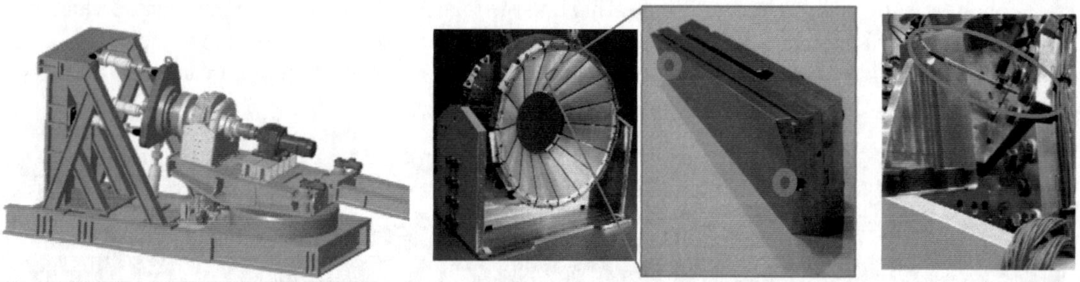

Figure 6. Test Setup; temperature measurement points, displacement transducer

In the test setup the bearing is equipped with a variety of different sensors to investigate the load taking capability under operation. Two temperature sensors are applied just behind the sliding surface of each pad. One sensor at the outer ring and one on the inner ring. This ensures the ability to detect critical thermal situations and gives information about the load distribution over the whole bearing.
In addition six displacement transducers are positioned at six upper pads to measure the deformation of the FlexPad arm under load.
Moreover the bearing is equipped with a torque sensor between the servomotor and the bearing to investigate the friction in the bearing.
During operation force and travel of each hydraulic cylinders is monitored by load cells and position sensors.
The bearing is lubricated with synthetic oil (PAO) of the viscosity index 320. The bottom pads are below the oil level (immersion lubrication) and some of the upper pads are additionally supplied by an external oil supply (pressure less).

IOP Conf. Series: Journal of Physics: Conf. Series **1222** (2019) 012026 doi:10.1088/1742-6596/1222/1/012026

3.2. *Load-bearing performance*

As this bearing design is new and only pure radial sliding bearings or pure axial sliding bearings are common in industry, the characteristics of the load-bearing behaviour are of major interest. Therefore, multiple tests from synthetic uniaxial loads (static) to realistic multiaxial load sequences (static and dynamic) have been conducted.
The temperature sensors, of which two are installed per pad, are best suited to evaluate the load zone. Also under pure hydrodynamic conditions the gap between the shaft and the pad is getting smaller in the loaded zone, which results in more liquid friction and a higher heat input into the pad.
Of course also critical operating conditions due to mixed friction or edge wear can be detected at an early stage.

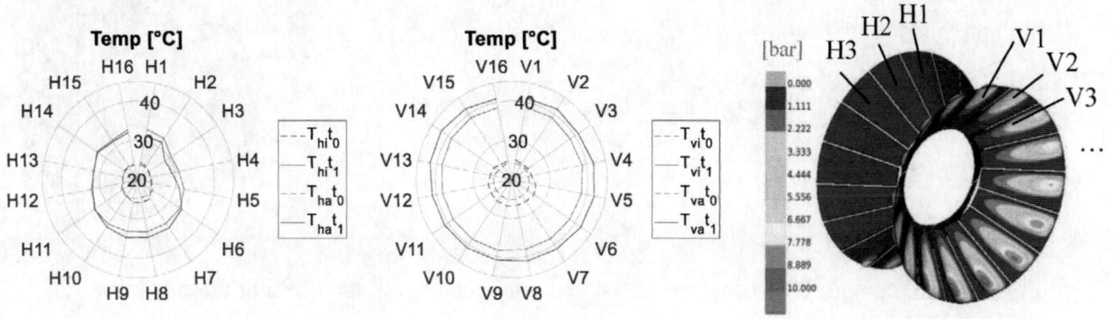

Figure 7. Temperature distribution during uniaxial thrust test (left: rear side of the bearing with pad #H1-H16, right: front side of the bearing with pad #V1-V16, simulation result

For a better understanding of how the load is distributed over the bearing, simple synthetic loads are investigated at nominal rotational speed (28 rpm). In the case of pure thrust the conical bearing takes the load with the front cone.
The thrust is ramped up in 7 steps up to 101 kN, whereby every step is hold for 10 minutes of operation. Figure 7 shows the temperature distribution, in which the dotted lines indicate the temperatures of each pad at the beginning of the tests (t_0 ; 0 kN thrust) and the continuous lines the temperatures at the end of the last load step (t_1 ; 101 kN thrust). The blue line indicates the temperatures at the outer ring measurement points and the red line shows the inner ring measurement points.

IOP Conf. Series: Journal of Physics: Conf. Series **1222** (2019) 012026 doi:10.1088/1742-6596/1222/1/012026

The measurements proof the expectation, that the front pads become warmer as they take the load. Figure 7 shows also that the pads are very equally loaded due to the very similar temperature distribution, which is an indicator for proper position tolerances of the bearing pads.
The outer ring measurement points generally become a bit warmer (~2-3°C), as the bearing diameter is larger at this location resulting in a higher circumferential speed.
Visible is also the gravity influence on the 1700 kg heavy shaft. In the measurements as well as in the simulation the rear bottom pads become slightly warmer.

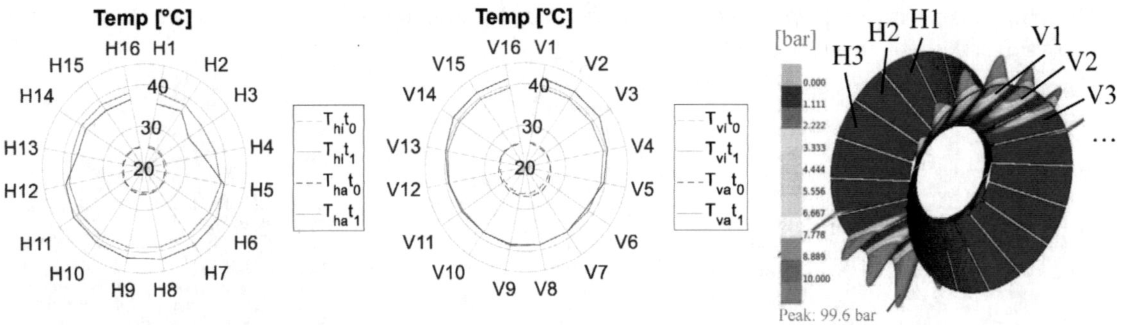

Figure 8. Temperature Distribution during uniaxial bending moment, Simulation result

Figure 8 shows the next simple synthetic load as pure bending moment is applied to the bearing. The bending moment is ramped up in 11 steps up to 100 kNm, whereby every step is hold for 10 minutes of operation. The continuous lines again show the temperatures of the different measurement points at the end of the last step.
The loaded zone changes to top front and bottom rear as expected and predicted by the simulation model. After completing the uniaxial tests, more tests with combined loads were conducted in order to cautious approach to the maximum design load values as follows:

- $M_y + F_x$ (Bending moment +Thrust)
- $M_y + F_z$ (Bending moment + Radial force)
- $M_y + F_x + F_z$ (Bending moment + Thrust + Radial force)
- $M_y + M_z + F_x + F_z$ (static design load)

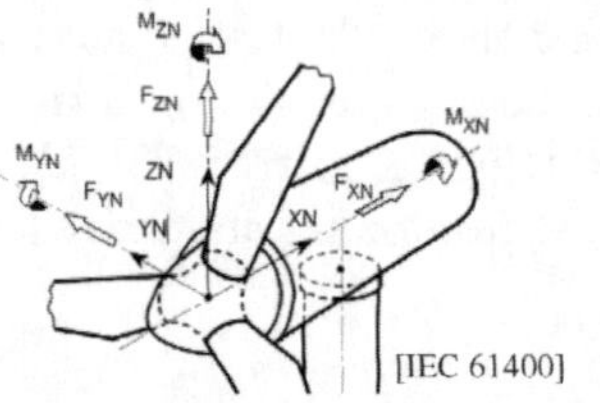

The measurements showed that combined loads lead to a combined contact pattern. All tests had in common that the load zone became larger at higher loads, but edge-wear never took place during all the tests as neither a critical temperature rise was measured nor any wear was detectable after the tests. Also the measured friction torque increased only slightly in each additional load step, whereas this is explainable with the smaller bearing gap and higher liquid friction due to the higher load.

IOP Conf. Series: Journal of Physics: Conf. Series **1222** (2019) 012026 doi:10.1088/1742-6596/1222/1/012026

3.3. *Deformation of FlexPad Arms*

Since the load-bearing performance under all relevant load scenarios is proofed successfully and edge wear is effectively prevented by means of the flexible support structure, now attention is to be payed to the deformation of the flexible structure.
Two exemplary load conditions are selected for this purpose: The pure bending moment M_y (left column) is selected, because in this case the minimum number of pads is loaded and the maximum absolute deformation is reached. Even combined loads do not reach this maximum deformation, as the additional load is always distributed over more pads.
The second regarded load condition is a realistic highly dynamic load. Here predictions about the adaptability to changing conditions can be made.
As shown in Figure 9 the deformation of the FlexPad arms is measured by LVDT displacement sensors on the back of six upper pads

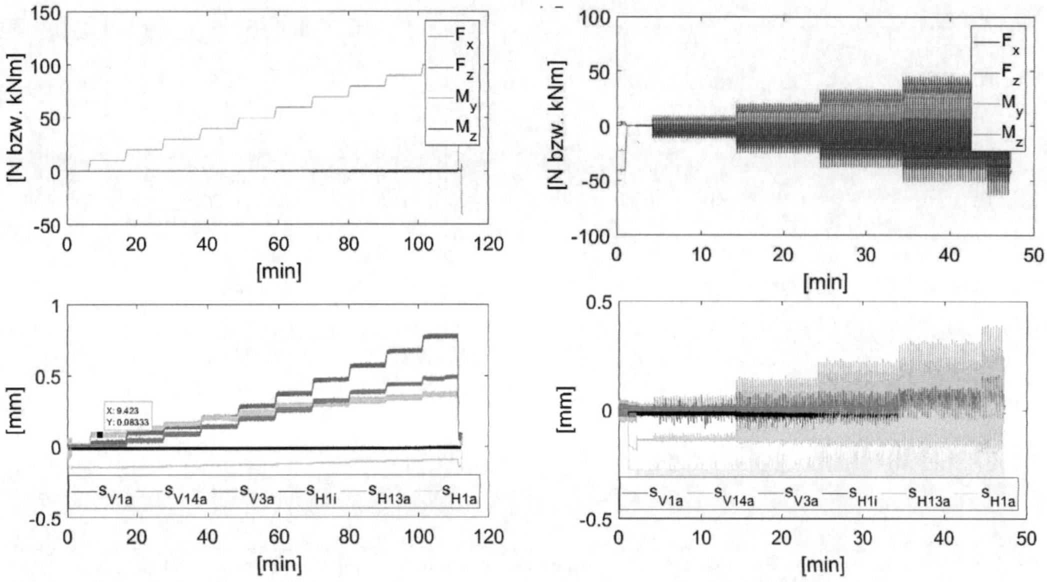

Figure 9. Deformation of FlexPad arm under pure M_y bending moment (left) and realistic highly dynamic load (right)

The results at pure bending moment show a deformation of 0.7 mm at maximum load measured at the front 12 o´clock pad position (Sensor S_{V1a}). The Sensors at front 10 and 2 o'clock position (S_{V3a} & S_{V14a}) measure with 0.5 mm and 0.4 mm a lower deformation corresponding to the load zone.
According to an FE-calculation a deformation of 1 mm leads to a von Mises stress of 190 N/mm^2 at the most stressed area of the pad supporting structure, which is about 20 % of the yield strength, hence the measured deformation is uncritical and fatigue issues should not occur. Nevertheless, further investigations will be carried out on this topic.
During static operational points the sensors detect also some slight rotational speed-dependent movement (~30 μm), which indicates that the FlexPad bearing adapts also manufacturing tolerances like roundness deviations of the shaft. Especially in comparison to roller bearings where this results in constraining forces between the rollers and the races, this is a big benefit.
The rear sensors (S_{H1i}, S_{H13a}, S_{H1a}) do not measure any deformation as they are placed on completely unloaded segments.
The measurement under dynamic loads shows that the bearing adapts quickly to new load situations, which is important for the application in wind turbines. Even though calculations concerning the fatigue under long term dynamic loading are yet to be made. The dynamic loads couldn't be run up to the highest

loads, because the used hydraulic pressure unit wasn't able to deliver the requested amplitudes of up to 85 kN in combination with the requested frequencies of up to 1.25 Hz.
During all measurements the FlexPad arms show a full elastic recovery and entirely uncritical load-bearing and temperature characteristics.

3.4. *Mixed Friction limit*

Compared to rolling bearings, low rotational speeds are particularly critical for sliding bearings, as the hydrodynamic separation of the surfaces depends on the circumferential speed. Therefore investigations on the mixed friction limit have been conducted at 60 % synthetic thrust (F_x) and bending moment (M_y). This represents a normal operation point of the considered wind turbine.
During the test the rotational speed is reduced from nominal speed (28 rpm) to a stepwise decreased minimum value. In each of the ten steps the minimum value is hold for 10 seconds.

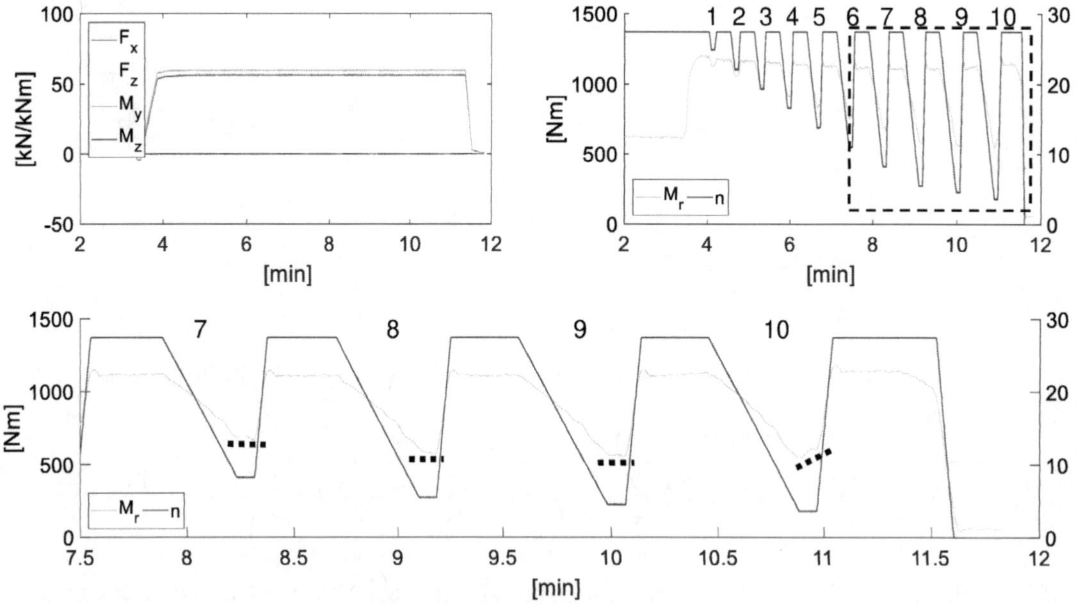

Figure 10. Dropping rotational speed and friction torque

Figure 10 shows the results of this investigations. In the detailed view of the last four steps in the bottom graph it is visible that the measured friction torque drops uniformly with the dropping rotational speed. This is explainable by the lower liquid friction at lower rotational speeds. From step 1 to 9 the lowest friction measured is still getting lower each step but in step 10. In step 10 the rotational speed is ramped down from nominal rotational speed (28 rpm) to 3.6 rpm. Despite the lower rotational speed, the friction torque during the holding phase is not lower than in the previous phase and it also rises during the holding phase, indicated by the dotted black lines.
This is an indication that the hydrodynamic pressure build-up is no longer able to completely separate the surfaces, so that mixed friction occurs. At the mean diameter of the conical shaft this speed corresponds to a circumferential speed of 0.127 m/s.
Normal operation points are far away from this rotational speed so that full lubrication is ensured under nominal conditions.
According to the test setup, there is a roller bearing inside of the load application unit, whose friction torque is also measured by the torque sensor, so that the absolute values shown in Figure 10 are overestimated.

3.5. *Wear during WT start ups*

Despite a very early transition to hydrodynamic lubrication, wind turbines pass through the mixed friction region during start-ups. According to a SCADA-data analysis of eight 1.5 MW on shore wind turbines over two years, each turbine had about 1000 start-ups per year counted by means of the closing operations of the circuit-breaker.

In order to be prepared for this circumstances, which are incidentally quite harsh for sliding bearings, special sliding materials were developed and investigated [3, 4, 6, 7]. New thermal spraying coatings have been developed which are featured, for example, by solid lubricants incorporated into the material matrix. Besides this technology enables the coating of complex surfaces, like the conical surface of the sliding pads.

To investigate the properties of the bearing and the coating during the start-ups, this conditions were simulated on the 1 MW system test bench. Start-up loads and the rotational speed ramp according to the load simulation were applied to the bearing in repetitive. The cycle is truncated at 7.5 rpm due to cycle time reduction, as it was proofed before, that a higher rotational speed is uncritical.

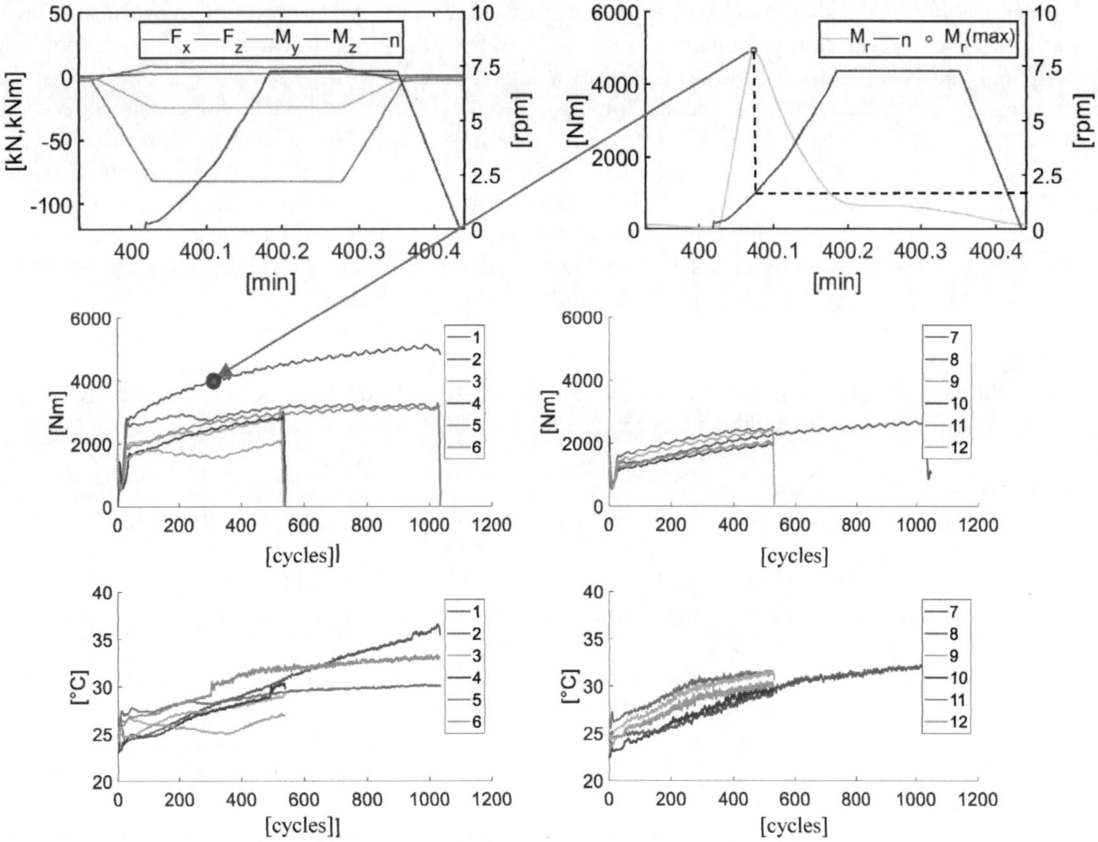

Figure 11. Start-up loads (top left), exemplary cycle showing friction-torque and the rotational speed (top right); oil sump temperatures during all 12 tests (bottom)

In the top right graph in Figure 11 one exemplary cycle showing friction-torque and the rotational speed ramp. As soon as the rotational speed ramp starts, the measured friction torque increases rapidly. In this phase a higher speed results in more friction by reason of asperity contact. From a speed of 1.6 rpm, the hydrodynamic separation of the surfaces begins and the friction torque drops again to the level of nominal operational points.

The lower 4 graphs show the series of maximum values of each cycle for the 12 conducted tests, which were run either with 1000 or 500 Cycles in a row. For a better perceptibility the first six and the last six tests are shown in different graphs.
It can be seen that the peak values of the mixed friction are very similar over the test cycles apart from test one and the beginning of test 2. In this two cycles probably running-in and surface adaption processes have taken place. In following ten cycles, the difference in the curves can be explained by different ambient conditions. E.g. in the bottom graphs, the oil sump temperature at the beginning of the tests is slightly different or in test six the gate to the test hall have been opened for a longer duration (winter, 5°C outside temperature). This directly correlates with the measured friction values. In total 8000 start-up cycles have been completed, which correspondents with about 40 % of the lifetime start-ups of a wind turbine. After the tests the bearing has been disassembled and the optical impression of the bearing surfaces corresponds with the measurement results. The highest loaded bearing pads showed a slight smoothing of the surfaces but without excessive wear.

4. Summary

This contribution shows an innovative approach concerning a new type of rotor main bearings for wind turbines. The conical sliding bearing design with its flexible support structure of the bearing pads combines the benefits of established double row tapered roller bearings with the benefits of sliding bearings. When a bearing fault occurs, for example, the bearing pads are replaceable individually without the need of disassembling the drivetrain, in contrast to roller main bearing applications.
In order to investigate the new bearing concept under real conditions, an extensive measuring campaign has been carried out on a full-scale 1 MW system test bench. Among other matters the load-bearing properties, the mixed fiction limit and the start-up behaviour have been investigated. All tests carried out delivered very promising results. Furthermore the described problem of edge wear is successfully solved by the innovative flexible support structure of the bearing pads.

Acknowledgments
The authors would like to thank the German Federal Ministry of Education for the financial funding and the possibility to do research on this topic.

References
[1] E. Hau: Windkraftanlagen: Grundlagen, Technik, Einsatz, Wirtschaftlichkeit. 4. Auflage, Springer 2008
[2] J. Wenske, H. Kyling: Abschlussbericht „ELAN – Entwicklung innovativer elektrischer Antriebskonzepte für Windenergieanlagen“ Fraunhofer Institute for Wind Energy Systems IWES; ResearchGate 2016
[3] T. Schröder, G. Jacobs, D. Bosse: Sliding moment bearing as a main bearing in wind turbine generators; Conference for Wind Power Drives 2017
[4] K. Bobzin, M. Öte, T. Königstein, W. Wietheger, T. Schröder, G. Jacobs, D. Bosse: Thermally Sprayed Hydrodynamic Main Bearings for Wind Turbines; International Thermal Spray Conference (ITSC) 2017
[5] T. Schröder, G. Jacobs, D. Bosse: EHD-Betrachtung eines neuartigen Gleitlagerkonzeptes für das Hauptlager einer Windenergieanlage, First-User-Meeting 2017
[6] J. Hölzl, T. Schröder, A. Rolink: Innovative Lagerkonzepte und neue Gleitlagerwerkstoffe für den Antriebsstrang von Windenergieanlagen; 3. VDI-Fachkonferenz: Schadensmechanismen an Lagern 2018
[7] K. Bobzin, M. Öte, T. Königstein, W. Wietheger, T. Schröder, G. Jacobs, D. Bosse: Thermisch gespritzte Gleitlager für die Hauptlagerung von Windenergieanlagen, ATK 2017-Antriebstechnisches Kolloquium 2017

IOP Conf. Series: Journal of Physics: Conf. Series **1222** (2019) 012027 doi:10.1088/1742-6596/1222/1/012027

Comparative Concept Design Study of Laterally Loaded Monopiles

K Kaltekis[1], S Panagoulias[2], B F J van Dijk[3,4], R B J Brinkgreve[5] and M Ramos da Silva[6]

[1] Geotechnical Engineer, Fugro, Nootdorp, the Netherlands
[2] Researcher, Plaxis, Delft, the Netherlands
[3] Senior Geotechnical Advisor, Arcadis, Amersfoort, the Netherlands
[4] Principal Engineer, Fugro, Nootdorp, the Netherlands (formerly)
[5] Associate Professor, Delft University of Technology, Delft, the Netherlands
[6] Senior Engineer, Fugro, Brussels, Belgium

Email: k.kaltekis@fugro.com

Abstract. Offshore wind turbine generators (WTG) are commonly founded on single large diameter piles, named monopiles. These monopiles are subjected to significant lateral loads and thereby sizeable overturning bending moments mainly due to action of wind and wave forces; thus the critical geotechnical design situation for monopiles supporting WTGs is often related to lateral loading conditions. The Pile Soil Analysis (PISA) joint industry research project [1] has recently proposed a monopile design method which encompasses finite element (FE) calculations under a specific design framework. Soil reaction curves that are crucial for monopile design (i.e. lateral force and moment reactions along the shaft and at the base of the pile) are derived from FE calculations, subsequently calibrated and entered into a 1D model which is then used for design optimisation. This method is implemented within the PLAXIS MoDeTo (Monopile Design Tool) software. This paper presents results of a concept monopile design study under lateral monotonic loading with the use of the PLAXIS MoDeTo method.

1. Introduction

Planning and development of offshore wind farm parks is booming during the past few years, as they are considered a key element in meeting renewable energy and carbon emission targets. The majority of both existing and planned offshore WTGs are supported by monopiles, which are single driven foundation piles of large diameter. Monopiles supporting WTGs are subjected to significant lateral forces and overturning moments from, for example, vessel impact and environmental loads induced by wind, waves and currents. In principle, the critical geotechnical design situation for wind turbine monopile foundations is related to lateral loading conditions.

Current design standards for long slender piles [2] recommend the p-y approach for assessing pile lateral response. In this method, based on the Winkler assumption [3], the soil surrounding the pile is modelled as a set of uncoupled non-linear elasto-plastic springs, which define the lateral pressure (p) applied to the pile at a given depth, as a function of the lateral pile displacement (y). The p-y method was developed and empirically validated for piles with high aspect ratios (length L over diameter D) exceeding 15. However, monopiles usually have a low L/D aspect ratio of less than 8, and are therefore intermediate foundations, for which distributed moment along the shaft (m), base shear (H_B) and base moment (M_B) will also influence the lateral behaviour of the monopile. Historic industry practice for concept monopile design was to modify the p-y curve formulations to account for large diameter effects or to derive them from finite element (FE) calculations.

The Pile Soil Analysis (PISA) joint industry project [1] has developed and proposed a new design framework for concept monopile design. The soil reaction curves (p-y, m-ψ, H_B-y and M_B-ψ, where ψ is rotation) are extracted from 3D FE analyses and used in a 1D Timoshenko beam model representing the monopile (Figure 1). A software tool called PLAXIS MoDeTo (Monopile Design Tool) has been developed for the purpose of monopile foundation design for WTGs aiming at implementing the PISA monopile design framework for practical application [4,5].

This paper illustrates the applicability of the PISA method in engineering practice via the use of PLAXIS MoDeTo. It presents results of a concept design for monopile foundations in a stiff overconsolidated clay profile, considering monotonic loading conditions.

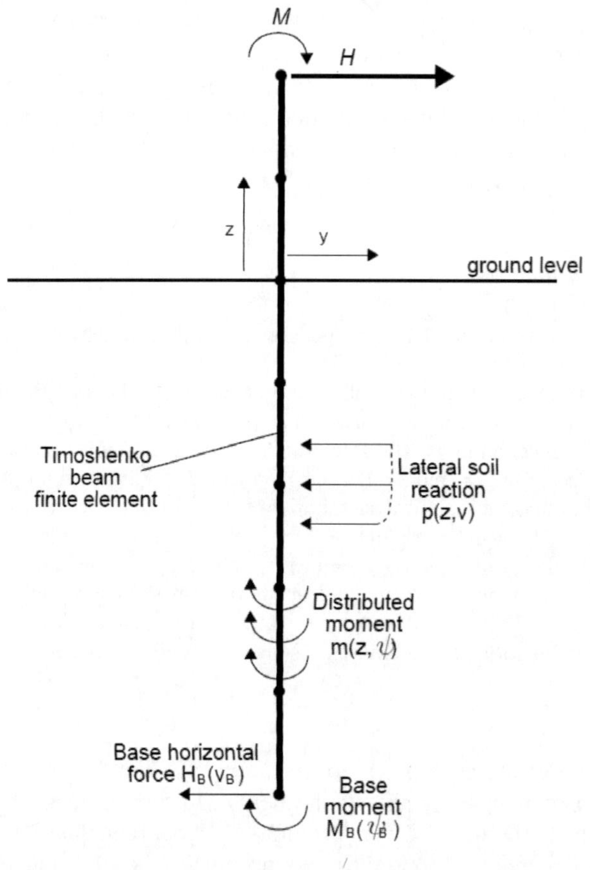

Figure 1. 1D FE model as developed in PISA project (as depicted in [4]). Note that notation v in the figure corresponds to notation y in the main text.

2. Ground Conditions

Table 1 presents the selected ground profile and soil design parameters. Profile selection included consideration of actual soil site specific investigation data including data from seismic in-situ tests and advanced laboratory tests (i.e. triaxial and direct simple shear tests with bender elements). The soil profile is idealised and does not represent or is representative of any specific location in the North Sea.

Table 1. Summary of soil parameters.

Depth	Effective unit weight	Undrained shear strength	Small strain shear modulus	Coefficient of lateral earth pressure at rest	Axial strain at 50% deviatoric stress
	(γ')	(s_u)	(G_0)	(K_0)	(ε_{50})
[m BSF]	[kN/m^3]	[kPa]	[MPa]	[-]	[%]
0-8	7.6	75	70	1.4	0.7
8-21	8.6	85	105	1.15	0.7
21-28	8.6	120	125	1	0.5
28-50	10.2	140	145	0.9	0.5

- BSF: below seafloor
- G_0 and K_0 are only used in the PLAXIS MoDeTo method
- ε_{50} is only used in the p-y method

3. Scope of the study

This paper considers required minimum monopile installation depth for ULS and SLS design criteria using the following methods:

- PLAXIS MoDeTo method following the PISA design framework;
- p-y method following the method by Stevens and Audibert [6].

The selected monopile (Figure 2) is modelled as an elastic beam, with an outer diameter of 9 m, wall thickness of 100 mm, to be installed by impact driving. The (unfactored) design lateral load is 9 MN resulting in an overturning moment of 594 MN at seafloor for a load eccentricity of 66 m. Table 2 presents the considered limit states and associated design criteria. Note that the presented study excludes specific considerations for scour management and the possible formation of a gap around the pile close to seafloor.

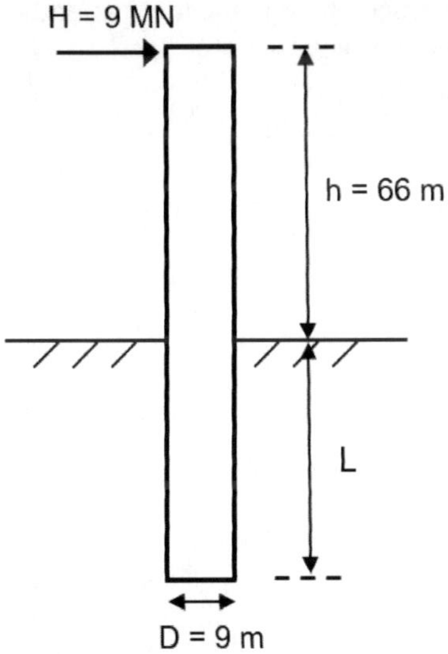

Figure 2. Schematic representation of the monopile design.

Table 2. Overview of limit states and concept design criteria.

Limit state	Design criteria
ULS	Working Stress Design (WSD) approach with a global safety factor of 1.5
SLS	Horizontal permanent rotation tolerance at seafloor of 0.25 degrees

- Note that the SLS design criterion refers to monotonic loading conditions for which the monopile rotation can be partially reversible. This is cautiously not considered for this study.

3.1. PLAXIS MoDeTo method

The PLAXIS MoDeTo method is following the numerical-based design philosophy from the PISA project [7] and thus it entails a step-by-step procedure to be followed rather than a prescription of formulas as is the case for p-y methods. PLAXIS 3D [8] is used to define and run the 3D FE monopile models which are necessary for calibrating the soil reaction curves used in the 1D FE model. The design procedure is as follows:

- Soil stratigraphy and parameter selection;
- Definition of geometrical parameter space for calibration of soil reaction curves;
- Calculation of the 3D FE (calibration) monopile models;
- Calibration of the soil reaction curves (i.e. p-y, m-ψ, H_B-y and M_B-ψ) extracted from the 3D FE calculations;
- Run of the 1D FE model with the calibrated (site-specific) soil reaction curves;
- Optimisation of the geometry of the 1D monopile model based on ULS and SLS design criteria;
- Accuracy check of the final design with a (geometrically) equivalent 3D FE model.

3.1.1. Soil parameter selection. Soil parameters are derived to be used as input in PLAXIS 3D constitutive models. The default constitutive model for modelling undrained clay behavior (total stress analysis) within PLAXIS MoDeTo is the NGI-ADP model [9]. The more advanced parameters of the NGI-ADP model are automatically derived using empirical correlations [4] and the user-input of basic soil parameters (Table 1). This study excludes verification of the suitability of the default selection of the advanced parameters and calibration of the constitutive model.

3.1.2. Calibration parameter space. A series of 3D FE monopile models, termed calibration models, were defined in order to span the design geometrical parameters, namely the embedded pile length L, the pile outer diameter D, the pile wall thickness t and the load eccentricity h. These geometrical parameters (Table 3 and Figure 3) were selected in such a way as to ensure that the final monopile design falls within the defined design space. Reliability of the 1D model is otherwise doubtful. An initial indication of the required monopile geometry can be obtained from p-y calculations. Four, eight and twelve 3D FE models were used for calibration to investigate the influence of the number of calibration models on the outcome of the calibration of the extracted soil reaction curves (see also Section 3.1.3) and consequently on the accuracy of the 1D model. The outcome of this sensitivity analysis is presented in Section 4.1. Figure 3 also shows the final geometry of the 1D model along with the equivalent PLAXIS 3D model used for quality check.

Table 3. Summary of 3D FE calibration models.

Calibration models	Length (L) [m BSF]	Outer diameter (D) [m]	Load eccentricity (h) [m ASF]	Wall thickness (t) [mm]
GeoDS_1	20	8	48	80
GeoDS_2	42.5	8.5	76.5	90
GeoDS_3	23.75	9.5	85.5	100
GeoDS_4	50	10	60	110
GeoDS_5	42	8.4	63	85
GeoDS_6	33	8.8	79.2	90
GeoDS_7	23	9.2	69	95
GeoDS_8	36	9.6	57.6	100
GeoDS_9	64	8	72	90
GeoDS_10	80	10	60	110
GeoDS_11	40	8	96	90
GeoDS_12	25	10	120	100

- BSF: below seafloor
- ASF: above seafloor

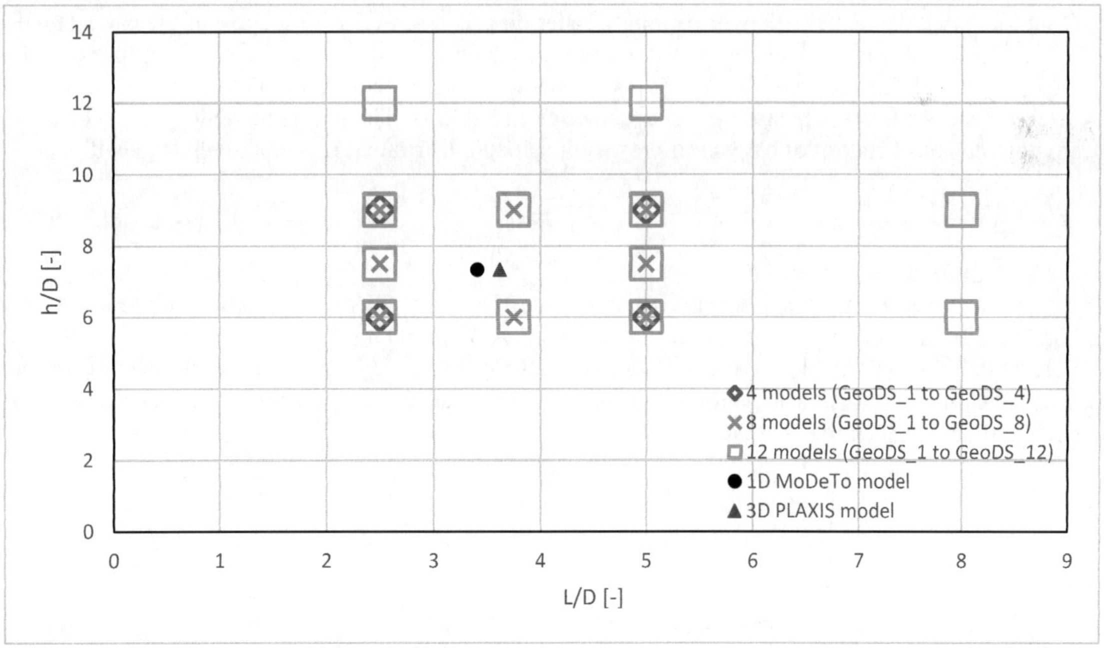

Figure 3. Calibration parameter space including the 3D calibration models, the final (optimised) 1D model and the final PLAXIS 3D model (h/D: load eccentricity ratio, L/D: aspect ratio, h: height above seafloor, L: monopile length below seafloor, D: monopile outer diameter).

The calculation of the 3D FE models is displacement-controlled; hence the prescribed value of pile head displacement should be large enough to ensure that nominal failure has occurred for all 3D models included in the calibration process. Nominal failure is presumed for pile lateral deflections at seafloor of at least 0.1D, where D is the monopile diameter [10].

3.1.3. Soil reaction curves. The soil reaction curves are extracted from the 3D FE calculations, normalised and subsequently used to calibrate the mathematical functions that will approximate the soil reaction curves in the 1D model [4]. Those functions are inherently dependent on the geometrical variations of the calibration models.

3.1.4. Design optimisation. The monopile geometry is optimised with regards to meeting the design criteria for ULS and SLS. The design parameter that has the primary focus for optimisation in this study is the embedded monopile length, since the selection of monopile diameter and wall thickness is typically driven by fatigue, buckling and natural frequency analysis. This is an iterative but fast procedure in which the embedded length of the 1D monopile model is modified (within the limits of the calibration parameter space, as presented in Figure 3) and the 1D model is ran until both ULS and SLS conditions are satisfied.

3.1.5. Accuracy check. Once the optimal monopile design is selected, an accuracy check is performed to confirm the robustness of the 1D calculation. An equivalent 3D model, of the same geometry as the final 1D model, is ran and the fitting of the resulting lateral load-deflection curves at seafloor is compared. PLAXIS MoDeTo computes an accuracy metric, termed η, which is recommended to be above 90 % [4].

3.2. p-y method

ISO suggests that the lateral behavior of long slender piles is assessed with the use of the p-y method, with the warning that large diameter piles with limited penetration may require a different formulation for the p-y relationships [2]. For this study, p-y curves were calculated according to a method proposed by Stevens and Audibert [6], based on database of pile load tests. The pile is modelled as a 1D elastic beam with no consideration of base shear, base moment and distributed moment along the shaft.

4. Results

4.1. Calibration parameter space

Figure 4 illustrates results from a sensitivity analysis carried out to check the accuracy of the 1D model calibrated with different number of 3D calibration models. Four 3D models were sufficient to calibrate the 1D model with reasonable accuracy. It was also observed that varying the pile wall thickness or the size of the calibration space had negligible influence on the calibration, as long as the final design was encompassed by the calibration space.

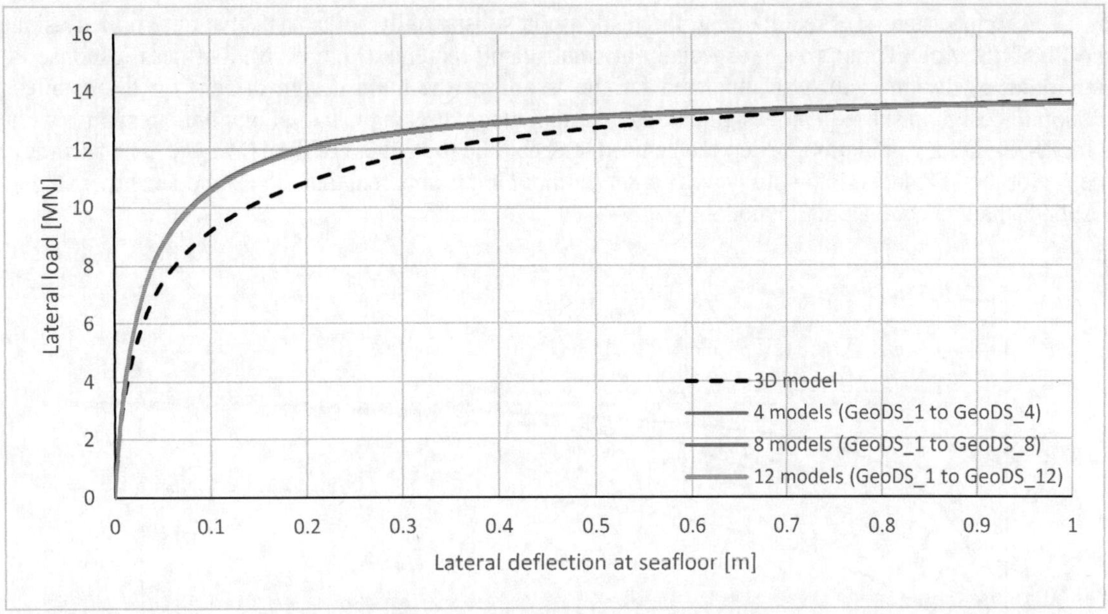

Figure 4. Comparison of resulting load-deflection curves for 1D models calibrated with different number of calibration models (see Table 3). The black dashed line represents the equivalent PLAXIS 3D model.

4.2. Concept monopile design

The required monopile design length according to the PLAXIS MoDeTo method is presented in Table 4. In comparison, the required monopile design lengths based on employment of a 3D FE model and the Stevens and Audibert p-y method are also displayed.

Table 4. Summary of required monopile lengths for all considered methods. The differences with the length predicted from the PLAXIS MoDeTo method (reference case) are also displayed.

Design method	Load case	Required monopile length [m BSF]	Aspect ratio [-]	Governing case	Difference
PLAXIS MoDeTo method (1D model)	ULS	30.7	3.41	✓	
	SLS	30	3.33		
PLAXIS 3D (equivalent) model	ULS	30.7	3.41		
	SLS	32.6	3.62	✓	+6%
Stevens and Audibert p-y method	ULS	34.6	3.84		
	SLS	39	4.33	✓	+27%

4.2.1. ULS. Figure 5 compares predictions of the monopile response for the ULS. For comparison purposes, load-deflection curves are also plotted for an installation depth of 30.7 m with the Stevens and Audibert p-y method. It is evident from the plots that the p-y method predicts a much lower ultimate pile capacity and considerably softer pile response compared to the other two methods. The pile would need to be 27 % longer than what the PLAXIS MoDeTo method predicts in order for the concept design to meet the ULS and SLS requirements. Furthermore, even by increasing the monopile installation depth by 13 % to meet the ULS requirement, the response is substantially softer than that determined using the PLAXIS MoDeTo approach, especially at small displacements (Figure 5b). Stiffness response is important, as design against accumulated fatigue is one of the main design drivers for the detailed monopile design in subsequent design phases. Figure 5 also presents the lateral monopile response with a PLAXIS 3D FE calculation, which can be used to check the robustness of the 1D model. The accuracy metric for the 1D model in this study was 96.8 %, although it is apparent that the 1D model shows stiffer response than its equivalent 3D model.

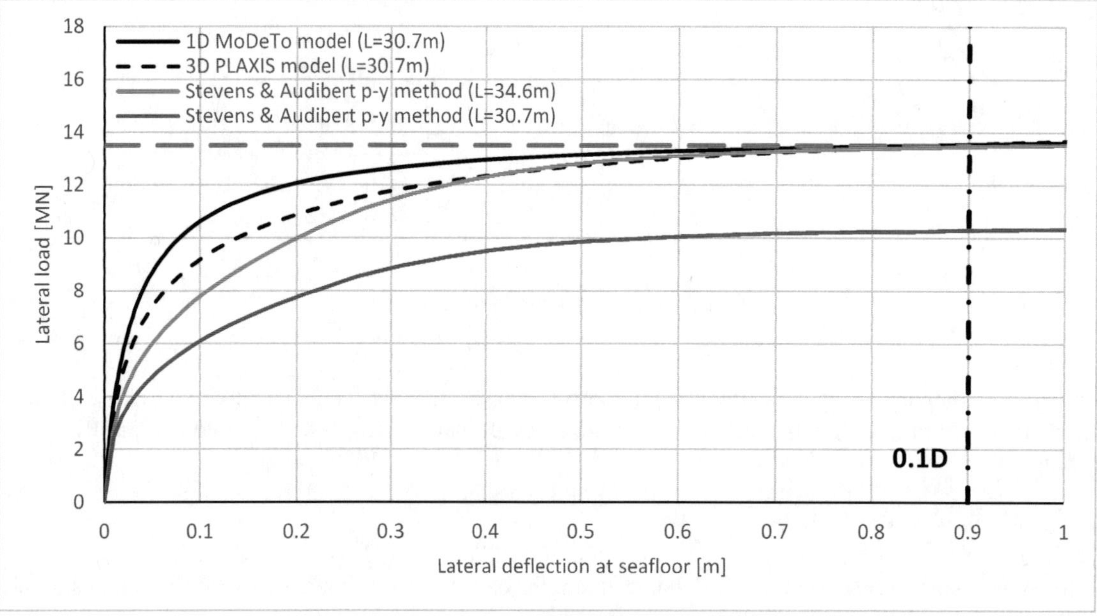

(a) Response to large displacements

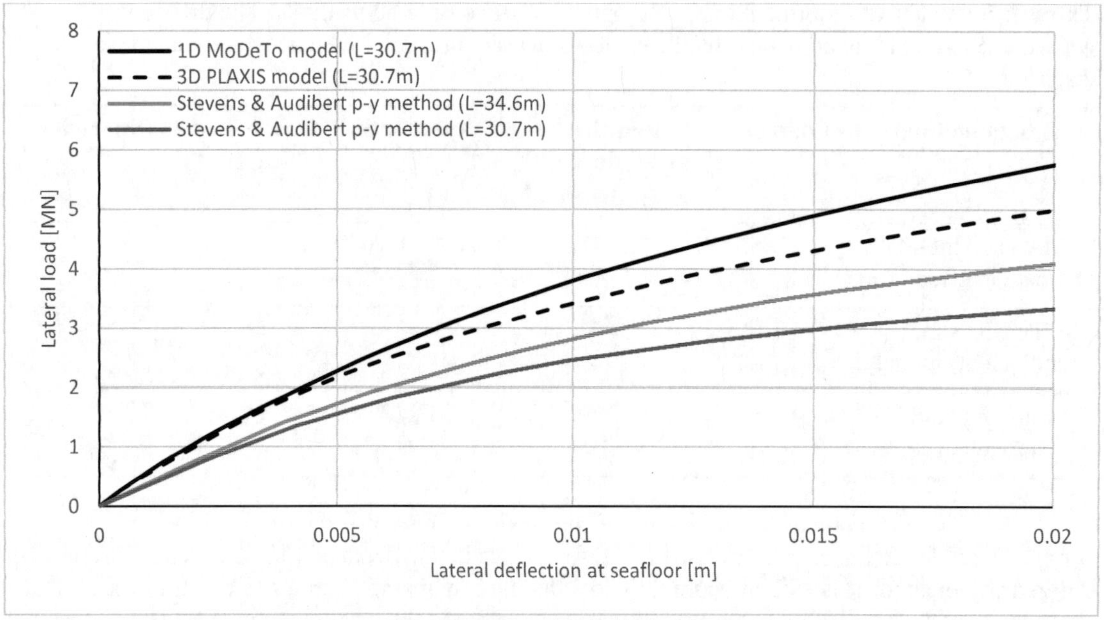

(b) Response to small displacements

Figure 5. Monopile response in ULS.

4.2.2. SLS. The pile response for the PLAXIS MoDeTo calculation shows a stiffer behaviour than both the 3D FE calculation and the p-y method (Figure 6).

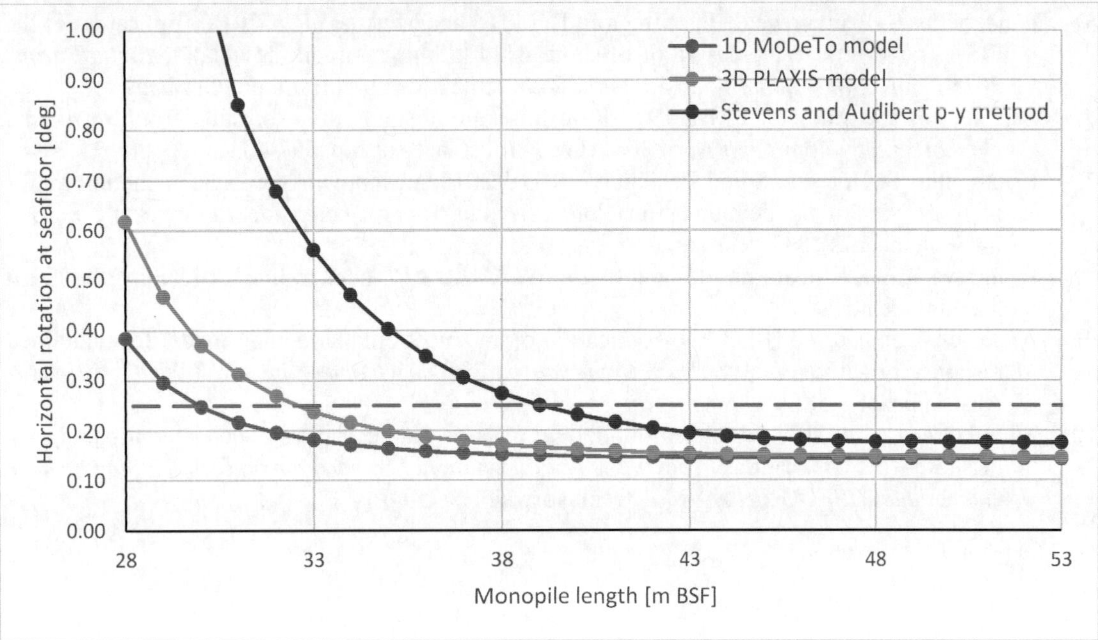

Figure 6. Horizontal monopile rotation at seafloor versus monopile length (SLS).

5. Conclusions

The PLAXIS MoDeTo method is a straightforward and easily applicable method for concept design of monopiles based on the PISA design framework. It provides a realistic representation of a typical large diameter monopile capturing the key elements of its behavior when subjected to lateral monotonic loading.

The quality check of the calibrated 1D model against its equivalent 3D model is within tolerable margins according to the accuracy metric, although in this study the calibrated 1D model was stiffer than its equivalent 3D model. This resulted in the 1D model showing a 6 % shorter monopile than the 3D model. Furthermore, the size of the calibration space did not seem to influence the calibration accuracy provided that the final design is within the defined calibration space. The MoDeTo team is working on further optimisation of the calibration procedure (i.e. calibration of the mathematical functions) to better match the 1D results with the 3D FE model results.

It was also observed that only a small number of 3D FE models (i.e. four in this study) is required for calibration of the 1D model; hence the overall computation time when employing the PLAXIS MoDeTo method is relatively limited.

Making use of a conventional p-y method for concept monopile design results in a substantially softer response and lower ultimate capacity of the pile, as anticipated.

6. References

[1] Byrne B W et al 2017 PISA: New Design Methods For Offshore Wind Turbine Monopiles *Proc. of the 8th International Conf. on Offshore Site Investigation and Geotechnics (OSIG)* Vol. 1 12-14 September 2017 London UK pp. 142-161

[2] International Organization for Standardization 2016 *ISO 19901-4:2016 Petroleum and natural gas industries – Specific requirements for offshore structures – Part 4: Geotechnical and foundation design considerations* Geneva: ISO

IOP Conf. Series: Journal of Physics: Conf. Series **1222** (2019) 012027 doi:10.1088/1742-6596/1222/1/012027

[3] Winkler E 1867 *Die lehre von elasticitaet und festigkeit* Verlag Dominicus Prague
[4] Panagoulias S, Brinkgreve R B J and Zampich L 2018 *PLAXIS MoDeTo Manual 2018* Plaxis bv Delft the Netherlands
[5] Panagoulias S, Brinkgreve R B J, Minga E, Burd H J and McAdam R A 2018 Application of the PISA framework to the design of offshore wind turbine monopile foundations *WindEurope 2018 Conf. at the Global Wind Summit* 25-28 September 2018 Hamburg Germany
[6] Stevens J B and Audibert J M E 1979 Re-examination of p-y curve formulations *11th Annual Offshore Technology Conf* April 30 - May 3 Houston Texas pp. 397-403
[7] Panagoulias S, Hosseini S and Brinkgreve R B J 2018 An innovative design methodology for offshore wind monopile foundations *26th European Young Geotechnical Engineers Conf.* 11-14 September 2018 Graz Austria
[8] Bringreve R B J, Kumarswamy S and Swolfs W M 2018 *PLAXIS manual 2018* Plaxis bv Delft the Netherlands
[9] Andersen L and Jostad H P 1999 Application of an anisotropic hardening model for undrained response of saturated clay *Proc. Numerical models in Geomechanics (NUMOG) VII* Graz Austria pp. 581-585
[10] Zdravković L et al 2015 Numerical modelling of large diameter piles under lateral loading for offshore wind applications *Proc. of the 3rd International Symposium on Frontiers in Offshore Geotechnics (ISFOG 2015)* Vol. 1 Oslo Norway pp. 759-764

Comparison of inflow-turbulence and trailing-edge noise models with measurements of a 200-kW vertical axis wind turbine

E. Möllerström[1] and F. Ottermo[1]

[1] The Rydberg Laboratory for Applied Sciences, Halmstad University, PO Box 823, SE-301 18 Halmstad, Sweden

E-mail: erik.mollerstrom@hh.se

Abstract. Models of inflow-turbulence noise and turbulent-boundary-layer trailing-edge noise are compared to earlier measurements of a 200-kW vertical axis wind turbine so that conclusions regarding the origin of the aerodynamic noise can be drawn. The measurement campaigns, which aimed at establishing the noise emission value and locating the aerodynamic noise sources with a microphone array, are here both compared to further modified versions of the trailing-edge and inflow-turbulence models respectively. Unlike the case for horizontal axis wind turbine, inflow-turbulence noise is deemed as the prevailing noise mechanism. Reducing the self-induced turbulence could then be an effective way of lowering the noise levels for vertical axis wind turbines. Also, looking at the directivity of the inflow-turbulence noise model which indicate most noise in the cross-wind directions, a deviation from the standard downwind measurement position for measuring noise emission is suggested for the VAWT case.

1. Introduction

Wind turbines can be categorized by the orientation of their axis of rotation into two groups: horizontal axis wind turbines (HAWTs) and vertical axis wind turbines (VAWTs). From a historical perspective, different sub-concepts of both HAWTs and VAWTs have had individual threads of development in both Europe and North America [1-3]. It was mainly one of these sub-concepts, the upwind 3-bladed HAWT that established itself on the market in the 1970-80s in Denmark, where it thereafter has developed into the large and economically feasible wind turbines of today.

However, the wind turbine industry is today facing several challenges, for example, local environmental impact as well as service cost and reliability. These challenges match some of the possible advantages of the VAWT technology, which has led to a renewed interest in the concept. By using a vertical shaft to transfer the torque, VAWTs can have the generator and other key parts located at ground level which enables designing them with focus on performance and economy rather than size and mass. Furthermore, maintenance and modifications are made easier with these parts placed on the easily accessible tower base. Yaw motors are superfluous since VAWTs are omni-directional, which allows for a design with essentially only one moving part, including blades, struts, shaft and generator rotor which are all jointed. Of special interest here, the concept has shown potential for lower noise emissions, which enables establishment closer to inhabited areas [4, 5]. Moreover, in [6] it has been shown that the concept is more suitable for up-scaling to > 10 MW than the HAWT concept; it has also been suggested that, mainly due to its low center of mass, it is more suitable for floating off-shore platforms [7].

Noise is considered as one of the disadvantages with wind turbines and noise immissions at dwellings and other sensitive areas is generally regulated in national legislation which restrains potential locations when planning for wind power. Noise from operating wind turbines can be divided into mechanical and aerodynamic noise with the aerodynamic noise being dominant for modern turbines [8]. While the mechanisms behind aerodynamic noise from HAWTs have been thoroughly investigated, the origin of aerodynamic noise from VAWTs is still to a large extent an open question. The work in this paper is a continuation of earlier attempts [4, 5] to understand how VAWT noise is created.

When designing a wind farm, the noise immission are typically simulated based on reported noise emissions of the specific wind turbine model. The noise emission is measured according to the IEC standard [9], which specifies individual procedures for HAWTs and VAWTs respectively. The difference being that for a VAWT, the sound pressure level is measured at a downwind distance $R_h = H + D$, which is further away than the $R_h = H + D/2$ that is stated for HAWTs.

Table 1. Properties of the T1-turbine.

Rated power	200 kW
Turbine diameter	26 m
Tower height	38 m
Hub height[1]	41 m
Wing length	24 m
Swept area	624 m^2
Cut-in wind speed	4 m/s
Rated wind speed	12 m/s
Cut-out wind speed	25 m/s
Survival wind speed	60 m/s
Rotational speed	16-33 rpm
Operational tip speed ratio	3.8
Power regulation	Passive stall
Wing/strut material	Fiberglass composite
Tower material	Laminated Wood

Figure 1. The T1-turbine.

1.1. The 200-kW VAWT H-rotor

The VAWT we consider in this study is a so called H-rotor, which is a Darrieus-type turbine with straight fixed blades. A 200-kW VAWT (hereafter referred to as the T1-turbine) was erected in 2010 just outside Falkenberg on the Swedish west coast by the company Vertical Wind AB in collaboration with Uppsala University. Only one 200-kW turbine was ever built by Vertical Wind AB, making the T1-turbine unique. This particular VAWT has a direct-drive permanent-magnet synchronous generator which is mounted at the bottom of the tower and connected to the rotor by a steel shaft. The rotor consists of three 24 m long straight blades that are connected to the shaft by two struts each. Both blades and struts are made out of fiberglass. The turbine's blades have a fixed pitch, but the variable speed of operation allows control of the rotor through the stall effect. The aerodynamic power coefficient has been measured to 0.33 [10], and the noise characteristics of the turbine have also been studied [4, 5]. The T1-turbine has a tower made out of laminated wood covered by fiberglass laminate. From the start the tower was free standing, but after two years it was complemented with three guy wires. The construction may

[1] The hub height includes the tower, the small hill covering the generator and half the hub shaft that is carrying the struts.

therefore be described as semi-guy wired [11]. The guy wires were added because small fatigue cracks appeared in some of the glue joints attaching the steel flanges to the glue-laminated wood. A more detailed description of the 200-kW experimental turbine can be found in [12] while a review of Uppsala University's entire work on VAWTs can be found in [13].

2. Theory

Aerodynamic noise sources can be divided depending on causation, most notably airfoil self-noise and inflow-turbulence noise. Airfoil self-noise is caused by the interaction between the turbulence created by the blade and the airfoil itself [14]. The airfoil self-noise sub-group turbulent-boundary-layer trailing-edge (TBL-TE) noise is created when the turbulent boundary layer, that develops along the blade surface and convects downstream, passes the trailing edge. The sudden change when the blade surface disappears causes sound to be scattered as broadband noise. In [15], microphone array measurements showed that for a modern HAWT, TBL-TE noise is the dominant noise source. Other airfoil self-noise sub-groups include tip noise, blunt trailing-edge noise, separation-stall noise and laminar-boundary-layer vortex-shedding noise. For more information about these and other sub-groups, well explained descriptions can be found in [8].

The inflow-turbulence noise is created when upstream turbulence interacts with the leading edge of the blade [8]. The sound is caused when the turbulent eddies is scattered at the leading edge, creating noise of broadband character. This noise source is dependent on the amount of incoming turbulence, hence both the atmospheric conditions and the eventual presence of turbulence creating structures in the vicinity, such as other wind turbines, will be of importance. It is to some extent an open issue to what extent inflow-turbulence noise contributes to the overall noise level of a wind turbine [8], although it has been shown that it is not the dominant effect for the large-scale HAWT studied in [15]. However, the contribution of inflow-turbulence noise can be assumed to be greater for VAWTs than HAWTs as the blades are passing in the turbulent wake of the previous passing blades.

2.1. TBL-TE noise model

The TBL-TE noise contribution is estimated using the TBL-TE part of the empirical model for airfoil self-noise in [14]. For small VAWTs (low Reynold numbers), laminar-boundary-layer trailing edge noise has been observed to be important [16], but for the larger-sized turbine considered here the Reynolds number is of the order of 1×10^6, so the TBL-TE is expected to be dominant over the laminar-boundary-layer part. A complication is that the model in [14] assumes static angles of attack, whereas in this case the angle of attack is constantly changing. However, static angle-of-attack lift and drag data has been successfully used in prediction models of VAWT performance [17], so it is reasonable to expect that a static approach will be acceptable here as well. The model for the sound pressure level has the following main characteristics [14, 15]:

$$SPL_{\text{TBL-TE}} = 10\log_{10}\left(\frac{L\,M^5\delta^*}{R^2}\,\frac{\sin^2\Theta/2\sin^2\Phi}{(1+M\cos(\Theta+\alpha))^4}\right) + K_1(f), \qquad (1)$$

where L is the blade span, M is the local Mach number, δ^* is the displacement thickness of the boundary layer at the trailing edge, R is the distance from observer, Θ and Φ are angles specifying the direction of the observer relative to the blade (see [4]), α is the angle of attack and $K_1(f)$ is a frequency dependent constant. The local Mach number and the angle of attack are calculated from a simplified model where the flow is undisturbed by the turbine. The displacement thickness and the constant K_1 are calculated according to the prescription in [14]. Further information about the application of this model to the current geometry is found in [4].

2.2. Inflow-turbulence noise model

An important difference between VAWTs and HAWTs with respect to noise generation is that the blades of a VAWT, at the downstream half of the rotation, pass the wake of the blades at the upstream half.

IOP Conf. Series: Journal of Physics: Conf. Series **1222** (2019) 012028 doi:10.1088/1742-6596/1222/1/012028

The turbulence levels at the downstream half of a VAWT is then expected to be much larger than the turbulence of the flow ahead of the turbine, due to the upstream blade disturbing the flow. Inflow-turbulence has been proposed to be a significant source of noise even for HAWTs, depending on the turbulence properties of the incoming flow [8]. The flow of a small 2-bladed VAWT was mapped experimentally in [18] where it was observed that a significant amount of turbulence is generated by the blades and struts. The velocity fluctuations appears to be dominated by the tip vortices, but a significant amount of turbulence is also seen to be generated by the struts and the tower, apparently a lot more than the contribution from the trailing-edge vorticity. The turbulence generated by the tower and the struts is indeed observed to collide with the downstream blade path. This means that turbulent inflow might be a potential noise source for VAWTs, irrespective of the atmospheric properties. The model used here builds on [19, 20], and the adaptation to the current geometry is further described in [5]. The main characteristics of the model in the mid frequency range is captured by:

$$SPL_{\text{Turb}} = 10\log_{10}\left(\frac{L\,M^{10/3}I^2}{R^2\ell^{2/3}}\,\frac{\sin^2\Theta\sin^2\Phi}{(1+M\cos\Theta)^4}\right) + K_2(f), \qquad (2)$$

where L is the blade span along which noise is created, I is the turbulence intensity, ℓ is the turbulence length scale, and $K_2(f)$ is a frequency dependent constant, which may be calculated according to the framework given in [5]. Regarding the self-induced turbulence, the noise-generating span L, as well as the approximate size of the turbulent air stream that is proposed to be originating from the strut-blade joint, is estimated from the microphone-array measurements of [5]. Moreover, the turbulence intensity I is estimated to be of the order of 0.4 using the an energy extraction argument based on a comparison of C_p measurements for the current design and strutless designs, see [5]. The turbulence length scale is taken to 0.3 m since that is the length scale of the structure that is proposed to generate the turbulence. In addition to the self-induced turbulence, the current study also includes the ambient turbulence hitting the leading edge of the blades during the whole revolution. The turbulence length scale for this background inflow turbulence is taken to be the height above the ground (40 m), as is conventional [20]; this scale is also indicated by the long term wind measurement used for the noise emission measurement.

3. Method

Two separate noise measurements campaigns on the T1-turbine have previously been performed. One noise emission measurement to evaluate the noise levels, and one microphone array measurement aiming to localize aerodynamic noise sources; for details of the measurement setups and data processing, see [4, 5]. The positions for the array measurement setup as well as the emission measurement setup are indicated in Table 2 and Figure 2.

Table 2: Wind speed, measurement-position angle, and measured atmospheric turbulence intensity for the measurement positions. The wind direction angle is relative to the measurement-position direction. For the emission (long-term) measurement, the mean and standard deviations are given.

Position	Wind speed at hub height (m/s)	Wind speed at 10 m (m/s)	Measurement-position angle	Turbulence intensity
Array pos 1	4.7	3.7	31°	0.20
Array pos 2	6.0	4.7	126°	0.11
Array pos 3	5.4	4.2	218°	0.12
Array pos 4	5.0	3.9	303°	0.13
Emission meas	8.0 ± 1.7	6.3 ± 1.3	353° ± 10°	0.16

The angle is always relative to the momentary wind direction. Table 2 also displays the measured wind speeds and turbulence intensities. We note that the measured atmospheric turbulence intensities are much lower than the estimated self-induced turbulence intensity at 0.4, (furthermore with a much shorter length scale) as discussed above. Note also that the mean position during the noise emission measurement is close to, but not exactly, downwind (which would correspond to 360° or 0°).

In the previous works, results from the noise measurements were compared to a model of the TBL-TE noise and the array measurements were compared to a model of the inflow-turbulence noise. A combined analysis was however not performed. Here, both models are compared simultaneously to the noise emission measurements and the array measurement respectively. Further differences in this study, compared to [4, 5], are (i) that for the comparison of the models to the long-term noise emission measurement, during which time the wind direction fluctuated (quantified in Table 2), the models used here also include averaging the level over a range of wind directions, and (ii) the effect of the atmospheric inflow turbulence is included.

4. Results and discussion

To investigate which noise source is the most important for the T1-turbine, models for the TBL-TE and inflow-turbulence noise are compared with measured results from the noise emission and microphone array campaigns. When comparing the measured results with the models for a certain recording position, the difference in directivity of the two noise contributions will alter the comparison depending on position. Therefore, the directivity of the two models must be considered before drawing conclusions. The directivities as well as marked out recording positions can be seen in Figure 2. Most noise is emitted when the blade is moving toward the wind for both models, and both models show a directivity that, for each blade position, suppresses noise if the observer is behind the blade and precisely in line with the blade chord. (For the inflow-turbulence model, noise is suppressed also at the corresponding position ahead of the blade, see Eq (2).) Summing up the contributions over the whole revolution, however, the inflow-turbulence model shows a sharper directivity pattern, due to the narrow region where the blade interact with self-induced turbulence (while noise is created more evenly during the entire revolution for the TBL-TE model). The dip in the noise level from the inflow-turbulence model is at the position where the observer is standing in line with the emitting blade.

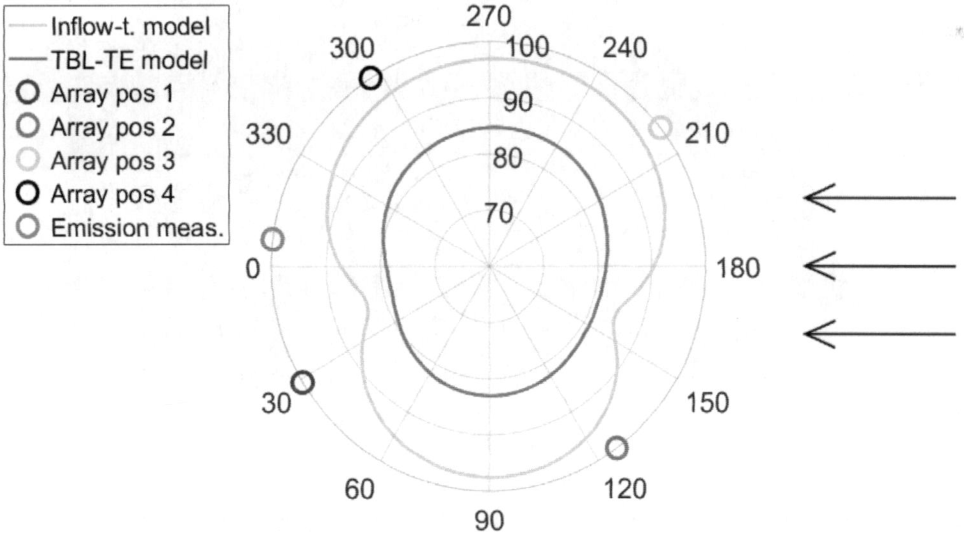

Figure 2. Directivity of the inflow-turbulence model (green) and TBL-TE model (red) as seen from above with the wind coming in from the right. The comparison is for a distance of 40 m from the turbine base, a wind speed (hub height) of 8.0 m/s (6.3 m/s at 10 m height) and n=22 rpm. The positions (measurement angles) for microphone array measurements and noise emission measurement are marked in the figure. Units in dBA and degrees counted counterclockwise from downwind position.

The noise emission measurement is compared to the models in Table 3 and Figure 3. For both the emission levels and 1/3-octave bands, it is clear that although the levels from both models are substantially lower than the measured levels, the inflow-turbulence model is closer than the TBL-TE model. The still lower levels of the inflow-turbulence model compared to the measured levels could be explained by directivity issues of the models; the directivity seem to have a sharper directivity in the model than in reality (the downwind position is close to where the least noise is expected, see Figure 2). Note that the values in Table 3 relates to integer wind speeds specified at the height 10 m, since noise emission from wind turbines are typically specified at wind speeds at 10 m height. Figure 2 and 3, on the other hand, relates to 8.0 m/s wind speed at hub height (which was the mean wind speed encountered during the noise emission measurement), corresponding to 6.3 m/s at 10 m.

Table 3: Measured noise emission and equivalent model values for 40 m downwind of the turbine. The models include averaging of a wind direction range with standard deviation 10°.

Wind speed at 10 m height (m/s)	5	6	7	8
Tip-speed ratio	3.8	3.8	3.4	2.9
Measurement (dBA)	93.1	94.1	95.4	96.2
TBL-TE model (dBA)	73.0	77.8	78.7	78.0
Inflow-turbulence model (dBA)	82.9	88.1	90.7	92.1

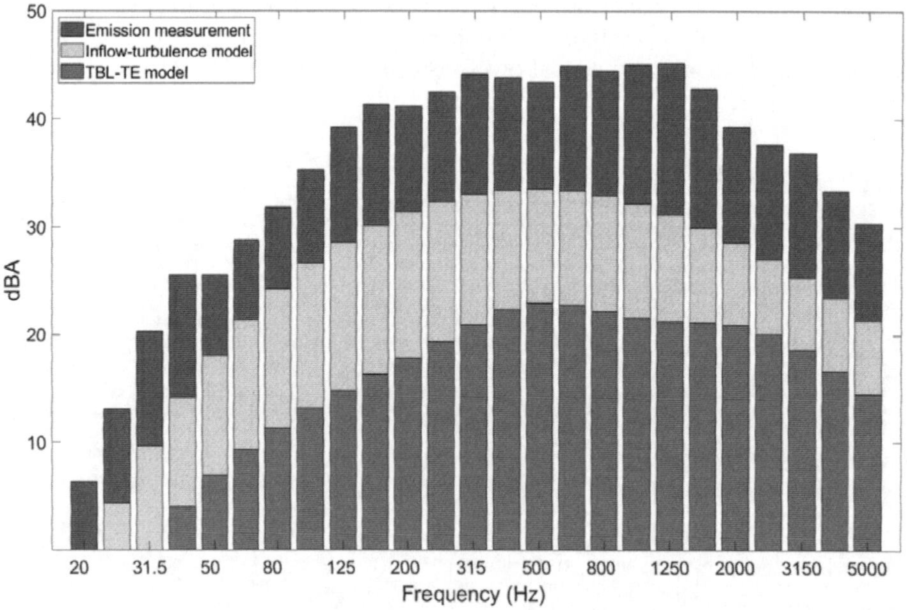

Figure 3. Comparison of 1/3-octave band of T1-turbine sound pressure level from noise emission measurements (blue), inflow-turbulence model (green) and TBL-TE model (red). All cases for a wind speed (hub height) of 8.0 m/s (6.3 m/s at 10 m height) and n = 22 rpm. The models include averaging of a wind direction range with standard deviation 10°.

The microphone array measurements are compared to the models in Figure 4, in which the experimental 1/3 octave bands and the corresponding bands from the models can be seen for the four recording positions. For each of the four positions, the wind speed and tip-speed ratio for that specific array recording has been implemented in the models. Overall, the inflow-turbulence model seems to fit better although, for most of the bands, the measured values lie between the two models. The directivity of both models (see Figure 2) results in low levels for position 1. Although the measured spectrum shows the lowest levels for position 1 (partly explained by the low wind speed during the position 1 recording), the difference compared to the other positions is smaller (apart from the 1000 Hz band). However, the uncertainty in this angle is large, due to varying wind direction and the fact that the wind direction was measured at a mast located 100 m away from the turbine. Moreover, even though it is evident from the results in [5] that most of the noise is created in the exact region handled by the current model, it is clear that some self-induced turbulence exists also behind the hub and the upwind blades in general, which hits the down-wind blade. The lower local wind speed (both where the turbulence is created and where it hits the downwind blade) will produce lower noise levels, but the directivity will be different to the modeled contribution and could possibly be of relevance when the directivity suppresses the dominant contribution. The ambient atmospheric turbulence, which is included in the current model, has a less prominent directivity as noise is created during the whole revolution. However, due mainly to the much longer length scale, this contribution only accounts for some 3 dB at position 1 (and much less for the other positions). It is clear that the inflow-turbulence model, as formulated here, needs to be extended to be able to adequately handle downwind positions.

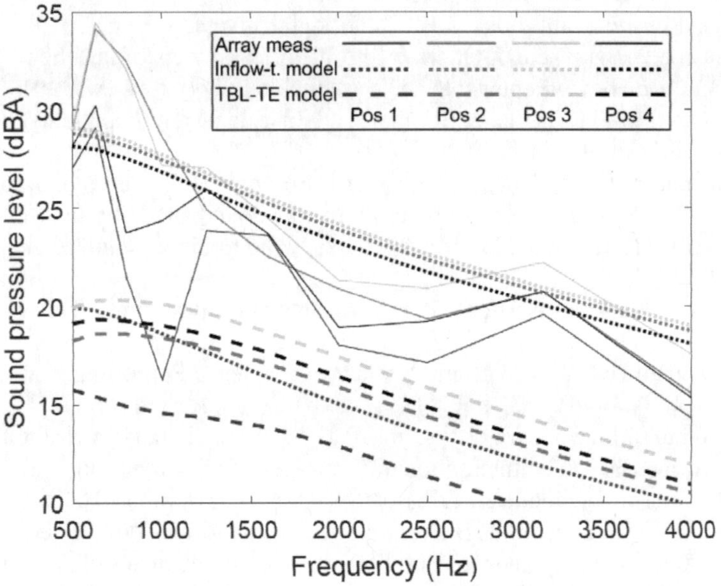

Figure 4. Comparison of 1/3 octave band of T1-turbine sound pressure level from inflow-turbulence model (dotted) and TBL-TE model (dashed) as well as the microphone array measurement (solid). For each position, the wind speed used for modelling was the same as during microphone array recording. No averaging over wind directions are done here since this is short-term measurement.

The comparison between the two models, the noise emission measurement and the microphone array measurements indicate that inflow-turbulence noise is the prevailing noise source for the T1-turbine. It seems like the noise produced by the T1 turbine is due to turbulent inflow that is created by the turbine itself, and creating this turbulence in turn degrades the power-conversion performance of the turbine, as discussed in [5]. If it is possible to mitigate the turbulence production in the blade-strut system, this is likely to both enhance the power factor and lower the overall noise level.

The IEC61400-11 standard implies measuring noise emission from a VAWT at a downwind distance $R_h = H + D$, which is a longer distance than the $R_h = H + D/2$ that is implied for HAWTs. This is probably the case because of the distance to the blade at its closest position will be approximately the same in both cases. However, as the most noise seem to be generated when the oncoming blade moves in the opposite wind direction and thus is at a tower distance from the downwind observer, using the same recording distance for VAWTs as for HAWTs seem to be more reasonable. Additionally, the indicated noise directivity with relatively low levels for the downwind position suggests that a cross-wind direction would be more appropriate when measuring noise emission from VAWTs. Specifying a crosswind direction, it is reasonable to actually keep the longer distance $R_h = H + D$, provided that the measurement is done on the side where the blade moves toward the wind.

5. Conclusions

Comparing results from the noise emission and the microphone array measurements with models of TBL-TE and inflow-turbulence noise respectively, it is suggested that the inflow-turbulence noise is the prevailing noise mechanism. Reducing the self-induced turbulence could then be an effective way of lowering the noise levels for VAWTs. Regarding the standard for measuring noise emission, it is suggested that the noise should be measured at the cross-wind direction where the blade move toward the wind.

References

[1] Gipe P. Wind Energy for the Rest of Us: A Comprehensive Guide to Wind Power and How to Use It. Bakersfield, CA, USA: Wind-Works.Org, 2016.
[2] Musgrove P. Wind Power. Cambridge, UK: Cambridge University Press, 2010.
[3] Möllerström E, Gipe P, Beurskens J, Ottermo F. A historical review of installed vertical axis wind turbines rated 100 kW and above. Renewable and Sustainable Energy Reviews. 2019;105:1-13.
[4] Möllerström E, Ottermo F, Hylander J, Bernhoff H. Noise emission of a 200 kW vertical axis wind turbine. Energies. 2016;9(1):19.
[5] Ottermo F, Möllerström E, Nordborg A, Hylander J, Bernhoff H. Location of aerodynamic noise sources from a 200kW vertical-axis wind turbine. Journal of Sound and Vibration. 2017;400:154-66.
[6] Ottermo F, Bernhoff H. An upper size of vertical axis wind turbines. Wind Energy. 2013;17(10):1623-9.
[7] Borg M, Shires A, Collu M. Offshore floating vertical axis wind turbines, dynamics modelling state of the art. Part I: aerodynamics. Renewable and Sustainable Energy Reviews. 2014;39:1214-25.
[8] Bowdler D, Leventhall HG. Wind Turbine Noise: Multi Science Publishing Company, Limited, 2011.
[9] IEC 61400-11: Wind turbine generator systems–Part 11: Acoustic noise measurement techniques, Edition 3. Geneva, Switzerland: The International Electrotechnical Commission (IEC); 2012.
[10] Möllerström E, Ottermo F, Goude A, Eriksson S, Hylander J, Bernhoff H. Turbulence influence on wind energy extraction for a medium size vertical axis wind turbine. Wind Energy. 2016.
[11] Möllerström E, Ottermo F, Hylander J, Bernhoff H. Eigen Frequencies of A Vertical Axis Wind Turbine Tower Made of Laminated Wood and the Effect Upon Attaching Guy Wires. Wind Engineering. 2014;38(3):277-90.
[12] Möllerström E. Noise, eigenfrequencies and turbulence behavior of a 200 kW H-rotor vertical axis wind turbine [Ph.D. thesis]: Uppsala University, 2017.
[13] Apelfröjd S, Eriksson S, Bernhoff H. A Review of Research on Large Scale Modern Vertical Axis Wind Turbines at Uppsala University. Energies. 2016;9(7):570.
[14] Brooks TF, Pope DS, Marcolini MA. Airfoil self-noise and prediction: National Aeronautics and Space Administration, Office of Management, Scientific and Technical Information Division, 1989.
[15] Oerlemans S, Sijtsma P, Méndez López B. Location and quantification of noise sources on a wind turbine. Journal of Sound and Vibration. 2007;299(4–5):869-83.
[16] Pearson C. Vertical axis wind turbine acoustics [Ph.D. thesis]: Cambridge University, 2014.
[17] Paraschivoiu I. Wind turbine design: with emphasis on Darrieus concept. Montreal, Canada: Presses inter Polytechnique, 2002.
[18] Tescione G, Ragni D, He C, Ferreira CS, Van Bussel G. Near wake flow analysis of a vertical axis wind turbine by stereoscopic particle image velocimetry. Renewable Energy. 2014;70:47-61.

[19] Paterson R, Amiet R. Acoustic radiation and surface pressure characteristics of an airfoil due to incident turbulence. 3rd Aeroacoustics Conference1976.
[20] Buck S, Oerlemans S, Palo S. Experimental validation of a wind turbine turbulent inflow noise prediction code. 22nd AIAA/CEAS Aeroacoustics Conference. Lyon, France2016.

IOP Conf. Series: Journal of Physics: Conf. Series **1222** (2019) 012029 doi:10.1088/1742-6596/1222/1/012029

Investigation of Modelling Depths for an Electromechanical Simulation of a Direct-Drive Generator Considering Parasitic Airgap Forces and External Loads

Tobias Duda*[1]; Georg Jacobs[1]; Dennis Bosse[1]

[1]Center for Wind Power Drives - RWTH Aachen University
Campus Boulevard 61, 52074 Aachen , Germany
*Corresponding author tobias.duda@cwd.rwth-aachen.de

Abstract. This paper shows a multiphysical simulation approach for direct - drive generators, including electromagnetic force calculations and multibody simulation. The objective of this method is to determine the airgap width distribution of the generator during operation.

In this investigation different modelling depths of the electromechanical model are discussed. Especially the modelling of the electromagnetic force application and integration of flexible structural elements is discussed.

1. Introduction

Renewable energy production must become competitive to fossil and nuclear energy sources. One way to achieve this is to keep maintenance and investment costs as low as possible. Therefore, novel and efficient simulation models are needed for improving the turbine's design. This paper shows an electromechanical simulation method for direct-drive generators, including the parasitic electromagnetic forces in the airgap of the generator in a multi-body simulation model. These parasitic forces aim to close the airgap resulting in a fatal turbine damage. To prevent collision of rotor and stator, the drivetrain is designed very stiff [1], which leads to an increased use of material. Most of recent research regarding direct-drive design addresses the matter of airgap stability, e.g. incorporating flexible couplings between hub and generator.[7] However, an accurate determination method of the airgap width distribution during operation is needed, which can lead to a reduction in the use of materials and thus be cost-reducing. [2]

2. Modelling Approach

Figure 1 shows the approach for the electromechanical simulation. First the geometry of the stator and rotor tooth and slots are modeled by permeance functions by means of conformal mapping [3]. This step is a preprocessor step and can be calculated in advance independently of the other simulations steps. After that a lumped parameter model of the generator calculates the ideal flux density distribution from the excitation voltage. The ideal flux density multiplied with the permeance functions of rotor and stator results in the airgap flux density distribution.

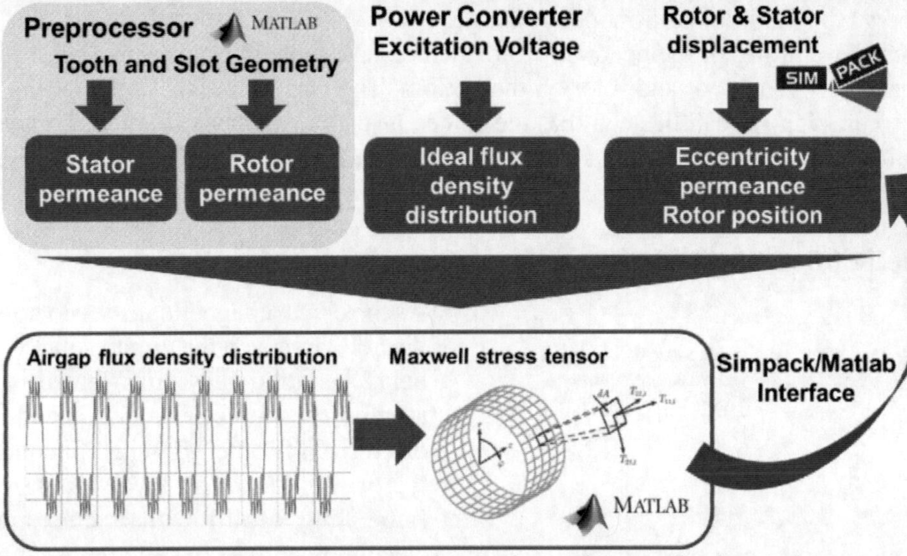

Figure 1: Approach for hybrid electromechanical simulation of a direct-drive generator.

From the airgap flux density distribution the Maxwell stresses on the airgap surface can be calculated, which results in tangential, radial and axial electromagnetic force densities [4]. The results of this electromagnetic model are handed over to force-elements in a multibody simulation (MBS) model of the turbine, which can also apply mechanical rotor loads and weight forces on the structural components of the generator. These forces, either mechanical or electromagnetic, result in a displacement of rotor and stator or lead to deformation in the structural elements of the turbine. Consequently the airgap permeance changes, which will be accounted in the electromagnetic field calculation iteratively. [2]

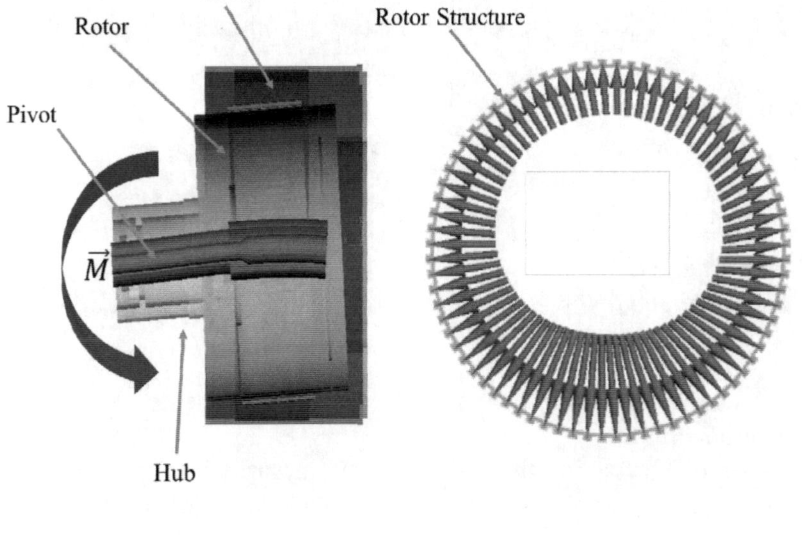

Figure 2: Bending moment application and resulting radial airgap forces.

Figure 2 shows the flexible MBS model of the generator. The applied bending moment on the rotor hub, causes a tilting of the generator rotor, which leads to a non-uniform airgap with the smallest space between stator and rotor on the bottom. The corresponding radial airgap forces are shown to the right, it can be seen that the forces near the smallest airgap width are much larger, resulting in the so called unbalanced magnetic pull [5], which will lead to increased loads on the generator structure.

3. Magnetic Force Application

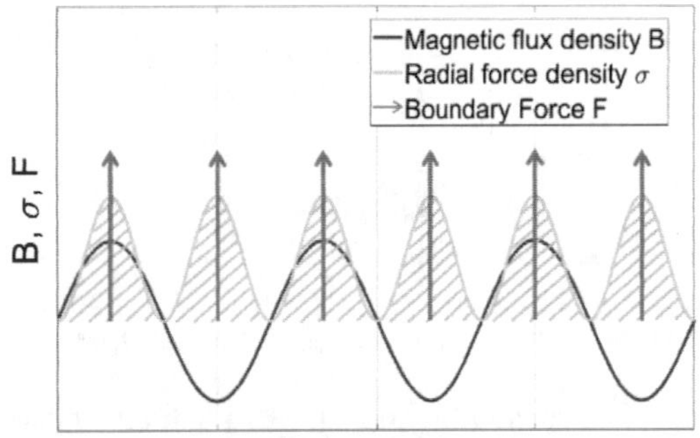

Figure 3: Schematic calculation of discrete force vectors from airgap flux density distribution.

As the electromagnetic forces do not act on discrete points on the airgap surface, rather being continuously distributed in the airgap, the force elements need to be discretized for the MBS [2]. In conclusion there exist different level of abstraction (modelling depth) with a corresponding calculation time. A simple method for discretization is shown in Figure 3.
For modelling generator tilting, the corresponding force elements are distributed axial on the permanent magnets. To investigate the model behaviour, the number of electromagnetic force elements on the permanent magnets were differed in axial direction, see Figure

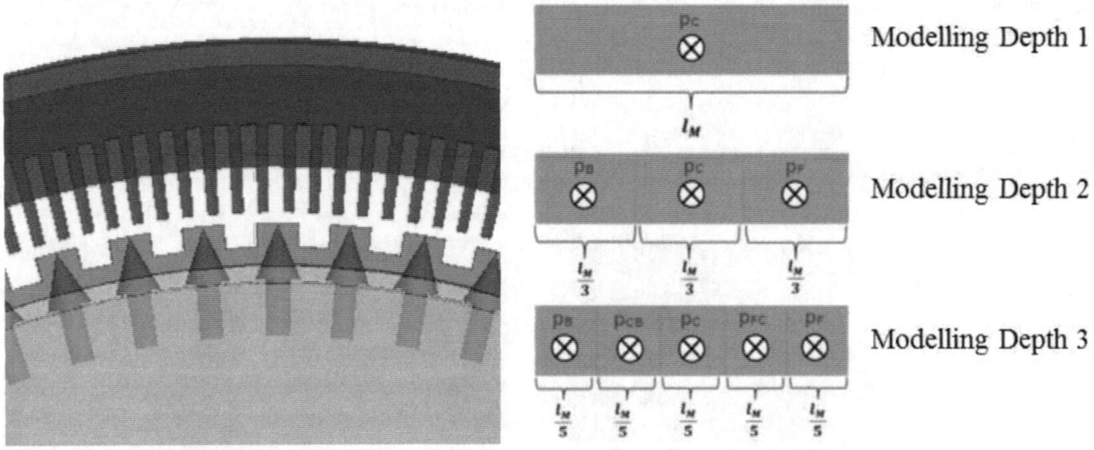

Figure 4: Electromagnetic force elements on rotor structure (3D model left; schematic right).

4. This leads to a change in deformation and movement of the generator rotor due to bending moments at the turbine hub.
With different level of abstraction the results for the airgap widths distribution will change. This is shown in Figure 5, where modelling depth 1 considers just one force element on the

rotor and stator structure per permanent magnet. Modelling depth 2 includes the parasitic airgap forces, by applying three force elements per magnet on the structure. Modelling Depth 3 considers the force distribution of five force-elements per magnet. As a central result, it should be noted that the additional loads caused by magnetic forces are in a significant range. For the most detailed possible consideration of the air gap width in generators, these should be taken into account in simulations.

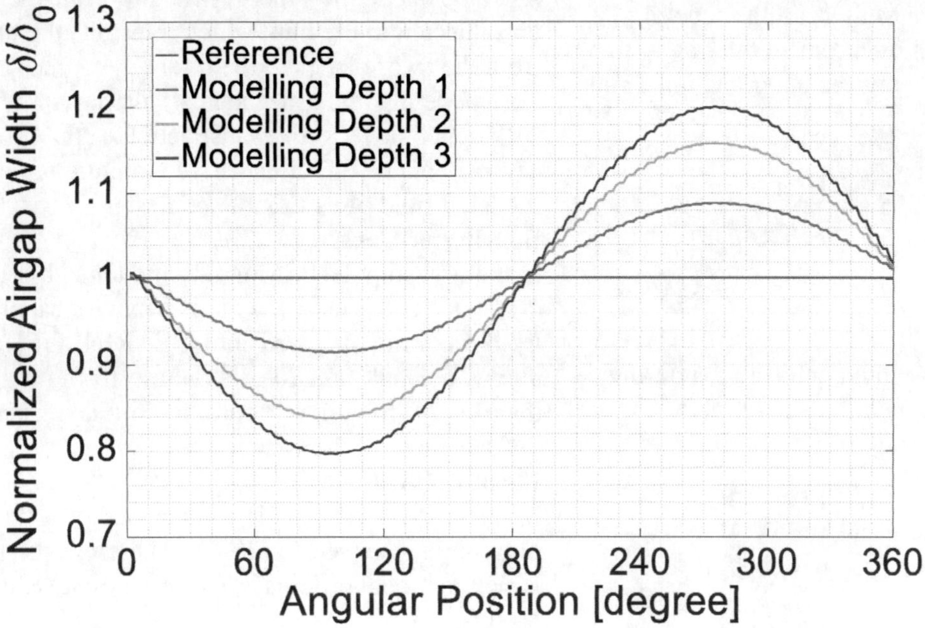

Figure 5: Resulting airgap width distribution due to extreme bending loads, modelled with different levels of abstraction of electromagnetic force excitation.

IOP Conf. Series: Journal of Physics: Conf. Series **1222** (2019) 012029 doi:10.1088/1742-6596/1222/1/012029

4. Structural Modelling

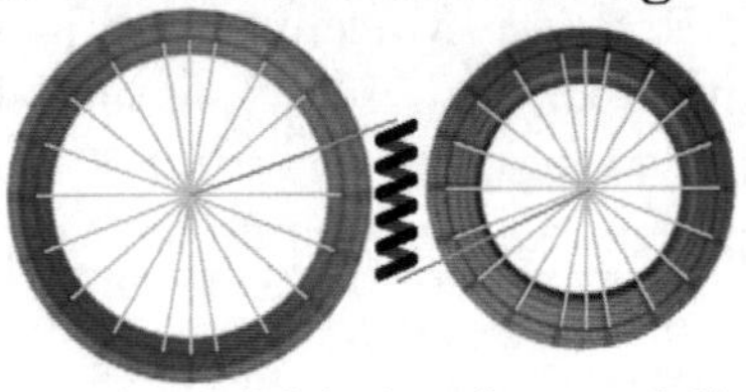

Figure 6: Main bearing modeling approach by means of bearing stiffness between outer and inner ring.

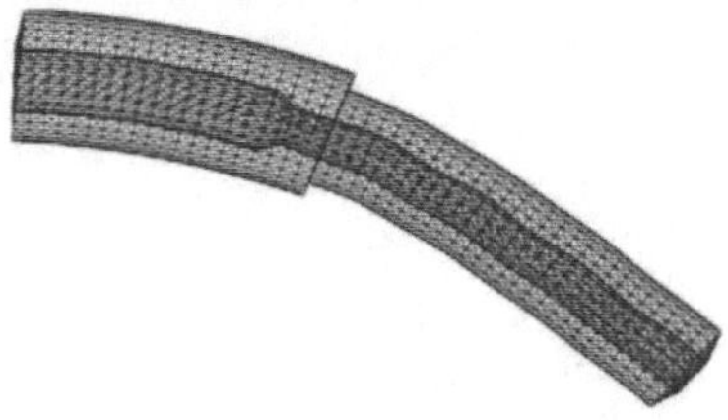

Figure 7: Pivot deformation during normal operation due to modelling as flexible body (factor 1000).

Figure 8: Rotor structure deformation during normal operation due to modelling as flexible body (factor 5000).

Figure 9: Stator deformation during normal operation due to modelling as flexible body (factor 10.000).

Figure 2 shows the drivetrain structure of the direct-drive generator as a 3D multibody model. The pivot is connected via a pressfit with the flange of the support structure, which in turn is attached to the foundation structure. The stator of the generator is also attached to the support structure while the rotor is connected to the hub, which transmits the torque from the rotor blades to the rotor.

For modeling the (mechanical) drivetrain with a sufficient level of abstraction, the major components of the direct-drive structure are either modeled rigid or as flexible bodies.

Bearing Stiffness

In the first step the bearings are modelled by spring systems (see Figure 6), with its behavior calculated by the commonly used tool BEARINX [6]. This allows the hub and rotor structure to move in all degrees of freedom according to the respective displacement of the bearing structure.

Flexible Pivot

In the next step the model is extended by a flexible pivot structure (see Figure 7). Therefore the pivot structure modal order is reduced via Finite Element Method.

Flexible Rotor

The model is extended by a flexible rotor support structure (Figure 8). This support structure carries the permanent magnets of the generator and is directly exposed to the Maxwell stresses.

Flexible Stator

Furthermore the MBS model is extended by a flexible stator structure (Figure 9). The stator windings of the generator are inserted in the stator structure. However these are not included in the model as they have no influence on the stator radial stiffness behavior.

Figure 10 shows the influence of the mechanical modeling depths on the airgap width distribution. It can be seen, that the addition of the flexibility of the pivot has the biggest influence on the airgap width. However each additional modelling depths has an influence on the airgap width distribution.

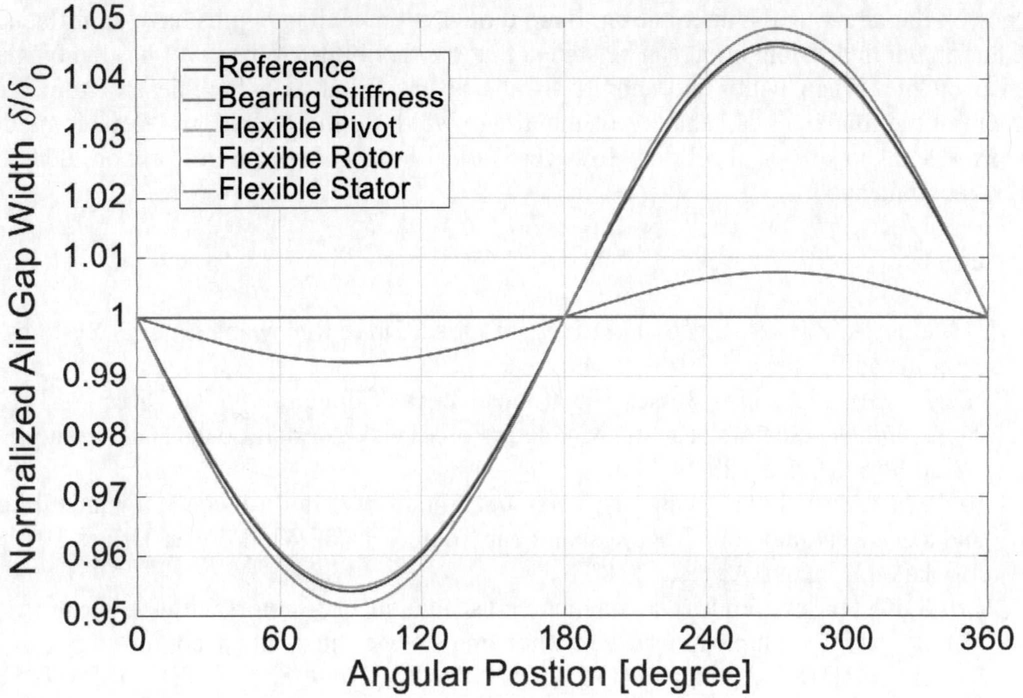

Figure 10: Resulting airgap width distribution due to normal bending loads, modelled with different levels of abstraction of electromagnetic force excitation.

Depending on the objective of the simulation model, different modelling approach are sufficient. In the case of fundamental design studies of the drivetrain behaviour (e.g. to ensure that the airgap clearance is maintained during turbine operation) modelling depth 2 seems sufficient.

On the other hand, if the objective of the model is to simulate the drivetrain in higher frequency domains with focus on local force distribution and deformation (e.g. to model the acoustic behaviour of the drivetrain) a higher modelling depths is recommended.

5. Conclusion

This investigation shows a modeling approach for direct-drive generators, considering the interaction of electromagnetic airgap forces and external loads. During this research a sensitivity analysis of the structural components and modeling depths of the electromagnetic force excitation were conducted.
It is shown that the axial distribution of electromagnetic force-elements influences the airgap width distribution. The biggest deviation in airgap width change is between the modeling with one force-element per magnet compared with three force-elements per magnet. Still the results between the model with five and the model with three force-elements per magnet differ significantly. For a design estimation, also considering model design effort, the modeling approach with three magnets seems sufficient, as it models rotor tilting in a more realistic way compared to one force-element, but leads to valuable results.

Furthermore it is shown, that the different ways of modeling the structural components have an influence on the airgap width distribution. Even though all modeling depths have an influence on the airgap width distribution, it can be stated, that the flexibility of the pivot has the biggest influence on the airgap width and should always be modelled as a flexible structure. For investigating the fundamental behavior of the airgap width during operation a flexible model of the pivot leads to sufficient results. However if local forces are of interest a more detailed model is recommended.

References

[1] M. Mueller; A. Zavvos; Electrical Drives for Direct Drive Renewable Energy Systems. Woodhead Publishing, 2013

[2] T. Duda, G. Jacobs, D. Bosse: Electromechanical Simulation of a Direct – Drive Generator Considering parasitic Magnetic Forces and External Loads , Conference for Wind Power Drives, 2019.

[3] S. Rick; M. Oberneder; K. Hameyer; Torsional vibrations in multi-megawatt wind turbine induction generators; In: Tagungsband zur Conference of Wind Power Drives 2017; Books on Demand, Aachen, 2017

[4] Oberretl, K.: Losses, torques and magnetic noise in induction motors with static converter supply, taking multiple armature reaction and slot openings into account.
In: Electric Power Applications, IET 1 (2007), Juli, Nr. 4, S. 517–531. –ISSN 1751–8660

[5] A. Ruf; M. Schroeder; A. Putri; D. Franck; K. Hameyer; Analysis and determination of mechanical bearing load caused by unbalanced magnetic pull
In: 17th International Conference on Electrical Machines and Systems (ICEMS), Hangzhou,China, 2014

[6] Schaeffler Technlogies GmbH & Co. KG, BEARINX Lagerauslegung auf hohem Niveau
Address of the website
(https://www.schaeffler.com/remotemedien/media/_shared_media/08_media_library/01_publications/schaeffler_2/brochure/downloads_1/pbb_de_de.pdf)
(01.10.2018)

[7] GE Renewable Energy – Pure Torque* Technology Haliade* 150-6MW Offshore Wind Turbine
Address of the website (https://www.ge.com/renewableenergy/wind-energy/technology/pure-torque)
(22.03.2019)

IOP Conf. Series: Journal of Physics: Conf. Series **1222** (2019) 012030 doi:10.1088/1742-6596/1222/1/012030

Determination of Wind Turbine Main Bearing Load Distribution

S. Kock[1], G. Jacobs[1] and D. Bosse[1]
[1]Chair for Wind Power Drives, Campus-Boulevard 61, 52074 Aachen, Germany

Abstract. The presented paper introduces simulative method for the determination of load and pressure distribution in the main roller bearings in multi-MW wind turbines. This method has been developed using the example of a generic 6 MW research wind turbine (WT). The investigated drive train comprises of a main gearbox and is mounted with a four point suspension. The fixed rotor bearing is a spherical roller bearing and the floating rotor bearing is a cylindrical roller bearing. The load and pressure distribution is determined by finite element (FE) simulation under consideration of real hub loads and elastic surroundings. The hub loads have been calculated using Multibody Simulation. Three different rotor bearing models are introduced, which consider FE contact conditions, macro-geometry, stiffness and rolling element profile. The developed method is transferable to different rolling bearing types and can be used to improve bearing load and pressure distribution in multi-MW WTs under consideration of elastic surroundings (main shaft, frame and bearing housing).

1. Introduction

1.1. Motivation

The nominal power of installed onshore and offshore wind turbines is growing continuously [1]. Nowadays, 10 MW offshore wind turbines (WTs) equipped with rotors reaching up to 164 m in diameter are already commercially available. The main bearings of this multi MW WTs are exposed to high dynamic axial and radial forces as well as bending moments. These high dynamic loads can lead to main shaft and bearing housing displacements to be in the mm range [2]. Consideration of elastic structure and deformations during the product development process is becoming ever more important in order to cope with increasing main bearing dimensions.

Additionally, the aforementioned loads and displacements can cause unequal bearing load and pressure distribution and also change the kinematic conditions of the main bearing. Such local contact conditions when subject to simultaneously slow rotational speeds, mixed friction and slippage can induce main bearing failures, such as micro-pitting, classical wear or cage fractures [3]. Consequently, the main bearings do not usually reach the calculated lifetime of 20 years and have to be replaced on average after seven years of field operation [4], [5]. These main bearing failures lead to high repair and replacement costs as well as long downtime of the affected wind turbine [6] while reducing the competitiveness of wind energy compared to conventional energy sources. Therefore, there is a need for a reduction of the main bearing failures.

Besides the testing methods, modern simulation tools contribute to a better understanding of the main bearing behaviour and reduction of the prevailing failures. This simulation tools are used to design and optimize the main bearings. One of the most important parameters for the optimization of main bearings is the knowledge of the main bearing load distribution.

1.2. Objective

The objective of this paper is to quantify the influence of the elastic surroundings (bearing housing, main shaft, main frame) on the main bearing load distribution on the example of a generic 6 MW research WT (4 point suspension). Furthermore, this paper introduces a method for determination of the bearing load distribution by the Finite Element Method (FEM). By knowing the bearing load distribution it is possible to identify edge loading or unequal pressure distribution. Subsequently, this unfavourable load conditions can be changed by adjusting of the rolling element profile or stiffness of the elastic surroundings.

2. Approach

The selected approach to achieve the objectives can be divided into two main steps: determination of the hub loads and determination of the main bearing load distribution, see Figure 1. The presented paper focuses on the second step.

The determination of the prevailing hub loads have been executed under consideration of the IEC wind class IIb: averaged medium wind speed of 8.5 m/s and lower turbulence up to 14%. The prevailing loads on the hub of the 6 MW research WT have been calculated by means of Multibody Simulation (MBS) model under start-up operation [7]. This MBS model considers all relevant structural components of the wind turbine as flexible bodies (e.g. blades, hub, main frame and shaft and gearbox housing) leading to a realistic calculation of the hub loads. The maximal hub loads for the specific time step (thrust up to 1.4 MN, radial force up to 1.1 MN and bending moments up to 900 kNm) during the start-up operation have been selected for the determination of the main bearing load distribution, see Figure 2. The prevailing torque under this conditions is 4407 kNm. Subsequently, the hub loads and torque are transferred to the FEM model for the determination of the global load distribution, the load distribution over the rolling element width and the contact pressure considering variations of

- elastic surroundings, clearance,
- rolling element profile, and hub loads.

The FEM model considers the detailed three-dimensional geometry of the elastic surroundings and of the main bearing. The FEM model includes the contact between housing and inner parts of the main bearing as well as the bolt connection to the main frame and is capable of simulating this nonlinear connections.

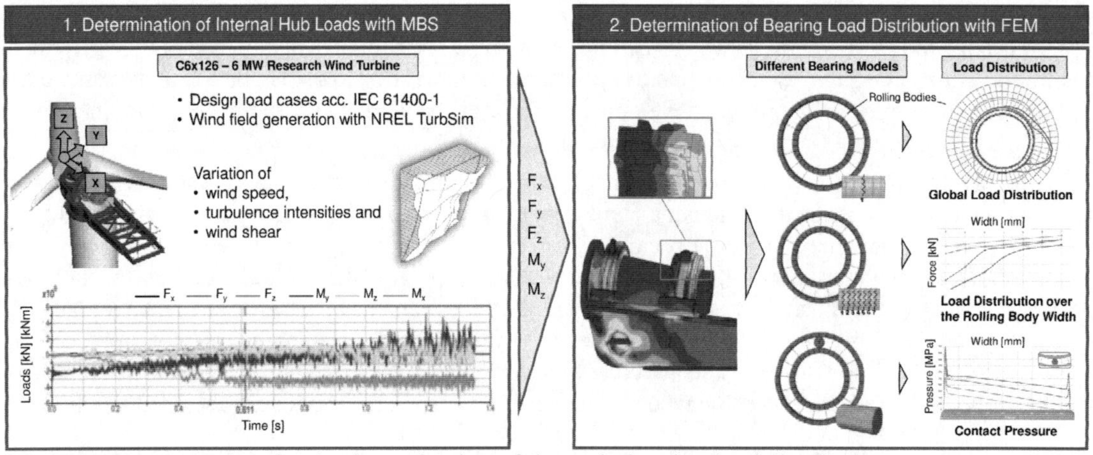

Figure 1: Approach for the determination of the main bearing load distribution

The core element of the FEM simulation is the modelling of the main shaft suspension. The main shaft suspension comprises of a fixed bearing (spherical bearing) and floating bearing (cylindrical bearing). The presented paper discusses the simulation results of the cylindrical bearing. Three bearing models with different numbers of parameters were used for the modelling, see Figure 1. All models consider the elasticity of the bearing rings as well as contacts between ring and housing as well as ring and shaft.

Furthermore, the stiffness of each rolling element is modelled. The first model considers rolling elements as a single stiffness and is used to describe global bearing load distribution so that each rolling element force can be calculated. The second model considers each rolling element with multiple stiffness's and is used for the calculation of force distribution over the rolling element width. This model is useful for the determination of tilting effects caused by deformations. The third model comprises of the 3D geometry of the rolling element, which has maximum load. This model is used for the calculation of the pressure distribution and optimisation of the roller profile.

3. Results

The next subsection comprises of the generated results. The behaviour of the bearing is described by calculation of the global load distribution, force distribution over the roller element width as well as the pressure distribution in the rolling contact. The sensitivity of these results to variations of elastic surroundings, hub loads and clearance of the main bearing has been analysed.

3.1. Influence of elastic surroundings on bearing loads

Figure 2 shows the results regarding the influence of the elastic surroundings (main shaft, bearing housing and main frame) on the global main bearing load distribution. It can be seen that with the consideration of the elastic surroundings more rolling elements are loaded. The main bearing load distribution becomes more irregular under simultaneously decreasing of the rolling element forces. The elastic surroundings determine the amount of the rolling elements under load, the highest force on the rolling element and general load distribution.

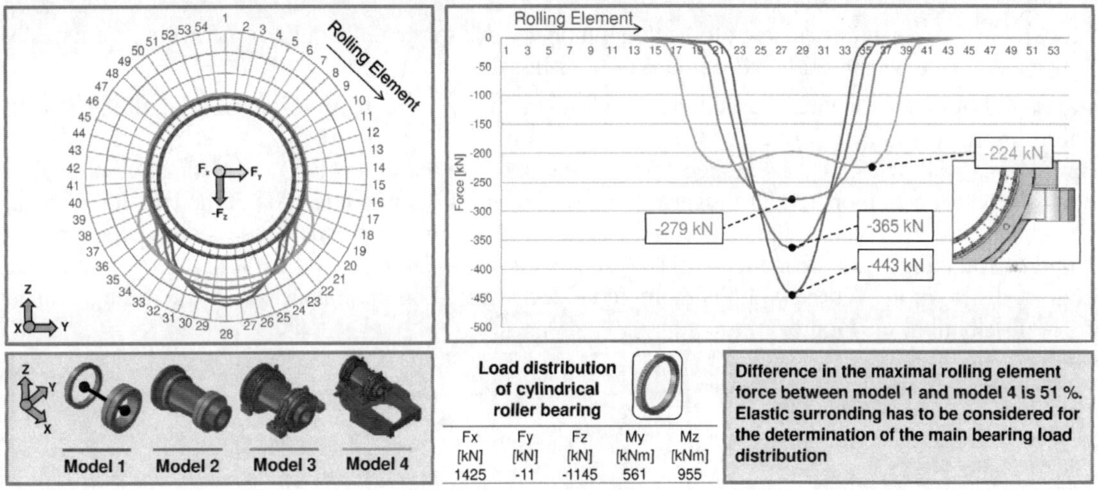

Figure 2: Influence of elastic surroundings on global load distribution of main bearing (maximum rolling element force is in the black rectangle)

Figure 3 shows the influence of elastic surroundings on the local load distribution over the rolling element width. The rolling element (without profiling) is simulated by seven equal springs which are responsible for the description of the elastic behaviour. The local load distribution is unequal and has almost linear gradient due to tilting and deformation of the main shaft.

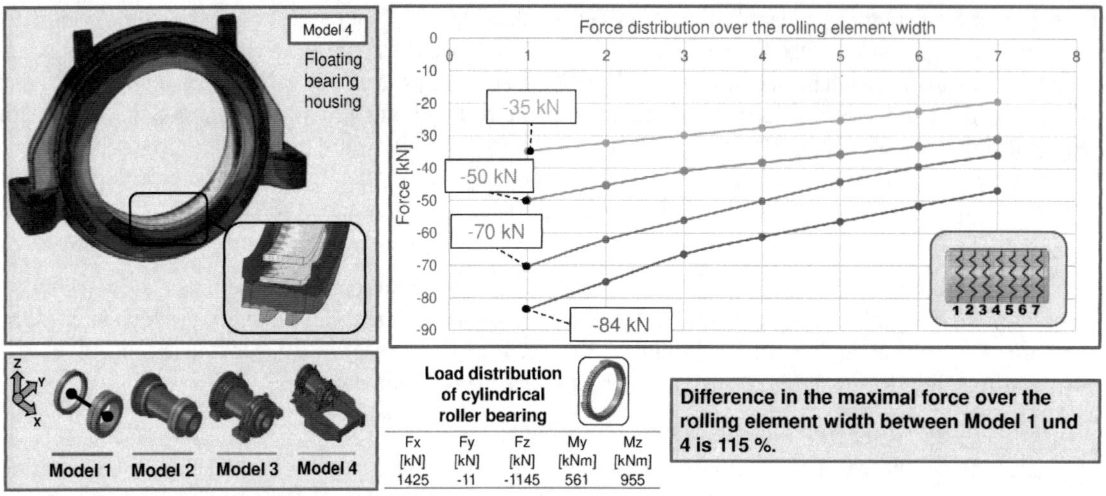

Figure 3: Influence of elastic surroundings on local load distribution over the rolling element width (maximum force over the rolling element width is in the black rectangle)

The upwind side of the rolling element is loaded higher than the downwind side. The difference in maximal rolling element force over the width between the model 1 (rigid surroundings) and the model 4 (elastic surroundings) is 115 %.

Model 4 has lower rolling element forces than model 1 due to the unloading of the rolling elements by deformation of the main shaft and frame as well as housing

Figure 4 shows the influence of elastic surroundings on the pressure distribution over the rolling element width. In this case the rolling element has a logarithmic profile according to Lundberg [8]. The three-dimensional rolling element is meshed finely at the contact area to resolve the contact pressure values and the contact area properly. The logarithmic profile avoids the pressure peak at the end of the rolling elements. The difference in the contact pressure over the rolling element width between the model 1 (rigid surroundings) and the model 4 (elastic surroundings) is 28 %. Model 1 has higher pressures. The contact distribution over the width is more equal due to profiling in comparison to the rolling element force distribution, see Figure 3.

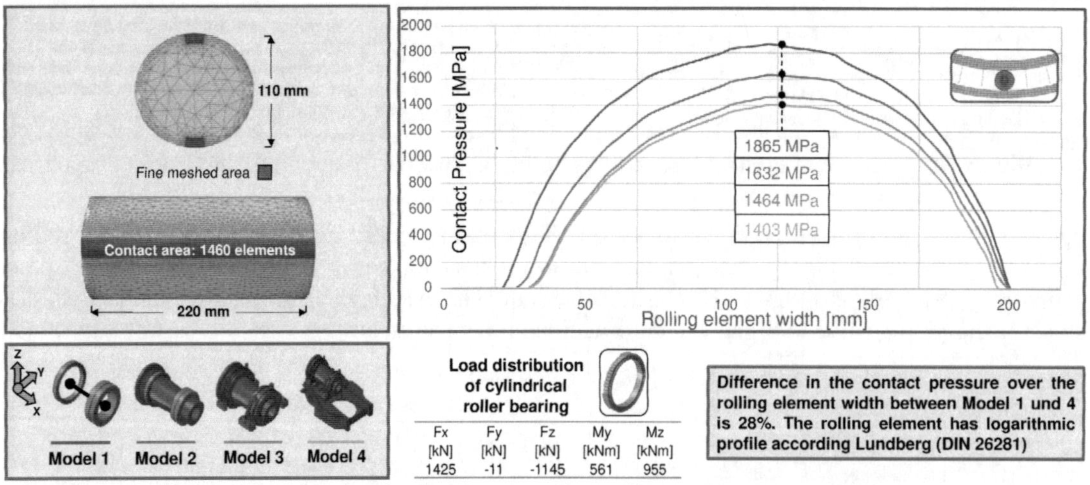

Figure 4: Influence of elastic surroundings on pressure load distribution of rolling element (maximum contact pressure is in the black rectangle)

The results of model 1 (rigid surroundings) have been validated by the bearing manufacturer software BEARINX. In case of the global load distribution there is a deviation of 2 % in comparison to the FEM. In case of the pressure load distribution there is a deviation of 8 % in comparison to the FEM. These

IOP Conf. Series: Journal of Physics: Conf. Series **1222** (2019) 012030 doi:10.1088/1742-6596/1222/1/012030

deviations can be attributed to the unknown material properties and unknown detailed geometry of the real bearing.

3.2. Influence of hub load directions

Figure 5 shows the results regarding the influence of direction of the hub loads on main bearing load distribution. Direction of the hub loads has an impact on the shape of the bearing load distribution and value of the maximal rolling element force. This can be attributed to the fact that bearing housing has fluctuating stiffness over circumference. If the hub loads are acting in the direction of area with high stiffness, fewer rolling elements are preloaded. In this case the rolling element forces are higher. Maximal rolling element forces can fluctuate up to 50% under variation of hub load direction. This fact underlines the importance of the consideration of the elastic surroundings during the stress assessment of the main bearing.

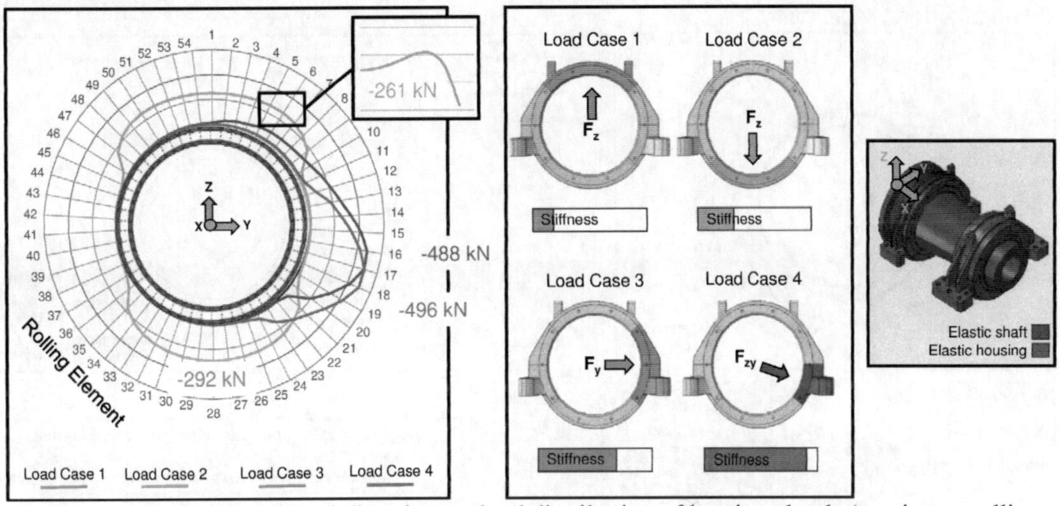

Figure 5: Influence of load direction on load distribution of bearings loads (maximum rolling element forces are shown with the corresponding color)

3.3. Influence of hub load increase

The main bearings are exposed to highly dynamic and fluctuating loads, which effect the behaviour of the main bearing. Figure 6 shows the influence of load increase on global load distribution of the main bearing. The hub loads, see Figure 6, have been calculated for start-up operation under consideration of IEC IIb wind field with a mean wind speed between 4 m/s and 5 m/s and a turbulence intensity of 0.2.

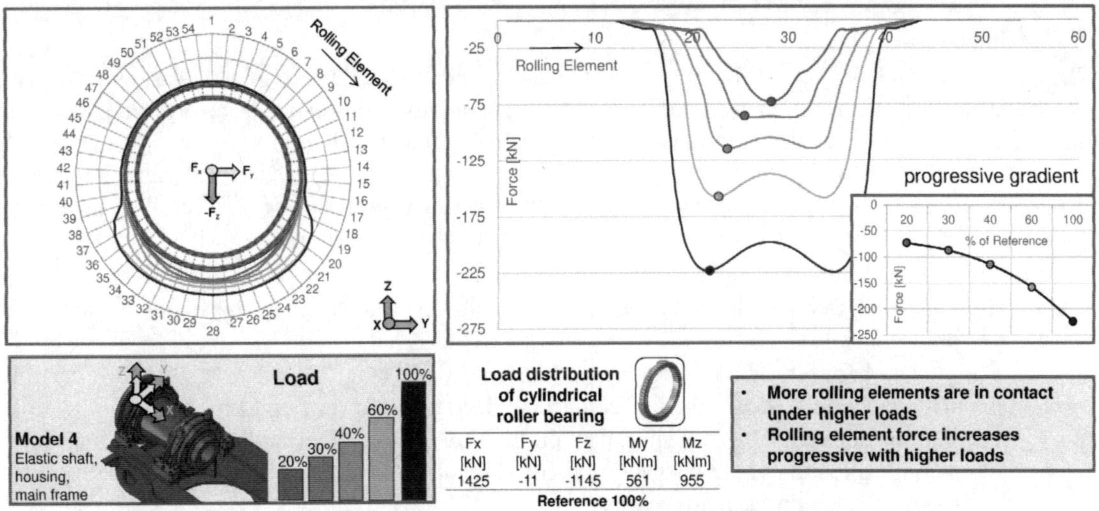

Figure 6: Influence of load increase on global load distribution of main bearing

It can be seen that the load increase has an effect on load distribution shape, maximal rolling element force as well as number of loaded rolling elements. Under high hub loads the influence of housing high stiffness in the area of the main frame connection can be seen clearly, see Figure 6 right hand graph. In this case the rolling force is higher and reaches 225 kN. With an increase of the hub loads more rolling elements are loaded. The maximal rolling element force has a progressive gradient if the hub loads are increased.

3.4. Influence of bearing clearance

The main bearings have an operational clearance, which depends on the manufacturing tolerances, mounting condition, temperature distribution of the bearing and on the hub loads. The effect of different clearances on the global load distribution have been carried out with the same load conditions as in the previous chapter. Figure 7 compiles the results and shows the influence of clearance on the main bearing load distribution.

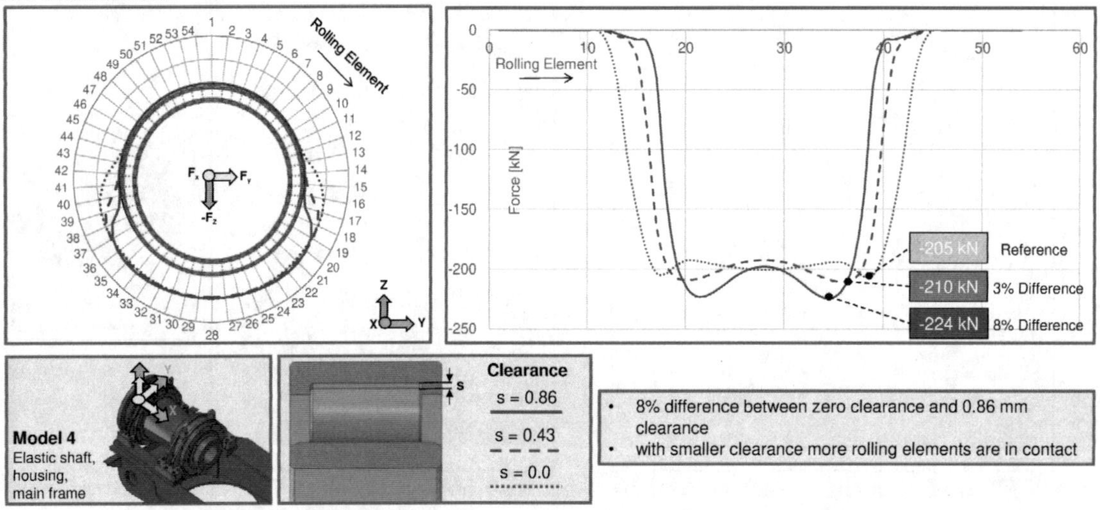

Figure 7: Influence of clearance on global load distribution of main (maximum rolling element force is in the black rectangle)

The clearance has been varied under the same load conditions between zero and 0.86 mm. It can be seen that clearance especially effects the amount of preloaded rolling elements. With less clearance, more rolling elements are in contact and the rolling element forces are smaller. The difference in maximal rolling element force can be up to 8% between zero clearance (205 kN) and 0.86 mm clearance (224 kN).

4. Conclusion

The elastic surroundings have a high impact on the global and the local bearing load distribution as well as on the pressure distribution over the rolling element width and should be always considered in development and fatigue life determination of multi-MW main bearings. In the case of the 6 MW research wind turbine with four point suspension of the main shaft, the elastic surroundings lead to a lower loading of the main bearing, but more unequal load distribution. The difference in prevailing maximal rolling element force between rigid and elastic surrounding can be up to 51%.

The project generated further insights as there are:

- High stiffness fluctuation of the main bearing housing can lead to high loads on the main bearing.
- The contact between bearing housing and bearing outer ring as well as shaft and inner ring should be modelled. Neglecting this contact can lead to an error of up to 14 %.
- The hub load direction has a high impact on the bearing load distribution. Resultant hub load, acting in the direction of the connection point of the main bearing housing and the main frame, leads to high loads on the main bearing.

- Gradual increase of the hub load leads to a non-linear, progressive increase of the maximal rolling element force.
- Gradual increase of the bearing clearance leads to higher loads on the main bearing.
- Validation of the simulation method has been done by measurement of main housing strains of the FVA-nacelle [9], [10]. In this case the deviation between simulation and measurement results is between 3 % and 20 %.

5. Bibliography

[1] C. Hernandez, T. Telsnig, A. Pradas, *JRS Wind Energy Status Report: market, technology and regulatory aspects of wind energy,* 2016, DOI: 10.2760/332535

[2] A. Heege, J. Hemmelmann, "*Matching experimental and numerical data of dynamic power train loads by modeling of defects*", Wind Energ., vol. 18, no. 3, pp. 541–558, 2015, DOI: https://doi.org/10.1002/we.1711.

[3] M. N. Kotzalas and G. L. Doll, *"Tribological advancements for reliable wind turbine performance",* Philosophical transactions. Series A, Mathematical, physical, and engineering sciences, vol. 368, no. 1929, pp. 4829–4850, 2010, DOI: 10.1098/rsta.2010.0194.

[4] Richard Brooks, *"Lessons Learned: Common Wind Turbine Bearing Failures, Cost & Solutions, Wind O&M Canada Toronto - Conference & Expo",* 2016.

[5] Konstruktion Zeitschrift, *"Titelthema: Wälz- und Gleitlager",* no. 9, 2014.

[6] Windstats Reports, *"Volume 19, 22, 25",* 2006, 2009, 2012.

[7] U. Jassmann *et al., "Model predictive control of a wind turbine modelled in Simpack",* J. Phys.: Conf. Ser., vol. 524, p. 12047, 2014, DOI: https://doi.org/10.1088/1742-6596/524/1/012047.

[8] DIN 26281, *Wälzlager - Dynamische Tragzahlen und nominelle Lebensdauer - Berechnung der modifizierten nominellen Referenu-Lebensdauer für Wälzlager,* 2010,

[9] D. Matzke *et al., "Coupling of electromagnetic and structural dynamics for a wind turbine generator", J. Phys.: Conf. Ser.*, vol. 753, p. 82034, 2016, DOI: https://doi.org/10.1088/1742-6596/753/8/082034.

[10] Reisch, G. Jacobs, D. Bosse, D. Matzke, *"Challenges and opportunities of full size nacelle testing of wind turbine generators", MPT,* vol. 2017, no. 0, 10-01, 2017, DOI: 10.1299/jsmeimpt.2017.10-01.

IOP Conf. Series: Journal of Physics: Conf. Series **1222** (2019) 012031 doi:10.1088/1742-6596/1222/1/012031

Parametric study of vane-type vortex generators under adverse pressure gradient by source term modelling in OpenFOAM

Iñigo Errasti, Unai Fernández-Gamiz, Pablo Martínez-Filgueira and Jesús María Blanco

inigo.errasti@ehu.eus
Corresponding author: unai.fernandez@ehu.eus
pablo.martinez@ehu.eus
jesusmaria.blanco@ehu.eus

Abstract. The aim of the study is to analyse and characterize the primary vortices produced by single low profile vane-type vortex generators (VGs) of different heights positioned on a flat plate with a backward-facing ramp and adverse gradient pressure for an incident angle of 15°. The effect of the vortex generator is implemented by using a source term in the Navier-Stokes equations according to the so-called jBAY source term model. In order to carry out the parametric study of the primary vortex, Computational Fluid Dynamics (CFD) simulations have been performed for different VG heights using the Navier-Stokes equations at a Reynolds number of Re θ=9100 based on the local boundary layer momentum thickness θ in open-source code OpenFOAM. As a preliminary result, the jBAY model reproduces relatively well the streamwise pressure coefficient distributions on the flat plate floor. Finally, the advantage of using this model over a fully mesh-resolved vortex generator model for certain cases must be remarked because a lower number of cells is needed in the model domain with a saving of computational time and resources.

1. Introduction

Flow separation control and the energy losses associated with the boundary layer (BL) have emerged as a key point in certain industrial fluid dynamics applications. The flow separation from a continuous surface is governed by the adverse pressure gradient and the viscosity. If the flow must remain attached to the wall, the stream should have enough energy to overcome the adverse pressure gradient, the viscous dissipation along the flow path and the energy loss caused by the modification in momentum. If the loss is such that further advancement of the fluid is no longer possible, then the flow separates from the surface.

Vortex generators (VGs) are passive devices to control flow which are able to change the motion performance of the fluid in the BL region. VGs are small vanes not aligned with the oncoming flow and they act by exchanging momentum from the distant flow region to the wall-closed inner region. Triangular and rectangular conventional VGs have been implemented onto wings of airplanes (**Figure**

1) for flow control to efficiently improve mixing of the BL and transfer momentum nearby the wall by delaying or suppressing the flow separation. In the experimental work performed by Velte [1] and the associated simulations carried out by Fernandez-Gamiz et al. [2] and Urkiola *et al* [3] demonstrated that the primary vortex produced by a rectangular VG mounted on a flat plate showed self-similarity for axial and azimuthal velocities. In most applications, VGs are designed with their height h similar to the local BL thickness δ and mounted normal to the surface with an incident angle to the flow to produce streamwise vortices. However, the residual drag related to these δ-scale conventional VGs might be relatively large in some flow-control application.

The generation of streamwise vortices by means of the implementation of vane-type devices with reduced height is a simple method to enhance their efficiency. Lin et al. [4] proved that when decreasing the height of standard VGs to a value lower than the local BL thickness, the momentum transfer keeps being large enough to avoid or delay flow separation downstream of the VGs. These so-called low-profile VGs were mounted on multi-element high-lift airfoils with the aim of controlling the flow separation on the flap. Ashill et al. [5] in a conceptual and experimental study showed a successful delay of the shock-induced separation on a transonic profile by mounting Sub Boundary Layer Vortex Generators (SBLVG). According to Lin [6], the implementation of these low-profile VG devices could be considered as a feasible option to be implemented when the flow-separation positions are relatively fixed and the vanes can be implemented upstream relatively nearly the flow separation. The European FP7 AVATAR project [7] and [8] has compiled a large amount of experimental and numerical work on VGs on flat plates, airfoils for 2D and 3D flows and their corresponding advances.

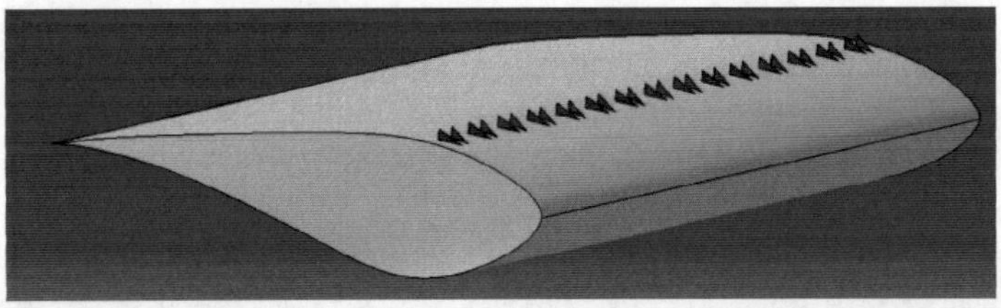

Figure 1. Sketch of an airfoil with a set of pairs of triangular vortex generators (VGs) placed on its upper surface.

In order to design and optimize the position of the vortex generator on a wind turbine blade, CFD tools can be used as for instance Fernández-Gamiz et al. [9] do. However, modelling the fully-meshed VGs on a full rotor computation becomes expensive. A different option of modelling VGs in CFD is to model its influence on flow using body forces. Bender et al. [10] developed a source term model called the BAY model. The underlying main of this model is based on replacing the vane-type VG by a local source term. The results when applying this model are quite satisfactory as shown for instance by Florentie et al. [11].

The main goal of this study is firstly the implementation of the so-called jBAY Source Term Model proposed by Jirasek [12] into OpenFOAM code [13] and secondly to investigate how well the open source CFD simulations are able to mimic the physics of the flow behind two modelled rectangular conventional and sub-boundary layer vortex generators (VGs) mounted on a three-dimensional flat plate with a backward-facing ramp under adverse pressure gradient conditions for an incident angle of 15°. This study is based on Errasti et al [14].

2. Theoretical background

The model consists of a single rectangular VG positioned on a flat plate with a backward-facing ramp. The three-dimensional computational domain modelled in OpenFOAM represents the extended test section of the wind tunnel experiment performed by Lin [15] and is shown in the illustration of **Figure 2**. Note that the single VG is placed upstream the backward-facing ramp. The domain has been designed following a previous simulation study by Konig et al [16].

The geometry dimensions of the vane-type VG determined by a length L two times the VG height H are depicted in **Figure 2** for two different heights H1=0.8δ and H2=0.2δ where δ is the local boundary layer thickness just upstream edge of the ramp. The vortex generators respectively are positioned at distances of 5δ and 2.5δ from the upstream edge of the ramp.

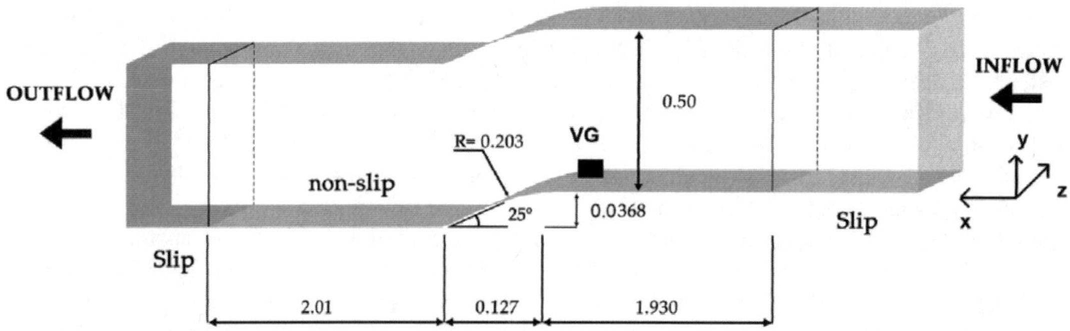

Figure 2. Description of the computational domain. The domain dimensions are expressed in meters.

In the current work, the jBAY source term model by Jirasek [12] based on the BAY model developed by Bender et al. [10] is implemented. The BAY model was designed for simulating vane-type vortex generators into finite volume CFD codes and allows substituting the VG geometry by a subdomain of similar size at the original VG location where a specific body force distribution is then applied. In the **Figure 3**, the selected cells where the body force will be applied are indicated. According to Bender et al. [10] a new term the final lift force on a single cell L_{cell} is given by the following expression:

$$\vec{L}_{cell} = C_{VG}\rho(\vec{u}\times\vec{b})(\vec{u}\cdot\vec{n})\frac{\vec{u}\cdot\vec{t}}{\|\vec{u}\|}S_{VG}\frac{V_{cell}}{V_S} \qquad (1)$$

where ρ *is* the local density, $\vec{u}$ the local velocity, $\vec{b}$ the unit vector defined as $\vec{b} = \vec{n} \times \vec{t}$ with $\vec{n}$ and $\vec{t}$ the unit vectors normal and tangential to the VG respectively, S_{VG} the vortex generator area, C_{VG} a relaxation constant parameter of value C_{VG}=10 (see Jirasek [12]), V_{cell} the volume corresponding to a single cell and V_s the total volume of the grid cells where the model is applied. The model is incorporated into the CFD code as a source term in the Navier-Stokes momentum and energy equations.

IOP Conf. Series: Journal of Physics: Conf. Series **1222** (2019) 012031 doi:10.1088/1742-6596/1222/1/012031

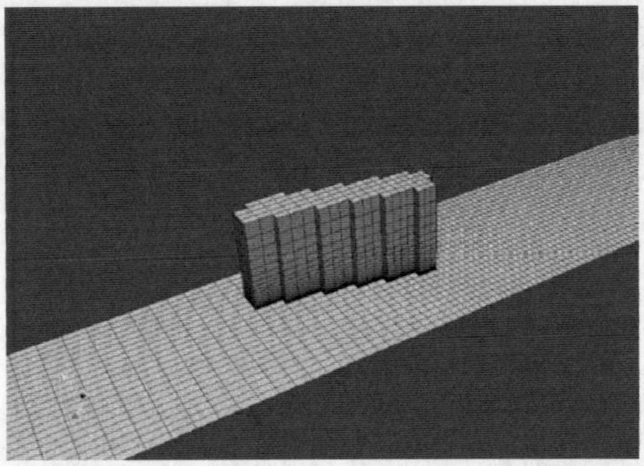

Figure 3. Selected cells representing a vane-type vortex generator (VG) of height H1=0.8 δ placed on a flat plate upstream the backward-facing ramp.

3. Experimental data

Experimental data for this study has been taken from the experiments carried out by Lin et al. [15]. The current computational domain partially based on a previous simulation study performed by Konig et al. [16] consists of a flat plate with a backward-facing ramp and adverse pressure gradient where selected cells corresponding to a single rectangular vortex generator according to the jBAY model and positioned at a point where the VG height is 20% and 80% the local boundary layer thickness δ at the upstream edge of the ramp for the cases H1 and H2 respectively. The angle of incidence of the vane with respect to the oncoming flow is β=15° for the two VG cases.

4. Computational configuration

Open source code OpenFOAM [13] is used for reproducing the primary vortex. This open source CFD code is an object-oriented library written in C++ to solve computational continuum mechanics problems. In the OpenFOAM CFD code, the user has only to specify the following three parameters for a specific VG: the grid cells where the source term model will be applied, the VG area and the angle of incidence β of the main flow direction with respect to the VG orientation. Steady-state, incompressible and turbulent flow is assumed and Reynolds Average Navier-Stokes (RANS) turbulence modelling according to the Shear Stress Transport (SST) turbulence model proposed by Menter [17] is applied in the simulations. The computational domain was discretized with a structured-type mesh and hexahedral faces of around 5×10^5 cells for the H1 = 0.8δ vane-type VG case and 2×10^6 cells for the H2 = 0.2δ vane-type VG case. Full second order linear-upwind schemes for the discretization were implemented for all the simulations. Two quality mesh parameters such as mesh orthogonality and mesh skewness for the two VG cases were studied in order to analyse the mesh quality. According to Richardson and Gaunt [18] Extrapolation Method, a grid dependency study was performed for three different mesh resolutions: coarse, medium and fine. The iterative solution process was carried out in the simulations of the two VG cases until the residual errors dropped below 10^{-4} for the pressure p and 10^{-5} for the velocities Ux, Uy and Uz and the turbulence quantities used, e.g., the turbulent kinetic energy k and the specific dissipation rate *w*. In the two meshes corresponding to the two VG cases analysed, the value for the dimensionless distance of the first wall cell y+ is lower than 1 as required by the turbulence model adopted.

5. Results

IOP Conf. Series: Journal of Physics: Conf. Series **1222** (2019) 012031 doi:10.1088/1742-6596/1222/1/012031

Figure 4 shows the vortex visualization based on the velocity field distribution at four planes normal to the streamwise direction and located at normalized distances x/δ respect to the boundary layer thickness δ from the backward-facing ramp where the domain origin is placed. This visualization allows just checking in a non numerical approach the vortex formation past the modelled VG.

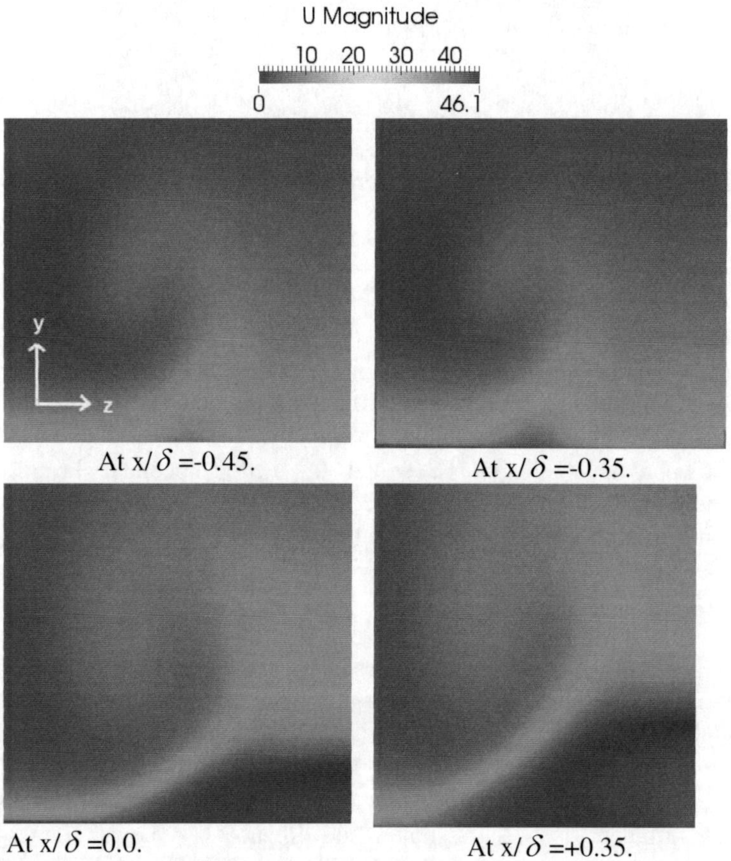

Figure 4. Velocity distribution (m s^{-1}) at planes normal to the streamwise direction and located at normalized distances x/δ from the backward-facing ramp where the origin is placed. H1=0.8 δ vane-type VG case.

As a preliminary result, he effect generated by the modelled VG of height H1=0.8 δ placed at x=-5 δ upstream the ramp on the streamwise pressure coefficient distribution c_p along a measurement line on the flat plate floor past the passive device is shown in **Figure 5**.

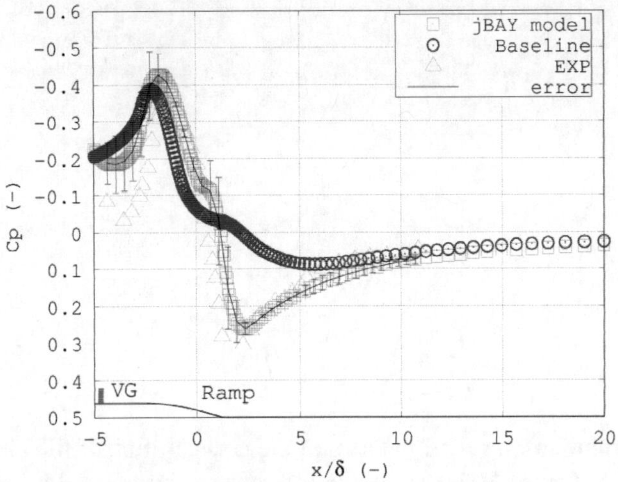

Figure 5. Streamwise pressure coefficient distribution cp along a measurement line on the flat plate floor past the VG. The red rectangles represent the simulated data (jBAY model), the blue circles the distribution data (Baseline) obtained from the baseline simulation and the green triangles the experimental (EXP) distribution data. Plots corresponding to the VG of height H1=0.8δ.

The path or trajectory of the vortex in the vertical (y) and lateral (z) directions can be determined by computing the location of the vortex center generated by the VG all along the downstream axis x. The modelled vertical and lateral paths are shown in Figure 6 are in concordance with the experimental vortex lateral and vertical paths.

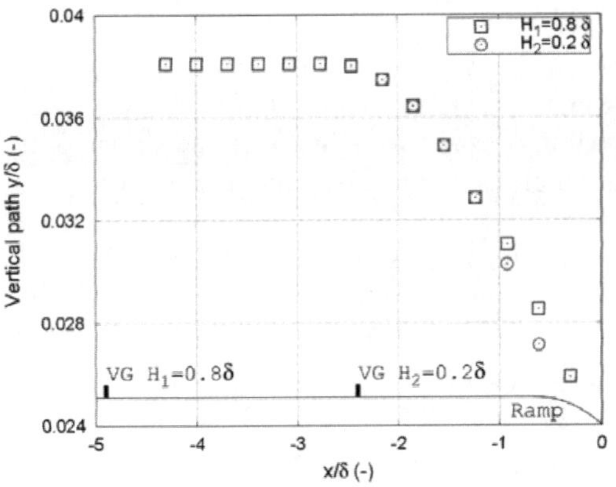

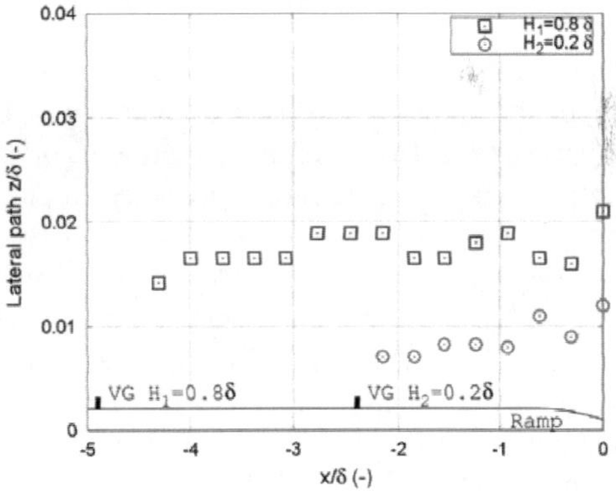

Figure 6. Normalized vertical (a) and lateral (b) paths function of the normalized distance x/δ from the backward-facing ramp for the H1=0.8δ (blue rectangles) and H2=0.2δ (red circles) vane-type VG cases. VGs, flat plate and ramp are also shown (not at scale).

A way to study the vortex size is by means of the half-life radius $R_{0.5}$ defined as the radial distance from the vortex center to the point where the vorticity is half the peak vorticity w_{max} captured in a cross-stream plane. **Figure 7** shows the vortex size evolution expressed in terms of the normalized half-life radius $R_{0.5}$ for the H1=0.8δ and H2=0.2δ vane-type VG cases at different locations x/δ past the VG.

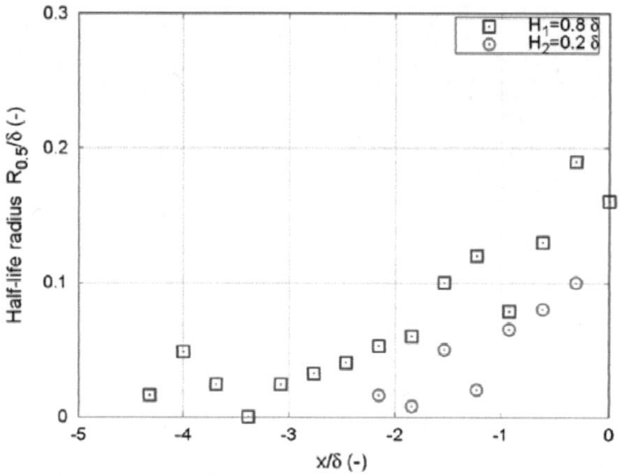

Figure 7. Vortex size evolution expressed in terms of the normalized half-life radius $R_{0.5}$ at difference normalized distances x/δ for the H1=0.8δ (blue rectangles) and H2=0.2δ (red circles) vane-type VG cases.

6. Conclusions

The generation of vortices and their effects by single low profile vane-type vortex generators of different heights positioned on a flat plate with a backward-facing ramp and adverse gradient pressure has been successfully carried out by CFD simulations using the OpenFOAM code in a preliminary approach. The influence of these vortex generators (VGs) on the computational domain flow is implemented by using a source term in the corresponding Navier-Stokes equations according to the so-called jBAY source term model.

The preliminary results obtained for the two vane-type vortex generators for the JBAY modelled pressure distributions are in concordance with the experiments where the influence of these VG devices is observed. However, the streamwise pressure distribution has been slightly underestimated in certain regions far from the ramp and overestimated in other regions by the jBAY model.

The implementation of the jBAY model represents an advantage over a fully mesh-resolved Vortex generator model for certain cases due to a meaningful decrease in the cell number of the computational domain in the CFD simulations. This implies saving of computational time and resources.

Ackowledgements: The authors are grateful to the Basque Government, which funded this research through project IT781-13 and IT1314-19.

References

[1] C. M. Velte. Vortex generator flow model based on self-similarity. AIAA J. 51(2), pp. 526-529. **2013**. DOI: 10.2514/1.J051865.

[2] U. Fernandez-Gamiz, C. M. Velte, P. Réthoré, N. N. Sørensen and E. Egusquiza. Testing of self-similarity and helical symmetry in vortex generator flow simulations. Wind Energy 19(6), pp. 1043-1052. **2016**. DOI: 10.1002/we.1882.

[3] A. Urkiola., U. Fernandez-Gamiz, I. Errasti, E. Zulueta. Computational characterization of the vortex generated by a Vortex Generator on a flat plate for different vane angles. Aerospace Science and Technology. **2017**. DOI: 10.1016/j.ast.2017.02.008.

[4] J. C. Lin, F. G. Howard and G. V. Selby. Small submerged vortex generators for turbulent-flow separation control. J. Spacecraft Rockets 27(5), pp. 503-507. **1990**.

[5] P. Ashill, J. Fulker and K. Hackett. Research at DERA on sub boundary layer vortex generators (SBVGs). Presented at 39th AIAA Aerospace Sciences Meeting and Exhibit, AIAA Paper 2001-0887. **2001**.

[6] J. C. Lin. Review of research on low-profile vortex generators to control boundary-layer separation. Prog. Aerospace Sci. 38(4-5), pp. 389-402. **2002**. DOI: 10.1016/S0376-0421(02)00010-6.

[7] Baldacchino, D. Experimental Data Description for Avatar Task 3.1: Experimental Investigation of Low-Profile Vortex Generators in a Boundary LayerWind Tunnel. Available online: http://www.eera-avatar.eu (accessed on 14 March 2019).

[8] Manolesos Marinos, J.P. Advanced Aerodynamic Tools for Large Rotors (AVATAR), Workpackage 3, Task 3.1: CFD and Experimental Database of Flow Devices, Comparison; FP7-Energy-2013-1/no 608396; European Project FP7-Energy-2013-1: Trondheim, Norway, 18 February 2015.

[9] Fernández-Gamiz U., Errasti I., Gutierrez-Amo R., Boyano A. and Barambones O. Computational modelling of rectangular sub-boundary-layer vortex generators. Applied Sciences. **2018**, 8(1), 138, 142-149, DOI: 10.3390/app8010138.

[10] Bender E.E., Anderson B.H. and Yagle, P.J. Vortex Generator Modelling for Navier–Stokes Codes. Proceedings of the 3rd ASME/JSME Joint Fluids Engineering Conference. **1999**, Paper FEDSM99-6929.

[11] Florentie L., Hulshoff S. J. and van Zuijlen A. H. Adjoint-based optimization of a source-term representation of vortex generators. Computers and Fluids. **2018**, 162, 139–151. DOI: 10.1016/j.compfluid.2017.12.009.

[12] Jirasek A. Vortex-Generator Model and Its Application to Flow Control. Journal of Aircraft. **2005**, 42(6), 1486-1491, DOI:10.2514/1.12220.

[13] OpenFOAM. https://openfoam.org. Accessed in October 1, **2018.**

[14] Errasti I., Fernández-Gamiz U., Martinez-Filgueira P. and Blanco J.M**.** Source term modelling of vane-type vortex generators under adverse pressure gradient in OpenFOAM". *Energies*. ISSN 1996-1073. **2019.** Accepted.

[15] Lin J.C. Control of Turbulent Boundary-Layer Separation using Micro-Vortex Generators. American Institute of Aeronautics and Astronautics. **1999**, 1-16, DOI: 2514/6.1999-3404.

[16] Konig B., Fares E. and Nolting S. Fully-Resolved Lattice-Boltzmann Simulation of Vane-Type Vortex Generators. 7th AIAA Flow Control Conference. **2014**, AIAA 2014-2795, 2795-3008. DOI: 10.2514/6.2014-2795.

[17] Menter, F.R. Zonal two equation k-w turbulence model for aerodynamic flows. AIAA J. **1993**, 93, 2906, DOI:10.2514/ 6.1993-2906.

[18] Richardson, L.F. and Gaunt, J.A. The deferred approach to the limit. Part I. Single lattice. Part II. Interpenetrating lattices. Philos. Trans. R. Soc. Lond. Ser. A **1927**, 226, 299–361

IOP Conf. Series: Journal of Physics: Conf. Series **1222** (2019) 012032 doi:10.1088/1742-6596/1222/1/012032

From SCADA to lifetime assessment and performance optimization: how to use models and machine learning to extract useful insights from limited data

Nikolay Dimitrov[1,2], **Anand Natarajan**[1]

[1] DTU Wind Energy, Technical University of Denmark, 4000 Roskilde, Denmark

[2] nkdi@dtu.dk

Abstract. A common challenge in the decision making process regarding operation and life extension of existing wind farms is the lack of accurate information about the actual dynamic states of the turbines in terms of its operation from inception. SCADA records normally contain limited number of channels, and are not necessarily kept for the entire operating period of the wind farm; design and site data may be outdated or inaccessible. Nevertheless, as long as a minimum amount of information is available, statistical analysis and augmentation with artificial intelligence based simulation can be used to supplement the information. In the present study, we delineate a combination of data analysis, physical modelling and machine learning, that produces a detailed assessment of the operating conditions experienced by a wind farm and establishes the corresponding power performance, loads and fatigue damage accumulation.

1. Introduction and purpose

The decision making process regarding operating strategy or end-of-life considerations requires characterizing the current state of the wind turbine structures, potentially combining several sources of information: 1) an overview of the wind turbine design assumptions and limitations, 2) recovering the operating and loading history of each turbine, and 3) visual or other form of non-destructive testing to determine the presence of other failure modes not captured by the standard modelling approaches used in the design calculations. In particular, the operating history in terms of dynamic states and loading history is necessary in order to determine whether any of the design assumptions has been violated during the wind farm lifetime. A common challenge in this process is the incompleteness of historical information. SCADA records [1] normally contain limited number of channels usually dealing with power production, rotor speed and other such system information, and are not necessarily kept for the entire operating period of the wind farm, but they have nevertheless been used for various applications. A commonly reported use of SCADA data is for predictive applications such as failure detections [1],[2],[3],[4]. Other common use is power output forecasting as shown in [5],[6]. In [7], SCADA data are used in combination with wake models for determining wind direction, wind speed, and for power output prediction. Operational wind farm data rarely include load measurements, therefore studies with such analysis are scarce and typically limited to a few turbines over a short-term load measurement campaign, as in e.g. [8],[9]. In a scientific field that previously has been dominated by power and fault prediction analysis, the present study aims at demonstrating how detailed fatigue load estimation can be obtained for each individual turbine in a wind farm, based on SCADA measurements supplemented with an aeroelastic model. The use of a combination of limited SCADA based measurements and wind turbine/farm simulations using a verified design basis of the turbines is very appropriate to train an artificial intelligence system to forecast the future performance of the wind turbine and the fatigue life consumption of its components. The following sections elaborate on the process followed to establish this method.

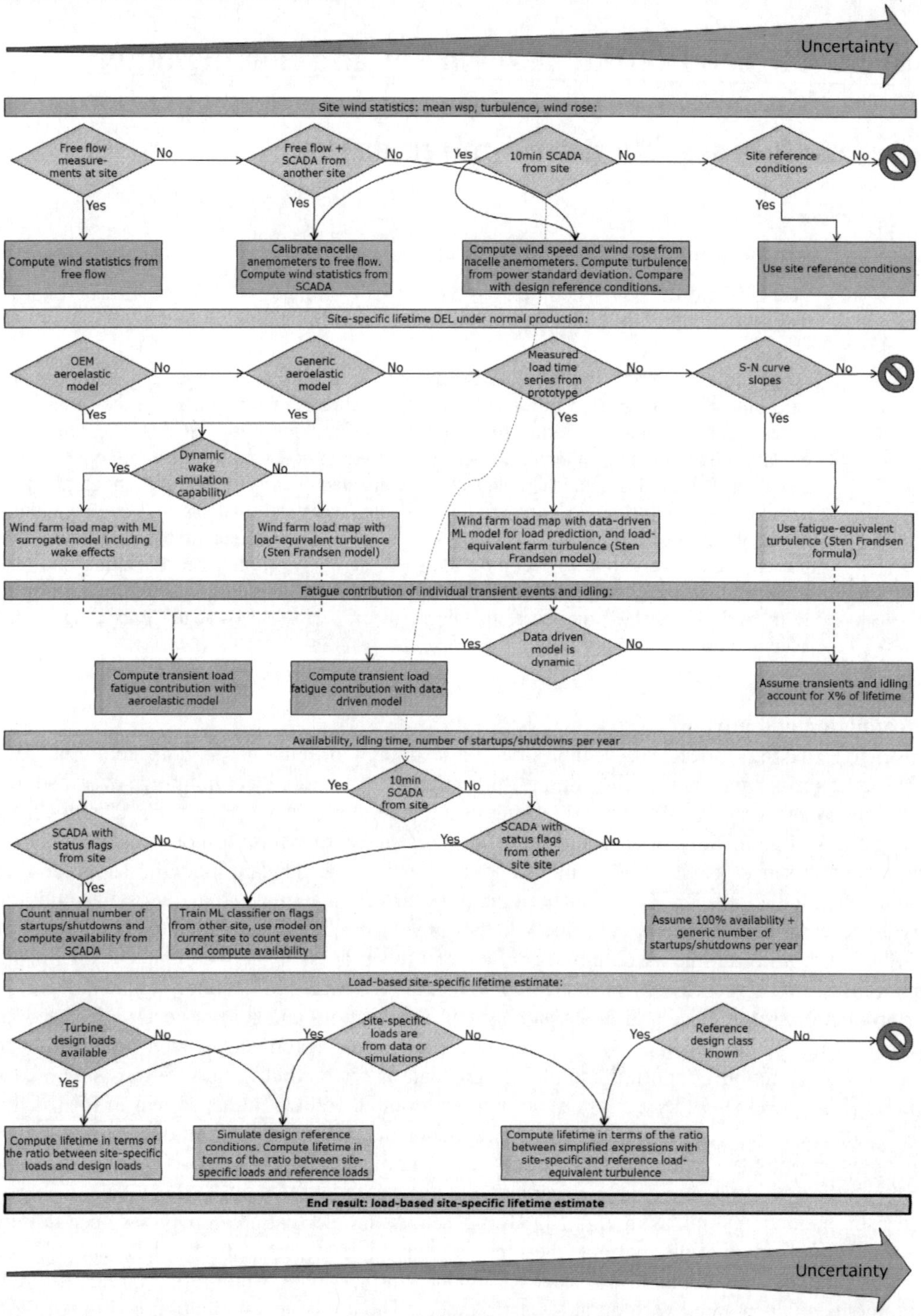

Figure 1. Chart describing a procedure for estimating load-based site-specific remaining lifetime, with different process alternatives depending on the type of information available.

2. Methods

Practically every wind farm in the world is unique as a project due to the varying environmental conditions, wind turbine technologies, project developers, owners, and operating history. As a result it is very likely that each project also comes with a unique combination of available data. Due to this data heterogeneity it is largely infeasible to define a fixed procedure for establishing the dynamic history of wind farms. Instead, we suggest a flexible procedure as shown on Fig.1. The procedure requires several predefined categories of information; however the numbers within each category can be obtained using a choice of different methods depending on the type of data and design information available. It is assumed that more detailed information will in general lead to lower uncertainty as indicated in Fig.1.

The SCADA data recorded by wind turbines typically contain information recorded by sensors mounted on the turbines such as nacelle anemometers. The wind speed measurements from a nacelle anemometer are potentially biased as the nacelle anemometers are operating in a disturbed flow downwind from the rotor. Therefore, in order to utilize the information available in SCADA, a transfer function between true wind speed and turbulence and readings from a nacelle anemometer needs to be established. It is expected that a few months of data on a site with met mast should be sufficient to establish the transfer function. Alternatively, the statistics of electrical power output (mean and standard deviation) can be used as proxies to rotor-equivalent wind speed and turbulence. The transfer function is calibrated by means of a machine learning (ML) regression model. Then, free wind speed and turbulence at farm level can be estimated by 1) correcting the nacelle anemometer readings using the abovementioned transfer function, and 2) averaging the corrected readings from turbines operating in free wind. Among the multiple potential choices for regression models, in the present study we use Artificial Neural Networks (ANNs)[10], due to their computational efficiency and versatility. By applying a series of simple operations (a weighted multiplication or convolution followed by a nonlinear activation function) ANNs provide a nonlinear mapping between inputs and outputs, which, given a network of sufficient size, can theoretically approximate any real-valued function [11]. Once the model has been trained, neural networks can make thousands of forward propagations (predictions) per second to simulate wind farm operational states that yield valuable information on fatigue lifetime. ANNs can also be trained to represent the dynamic behavior of a dynamic system where the approach is taken to map input-output responses that are a function of time. Response surface methodology and neural networks have been adopted in literature for nearly two decades to represent several types of aeroelastic systems[12],[13].

SCADA signals measured from wind turbines such as the rotor speed, blade pitch angle etc. can be mapped to corresponding mechanical loads on the turbine components during operation either using load measurements or aeroelastic simulations. If measured loads on the wind turbine are unavailable, then an aeroelastic design basis of the wind turbine can be set-up so as to successfully reproduce the operating conditions of the wind turbine and simulate the exact rotor speed, power output, blade pitch angle as measured on the actual turbine. Consequently a neural network can be trained to reproduce the loads on the blade root, tower top, tower base using a set of representative SCADA signals as the input.

In the following, we demonstrate examples of several of the potential solutions outlined in Fig.1 for specific combinations of real-world data. In Section 3, we show how the lifetime-equivalent fatigue loads under normal operation at the Horns Rev 1 wind farm can be estimated based on approx. 3 years of basic SCADA (nacelle wind speed, yaw direction, power and status flags) supplemented with load simulations as well as mast measurements synchronized with SCADA from a single turbine. In Section 4, two additional solutions are shown for situations where 1) some load measurements are available for at least one turbine, e.g. a prototype, and 2) if the wind turbine operating status flags are not available in the wind farm SCADA, but may be available on a smaller data set, e.g. a prototype of from a dedicated campaign.

3. Case study: lifetime estimation for the Horns Rev 1 wind farm

3.1. Overview of available data

In this example, we use the Horns Rev offshore wind farm in western Denmark (Figure 2) consisting of 80 Vestas V80 turbines with 2MW power rating. The following data types are available:

- Ten-minute SCADA statistics (power, nacelle wind speed and yaw direction, including mean values, standard deviations and status flags) for 3 years;
- Wind speed, turbulence and wind direction at close to hub height (70m) from 3 met masts;
- A generic aeroelastic model of the V80 turbine using the DTU Wind Energy controller [14].

As no load measurements from the wind farm are available, we need to use a load model to estimate the accumulated fatigue damage. The aeroelastic load model requires inputs in terms of the free wind conditions, and this information can be obtained from the SCADA. As discussed earlier, the SCADA data are not guaranteed to be representative of the free wind conditions due to various disturbances. However, as data from a nearby met mast are also available, they are used to calibrate transfer functions which map the SCADA readings to the reference free wind conditions. The transfer functions are calibrated based on comparing data from the met mast and from a single turbine in the farm. With this approach we aim at also covering a scenario where there is no mast at the wind farm location but measurements could be available from another site with e.g. a single prototype turbine. Prior to analysis, the SCADA and met mast data are filtered based on the status signals. For a ten-minute period to be considered valid for analysis, at least one turbine in an outer row facing the incoming wind direction needs to be available and producing power. To enable comparison with met mast data, an additional condition is that mast wind speed and wind direction have to be available. This results in approximately 150,000 data points valid for analysis.

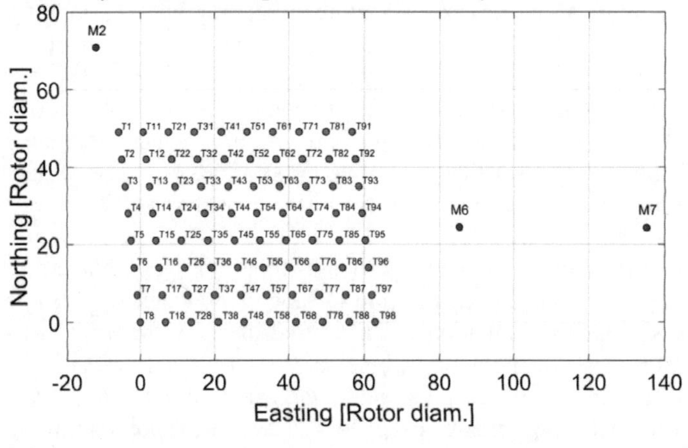

Figure 2. Horns Rev 1 wind farm layout and met mast locations.

3.2. Computing wind direction, ambient wind speed and ambient turbulence from SCADA

The yaw records of individual turbines can be unreliable – for example, in the SCADA data set used in the present study, the availability of yaw information for some turbines is as low as 30%. It is therefore not sufficient to calculate wind direction statistics for individual turbines. Instead, farm-average wind direction is calculated based on the yaw directions of turbines operating in free wind using the following algorithm:

- For a given 10-minute period, a preliminary average direction is estimated as the median of the yaw angle readings of all operational turbines in the farm;
- The preliminary wind direction combined with the farm geometry is used to estimate which turbines are operating in free wind conditions (no turbines upwind within a ± 20deg sector);

- Final wind direction is estimated as the median of valid yaw readings from the turbines operated in free wind conditions.

In the above, the median is preferred instead of the mean, because the mean would be affected by outliers such as erroneous yaw readings. Alternatively, the mean can be used together with a filter which eliminates all yaw readings which are certain number of standard deviations away from the mean [7]. The SCADA-based estimate of the wind direction computed with the above procedure shows very high correlation ($r^2 = 0.973$, Table 1) with the wind direction measured by met mast 6 (M6). There is a constant bias of approx. 9deg, which can be due to bias in the reference (true north) orientation of either the met mast or the wind turbines. When the bias is subtracted from the wind direction estimated from SCADA, the resulting time series show excellent agreement with the wind direction measured from the mast, Fig.3. This also results in a close resemblance between the wind roses estimated from the mast and from SCADA, as shown on Fig.4.

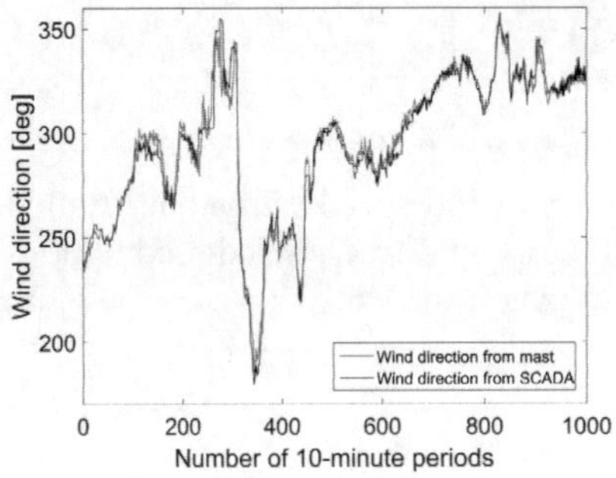

Figure 3. Comparison of the wind direction prediction from SCADA data with measurements from met mast M6.

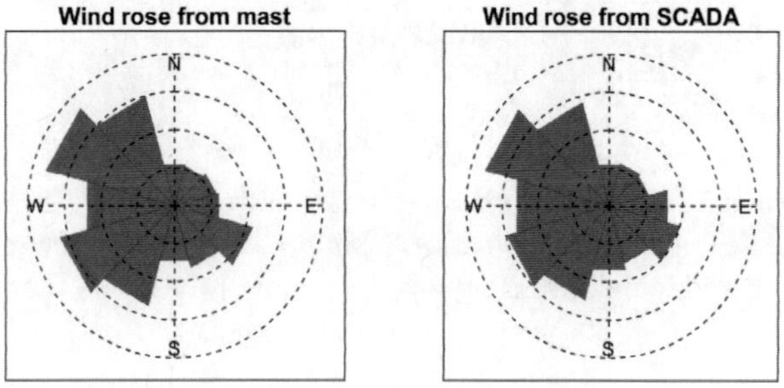

Figure 4. Comparison of the wind roses at Horns Rev 1 derived from M6 mast measurements and from SCADA.

Having determined the wind direction, it is straightforward to identify the turbines which are operating in free wind conditions. The criterion applied is that there should be no wind turbines upwind within a sector of ± 20deg from the mean wind direction. Using the same approach as with the wind direction, the free wind speed and turbulence can be estimated as the median of the nacelle wind speed and turbulence recorded by the turbines in free wind conditions. However, due to the influence of the rotor, the nacelle wind speed readings show a nonlinear behavior with respect to the free wind speed, with a change of slope at around rated wind speed. On Figures 5 and 6 the raw nacelle wind speed readings are plotted against the readings from M6 (black circles), for a single turbine and for the average from free-wind turbines, respectively. The change of slope around rated wind speed can be best observed on Figure 6 where the raw nacelle wind speed is lower than the mast wind speed for wind speeds above rated. In order to eliminate some of the nonlinearity, a transfer function is calibrated between the nacelle wind speed reading of a single turbine and the wind speed measured by M6. The transfer function used is a simple ANN regression model with a single hidden layer with 5 neurons and hyperbolic tangent (tanh) activation functions. The transfer function is calibrated for wind turbine 81 which is in the northeast corner of the wind farm. This choice maximizes the availability of data where both the turbine and the met mast are in free wind conditions. The transfer function is then applied on the nacelle wind speed readings of all wind turbines in free wind, and a corrected farm wind speed is estimated. Shown with blue dots on Figures 5 and 6, the correction almost completely eliminates the bias for the turbine where the function is trained on, while for the wind farm there is a noticeable reduction in bias while the coefficient of determination remains practically unchanged. A similar corrective approach is also applied to the turbulence (standard deviation of wind speed) from the nacelle anemometers, however with four input variables to the neural network model: nacelle wind speed, nacelle turbulence, mean power production, and standard deviation of power production. Due to the larger number of inputs the model used is also slightly larger, with 10 neurons in the hidden layer. The correlation between the nacelle readings and mast data is shown on Figure 7 for a single turbine and on Figure 8 for all turbines in free wind. There is significantly more uncertainty than with the wind speed estimation; nevertheless applying the correction noticeably improves the correlation.

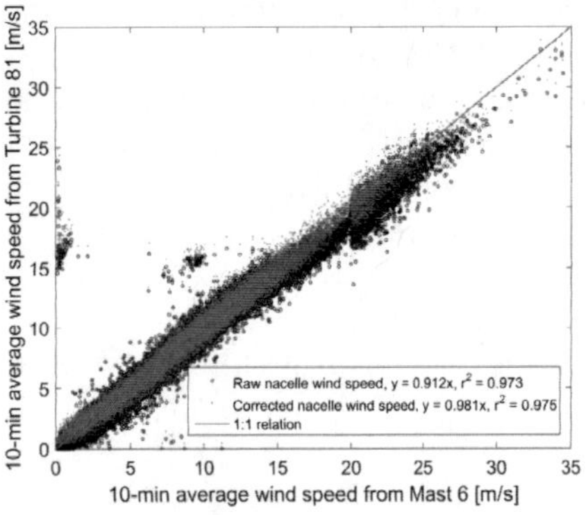

Figure 5. Mean wind speed estimation from the nacelle anemometer of a single turbine, compared to data from M6.

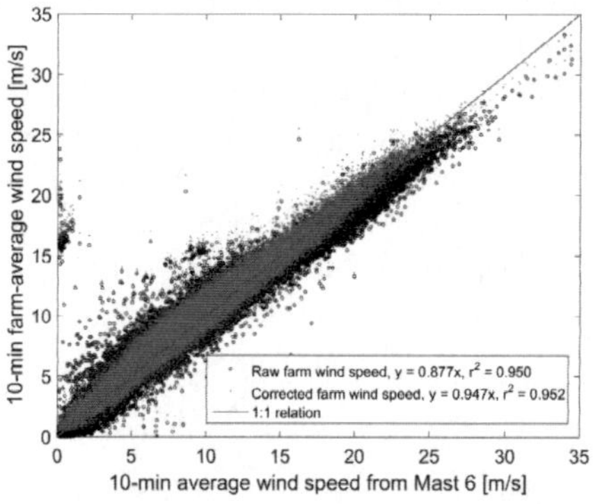

Figure 6. Mean wind speed estimation from all wind turbines in free wind conditions, compared to data from M6.

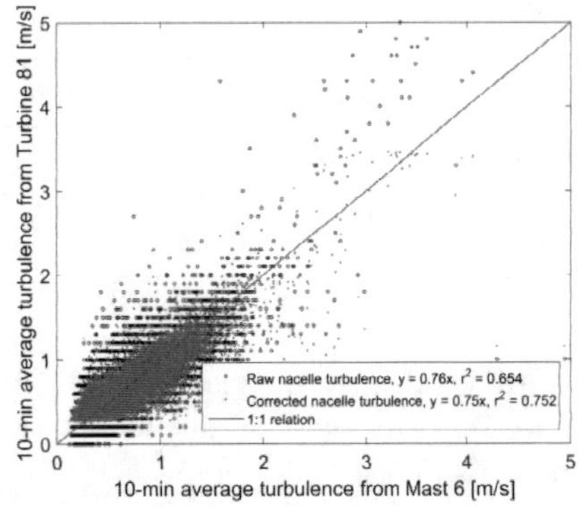

Figure 7. Turbulence estimation from the nacelle anemometer of a single turbine, compared to data from M6.

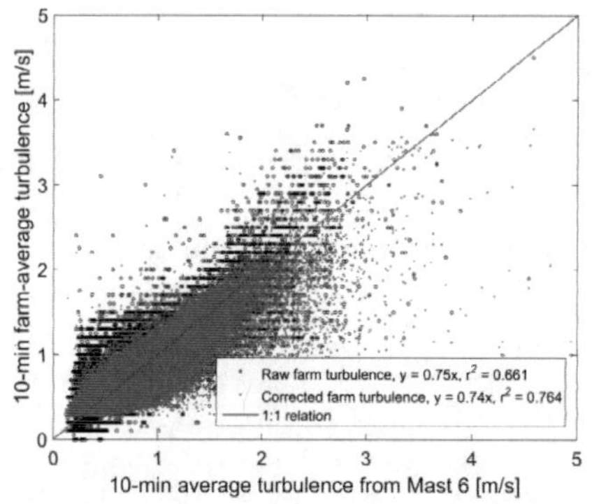

Figure 8. Turbulence estimation from all wind turbines in free wind conditions, compared to data from M6.

3.3. Load prediction including wake-induced effects

The SCADA records in existing wind farms usually do not include reliable and extensive load measurements. This is also the case for Horns Rev 1. Instead, a model can be trained that maps the mechanical loads on major mechanical components as function of the inflow conditions. If a certain amount of SCADA measurements (e.g. a few months on a turbine prototype) in terms of measured power production, wind speed and rotor speed, are available, the load mapping can be realized by training a data-driven regression model using e.g. artificial neural networks (ANN). This is feasible both in terms of load time series and in terms of 10-minute load statistics. An alternative is to use an aeroelastic model of the turbine which can be used to generate a synthetic data set, and the ML model is then trained on the synthetic data set [15],[16]. The use of the aeroelastic model can be beneficial in most cases also as a supplement to measured loads, as it is unlikely that a single load measurement campaign can provide enough information about the full range of possible operating conditions. In the present study, we use a generic aeroelastic model of the Vestas V80 wind turbine developed in the Hawc2 aeroelastic simulation software [17]. As the OEM-installed turbine controller is confidential, the model features the DTU Wind Energy controller [18] tuned to reproduce the known power, pitch and rpm curves of the V80 turbine. The model includes a monopile foundation with variable water depth between 7 and 15m. The model mass and stiffness are tuned so that it reproduces the natural frequencies of the real structure, including the variation of the frequency with water depth.

Wake effects are a major load driver in wind farms and a fatigue life prediction model needs to take wakes into account. In the present study, we use the approach described in [16] and [19], where the wake-induced effects are accounted for by introducing three variables describing the relative location of upwind turbines: R_D, the distance in rotor diameters to the closest turbine upwind, θ, the relative angle between the wind direction and the direction of the upwind turbine (or row of turbines), and N_{rows}, the number of turbines upwind in case there is an aligned row of turbines. A random sample of 30,000 combinations of these variables with other external conditions (wind speed, turbulence, wind shear, and water depth) is generated, and aeroelastic load simulations are carried out using the Hawc2 aeroelastic simulation tool[17], where wake effects are simulated with the Dynamic Wake Meandering model [20]. In order to reduce the statistical (realization-to-realization) uncertainty due to using random turbulent realizations, six simulations with random turbulence seeds are carried out for each sample point, resulting in a total of 180,000 simulations. A regression model is trained on the results using an ANN with 3 hidden layers with 24 neurons each, and with tanh activation functions. The process of training the model and its performance are discussed in details in[16]. As there are no load data available, it is not possible to directly validate the load predictions with measurements. However, the same ANN model can be used to predict the power output, and the results can be directly compared to the SCADA records. The prediction is carried out on each individual turbine, and the values are summed to also estimate the total power output of the wind farm. Figure 9 shows a comparison of the predicted total power output time series vs. the one recorded from SCADA, while Figure 10 shows the correlation between measured and predicted power outputs at individual turbines. Both the total and individual turbine power output predictions show high correlation with measurements. From some differences visible on Figure 9 it may be argued that the power curve of the aeroelastic model may not be fully accurate at close to rated wind speeds. This is to some extent expected as the load and power predictions are based on a generic model and not on one provided by the OEM.

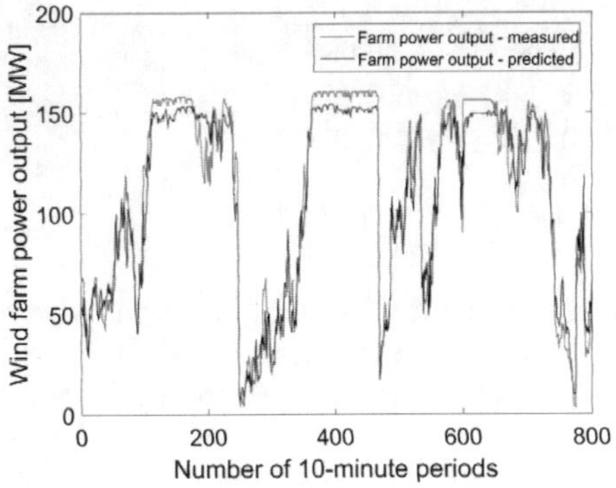

Figure 9. Comparison of measured and predicted time series of total electrical output from the Horns Rev 1 wind farm.

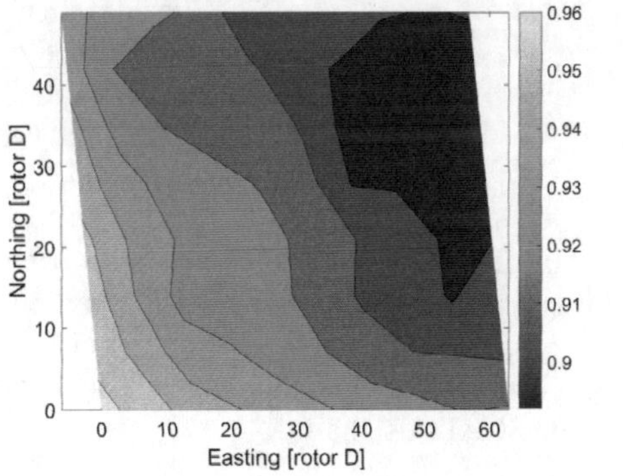

Figure 10. Correlation (r-square) between measured and predicted power output at individual turbines.

Damage-equivalent fatigue loads (DEL) are predicted simultaneously with the power predictions, based on the same input data. Based on these results, ten-minute fatigue damage increments are computed for individual turbines, thus establishing their fatigue damage accumulation history. The fatigue damage is computed relative to an assumed 20-year design lifetime under IEC 1A reference conditions. Figure 11 shows the estimated damage accumulation history for main shaft torsion, while Figure 12 provides a map of the accumulated blade root flapwise fatigue damage over the wind farm based on approximately 3 years of operation. These estimates are based on directly applying the load prediction model on the free wind time series obtained from SCADA and the prediction length is limited to the amount of data available. In order to estimate the damage accumulation over the entire operating period, inflow statistics (wind rose, wind speed and turbulence probability distributions) need to be established from the data. Afterwards the lifetime-equivalent loads can be estimated by numerically integrating the load model outputs over the joint distribution of inflow conditions.

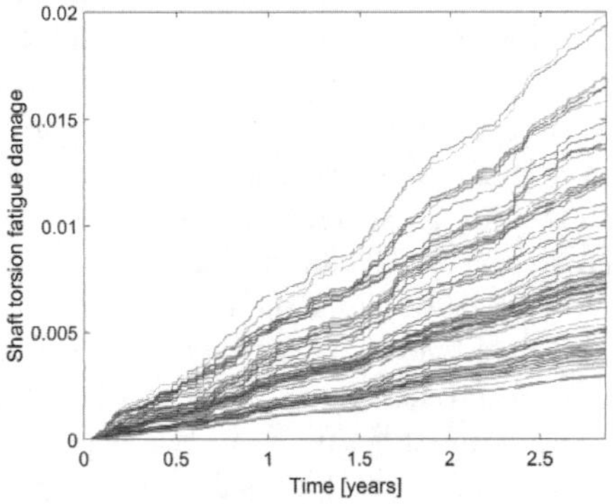

Figure 11. Time series of accumulated shaft torsion fatigue damage for all individual turbines in Horns Rev 1.

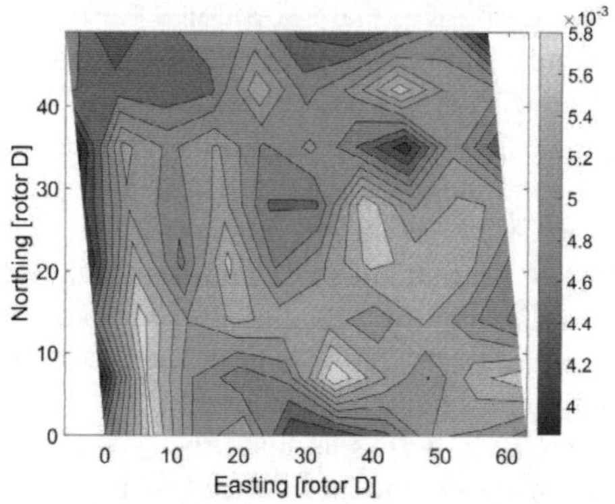

Figure 12. Map of blade root flapwise lifetime fatigue damage estimates for all individual turbines in Horns Rev 1.

Finally, Table 1 provides a summary of the accuracy of the various ML models presented above. The measures used are the coefficient of determination (R-square), and normalized root mean squared error (NRMSE).

Table 1. Overview of the accuracy of the estimations based on SCADA and ML models.

Variable names	R-square	NRMSE
Wind direction	0.973	0.075
Free wind speed	0.952	0.085
Ambient turbulence	0.764	0.256
Individual turbine power output	0.919	0.229
Total wind farm power output	0.956	0.188

4. Examples of alternative analysis approaches

In the previous section we showed a particular case study for the Horns Rev 1 wind farm, where a generic load model, and a certain amount of mast and SCADA data were available. In the following, we present further alternative steps which can be implemented when more information is available from the SCADA signals.

4.1. Load time series

While the last section described the ability to successfully reproduce mean wind farm performance using neural networks, similar models will be utilized here to reproduce component loads time series given SCADA based measurement time series. Based on measured time series from the SCADA system of electrical power, wind speed, blade pitch angle, tower top accelerations and rotor speed, a 3-layer feed-forward neural network is trained to output the corresponding time series of blade root flap moment and tower base fore-aft moment. The training is based on a selected sample of measured loads on a selected turbine within a wind farm over various mean wind speeds and time periods. The trained neural network is subsequently simulated using a different set of measured SCADA inputs not used in the training process. The result of simulation of loads using this neural network for such an arbitrary

10-minute SCADA input is shown in Figure 13, which demonstrates that a reasonable fidelity of the structural loading can be reproduced using measured SCADA signals.

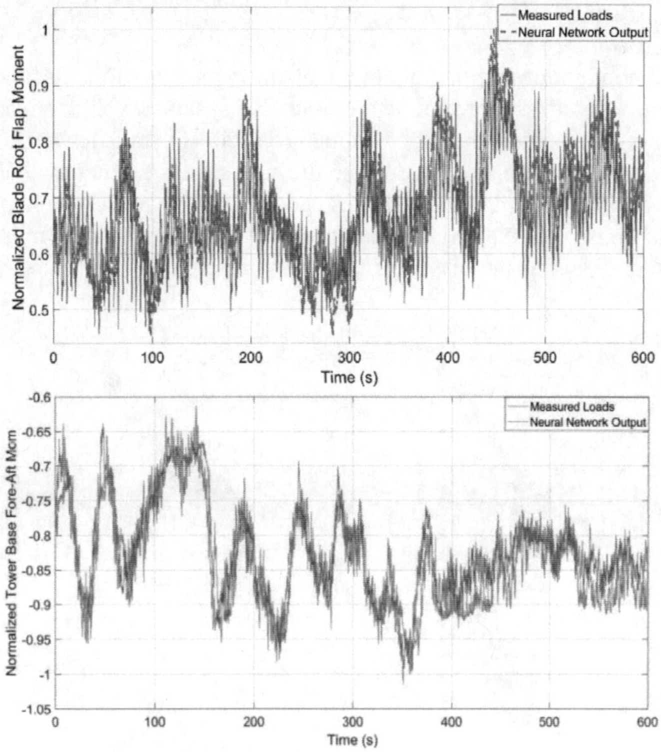

Figure 13. Mapping of wind turbine load time series to wind speed, rotor rpm and power by means of a Neural Network model.

Based on the neural network load simulation shown in Fig. 13, the corresponding load time series can be processed to yield damage equivalent loads or damage. The ANN model appears to act as a low-pass filter. However, due to the exponential relationship between load amplitude and fatigue damage, the large load cycles are predominant and the ANN model is capable of tracking the fatigue life consumption of the corresponding wind turbine structure within a wind farm. A comparison of the normalized damage equivalent blade root flap moments between measurements and simulations shows that difference is of the order of 10%-15% for some of the operating wind speeds as shown in Fig. 14.

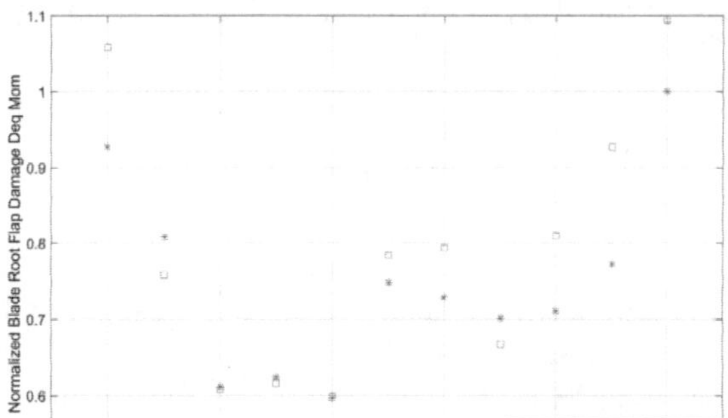

Figure 14: Mean flapwise damage equivalent moments on the blade root.

It is therefore possible to forecast a large fraction of the damage in the future based on a sufficient database of neural network simulation of loads based on SCADA inputs, which enables wind farm control to either diminish fatigue damage using de-rating or increase production.

4.2. Turbine operating status

SCADA data will often contain information about the operating status of the turbine, in terms of status flags (e.g. grid availability, available for production) and also system level information such as power, rotational speed etc. This information should be enough to establish statistics regarding the prevalence of different operating regimes of the turbine. In case the status flags are not available, a similar result can be obtained using basic operating data variables (power, rpm, nacelle wind speed and pitch low-order statistics), by training a ML classifier which predicts the turbine operating status and uses the basic variables as inputs. Training the classifier requires a reference data set with status flags for at least one turbine.

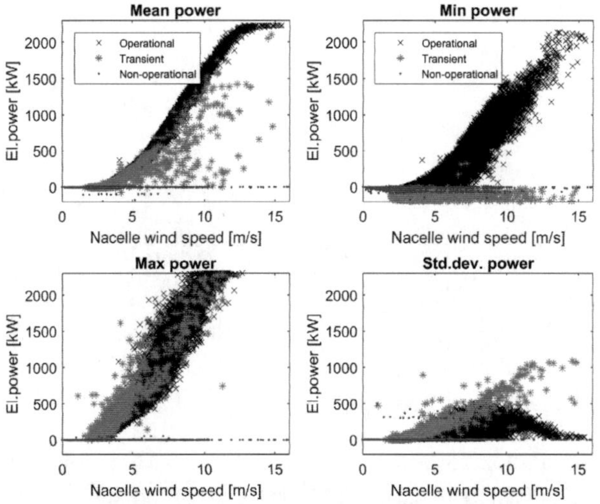

Figure 15. Illustration of the effect of turbine operational status on 10-minute power statistics from SCADA. Top left: mean power, top right: minimum power, bottom left: maximum power, bottom right: standard deviation of power.

Figure 15 shows the outputs of a ML classification model based on the random forest algorithm, for predicting the operating status of an onshore wind turbine. For training the model, 2 months of SCADA records on a single turbine including operational status flags (grid availability, turbine availability) are used. The trained model has an average precision score of 99.8%, with just 21 out of 10306 points being mislabeled. The outcome is a classification tool which can distinguish between three classes of operating conditions (operational, non-operational and transient) based on information about power, rotor rpm, nacelle wind speed, and blade pitch angle – all of which are basic variables typically present in SCADA records.

5. Conclusions.

The present study charted a flexible procedure for obtaining a load-based site-specific remaining lifetime estimate, where different alternatives are suggested based on the type and amount of information available. A particular case study on the Horns Rev 1 wind farm was shown, where free wind conditions were derived from SCADA, and the results were fed to a load prediction surrogate model trained on load simulations. The model predictions were validated against available SCADA data. It can be concluded that SCADA data can be used to derive an estimate of free wind conditions which is sufficiently good for carrying out load predictions. Furthermore, it was shown that a simulation-based surrogate model where wake-induced effects are taken into account can perform well in predicting power output at wind farm as well as at wind turbine level.

Given measured load time series of rotor speed, power production, wind speed, pitch angles and tower top accelerations, it was determined that a neural network can reproduce the blade root and tower base load time series sufficiently well. If only 10-minute statistics of the SCADA signals are available, then an aeroelastic design basis of the wind turbine can be used to reproduce the 10-minute measured statistics and subsequently use the simulated time series to train the neural network.

Acknowledgements
The work on this paper has been supported by the Danish Energy Agency through the EUDP LifeWind project (Grant Number 64017-05114). The support is greatly appreciated.

References
[1] Kim, K., Partahasarathy, G., et al., "Use of SCADA data for Failure detection in wind turbines", Proceedings of 2011 Energy Sustainability Conference and Fuel Cell Conference August 7-10, 2011, Washington DC, USA
[2] Tautz-Weinert, J. & Watson, S. J. Using SCADA data for wind turbine condition monitoring – a review. *IET Renew. Power Gener.* **11,** 382–394 (2016).
[3] Kusiak, A. & Li, W. The prediction and diagnosis of wind turbine faults. *Renew. Energy* **36,** 16–23 (2011).
[4] Colone, L., Reder, M., Dimitrov, N. & Straub, D. Assessing the Utility of Early Warning Systems for Detecting Failures in Major Wind Turbine Components. *J. Phys. Conf. Ser.* **1037,** 032005 (2018).
[5] Kusiak, A., Zheng, H. & Song, Z. Wind farm power prediction: a data-mining approach. *Wind Energy* **12,** 275–293 (2009).
[6] Göçmen, T. & Giebel, G. Data-driven Wake Modelling for Reduced Uncertainties in short-term Possible Power Estimation. *J. Phys. Conf. Ser.* **1037,** 0–10 (2018).
[7] Mittelmeier, N., Blodau, T. & Kühn, M. Monitoring offshore wind farm power performance with SCADA data and advanced wake model. *Wind Energy Sci. Discuss.* 1–25 (2016). doi:10.5194/wes-2016-16
[8] Larsen, T. J., Madsen, H. A., Larsen, G. C. & Hansen, K. S. Validation of the dynamic wake meander model for loads and power production in the Egmond aan Zee wind farm. *Wind Energy* **16,** 605–624 (2013).
[9] Larsen, T. J., Larsen, G. C., Pedersen, M. M., Enevoldsen, K. & Madsen, H. A. Validation of the Dynamic Wake Meander model with focus on tower loads. *J. Phys. Conf. Ser.* **854,** (2017).
[10] Goodfellow, I., Bengio, Y. and Courville, A. (2016) *Deep Learning*. MIT Press
[11] Cybenko, G. (1989) Approximation by superpositions of a sigmoidal function. *Mathematics of control, signals and systems 2*(4), pp.303–314
[12] Scott, R.C. and Pado, L.E., "Active control of wind-tunnel model aeroelastic response using neural networks", Journal of Guidance, Control, and Dynamics, Vol. 23, No. 6, 2000, 1100-1108.
[13] Bernelli-Sazzera, F., Mantegazza, P., Mazzoni, G. and Rendina, M., "Active flutter suppression using recurrent neural networks", Journal of Guidance, Control, and Dynamics, Vol. 23, No. 6, 2000, 1030-1036.
[14] Hansen, M. H., Henriksen, L. C. (2013) Basic DTU Wind Energy Controller, *DTU Wind Energy E-0028*. DTU Wind Energy, Technical University of Denmark.
[15] Dimitrov N., Kelly M., Vignaroli A., Berg J. (2018) From wind to loads: wind turbine site-specific load estimation with surrogate models trained on high-fidelity load databases. *Wind Energy Science 3*, 767-790, https://doi.org/10.5194/wes-3-767-2018
[16] Dimitrov, N. (2019) Surrogate models for parameterized representation of wake-induced loads

in wind farms, *under review for Wind Energy*.

[17] Larsen T. J. and Hansen A. M. (2014). How 2 HAWC2, the user's manual, *Risø-R-1597(ver.4-5)(EN)*. Risø National Laboratory, Technical University of Denmark.

[18] Hansen, M. H. & Henriksen, L. C. (2013) Basic DTU Wind Energy controller. *DTU Wind Energy Report E-0028*.

[19] Galinos C., Dimitrov N., Larsen T. J., Natarajan A., Hansen K. S.(2016) Mapping Wind Farm Loads and Power Production - A Case Study on Horns Rev 1. *Journal of Physics: Conference Series 753*(3), article number 032010.

[20] Larsen G. C., Madsen H. A., Thomsen, K. and Larsen T. J. (2008). Wake meandering - a pragmatic approach. *Wind Energy 11*, pp. 377–395.

Lifetime Extension, Repowering or Decommissioning? Decision Support for Operators of Ageing Wind Turbines

JH Piel[1,2], C Stetter[1,2], M Heumann[1], M Westbomke[2,3] and M H Breitner[1]

[1] Institut für Wirtschaftsinformatik, Leibniz Universität Hannover, Königsworther Platz 1, 30167 Hannover, Germany

[2] Nefino GmbH, c/o Leibniz Universität Hannover, Königsworther Platz 1, 30167 Hannover, Germany

[3] Institut für Integrierte Produktion, Leibniz Universität Hannover, Hollerithallee 6, 30419 Hannover, Germany

E-mail: jan-hendrik.piel@nefino.de, stetter@iwi.uni-hannover.de, heumann@iwi.uni-hannover.de, westbomke@iph-hannover.de, breitner@iwi.uni-hannover.de

Abstract. In Germany, more than one third of the installed wind energy capacity will leave the feed-in tariff funding between 2021 and 2025. Operators of affected turbines are therefore increasingly concerned with the design of profitable end-of-funding strategies. This requires feasibility analyses of both lifetime extension and repowering options and entails the subsequent challenge to determine the optimal lifetime extension and corresponding repowering timing. To support operators and other stakeholders dealing with wind turbines' end-of-life issues, this study presents a geographic information system that permits evaluating optimal end-of-funding strategies at different spatial scales reaching down to detailed analyses on individual turbine level. The decision support system processes topographic, wind, turbine, and finance data in an integrated system of resource simulations, spatial planning analyses and economic viability assessments. Case-study results show that a uniform end-of-funding strategy cannot be applied to all ageing turbines. Conducted sensitivity analyses rather indicate that the best strategy highly depends on various turbine-specific aspects, especially the location, type and maintenance costs as well as exogenous factors, including the developments of electricity spot market prices and tendered feed-in premiums. In light of latest trends regarding the exogenous factors, lifetime extension and repowering potentials increase. However, the results also indicate that dismantling, disposal and recycling of numerous ageing turbines will become a major challenge for the wind energy sector in the next decade.

1. Introduction

At the end of 2020, more than 5,000 wind turbines ($\approx$ 3.9GW) will reach the end of the feed-in tariff funding period according to the German Renewable Energy Sources Act (EEG). More than 8,000 turbines ($\approx$ 12.5GW) will follow by the end of 2025. If a lifetime extension beyond the funding period is technically viable, operators of affected turbines need to find alternative sales models for the generated electricity once the feed-in tariff expires. The most popular sales model is trading the generated electricity directly or via contracts with trading companies on the

IOP Conf. Series: Journal of Physics: Conf. Series **1222** (2019) 012033 doi:10.1088/1742-6596/1222/1/012033

European Energy Exchange (EEX). An alternative is the signing of power purchase agreements (PPAs) with industrial partners, which is already a widespread sales model in many countries worldwide and is becoming increasingly established throughout Europe. In Germany, the first PPAs with the purpose of ensuring a lifetime extension of turbines beyond 2020 were concluded in 2018. Nevertheless, regardless of the sales model chosen, the future sales price per unit of electricity is likely to be significantly lower than the previous feed-in tariff, which raises the question in how far a lifetime extension beyond the EEG funding period is economically viable.

Consequently, an alternative to a lifetime extension could be a repowering of one or multiple old turbines by a modern turbine. In Germany, this would require a successful participation in the renewable energy auctions implemented with the 2017 EEG amendment to determine the level of a feed-in premium, which would then be guaranteed to the repowering project for another twenty years of operation. However, whether the repowering option can be exercised also depends on various spatial aspects in the immediate vicinity of the location of the old turbines. Increasing overall heights of modern turbines in combination with restrictive regulations regarding required minimum distances to settlements and other areas impede a repowering for many old turbines. If both lifetime extension and repowering are not viable, the only remaining option is a permanent shut-down of wind turbine operation at the affected location.

Since the viability of lifetime extension and repowering depends on various endogenous and exogenous factors, selecting and designing the optimal end-of-funding (i.e., expiry of feed-in tariff funding) strategy for a specific turbine are challenging tasks. To provide decision support to operators and other stakeholders, this study presents a geographic information system (GIS) that permits systematically evaluating the optimal choice between lifetime extension, repowering and decommissioning for operational wind turbines in Germany at different scales reaching down to detailed analyses on individual turbine level. Current research tends to investigate lifetime extension (e.g., [1–3]) and repowering (e.g., [4–6]) separately. Studies combining both options to decide on the optimal repowering timing are rare (e.g., [7–10]). Furthermore, to the knowledge of the authors, no study has so far been published combining resource, spatial and economic analyses to investigate both lifetime extension and repowering potentials on a macro scale. Consequently, the focus of this study is on a macro-scale analysis of end-of-funding strategies for ageing wind turbines located in the German federal state of Lower Saxony.

2. Methodology and Data

To permit the macro-scale analysis, the GIS handles extensive topographic, wind, turbine, and finance data in an integrated system of resource simulations, spatial planning analyses and economic viability assessments. Figure 1 shows the system architecture of the GIS. Firstly, the spatial analysis is applied to the coordinates of all operational turbines under investigation to determine whether the corresponding location allows a repowering with respect to the spatial planning regulations or whether a lifetime extension of the old turbine is the only remaining option. Secondly, based on the results of the spatial analysis, the economic analysis is applied to each turbine to determine the optimal end-of-funding strategy.

2.1. Spatial Analysis

When it comes to the repowering of a wind turbine in Germany, the same legal regulations apply as for the construction of a turbine at a greenfield site. Whether or not a construction permit is granted depends in particular on a wide variety of spatial aspects. In general, the protection of the habitats of man and nature is in the foreground of legislative measures. By preventing the construction of turbines, harmful impacts on the environment and life are to be limited. Correspondingly relevant guidelines are protection laws such as the Federal Nature Conservation Act, Landscape Protection Act and Immision Protection Act. These regulations result in legally conditioned exclusion areas for wind energy operation. Based on these exclusion areas, which also

IOP Conf. Series: Journal of Physics: Conf. Series **1222** (2019) 012033 doi:10.1088/1742-6596/1222/1/012033

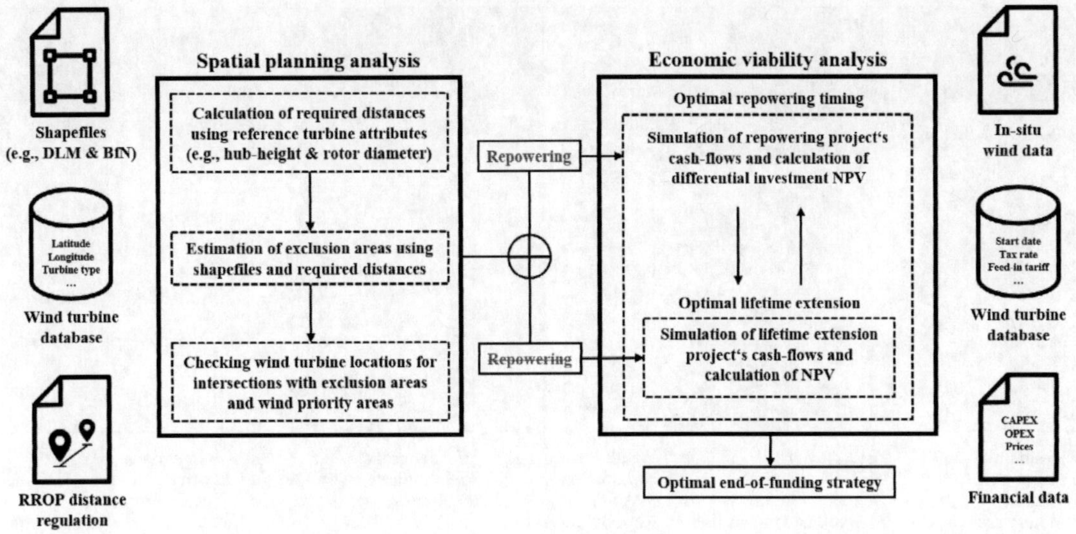

Figure 1. System architecture of the decision support system and required data inputs.

consider specific distance regulations, the spatial planning laws (RROP) resulting from federal law feature wind priority areas at municipal level, which eventually represent the areas legally intended for the construction of wind turbines. These priority areas are based on applicable law but offer municipal legislators the possibility of supplementary individual regulations to enable more granular political action. The approval of a repowering project within such designated areas is extremely probable, whereas outside it is rather unlikely, making priority areas well suited as indicators for a possible project approval. However, due to increasing turbine heights and more restrictive distance regulations, it is highly uncertain for many old turbines whether the repowering option can be exercised, since many old turbines were built very close to settlements or other protection areas. Consequently, the repowering option is especially dependent on the feasibility of an old turbine's location to be designated as a wind priority area.

To determine viable locations, various available datasets on protected areas, infrastructural exclusion areas and others were implemented in the GIS and applied to the locations of all turbines under investigation (see Figure 2). Necessary data was imported by vector-based shapefiles, converted into tabular form and imported or graphically created. The resulting geospatial datasets were stored in a local database. Based on the datasets relevant distance criteria is applied by generating buffer zones around each affected object. Turbine-specific distances are generated on the basis of the attributes of a preselected reference repowering turbine: a Vestas V150-4.0 with a rotor diameter of 150m, an assumed hub-height of 120m and a total turbine height of 195m. The choice of the reference turbine is based on the determined average wind speed in Lower-Saxony. The respective distance regulation relevant for Lower Saxony can be seen in Table 1. Further protected areas without distance regulation comprise industrial areas provided by DLM, nature reserves, national parks, nature parks, landscape reserves, protected forests, special protection areas, flora-fauna habitats, biosphere reserves and wetlands of international importance provided by BfN as well as military zones.

Based on the regulation, the exclusion areas are generated and matched with the old turbines' locations. By applying an intersect function, all locations are checked for an overlap with exclusion areas, deriving repowerable and non-repowerable turbines. In addition, a distinction is made between exclusion from currently designated priority areas. In this way a differentiated statement can be made as to how likely a repowering is at a certain location.

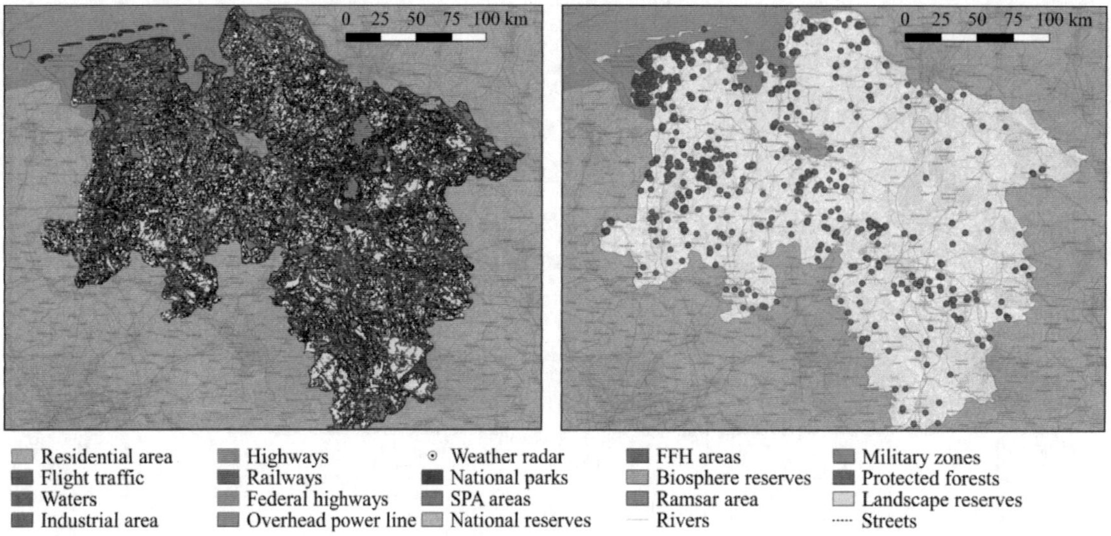

Figure 2. Relevant protection areas (left) and locations of investigated turbines (right).

Table 1. Relevant distance regulations in Lower Saxony and corresponding layer sources.

Protected areas	Distance regulation	Source of layer*
Residential areas	2 × turbine height	DLM
Waters	50 meters	DLM
Weather radar locations	5,000 meters	DWD
Overhead power lines	3 × rotor diameter (RD)	DLM
Railway	1.5 × (RD + hub-height)	DLM
Highway, federal highway, county roads	40, 20, 20 meters	DLM
Airports	5,000 meters	LuftVG

* Digitales Landschaftsmodell (DLM), Deutscher Wetterdienst (DWD), Luftverkehrsgesetz (LuftVG), Bundesamt für Naturschutz (BfN)

2.2. Economic Analysis

To determine the optimal end-of-funding strategy for a turbine, an economic analysis is applied based on the results of the spatial analysis. In the economic analysis, turbines not qualified for repowering according to the spatial analysis are examined for the economic viability of a lifetime extension, whereas all remaining turbines are also being investigated for the economic viability of a repowering. In both cases, a discounted cash-flow (DCF) model is applied to determine the economic viability. The net present value (NPV) is used as the decision criterion: a non-negative NPV indicates economic viability. Only if both options, lifetime extension and repowering, feature a negative NPV, the permanent shut-down of wind turbine operation at the respective site remains as the only viable option at the end of the funding period.

The general equation for determining the NPV is:

$$NPV = -I + \sum_{t=1}^{T} \frac{Z_t}{(1+r)^t}, \tag{1}$$

IOP Conf. Series: Journal of Physics: Conf. Series **1222** (2019) 012033 doi:10.1088/1742-6596/1222/1/012033

where I represents the investment and Z_t the net cash-flows. The latter are discounted using the discount rate r and summed over the lifetime T. To calculate the NPV of the lifetime extension option, expected net cash-flows generated over the expected lifetime extension period are compared with the mandatory investment in a technical lifetime extension assessment. Instead, the NPV for the repowering option is calculated on the basis of a standard project valuation in which the capital expenditures are compared with the expected net cash-flows over a project lifetime of 20 years. To determine the optimal lifetime extension period in addition to the economic viability of both options, a comparison between the options is necessary. For this purpose, a differential investment analysis following [10] is applied based on the cash-flow analysis of both options. A differential investment analysis allows to apply DCF models to compare investment alternatives with different characteristics, such as capital expenditures or horizons. In this study, the differential investment analysis permits evaluating the economic viability of the repowering option and the optimal lifetime extension period simultaneously. The end of the optimal lifetime extension period also represents the optimal repowering timing, which determines the commissioning of the repowering project. Since the resulting optimal lifetime extension period can feature negative values, it is also possible that the optimal repowering timing occurs already before the end of the regular 20-year lifetime of the old turbine.

The aim of the differential investment analysis is to constantly compare the profitability of the repowering and lifetime extension options in order to maximize the NPV of the combined project, which represents a hypothetical investment reflecting the difference between the cash-flow streams of the repowering and lifetime extension projects. The NPV of the differential investment also consists of an investment and net cash-flows and is calculated as follows:

$$NPV_{diff,t} = -C_t + R_t, \qquad (2)$$

where C represents costs and R revenues arising from postponing the start of the repowering project to a later period t and continuing the operation of the old turbine. The costs mainly refer to the additional discounting effect that reduces the present value of the repowering project. The net cash-flows, which are compared to this hypothetical investment, mainly comprise the additional revenues arising from the lifetime extension of the old turbine. Nontheless, both costs and revenues can also be influenced by other aspects, such as technological advances or the development of electricity spot market prices and feed-in premium levels in future auctions, which can increase or decrease the profitability of both lifetime extension and repowering over time. Consequently, it can be drawn from Eq. 2 that the optimal repowering timing is reached when the NPV of the differential investment is equal to or smaller than zero, as this implies that a further lifetime extension of the old turbine is no longer economically viable. The GIS determines the optimal repowering timing by solving an optimal stopping problem: Once the costs of the differential investment exceed the revenues, the old turbine should be dismantled and the repowering project implemented. Accordingly, the planning process for the repowering project must be initiated in advance of the optimal repowering timing.

The economic analysis is performed for any single turbine of the dataset and, hence, the in-situ wind resources are necessary for unbiased results. The wind speeds drive the electricity yield, which by means of the resulting revenues determine the net cash-flows. To estimate bias-corrected wind speeds for an arbitrary location, spatial statistical downscaling following [11] is applied to reanalysis wind data. The reanalysis data was obtained from NASA's Modern-Era Retrospective analysis for Research and Applications (MERRA-2) dataset following the virtual wind farm model [12], which is processed as follows: (1) acquisition of hourly wind speeds at 10m and 50m above ground at each of the twelve MERRA-2 grid points closest to the selected location; (2) utilization of LOESS regression [13] for the spatial interpolation of hourly wind speeds to the geographic coordinates of the selected location. Based on steps (1) and (2) the spatial statistical downscaling is applied as follows: (3) collection of wind speed distributions for the same heights and the closest available grid point of the Global Wind Atlas (GWA), which features micro-scale information; (4) adjustment of the interpolated hourly wind speeds estimated from the MERRA-2 dataset to feature the local wind characteristics as captured by the GWA data using the in-situ roughness data from ESA's Global Land Cover Map. To calculate the expected electricity yield, the turbine-specific power curve is applied to the hourly wind speed time series. In addition, a turbine efficiency level is considered which covers wake effects, turbine availability and technical efficiency. The efficiency of a turbine in the first year of operation is set at 88.5% and an annual degradation of 0.5% is assumed.

Based on the expected electricity yield, the site quality according to the EEG is calculated for each location of the old turbines, which is then used to derive capital expenditures (CAPEX) and operating expenditures (OPEX) for both lifetime extension and repowering projects using the data shown in Table 2. The CAPEX of the lifetime extension project include the mandatory technical assessment at costs of 25,000 € [14], which are linearly depreciated over the maximum lifetime extension period of five years. The technical assessment includes the inspection and certification costs, but excludes costs for potential repairs needed to obtain permission. The site quality and the commissioning date of the old turbines are used to estimate the current feed-in tariff according to the corresponding EEG amendment. In addition, assumptions were made with regard to the development of the electricity spot market prices (3.86 ct/kWh in 2018 and 2.5% annual increase) and the CAPEX and OPEX (1.5% annual decrease) of a repowering project. The development of the tendered feed-in premiums (6.28 ct/kWh for a site quality of 100% in 2018) follows the same development as the cost parameters. The turbine-specific corporate tax rate was calculated according to the German tax legislation using a location-dependent municipalized tax multiplier. For the calculation of the discount rate r the weighted average cost of capital approach was used based on cost of equity of 5%, cost of debt of 3.5% and variable debt-to-equity ratios. Whereas the lifetime extension project was assumed to be entirely equity financed, the optimal debt-to-equity ratio was calculated individually for the repowering project based on turbine-specific cash-flows using a debt sculpting approach [15].

Table 2. CAPEX and OPEX data oriented towards recent cost studies [14, 16, 17].

Site quality	60%	70%	80%	90%	100%	110%	120%	130%	140%
CAPEX (€/kW)	1,355	1,355	1,308	1,308	1,308	1,216	1,216	1,216	1,216
Initial OPEX* (ct/kWh)	2.45	2.35	2.26	2.17	2.17	2.07	2.07	1.98	1.98
Basic OPEX (ct/kWh)	2.73	2.64	2.54	2.45	2.35	2.35	2.26	2.26	2.17

* OPEX in the first ten years of operation.

3. Results and Discussion

Due to the superior wind conditions, Lower Saxony is the federal state with the most and oldest wind turbines in Germany. More than 26.5% of the about 13,200 wind turbines that reach the end of the EEG funding period by the end of 2025 are operated in Lower Saxony. In this study, the GIS is used to investigate optimal end-of-funding strategies for the 1,645 turbines located in Lower Saxony reaching the end of their funding period at the end of 2020.

3.1. Spatial Planning Analysis

Initially, the spatial planning analysis is applied to the 1,645 locations using the attributes of the reference turbine. Afterwards, an additional sensitivity analysis is conducted based on changes

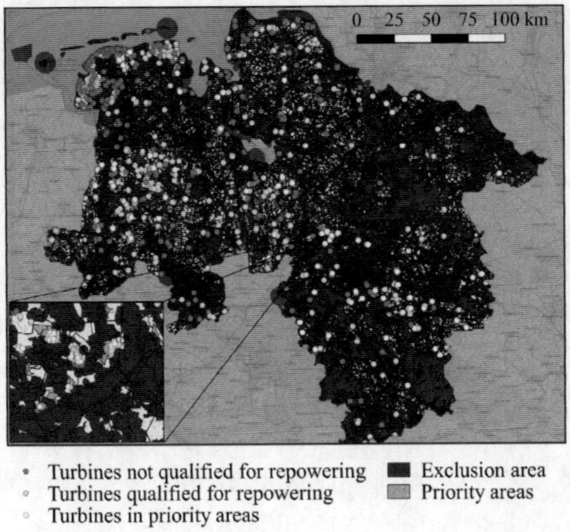

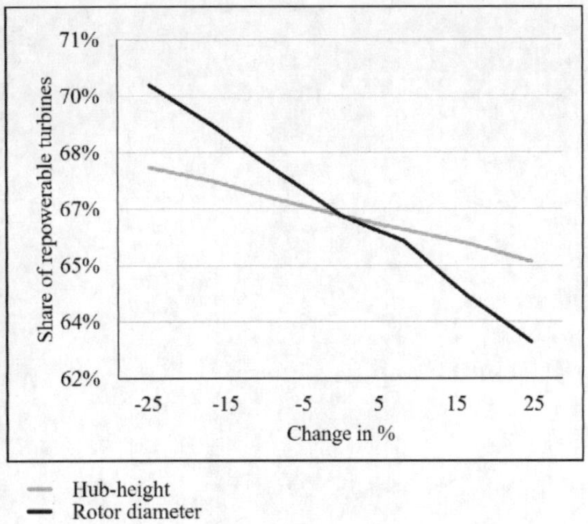

Figure 3. Results of spatial planning analysis and corresponding sensitivity analysis.

in the hub-height and rotor diameter of the reference turbine, since the criteria catalogue of Lower Saxony includes spatial restrictions that depend on the total turbine height. This enables an estimation of the extent to which the turbine attributes influence the spatial repowering potential. Figure 3 shows the estimated exclusion areas calculated based on the attributes of the reference turbine and the investigated locations. 547 locations (33.25%) intersect with the estimated exclusion areas and are thus not qualified for repowering. Accordingly, the remaining 1,098 locations (66.75%) do not violate the distance regulation and are thus qualified for a repowering by the reference turbine. As shown in Figure 3, the results of the sensitivity analysis indicate that the determined repowering potentials are robust against changes in the attributes of the reference turbine. Even a simultaneous increase in both hub-height and rotor diameter to 150 meters reduces the number of turbines qualified for a repowering by only 9.74%.

To estimate the current regulatory repowering potential in addition to the calculated repowering potential, Figure 3 also includes the current wind priority areas specified by the RROP of Lower-Saxony. The corresponding analysis indicates that only 317 locations (19.27%) intersect with current wind priority areas. For all remaining turbines the feasibility of a repowering is therefore highly uncertain, although 58.81% of the corresponding locations would allow a repowering given the current distance regulation as well as the attributes of the selected reference turbine. Consequently, the current RROP only partially exploits the calculated repowering potential given the estimated exclusion areas. A typical repowering ratio is to replace two to three old turbines by one modern turbine. Since the RROP only allows the replacement of one in five turbines, the results indicate a strong need for action on the part of the responsible authorities to increase the incentives for repowering projects in Lower Saxony.

3.2. Economic Viability Analysis

The results of the spatial analysis are explicitly considered in the economic analysis. For the 546 turbines that cannot be repowered due to distance regulation, only the economic viability of the lifetime extension option is assessed. All remaining turbines are also examined with regard to their economic repowering potential as well as their optimal lifetime extension and corresponding repowering timing by means of the differential investment analysis. Figure 4 shows the results of the economic viability analysis. The results indicate a high lifetime extension potential for the investigated wind fleet. 917 old turbines (55.75%) show a positive economic viability for a lifetime

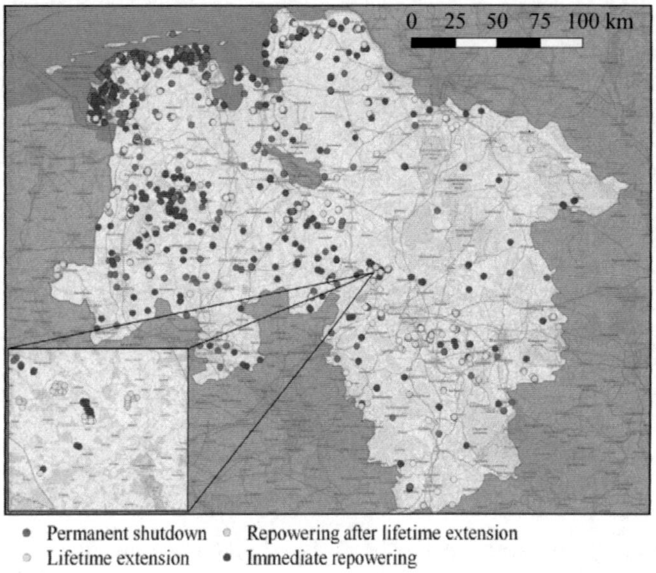

Figure 4. Optimal end-of-funding strategy for the 1,645 wind turbines under investigation.

extension beyond 2020, although this is not necessarily the optimal end-of-funding strategy for these turbines. An immediate repowering in 2021 without a preceding lifetime extension is the optimal strategy for 543 old turbines (33.01%), whereas further 469 turbines (28.51%) should be operated temporarily beyond 2020 before a repowering project is implemented in the following years. In contrast, a lifetime extension without a subsequent repowering is the optimal strategy for 281 (17.08%) turbines, while it must be taken into account that this figure increases significantly if the regulatory repowering potential based on the wind priority areas is considered. Consequently, 352 turbines (21.4%) and their corresponding locations are neither viable for a lifetime extension nor for repowering. At these locations the permanent shutdown of wind energy operation remains the only option, while the operators of affected turbines will then face further end-of-life challenges regarding the dismantling, disposal, recycling and/or resale of the old turbine. However, these challenges are also relevant in the case of a repowering project and are only postponed if a lifetime extension is carried out.

As the results of the economic viability analysis are based on assumptions for several inputs, a sensitivity analyses are conducted for a representative wind turbine with a site quality of 100% according to the reference yield model of the latest EEG amendment. On the one hand, these sensitivity analyses include calculations for the economic viability of a lifetime extension given changes in the electricity spot market price, the electricity yield (i.e., site quality), the OPEX as well as the lifetime extension investment and duration (see Figure 5). On the other hand, the sensitivity of the optimal repowering timing was evaluated by means of the estimated NPV of the differential investment at the end of funding period given changes in the costs (OPEX or CAPEX), feed-in premium and electricity yield of the repowering project (see Figure 5).

The results show that the economic viability of a lifetime extension highly depends on the OPEX and electricity yield of the old turbine as well as the development of the electricity spot market prices and/or related PPA prices. For the representative turbine, a drop of 23% in spot market prices would make a lifetime extension unprofitable. The same applies if the OPEX raise by 20%. With regard to the optimal repowering timing the results show that both OPEX and CAPEX of the repowering project as well as the level of the tendered feed-in premium have the highest impacts. The impact of the electricity yield is considerably lower due to the reference yield model implemented in the German auction mechanism. A change in the expected electricity yield also changes the site quality, which in turn adjusts the level of the

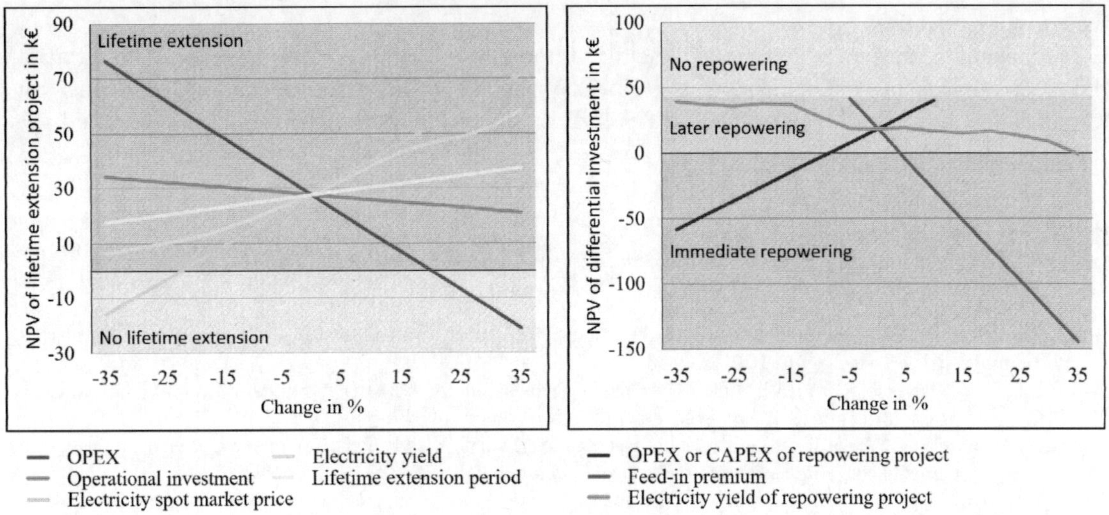

Figure 5. Sensitivity analyses of lifetime extension (left) and repowering (right) options.

feed-in premium, such that the opposite effects largely cancel each other out. The results further indicate that a repowering is no longer economically viable if the OPEX or CAPEX increase by 12% or the feed-in premium drops by 5%. Furthermore, they confirm the results of macro-scale analyses that an immediate repowering is the most likely optimal-end-of-funding strategy given that no spatial restrictions prevent a repowering at the corresponding location.

4. Conclusions and Outlook

In this study, the selection and design of optimal end-of-funding strategies for the ageing German wind fleet were under investigation. For this purpose, a GIS was developed integrating wind resource simulations with spatial planning and economic viability analyses. In a case study of the German federal state of Lower-Saxony, the GIS was applied to the 1,645 turbines reaching the post-EEG era at the end of 2020. Results show that increasing electricity spot market prices and tendered feed-in premiums in 2018, have a positive effect on the economic viability of both lifetime extension and repowering options. The highly limited availability of wind priority areas however prevents repowering projects in many locations, even though it would be an economically viable option. Nevertheless, a lifetime extension of the old turbine beyond 2020 will be an economically viable option at many of these locations. Consequently, in light of latest trends, the lifetime extension option is becoming more attractive, which could result in a smoothing effect on the entire German dismantling volume in the next decade.

In order to address related issues, future research will focus on analyzing the entire German wind fleet. First results indicate that the lifetime extension and repowering potentials are significantly lower in most of the remaining federal states. The relatively high lifetime extension potential in Lower-Saxony is due to the high average wind resources in this region that are well above the German average, whereas the relatively high repowering potential can be explained by spatial planning aspects, such as a moderate distance regulation and a relatively low population density. Consequently, despite the manifold results, the analyses indicate that the German wind energy sector is facing a large quantity of ageing turbines to be dismantled in the upcoming years. As other countries will also face this and other challenges related to the end-of-life of wind turbines in the next decade, future research will further focus on extending the scope of the GIS to other regions. In the course of the next decade, worldwide around 180 GW of installed wind energy capacity will reach the end of their originally planned operating lifetime of 20 years.

IOP Conf. Series: Journal of Physics: Conf. Series **1222** (2019) 012033 doi:10.1088/1742-6596/1222/1/012033

Acknowledgments

We gratefully acknowledge the financial support of the German Research Foundation (DFG) (research project "Entwicklung eines Wirkmodells für eine effiziente Gestaltung von Demontagenetzwerken für XXL-Produkte" (UL 419/6-1) funded by the DFG) that enabled this work.

References

[1] Ziegler L, Gonzalez E, Rubert T, Smolka U and Melero J J 2018 Lifetime extension of onshore wind turbines: A review covering Germany, Spain, Denmark, and the UK *Renewable and Sustainable Energy Reviews* **82** 1261–1271

[2] Rubert T, McMillan D and Niewczas P 2018 A decision support tool to assist with lifetime extension of wind turbines *Renewable Energy* **120** 423–433

[3] Rubert T, Niewczas P and McMillan D 2016 Life extension for wind turbine structures and foundations *Proceedings of International Conference on Offshore Renewable Energy 2016*

[4] Serri L, Lembo E, Airoldi D, Gelli C and Beccarello M 2018 Wind energy plants repowering potential in Italy: technical-economic assessment *Renewable Energy* **115** 382–390

[5] Villena-Ruiz R, Ramirez F J, Honrubia-Escribano A and Gomez-Lazaro E 2018 A techno-economic analysis of a real wind farm repowering experience: The Malpica case *Energy Conversion and Management* **172** 182–199

[6] Colmenar-Santos A, Campíñez-Romero S, Pérez-Molina C and Mur-Pérez F 2015 Repowering: An actual possibility for wind energy in Spain in a new scenario without feed-in-tariffs *Renewable and Sustainable Energy Reviews* **41** 319–337

[7] Ziegler L, Lange J, Smolka U and Muskulus M 2016 The decision on the time to switch from lifetime extension to repowering *Proceedings of Wind Europe Summit 2016*

[8] de Simón-Martín M, de la Puente-Gil Á, Borge-Diez D, Ciria-Garcés T and González-Martínez A 2019 Wind energy planning for a sustainable transition to a decarbonized generation scenario based on the opportunity cost of the wind energy: Spanish Iberian Peninsula as case study *Energy Procedia* **157** 1144–1163

[9] Himpler S and Madlener R 2014 Optimal timing of wind farm repowering: A two-factor real options analysis *Journal of Energy Markets* **7** No. 3

[10] Madlener R and Schumacher M 2011 Ökonomische Bewertung des Repowering von Onshore-Windenergieanlagen in Deutschland *Zeitschrift für Energiewirtschaft* **35** 297–320

[11] González-Aparicio I, Monforti F, Volker P, Zucker A, Careri F, Huld T and Badger J 2017 Simulating European wind power generation applying statistical downscaling to reanalysis data *Applied Energy* **199** 155–168

[12] Staffell I and Green R 2014 How does wind farm performance decline with age? *Renewable Energy* **66** 775–786

[13] Staffell I and Pfenninger S 2016 Using bias-corrected reanalysis to simulate current and future wind power output *Energy* **114** 1224–1239

[14] Wallasch A K, Lüers S, Rehfeldt K and Vogelsang K 2017 Perspektiven für den Weiterbetrieb von Windenergieanlagen nach 2020 *Deutsche WindGuard Technical Report*

[15] Bodmer E 2014 *Corporate and project finance modeling: theory and practice* (John Wiley & Sons)

[16] Lüers S, Wallasch A K and Rehfeldt K 2015 Kostensituation der Windenergie an Land in Deutschland–Update *Deutsche WindGuard Technical Report*

[17] Wallasch A K, Lüers S and Rehfeldt K 2016 Weiterbetrieb von Windenergieanlagen nach 2020 *Deutsche WindGuard Technical Report*

IOP Conf. Series: Journal of Physics: Conf. Series **1222** (2019) 012034 doi:10.1088/1742-6596/1222/1/012034

Farm-wide assessment of wind turbine lifetime extension using detailed tower model and actual operational history

A. Kazemi Amiri, R. Kazacoks, D. McMillan, J. Feuchtwang, W. Leithead
Wind Energy and Control Centre, University of Strathclyde, Glasgow, UK

Abbas. Kazemi-amiri@strath.ac.uk

Abstract. Lifetime extension is receiving increasing attention because of ageing asset bases, the need for efficient use of capital budgets, and the optimistic lifetime assumptions used at the project design stage. Based on the industrial attention and the overall observable reduction in subsidies for new investments and repowering, especially in onshore wind energy, lifetime extension is expected to become essential in the future. This contribution presents a methodology for life extension assessment of individual onshore wind turbine towers, as the key structural components, using the joint aeroeastic-finite element analysis and taking account of wind directionality and stress magnification around the tower door. The results demonstrate that the spread in the wind rose provide the potential for tower lifetime extension, however, the stress concentration around the tower door and site-wide variations of wind characteristics have to be also taken into account. The outcomes of this paper indicates that the wind rose dispersion can also provide a lifetime extension potential in addition to a more benign weather and operational conditions.

1. Introduction

Over the past years, wind turbine lifetime extension has been receiving increasingly attention. This is due to the fact that the European wind turbine fleet is aging and there are many farms with turbines close to the end of the design life. Moreover, the new subsidies and investments in wind energy has been shifted towards harvesting the offshore wind energy. Therefore, prolonging the service life of onshore wind turbines can be significantly profitable as an alternative for decommission. A useful review of the lifetime extension of the onshore wind turbines across Germany, Spain, Denmark and the UK can be found in [1]. The levelised cost of energy of onshore wind turbines and investment decision making for lifetime extension has been studied in [2]. Although the offshore fleet is comparatively younger than its onshore counterpart, nevertheless the lifetime extension studies are not limited to the onshore sites [3],[4].

Wind turbine lifetime extension is a very complex task and numerous factors and components have to be taken into account. The details of the life extension assessment procedure may differ from one component to another, but nevertheless, the general approach is the comparison between the estimated life under design loads with that under the site specific and operational loads. Several guidelines are available for the purpose of lifetime extension, which are mostly recommended by the industrial firms [5]–[7], but there is no commonly accepted procedure and a published standard, to date.

The existing lifetime extension guidelines propose a procedure, which is based on the assessment of the operational damage using the aeroelastic model [5]–[7]. The design reference fatigue evaluation by IEC61400-1 also recommends the same methodology. Such methodology as far as can be determined from the literature is based on unidirectional wind that leads to a conservative design. Consequently, the relation between wind directionality and life extension requires further investigations. This study applies a joint aeroelastic-finite element analysis and demonstrate the potential for lifetime extension due to the wind direction dispersion. Therefore, this study demonstrate that the lifetime extension is not limited only to a more benign operating and environmental conditions as mentioned by the existing lifetime extension guidelines. Moreover, this contribution shows that this lifetime extension potential also depends on the towers' own wind rose properties and tower door orientation relative to the prevailing wind. This is because the stresses around the tower door opening are magnified due to stress concentration effect, which can substantially increase the fatigue damage. This contribution develops and implements a methodology to evaluate the lifetime extension of wind turbines using the turbines' own SCADA data, taking account of the wind directionality and tower door opening stress concertation effect. According to the results, there is generally the potential for tower life extension under multi-directional wind as opposed to the unidirectional design assumption. However, this potential depends on the spread of the wind rose and also the tower door alignment relative to the wind seen by the individual turbine. Furthermore, the variations of wind properties across the wind farm and its influence on the lifetime extension potential is demonstrated.

2. Analysis methodology

Aero-elastic modelling is a powerful means for wind turbine response analysis under different load cases. The aero-elastic modelling is not intended for stress analysis, while finite element analysis is normally used for this purpose. As such, a joint aero elastic–finite element analysis is a reasonable analysis solution from the level of wind turbine load simulation up to stress analysis of different components under dynamic response. Figure 1 presents the flowchart of the applied procedure for tower lifetime extension evaluation in this study.

According to the flowchart, a validated aeroelastic model is required to analyse the hub loads under various fatigue load cases corresponding to different operational conditions (box 1). The hub loads are applied on the full-scale tower finite element model for dynamic response analysis, followed by stress time history simulation at different critical stress locations at tower base and around the door opening (box 2). The weather conditions information, including mean wind speed, wind direction (nacelle position), turbulence intensity, etc. can be obtained from towers' own SCADA data to evaluate the turbines' operational history (box 3). The tower fatigue life is normally controlled based on the largest damage of critical stress points under a unidirectional wind assumption [8]. As such, this is selected as the design reference damage. However, wind blows from different directions in reality, which implies that a critical stress location would be loaded at lower levels in contrast to the unidirectional design assumption. As such, the critical stress point with largest cumulative damage among various points on tower circumference and around the tower door under stress concentration effects should be identified. Finally, the damage indicator as a ratio between the reference and largest operational damage is utilised to evaluate the life extension potential (box 4). This procedure is repeated for other turbines within the wind farm. Through this procedure, the candidate turbines with the highest potential can be selected for the further detailed analysis and decision making process for the likely lifetime extension (box 5).

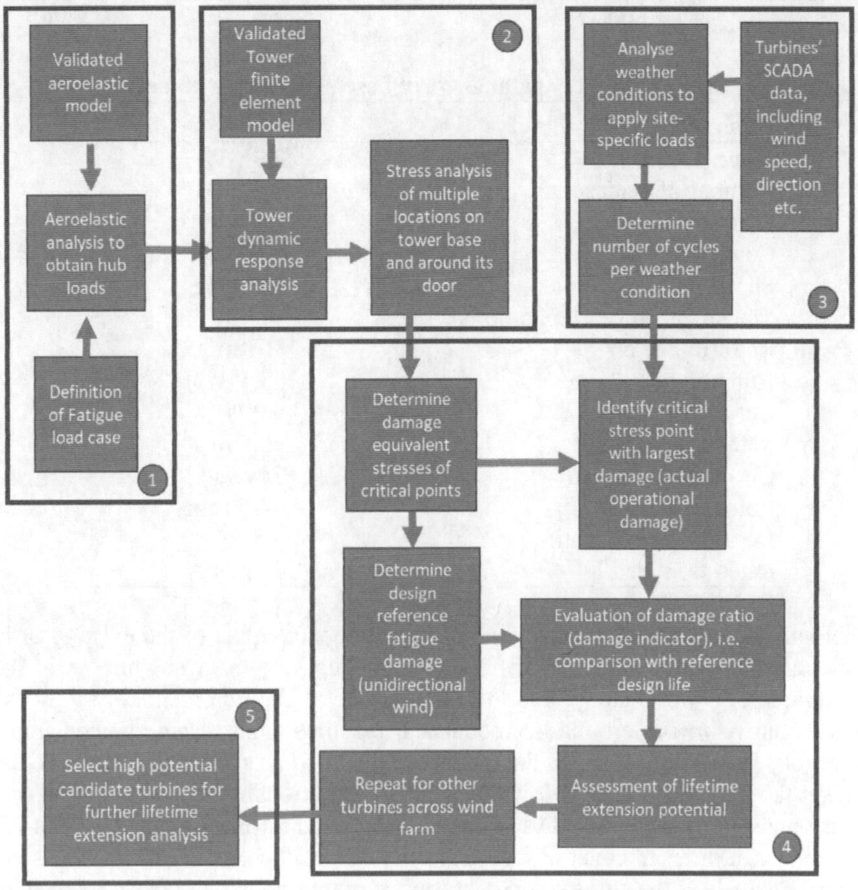

Figure 1. Step by step procedure of the applied of the applied methodology

2.1. *Wind turbine Aeroelastic model and operational loads simulation*

According to IEC 61400-1 standards [8], there are five design load cases (DLC) for evaluation of fatigue wind turbine fatigue damage and design life. These five DLCs correspond to design situation for power production without/with fault, i.e. (DLC 1.2) or (DLC 2.4), starts up (DLC 3.1), shut downs (DLC 4.1) and idling (DLC 6.4). In this study, the normal power production (DLC1.2) without any fault is used to produce time series. According to the base model of this study, the normal turbulence model (NTM) of IIB wind class is considered for the normal power production. DLC 1.2 characterises the wind turbine in its power production range and connected to the electrical load with normal turbulence model (NTM), where wind speed in the range from 4 m/s to 24 m/s with 2 m/s increments.

This study aims to replicate the power production loads of the actual wind turbines within an existing wind farm, to be able to evaluate the lifetime extension potential by use of the available operational data. Therefore, the base model of this study is an existing wind turbine, which is commonly used around the world. The aero-elastic model of base wind turbine model was developed by down scaling the generic upwind, variable speed and pitch regulated 3MW wind turbine model [9] by using the scaling rule with similarities [10]–[12]. The aerodynamic properties of down scaled model were adjusted according to the report developed by Carlen [13]. He performed the reverse engineering for the existing turbine to calculate the aerodynamic properties of blades of this turbine. His report includes the type of aerofoils,

IOP Conf. Series: Journal of Physics: Conf. Series **1222** (2019) 012034 doi:10.1088/1742-6596/1222/1/012034

used in the blades of the existing turbine, and other aerodynamic properties. The key parameters of the base model are summarized in *Table 1*.

Table 1. The key parameters of base aero-elastic model

Parameter	Value
Rated power	2.3MW
Number of blades	3 blades
Rotor position	Up-wind
Control	Collective pitch
Speed type	Variable
Transmission gearbox ratio	91
Rotor diameter	101 m
Hub height	73.5 m
Cut-in wind speed	4 m/s
Rated wind speed	10.5 m/s
Cut-out wind speed	25 m/s
Rotor speed	6-16 rpm
Design tip speed ratio	9.5
Rated tip speed	75 m/s

The full model validation requires the envelope of high frequency data of the existing turbine, to see harmonic peaks and structural modes of different components of the wind turbine at the different operational conditions. Unfortunately, this information due to the confidentiality reasons cannot be published. The comparison of aero-elastic model and existing wind turbine demonstrates adequate similarities until 6P. This is sufficient for the model to be a good representation of the actual machine, as the significant fatigue loads occur at low frequencies. Additionally, the output of developed aero-elastic model is validated by using SCADA of the existing wind turbine model. Two full years of data is used for modelling in order to cover any possible weather conditions in different seasons. Figure 2 depicts two graphs that show the comparison between the output power and mean pitch angle from the actual wind turbine SCADA and the so-called artificial SCADA. The artificial SCADA is derived from the base aero-elastic model for the full envelope power production (DLC 1.2) wind speeds (from 4 to 25 m/s with 1 m/s increments) with 6 seeds per wind speed as specified in IEC standards. The artificial SCADA is obtained from the time series simulation outputs of aero-elastic model. The mean values of 10-minute time series are used for the artificial SCADA in Figure 2. The existing wind turbine is designed for the wind turbine class of IIB, which corresponds to 0.14 reference turbulence intensity and 0.2 wind shear coefficient. However, the SCADA data provided by the operator shows that the turbines also experience much higher turbulence intensities. Therefore, the simulation results of turbine operation under different turbulence intensities are also included in artificial SCADA data as shown in Figure 2. According to this figure, firstly, the operation under simulated SCADA data lies within the envelope of the turbine's power curve and secondly, the broad scattering due to the large turbulence intensity of the site is also reflected in the simulated data. As such, the developed model consistently represents the actual wind turbine.

IOP Conf. Series: Journal of Physics: Conf. Series **1222** (2019) 012034 doi:10.1088/1742-6596/1222/1/012034

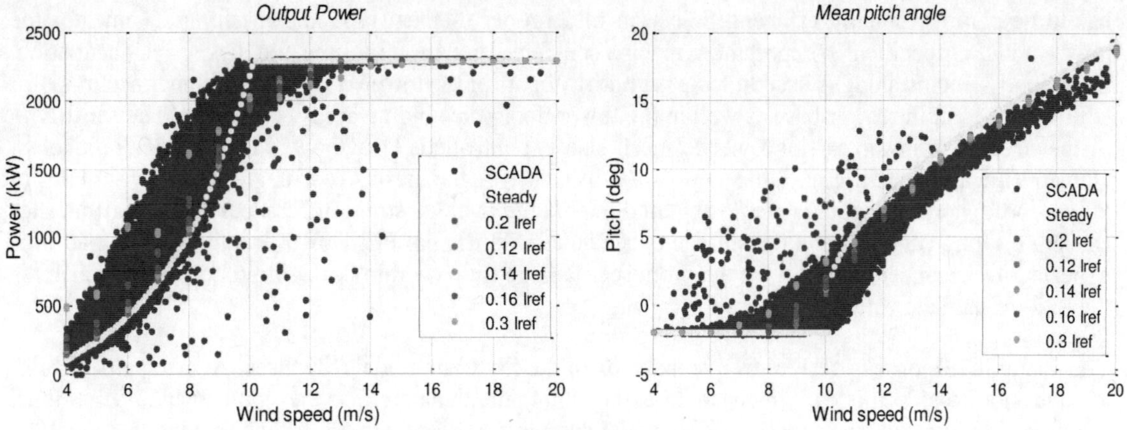

Figure 2. Power production (a) and mean pitch angle (b) against wind speed, comparison between the artificial SCADA of developed base model with SCADA of the actual wind turbine

2.2. *Tower stress analysis*

The wind turbine tower finite element (FE) model is used for tower dynamic response analysis under wind turbine loads, which is then followed by tower stress analysis for fatigue damage estimation. The tower loads are the power production turbine hub loads in different wind speeds, taking into account the tower door opening and wind directionality through rotation of hub load around tower vertical axis. The full-scale FE model of the wind turbine tower with the height of 71m is developed in ANSYS[14], based on the partially available information Figure 3. The door opening and its stiffener flange are modelled according to the detailed dimensions obtained from in situ measurements. There are some uncertainties with regard to tower FE modelling, since precise values for some parameters including rotor and nacelle mass properties, tower wall radius and tower wall thickness variations and actual mechanical properties of tower material are due to data protection policy of the manufacturers unavailable.

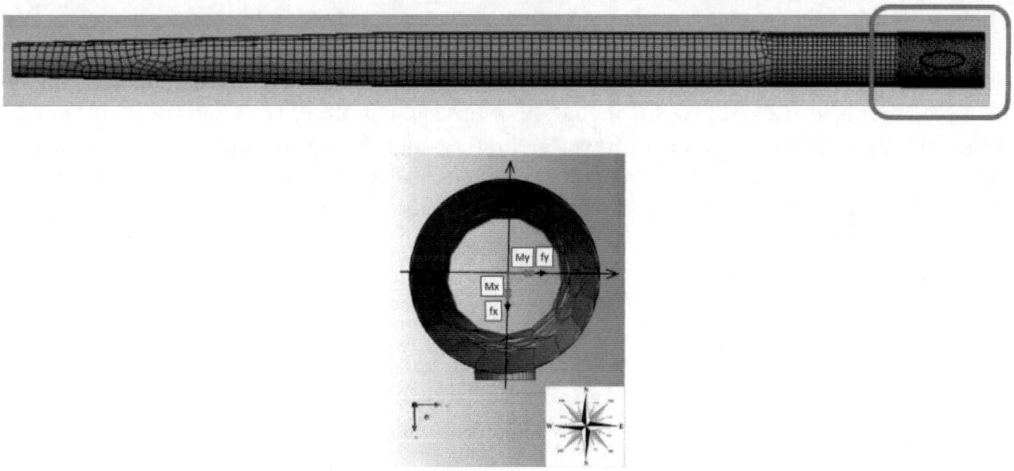

Figure 3. Tower finite element model (above), wind turbine hub loads and analysis coordinate system (below)

The tower FE model consists of about 13500 nodes and 7700 solid-shell elements (SOLSH190). Owing to the substantially large dimensions of the wind turbine model, high sampling rate of dynamic response analysis (25 samples/Sec) and long analysis time-windows (10 minutes), the finite element modelling has to be computationally efficient. Selection of a proper element type and meshing scheme for the tower model is specifically important to reach a sensible simulation runtime and disk space occupation. Sufficiently fine meshing is used in lower cylindrical part of the tower body, where higher accuracy for stress analysis at the tower base as well in the tower door surrounding areas is required. The sensitivity of the stresses in lower part of tower to mesh size is controlled. This showed that for this number of elements the average difference in stress values by reducing the mesh size is negligible. The samples of 20-second dynamic responses were simulated to find the average stress differences. Too small meshes caused too long runtimes and the storage exceedance of a normal PC. However, any approximation is expected to be cancelled out, since the proposed damage ratio (within box 4 in procedure flowchart) is the ratio of damage values.

The dynamic response is derived from expansion of the first four modal responses, i.e. two pairs of fore-aft and side-to-side modes, which capture the dominant dynamics of the wind turbine vibrations, required for fatigue damage analysis. The modal damping ratio is set to 5% for the first mode, to include the additional rotor aerodynamic damping, and 1% for the other three modes. The validity of the tower finite element model against the aeroelastic model is also checked. Figure 4 represents a sample of tower top total displacements derived from aeroelastic and finite element model at 15 m/s mean wind speed. The tower top displacement of the finite element model corresponds to the node on the tower top cross section that undergoes the largest deflection. The displacement time series as well as the power spectrum graph show adequate agreement between two displacements. The power spectrum graph demonstrates that for the fatigue damage analysis purpose, the dominant frequencies match sufficiently well, although two models are from different sources with different levels of complexity and details.

3. Results

For the estimation of the remaining useful life of the tower, the critical stress points that experience the highest fatigue damage have to be identified. According to the linear damage theory, Miner's rule [14], the cumulative damage is linearly proportional to the number of cycles, whereas it is exponentially proportional to stress levels. Consequently, it is required to check the damage at critical stress points, including the stress concentration points around the tower door, in addition to the points on the tower circumference. In order to evaluate the level of damage increase at the points around the door as compared to that at tower based, the concept of damage equivalent loads can similarly be applied to the stress time history to obtain an equivalent stress level with a constant amplitude (damage equivalent stress) [15], using the rainflow counting method with Woehler coefficient of 4. The number of cycles for the calculation of the damage equivalent loads is arbitrarily set to 600 cycles. The damage equivalent stresses are computed in 12 cardinal directions (30° interval) with the mean wind speeds from 4 to 24 m/s in each direction. Note that since the tower section has an axis of symmetry and due to the central symmetricity of the tower section, the damage equivalent stresses can be computed only for six cardinal directions and the Goodman's mean stress correction [14],[15] is conservatively applied where there is the tensile mean stress.

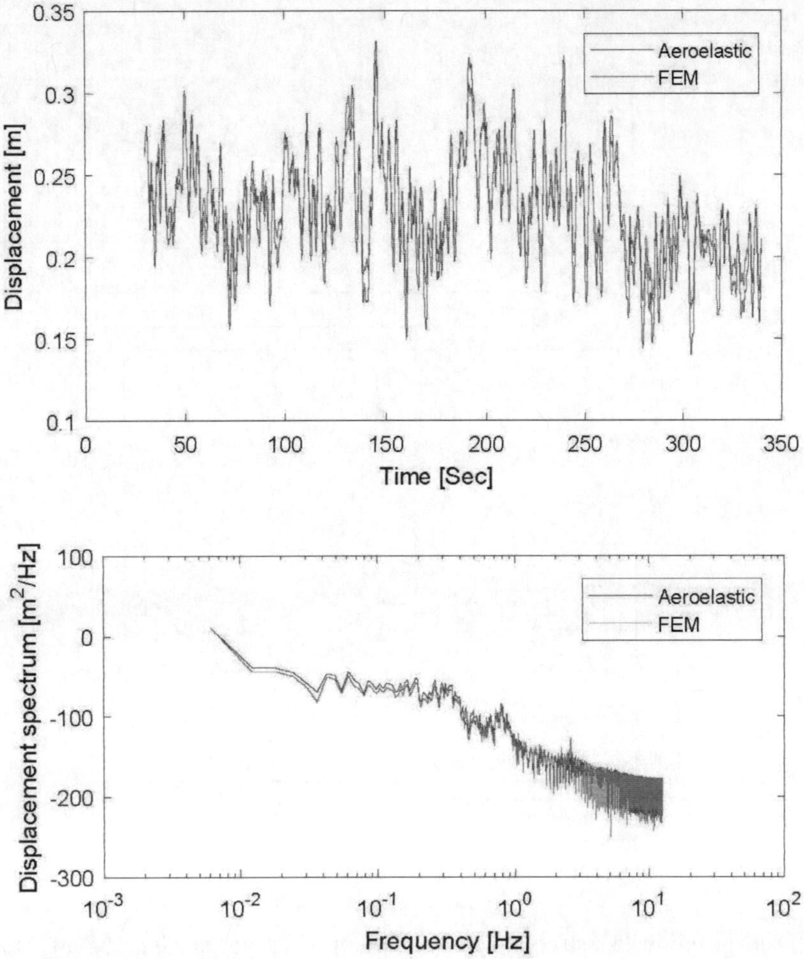

Figure 4. Comparison of aeroelastic model's tower top total displcament with that of the finite element model, time series (above) and power spctrum (below)

Through running a set of trial dynamic response analysis under wind blowing from each of the compass directions, four critical points around the tower door are detected to have stress values higher than that at the tower base within six wind directions. Two of those points are located above (DOA) and under (DOU) the door outside the tower, whereas the other two points lie inside the tower on the left (DIL) and right-hand (DIR) side of the door. Figure 5 represents the 10-minute damage equivalent stress at the critical stress points under southerly wind when the tower door faces south with and without mean stress correction. This figure compares the maximum damage equivalent stresses at tower base and other points around the door.

IOP Conf. Series: Journal of Physics: Conf. Series **1222** (2019) 012034 doi:10.1088/1742-6596/1222/1/012034

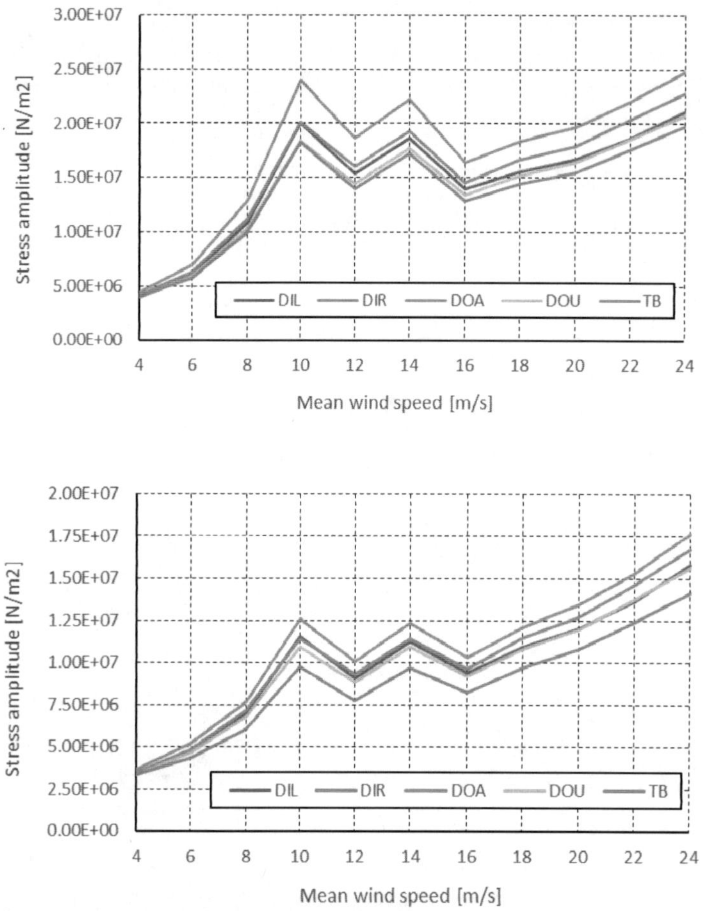

Figure 5. Damage equivalent stresses across different wind speeds with (above) and without (below) mean stress correction

This figure reveals the damage increase due to higher stresses at the points around the door as compared to that at tower base at each wind speed. The damage equivalent stresses are also higher where the mean stress correction is considered. This figure shows that DOA experiences the highest damage, while the lowest damage occurs at tower base. Nevertheless, the location of the point that experiences the highest damage will depend on the wind speed and direction distribution on site. Therefore, by changing the wind direction and the associated hub load on the tower top one can find a good picture of damage variations across different wind directions. Such an analysis helps to stablish the rosette plots of damage equivalent stresses at critical stress points around the tower door and at the tower base. The examples of theses rosette plot at 8 and 18 m/s mean wind speeds are presented in Figure 6. In the same figure, the tower top view with the red dots next to the rosette plots shows the location of the critical stress points, whose associated damage equivalent stresses are plotted. Note that the location of the point at the tower base is situated on the east compass axis on the tower circumference.

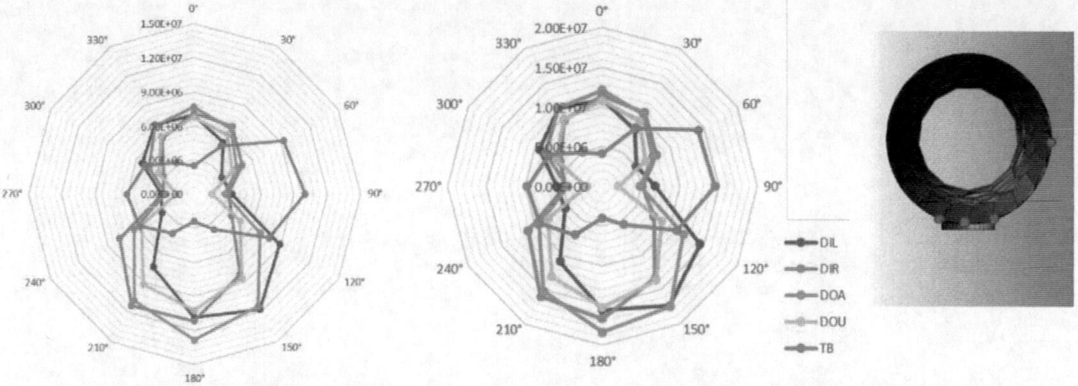

Figure 6. Samples of damage equivalent stress rosette plots at wind speeds of 8 (left) and 18 m/s (middle) together with the tower top view and representation of critical stress points (right)

Figure 7 shows the damage increase due to the stress concentration effect. This figure represents the ratio between damage equivalent stresses at DOA above the tower door opening (under multidirectional wind) and tower base (under unidirectional wind). Since the tower door has a fix orientation, it experiences different levels of damage in different wind direction. The plots with and without mean stress correction demonstrate that the damage at DOA is higher in six and three wind directions, respectively. Consequently, if the prevailing wind direction falls within these six or three directions the damage at DOA will be higher than that of tower base due to the stress concertation effect. Fatigue damage can significantly increase due to the stress concentrations, as it is exponentially proportional to the damage equivalent stress through the Woehler coefficient. Note that in case without mean stress correction the equivalent stress rosette plots becomes central symmetric, i.e. there is no differentiation between compressive or tensile stress.

The damage corresponding to the critical stress points on tower circumference and around the door opening is accumulated based on the procedure flowchart. The point with the highest damage is the driver for tower lifetime determination. An indicator is defined to be able to evaluate the lifetime extension potential (damage indicator). The damage indicator is derived by scaling the damage of the under multidirectional wind (operational condition) to the unidirectional wind along the door normal axis (design reference damage). It should be noted that according to the design guidelines the tower has to be designed for the highest possible damage, and selection of this point for the design reference is due to this fact. Hence, the damage at the point above the tower door under unidirectional wind along the door normal axis, which causes the highest damage above tower door (DOA) is assumed to be the fatigue design basis. The damage indicator indicates the amount of damage that has been spent under operational conditions. This means that the higher the damage indicator is the lower will be the lifetime extension potential.

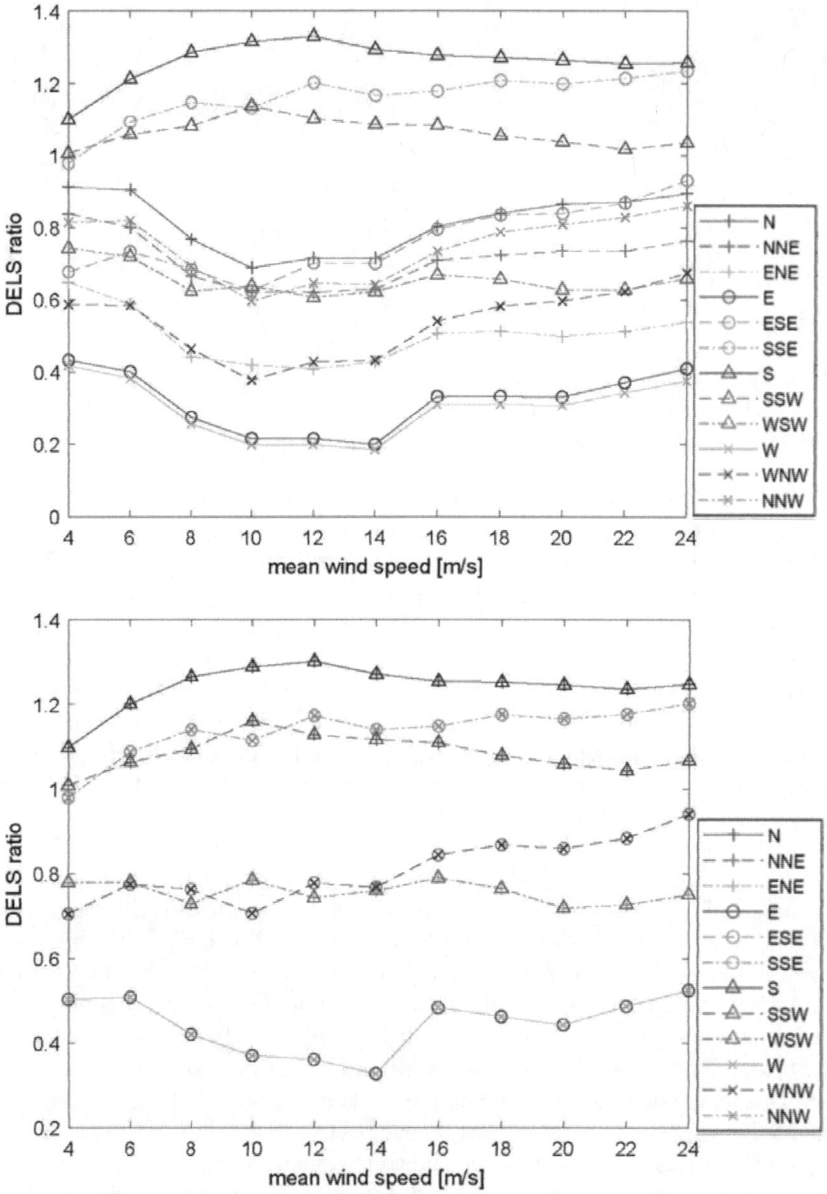

Figure 7. Damage equivalent stress magnification due to stress concentration around tower door opening, with (above) and without (below) mean stress correction

At this stage the proposed methodology for lifetime extension assessment is implemented. Firstly, the implementation results for two wind turbines are inspected in details. These two wind turbines, named T2" and T4 here, are picked from two different locations within an actual wind-farm. The damage indicators are evaluated with and without consideration of the mean stress correction, using the one-year weather conditions history from the turbines' own SCADA data. Inspection of the wind roses of these two turbines in Figure 8 indicates the wind speed properties and prevailing wind direction variations.
The damage indicators for turbine T2 and T4, applying the mean stress correction, are respectively 18% and 9%. These values correspond to critical tensile stresses of the points of TB2 and TB1, which expectedly fall on the axes of the turbines' prevailing wind directions. These values show that the tower has just spent 18% and 9% of its fatigue design capacity. The turbine towers' damage indicators for the case without mean stress correction are 60% and 37%, both relating to DOA. In this case the damage

magnification due to stress concentration around door opening is the driver for lifetime extension. Whether or not the mean stress correction should be considered for lifetime extension evaluation depends on the manufacture's methodology for fatigue analysis. Since the authors are not aware of the methodology used by the turbine manufacturer for fatigue reference design, the life extension potential results, considering both with and without mean stress correction are provided.

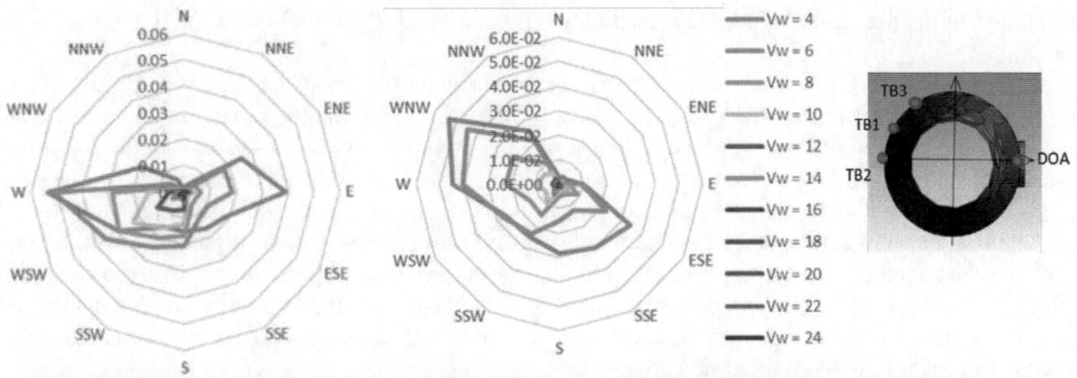

Figure 8. Wind rose plots of turbines T2 (left) and T4 (middle) and location of critical stress points (right) for damage estimation, wind speeds in m/s

In order to evaluate the variations of the lifetime extension potential across the wind farm, the proposed methodology is applied to five more wind turbines within a wind farm. Figure 9 compares the lifetime extension potential of all seven wind turbines. The wind farm has three main arrays and there are 3 rows of turbine in the first array, whereas the rest of arrays consist of two rows of turbine. The wind farm's layout cannot be displayed due to confidentiality issues. The turbines are purposely selected from different arrays and rows to reflect the wind speed properties variations in the results. The location of the stress points with highest damage are already shown in Figure 8. According to Figure 9, turbine T1 and T2 have undergone the highest operational fatigue damage. The wind roses of T1 and T2 are similar and both stretched along the tower door normal axis with a small spread. Turbines T4 and T5 have experienced significantly less damage in comparison to the other turbines, with and without mean stress corrections. Therefore, these two turbines have the highest potential for lifetime extension.

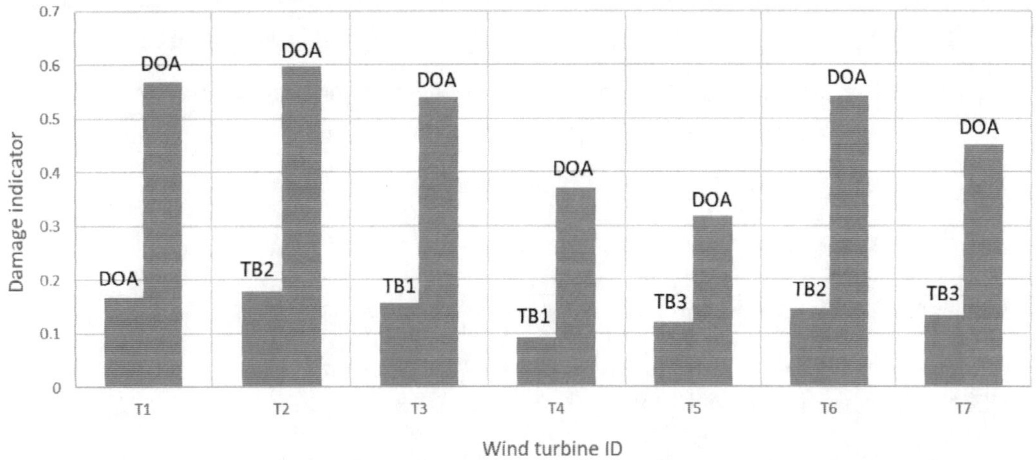

Figure 9. Comparison of lifetime extension potential of different turbines

4. Conclusions

Variation of wind direction implies the fact that the location of the maximum fatigue damage on the tower circumference is not fixed over the wind turbine life. Therefore, the point that undergoes the largest damage over the longest period of time under turbine's operational life will govern the tower lifetime. The shape of the wind rose and its spread will determine the potential for lifetime extension, in contrast to the design fatigue damage analysis that assumes a unidirectional wind direction. The damage equivalent stresses around the tower door are higher than those at tower base. Hence, the relative orientation of the prevailing wind direction and the tower door can also be limiting to the tower fatigue life.

The lifetime extension potential as a consequence of the wind direction dispersion holds true for all conventional large turbines. This is due to the fact that wind directionality distributes the operational loads over multiple points at tower base and around door opening, which causes distribution of fatigue damage as well. According to the results presented here, there is generally the potential for tower life extension under multi-directional wind as opposed to the unidirectional design assumption. However, this potential depends on the spread of the wind rose and also the tower door alignment relative to the wind seen by the individual turbine. Furthermore, the variations of wind properties across the wind farm influence the lifetime extension potential. Through the proposed methodology in this paper, the candidate turbines with the highest potential can be selected for the further detailed analysis and decision making process for the likely lifetime extension.

Acknowledgements

The authors would like to greatly appreciate the support of SSE and SPR Scottish power through the technology and innovation centre, Glasgow, for the work presented here.

References

[1] Ziegler L, Gonzalez E, Rubert T, Smolka U, Melero JJ. Lifetime extension of onshore wind turbines : A review covering Germany ,. Renew. Sustain Energy Rev. 2018;82 (January 2017):1261–71.

[2] Rubert T, Mcmillan D, Niewczas P. A decision support tool to assist with lifetime extension of wind turbines. Renew. Energy. 2018;120:423–33.

[3] Lisa Z, Muskulus M. Fatigue reassessment for lifetime extension of offshore wind monopile substructures.

[4] Bouty C, Sebastian S, Lisa Z, Michael M. Lifetime extension for large offshore wind farms: Is it enough to reassess fatigue for selected design positions? Energy Procedia. 2017;137:523–30.

[5] DNVGL. DNVGL-ST-0262 Standard: Lifetime extension of wind turbines. 2016.

[6] Megavind. Report from Megavind: Useful Lifetime of a Wind Turbine. 2016.

[7] Guidelines for Wind turbines life time extension, Bureau Veritas.

[8] International Electrotechnical Commission. Wind Turbine—Part 1: Design Requirements, IEC 61400-1. Geneva, Switzerland; 2005.

[9] Kazacoks R, Jamieson P. A Generic Evaluation of Loads in Horizontal Axis wind Turbines. Strathclyde Universirt; 2017.

[10] Ashuri T. Beyond Classical Upscaling : Integrated Aeroservoelastic Design and Optimization of Large Offshore Wind Turbines. Delft University of Technology; 2012.

[11] Manwell JF, McGowan JG, Rogers AL. Wind Energy Explained: Theory, Design and Application. Wind Engineering. 2002. vii, 577 p.

[12] Jamieson P. Innovation in Wind Turbine Design. John Wiley & Sons; 2011. 298 p.

[13] Carlen I. Aerodynamic characteristics for the Siemens-2.3-93 rotor. 2005.

[14] ANSYS I. ANSYS 17.1. Canonsburg, PA, USA; 2016.

IOP Conf. Series: Journal of Physics: Conf. Series **1222** (2019) 012035 doi:10.1088/1742-6596/1222/1/012035

Challenges of decommissioning offshore wind farms: Overview of the European experience

Eva Topham[1,2]*, Elena Gonzalez[3], David McMillan[4], Elsa João[5]

[1] DNV GL, Renewables Advisory, Zaragoza, Spain.
[2] Orseide Renewable Energy, Madrid, Spain.
[3] Wind and Marine Energy Systems CDT, University of Strathclyde, Glasgow, UK.
[4] Electronic and Electrical Engineering, University of Strathclyde, Glasgow, UK.
[5] Civil and Environmental Engineering, University of Strathclyde, Glasgow, UK.

E-mail: eva.topham@dnvgl.com

Abstract. In the coming years, an important number of offshore wind turbines will reach the end of their initially planned service life. In a wind turbine end-of-life scenario, owners can decide between extending the life of the asset, repowering the site or decommissioning. This decision- making process is affected by important sources of uncertainty, especially in offshore environments. The limited experience makes the decommissioning procedure challenging, as it is still largely unexplored. This work assesses the current state of knowledge about the challenges surrounding the decommissioning process of an offshore wind farm. The four main challenges encountered are identified and analysed in detail, namely the regulatory framework, the overall planning of the process, the logistics and vessels' availability, and the environmental impacts confronted. Ultimately, this paper aims at stimulating the dialogue among stakeholders and raising the awareness of adequately regulating and preparing the upcoming decommissioning of offshore wind farms in Europe.
Keywords: Offshore wind, End-of-life strategies, Decommissioning, Challenges.

1. Introduction

Since the commissioning of the first offshore wind f arm i n 1 991, the offshore wind industry has successfully evolved, and a total installed capacity of 18.5 GW can be now found in Europe [1]. While the main problems previously encountered were related to the planning and construction [2], the new concerns emerge regarding the ageing of the fleet [3], which are shared with the onshore wind industry. Nevertheless, the required decisions in a wind turbine end-of-life scenario are more critical offshore, due to the natural difficulties and constraints posed by the environment.

Wind turbine end-of-life has recently emerged as a topic of major interest. Almost 30% of the total installed wind turbine capacity in Europe will be older than 15 years by 2020 [3]. Given the wind turbine service life is of 20 years, there is an urgent need to thoroughly prepare for the different end-of-life possible scenarios. By the end of this 20-year service life, wind farm operators need to decide between the extension of the asset lifetime, repowering or decommissioning the site [4]. The two first options may be explored under the right conditions, with respect to the technical, economic, regulatory and environmental aspects. Nevertheless, regardless the final service life of a wind turbine, decommissioning will always happen [5]. Despite the important challenges, this topic is still largely unexplored mainly due to the lack of experience [6].

IOP Conf. Series: Journal of Physics: Conf. Series **1222** (2019) 012035 doi:10.1088/1742-6596/1222/1/012035

The present work assesses the current state of knowledge about the challenges surrounding the decommissioning process of an offshore wind farm. A thorough review of the current state-of-the-art has been conducted, based on the available scientific literature, together with the information documenting the few actual experiences of offshore wind farm decommissioning. As a result, four key challenges have been identified and studied in detail, i.e. the regulatory framework, the planning of the decommissioning procedures, the logistics and vessels' availability, and the environmental impacts involved. Ultimately, this paper aims at stimulating the dialogue among stakeholders and raising the awareness of adequately regulating and preparing the upcoming decommissioning of offshore wind farms in Europe.

The paper is organised as follows. Section 2 describes the research methodology combining literature review and media information documenting the existing experiences of already decommissioned projects in Europe. These experiences are described in detail in Section 3 together with background information on the offshore wind assets' life in Europe. Section 4 presents results of the challenges surrounding the decommissioning of offshore wind farms as an outcome of the review of scientific literature; the four identified challenges are discussed. Conclusions are presented in Section 5.

2. Methodology

As previously stated, the decommissioning of offshore wind farms is still a largely unexplored topic. Consequently, the available scientific contributions contain limited information on the current practice within the offshore wind industry. To overcome this shortcoming, further data was gathered through the documentation of the limited experience of already decommissioned wind farms, which only refer to Europe and are mainly available through press articles.

From the review of all the collected information, a critical analysis of the procedures involved in the decommissioning of offshore wind farms was performed. Additionally, the major aspects that condition the process were assessed together with their interaction. As a result, four main challenges have been identified and covered in detail. Further study of these should lead to the design of adequate decommissioning procedures, such as a guideline elaboration on recommended practices for an appropriate offshore wind farms dismantling.

3. The European experiences of offshore wind farm decommissioning

With a total installed capacity of 18.5 GW, Europe is the world's leader in the offshore wind industry [1]. Nevertheless, the development of wind energy in the offshore sector came later than in the onshore environment. As a result, the offshore wind fleet is still predominantly young. This situation will dramatically change in the coming years, as more and more offshore wind turbines will reach the end of their 20-year design life. The annual number of offshore wind turbines that will reach the end of their planned service life in Europe is illustrated in Figure 1. Between 2020 and 2030, decisions between lifetime extension, repowering or decommissioning will be needed for over 1,800 offshore wind turbines. From 2030 to 2040, almost 20,000 offshore wind turbines will be facing these end-of-life scenarios in Europe.

Although the importance of the required end-of-life decisions for offshore wind turbines might be seen as a future problem, Europe has already witnessed the end of the service life and decommissioning of several projects. A summary of the decommissioned offshore wind projects is presented in Table 1. For the two last listed projects, Blyth and Beatrice Demo, decommissioning has not yet ocurred although it has already been planned as their operation has ceased. Prototype projects, such as Nogersund (Sweden, 0.22MW), Windfloat (Portugal, 2MW) or Hooksiel (Germany, 5MW), have been discarded in this study.

IOP Conf. Series: Journal of Physics: Conf. Series **1222** (2019) 012035 doi:10.1088/1742-6596/1222/1/012035

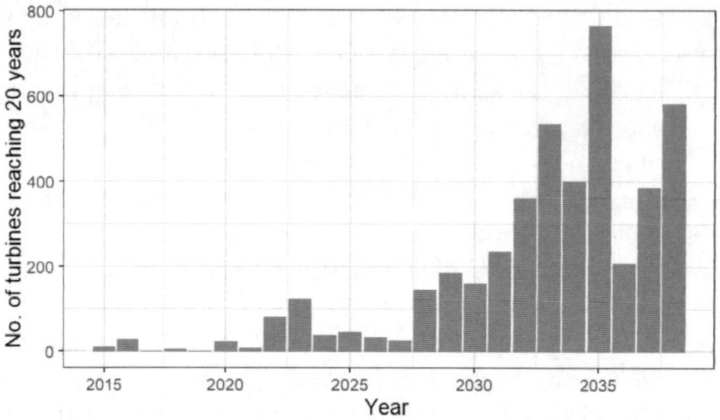

Figure 1. Number of offshore wind turbines reaching 20-years of operation annually in Europe. Source: [7].

Table 1. Offshore wind farms decommissioned to date. Source: [7].

Wind farm	Country	Capacity and nb. of WTs (MW)	Foundation type	Years of operation	Decommissioning year
Yttre Stengrund	Sweden	10 (5 x 2MW)	Monopiles	15 (2001-2015)	2015
Lely	Netherlands	2 (4 x 0.5MW)	Monopiles	20 (1994-2014)	2016
Vindeby	Denmark	4.95 (11 x 0.45MW)	Monopiles	26 (1991-2017)	2017
Utgrunden I	Sweden	10.5 (7 x 1.5MW)	Monopiles	18 (2000-2018)	2018
Blyth	UK	4 (2 x 2MW)	Monopiles	13 (2000-2013)	2019
Beatrice Demo	UK	10 (2 x 5MW)	Jacket	8 (2007-2015)	2024-2027

From Table 1 it can be noted that only two wind farms reached their initially planned service life: the projects Lely and Vindeby [8, 9]. In the case of Vindeby, the lifetime of the asset was even extended until 26 years of operation. For the rest of the cases, decommissioning occurred or is expected to happen before reaching the end of their planned service life. This highlights the high uncertainty related to the wind turbine end-of-life scenario for the offshore environment.

From a more generic point of view, The necessary activities to be performed are detailed in several contributions [5, 10, 11, 12, 13, 14, 15, 16, 17]. In general, the decommissioning of offshore wind projects can be divided into three different phases; firstly, the preliminary work to plan the programme and achieve the required permits; this is then followed by the process itself which corresponds to the removal of the components (turbines, foundations, etc.); and finally the monitoring phase that will check that the site is left as it should [14]. While the older publications [13, 17] offered a theoretical description of the process from a technical point of view, recent contributions have been able to extract conclusions from the comparison to the actual experiences in Europe. This way, Topham and McMillan concluded that the initially planned duration and costs are not adequately captured as part of the planning activities, and recommend considering the decommissioning process from the initial development stage of an offshore wind project [5]. Furthermore, complications can also be the cause for radical changes to the levelised cost of energy and for the owner's business case [12]. Finally, Jensen highlights the uncertainty surrounding the process due to the differences between sites in terms of turbine size, location, foundation type, water depth or distance to shore, where again, planning is found to be crucial for an optimised outcome [11]. It is therefore clear that the complexity of the offshore environment, the higher variability between projects, and the lack of experience makes planning and risk identification essential for the successful decommissioning of offshore wind farms.

4. Challenges encountered in the decommissioning of offshore wind farms

From the few existing experiences, the decommissioning process is found to be affected by high uncertainties and many unexpected challenges. Based on the literature available, the most important challenges have been reviewed and gathered into four main aspects: the regulatory framework, the planning of the decommissioning process, the logistics and vessels' availability, and the environmental impact. The high variability between projects might difficult the preparation of a general guideline for the decommissioning of offshore wind farms. Nevertheless, regrouping the challenges encountered in a set of key aspects to be tackled may ease a draft of recommended practices. The identified major challenges are summarised in Figure 2 and are detailed in the following sections.

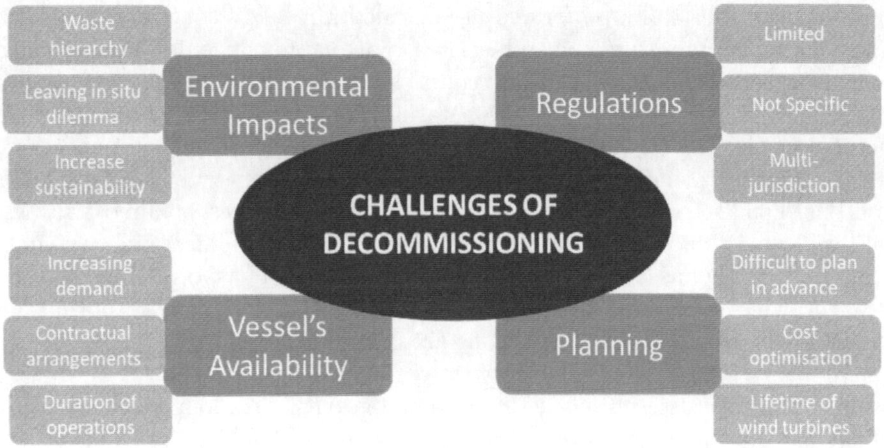

Figure 2. Summary of the major challenges encountered during the decommissioning of an offshore wind farm.

4.1. Regulatory framework

The decommissioning process for the whole offshore industry is currently insufficiently regulated and lacks relevant guidelines for recommended practices. This problem is not unique for the offshore wind industry but does also affect other sectors, such as the oil and gas. Moreover, it has to be noted that this limited and vague current available practices are mainly based on oil and gas projects due to the synergies encountered, and often do not apply to renewable energy projects.

A first overview of the regulations concerning the decommissioning of offshore wind farms can be found in [18]. The authors highlight the importance of guideline elaboration as part of the European Maritime Policy, and recommend the total removal of any offshore installation, unless there are strong reasons not to do so. The obligations of offshore wind farm owners in the UK regarding the decommissioning of the installations are contained in [19]. From these regulations, a decommissioning programme is required in the UK from developers to gain their construction approval. This plan includes a description of the operations that would need to take place together with their cost implications. Nonetheless, the detailed programmes are found to be rather simplistic and the costs underestimated [5]. Both aspects may lead to lower rates of return and to non-sustainable decommissioning solutions. Additionally, no prescription is indicated on what needs to be removed; the decision between partial or total removal would be made based only on economic reasons. Kaiser and Snyder elaborate on the bonding requirements for offshore wind farms and on the owner's liability in [17, 20], and also highlight the lack of specific regulations.

It can be easily concluded that the decommissioning phase of offshore wind farms currently lack a specific regulatory framework. Apart from limited existing regulations, the polluter pays' principle is the only strong guiding principle, where any damage done to the environment will need to be remediated by the owner [5]. Thus, the elaboration of specific guidelines is strongly recommended including precise liabilities for the owners.

4.2. Process planning

As previously mentioned, the decommissioning procedures should be planned within the design phase of a project. The process management should allow owners to answer what needs to be done and its duration. Implicitly, decommissioning planning is strongly related to costs.

The necessary activities to be performed are detailed in several contributions [5, 10, 11, 12, 13, 17, 21, 22]. From the literature, it can be noted that planning the decommissioning phase is far from being easy as there is no a unique option. The decisions are determined by the project's characteristics, such as the distance to the operating ports, the water depth, the turbine size or the type of foundation. Nevertheless, other aspects such as duration of the working time offshore, the environmental impacts and the total dismantling costs will also impact the decision significantly. Based on [21], the selection of a specific dismantling method almost always will lead to technical problems.. The number of planned vessels might also affect the cost and duration of the decommissioning phase [22], as well as other unplanned aspects such as the weather. Additionally, the wide spectrum of decommissioning possibilities includes options for full or partial removal of the asset. As no regulations are currently applicable, the decision is mainly driven by economic reasons.

The options for the electric infrastructure and the gravity base foundations are also largely unexplored. While wind turbines have a designed life time of 20-25 years, cables could last for 50 years [23] and gravity foundations for 100 years [12], although there is little experience to confirm these numbers. The importance of this part of the infrastructure may even affect the operation and decommissioning of the whole asset. For instance, the Nysted project that was commissioned in Denmark in 2003, considered an estimated life for both the foundations and the electrical system of 50 years [24]. This is a real example of adequate planning, since for this project the seabed can be leased for twice as long the usual (40 or 50 years instead of 20 to 25) [25]. This approach is currently observed in many decommissioning programmes in the UK.

From an economic point of view, several life cycle models can be found in the literature [26, 27]. Both highlight the high uncertainty affecting the planning and cost estimation due to the lack of experience and real data, and hence, the impossibility of validating assumptions and models. Indeed, based on the few experiences, the process duration and costs do not seem to be effectively captured in planning activities [5]. This uncertainty is slightly considered in [26] by following a stochastic approach for the lifecycle model, including several sources of uncertainty such as the vessels or technician's availability.

Another important aspect is the moment when the decommissioning process will happen. Since its planning should be made a decade in advance, substantial changes regarding the technical feasibility may arise within that timeframe. On the one hand, decommissioning might occur earlier than planned as a consequence of the lack of spare parts required for continuing operation. This was the case at the Lely wind farm, that had to be decommissioned since repowering was discarded due technical obsolescence [9]. Similarly, site dismantling can also happen later than expected if lifetime extension is found to be profitable and technically feasible by the end of the 20-year planned service life [3]. On the other hand, the technical feasibility of the decommissioning process may also be affected by time. Certainly, the equipment, technologies

IOP Conf. Series: Journal of Physics: Conf. Series **1222** (2019) 012035 doi:10.1088/1742-6596/1222/1/012035

or procedures that were initially planned can become outdated as the time to proceed arrives; conversely, new techniques can also arise and hence ease the general process. As a result, the initial plan for the decommissioning process should be reviewed every few years to keep it technically updated and to include any modifications made to the project that could affect its implementation [16].

Finally, some recent attempts have been made to plan and improve the whole process. An optimisation model is proposed in [10]. The model aims to enhance the decommissioning scheduling in order to minimise the costs, considering inputs such as the tasks to be performed, the vessels' availability, duration or the required expenses. Similarly, a framework for identifying the most appropriate decommissioning method for a given offshore wind farm is presented in [28]. These recent works set out the basis for further work to facilitate the decommissioning planning for practitioners in the field.

4.3. Vessels' availability

Specialised vessels with heavy lifting and specific stability characteristics are required for the performance of the decommissioning operations. However, the vessels also need to adapt to the site conditions. The number of turbines, the foundation type, the water depth, the distance to the operating ports and the seabed type need to be considered, as not all the vessels work under the same conditions nor have the same speed. Even though there are numerous vessels that can be chartered [20], their availability can be compromised due to the demand forecasting of new offshore installations expected in the upcoming years [29], operation and maintenance procedures within already operational projects, and the decommissioning oil and gas facilities as a consequence of the sector synergies found. To avoid this potential bottleneck, contractual arrangements must be ensured largely in advance. Recent contributions also suggest the use of new vessels able to handle very large platforms where the inherent risk of cranes will be removed [30]. Nevertheless, this will not eliminate the competition between installation and decommissioning activities.

Concerning the costs, the vessels' availability accounts for a very large part of the total decommissioning costs. In general, the deck capacity, the working versatility, or the speed will determine the price. To reduce extra costs, the duration of the operations must be meticulously studied. Vessel's operations are not only impacted from high daily rates but also depend on important sources of uncertainty, such as the equipment used, the weather, and the market. Therefore, a best practice would involve a strategy with the least amount of offshore lifting and operating time, leading to reduced costs and risks. This practice is supported in [22], where the total duration and consequently costs, can be reduced by a 6% only by adapting the number of vessels.

4.4. Environmental impact

One of the most controversial topics around the decommissioning of an offshore wind farm is the impact in the marine environment, as detailed in [31]. Two important aspects need to be considered in this regard.

First, the big debate between the need of a total or partial removal. While there seems to exist the belief that leaving the site practically as it was before the project was implemented, recent contributions emphasise the environmental benefits of only partially dismantling an offshore wind farm. Concerning the electric infrastructure, the subsea cables are usually buried between 1 to 2 meters [32], and therefore could be left in situ. Their complete removal would require excavation and pulling out of the trenches and given their extensive length, this would produce an important marine disruption, as well as notable costs. Similar arguments are found for the substructures. A 'renewables-to-reefs' programme is proposed in [13], where the environmental and economic benefits of a partial removal are shown, as opposed to complete removal, especially if the habitat created on the left structures has conservation or commercial value. Substructures could definitely become habitats for marine wildlife, such as fish or crustaceans, as presented in [33]. Oudman also suggests partial removal as more beneficial than complete, and strengthens the idea of turning already existing jackets in artificial reefs, as an example of a green decommissioning. Similar conclusions are provided in [34, 35] where biodiversity enhancement, provision of reef habitats, and protection from bottom trawling are negatively affected by the complete removal of offshore wind farms. All the contributions also highlight the strong link between the regulatory framework and the environmental impacts.

Secondly, to perform a decommissioning strategy as sustainable as possible, this should involve reusing (if possible) or recycling measures. Wind turbines are mainly fabricated with metals, so about 95% can be recycled (mainly from the tower, the gearbox, the main shaft, the generator, castings, bearings and parts from the nacelle and hub) [36]. While there is extensive experience in its recycling, an optimization of the transportation to the recycling plants could enable a cost reduction [36]. The remaining 5% involves power electronics, lubricant and cooling substances, and polymers, which mainly come from the blades and are currently non-recyclable [37]. Blades are certainly the biggest challenge encountered within its material recyclability and transportation logistics perspectives [36]. Any interested readers in the topic are referred to [11, 36] for more detailed reviews of recycling offshore wind farms.

Finally, as concerns the evolution of offshore wind turbines recyclability, it is likely to be negatively affected by the growing size of wind turbines. Indeed, the amount of raw materials required for two small wind turbines is found to be lower than a bigger turbine, equivalent in terms of rated capacity [36]. As a result, the current trend towards installing larger offshore wind turbines will imply a much more important use of raw materials, compromising the sustainability of the subsequent decommissioning process.

5. Conclusions

This paper explores the state of the knowledge regarding the decommissioning process of offshore wind farms in Europe. From the limited existing experiences, it is clear that wind turbine end-of-life scenarios and particularly, decommissioning, are affected by high uncertainties. In addition, the complexity of the offshore environment, the high variability between projects, and lack of experience makes planning and risk identification essential for the successful decommissioning of these projects. As a result, the major challenges encountered during this process are identified and individually analysed, namely the regulations, the planning of the process, the vessels' availability and the environmental impacts.

The decommissioning phase of offshore wind farms currently lack a specific regulatory framework. This negatively affects concepts such as liabilities, but strongly impacts other fields like the planning of the process and the resulting environmental impacts. Offshore wind farm owners lack specific guidelines that could help ease their decommissioning planning, that should ideally be included as part of the development phase. With regards to the environmental impacts, the absence or limited dedicated maritime regulations may lead to decisions that can harmfully disturb the marine environment. As a result, the elaboration of specific regulations and guidelines is strongly recommended including precise liabilities for the owners. Concerning the planning, the ideal guidelines should help the owners with the decision about what needs to be done, estimating the decommissioning costs and ensuring that potential technical modifications arisen over time are adequately monitored and considered in the initial plan. The availability of appropriate vessels is also found as a potential bottleneck; decommissioning programmes might compete with new installation activities, operation and maintenance procedures or decommissioning oil and gas facilities, which all require similar vessel characteristics. Again, anticipation is key to ensure adherence to planned duration and costs, together with reduced risks. Finally, the environmental impacts of a partial removal of the asset together with a proper recycling strategy could be the ideal combination for achieving a sustainable decommissioning. Approaches for blade recycling could as well diminish the wind energy carbon footprint and make the process even more sustainable.

Further research should aim at tackling the challenges identified. The assessment of the uncertainty affecting the whole decommissioning process should also become a priority, to allow a proper scheduling, to minimise expenditures and to mitigate risks. Certainly, the decommissioning phase, such as other wind turbine end-of-life scenarios, is affected by important sources of uncertainty. Nevertheless, the complexity of the offshore environment, the diversity of the site characteristics and the limited experience compared to onshore, makes the decommissioning of offshore wind farms more critical. Furthermore, possibilities for blade recycling need to be explored to achieve more sustainable alternatives.

Acknowledgements
This work was supported by DNV GL Renewables Advisory (Zaragoza, Spain) and the Engineering and Physical Sciences Research Council (EPSRC) with grant number EP/L016680/1.

References

[1] Wind Europe, "Offshore wind in Europe. Key trends and statistics 2018," tech. rep., 2019.
[2] F. Olsen and K. Dyre, "Vindeby Off-Shore Wind Farm - Construction and Operation," *Wind Engineering*, vol. 17, no. 3, pp. 120–128, 1993.
[3] L. Ziegler, E. Gonzalez, T. Rubert, U. Smolka, and J. J. Melero, "Lifetime extension of onshore wind turbines: A review covering Germany, Spain, Denmark, and the UK," *Renewable and Sustainable Energy Reviews*, vol. 82, pp. 1261–1271, feb 2018.
[4] M. Luengo and A. Kolios, "Failure Mode Identification and End of Life Scenarios of Offshore Wind Turbines: A Review," *Energies*, vol. 8, pp. 8339–8354, aug 2015.
[5] E. Topham and D. McMillan, "Sustainable decommissioning of an offshore wind farm," *Renewable Energy*, vol. 102, pp. 470–480, mar 2017.
[6] K. Ortegon, L. F. Nies, and J. W. Sutherland, "Preparing for end of service life of wind turbines," *Journal of Cleaner Production*, vol. 39, pp. 191–199, jan 2013.
[7] 4C Offshore, "4C Offshore Database."
[8] D. Weston, "Dong announces Vindeby decommissioning," *Windpower Offshore*, 2016.
[9] J. Dodd, "Decommissioning - Should they stay or should they go?," *Windpower Monthly*, 2015.
[10] C. A. Irawan, G. Wall, and D. Jones, "An optimisation model for scheduling the decommissioning of an offshore wind farm," *OR Spectrum*, feb 2019.
[11] J. P. Jensen, "Evaluating the environmental impacts of recycling wind turbines," *Wind Energy*, vol. 22, pp. 316–326, feb 2019.
[12] P. Hou, P. Enevoldsen, W. Hu, C. Chen, and Z. Chen, "Offshore wind farm repowering optimization," *Applied Energy*, vol. 208, pp. 834–844, dec 2017.
[13] K. Smyth, N. Christie, D. Burdon, J. P. Atkins, R. Barnes, and M. Elliott, "Renewables-to-reefs? – Decommissioning options for the offshore wind power industry," *Marine Pollution Bulletin*, vol. 90, pp. 247–258, jan 2015.
[14] G. Smith and G. Lamont, "Decommissioning of Offshore Wind Installations - What we can Learn - What we can Learn," in *Offshore Wind Energy 2017*, 2017.
[15] G. Smith, D. Michael, R. Paul, and W. Arlene, "Assessment of Offshore Wind Farm Decommissioning Requirements," Tech. Rep. May, Ontario Ministry of the Environment and Climate Change, 2016.
[16] G. Smith, C. Garrett, and G. Gibberd, "Logistics and Cost Reduction of Decommissioning Offshore Wind Farms," in *EWEA Offshore 2015*, 2015.
[17] M. J. Kaiser and B. Snyder, "Offshore wind decommissioning regulations and workflows in the Outer Continental Shelf United States," *Marine Policy*, vol. 36, pp. 113–121, jan 2012.
[18] C. Januário, S. Semino, and M. Bell, "Offshore Windfarm Decommissioning : A proposal for guidelines to be included in the European Maritime Policy," in *EWEC 2007*, no. April 2007, (Milan), 2007.
[19] Department of Energy & Climate Change - British Government, "Decommissioning of offshore renewable energy installations under the Energy Act 2004," Tech. Rep. January, 2011.
[20] M. J. Kaiser and B. F. Snyder, *Offshore Wind Energy Cost Modeling*. Green Energy and Technology, London: Springer London, 2012.

[21] N. Hinzmann, P. Stein, and J. Gattermann, "Decommissioning of Offshore Monopiles, Occuring Problems and Alternative Solutions," in *Volume 9: Offshore Geotechnics; Honoring Symposium for Professor Bernard Molin on Marine and Offshore Hydrodynamics*, p. V009T10A020, ASME, jun 2018.

[22] T. Beinke, A. Ait Alla, and M. Freitag, "Decommissioning of Offshore Wind Farms: A Simulation-Based Study of Economic Aspects," in *Dynamics in Logistics* (M. Frietag, H. Kotzab, and J. Pannek, eds.), pp. 216–222, Springer, 2018.

[23] C. Birkeland, *Assessing the life cycle environmental impacts of offshore wind power generation and power transmission in the North Sea. MSc.* PhD thesis, 2011.

[24] S. Properzi and H. Herk-Hansen, "Life cycle assessment of a 150 MW offshore wind turbine farm at Nysted/Roedsand, Denmark," *International Journal of Environment and Sustainable Development*, vol. 1, no. 2, p. 113, 2005.

[25] E. Lantz, M. Leventhal, and I. Baring-gould, "Wind Power Project Repowering : Financial Feasibility, Decision Drivers, and Supply Chain Effects," *National Renewable Energy Laboratory (NREL)*, no. December, 2013.

[26] F. Judge, F. D. McAuliffe, I. B. Sperstad, R. Chester, B. Flannery, K. Lynch, and J. Murphy, "A lifecycle financial analysis model for offshore wind farms," *Renewable and Sustainable Energy Reviews*, vol. 103, pp. 370–383, apr 2019.

[27] M. Shafiee, F. Brennan, and I. A. Espinosa, "A parametric whole life cost model for offshore wind farms," *The International Journal of Life Cycle Assessment*, vol. 21, pp. 961–975, jul 2016.

[28] H. Kerkvliet and H. Polatidis, "Offshore wind farms' decommissioning: a semi quantitative Multi-Criteria Decision Aid framework," *Sustainable Energy Technologies and Assessments*, vol. 18, pp. 69–79, dec 2016.

[29] D. Mcmillan and I. Dinwoodie, "Forecasting Long Term Jack up Vessel Demand for Offshore Wind," in *European Safety and Reliability Conference 2013*, 2013.

[30] A. Choudhary and M. Howes-roberts, "Crane-Less, Diver-Less Platform Decommissioning Vessel to Reduce Cost and Enhance Safety and Efficiency," in *SPE Symposium: Decommissioning and Abandonment*, (Kuala Lumpur, Malaysia), 2018.

[31] J. Kaldellis, D. Apostolou, M. Kapsali, and E. Kondili, "Environmental and social footprint of offshore wind energy. Comparison with onshore counterpart," *Renewable Energy*, vol. 92, pp. 543–556, jul 2016.

[32] Statoil, "Sheringham Offshore Wind Farm Decommissioning Programme," tech. rep., 2014.

[33] V. Gartman, L. Bulling, M. Dahmen, G. Geißler, and J. Köppel, "Mitigation Measures for Wildlife in Wind Energy Development, Consolidating the State of Knowledge — Part 2: Operation, Decommissioning," *Journal of Environmental Assessment Policy and Management*, vol. 18, p. 1650014, sep 2016.

[34] A. M. Fowler, A.-M. Jørgensen, J. C. Svendsen, P. I. Macreadie, D. O. Jones, A. R. Boon, D. J. Booth, R. Brabant, E. Callahan, J. T. Claisse, T. G. Dahlgren, S. Degraer, Q. R. Dokken, A. B. Gill, D. G. Johns, R. J. Leewis, H. J. Lindeboom, O. Linden, R. May, A. J. Murk, G. Ottersen, D. M. Schroeder, S. M. Shastri, J. Teilmann, V. Todd, G. Van Hoey, J. Vanaverbeke, and J. W. Coolen, "Environmental benefits of leaving offshore infrastructure in the ocean," *Frontiers in Ecology and the Environment*, vol. 16, pp. 571–578, dec 2018.

[35] S. Salvador, L. Gimeno, and F. J. Sanz Larruga, "The influence of regulatory framework on environmental impact assessment in the development of offshore wind farms in Spain: Issues, challenges and solutions," *Ocean & Coastal Management*, vol. 161, pp. 165–176, jul 2018.

[36] E. Topham, D. McMillan, S. Bradley, and E. Hart, "Recycling offshore wind farms at decommissioning stage," *Energy Policy*, vol. 129, pp. 698–709, jun 2019.

[37] E. Gonzalez, A. Ortego, E. Topham, and A. Valero, "Is the future development of wind energy compromised by the availability of raw materials?," *Journal of Physics: Conference Series*, vol. 1102, p. 012028, oct 2018.

The Portuguese Offshore Energy SWOT Analysis

Ana Cristina Pego

CICS Nova- Universidade Nova de Lisboa
Avenida de Berna, nº 36 C, 1069-061 Lisboa

anapego@campus.fcsh.unl.pt

Abstract. Renewable energy has gained importance in recent years because of its cooperation in ecosystems, welfare and the circular economy. The new business model comprises new values for organizational sustainability. Moreover, renewable energy such as offshore energy is associated with innovation systems, knowledge and collaboration between organizations. The participation of the offshore energy sector has increased in the Portuguese renewable energy supply. It is known that renewable offshore energy suppliers will cover economic necessities. One of the strategies was to set up a new business model based on supplier organization networking for the construction of an offshore wind farm. This will contribute positively to lower fuel dependency, fewer imports, control of energy prices and market regulation in terms of supply/demand. The methodology used was an electronic questionnaire applied to emergent offshore energy cluster stakeholders. The SWOT analysis of the offshore energy sector contributes to an understanding of the offshore energy sector in Portugal and reveals the business model with which the stakeholders are faced. Moreover, it draws up a business strategy oriented towards competitiveness and cooperation with other energy sectors.

1. Introduction

The RES (renewable energy sector) and in particular the offshore energy sector in Europe and in the Portuguese economy has taken on new importance due to the new European strategy for the energy sector.

The RES has particular importance due to the interest in its implementation and discussion about its links to the pentagonal problem discussed by [1] showed that the RES is responsible for a new concept of energy production. As such, the offshore energy sector and its clustering have begun to build a space in economic development and create a positive flux in the economy. This implies a new assumption that the business model adopted by organizations should change towards greater collaboration and competitiveness with other energy sectors. This article shows the importance of the offshore energy sector in Portugal, and presents a new market strategy based on SWOT cluster analysis.

The article is divided into three main parts. Firstly, it explains the Portuguese maritime cluster, secondly the offshore energy cluster and thirdly, it demonstrates the business model for the offshore energy cluster based on stakeholder opinions. The management model and its applicability to the offshore energy sector can be achieved if there is also vertical cooperation in the value chain, like the participation of other institutions towards a *profit* network. Innovation, investment, human skills and production need to be linked towards a better decision-making by managers. This networking is the basis for a business model with competitiveness and cooperation at regional, national or European levels.

2. The case of the Portuguese Maritime Cluster

[2:21-23] defines the maritime economy as: "all the sectorial and cross-sectorial economic activities related to the ocean, sea, and coast. This includes the closest direct and indirect supporting activities necessary for the functioning of these economic sectors, which can be located anywhere, including in landlocked countries".

The concept of ocean economy presented by [3], cited in [2], refers to "economic activities that take place in the ocean, receive outputs from the ocean, and provide goods and services to the ocean". The marine economy is related to emerging ocean industries and established ocean-based industries. Maritime sectors comprise activities linked to the sea; the link between activities and the sea may be explained by the use of marine resources, marine areas or by the vicinity of these areas. These linkages between sectors can follow the cluster concept when they have some direct or indirect economic flux ([4];[5]). According to [6], maritime clusters can derive benefit from knowledge sharing and knowledge transfer from research, encouraging joint innovation exercises (e.g. product development), and ensuring availability of know-how (e.g.joint training programmes) or innovative organisation methods covering a group of enterprises (e.g. common procurement or distribution).

The marine cluster project was launched by the Marche region in October 2004 with the goal of establishing a new, technically specialised yacht and shipbuilding system in Ancona. the initiative promoted a system of goods and services related to navigation, namely shipbuilding, pleasure craft, accessories and infrastructures for tourism and commerce and was horizontally connected to the other main regional sectors with the aim of further integrating the different clusters [7]. By definition, all maritime cluster organizations capture more than one maritime sector. Maritime cluster organizations represent almost every traditional maritime sector; although in practice this does not (yet) always seem the case– except the sectors that only answer a broader definition of maritime sectors, such as the Navy and coastguard, inland navigation and maritime works.

The fisheries and coastal (and marine) tourism and recreation sectors are sometimes represented by the national cluster organization, although this is less frequent than the traditional maritime sectors [8]. [9] argue that studies on a maritime cluster or other clusters are mainly concerned with the relationships between ports, the shipping industry, or services from the shipbuilding industry.

[10] argue that Michel Porter´s theories' can be applied to the marine sector, as the Norwegian maritime cluster presents the majority of characteristics that one can find in large industrial groups, including strong inter-sectoral linkages, economic diversity and competitive rivalry. [7] presents maritime clusters as relationships between competences, entrepreneurship and experiences, both inter- and intra-industry. This means that the marine cluster is related to organizations and services which comprise intra- and inter-sector relationships.

In particular, [7] identifies those sectors which have more impact (direct or indirect) on the marine economy. From [7] point of view, the engines and propulsion systems, tourism, services and manufacturing constitute indirect impacts on the marine cluster with the other services being the direct impact. [6] identifies the relationship between different levels of clusters (regions, Member State, European Union, international organization); these relationships are based on similar sectors and activities, such as transport, resources, public sector, other services, research and leisure and tourism. The dynamic between them comes from the interaction between their activities from different clusters. The positive aspect is complementarity and cooperation of goods and services between all levels of clusters.

Maritime cluster organizations exist at different geographical levels, focusing on international and/or European, national, regional (incl. cross-border) or local levels. Policy (initiatives) and actions are consequently translated to European national and regional levels, although not in a uniform manner [8]. The "*Cluster do Mar Português / Portuguese Maritime Cluster*",[1] is promoted by FórumOceano – Associação da Economia do Mar, and was recognized as a Competitiveness Cluster Strategy by the Ministers of Economy, Maritime, Planning and Infrastructures, and National Defence. Fórum Oceano is the merger by incorporation of two Portuguese associations, Oceano XXI – Association for the Knowledge and Economy of the Sea (established in 2009) and AFEM – Association Business Forum of the Sea Economy (established in 2010), which occurred in July 2015.

[1]http://www.forumoceano.pt

VISION– Maritime activities are dynamic and competitive segments of the economy, enhancing the sea and its resources, generating value, promoting high-value jobs, and contributing to strengthening Portugal's position in the global economy. *MISSION–*To strength the dynamics of strategic cooperation between players; to promote the competitiveness of the main value chains that utilize the sea and marine resources as the central elements of its activity; to contribute, in a sustainable way, to economic growth and the growth of exports and employment; to increase the maritime economy's importance to the national economy.

The association has more than 100 members connected to many sectors of the sea economy, including *traditional* ones, such as: seafood conservation, processing and commercialization; shipbuilding, repair and maintenance; nautical activities and marine tourism; maritime construction; fisheries and aquaculture; ports, transport and logistics, *or complementary* areas, like: maritime culture; maritime defense and security; services; or *emerging* ones, such as: marine bio resources and biotechnology; marine renewable energy; and offshore and oil &gas. The Portuguese sea cluster has three main objectives: to increase the value added and the consolidated business from the sectors which comprise the cluster; to promote emergent activities and to increase international market networks and investments. These objectives are based on entrepreneurship, innovation and modernization, sea technologies and knowledge, internationalization and strategic information.

3- The offshore energy sector in Portugal

Wind and wave offshore energy are the future of energy supply in Portugal. The key challenge of offshore suppliers is to compete with other fossil energy production. Offshore energy is directly related to expert employment, technology and direct investment in electric equipment, dry docks, maritime transport support, engineering and consultancy studies and expert knowledge.

There are a few aspects with which this sector deals in a negative way – the spatial needs for energy sectors –which means that the energy and economic infrastructures do not meet regional needs. According to [11], exploration, fixed installation, and decommissioning are the fragile points with which countries need to deal concerning environmental and use impacts. In order to achieve the target proposed by the European Union in the 2020 programme, every country has to design different strategies to accomplish the general goal. The development of renewable energy has an outstanding effect on sustainable economic growth, for the harmonious coexistence of humans and the environment as well as for sustainable development. Renewable energy usually refers to that energy that does not pollute the environment and could be recycled in nature [12].

Offshore renewable energy faces similar changes in most countries, such as the location of offshore wind farms, port facilities, underwater cables, operation and maintenance and decommissioning projects. On the other hand, the environmental impact and spatial competition between energy and other maritime activities should be analyzed. The environmental impacts are normally related to noise, infrastructure, and decommissioning projects, this means that the impact on sea species and birds, as well as, oil spill should be taken in account by stakeholders. Moreover, benefits from energy sector on economy (spatial synergies from wave-tidal-wind; certainty and consistency (planning certainly from stakeholders towards investments in this sector); grid development and grid interconnection within other sectors; cross border cooperation in knowledge, skills, development and experience towards a market corridors for regional energy market integration; data collection from energy sector) should be analysed. The objective is to reduce the cost of installation for each type of energy and guarantee a better performance, in technology innovation and learning by doing must be translated into a comprehensive cost-reduction pathway if wave and tidal energy technologies are to achieve cost competitiveness on commercial markets [13].

From the following tables, it is possible to conclude about the use of RES in Europe and Portugal. Few factors can be pointed out to explain the data, the demand, the green economy, and the ecosystems. The data shows that the European market has increased its consumption and primary energy uses. Clearly, there is a possibility to change the market energy supply, whereas the use of RES will be in majority of energy. This feature also explains why a different model should be adopted by organizations. The market is changing, as well as attitudes about the environment, green economy, circular economy and welfare.

Table 1–EU Renewable Gross Electricity Production (wind and tidal, wave and ocean), 2018, 1990-2013 (GWH)

	1990	**1995**	**2000**	**2005**	**2011**	**2012**	**2013**	**Growth rate (1990/2013)**
Wind	778	4068	22254	70455	179669	205980	235012	301.1%
Tide, wave and ocean	503	507	507	481	478	462	420	-16.5%

Source: [14]

Table 2–Share of Renewable Sources in Gross Final Consumption of Energy (%)

	2011	**2012**	**2013**	**TARGET 2020**
EU-28	12.9	14.3	15.0	20
Portugal	24.7	25.0	25.7	31

Source: [14]

Table 3–Renewable Sources as a Percentage of Total Primary Energy Supply

	1970	**1990**	**2010**	**2014**
EU-28	21.6%	19.5%	23.2%	24.6%
Portugal	----	4.3%	9.8%	-----
OECD	-----	5.9%	7.8%	9.2%

Source: [2]

Moreover, over the course of recent years, Portugal has created a series of financial and fiscal measures to support investment in renewable energy. These measures have been boosted with the creation of differentiated tariffs for electricity produced in renewable plants, feed-in tariffs (FIT), according to the degree of maturity of the various technologies available in the national market.

These measures have helped to achieve the overall objectives of the national energy policy. Renewable energy has become increasingly important and visible in the national strategy. The latest National Energy Strategy (NES 2020)[2] continues to attribute a pivotal role to renewable energy and the targets that have been delineated for this sector, with a very significant impact on the Portuguese economy.NES 2020 incorporates the objectives of the energy policy and seeks to maintain Portugal at the fore front of the energy revolution with regard to the use of renewable energy (increase of installed hydro and wind energy capacity). Wave energy will be very important in the future: specific instruments have been envisaged to promote it and to develop the necessary technologies. Amongst

[2]Approved by Government Resolution No. 29/2010, of 15 April

these measures, it is especially important to note the implementation of the pilot zone created by Decree-Law No. 5/2008 to test technologies that harness wave energy.

4- Uploading a Portuguese Offshore Energy SWOT analysis

The methodology used to discuss a business model for Portuguese offshore energy was a SWOT analysis. An electronic questionnaire was applied to 14 organizations which belong to the Portuguese offshore energy sector (NGO, government, research centers, and suppliers). It tested opinions about the future of the offshore energy cluster in Portugal and how future organisations are involved in the cluster. The questionnaire was sent to suppliers, government entities and others which can influence positively the potential energy cluster in Portugal. The total of inquires was 14 organisations, divided by private organisations (60%), universities (13%), research centres, public organisations, the government and NGOs (7%). Some results are pointed out to discuss the cluster analysis, such as the relationship within the sectors: The results showed that 36% of the companies were independent or included in an international group; 21% were included in a national group and just 7% were considered as "other", such as research centres and universities; according to the same questionnaire, the cooperation and collaboration showed that services are mainly inside the sector, main services and goods supply from the sector, showed that the interviewees are mainly supplied by EU organisations, the organisations cooperated with other organisations (75%), technological cooperation, the interviewees stated that with organisations, technological cooperation is in process production and in new products, and in financial terms all organizations agreed that Legislation, policies, financial incentives and taxation are part of this emergent cluster, namely Portugal 2020, Interface Program and Blue Fund.

The study comprises the study of the offshore energy cluster in Portugal [15].The SWOT analysis made on this cluster contributes to an understanding of the factors which influence the business model and the principal features to exploit in the market in the near future.

SWOT analysis is the beginning of a business strategy in any market; therefore, in order to survive, any organization needs to accept and evaluate products, to be competitive. In this outlook it is important to analyse the market position in comparison with other organizations with the same market segmentation. Moreover, SWOT analysis completes how clusters deal with their strengths, weaknesses, opportunities, and threats. [16] define strategic planning as an extended tool for regional development and territorial structure. Consequently, SWOT analysis is an instrument which comprises internal and external factors associated with favourable and unfavourable outcomes [17]. In this point of view, there are four main fields in SWOT analysis:"Strengths": internal positive characteristics that the organization can exploit to achieve its goals;"Weaknesses": internal characteristics that may inhibit the organization from achieving its goals; "Threats": aspects of the external environment that may prevent the organization from achieving its strategic goals; and "Opportunities", which are described as the characteristics of the external environment that have the potential to help the organization to achieve or exceed its strategic goals [18]. SWOT analysis is a methodology for setting up a business plan based on hierarchical, quantitative and consistent objectives, and with real goals [19]. SWOT analysis presents some limitations, such as an inadequate definition or lack of prioritization of factors; over-subjectivity in the generation of factors: compiler bias [20].

[16], [21], [22] and [23], point out the importance of using SWOT analysis to face problems in a renewable energy industry, namely, suitable strategies that could overcome such problems, and that it also summarizes key issues of the business environment, including the organization's capability for strategy [18]. Not surprisingly, clusters can be related to SWOT analysis. Regional planning and regional strategy are both concepts which comprise better cluster performance. SWOT analysis confirms the plans for better utility of resources.

The general balance of both from internal or external factors quickly and economically exposed so that an organization can work via a business plan towards facing problems which emerge in its relationship with the environment. [16] argues that is possible to present the most important points to the SWOT analysis renewable wind energy sector:

IOP Conf. Series: Journal of Physics: Conf. Series **1222** (2019) 012036 doi:10.1088/1742-6596/1222/1/012036

1- Strengths: Domestic source; new business opportunity; resources are environmentally friendly; wind energy is renewed on a daily basis; it provides a central control for each government; the supply of sources depends on the needs; well-planned transport system can reduce costs in the supply chain; contribution to long-term energy demand; contribution to energy diversification and energy competitiveness; 2-Weaknesses: Energy share still small in the world market, although this share will increase but it will not surpass fossil fuel till 2030. Governments are required to promote this type of energy otherwise it will fall behind fossil fuel. The initial investment is high and the production does not occur at a stable level; 3 - Opportunities: Development of new technologies; market size will increase with increasing effectiveness; SMEs technologies; 4 - Threats: Few grants; tax exemptions need to be provided by governments, otherwise the prices will be higher than fuel energy; some initial setup costs are high, with negative effects on market entry.

The SWOT analysis presented by [24] for central Europe (Austria, Czech Republic, Hungary, Slovakia, Italy, Poland) on energy and the environment presents the following strengths: better energy promotion and environment industry; transfer of knowledge and know-how; internationalization of clusters, sharing research and testing facilities, and developing new and better services to clustered firms; ability to innovate, common projects with real business value; generates market for others products; sharing best practices; influence of regional development policy; boosting interests of potential members due to access to European Union funding. The threats are :economic crisis and lack of sources of financing; decrease in clusters member activity; cultural differences; lack of trust among clusters; differences among the meta cluster members (national/regional, priority differences); the meta clusters can become a platform without any real content.

From the Portuguese perspective the first study to present the swot analysis about the energy sector in Portugal was [25]. [25] developed a SWOT energy sector analysis in Portugal, referring to the characteristics of the Portuguese energy system and the internal features of companies presently operating in the market and project investments.

The cluster SWOT analysis made from the questionnaire enabled conclusions about strengths, weakness, opportunities and threats that the cluster will face in the near future in Portugal. The emergent Portuguese offshore energy cluster has just started to take the first steps towards a competitive European energy offshore cluster. Nevertheless, from the analysis, it is possible to come to conclusions about its performance and capacity in the market and the behaviour in the economy.

The SWOT analysis leads to an innovation sector with a high level of research and development activities, it based on expertise knowledge and is environment friendly. The strengths will bring the sector into high competiveness within the Portuguese energy sector in terms of innovation and performance. However, the sector faces points of weakness related to its strategic planning to set up the links between organisations. The inexistence of a road map to deliver a strategy and regulation within the sector, the lack of communication between the stakeholders, and high level of investment are other negative points which the sector will face when in the market.

The SWOT analysis also refers to the opportunity for new investments and R&D activities, such as economies of scale and emerging markets (opportunities).Additionally, the sector faces threats related to its capacity to compete with other energy sources, and the direct costs related to its production, as well as funding levels and priorities. The SWOT analysis demonstrates a sector with a high capacity for expert employment, development of technologies and R&D activities. However, it is an emergent sector in the Portuguese economy and will face competitiveness with other energy sources, cooperation between suppliers and a high level of investment.

The SWOT analysis showed also that this sector has the capacity to be a cluster if the stakeholders and others commit to cooperating in the long run in technology, expertise and knowledge. Moreover, an internationalisation of energy production is also expected in this sector which will lead to a better relationship with other European partners.

The SWOT analysis allowed a description of stakeholders 'expectations about the potential cluster itself and its relationship within the energy market. Therefore, the capacity for staying in the market depends on its capacity to promote renewable energy with lower production costs and short-term return on investment. Also, human resources specialization, knowledge transfer and innovation constitute important decision variables for competitiveness in the sector.

To conclude, the offshore business model from the SWOT analysis is based on the potential in human resources, knowledge transfer and investment. However, vertical and horizontal decisions need to be consistent with a competitive sector based on key initiatives, such as standardization and certification, minimizing the environmental impact of offshore wind farms, grid integration, and offshore development based on collaboration and cooperation. On the other hand, the business model is also linked with an ecosystem, welfare and the green economy.

Table 4–Offshore Energy Cluster in Portugal: SWOT Analysis

	Most positive points (>50% of responses)	Less positive points (<50% of responses)		Most positive points (>50% of responses)	Less positive points (<50% of responses)
STRENGTHS	Innovation; extensive R&D development activities; environmentally friendly and expert knowledge	public interest; efficient government support, return on investment, and geographic diversification; network investments and human resources; good natural resources	WEAKNESSES	lack of strategic plan;	No road map; high level of investment and communication between stakeholders; security, environmentally unfriendly, data challenges;
OPPORTUNITIES	R&D activities; Investment opportunities;	Economies of scale; emerging market;	THREATS	Competition with other energy sources; price of energy from offshore energy production; funding levels and priorities	Environmental regulations; environmental risks.

Source: Pego, A. (2019:203)

5- Conclusions

It is clearly understandable that the offshore energy sector is an emergent sector in Portugal. This emergent sector (cluster) is a concern with an innovative cluster based on competitiveness, cooperation, and collaboration with other sectors. Apart from this, the offshore energy sector comprises a new business model based on environmentally friendly technology and welfare, which will draw in a new production model based on demand (*INOVGRIDEDP*).

This SWOT analysis will contribute to positive benefits in the sector related to the use of innovation and technology, energy potential, green technology potential, investment in the sector and knowledge transfer; however, according to the research questionnaire, there is a lack of connection between the stakeholders and the government, as well as deficient regulatory laws and lack of a roadmap. Therefore, the business strategy begins with an attitude towards more cooperation within the sector. The interest in the future is to draw a model where it can be performed the relationship with other sectors in the market, the renewable road map and the strategy for more cooperation and collaboration within the sector.

References

[1] Pego A 2018a The pentagonal problem and the offshore energy sector in Portugal. Why does it matter? *Entrepreneurship along the Industry Life Cycle: The changing role of Entrepreneurial Activities and Knowledge Competencies* ed by S Cubico G Favretto J Leitão and U Cantner (Springer Book) pp 313-27

[2] Organisation for Economic Co-operation and Development Staff 2016 *OECD factbooks 2015-2016: economic, environmental and social statistics* (Paris: OECD Publishing)

[3] Park D Seo K Kildow D Judith T 2014 Rebuilding the classification system of the ocean economy *J Ocean and Coastal Economics* **1** 4

[4] Salvador R 2015 Maritime clusters evolution The (not so) strange case of the Portuguese maritime cluster *J Maritime Research* 11 **1** 53-9

[5] Simões A D S 2013 *O e-planning como instrumento na construção do cluster do mar em Portugal - quantificação, participação e envolvimento dos setores socioeconómicos nacionais.* PHD thesis (Lisboa: Universidade Nova de Lisboa)

[6] Wijnolst N 2006 *Dynamic European Maritime Clusters* **30** (IOS Press)

[7] Cooke P 2011 *Resilience and Potential in Maritime Clusters* 2011-9 (CIEO-Research Centre for Spatial and Organizational Dynamics, University of Algarve)

[8] Peeters C 2008 The role of maritime clusters to enhance the strength and development of maritime sectors *European Commission DG Fisheries Maritime Affaire*

[9] Benito G R Berger E de La Forest M Shum J 2003 A cluster analysis of the maritime sector in Norway *International J Transport Management* 1 **4** 203-15

[10] Salvador R Simões A Soares R G 2015 *Features of the European Maritime Clusters* (ERSA-European Regional Science Association)

[11] European Commission 2015b *Energy Sectors and the implementation of the Maritime Spatial Planning Directive* (Luxemburg: Publications Office of the European Union)

[12] Zhang W Lam J S L 2013 Maritime cluster evolution based on symbiosis theory and Lotka–Volterra model *Maritime Policy and Management* 40 **2** 161-76

[13] Badcock-Broe A Flynn R George S Gruet R Medic N 2014 *Wave and tidal energy market deployment strategy for Europe*

[14] EUROSTAT 2015 *Energy, transport and environment indicators* Statistical Book (European Commission)

[15] Terrados J Almonacid G Hontoria L 2007 Regional energy planning through SWOT analysis and strategic planning tools: Impact on renewables development *Renewable and Sustainable Energy Reviews* 11 **6** 1275-87

[16] Valentin E K 2001 SWOT analysis from a resource-based view. *J Marketing Theory and Practice* 9 **2** 54-69.

[17] Kartakoullis N L Karlis G 2002 Developing Cyprus as a sport tourism destination: the results of a SWOT analysis *J Sport Tourism* 7 **4** 3-17

[18] Kotler P Keller K L 2007 *Administração de Marketing* 12 editions (São Paulo: Pearson Prentice Hall)

[19] Pickton D W Wright S 1998 What's swot in strategic analysis? *Strategic Change* 7 **2** 101-09

[20] Markovska N Taseska V Pop-Jordanov J 2009 SWOT analyses of the national energy sector for sustainable energy development *Energy 34* **6** 752-6

[21] Aydin B 2014 SWOT analysis of renewable energy In *Green Energy for Sustainable Development (ICUE) 2014 International Conf. and Utility Exhibition on* (1-7).IEEE.

[22] Chen W M Kim H Yamaguchi H 2014 Renewable energy in eastern Asia: Renewable energy policy review and comparative SWOT analysis for promoting renewable energy in Japan, South Korea, and Taiwan *Energy Policy* 74 319-29

[23]EUROPEAN COMMISSION 2013 http://www.central2013.eu/fileadmin/user_upload/Downloads/outputlib/CLUSTERS_CORD_Energy_Environment_strategic_plan.pdf

[24]Pego A 2019 The offshore energy cluster in Portugal. A cluster perspective PHD Thesis (Lisbon: FCSH Universidade Nova de Lisboa) (to be published in May 2019).

[25]Ferreira P V Vieira F D 2010 November Evaluation of an offshore wind power project: economic, strategic and environmental value. *In International Conf. on Environmental Systems Engineering and Technology (WASET)*

IOP Conf. Series: Journal of Physics: Conf. Series **1222** (2019) 012037 doi:10.1088/1742-6596/1222/1/012037

Deep Learning is blowing in the wind. Deep models applied to wind prediction at turbine level

Jaume Manero[1], Javier Béjar[1], Ulises Cortés[1,2]

[1]Knowledge and Machine Learning Group - Computer Science Department.Universitat Politècnica de Catalunya - BarcelonaTECH. Barcelona

[2]Barcelona SuperComputing Center Barcelona

E-mail: `jaume.manero@upc.edu, bejar@cs.upc.edu, ia@cs.upc.edu`

Abstract. Wind Energy generation depends on the existence of wind, a meteorological phenomena intermittent by nature, with the consequence of generating uncertainty on the availability of wind energy in the future. The grid stability processes require continuous forecasting of wind energy generated. Forecasting wind energy can be performed either by using weather forecast data or by projecting (or regressing) the past time-series data observations into the future. This last method is the statistical or time series approach. Wind Time Series show non-linearity and non-stationarity properties, and these two properties increase the complexity of the forecasting task using statistical methodologies. In this paper we explore the use of deep learning techniques, which can represent non-linearity, to the wind speed prediction using the largest public wind dataset available, the Wind Toolkit from the National Renewable Laboratory of the US. Several deep network architectures like Multi Layer Perceptrons, Convolutional Networks or Recurrent Networks have been tested on the 126,692 wind-sites and with the results obtained valuable comparisons and conclusions have been obtained. The distribution of the wind sites across the North American Geography has allowed to include in the analysis relationships between terrain, wind forecast complexity and deep methods.With the developed testing workbench and with the availability of the Barcelona Supercomputing Center new architectures are being developed. This work concludes with the feasibility of deep learning architectures for the wind and energy forecasting.

1. Introduction

The energy industry is changing its dependence on fossil fuels by investing heavily in the generation using renewable sources. This transformation is a consequence of the Climate Change impact in our economies, and is strongly pushed by international organizations and agreements like the Paris Climate Agreement in 2015 [1].

This transformation has many challenges, being one of the them to cope with the inherent intermittence of the renewable sources which it is related to the inner nature of the natural phenomena that creates them (sun, wind, etc.) as they appear and disappear depending on actual weather, day-night cycles, combination of temperature, clouds and humidity. To cope with this intermittence it is required to develop forecasting strategies that will allow to understand the future patterns of energy generation and to develop mitigating strategies for the different output levels [2]. One specially important forecasting window is 12-24 hours, as it has intrinsically high value (it is used as load confirmation from most network operators), but this forecasting horizon

IOP Conf. Series: Journal of Physics: Conf. Series **1222** (2019) 012037 doi:10.1088/1742-6596/1222/1/012037

has a high level or uncertainty due to the physical complexity of the wind generation dynamics at local level [3].

Research and implementation of Wind forecasting methods has been extensively developed in the industry, as the requirement has been around for the last couple of decades and massive investments have been done in developing the right tools for the problem [4], however, the use of Machine Learning and Deep Learning techniques is still nascent in this area [5][6][7].

The application of Machine Learning to Time Series forecasting has shown good results in some applications [8], and is currently a topic of research. The representation capabilities of these techniques for non-linear modelling are being used with series that can't be modelled with traditional statistical models [9].

In this work deep learning models are applied to the U.S. National Renewable Laboratory (NREL) Wind Toolkit, the largest publicly available wind dataset. This dataset contains 126,692 wind sites [10] with wind speed, humidity, pressure, wind direction measures sampled every 5 minutes during 7 years (2007-2013). This work has developed experiments with Several deep learning architectures and applied them to all the wind sites, which has allowed to obtain conclusions from a very large resource of data. This wealth of data has allowed to develop geographical interpretations of the results.

This article is organized as follows. First the Time Series Forecasting task is presented as a Regression problem, then the deep learning families of algorithms developed are analyzed, and finally the experimentation is described in some detail. The final conclusions and future work actually in place is described in the last section.

2. Wind Time Series and Forecasting

A wind time series is a set of several observations (wind speed at different heights, wind direction, temperature, pressure, etc.) recorded in a geographical point or site. Usually they are obtained from the extensively sensorized turbines that generate the wind energy. A wind can series is then composed of observations, defined formally as an observation tuple O_i

$$O_i = \langle O_{temperature}, O_{wind-direction}, O_{pressure,\cdots} \rangle \tag{1}$$

There can be many elements in a observation tuple in one time stamp, like wind speed at different heights, humidity, sun radiation and others. Then a wind time series $\mathbf{X}$ can be described as a sequence of n observation tuples.

$$\mathbf{X} = \langle O_1, O_2, \cdots, O_n \rangle \; or \; \langle \mathbf{x}_1, \mathbf{x}_2, \cdots, \mathbf{x}_n \rangle \tag{2}$$

A forecast $\hat{\mathbf{Y}}$ is a synthetic generated time series obtained by using inference methods in the past observations or by integrating information from other sources (like Weather Meteorological Data for instance). A forecast $\hat{\mathbf{Y}}$ will have two dimensions, an horizon H and a number of predictions. Horizon defines a point in the future H steps ahead. A forecast can have be a single step prediction (only for the horizon) $\hat{\mathbf{Y}} = \hat{\mathbf{y}}_{\mathbf{n+H}}$ or multi-step where a prediction is obtained for each or all the future steps inside the horizon interval $\hat{\mathbf{Y}} = \langle \hat{\mathbf{y}}_{\mathbf{n+1}}, \hat{\mathbf{y}}_{\mathbf{n+2}}, \cdots, \hat{\mathbf{y}}_{\mathbf{n+H}} \rangle$.

The wind forecast problem can be considered as a Regression learning problem. A model $\mathfrak{M}$ is developed that, using the most recent observations $\mathbf{X}$ as an input generates predictions (with an horizon H) as an output $\langle \hat{y}_{n+1}, \hat{y}_{n+2}, \cdots, \hat{y}_{n+H} \rangle$. The objective is to develop a function (or model) $f : \mathbb{R}^n \rightarrow \mathbb{R}$ that generates predictions from the existing observed data.

In Deep learning the regression function is obtained by learning the mapping $f : \mathbb{R}^n \rightarrow \mathbb{R}$ from examples, the learning is acquired by the model in the training phase of the algorithm. Training a Network consists in presenting a large set of data and the network, through its internal optimization mechanisms, learns the relationships between inputs and outputs. Training in a network has to converge to an optimum value, and this convergence is a key element in the

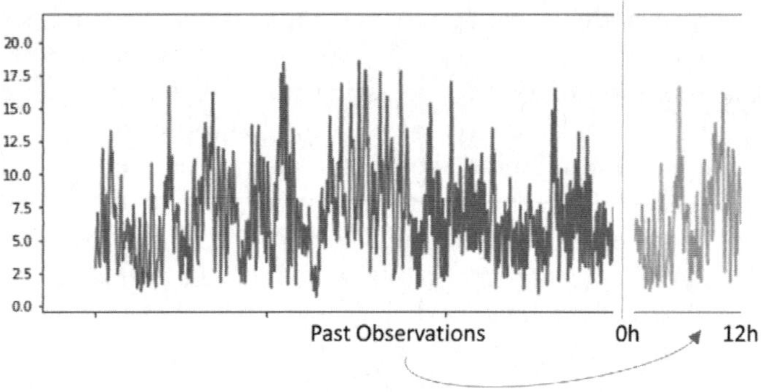

Figure 1. wind time series and prediction

definition and development of network models. There are many optimization algorithms and techniques for Network training [11]. In this work the optimizer used has been Adamax and elements like early-stopping have been included in the architectures [12].

This model is then applied to real data and generates a prediction result that needs to be rated based on the accuracy observed, and in our case we have chosen the R^2 or coefficient of determination which is a very informative measure for linear and non-linear regression problems [13].

$$R^2 = 1 - \frac{\sum_{i=1}^{N}(\hat{y}_i - y_i)^2}{\sum_{i=1}^{N}(y_i - \bar{y})^2} \tag{3}$$

For linear models R^2 can be interpreted as the explained variance (of the regressed function) divided by the existing variance. A value of 1 shows a very good fit, but can become negative if the resulting function does not have good results.

The purpose of this work is to predict energy, but the kinetic energy of wind is transformed to Energy through the energy formula $E = \frac{1}{2}\rho A t v^3$, however this transformation depends on many factors, which make the transformation a more complex function like $f : \mathbb{R}^n \to \mathbb{R}$ where the inputs are wind speed, direction, turbine model, technical loses, etc. There is a theoretical Power curve developed by the manufacturer but practically the transformation is not linear and has a level of uncertainty [14] [15]. There are some works that support the view of separating the wind-speed problem from the energy conversion, problems that are of different nature [4]. In this article, we have focused only in the wind speed forecast problem as we have considered it as a more general. To produce the conversion from wind speed into energy the specifications of the sites and turbines need to be used and this conversion requires an approach based on the real power curve function of the site. This specialized approach would use much more local information about the site (geographically, turbine models used, transformation infrastructure, technical loses, etc.). From now on this article when we refer to forecasting we understand forecasting wind speed.

Wind Time series forecasting is a specific breed of general Time series prediction due to the inner nature of the wind time series which are dependent on the physical characteristics of the wind formation, like weather, terrain or even the application of the Coriolis force on air [16], [3]. Linearity is the property of a time series defined by the possibility to adjust the time series by a linear model like: $\mathbf{X}_t = \alpha_1\mathbf{X}_{t-1} + \alpha_2\mathbf{X}_{t-2} \cdots \alpha_p\mathbf{X}_{t-p} + \epsilon_t$ Stationarity is understood as the statistically property defined by observing a constant mean, variance or covariance in a time series.

Wind Time Series present non-linearity and non-stationarity properties this added to the fact that linear models applied to non-linear time series offer poor results, makes mandatory the use of non-linear modelling for wind time series. There are many non-linear models used for wind time series in the literature [17], but there is limited application of Machine Learning or Deep Learning models to the wind forecasting area [5].

Deep Learning and Machine Learning methods are able to represent non-linear relationships and intuitively appear to be a good alternative for Wind Time Series Forecasting, defining the motivation for this work.

3. Forecasting wind with Deep Learning

One practical example of Deep Learning application is the image classification task, where the algorithm is trained with a set of images or examples. In this way the algorithm learns patterns that exists in the images and constructs a model for the classification of new images. With the Deep Learning techniques the models built have shown accuracy levels superior to the human ability classifying images, specifically when the image is clear, [18], [19].

Regression can be considered a variation of classification, in this case the task consists in adjusting a function or a model developed by learning all the examples in a training dataset. The resulting model will apply the learned structures to new data in order to perform a forecast or a regression of the curve (wind speed in this application).

The Deep Learning algorithm will require to be presented with a set of examples Z where each example will be a series of length *lag*, $Z : (Z_1, Z_2, \cdot, Z_n)$. Each example will be a tuple with the main variable (wind speed) and all the auxiliary variables that the dataset provides us (wind direction, temperature, humidity, density), $Z_i = (Z(\text{wind speed})_i, Z(\text{T})_i, Z(\rho)_i, \cdots,)$. All the algorithms described below use this Examples dataset for its training.

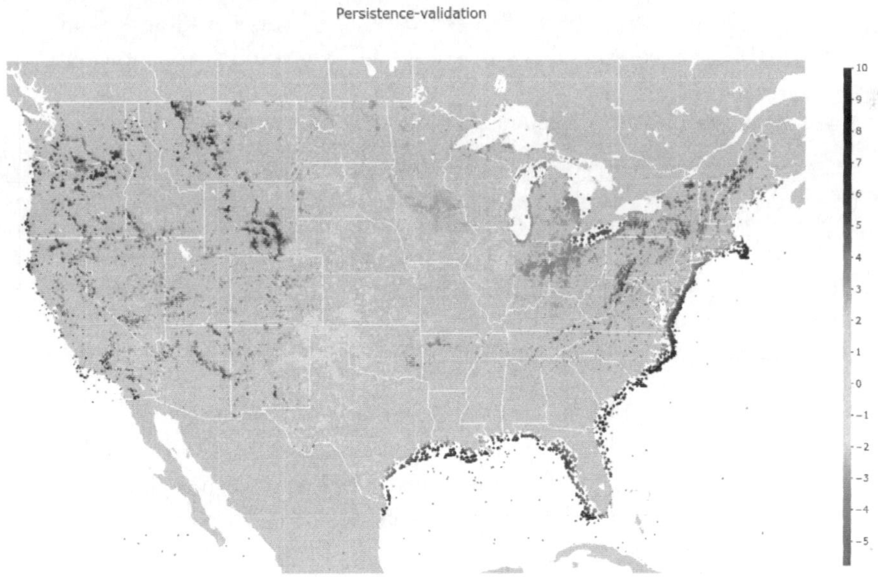

Figure 2. Persistence method applied to all the wind sites. The worse rated sites (yellow-red) mark high wind variability

3.1. Multi Layer Perceptron MLP

Multi Layer Perceptron or Feed Forward Network can be considered as the traditional Neural Network. All the neurons are fully connected to the next layer and the information flows in one direction. The activation function most widely used is the ReLu function [11]. In order to use the MLP architecture two possible approaches can be used. A Direct Regression approach, where the network forecasts a single data point in one execution. Or a sequence to sequence approach where the network forecasts a time series from $n+1$ to k where k is the horizon in one execution. The Direct Regression approach is very expensive computationally wise, as multiplies k times the number of executions and impacts training as well, some preliminary comparisons between direct regression approach and multi-step approach have been performed and the multi-step has shown better accuracy. With this conclusion all the models have been developed with multi-step regression or sequence to sequence. The sequence to sequence approach was defined for Recurrent Neural Networks (RNN) initially and has proven to be very effective for language and voice data [20]. The network is presented with a sequence of values and predicts a sequence of outputs, which is the predicted time series of length 12 (all the work has been done with 12 hours ahead prediction).

The training is performed using backpropagation and standard Gradient Descent with ADAMAX as optimizer [12].

3.2. Convolutional Network CNN

Convolutional Networks are extensively used in the image recognition field. Their main capacity is the application of filters to matrixes of data that can specialize in small patterns in the image.

There is some, but limited, experience on the application of CNN to time Series like [21] or [22], which show applications of convolutional networks to time series with good results.

CNN are feed forward networks in the sense that information flows in only one direction. The structure used in the experiments is a seq2seq CNN where the examples are presented with one dimensional data that is the sequence to be trained on. The CNN generates a sequence as an output which represents the sequence of horizon H to be predicted.

3.3. Recurrent Neural Network RNN

Recurrent Neural Networks are an evolution of feed forward networks, in this case the information is able to flow backward and forward, allowing to process history or long sequences of data. At each time step, RNN will take the input x_i and the output of the previous node h_{i-1}, producing an output h_i.

$$\begin{aligned} h_t &= f(x_i + h_{i-1}) \\ \hat{y}_t &= f(h_t) \end{aligned} \tag{4}$$

RNN can be feed with sequence to sequence like MLP and Convolutional and with encoder-decoder algorithms. An encoder-decoder architecture will take a sequence as an input and a generate a sequence as an output.

There are several variants of RNN networks depending on how the function f behaves, which inputs will process and how to gate them into the function. The two major structures in the neuron are the Long Short Term Memory (LSTM) [20] and the Gated Recurrent Units (GRU) cells [23].

4. Experimental application of Deep Learning to the NREL wind Dataset

The contribution of this work consists in the application of new deep learning architectures on a very large wind dataset. The architecture models described in the previous section have been used in many applications, but there are limited experiences of their use with time-series or with wind data. Regarding data there is not available a study on more than a handful of wind sites using deep learning, for this reason using the NREL database with over 126,692 data sites in North-America [10]

There are some works in the literature using this dataset [24], but, probably due to computing resource availability there are not large exercises available yet. We have used the Super Computing Center in Barcelona [25] as a key computing resource that has made the experiments possible.

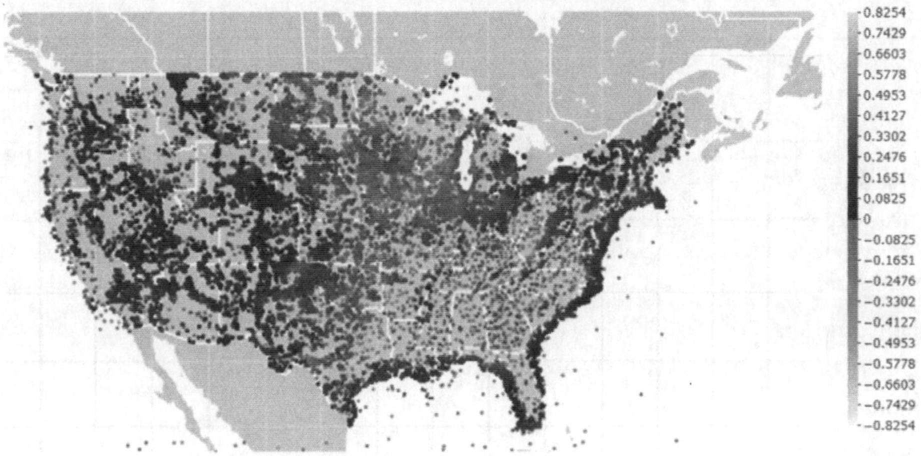

Figure 3. Map of US with the site comparison of MLP vs RNN comparison (blue better MLP, red better RNN)

To perform the experiments we have developed a software framework that allows the application of each architecture to all wind sites under the same conditions in order to obtain comparable homogeneous results.

The algorithms try to learn the wind speed patterns by presentation of different examples, which is accomplished by an sliding window technique.

$$\mathbf{X} = \langle \mathbf{x}_1, \mathbf{x}_2, \cdots, \mathbf{x}_n \rangle \tag{5}$$

The Wind Series in (2) is then decomposed in multiple examples of width lag

$$\begin{aligned} \mathbf{X_1} &= \langle \mathbf{x}_1, \mathbf{x}_2, \cdots, \mathbf{x}_{lag} \rangle \\ \mathbf{X_2} &= \langle \mathbf{x}_2, \mathbf{x}_2, \cdots, \mathbf{x}_{lag+1} \rangle \\ \mathbf{X_3} &= \langle \mathbf{x}_3, \mathbf{x}_4, \cdots, \mathbf{x}_{lag+2} \rangle \quad \cdots \end{aligned} \tag{6}$$

There are n $\mathbf{X_i}$ possible windows in the series being each one of length lag which are the examples that will be used in the training process by the different architectures. Each model generates a prediction series from n to n+12 steps (hours). The models tested can be seen in table §1.

The data is z-normalized as a pre-process for training so the value obtained for the MSE is also normalized, as a consequence $R^2 = 1 - MSE$, in this sense MSE can be considered an error measure and then R^2 is an accuracy measure, where the closer to 1 the more accurate is the result.

IOP Conf. Series: Journal of Physics: Conf. Series **1222** (2019) 012037 doi:10.1088/1742-6596/1222/1/012037

Table 1. Deep Models tested in Experiments

Model	Variant	Description
Multi Layer Perceptron	seq2seq	MLP sequence to sequence
Multi Layer Perceptron	Direct	MLP with Direct Regression
Convolutional Network	seq2seq	CNN sequence to sequence
Recurrent Neural Network	seq2seq	RNN seq2seq
Recurrent Neural Network	encoder decoder	RNN encoder decoder
KNN	KNN neighbors	trained like the Networks

A z-normalization consists on adjusting the scale of the the data obtaining mean of zero and variance (σ^2) of one. This is a usual data transformation process performed if working with neural networks [11] [26].

The accuracy of each method has been calculated using the coefficient of determination $\mathbf{R^2}$. Each one of the models has been applied to the 126,692 wind sites. The experimentation process has followed several steps.

1 Baseline execution

2 MLP Direct Regression vs MLP seq2seq model

3 CNN model

4 RNN models

5 Comparisons and results analysis

The first step has been to perform the baseline models for comparison with the deep learning architectures that will be executed later. The two baseline models chosen have been persistence and k-Nearest Neighbor or k-NN method. Persistence consists in using the x_n value as $\langle \hat{y}_{n+1} = x_n, \hat{y}_{n+2} = x_n, \cdots, \hat{y}_{n+H} = x_n \rangle$ this method obtains reasonable results up to 3 hours and then it deteriorates and in its R^2 becomes highly negative. An interesting finding with this method is that the accuracy of the method marks the wind variability in a site, and the location of these wind variability across the US geography is mostly located in areas with high wind resources and variability like the west part of the central plains (see fig. 2).

The k-NN method is trained with examples 18hours long. The algorithm identifies the nearest neighbors using Euclidean distance. With the 13 nearest neighbors selects the averaged next 12 hours. This algorithm obtains as a sum of the R^2 for the 12 steps ahead a value of 4.9, much higher than persistence which is below 2.5, but lower than all the Deep Learning models which are above 6.

The next step has been to determine which is the best regression strategy, and for this reason an experiment with the MLP has been designed. A model with direct regression (one MLP for each step) and a model for MLP seq2seq (input and output are sequences). Both models are executed in a selected number of sites and compared. The comparison confirms the better accuracy of the MLP seq2seq and the conclusion is to use seq2seq for all the experiments.

The third step has been to test each architecture.Each algorithm has been tested to identify the best parameters for each architecture, and then it has been executed with the same set of parameters with the whole dataset. The data has been divided in three sets, 5 years of data for training (2007-2011), one year for testing (2012) and another year for validation (2013). Each single execution on each site has been stored and used for the analysis.

From the comparison between the different methods we obtain valuable conclusions. The comparisons between the different methods shows a slightly better accuracy of the MLP seq2seq model than both RNN approaches and the CNN, however the four methods obtain very similar ratings. Another conclusion is that there is not something like the best model everywhere. We find sites where the best model is CNN and others where the best is RNN or MLP. In figure 2 we can see the result comparison between MLP and RNN. MLP is better in most sites, but RNN has areas from the Plains to the West Coast where it performs better, there is even a small part in southern Florida where RNN is the best method.

Analyzing the boxplot of each experiment (see 4 we can analyze the performance of the methods comparing the results between validation and test. Two observations can be made, one is the differences between the two sets of data, which probably implies that the model should be retrained in order to cope with the wind regime changes over time, and the second is the difference of accuracy comparison hour over hour. Could it be that there are methods that perform better in shorter time frames than others?

The results from the experiments can be summarized as follows:

(i) Persistence obtains poor results for the horizon of 12 h.(as it was expected).
(ii) Deep learning models are better than statistical baselines (k-NN).
(iii) MLP direct regression has worse results than the seq2seq regression approach.
(iv) MLP with sequence to sequence, as an average, has better accuracy than RNN or CNN models
(v) The best model in a site depends on some characteristics of the site (terrain complexity or wind regimes) showing a dependence between terrain characteristics and the most accurate model (see figure §3).
(vi) There are some differences in forecasting year 6 or year 7, which explains changes in the wind resource year on year. The conclusion is that the models need to be retrained, as the wind is not constant over time.

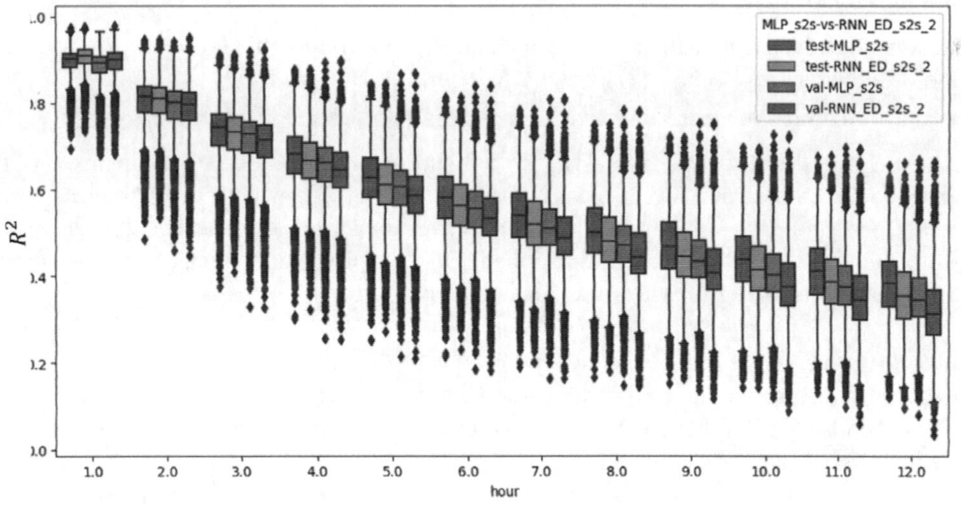

Figure 4. MLP vs RNN boxplot for validation and test datasets

There are narrow differences between the deep learning architectures, and we believe there is room for tuning the deep model structures in order to obtain better results.

5. Conclusions and future Work

The deep learning algorithms tried on the wind toolkit show better accuracy for 12 hour forecast ahead compared to traditional models. The experiments have shown dependency between

wind site characteristics and the best performing method. In areas with high and variable winds (Southern plains, Gulf of Mexico and West Coast) the best method is the recurrent neural network, while in other more stable areas the best method is MLP and in some points CNN. This relationship between best method and site characteristics is an objective of future experimentation.
As the best models differ site to site, and there is some model accuracy drifting between the 12 hour steps, the added accuracy of using different models could improve the overall accuracy, with this finding, we can explore the combination of methods in ensembles in order to improve accuracy.
It is common, with the statistical methods, to adjust the forecasting models by heavy manual tuning, as the forecast is usually made at wind park level and not at turbine like in this work. Defining several models and adjusting the best model to each terrain characteristics may generate an improvement in accuracy, and using an ensemble approach will improve the overall results of the architectures tried. Deep Learning ensemble methods applied to the NREL dataset is another area of further work.
The experiments have been made only using past observations. The variables have different levels of information. Wind speed is the most relevant, while temperature or humidity have less information. From another point of view, temperature or humidity are low resolution variables which can be predicted with great accuracy from weather forecasts. We propose a hybrid model where the past observations will be combined with some low resolution variables from weather forecast to improve the overall accuracy.
Finally, even though the 7 years of measures sampled every 5 minutes looks like a lot of data, the experience shows us that Deep Learning exercises require much larger datasets to obtain the best results. It would be desirable to have alrger datasets, and this could be accomplished by using longer time series, but from the experimentation it has been observed that the use longer series presents some challenges, like:

(i) Winds change over time (due to climate change impact) and this fact compromises the the value of longer series of data [27].
(ii) There are limited sources of sites with longer series of data

Due to these challenges it will be interesting to use synthetic data, and one of the areas of future work is how to use Generative Adversarial Networks to develop synthetic data to enrich training and forecast processes like it is already made in other areas and applications[28]. In summary, the authors believe that Deep Learning applied to wind prediction opens a whole new set of tools to improve forecasting for the wind renewable industry.

Acknowledgments
We thank the Barcelona Supercomputing Center for the extensive use of their infrastructure in this project, and the United States National Renewable Energy Laboratory (NREL) for the use of their wind energy datasets.
This work is partially supported by the Joint Study Agreement no. W156463 under the IBM/BSC Deep Learning Center agreement, by the Spanish Government through Programa Severo Ochoa (SEV-2015-0493), by the Spanish Ministry of Science and Technology through TIN2015-65316-P project, and by the Generalitat de Catalunya (contracts 2014-SGR-1051)

References

[1] United Nations Framework Convention on Climate Change (UNFCCC) Secretariat 2015 Report of the conference of the parties on its twenty-first session, held in Paris from 30 November to 13 December 2015. addendum. part two: Action taken by the conference of the parties at its twenty-first session united Nations Office at Geneva
[2] Fares R 2015 *Scientific American* – Nature Publishing Group

[3] Landberg L 2015 *Meteorology for Wind Energy* (John Wiley & Sons Ltd) ISBN 9781118913444
[4] Giebel G, Brownsword R, Kariniotakis G, Denhard M and Draxl C 2011 *The State-Of-The-Art in Short-Term Prediction of Wind Power: A Literature Overview, 2nd edition* (ANEMOS.plus) project funded by the European Commission under the 6th Framework Program, Priority 6.1: Sustainable Energy Systems
[5] Manero J, Béjar J and Cortés U 2018 *Lecture Notes in Computer Science (including subseries Lecture Notes in Artificial Intelligence and Lecture Notes in Bioinformatics)* **11314 LNCS** 89–98
[6] Hervás-Martínez C, Salcedo-Sanz S, Gutiérrez P A, Ortiz-García E G and Prieto L 2012 *Neural Computing and Applications* **21** 993–1005 ISSN 1433-3058 URL `https://doi.org/10.1007/s00521-011-0582-x`
[7] Torres J M and Aguilar R M 2018 *Complexity* **2018** 10 URL `https://doi.org/10.1155/2018/9327536`
[8] Ahmed N K, Atiya A F, Gayar N E and El-Shishiny H 2010 *Econometric Reviews* **29** 594–621
[9] Ben Taieb S, Sorjamaa A and Bontempi G 2010 *Neurocomput.* **73** 1950–1957 ISSN 0925-2312 URL `http://dx.doi.org/10.1016/j.neucom.2009.11.030`
[10] Draxl C, Clifton A, Hodge B M and McCaa J 2015 *Applied Energy* **151** 355–366
[11] Goodfellow I, Bengio Y and Courville A 2016 *Deep Learning* (MIT Press) ISBN 978-0262035613
[12] Abadi M, Agarwal A, Barham P, Brevdo E, Chen Z, Citro C, Corrado G S, Davis A, Dean J, Devin M, Ghemawat S, Goodfellow I, Harp A, Irving G, Isard M, Jia Y, Jozefowicz R, Kaiser L, Kudlur M, Levenberg J, Mané D, Monga R, Moore S, Murray D, Olah C, Schuster M, Shlens J, Steiner B, Sutskever I, Talwar K, Tucker P, Vanhoucke V, Vasudevan V, Viégas F, Vinyals O, Warden P, Wattenberg M, Wicke M, Yu Y and Zheng X 2015 TensorFlow: Large-scale machine learning on heterogeneous systems software available from tensorflow.org URL `http://tensorflow.org/`
[13] Schneider A, Hommel G and Blettner M 2010 *Dtsch Arztebl Int* **107** 776–782 ISSN 1866-0452 21116397[pmid]
[14] Sohoni V, Gupta S C and Nema R K 2016 *Journal of Energy* 18 ISSN 8519785
[15] Shetty R P, Sathyabhama A and Pai P S 2018 *Frontiers in Energy* ISSN 2095-1698 URL `https://doi.org/10.1007/s11708-018-0553-3`
[16] Abkar M and Porté-Agel F 2016 *Phys. Rev. Fluids* **1**(6) 063701 URL `https://link.aps.org/doi/10.1103/PhysRevFluids.1.063701`
[17] Okumus I and Dinler A 2016 *Energy Conversion and Management* **123** 362–371
[18] Geirhos R, Janssen D H J, Schütt H H, Rauber J, Bethge M and Wichmann F A 2017 *CoRR* **abs/1706.06969** URL `http://arxiv.org/abs/1706.06969`
[19] He K, Zhang X, Ren S and Sun J 2015 *CoRR* **abs/1502.01852** (*Preprint* 1502.01852) URL `http://arxiv.org/abs/1502.01852`
[20] Sutskever I, Vinyals O and Le Q V 2014 *CoRR* **abs/1409.3215** (*Preprint* 1409.3215) URL `http://arxiv.org/abs/1409.3215`
[21] Qiu X, Zhang L, Ren Y, Suganthan P N and Amaratunga G 2014 Ensemble deep learning for regression and time series forecasting *2014 IEEE Symposium on Computational Intelligence in Ensemble Learning (CIEL)* pp 1–6
[22] Borovykh A, Bohte S and Oosterlee C W 2017 *arXiv e-prints* arXiv:1703.04691
[23] Chung J, Gulcehre C, Cho K and Bengio Y 2014 *arXiv e-prints* arXiv:1412.3555 (*Preprint* 1412.3555)
[24] Heinermann J and Kramer O 2016 *Renewable Energy* **89** 671 – 679 ISSN 0960-1481 URL `http://www.sciencedirect.com/science/article/pii/S0960148115304894`
[25] Martorell J M 2016 *Contributions to Science* **12** 5–11 ISSN 0960-1481
[26] Ioffe S and Szegedy C 2015 *CoRR* **abs/1502.03167** (*Preprint* 1502.03167) URL `http://arxiv.org/abs/1502.03167`
[27] Pryor S and Barthelmie R 2010 *Renewable and Sustainable Energy Reviews* **14** 430 – 437 ISSN 1364-0321
[28] Esteban C, Hyland S L and Rätsch G 2017 *ArXiv e-prints* (*Preprint* 1706.02633)

Increasing the Value of Offshore Wind by Integrating On-Board Energy Storage

D. Buhagiar[1*], T. Sant[1,2], R. N. Farrugia[1], D. Farrugia[1]

[1] Department of Mechanical Engineering, University of Malta, Malta
[2] Institute for Sustainable Energy, University of Malta, Malta

*corresponding author: *daniel.buhagiar@um.edu.mt*

Abstract. Energy storage technologies are considered a promising solution for overcoming one of the most pertinent hurdles to high renewable energy penetration: the mismatch between energy supply and consumer demand. The intermittent nature of variable renewable energy technologies at high penetration rates leads to a loss of value for each unit of energy produced. Generation-side energy storage can allow wind turbines to alter their generation strategies and derive additional value through improved market participation. On-board storage leads to more efficient use of space and a potential for cost reductions. In the present work, a brief review of existing work on these aspects was undertaken, followed by a time-series analysis of an offshore 6 MW wind turbine coupled to an energy storage system. The performance of the *wind+storage* system was simulated using one year of data from the *Egmond aan Zee* offshore wind farm site. A statistical analysis was undertaken to estimate the required charge/discharge cycles and establish the required storage capacity under different operating conditions. A lithium-ion battery was then considered as the competing energy storage technology, and a cumulative damage model was applied based on the depth-of-discharge characteristics. Findings indicate that despite their competitive capital costs, battery technologies would have a limited lifetime resulting from high charging/discharging cycles. A more viable approach in the long-term could be to opt for technologies that are less dependent on charge/discharge cycles and which have a lifetime that can match that of the wind turbine itself.

1. Introduction

After long disregard, energy storage is now receiving significant attention by the energy industry; and particularly in the renewables sector, thanks to an increasing requirement for flexibility, intermittency regulation and the provision of an uninterruptable power supply [1]. According to published estimates [2], 50-90 GW of deployable electrical power from energy storage is required by the year 2050, considering a net output variation in wind power of 15-30%.

Integrated energy storage solutions can significantly increase revenue from a renewable energy system such as wind farms due to their capabilities when it comes to grid stability, load shifting, operational support, as well as increasing overall power quality and reliability [3]. It also allows the wind farm to provide Energy Storage as a Service (ESaaS) [4].

In the present work, the potential value-adding benefits of integrating energy storage at the point of generation are investigated, along with caveats of existing battery technologies due to the harsh nature of charging cycles associated with the reduction of intermittency from wind power generation.

2. Background: Increasing the Value of Wind Energy

One unit of electricity from whichever source, at any point in time, should have the same intrinsic monetary value. However, from a supply-demand perspective, 1 MWh of energy generated during times of high demand can attain a higher market value than the same 1 MWh generated during times of lower demand. This basic principle leads to what has been referred to as the "self-cannibalisation effect" of Variable Renewable Energy (VRE) generators [5]. The issue arises from the fact that the abundance of the VRE itself tends to depress market prices during periods of high availability. This can be observed in Figure 1. VRE technologies therefore lose market value [6], since their variability makes it more difficult for an accurate and strategic supply-demand match [7].

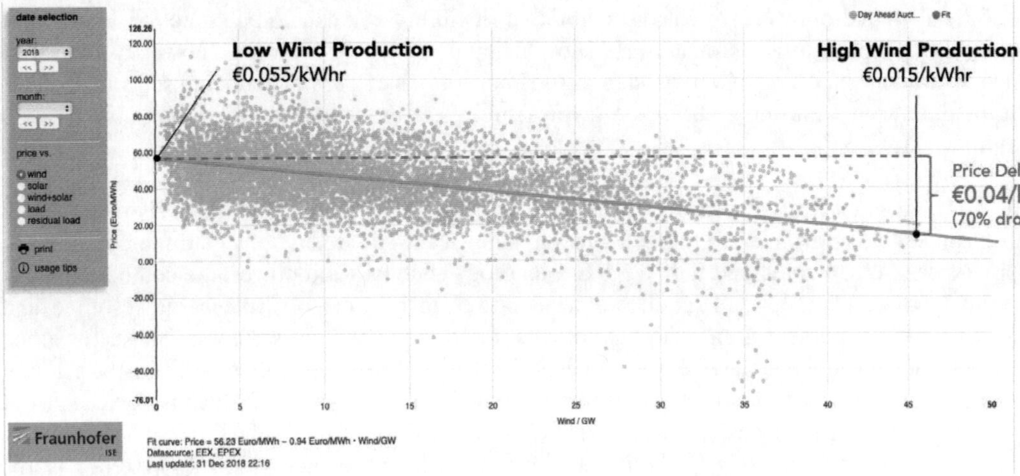

Figure 1: Spot prices vs. wind power in Germany during 2018 (source: [8])

Wind power exhibits intermittency on various time scales, often resulting in a supply variable that is disproportionate to the typical demand variability [9]. The variable nature of wind energy coupled with zero fuel costs implies that at times price suppression can lead to even negative prices [10]. This results from excess production, typically at times when prices are already low and wind power generation is particularly high. Hirth and Müller [5] state that this value drop can be significant, such that with 30% wind penetration its variable input can be worth 20-50% less than electricity from a steady production source. This impact on market value is detrimental to the competitiveness of VREs, and therefore to the decarbonisation of the energy system.

It is evident that the deployment of large quantities of wind power in the grid calls for suitable grid integration measures and an increase in flexibility. Some grid integration measures related to *generation strategies* and means of adding flexibility through *energy storage* are briefly discussed in this section.

2.1. Generation Strategies

Before attaining the high penetration rates when energy storage becomes a fundamental element for VRE integration, the VRE generators themselves can contribute through strategic generation and alternative market participation. Mills and Wiser [11] derive the economic value of wind as the sum of the *energy (market) value*, *capacity value*, *ancillary services value* and *forecast error value*. The *forecast error* is the result of deviating from the day-ahead generation schedule, and is typically *negative* if wind is traded passively on the day-ahead market (without selling regulating power). For the sake of brevity, a thorough overview of generation strategies is omitted in the present work, which instead expands on energy storage as a means of increasing the value of wind power.

IOP Conf. Series: Journal of Physics: Conf. Series **1222** (2019) 012038 doi:10.1088/1742-6596/1222/1/012038

2.2. Energy Storage

Beyond penetration levels where generation strategies can mitigate intermittency issues, energy storage becomes a key enabling technology for the wide-spread use of VREs [12]. But even before this happens, renewables with schedulable output (such as hydropower), can already bring about a net increase in the value of wind energy. Hydropower is essentially a form of generation integrated energy storage [13], since the reservoir acts as a storage system for the natural resource prior to generation. Of course, it can also operate as a fully-fledged electrical energy storage system through the use of reversible pump-turbines (Pumped Hydro Storage – PHS). An interesting scenario is illustrated by Hirth [14], where it is shown that at 30% wind penetration, 1 unit of wind energy in Sweden can be worth 12-29% more than the same unit in Germany. This higher value results from the fact that Swedish grids have a significant quantity of installed hydropower, which adds additional flexibility that can compensate for wind power output variability. Pérez-Díaz and Jiménez [15] showed that particularly for isolated power systems with high wind penetration, such as the Canary Islands, the implications of PHS on reducing scheduling costs can result in significant wind-integration cost savings, and therefore an overall increase in the per-unit value of wind energy.

Apart from geographically-restrictive PHS, lithium ion (li-ion) batteries have also become a formidable contender for grid-scale storage thanks to significant cost reductions, and the prospects of further price drops. When combined with a VRE technology such as wind power, the combination can form a virtual power plant [16]. This allows wind power to become a dispatchable resource and participate in real-time markets along with the potential to provide ancillary services. One such service is Frequency Containment Reserve (FCR), which is particularly lucrative for batteries, and is the primary revenue source in current market conditions. Batteries are suitable contenders for such a service since it typically requires low Depth-of-Discharge (DoD) micro-cycles, which prolong the lifetime of the battery system [17]. When it comes to *energy-shifting* applications, where the energy storage system absorbs supply surpluses to then be used to supplement deficits, batteries tend to suffer due to the requirement for deeper DoD, which significantly reduces their lifetime. Li-ion battery lifetime is notoriously difficult to predict, and practical lifetimes have been shown to deviate significantly from manufacturer quoted values [18]. The actual lifetime not only affects the return on capital investment, but also the viability from an environmental stand-point, since li-ion batteries are particularly difficult to recycle [19]. As such, a life-cycle perspective is crucial to evaluate the viability of battery storage for energy-shifting in wind power applications [20].

3. Methodology

The present study focuses on the integration of energy storage at the point of generation of a wind energy production system. The following approach is adopted:

- A hypothetical 6 MW offshore wind turbine with an integrated energy storage system is considered.

- Open-source, year-long offshore wind data from the *Egmond aan Zee* site [21] is used to simulate the behaviour of the energy storage system as it operates to provide a range of services to the Energy Management System (EMS). The data corresponds to a mean wind speed of 9.06 ms^{-1} at 112 m above mean sea level, and a wind turbine capacity factor of 50.4%. Simulations use the 10-min average wind speed records. Wake losses are neglected since the scope is limited to a single turbine.

- Two different operational strategies requiring storage are considered (Figure 2):
  - (i) *Stepped Output*: a fixed power output is supplied over a specified time window (up to 6 hours), based on the forecast mean output during that window. Any deviations are compensated for by the energy storage system.
  - (ii) *Ramp Regulation*: the power delivered to the grid corresponds to a moving average over a specified window (up to 8 hours), such that sharp increases/decreases in wind turbine output are compensated for by the energy storage system.

- Statistical methods are applied to assess the implications of the observed performance attributes on the required technical characteristics of the storage system (capacity, power rating, response time, etc.).
- A cumulative damage model is applied to estimate the expected lifetime of existing battery-based solutions when applied to this context.

All numerical modelling, statistical operations and data visualisation is carried out using MATLAB®.

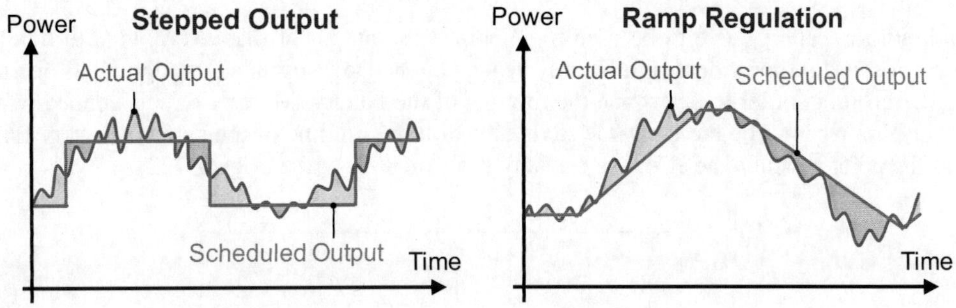

Figure 2: Output power for stepped output (Left) and ramp regulation (Right) operational strategies. Shaded blue regions correspond to deviations that are compensated for by the energy storage system.

3.1. EMS Operational Strategies
The EMS interfaces between the wind turbine generator and energy storage system use two distinct operational strategies (Figure 2).

(i) Stepped output: In the first case, the system delivers a fixed power output for specific number of hours (windows of up to 6 hours are considered). This output power is established by taking the arithmetic average wind power for that window, based on the wind speed time series and the wind turbine power curve. In this case, the wind time series is historical in nature, but in practice, forecasting methods can be adopted. Based on this average value, the EMS will direct energy flows to charge/discharge the energy storage system, and compensate for deviations between the actual wind turbine output and the defined average power output. In this case, the value of wind power is increased by converting it from an intermittent output into a schedulable, on-demand, supply.

(ii) Ramped output: In the second case, rather than deliver a fixed output the EMS operates to time-average the output. A moving average is adopted, taking into account the data points falling within the defined averaging window. Based on this principle, at each time-step, the EMS refers to the required power output as computed by the moving average filter, and compares this to the actual power output. Any deviations are compensated for by routing power to charge/discharge the energy storage system. Such an averaging process improves the characteristics of the output power by eliminating sharp (ramping) power fluctuations that can cause grid instability. Compensating for these fluctuations typically requires other grid services to be available at an additional cost.

In both cases, a hypothetical energy storage system with infinite capacity is considered. The state-of-charge is represented as statistical distribution in time (number of occurrences for binned values), which allows for determining the required storage capacity for each mode of operation.

3.2. Cumulative Damage Model

Miner's Law is a cumulative damage model [22] that assesses the fraction of product life consumed by exposure to a series of stress cycles at differing levels of severity. The rule states that if there are k different stress levels and if the number of cycles to failure at the ith stress S_i is N_i, then the damage fraction C is given by:

$$\frac{\sum_{i=1}^{k} n_i S_i}{N_i S_i} = C \tag{1}$$

where n_i is the number of cycles accumulated at S_i and C is 1 at the point of system failure.

If a lithium-ion battery were to be used to smoothen the output of the 6 MW wind turbine in the above scenarios, its lifetime would significantly depend on the DoD sustained throughout its operational lifetime. Therefore, in order to determine the lifetime of the battery, Miner's Law is applied, whereby the stress level is taken to be the DoD due to its effect on the lifetime of the battery. Battery failure is defined as the point at which the effective capacity drops to 80% of the original [22].

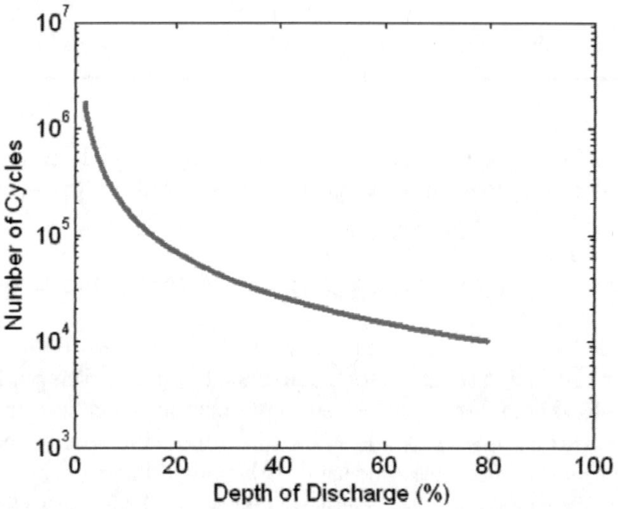

Figure 3: Typical *lifetime-DoD* curve at 25°C for a *SAFT* Li-Ion battery (source: [23])

Using a typical *lifetime-DoD* curve for a li-ion battery, such as that shown in Figure 3, one can see that, for example, the battery exhibits a lifetime of 8,000 cycles at a DoD of 100%. However, lower DoD cycles can result in much higher lifetimes, since the relationship is non-linear.

In the present work, the damage fraction sustained throughout the year under investigation can be determined by considering the various DoDs of the storage system as it operates to provide the desired mode of operation specified by the EMS, and taking their total sum. Therefore, the damage fraction sustained during a year of operation is given by:

$$C = \frac{\sum DoD}{100 \times L} \tag{2}$$

where $\sum DoD$ is the sum of the DoD timeseries for the one-year simulation and L is the lifetime (in number of cycles) of the energy storage system.

IOP Conf. Series: Journal of Physics: Conf. Series **1222** (2019) 012038 doi:10.1088/1742-6596/1222/1/012038

Since failure occurs at a damage fraction of 1, and assuming that subsequent years exhibit identical discharge behaviour, then the lifetime of the battery in years can then be determined. For this method to be consistent with the most standard universal definition of DoD [22], this value was computed based on instantaneous capacity at the end of a discharging period (at the time-step where the system stops discharging) and not the instantaneous level of charge at each time step, see Figure 4.

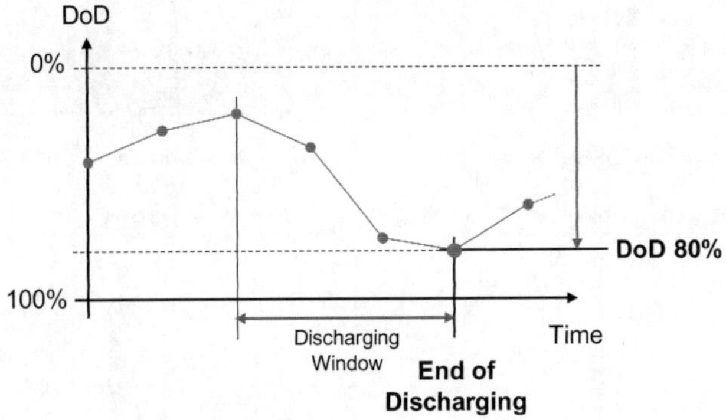

Figure 4: Illustration of how the DoD is established for the cumulative damage model.

For the purpose of establishing the DoD, and then applying the cumulative damage model, the storage capacity must be established *a priori*. This is based on the findings extracted from the statistical distribution of the time-series of instantaneous stored energy (state-of-charge), obtained for each of the operational strategies outlined in Section 3.1.

4. Results

For both (i) *stepped output* and (ii) *ramp regulation*, a statistical analysis is carried out on the time-series of the energy storage system's state-of-charge. This allows key results to be extracted, such as the required storage capacity for each mode of operation, and the estimated number of cycles over a 30-year operating window. Selected results for stepped output and ramp regulation operating modes are shown in Figure 5 and Figure 6, respectively.

These results allow for the determination of the required energy storage capacity, which is based on the maximum value of energy stored throughout the year. The number of energy storage cycles is also estimated by considering the number of occurrences that the system will switch from charging to discharging mode (Figure 4). Other definitions of a charge-discharge cycle can be applied, although technology-specific definitions have been avoided to make the findings applicable to a wider range of storage solutions. Extracted values are shown in Table 1. These values are conservative, since round trip energy losses are ignored, and the required capacity is based on the system handling 100% of the regulation requirements, which might not be the most practical or cost-effective approach.

Table 1: Storage system capacity and number of cycles over a 30-year operating window.

Stepped Output			Ramp Regulation		
Scheduling Window Hours	Storage Capacity MWh	Storage Cycles n_i	Averaging Window Hours	Storage Capacity MWh	Storage Cycles n_i
1	1.6	197,700	2	4	165,210
2	3.8	138,600	4	5	120,420
4	7.5	94,110	6	6	102,750
6	12	63,900	8	7	89,130

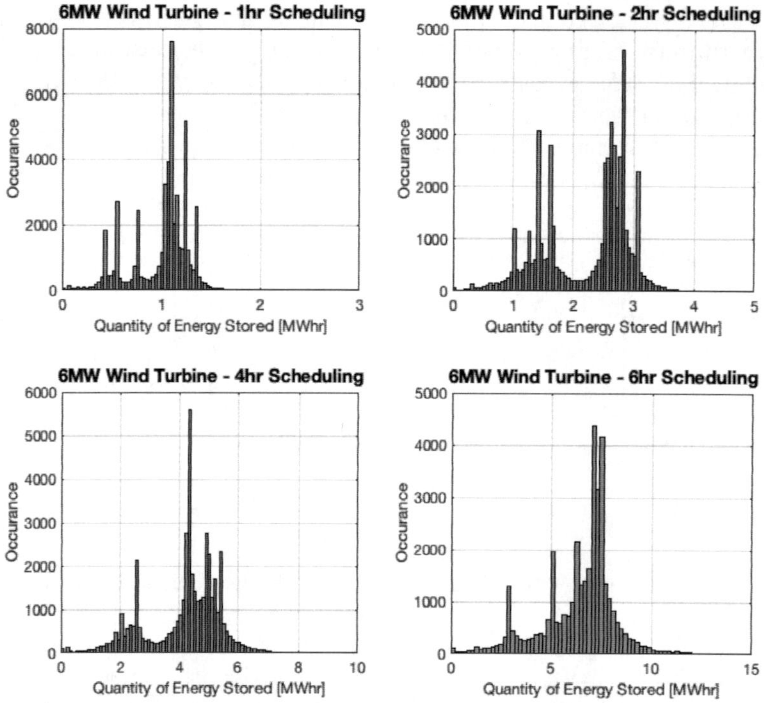

Figure 5: Distribution of energy stored during one year of operation, for scenarios during which the wind turbine output is fixed for 1, 2, 4 and 6 hours.

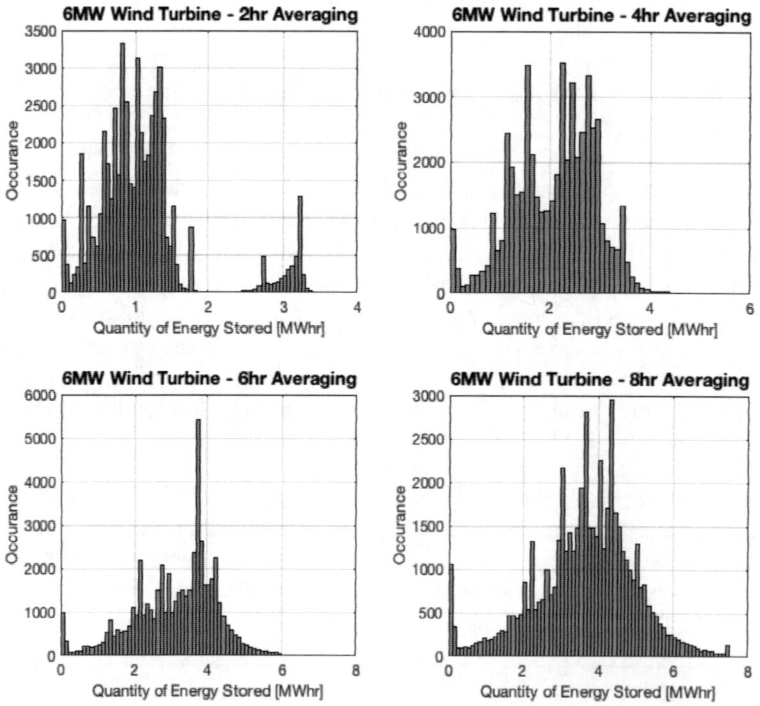

Figure 6: Distribution of energy stored during one year of operation, for scenarios during which the wind turbine output corresponds to the moving average for 2, 4, 6 and 8-hour averaging windows.

The time-series of stored energy can also be used to compute the instantaneous DoD of the energy storage system for the different operational strategies. This is taken to be the state-of-charge at which the system switches to charging mode following a period of discharging (Figure 4). DoD is fundamental when estimating the expected lifetime of the energy storage system. Lifetime is crucial to increasing the net value, since a cost-effective energy storage solution will be one that does not require frequent replacements, particularly in the context of a system deployed offshore.

For the purpose of establishing the effective lifetime of a battery through the application of Miner's Law, a hypothetical li-ion battery with a lifetime of 8,000 cycles at 100% DoD, having a capacity based on the mode of operation (Table 1) is considered. Based on these criteria, and using Miner's law, in the case of a 4-hr stepped output, the lifetime of the li-ion battery in this context was determined to be around 5.5 years. This relatively short lifetime is due to rapid charging/discharging cycles, and frequent occurrences of high DoD. The latter can be observed in Figure 7 where it is shown that the DoD is greater than 28% for 90% of the occurrences. Expected li-ion battery lifetime for the various operational strategies outlined in Table 1, are shown in Table 2.

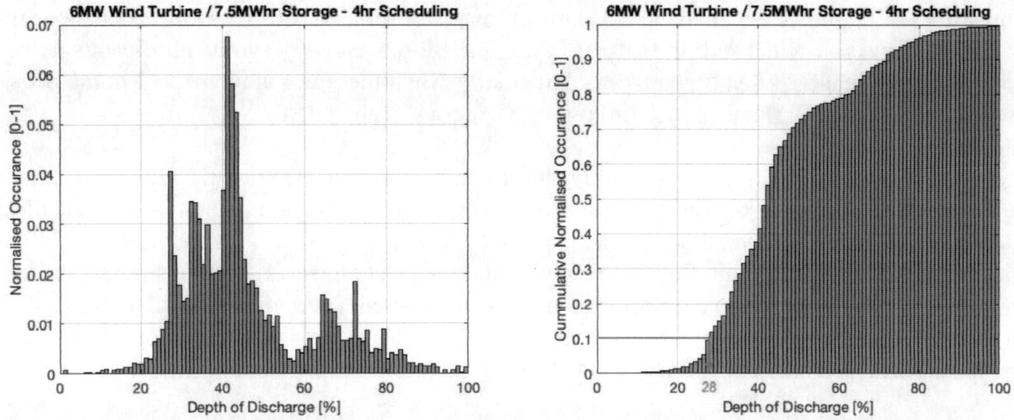

Figure 7: Distribution of DoD for one year of operation, with wind turbine output fixed for 4 hours.

From the lifetimes computed in Table 2 one can note that in all cases the lifetime is relatively short, compared to the 25-30 year operating window of typical offshore wind turbines. Moreover, it can be observed that the lifetime seems to increase as the scheduling/averaging window increases. This effect is observed since, with every increase in the hourly window, the storage capacity is also increasing. This implies that the storage system sustains lower DoD cycles, despite the longer time window. However, it must be noted that the capacity increase is significant. Assuming a fixed unit capital cost (€/kWh), in order to increase the lifetime from 3.14 years to 7.64 years (a factor of 2.4), and the scheduling window from 1 to 6 hours, the total cost would increases by a factor of 7.5.

Table 2: Effective lifetime for a li-ion battery determined by Miner's Law under different operational strategies. The battery has a specified lifetime of 8,000 cycles at 100% DoD.

Stepped Output			Ramp Regulation		
Scheduling Window Hours	Storage Capacity MWh	Effective Lifetime Years	Averaging Window Hours	Storage Capacity MWh	Effective Lifetime Years
1	1.6	3.14	2	4	2.03
2	3.8	3.41	4	5	3.37
4	7.5	5.51	6	6	4.27
6	12	7.64	8	7	5.07

IOP Conf. Series: Journal of Physics: Conf. Series **1222** (2019) 012038 doi:10.1088/1742-6596/1222/1/012038

5. Conclusion

Generation-side energy storage can add value to wind energy by mitigating price suppression resulting from congested energy grids, particularly during periods of high wind energy supply. Simulation results using one year of real wind data for a North Sea site and the corresponding operation of a 6 MW wind turbine operating in conjunction with an energy storage system were used to establish the storage capacity required under different modes of operation. The use of a li-ion battery system with a specified lifetime of 8,000 cycles at 100% DoD was evaluated using a cumulative damage model (Miner's Law). Results indicate that the battery system would reach end of life well before the wind turbine. In the best case scenario (12 MWh system for 6 hour scheduling), the lifetime was found to be 7.64 years.

One possible solution to increase lifetime is to oversize the battery bank, and therefore reduce occurrences of high DoD. However, this implies a significantly higher capital expenditure and hinders the net value delivered by energy storage to the wind farm. It also results in higher spatial requirements, making it unfeasible to integrate within existing wind turbine structures. Another approach is to utilise alternative technologies whose lifetime is less sensitive to charge/discharge cycles and which in effect have a lifetime that can match that of the wind turbine itself. Systems such as hydro-pneumatic accumulators [24] could be ideal, since they use infrastructure that is already well-understood in the offshore sector and can withstand up to 100,000 cycles. Moreover, they avoid significant safety and environmental challenges posed by batteries, particularly when these are implemented in the offshore context, due to issues with flammability and use of hazardous materials.

Integrating on-board energy storage could result in additional revenue streams and increase the net value for offshore wind farms. The key issue is to adopt storage technologies that can handle the intermittency associated with wind power and that can maintain operation throughout the lifetime of the turbine with minimal intervention. Environmental issues and safety must also be considered, especially if a storage system is to operate in the marine environment. As such, alternative technologies to batteries could be more suited due to the longer service life. The right energy storage technology could be a key enabler to increase the value of offshore renewables.

Acknowledgements

This work was carried out as part of Project FLASC (Grant R&I-2015-044-T) with the financial support of the Malta Council for Science and Technology through the FUSION R&I Technology Development Programme. The project is also being supported by the Research and Innovation Development Trust of the University of Malta and Malta Marittima. The authors would also like to acknowledge the use of the open source wind measurements from the Egmond aan Zee wind farm [21].

References

[1] E. M. G. Rodrigues, R. Godina, S. F. Santos, A. W. Bizuayehu, J. Contreras and J. P. S. Catalao, "Energy storage systems supporting increased penetration of renewables in islanded systems," *Energy,* vol. 75, no. 10, pp. 265-280, October 2014.

[2] G. L. Kyriakopoulos and G. Arabatzis, "Electrical energy storage systems in electricity generation: Energy policies, innovative technologies, and regulatory regimes," *Renewable and Sustainable Energy Reviews,* vol. 56, no. 4, pp. 1044-1067, April 2016.

[3] M. Y. Suberu, M. W. Mustafa and N. Bashir, "Energy storage systems for renewable energy power sector integration and mitigation of intermittency," *Renewable and Sustainable Energy Reviews,* vol. 35, no. 7, pp. 499-514, July 2014.

[4] F. Teng and G. Strbac, "Business cases for energy storage with multiple service provision," *Journal of Modern Power Systems and Clean Energy,* vol. 4, no. 4, pp. 615-625, 2016.

[5] L. Hirth and S. Müller, "System-friendly wind power: How advanced wind turbine design can increase the economic value of electricity generated through wind power," *Energy Economics,* vol. 56, no. 5, pp. 51-63, May 2016.

[6] P. Sorknæs, S. R. Djørup, H. Lund and J. Z. Thellufsen, "Quantifying the influence of wind power and photovoltaics on future electricity market prices," *Energy Conversion and Management,* vol. 180, no. 1, pp. 213-324, January 2019.

[7] W. Katzenstein and J. Apt, "The Cost of Wind Power Variability," *Energy Policy,* vol. 51, no. 12, pp. 233-243, 2012.

[8] Fraunhofer ISE, "Spot prices vs. wind in Germany in 2018," EEX, EPEX, 31 December 2018. [Online]. Available: https://www.energy-charts.de/price_scatter.htm?source=priceVSWind&year=2018 [Accessed 29 January 2019].

[9] M. Waite and V. Modi, "Impact of deep wind power penetration on variability at load centers," *Applied Energy,* vol. 235, no. 2, pp. 1048-1060, February 2019.

[10] K. Skytte and L. Bobo, "Increasing the value of wind: From passive to active actors in multiple power markets," *WIREs Energy and Environment,* vol. 328, September 2018.

[11] A. D. Mills and R. H. Wiser, "Changes in the economic value of wind energy and flexible resources at increasing penetration levels in the Rocky Mountain Power Area," *Wind Energy,* vol. 17, no. 11, pp. 1711-1726, November 2013.

[12] F. Cebulla, J. Haas, J. Eichman, W. Nowak and P. Mancarella, "How much electrical energy storage do we need? A synthesis for the U.S., Europe, and Germany," *Journal of Cleaner Production,* vol. 181, no. 4, pp. 449-459, April 2018.

[13] S. D. Garvey, P. C. Eames, J. H. Wang, A. J. Pimm, M. Waterson, R. S. MacKay, M. Giulietti, L. C. Flatley, M. Thomson, J. Barton, D. J. Evans, J. Busby and J. E. Garvey, "On generation-integrated energy storage," *Energy Policy,* vol. 86, no. 11, pp. 544-551, November 2015.

[14] L. Hirth, "The benefits of flexibility: The value of wind energy with hydropower," *Applied Energy,* vol. 181, no. 11, pp. 210-223, November 2016.

[15] J. I. Pérez-Díaz and J. Jiménez, "Contribution of a pumped-storage hydropower plant to reduce the scheduling costs of an isolated power system with high wind power penetration," *Energy,* vol. 109, no. 8, pp. 92-104, August 2016.

[16] F. J. Heredia, M. D. Cuadrado and C. Corchero, "On optimal participation in the electricity markets of wind power plants with battery energy storage systems," *Computers & Operations Research,* vol. 96, no. 8, pp. 316-329, August 2018.

[17] M. Sandelic, D. Stroe and F. Iov, "Battery Storage-Based Frequency Containment Reserves in Large Wind Penetrated Scenarios: A Practical Approach to Sizing," *Energies,* vol. 3065, no. 11, 2018.

[18] M. Dubarry, N. Qin and P. Brooker, "Calendar aging of commercial Li-ion cells of different chemistries – A review," *Current Opinion in Electrochemistry,* vol. 9, no. 6, pp. 106-113, 2018.

[19] C. J. Barnhart, M. Dale, A. R. Brandt and S. M. Benson, "The energetic implications of curtailing versus storing solar- and wind-generated electricity," *Energy & Environmental Science,* vol. 6, p. 2804–2810, 2013.

[20] B. Zakeri and S. Syri, "Electrical energy storage systems: A comparative life cycle cost analysis," *Renewable and Sustainable Energy Reviews,* vol. 42, no. 2, pp. 569-596, February 2015.

[21] NoordzeeWind, 2006. [Online]. Available: http://www.noordzeewind.nl/en/knowledge/reportsdata/.

[22] B. Xu, A. Oudalov, A. Ulbig, G. Andersson and D. S. Kirschen, "Modeling of Lithium-Ion Battery Degradation for Cell Life Assessment," *IEEE Transactions on Smart Grid,* vol. 9, no. 2, pp. 1131-1140, March 2018.

[23] J. Marcos, I. de la Parra, M. García and L. Marroyo, "Control Strategies to Smooth Short-term Power Fluctuations in Large Photovoltaic Plants Using Battery Storage Systems," *Energies,* vol. 7, pp. 6593-6619, 2014.

[24] D. Buhagiar and T. Sant, "Modelling of a novel hydro-pneumatic accumulator for large-scale offshore energy storage applications," *Journal of Energy Storage,* vol. 14(2), pp. 283-294, December 2017.

IOP Conf. Series: Journal of Physics: Conf. Series **1222** (2019) 012039 doi:10.1088/1742-6596/1222/1/012039

Optimized Energy Management Strategy for Wind Plants with Storage in Energy and Reserve Markets

Amaia González-Garrido[1,2], Igor Villarreal[1], Haizea Gaztañaga[1], Andoni Saez-de-Ibarra[1], Pablo Eguia[2]

[1] Energy Storage and Management Area, *IK4-IKERLAN Technology Research Centre* Pº. J. M.ª Arizmendiarrieta, 2, 20500, Arrasate-Mondragón, Gipuzkoa (Spain)

[2] Electrical Engineering Department, *University of the Basque Country*, Plaza Ingeniero Torres Quevedo 1, 48013 Bilbao, Spain

E-mail: amaia.gonzalez@ikerlan.es

Abstract. This paper addresses the joint operation of wind plants with energy storage systems in multiple markets to increase the value of wind energy from an economic and technical point of view. The development of an optimized energy management allows scheduling the wind generation in energy markets, as well as contributing to the system stability through the joint participation in frequency ancillary services. The market optimization maximizes market revenues considering overall storage costs, while avoiding energy imbalances and market penalties. Moreover, wind power fluctuations, forecast errors and real-time reserve requirements are controlled by the energy storage system and managed afterward through the participation in continuous intraday market. Furthermore, model predictive control approach enables a high compliance of reserve requirements and a huge reduction of energy imbalances in real-time operation. Different energy storage capacities are selected in order to evaluate their cost-effectiveness enhancing the wind plant operation under the considered study case.

1. Introduction

As a result of increasing penetration of wind energy in the electric power system, the uncertainty in wind energy forecast and the intermittency of power generation [1] could affect the grid stability.

Moreover, due to their power variability, wind plants (WP) have difficulties in participating and operating suitably under current electricity markets from a technical and economic point of view compared to traditional generators. However, Energy Storage Systems (ESS) are considered one of the key technologies [2], which enable more flexibility and controllability to their market operation [3], as well as the opportunity of providing ancillary services (e.g. frequency control). As a result, WP+ESS plants will participate in a more reliable and profitable way in electricity markets.

Until now, renewable energy sources (RES) participate mostly in energy markets. Firstly, ESS helps WP to deliver a steady output generation [4-5] for the Day-Ahead Market (DM), in which a certain hourly amount of energy is bidden beforehand and delivered later. In order to accommodate the wind production and their inherent forecast uncertainty, Intraday Market (IM) allows them to correct the DM scheduling, reduce large energy imbalances, and thus, maximize their market revenues [5-9].

The growth of variable renewable sources has increased the importance of efficient markets. For that reason, Continuous Intraday Market (CIM) was included in Spain since June 2018 through XBID project [11], which enables RES to manage and reschedule their generation each hour. While the Spanish IMs through six daily auctions is widely known and analyzed [8,10], recent studies take advantage of CIM in order to schedule industrial processes [12], make arbitrage [13], or operate a wind farm but incurring deliberately in energy imbalances when it is profitable [14]. Although some researches analyze the impact of the low liquidity in Nord Pool CIM [9,14,15], the current Spanish intraday trading volume through auction is relevant, around 20% of the total generation. Thus, the WP+ESS plant is supposed to participate in CIM with enough liquidity over time.

Furthermore, researches that are only limited to accommodate renewable energy or make arbitrage do not reflect the potential value of ESS with WP. With increasing renewable generation, large-scale wind farms and ESS could contribute to provide frequency ancillary services in the same way as conventional generators. In the case of Spain, the participation of RES in balancing services is allowed since 2017 [16]. In frequency ancillary markets, generators offer a power availability band the previous day in Spanish Secondary Reserve Market, known also as Frequency Restoration Reserve (FRR). In contrast to energy markets, FRR energy delivery requirements (known as, AGC dispatch signal) are uncertain until real-time operation. This issue has huge implications in their joint operation. Control techniques developed for energy markets or for conventional generation are not directly applicable for WP+ESS under ancillary market participation. Thus, the development of advanced Energy Management Strategies (EMS) is needed for their joint participation in multiple markets.

From the point of view of the system operator, the grid stability and reliability is improved by providing regulation control, while from the WP owner point of view, overall market profits increase. However, some authors do not consider the joint WP+ESS operation for FRR market [10], and thus, the WP do not participate actually in this ancillary service. Other authors show that the wind energy imbalances increase in case of FRR participation [10][17], which is contradictory with the objective of providing FRR. Regarding FRR capacity reservation, some strategies give priority of downward reserve bids (more advantageous for WP) [17], offer a small FRR band compared to the WP capacity [18], assure a low percentage of FRR compliance [19] or operate under the maximum available power [20]. After the operation, while some researches do not analyze clearly the level of FRR compliance [21], other analyses result in high reserve penalties [17-19], or otherwise, achieve a high technical FRR reliability because wind forecast errors are not really considered in the case study [18,20].

In order to fulfill the identified gaps from a global perspective, this paper presents a joint optimal scheduling and operation of DM, CIM and FRR markets in the Spanish market for a wind plant with ESS, similarly as the previous study carried out by the same research group for a solar plant with ESS [22]. Although the main objective of the proposed EMS is to maximize benefits while reducing overall system costs, energy imbalances and reserve penalties are almost reduced to zero, in order to guarantee high technical reliability regarding market scheduling and reserve requirements.

As well as the uncertainty in renewable generation forecast, the AGC dispatch signal at real-time is so variable and uncertain to be predicted that important implications arise for the optimization technique and, even more, on the real-time operation strategy that should be applied. While most researches address this problem from a non-deterministic approach, this paper proposes a mixed integer linear programming to calculate optimal market bids (energy schedule and reserve capacity bids in both directions according to the TSO requirements). Moreover, a model predictive approach enables making decisions closer to the real-time plant operation in order to accommodate wind forecast errors. Furthermore, there are other aspects that are hardly analyzed, such as ESS acquisition, degradation and replacement costs [17-21]. These costs should be analyzed together with market revenues in order to justify economically the installation of ESS with WPs.

This paper addresses the joint operation of WP+ESS in multiple markets to increase the value of wind energy from an economic and technical point of view. On the one hand, the market optimization maximizes WP+ESS market revenues while reduces energy imbalances and considers overall ESS costs. On the other hand, a high compliance of FRR is achieved thanks to the CIM participation and real-time operation based on a Model Predictive Control (MPC) approach in order to calculate the final grid output at each time step by following the AGC dispatch signal; and being aware of the current State Of Charge (SOC) of the ESS, the most recent wind forecast, and real generation.

2. Energy Management Strategy

In this section the proposed EMS to participate in the market is described. This strategy is divided into several blocks, as can be observed in Figure 1: scenario and market definition, optimization process, ESS degradation analysis, real-time operation and techno-economic analysis.

In the scenario definition, main design and operation variables are defined, such as: wind generation profile, energy storage capacity, market prices data, investment, operation and maintenance costs of wind plant and storage system, several operating conditions and other design parameters.

The electricity market design for this study is based on the Iberian electricity market, composed by:

- Day-ahead Market (DM): This market is the main energy trading market to meet demand of the following day. In Iberian Market, the gate closure is at 12:00 of the day before and hourly energy products are remunerated at marginal price (€/MWh) through an auction.
- Continuous Intraday Market (CIM): This market allows updating the scheduling generation in DM with a closer closure time. In this study, CIM market will be modeled because hourly re-scheduling greatly improves forecast accuracy and reduces energy deviations. CIM energy bids at pay-as-bid price should be matched at least one hour before the delivery hour.
- Imbalance settlement (IB): In Spanish framework, two-price system is established for imbalance settlement. This financial settlement mechanism, calculated after the operation, aims to charge and/or pay power plants for the energy imbalances from DM+CIM schedule.
- Frequency Restoration Reserve (FRR): The auction, at 17:30 of the day before, consists of a price for capacity reservation (€/MW) and two prices for energy product in real-time operation (known as AGC signal) (€/MWh), concerning to upward and downward energy products respectively.

MPC is applied for the analysed horizon, with a time step of 15 minutes (Δt = 0.25). Each sample corresponds with the subscripts k. In this study, the simulation period is one year, although the optimization only covers is carried out independently for each day.

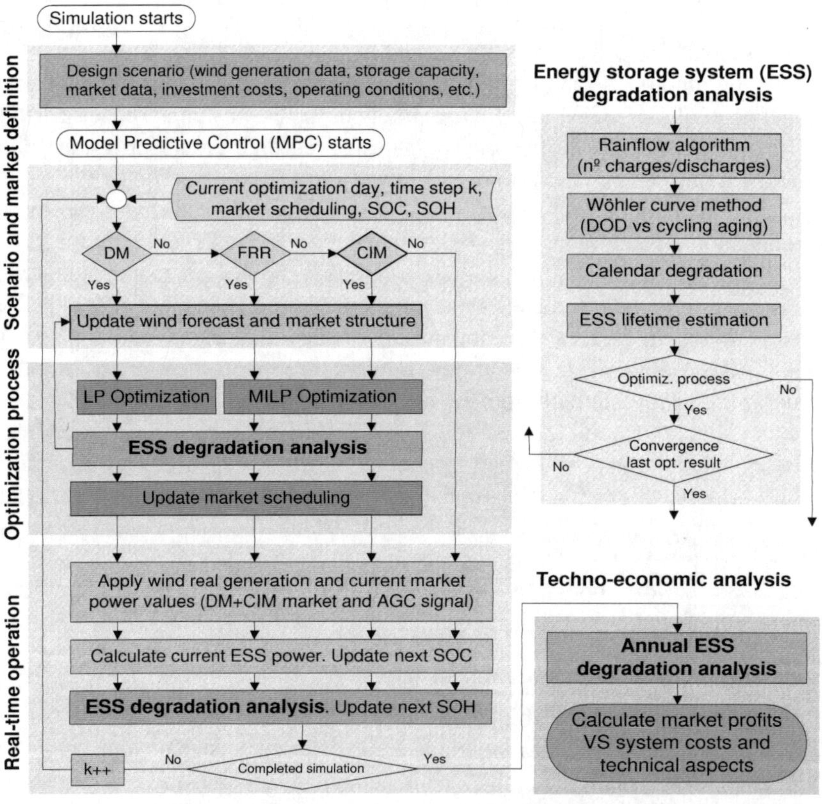

Figure 1. Proposed Energy Management Strategy.

After that, a market evaluation decides whether optimal market bids to DM, FRR or CIM can be calculated, or otherwise, real-time operation for next time step should be directly calculated.

Matlab software was used to model the entire EMS and carry out the simulation. In particular, *linprog* and *intlinprog* functions have been applied to calculate the optimal market operation. For DM optimization, a simplified linear model (LP) is applied, as described in [21], focused on DM profit maximization, including day-ahead generation forecast and hourly mean prices. By the time of FRR auction, the generation forecast improves and real daily DM prices are published. Taking advantage of this current information, a more complex objective function is maximized, through a Mix-Integer Linear Programming (MILP) optimization, described in (1). The first term considers converter and other installation costs (c^{MW}), battery system costs (c^{MWh}), converter power (P^{conv}), nominal ESS capacity (C^{ESS}) and their expected lifetime ($life^{conv}, life^{ESS}$). The second term controls the final daily energy value (E^{ESS}_{kend}) at the middle of the ESS capacity, increasing their weight in the objective function by the multiplier *M*. The third term calculates the hourly DM+CIM schedule (P_h^{mkt}) including estimated CIM prices (hourly prices to be highly matched under the pay-as-bid market process). The fourth and fifth terms minimize positive and negative energy imbalances (P_k^{imb+}, P_k^{imb-}) as much as possible, and only incurring them if the ESS is not able to operate and control the wind generation. The last three terms are related to the profits for hourly FRR capacity reservation band (P_h^{band}) and average upward/downward energy products, with their associated market prices (λ).

Apart from optimal market schedule, optimal daily ESS power and energy profiles (P_k^{ESS}, E_k^{ESS}) are calculated according to the operating limits and technical constraints including in the model.

$$max\begin{bmatrix} -(c^{MW}\cdot P^{conv}/life^{conv} + c^{ESS}\cdot C^{ESS}/life^{ESS})/365 + M^3\cdot E^{ESS}_{kend} \\ +\Sigma_h(\lambda_h^{mkt}\cdot P_h^{mkt}) - M\cdot\Delta t\cdot\Sigma_k(\lambda_h^{imb+}\cdot P_k^{imb+} + \lambda_h^{imb-}\cdot P_k^{imb-}) \\ +\Sigma_h(\lambda_h^{band}\cdot P_h^{band} + \lambda_h^{uw}\cdot\bar{E}_h^{uw} - \lambda_h^{dw}\cdot\bar{E}_h^{dw}) \end{bmatrix} \forall k,h \quad (1)$$

In real-time operation, ESS energy profile (E_k^{ESS}) varies from the optimal value, in order to fulfill the most recent market schedule and AGC power signal (P_k^{AGC}), taking into account real wind generation (WP_k^{real}). Due to these unpredictable changes, optimal WP+ESS plant operation is recalculated in each successive CIM taking into account previous energy deviations thanks to MPC approach, in order to maintain the state of charge (SOC_k^{ESS}) within the operating limits (5-95%) as well as reducing energy imbalances and FRR penalties almost to zero, as discussed in Section 3.

Due to MPC approach, the first values of the market optimization are applied to calculate the final grid output (P_k^{grid}), the charge and discharge power (P_k^{ESS}) and the current state of charge (SOC_k^{ESS}) at each sample *k*, according to (2)-(3). The ESS operation differs from the optimal calculated one, because real wind generation (WP_k^{real}) and real-time AGC power signal (P_k^{AGC}) are finally applied.

$$P_k^{ESS} = WP_k^{real} - P_k^{grid} = WP_k^{real} - [\,P_k^{mkt} + P_k^{imb+} + P_k^{imb-} + P_k^{AGC}\,]\ \forall k = 1 \quad (2)$$

$$SOC_k^{ESS} = [E_{k-1}^{ESS} + \Delta t\cdot(P_k^{ESS}\cdot(u_k\cdot\mu_{ch} + (1-u_k)/\mu_{dch})] \,/\, C^{ESS}\quad u_k = \{0,1\};\ \forall k = 1 \quad (3)$$

After each market optimization and at the end of each day, an ESS degradation analysis is carried out. In regard to cycling degradation, Wöhler curve-based ageing models are applied to evaluate the degradation of the ESS regarding to the number of cycles at a certain Depth of Discharge (DOD). Secondly, a theoretical calendar ageing is added. And finally, the lifetime of the ESS ($life^{ESS}$) is calculated based on both aging models and this term is updated in the objective function.

Finally, an annual ESS degradation analysis is conducted. Moreover, annual market profits are calculated, and annualized WP+ESS costs are calculated in order to assess the profitability of the proposed WP+ESS plant. Furthermore, other technical aspects are analyzed, such as, the traded energy, the level of FRR compliance by following P_k^{AGC} and the Full Equivalent Cycles (FECs).

IOP Conf. Series: Journal of Physics: Conf. Series **1222** (2019) 012039 doi:10.1088/1742-6596/1222/1/012039

3. Techno-economic results

The proposed EMS of the WP+ESS plant was validated for one year in different cost scenarios shown in Table 1. This EMS is applied for an annual wind generation profile based on [23], with available day-ahead and intraday updated forecasts. This profile was scaled for a WP of 30 MW resulting in 2230 Equivalent Full-Load Hours which match properly with a great location in Spain. An ESS capacity between 0.21 and 0.54 MWh_{ESS} / MW_{WP} is considered for the sensitivity analysis. Spanish market prices for 2017 were used for this study case, downloaded from the Spanish system operator website [24]. Investment costs of WP are considered in €/MW [25]. Investment costs of the ESS [26-28] are considered based in energy and power terms: investment energy-related term (in €/MWh), investment power-related term (€/MW) and replacement energy term according to the number of replacements needed (n) during the lifetime of the WP, whose value will be reduced.

The joint operation of a single day is shown in Figure 2. As can be observed, DM schedule is calculated considering the DM forecast. During the day, an evaluation of the current SOC, AGC signal required up to that time, most recent updated forecast and expected CIM prices is made. Based on this real-time information, the MILP optimization recalculates the optimal DM+CIM schedule for current conditions according to the objective function in eq. (1). For this particular day, the ESS is not fully charged and discharged and it is able to control the grid output at all times. Therefore, output power (P_k^{grid}) is equal to DM+CIM scheduling plus the AGC signal required for FRR markets. The power difference between P_k^{grid} and WP_k^{real} is provided by the ESS, according to eq. (2)-(3).

In order to evaluate the profitability of the WP+ESS, different energy storage capacities have been selected to compare their annual market benefits and annualized net profits (market benefits minus the annualized WP+ESS costs according to Table 1). Figure 3 summarizes all these economic results, in which the BC corresponds to annualized net profits of a WP without ESS participating in DM+CIM market under WP costs (1.25 mill.€/MW) reported in [25].

Table 1. WP+ESS costs for different cost scenarios.

WP+ESS costs	Base Case (BC)	CS1	CS2
WP investment costs (mill. €/MW)	1.25	1	0.75
ESS inv. energy costs (mill. €/MWh)	0.4	0.3	0.2
ESS inv. power costs (mill. €/MW)	0.5	0.4	0.3
ESS replacement costs (mill.€/MWh)	$\Sigma_{r=1}^{n} c_{ESS}(1) \cdot [1 - (0.8 \cdot e^{-0.11 \cdot yr(r)} + 0.13 \cdot e^{0.017 \cdot yr(r)})]$		

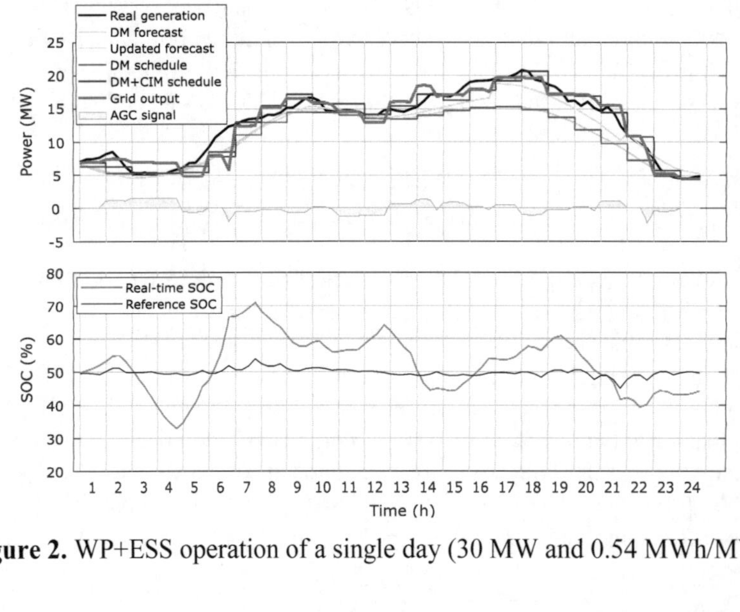

Figure 2. WP+ESS operation of a single day (30 MW and 0.54 MWh/MW).

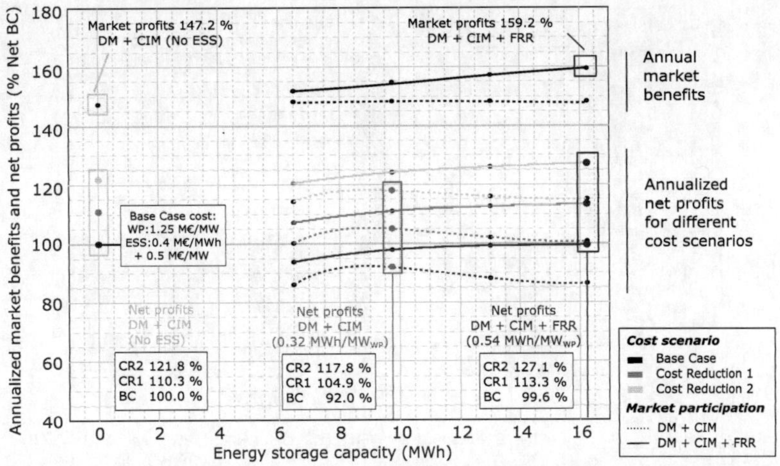

Figure 3. Annualized market profits and net profits (%) for a WP of 30 MW, considering different energy storage capacities (without ESS and 0.21-0.54 MWh/MW), considering two market participations (DM+CIM and DM+CIM+FRR) and three cost scenarios (CS).

On the one hand, the profitability of the WP+ESS under current cost scenario will be analyzed. Based on a 100% of net profits for the BC, additional costs of ESS for the DM+CIM participation are higher than annual market benefits obtained from the reduction of annual energy imbalances costs. For the optimal ESS sizing (0.32 MWh/MW, 9.75 MWh), net profits for this market participation (black dotted curve) are reduced 8% respect to the BC without ESS. In case of FRR participation, the optimal ESS capacity increases up to 0.54 MWh/MW (16.25 MWh). Net profits with this ESS sizing (99.6%) reach nearly the BC without ESS (100%), because annual market profits increase by 12% thanks to the proportional increment of capacity reservation in the FRR auction.

However, the Spanish WPs nowadays participate mainly in IM auctions every 4/6 hours, and therefore, their energy imbalances costs are quite higher than the ones obtained in this analysis. Thus, their current net profits could be less than the obtained ones with a DM+CIM+FRR participation.

On the other hand, if a reduction of considered WP and ESS costs (green curves), the economic results show that the joint operation of a WP+ESS will be more profitable than a WP without ESS when FRR is contemplated. In order to assure a more valuable option than without ESS, the minimum ESS capacity that should be installed will be 10 MWh in case of CR1 scenario and 7.5 MWh in case of CR2 scenario. The trends reveal that the best ESS capacity would be 16.25 MWh.

Focusing on all terms that composed net profits, they could be divided into benefits and/or costs related to: DM, CIM, imbalances, FRR band availability, upward AGC signal, downward AGC signal, AGC penalty, WP investment, ESS energy term (investment and replacements) and ESS power term.

The three optimal choices drawn from Figure 3 are compared below. As can be concluded from Figure 4, the usage of CIM increases when ESS is installed, because wind forecast errors are absorbed by the ESS, and then, this energy is shifted in the coming hours through higher CIM bids. Moreover, positive and negative imbalances terms are reduced hugely. Especially, energy imbalances are practically zero for DM+CIM+FRR participation with an ESS of 16.25 MWh.

Moreover, the FRR participation increases benefits by 10.65% for band availability, 5.22% for upward AGC signal, -2.66% for downward AGC signal and -0.006% for AGC penalty. On the contrary, ESS investment and replacement costs increase annualized costs by 7.57% for energy terms plus 4.88% for power terms, according to the third WP+ESS plant (darker bar) in Figure 4.

Annual traded energy with respect to the annual wind generation is analyzed in Figure 5, with similar conclusions drawn from Figure 4. As can be observed, the energy traded in DM corresponds mostly to the DM forecast, and the SOC initial values. Regarding energy imbalances, they are reduced significantly as the ESS sizing increases. Furthermore, the CIM participation increases in order to

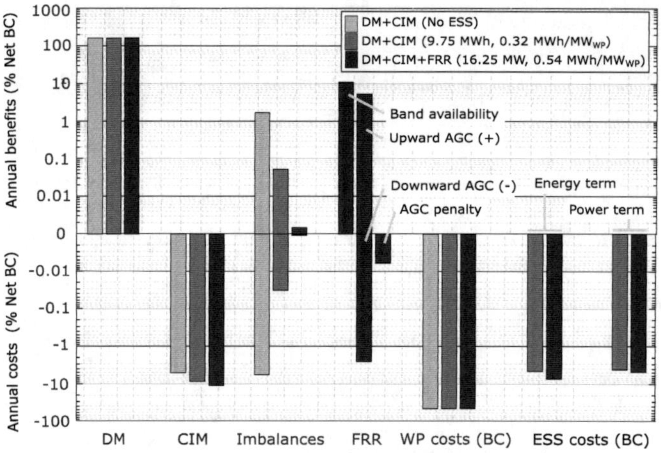

Figure 4. Annual benefits and costs of the three optimal plants under the Base Case cost scenario: 1) DM+CIM without ESS, 2) DM+CIM with 0.32 MWh/MW, 3) DM+CIM+FRR with 0.54 MWh/MW.

ESS capacity (MWh/MW_{WP})	DM°	CIM° (+)	CIM° (-)	Imbalances° (+)	Imbalances° (-)	AGC signal* (+)	AGC signal* (-)	AGC signal* pen.	ESS* losses	
0	105.9	4.26	8.15	1.48	3.48	0	0	0	0	DM + CIM participation
0.21	105.9	4.8	10.89	0.16	0.25	0	0	0	0.07	DM + CIM participation
0.32	106.2	4.65	11.19	0.05	0.02	0	0	0	0.07	DM + CIM participation
0.43	106.4	4.58	11.3	0.04	0.01	0	0	0	0.07	DM + CIM participation
0.54	106.4	4.58	11.31	0.04	0	0	0	0	0.07	DM + CIM participation
0.21	105.9	4.65	11.38	0.06	0.14	1.2	0.6	0.23	0.08	DM + CIM + FRR participation
0.32	106.2	4.65	11.38	0.01	0.02	2.05	1.13	0.08	0.09	DM + CIM + FRR participation
0.43	106.4	4.41	12.42	0	0.01	2.8	1.58	0.04	0.10	DM + CIM + FRR participation
0.54	106.4	4.45	12.82	0	0	3.49	1.99	0.04	0.11	DM + CIM + FRR participation

Resolution: ° Hourly net value, * For all simulation steps

Figure 5. Annual traded energy with respect to the annual wind generation (%)

manage hourly net imbalances from forecast errors, and real-time AGC power signal. Concerning FRR service, the hourly available FRR band and the energy delivered through the AGC signal increase proportionally to the ESS capacity, because the maximum hourly band is limited up to 20% of the C^{ESS}. As can be observed, AGC penalties are also reduced in energy terms when more ESS capacity is installed. Finally, ESS losses are calculated based on the ESS power profile applying the ESS efficiency considering all the conversion stages related to the energy storage system.

Other technical aspects that should be discussed are related to the ESS operation and the level of compliance of FRR service. Regarding annual Full Equivalent Cycles, FECs are reduced when ESS size is larger. In case of FRR participation, annual FECs increase due to a more demanding operation in order to follow AGC signal in real-time operation. Thus, annual FECs result from 501 FECs (0.21 MWh/MW) to 281 FECs (0.54 MWh/MW). According to these annual FECs, the lifetime of the ESS could be estimated between 8.6 years -181 FECs- and 13.3 years -501 FECs-, by applying Wöhler curve method and limiting the calendar lifetime up to 15 years. This means that one or two ESS replacements will be made during the WP lifetime depending of the ESS sizing.

Finally, the technical reliability of FRR participation is worth analyzing. Other studies are not focused on this parameter, while other ones address it [17-19] but not to the point that the power quality standards need. The developed EMS enables a huge reduction of energy imbalances and

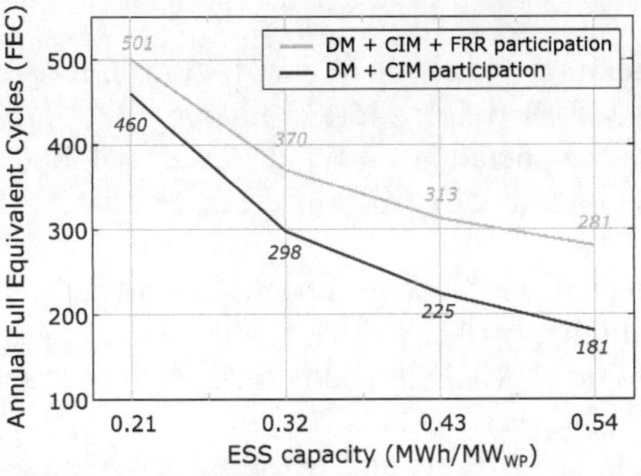

Figure 6. Annual Full Equivalent Cycles for each market participation and ESS capacity.

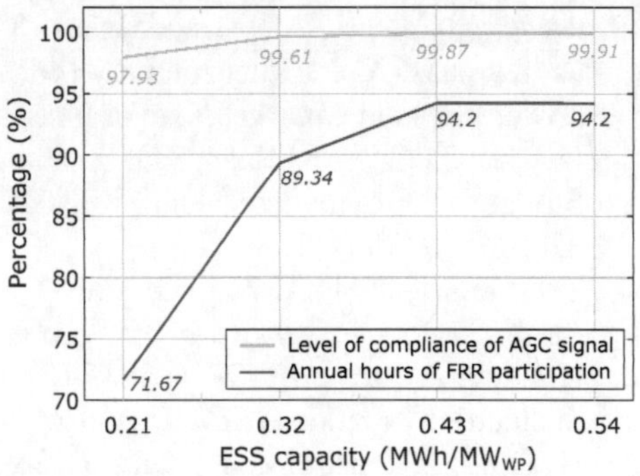

Figure 7. Level of compliance of AGC signal and annual hours of FRR participation.

achieves a high technical reliability of upward and downward FRR needs (from 97.9 % up to 99.9%).

Moreover, the daily number of hours of FRR participation increase with the ESS size, taking into account that an hourly minimum band must be 1 MW and a maximum band is limited up to 20% of the ESS capacity. Moreover, the WP+ESS plant participates daily a maximum of 23 hours, because the last hour is destined by design to manage the final daily SOC at the middle value. Thus, the mean daily hours increase from 17.2 h (with 6.5 MWh) to 22.6 h (with 13.25 MWh).

4. Conclusion

As can be concluded from the above results, this optimal EMS addresses the wind integration challenge to a great extent in the grid and in the electricity market. Through a MILP optimization, the daily DM+CIM schedule and hourly bands for FRR service are calculated in order to maximize the overall market benefits, taking into account the lifetime of the ESS and their associated costs. Moreover, the implementation of MPC approach enables the avoidance of annual energy imbalances and achieves a high technical reliability of FRR needs (up to 99.9%) with an ESS sizing of 0.54MWh per MW of wind power capacity installed for the study case. This analysis calculates the ESS degradation according to the annual resultant operation of the ESS.

This fact results in less annual Full Equivalent Cycles for higher ESS sizing and only one ESS replacement in the twelfth year. Therefore, the additional increase in market benefits due to the FRR participation results in more annualized net profits, compensating the incremental ESS costs. In the considered scenarios, under a light reduction of these ESS costs (corresponding to the CR1 or CR2), the joint operation of a WP+ESS plant could be more profitable than a WP without ESS. These obtained economic results are simulated and validated based on Spanish market framework and market prices of 2017, and under no feed-in-tariff scheme. Therefore, this EMS enables a high replicability to evaluate the WP+ESS cost-effectiveness under other markets, prices and wind profiles.

Acknowledgments
This work was partially supported by the Basque Government under Project Road2DC (ELKARTEK Research Program KK-2018/00083).

References

[1] United States Department of the Energy DOE 2013 *Grid Energy Storage* pp 1–67
[2] EASE and EERA 2017 *European Energy Storage Technology Development Roadmap* pp 1–128
[3] M. McPherson, S. Tahseen 2018 Deploying storage assets to facilitate variable renewable energy integration: The impacts of grid flexibility, renewable penetration, and market structure. *Energy* **145** 856–870
[4] P. Pinson, G. Papaefthymiou, B. Klockl, and J. Verboomen 2009 Dynamic sizing of energy storage for hedging wind power forecast uncertainty *IEEE Power and Energy Society General*

Meeting (Calgary, AB, Canada) pp 1–8
[5] M. Korpaas, A. T. Holen, and R. Hildrum 2003 Operation and sizing of energy storage for wind power plants in a market system, I*nt. J. Electr. Power Energy Syst.* **25**(8) 599–606
[6] G. He, Q. Chen, C. Kang, Q. Xia, and K. Poolla, Cooperation of Wind Power and Battery Storage to Provide Frequency Regulation in Power Markets, *IEEE Trans. Power Syst.*, 2017; **32**(5):3559–3568.
[7] J. Usaola and M. Á. Moreno 2009 Optimal bidding of wind energy in intraday markets, *Int. Conf. European Energy Market EEM* (Leuven, Belgium) pp 1–7
[8] H. Ding, Z. Hu, and Y. Song 2015 Rolling optimization of wind farm and energy storage system in electricity markets *IEEE Trans. on Power Systems* **30**(5) 2676–2684
[9] H. Ding, P. Pinson, Z. Hu, and Y. Song, 2016 Optimal Offering and Operating Strategies for Wind-Storage Systems With Linear Decision Rules *IEEE Trans. Power Syst.* **31**(6) 4755–4764
[10] F. J. Heredia, M. D. Cuadrado, and C. Corchero 2018 On optimal participation in the electricity markets of wind power plants with battery energy storage systems *Computers & Operations Research* **96** 316–329
[11] XBID: Cross-border intraday market project. [Online]. Available: [Accessed: 11-Sept-2018] https://www.epexspot.com/en/market-coupling/xbid_cross_border_intraday_market_project
[12] D. Metz and J. Tomé 2018 Use of battery storage systems for price arbitrage operations in the 15- and 60-min German intraday markets *Electric Power Systems Research* **160** 27–36.
[13] J. Meese, B. Dahlmann, M. Zdrallek and A. Voelschow 2017 Intraday Redispatch - Optimal Scheduling of industrial processes at day-ahead and continuous intraday market *International ETG Congress* (Bonn, Germany) pp 1–6
[14] A. Skajaa, K. Edlund and J. M. Morales 2015 Intraday Trading of Wind Energy *IEEE Transactions on Power Systems* **30**(6) 3181–3189
[15] B. Zakeri and S. Syri Value of energy storage in the Nordic Power market - benefits from price arbitrage and ancillary services, *Int. Conf. Eur. Energy Mark.,* Porto, Portugal, 2016; 1–5.
[16] Spanish Ministry of Industry, Energy and Tourism 2015 *Criteria for participation in ancillary services (Spanish law: BOE-A-2015-13875)* pp 119723–119944
[17] T. Rodrigues, P. J. Ramírez, and G. Strbac 2016 The value of storage for a wind farm offering energy and spinning reserve *IET Int. Conf. Renew. Power Gener* (London) pp 1–6
[18] J. Tan and Y. Zhang 2017 Coordinated Control Strategy of a Battery Energy Storage System to Support a Wind Power Plant Providing Multi-Timescale Frequency Ancillary Services," in *IEEE Trans. on Sustainable Energy* **8**(3) 1140-1153
[19] J. Wang, H. Zhong, Q. Xia, Z. Ma et al. 2017 Robust bidding strategy for microgrids in joint energy, reserve and regulation markets *IEEE PES General Meeting* (Chicago, IL, USA) pp 1–5
[20] L. Johnston, F. Díaz-González, O. Gomis-Bellmunt, C. Corchero-García, and M. Cruz-Zambrano 2015 Methodology for the economic optimisation of energy storage systems for frequency support in wind power plants *Appl. Energy* **137** 660–669
[21] I. Andri, A. Pina, P. Ferrão, J. Fournier, B. Lacarrière, and O. Le Corre 2017 Evaluation of the uncertainty in the scheduling of a wind and storage power plant participating in day-ahead and reserve markets *Energy Procedia* **136** 73–78
[22] A. González-Garrido, A. Saez-de-Ibarra, H. Gaztañaga et al. 2018 Annual Optimized Bidding and Operation Strategy in Energy and Secondary Reserve Markets for Solar Plants with Storage Systems *IEEE Transactions on Power Systems* 1-10
[23] Wind-Power Generation Data. [Online]. Available: http://www.elia.be [Accessed: 15-Oct-2018]
[24] ESIOS [Online]. Available: http://www.esios.ree.es [Accessed: 31-dic-2018]
[25] IRENA 2018 *Renewable Power Generation Costs* pp 93
[26] Lazard 2018 *Lazard's Levelized Cost of Storage Analysis* pp 1–60
[27] EPRI 2016 *Batteries and Energy Storage* pp 1–17
[28] EPRI 2016 *Energy Storage Cost Summary for Utility Planning* pp 1–24

Visualisation of Probabilistic Access Forecasts for Offshore Operations

C Gilbert, J Browell and D McMillan
University of Strathclyde, 16 Richmond St, Glasgow G1 1XQ, UK
E-mail: ciaran.gilbert@strath.ac.uk

Abstract. Access forecasting aims to predict the quality of transfer of maintenance technicians to/from vessels and the constituent offshore structures at the wind farm. This is highly dependent on sea state conditions as well as other environmental factors such as visibility. Typically, scheduling or dispatch decisions are made on the basis of deterministic forecasts of significant wave height, often coupled with service contracts where transfers are expected to be attempted below a threshold significant wave height. However, there is always uncertainty in a weather prediction which can be accounted for by probabilistic forecasts. The aim of this work is to explore visualisation ideas for transforming vessel specific access forecasts into an interpretable and intuitive decision-support tool. Three simple methods are proposed based on a score out of 10, classification of transfer conditions, and a threshold score. Methods for summarising access conditions for 2–5 days ahead are also developed. This new forecasting and visualisation capability has significant implications for marine coordinators and skippers who will be able to make better informed safety-critical decisions, with the potential for reductions in the cost-of-energy offshore.

1. Introduction

Access forecasting for offshore wind farm operations is critical to ensure safe transfer of maintenance technicians to and from turbines and to maximise turbine availability. Improving access can reduce the cost of operations and maintenance (O&M) as well as construction costs. O&M is estimated to represent 25% of the levelised cost of offshore wind energy [1]. Independent studies show that that improved access to the turbine could comfortably shave £15m from the operating costs of an offshore wind farm [2]. Improved access in tandem with improved prognostic capability from condition monitoring has the potential to unlock even greater savings. However, the pressure to achieve increased access to the turbine implies a greater number of marginal-weather crew transfers, which potentially carry a greater safety risk. Improved access is especially valuable to the industry as the offshore fleet ages and life extension initiatives kick in, necessitating many more crew transfers for inspection and to deal with increased failures in late life.

In order to achieve O&M cost savings while maintaining high levels of safety, this project seeks to develop a new crew transfer access forecasting capability within the context of the wider decision-making processes which govern and incentivize behaviour on site. Information flows from assets, weather forecasts and data systems must be assimilated by the marine coordination team, who then form part of a crucial decision-making chain along with crew transfer vessel (CTV) skipper and crew, and finally maintenance technicians. Typically, these

IOP Conf. Series: Journal of Physics: Conf. Series **1222** (2019) 012040 doi:10.1088/1742-6596/1222/1/012040

scheduling/dispatch decisions are made on the basis of single-valued (deterministic) forecasts of significant wave height, often coupled with service contracts where transfers are expected to be attempted below a threshold significant wave height. However, there is always uncertainty in a weather prediction which can be accounted for by probabilistic forecasts. The economic benefit of probabilistic forecasting in O&M decision making has demonstrable value [3].

To forecast the quality of a CTV transfer given uncertainties in the ocean conditions requires vessel specific forecast scenarios of motion during transfer. Given this information content, the goal of this paper is to outline ways to best visualise forecasts for end-users. The communication of uncertainty forecasts is crucially important; one of the main barriers to uptake by industry (and therefore decision-makers) of uncertainty forecasts is the lack of clear communication of information content [4]. Stakeholders in this project have indicated that simplicity is key and that condensing complex information into the simplest form would help decision-makers uptake this information into their workflow.

Scenario forecasts [5] (i.e. trajectories) are required to inform time-dependent decision making [4], such as the case here. Consider that access to the turbine at one narrow point in time is not sufficient as technicians require to be dropped off, complete work, and picked up again before the end of the available work window. Since meteorological forecast errors can persist in time, if the forecast is incorrect at one time point then it is likely to be incorrect *in a similar way* at the adjacent time step. Therefore, uncertainty in the forecasts must be accounted for and linked between time horizons. This is the purpose of scenario forecasts and necessitates their inclusion in this study. Scenario forecasts can be compared to the more familiar physical ensemble forecasts which alternatively can be used in this type of application [4], although site specific ensemble re-calibration would equally be necessary [6]. The focus of this work is in transforming vessel specific scenario forecasts into a digestible format for decision-makers.

The site of interest is an east coast wind farm in the UK, from which we have historical forecasts from the European Centre for Medium Range Weather Forecasting (ECMWF), ocean measurements from a wave buoy, and vessel motion data including over 700 crew transfers.

This paper is organised as follows: Section 2 covers some background on how to obtain vessel specific scenario forecasts. Section 3 describes the problem and Section 4 details the proposed solutions. These are followed by discussion points in Section 5, and conclusions are drawn in Section 6.

2. Generating Vessel Specific Forecast Scenarios

For more information on this stage the reader is referred to related work by the authors [7]. The approach can be summarised in two stages: 1) post-processing of weather forecasts to generate site-specific forecasts; this involves mapping the relationship between concurrent measurements of the sea state from a measurement buoy at the site and the corresponding historical Numerical Weather Predictions (NWP) and 2) creating a crew transfer model which learns the relationship between the vessel motion during push-on and the corresponding met-ocean measurements from the wave buoy. Both stages involve statistical learning techniques to find patterns in the data and map the complex relationships. The block flow diagram in Figure 1 illustrates the modelling stages and how they interlink with each other, including the both the sources and streams of data used, for the training and operational stage of the decision-support tool.

Site specific post-processing of NWP data is known to significantly increase accuracy of the forecasts [8]. This involves using the historic NWP forecasts and observations (e.g. wave buoy or met mast) to identify systematic biases in the NWP and remove them from subsequent forecasts. In this study probabilistic forecasts are generated for each variable of interest (significant wave height and peak wave period) to quantify forecast uncertainty. Both quantile regression and parametric distribution regression are considered.

To refine the forecasting model, clustering is used to separate distinct sea states. In this

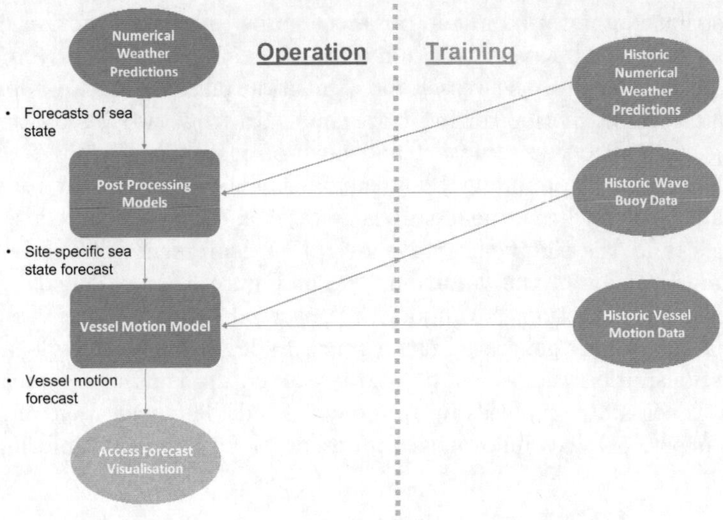

Figure 1. Summary of methodology to generate vessel specific scenarios

case, when either locally generated wind waves or swell from the North Sea dominate. Weather forecasts are used to predict the probability of cluster membership via logistic regression. This allows wave direction to enter the vessel motion modelling process in a straightforward manner.

An example of a probabilistic forecast of significant wave height for up to 5 days ahead is presented in Figure 2. The forecast has a high degree of confidence for the first 4 days for the forecast horizon, after which uncertainty increases significantly. The widest interval stipulates that there is a 90% chance that the observation will fall in this region at each time step.

To use post-processed forecasts as inputs to a vessel motion model, the joint distribution of each time horizon must be estimated using the marginal distributions (i.e. Figure 2) and a copula function [5]. Samples with the correct temporal dependency structure can then be drawn

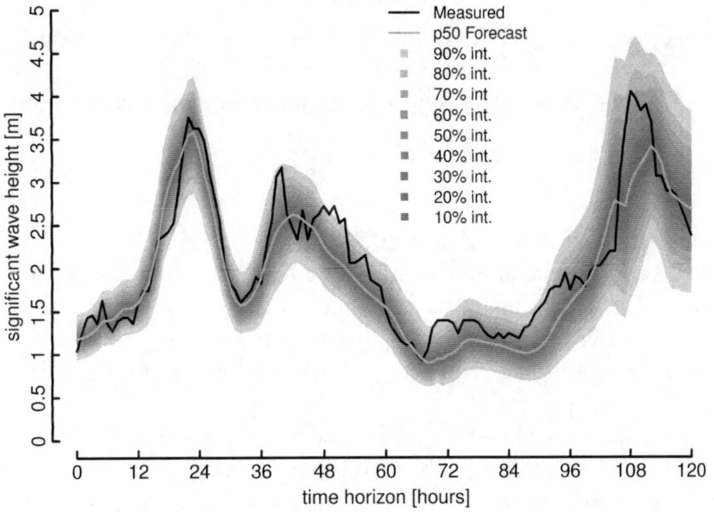

Figure 2. Example post-processed density forecast of significant wave height

from the joint distribution and used as temporal scenarios to a vessel motion model.

Accessibility is constrained by vessel motion during transfer, which is driven by the sea-state, and interactions between waves, the vessel and transition piece. Here we adopt a data-driven approach to estimate vessel motion during transfer. Data from two vessels with identical hull design comprising of detailed vessel motion and push-on instances (≈700) is available and has been analysed alongside corresponding buoy measurements and weather forecasts. There is a clear relationship between heave peak-to-peak vessel motion, which is a crucial metric for successful transfers, and the significant wave height measurements of the buoy — as shown in Figure 3. A large portion of the variation seen in Figure 3 is explained by other variables such as wave period and direction. To model the observed heave-to-peak versus observed sea-state distributional models are generated using general additive models for location, shape, and scale [9]. The relationship between sea state and vessel motion is used to forecast vessel motion during transfer from sea-state scenarios up to 5 days ahead. The next stage of this process, and the focus of this paper, involves innovative communication of forecast information content to end-users.

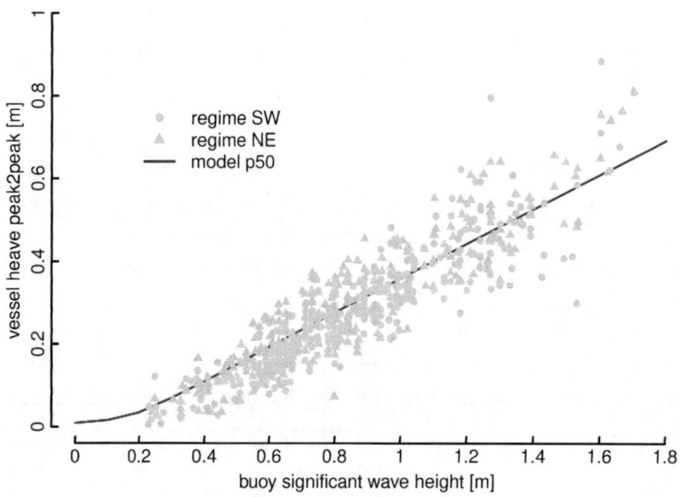

Figure 3. Vessel motion and measured significant wave height. Sea state is divided into two regimes defined on wave direction depending on whether to prevailing wave direction is from the South West (SW) or North East (NE). Blue line shows the median of the model fit in the south west regime.

3. Motivation for Forecast Visualisation Stage

The raw output of the vessel motion model driven by forecast scenarios of the sea-state is shown in Figure 4 and clearly illustrates the motivation behind this study. Although these equally-likely scenarios are useful for driving scheduling or dispatch tools, they present difficult information content for human interpretation. Note the similarity of the forecast profile compared to Figure 2.

Another important aspect to consider which motivates transformation of the forecasts is that often the forecast significant wave height is beyond the maximum measured wave height during transfer. This is understandable given that in reality a CTV skipper would not attempt a transfer in conditions which are unsafe. So beyond vessel heave peak to peak values of ≈0.9m access predictions should reflect that there is zero chance of a successful transfer. Forecasts must then be transformed to adhere to the physical reality of the problem.

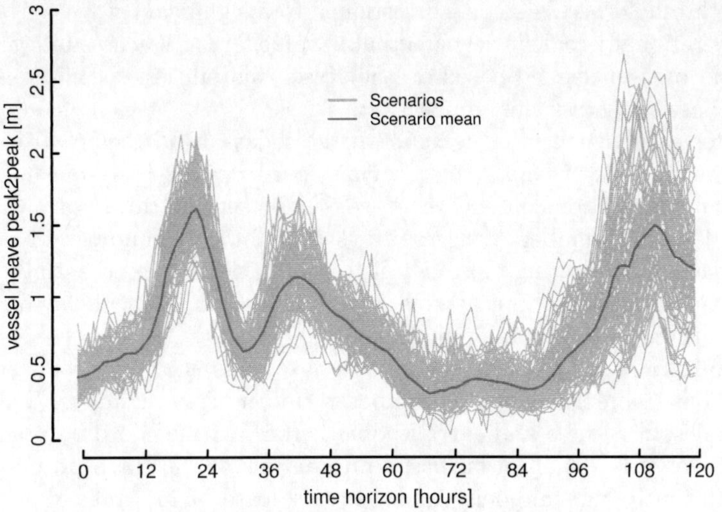

Figure 4. Scenario or trajectory forecasts contain temporal dependency structure that is important to consider in sequential decision making across extended periods of time; however, they are difficult to interpret in this form. Only 10% of the generated scenarios are plotted here.

4. Visualisation Options

This section proposes some methods to transform the vessel specific forecasts into more interpretable and realistic access scores. First it is important to consider what the end user requires; from discussions with industrial partners, marine coordinators desire simplicity from a decision support tool. To achieve this three simple functions are proposed to transfer the forecasts, based on a score out of 10, classification of access conditions, and a threshold score. The purpose of this paper is not to find the best method for communicating information content, it is to propose methods which can be built upon more rigorously in the future, and to introduce visualisation methods to the end-users for feedback. The plots described are implemented in `R` using the base graphics package and `ggplot2` [10, 11].

The user-defined functions for transforming the vessel scenarios into the score out of 10 and access classification are shown in Figures 5 and 6 respectively. The value used for the third

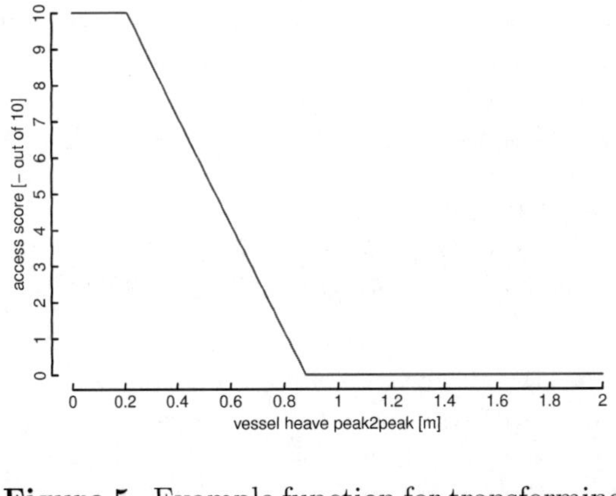

Figure 5. Example function for transforming scenarios into a continuous access score

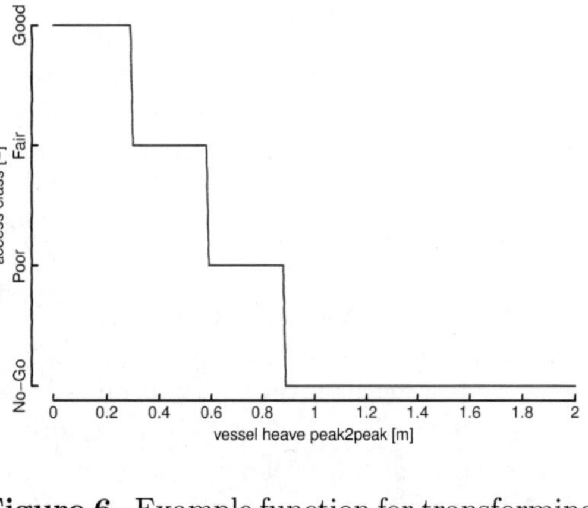

Figure 6. Example function for transforming scenarios into an access classification

method, i.e. the threshold access score, corresponds to a significant wave height value of 1.5m, which sites use operationally to define the maximum value for CTV accessibility. These scores are all flexible and can be user-defined based on the vessel capabilities, specific mission, experience of the site, and appetite for risk, for example.

The detailed forecast visualisation focuses on the 1-day-ahead region with further horizons summarised to the right of the main plot. This opens up the issue of summarising further horizons (2–5 days) information content; three ways of presenting this information are proposed. These are defined as the number of access days available, the number of available weather windows, and a panel plot showing more detailed information. This essentially gives 9 different propositions of visualising these forecasts as any combination of the 1-day-ahead forecast can be combined with three further horizon methods.

In Figure 7 an example of the continuous access score is shown in good access conditions, with further horizons summarised by the number of upcoming access days. This is the simplest of all the options described here and early feedback from marine coordinators have indicated a general preference for this visualisation based on analysis of the three proposed options. The advantages are obviously the simplicity and therefore ease of interpretation. Site and vessel experience are essential for defining the score function.

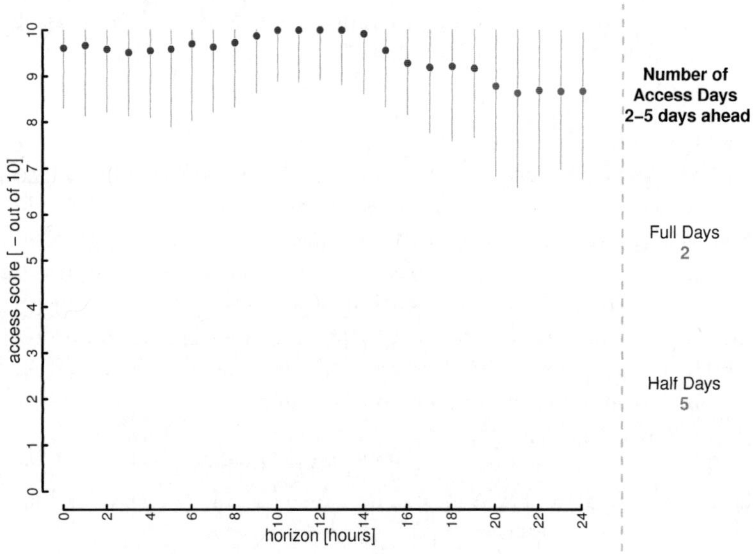

Figure 7. An example visualisation of a 'transfer quality' forecast. Large vessel motions and/or other factors deemed to degrade transfer quality result in a reduction in the score from a maximum of 10. The use of probabilistic forecast enables the uncertainty on this score to be quantified and communicated to the decision-maker. Longer-term forecasts of accessibility is shown for context: if this is the only chance of access for a long time, potential weather windows are more valuable. This plot indicates good access conditions.

An example of the access classification forecast is shown in Figure 8 in variable access conditions. Here the bars quantify the percentage of scenarios belonging to each class at every time step. Further horizon information is presented as the number of available weather windows for access. The advantages of this method is that every scenario produced here is accounted for and therefore the end user views a complete picture of the spread of possibilities at every time step. Additionally, the further horizon information would suit a more opportunistic operations scheduling method because the length of weather windows is flexible and user-defined. The disadvantages of this method is that it is not as immediately intuitive compared to the other visualisation tools.

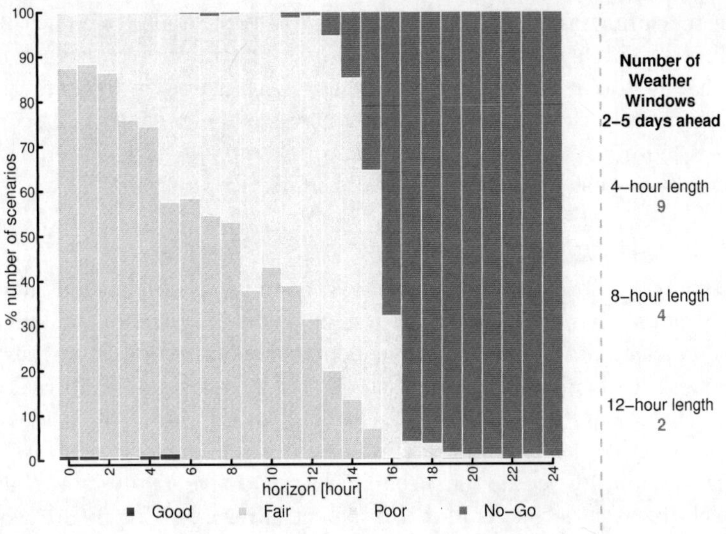

Figure 8. Similar to Figure 7 but scores are classified into descriptive categories. Note that all scenarios are visualised at each time step in this case. Slightly more detailed contextual information is provided in the left panel. This plot indicates variable access conditions.

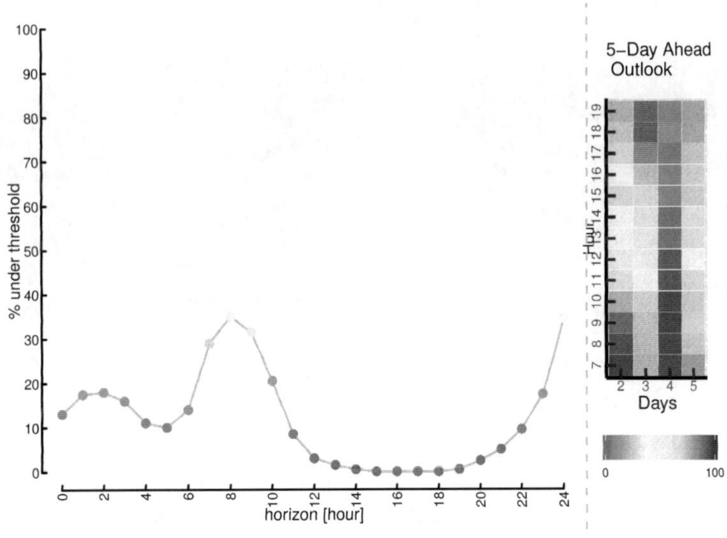

Figure 9. Threshold access score, where the limit is set approximately to the maximum observed vessel motion in 1.5m significant wave height. High values correspond to good access conditions. Further horizons shown in more detail via a panel plot where the colour of each panel corresponds to the percentage under threshold. This plot indicates poor access conditions.

Figure 9 is an example of the threshold score in bad access conditions. The plotted points quantify the percentage of scenarios under the defined limit; this means that high percentages are desirable for access. Details on conditions further into the horizon are shown with more detail via a panel plot where the colour of each panel indicates the threshold score over the next 4 work-days. However, the extra detail is clearly valuable. In Figure 9 good access conditions are clearly visible the following working morning and 4 days ahead which would enable more informed scheduling dispatch decisions immediately. The main advantages of the threshold access score are that it could be easily incorporated within existing contractual vessel access limits and that this visualisation tool is not as sensitive to the user-defined score function. The disadvantage of this score is that it doesn't inform of the spread of the scenarios, consider that 90% of scenarios could be just below the threshold limit.

5. Discussion & Future Work

On marginal access days the opportunity cost of lost revenue can be large if they are followed by days of windy conditions with poor access. It was envisioned at the outset of this visualisation work that access forecasts would be shown concurrently with power forecasts, as shown in Figure 10. Other methods were considered such as a cost/loss model and real-options decision tools. However, operators are currently understandably reluctant to consider power forecasts alongside access forecasting as it implies a potential trade-off between revenue and safety. Alternatively, this information could be used in retrospective analysis to better understand the impact of vessel dispatch decisions and this aspect is reserved for future work.

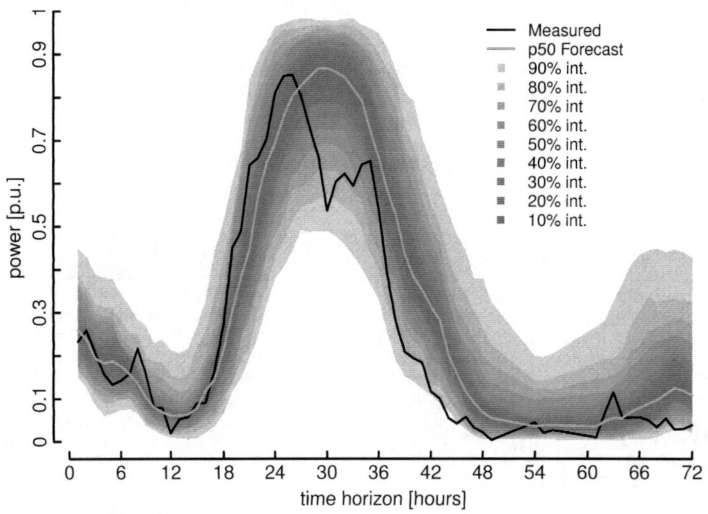

Figure 10. An example of a probabilistic power production forecast and out-turn

There are some crucial environmental factors for access quality which are also left for future work such as lightning risk, visibility, and wind speeds for craning. The intention of this work is to visualise the access forecast information content based on essential factors of wave height, period, and direction. For example, access scores in further work could simply be augmented to include lightning strike forecasts by simply over-riding the access scores to the minimum.

6. Conclusions

This work describes a visualisation strategy for forecasting access quality up to 5 days ahead considering vessel motion for offshore operations. Weather predictions are tuned via site-specific post-processing and forecast uncertainty is quantified using widely available historic data and

IOP Conf. Series: Journal of Physics: Conf. Series **1222** (2019) 012040 doi:10.1088/1742-6596/1222/1/012040

machine learning; please refer to [7] for more detail on the forecasting methodology. The raw output of the vessel motion model driven by forecast scenarios of the sea-state is transformed to better communicate information content. Three simple access forecasts are proposed based on a score out of 10, classification of transfer conditions, and a threshold score. The purpose of this paper is to propose methods which can be built upon more rigorously in the future, and to introduce visualisation methods to the end-users for feedback. This new forecasting and visualisation capability has significant implications for marine coordinators and skippers who will be able to make better-informed safety-critical decisions, with the potential for reductions in the cost of offshore wind energy.

Acknowledgments

This work is supported by the EPSRC Supergen Wind Hub project ORACLES, EP/L014106/1. The authors would like to thank Amine Hadjer, John Best, and Pete Leach at James Fisher Marine Services, Lars Johanning and Giovanni Rinaldi at Exeter University, Sally Shenton at Generating Better, and Christopher Briggs at SiemensGamesa for their excellent discussion points and feedback throughout this work. **Data Statement:** Due to confidentiality agreements with research collaborators, access to vessel motion data is restricted. Numerical weather predictions from ECMWF are available at `www.ecmwf.int` for qualifying research institutions or commercial customers. Wave buoy measurements from the Centre of Environment Fisheries and Aquaculture Science (CEFAS) are available at `http://wavenet.cefas.co.uk/`.

References

[1] Shafiee M, Brennan F, Espinosa IA. A parametric whole life cost model for offshore wind farms. International Journal of Life Cycle Assessment. 2016;.

[2] Catterson VM, McMillan D, Dinwoodie I, Revie M, Dowell J, Quigley J, et al. An economic impact metric for evaluating wave height forecasters for offshore wind maintenance access. Wind Energy. 2016 2;19(2):199–212.

[3] Taylor JW, Jeon J. Probabilistic forecasting of wave height for offshore wind turbine maintenance. European Journal of Operational Research. 2018;.

[4] Bessa R, Möhrlen C, Fundel V, Siefert M, Browell J, Haglund El Gaidi S, et al. Towards Improved Understanding of the Applicability of Uncertainty Forecasts in the Electric Power Industry. Energies. 2017;10(9):1402.

[5] Pinson P, Madsen H, Papaefthymiou G, Klöckl B. From Probabilistic Forecasts to Statistical Scenarios of Short-term Wind Power Production. Wind Energy. 2009;12(1):51–62.

[6] Pinson P, Girard R. Evaluating the quality of scenarios of short-term wind power generation. Applied Energy. 2012;96:12–20.

[7] Gilbert C, Browell J, McMillan D. Probabilistic Access Forecasting for Offshore Operations [in preparation]. 2019;.

[8] Giebel G, Kariniotakis G. Wind power forecasting-a review of the state of the art. In: Renewable Energy Forecasting: From Models to Applications. Elsevier; 2017. p. 59–109.

[9] Rigby RA, Stasinopoulos DM, Lane PW. Generalized additive models for location, scale and shape. Journal of the Royal Statistical Society Series C: Applied Statistics. 2005 6;54(3):507–554.

[10] R Core Team. R: A Language and Environment for Statistical Computing. Vienna, Austria; 2016. Available from: https://www.r-project.org/.

[11] Wickham H. ggplot2: Elegant Graphics for Data Analysis. Springer-Verlag New York; 2016. Available from: http://ggplot2.org.

Transferability of Operational Status Classification Models Among Different Wind Turbine Types

Z. Trstanova[1], A. Martinsson[1], C. Matthews[1], S. Jimenez[2], B. Leimkuhler[1], T. Van Delft[2], M. Wilkinson[2]
[1]University of Edinburgh, Edinburgh, United Kingdom
[2]DNV GL, Oslo, Norway

E-mail: zofia.trstanova@ed.ac.uk, michael.wilkinson@dnvgl.com

Abstract. A detailed understanding of wind turbine performance status classification can improve operations and maintenance in the wind energy industry. Due to different engineering properties of wind turbines, the standard supervised learning models used for classification do not generalize across data sets obtained from different wind sites. We propose two methods to deal with the transferability of the trained models: first, data normalization in the form of power curve alignment, and second, a robust method based on convolutional neural networks and feature-space extension. We demonstrate the success of our methods on real-world data sets with industrial applications.

Keywords: Machine learning, classification, generalization, CNN, wind turbine, wind energy

1. Introduction

Classification of operational status is an important step for performance analysis of wind farms from data of SCADA (Supervisory Control and Data Acquisition) type. The main goal of the analysis is future energy production assessment and the identification of performance issues. The analysis of a typical wind farm may consist of hundreds of millions of SCADA records. Until now, classification of under-performance and different operational states has been performed by a human analyst, automatization of this process can save many working hours and provide additional insight into wind turbine operations. Typically, companies possess historical data from various farms where the power generation status has already been analyzed. Understanding the operational status can explain under-performance, for example by indicating when the turbine may require maintenance. This under-performance may be a consequence of power derating, blade icing, or other conditions. In order to automate this task, we use the knowledge of these under-performing operations within a supervised learning model, which is then employed to classify unlabeled data from various turbine types and farms.

Machine learning has previously been applied to wind turbine SCADA data: (i) for example, time-series regression setting for power prediction [2, 12, 14], and (ii) for fault classification [6]. In this paper, we are interested in multi-class classification of

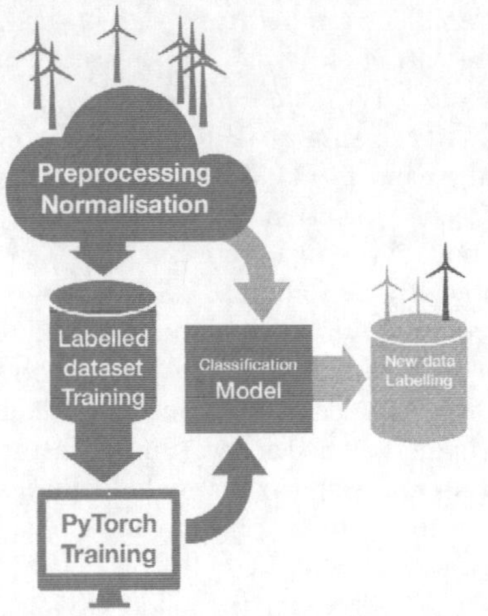

Figure 1. Diagram of the acwind software system for automatic classification of wind turbine performance.

the operational status of power generation for post-construction operational analysis. In a previous article, a classifier was applied to wind turbine fault analysis [6] using support vector machines. In that work, multiple binary classifiers were trained on separate test sets for each class. In real-world problems, the test set contains multiple classes, which makes such an approach difficult to implement in the industrial setting. Another challenge which impedes the industrial application of trained models among different data sets, is the variation between wind turbines: turbine manufacturer, model, geographic location, etc. [3, 6, 13]. This results in different nacelle anemometer power curves (various cut-in and cut-out wind speeds, rated powers, etc., see Figure 2). One of the main challenges in developing a robust method is, therefore, to determine how to handle the issue of generalization of the machine learning model between different farms.

To handle the transferability issue of classification of the operational state, we propose two solutions providing a robust method based on data normalization and feature-space extension using machine learning based on convolutional neural networks (CNN). We incorporate these techniques into a python software package "acwind" which is planned for release in open source mode (see Figure 1 for a diagram illustrating its functionality).

This paper is organized as follows: In Section 2, we describe the data sets and the classification challenge. In Section 3, we introduce baseline normalization in the preprocessing of industrial data sets and the proposed classification methods. Finally, in Section 4 we demonstrate the results of our method on real-world data.

2. Data sets

We use historical SCADA data provided by DNV GL‡. These standard operational 10-minute SCADA contain channels found in a typical modern wind turbine (e.g. wind

‡ www.dnvgl.com

speed, power, pitch, rotor speed, ambient temperature, and others). Each data point has been manually labeled as representing either normal operation or abnormal with a particular status type (such as *anemometer icing, unavailability, partial unavailability, derating, other performance issues*, for example). By a data set, we mean a collection of historical SCADA data from one wind farm. More precisely, these data sets relate to time series data for six farms (i.e. collections of wind turbines of the same type and location), with each time point classed as "normal", C_i (with $i = 1, \dots 4$) or "other". We denote by "other" a class which contains the remaining potential classes of the data set. The data sets contain time series over periods between 18 and 28 months, and with between five and fifteen individual turbines in each farm. Pairs of farms $(1, 2)$, $(3, 4)$ and $(5, 6)$ have similar classes but also different power curves (Figure 2). Note that the data sets include neither warning nor status data, although this information might have been used during the manual labeling process. As part of the preprocessing, the data sets have been anonymized by multiplication by a constant factor and min-max normalized to values in $[0, 1]$.

The wind speed power curve is a useful metric for identifying periods of relative under-performance of a wind turbine, and it can be learned from the data [4, 10]. To deal with the deviations among various data sets, we use a statistical method to learn the reference power curve for each turbine type from historical data. This is possible under the assumption that the turbine operates normally most of the time (in our data sets, around 80% of points were labeled as normal). The power curve baseline can then be obtained by two-dimensional histogram analysis of the operational data, providing for each discrete value of power the value of the corresponding wind speed with maximum frequency. Figure 2 shows the baselines of six farms inferred from the operational data.

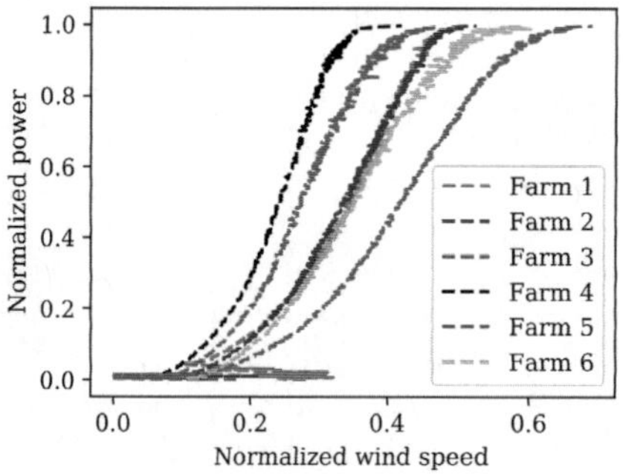

Figure 2. Characteristic power curve of data from six wind turbine farms, learned from the historical time-series of wind speed and power signals.

3. Methods

3.1. Baseline normalization

The goal is to train a supervised learning model on a reference data set, for which various operational states have been identified. As mentioned above, the main challenge is that different farms have a different underlying distribution of SCADA signal values, leading to very different normal behavior described by the baseline, see Figure 2. The aim of this section is to describe a method which allows for the generalization of the classification model to data sets with very different baselines.

The main idea is to preprocess each data set associated with a particular wind farm in such a way that the baselines among the farms are superimposed. In order to map the baseline of one data set onto another "reference" farm, we initially scale the power signal so that all entries lie between 0 and 1. We then apply a linear transformation to the wind speed signal, by multiplying it by a constant α and shifting by a constant β. The values of α and β are chosen to minimize the L1 norm (integrated absolute value) of the difference between the histogrammed distributions of wind speed and power, relative to those of the target wind farm. Standard optimization techniques can be used to achieve this minimization efficiently.

This transformation only requires finding two parameters and only need be done once. Even though nonlinear transformations can be applied to relate the two distributions, we have found that this simple linear transformation is sufficient to improve the generalization across training. Figure 3 illustrates such baseline normalization.

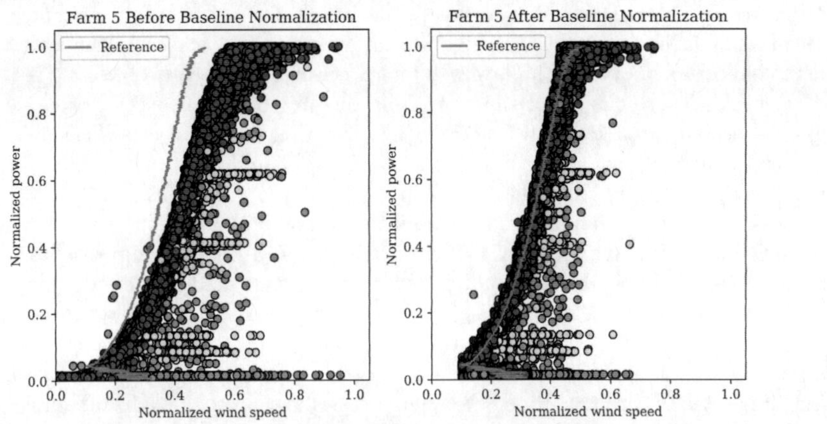

Figure 3. Normalization of data set from wind farm 5 with respect to the reference baseline of wind farm 1. Individual points are colored by the class type provided by four labels, dark blue represents the normal class.

3.2. Model accuracy

We consider a data set $\{(x_1, y_1), \ldots, (x_N, y_N)\}$ containing N data points x_i with associated class labels y_i, where $y_i \in \{1, 2, \ldots, n\}$. Each data point x_i is in $\mathbb{R}^p$ for p available signals. A supervised learning model parameterizes a function $f_\theta(\cdot)$, for parameter vector θ, such that $f_\theta(X) \approx Y$ [5]. The accuracy of a classifier can be estimated by the average over the correct predictions on the training data set,

$$\frac{1}{N}\sum_{i=1}^{N}\delta(y_i - f(x_i)), \tag{1}$$

where N is the number of data points in the set, and δ represents a Dirac delta function.

In order to prevent overfitting on the training data set, we measure the average score (1) over the test set. Due to a class imbalance problem, we estimate the average accuracy per class in the test set.

3.3. KNN and feed-forward neural network

K-nearest neigbors(KNN) algorithm [5] is a non-parametric method, which, for a particular point x estimates the conditional probability of belonging to a class C_j as the fraction of k-nearest neighbors of x that belong to the class $C_j, j = 1, \ldots M$. We use the Euclidean distance to define the set of k-nearest neighbors $S_K(x)$. More precisely, the probability of a point x belonging to the class C_j is

$$\mathbb{P}(y = C_j \mid X = x) = \frac{1}{K} \sum_{i \in S_K(x)} \delta(y_i - C_j).$$

The point x is then classified according to Bayes formula by the class with the largest estimated probability.

Feed-forward neural networks are nonlinear models with specific parameters and layered structure of nodes/neurons, historically motivated by neural science. The parameters of each layer l consist of weights and biases $\theta_l = (w_l, b_l)$ which are associated to connections between the layers. More precisely, the output of layer ℓ is given by $\sigma(w_\ell^T x + b_\ell)$ where x is either the input from the previous layer, or (for $\ell = 1$) the data itself, and σ is an activation function. The loss function for a classification task is the cross-entropy, which is based on maximum likelihood estimation of parameters, aiming at maximizing the probability to observe the labeled data set (x, y). It is possible to access the inferred probabilities of the classifier, however, they might be misleading in the estimation of model uncertainty and Bayesian approaches based on a sampling of the posterior should be favored for this task. For more details on neural networks, see for example [9]. One of the main advantages of neural networks is that, due to their structure and the backpropagation algorithm (based on the chain rule), it is possible to efficiently parameterize the networks even for many parameters and for large data sets. Libraries as PyTorch [11] or Tensorflow [1] provide a framework for efficient handling of distributed training on GPUs.

3.4. Convolutional neural networks

Convolutional neural networks (CNN's) are feed-forward neural networks that have had great success in image and speech recognition [7]. In addition to the fully connected layers, they incorporate convolutional and pooling layers, allowing for feature extraction from a data set, e.g an image, reducing the size of the representation and creating shift invariance.

Time-dependence is an important factor in many of the classes. Motivated by this idea, we extend the features of each data point by the local time series. As in [14], we first preprocess the data by arranging it into a matrix, with the principal data point (with timestamp t_0) located in the center. Figure 4 shows such an augmented data point, where t_k indicates a point at time $t_0 + k$ minutes.

Other options for this matrix are available, for example in the case of online classification we may wish to only consider previous data points. This could easily be used without other changes to the methodology we describe.

t_{-120}	t_{-110}	t_{-100}	t_{-90}	t_{-80}
t_{-70}	t_{-60}	t_{-50}	t_{-40}	t_{-30}
t_{-20}	t_{-10}	$\mathbf{t_0}$	t_{10}	t_{20}
t_{30}	t_{40}	t_{50}	t_{60}	t_{70}
t_{80}	t_{90}	t_{100}	t_{110}	t_{120}

Figure 4. An augmented data point from the local time series organized as a matrix: The central point determines the class label.

We use the following architecture of the feed-forward CNN: two convolutional layers (32 output signals and kernel with size 3) followed by a three-layer feed-forward neural network with ReLU activation functions. We use the cross-entropy as a cost function and we train the network using PyTorch [11] (Adam optimizer, learning rate 0.01). We are aware of the existence of neural networks specifically designed to handle time series as, for example, recurrent neural networks, however we opted to use CNNs for their translation-invariance property, and also motivated by the fact that the labeling process is mostly done by an analyst looking at feature space projections and very occasionally reviewing the time series. As we will show in the results section, extending the feature space in the local time series indeed improves the predictions.

4. Results and discussion

The results described in this section were obtained using our *acwind* [8], which is an open source software package for automatic classification of wind energy analytics. The package requires PyTorch and can be used to train CNN's to classify the operational status of SCADA data. We compare the methods proposed in Section 3: KNN or feed-forward neural network with baseline normalization and CNN. We demonstrate that a CNN does not require data normalization during preprocessing and that the model automatically generalizes among farms with various turbine types.

We do not expect high accuracy for predictions of the "other" class since it contains a different mixture of classes in every data set and the trained model should not relate them. Class C_2 is a very small class, which contains points of unknown labels, also denoted as "spurious". It is therefore difficult to learn and we do not expect to achieve high accuracies for this class, either. When the model is trained on the data set from Farm j, the accuracy is estimated on its test subset (splitting ratio 1/3). We then evaluate the model on Farm k, which we refer to as "test set" in the tables below.

In the first example, we preprocess the data sets with baseline normalization and use a KNN model or a feed-forward neural network. More precisely, after the baseline normalization of the data sets, we parameterize/train two supervised learning models: k-nearest neighbors (KNN) with $K = 50$ neighbors and a feed-forward neural network with three hidden layers with $(12, 6, 6)$ nodes respectively and rectified linear unit (ReLU) activation functions. We use the cross-entropy as cost function. The hyperparameters were chosen by cross-validation on various subsets of the data set. In Table 1 we show results using 3 signals. Table 2 shows equivalent results for a feed-forward neural network. In both cases, the normalization improves the generalization

Accuracy %

Test Set	Farm 1	Farm 3
Farm 1[a]	99(100, 99, 6, 94)	77(82, 97, 2, 37)
Farm 3[a]	73(69, 92, 2, 74)	97(99, 98, 18, 88)

[a] Training set.

Table 1. KNN with baseline normalization using wind-speed, power and pitch when classifying: Normal, C_1, C_2 and other

Accuracy %

Test Set	Farm 1	Farm 3
Farm 1[a]	98(99, 98.6, 31, 90)	72(82, 98, 11, 9)
Farm 3[a]	69(72, 87, 0, 15)	96(99, 93, 0, 79)

[a] Training set.

Table 2. Feed-forward NN with baseline normalization using wind-speed, power and pitch when classifying: Normal, C_1, C_2 and other

between the data sets.

Table 3 shows results using the CNN with four input signals: wind-speed, power, rotor-speed, and pitch. We observe an increase in average accuracy and improved generalization between different farms. In our data sets, only a few classes are present in all six data sets. For example, in the data sets used for results in Table 3, the C_4 class is not present in Farm 1 and 2 and conversely, C_3 is not present in either Farm 3 nor 4. In this example, we replace these missing classes by inserting a new column of "False" values. We can then train on six classes: "normal", C_1, C_2, C_3, C_4 and "other". The results in Table 3 show the robustness of the CNN in transferring the learning among various farms if there is enough representation of the target class in a data set.

We also train a CNN on a *mixed* data set, created by concatenating the data from Farms 1, 2 and 3 (after preprocessing them with the baseline normalization method). The results shown in Table 3 demonstrate that using a mixed data set can give better generalization than using any one data set. We ran experiments using other combinations of farms for the mixed data set and found qualitatively similar results.

We grouped the data sets such that there is a representation of the specific class (C_4 or C_3) in the test set. The results from Table 4 and 5 reinforce the conclusion that, with a good representation of a particular class, CNN can learn the classes very well.

Finally, we note that the machine learning model requires consistency in labeling the data sets used for training since in the absences of this lower accuracies are observed during cross-validation. This issue cannot be resolved by increasing the size of the training data set and can only be addressed by ensuring that the training set does not contain any ambiguities of operational status interpretation.

Accuracy %

Test Set	Farm 1	Farm 2
Farm 1[a]	98(99, 95, 98, 0, 10)	97(99, 80, 0, 99, 0, 0)
Farm 2[a]	94(94, 96, 98, 0, 0)	97(99, 90, 0, 99, 0, 0)
Farm 3[a]	60(61, 0, −, 94, 0, 9)	49(53, 0, 55, 89, 0, 29)
Farm 4[a]	81(87, 0, −, 94, 0, 0)	91(100, 0, 29, 91, 0, 0)
Mixed[a]	98(99, 94, −, 98, 29, 0)	98(99, 85, −, 99, 27, 0)
Test Set	**Farm 3**	**Farm 4**
Farm 1[a]	85(99, −, 0, 96, 0, 0)	94(99, −, 0, 95, 0, 0)
Farm 2[a]	85(99, −, 0, 96, 0, 0)	94(100, −, 0, 92, 0, 0)
Farm 3[a]	97(99, 90, 96, 1, 53)	92(92, 90, 93, 3, 0)
Farm 4[a]	94(100, 70, 95, 0, 0)	99(100, 96, 96, 0, 0)
Mixed[a]	96(98, −, 89, 96, 8, 0)	94(95, −, 75, 97, 57, 0)

[a] Training set.

Table 3. CNN with wind-speed, power, rotor-speed and pitch. Classification of 6 classes: normal, C_3, C_4, C_1, C_2, other

Accuracy %

Test Set	Farm 1	Farm 2
Farm 1[a]	98(100, 93, 99, 1, 3)	97(100, 72, 98, 12, 0)
Farm 2[a]	97(99, 87, 96, 0, 0)	97(100, 74, 97, 0, 0)

[a] Training set.

Table 4. CNN with wind-speed, power signal, rotor-speed and pitch when classifying: Normal, C_3, C_1, C_2 and other

Accuracy %

Test Set	Farm 3	Farm 4
Farm 3[a]	97(99, 89, 97, 3, 61)	96(97, 88, 93, 0, 1)
Farm 4[a]	94(100, 62, 97, 0, 0)	98(100, 85, 97, 0, 0)

[a] Training set.

Table 5. CNN with wind-speed, power signal, rotor-speed and pitch when classifying: Normal, C_4, C_1, C_2 and other

5. Conclusions

We have proposed two methods to deal with the generalization of classification models among different wind turbine types. In case relatively few signals are available, data normalization can improve the generalization as demonstrated on both KNN and feed-forward neural networks. The use of a CNN can provide a more robust method, which is invariant to the baseline differences and allows for the use of feature-space extension in the form of a local time series. If the class representation within the data set is sufficient, the accuracy is correspondingly high. This suggests that historical data from many farms can be combined to create a "mixed training set" with all the possible operational states. In order to combine SCADA data from various farms into the mixed data set, each of the data sets needs to be normalized in advance. A CNN based model provides robust predictions for the specific classes, thus providing a comprehensive, robust solution for demanding industrial applications.

References

[1] Martín Abadi, Paul Barham, Jianmin Chen, Zhifeng Chen, Andy Davis, Jeffrey Dean, Matthieu Devin, Sanjay Ghemawat, Geoffrey Irving, Michael Isard, and Others. Tensorflow: a system for large-scale machine learning. In *OSDI*, volume 16, pages 265–283, 2016.

[2] A Clifton, L Kilcher, J K Lundquist, and P Fleming. Using machine learning to predict wind turbine power output. *Environ. Res. Lett.*, 2013.

[3] Simon Gill, Bruce Stephen, and Stuart Galloway. Wind turbine condition assessment through power curve copula modeling. *IEEE Trans. Sustain. Energy*, 3(1):94–101, 2012.

[4] E Gonzalez, B Stephen, D Infield, and J J Melero. On the use of high-frequency SCADA data for improved wind turbine performance monitoring. In *J. Phys. Conf. Ser.*, volume 926, page 12009. IOP Publishing, 2017.

[5] Gareth James. *An Introduction to Statistical Learning.* Springer.

[6] Kevin Leahy, R Lily Hu, Ioannis C Konstantakopoulos, Costas J Spanos, and Alice M Agogino. Diagnosing Wind Turbine Faults Using Machine Learning Techniques Applied to Operational Data. *Progn. Heal. Manag. (ICPHM), 2016 IEEE Int. Conf.*, 2016.

[7] Yann Lecun, Yoshua Bengio, and Geoffrey Hinton. Deep learning. *Nature*, 2015.

[8] Anton Martinsson, Zofia Trstanova, and Charles Matthews. https://github.com/acwind-lib/acwind, 2019.

[9] Kevin P Murphy. *Machine learning : a probabilistic perspective.* MIT Press, Cambridge, Mass. [u.a.], 2013.

[10] Joon-Young Park, Jae-Kyung Lee, Ki-Yong Oh, and Jun-Shin Lee. Development of a novel power curve monitoring method for wind turbines and its field tests. *IEEE Trans. Energy Convers.*, 29(1):119–128, 2014.

[11] Adam Paszke, Sam Gross, Soumith Chintala, Gregory Chanan, Edward Yang, Zachary DeVito, Zeming Lin, Alban Desmaison, Luca Antiga, and Adam Lerer. Automatic differentiation in PyTorch. 2017.

[12] Alla Sapronova, Meissner Catherine, and Matteo Mana. Short time ahead wind power production forecast. *J. Phys. Conf. Ser.*, 2016.

[13] Onder Uluyol, Girija Parthasarathy, Wendy Foslien, and Kyusung Kim. Power curve analytic for wind turbine performance monitoring and prognostics. In *Annu. Conf. Progn. Heal. Manag. Soc.*, volume 2, pages 1–8, 2011.

[14] Anwen Zhu, Xiaohui Li, Zhiyong Mo, and Huaren Wu. Wind Power Prediction Based on a Convolutional Neural Network. *Int. Conf. Circuits, Devices Syst. Wind*, 2017.

IOP Conf. Series: Journal of Physics: Conf. Series **1222** (2019) 012042 doi:10.1088/1742-6596/1222/1/012042

Analysing Current Signature Data to Diagnose an In-Service Wind Turbine Generator

Estefania Artigao[1,*], Elena Gonzalez[2], Andres Honrubia-Escribano[1], Emilio Gomez-Lazaro[1]

[1] Renewable Energy Research Institute, DIEEAC-EDII-AB University of Castilla La Mancha, Campus Universitario s/n, 02071, Albacete, Spain.

[2] ORSEIDE Renewable Energy, Calle Dublin 62, 31621, Pamplona, Spain.

E-mail: `Estefania.Artigao@uclm.es`

Abstract. Condition-based maintenance (CBM) is currently seen as the preferred approach to optimise operation and maintenance (O&M) strategies by the early detection and diagnosis of critical failures for wind turbines (WT). WT reliability is highly affected by failures of the mechanical components from the drive train, although doubly-fed induction generators (DFIG) also contribute to high failure rates and downtime. As DFIG is the dominant technology employed for variable speed WTs, dedicated failure detection and diagnosis techniques are required. Current signature analysis (CSA) has emerged as a powerful tool for this purpose, despite some limitations. In this work, the variability of CSA through time by means of spectra waterfall and the wavelet transform are investigated to overcome these limitations. The analysis of a faulty in-service WT DFIG presented in this work raises the usefulness of the suggested approaches to identify patterns related to failure development through time.

1. Introduction

The global installed wind energy capacity continues to grow, reaching nearly 540 GW by the end of 2017 [1]. Of these, around 18 GW were located offshore, market led by the European countries [2]. Onshore availability has been found to be over 97%, achieved through appropriate O&M activities [3]. Different numbers are obtained for offshore sites, of about 85% availability; this significant difference is mainly due to inaccessibility issues and to the current uncertainty surrounding failure characteristics [4].

With larger and more modern wind turbine (WT) generators under continuous development, operation and maintenance (O&M) activities are key to improve reliability and availability of WTs while reducing costs [5], hence becoming the spotlight in the wind energy sector. Three maintenance strategies are commonly implemented for WT O&M: time-based maintenance (TBM), failure-based maintenance (FBM), and condition-based maintenance (CBM) [6]. TBM refers to scheduled actions that take place to ensure a certain performance level of the system; these interventions are regularly programmed disregarding the presence of failures in the system, such as the typical summer maintenance campaigns in offshore wind farms. Conversely, FBM is carried out once a functional failure is confirmed; these interventions are inevitably linked to higher WT downtime and costs. Conventional onshore O&M activities consist of a combination of TBM and FBM, such strategy has room for improvement and is unwise offshore. New trends are moving from these two towards CBM, where condition monitoring (CM) determines the optimum point between scheduled and corrective actions [7]. Necessary interventions can therefore be planned when needed, with sufficient time, leading to reduced downtime and costs [8].

WT components are unequally affected by failures, both in terms of occurrence and criticality, which is typically linked to the resulting downtime and repair or replacement cost. The mechanical drive train is among the top three contributors to failure rates and downtime of WTs, together with electric generators [9], where doubly-fed induction generator (DFIG) is the dominant technology employed for variable speed WTs [10]. Early detection of gear, bearing and generator faults is therefore crucial if CBM is to be implemented in order to reduce corrective and scheduled actions. Nonetheless, while commercial solutions are readily available and widely used for monitoring the mechanical condition of rotating equipment, mainly based on vibration analysis, there is no such an established approach for electric systems.

Data from real operating WTs over the period preceding a functional failure are rarely academically analysed since they are hardly ever made available. Particularly, there is a lack of electrically-based diagnosis techniques applied specifically to in-service WT equipped with DFIGs. Many efforts have been made through computer simulations and at laboratory test benches, but very few on in-service WT generators. Furthermore, the studied failure modes are commonly known to the authors before the analysis is carried out, and sometimes a baseline of the healthy-stage is also available, so that the deviations due to the developing failure symptoms are more easily identified. Hence, it is clear that further research applied to in-service WTs is required to validate experimental simulations and models, especially for machines under unknown health status.

In the present work, an in-service WT DFIG has been analysed for a period of eight months to assess the development of a failure. Such failure had been previously diagnosed using current signature analysis (CSA), which was confirmed through vibration analysis, see [11]. CSA for wind turbine monitoring is still at an early stage; current spectra are particularly difficult to be tracked over time. Unlike vibration analysis where specific features from the spectra can be traced over time and linked to specific failure modes, CSA is negatively affected by a higher complexity and a lack of maturity currently limiting their detection and diagnosis capabilities. In this work, we explore the variability of CSA over time by means of spectra waterfall and wavelet transform, and address its usefulness to identify patterns related to failure development as a first step towards improving the detection and diagnosis of WTs equipped with DFIGs.

2. Methodology

Condition monitoring approaches for induction machines, such as DFIGs, include temperature, chemical and wear, mechanical vibration, and electrically-based analysis [12]. The latter usually covers current, voltage, instantaneous power and flux analysis. Of all these, stator current analysis is the most widely applied, as it is cheap, non-intrusive, and can monitor both electrical and mechanical failures. DFIGs major failures are most often due to bearing (between 40% and 60%), stator (between 28% and 38%) and rotor failures (around 10%) [13].

2.1. Current Signature Analysis (CSA)

This work explores the possibility of detecting failure related symptoms revealed as variations in CSA over time. Indeed, CSA is based on the principle that each developing failure mode has its own effect on the current spectra [10]. The appearance and development of an incipient failure should be therefore translated into a variation of the current spectra. Several signal processing methods can be applied for the purpose of faulty state identification, typically classified into time-domain, frequency-domain and time-frequency domain [14]. CSA lies within the frequency-domain techniques, whose most obvious disadvantage is the loss of temporal information. To overcome this limitation, two approaches are explored in the present work, and applied to data from a real operating WT equipped with a DFIG for a period of eight months.

Firstly, frequency-domain analysis is applied to current signals from the stator via Fast Fourier Transform (FFT). This technique transforms the waveform signal from the time domain to the frequency domain. It is widely used for fault diagnosis because the variations of certain harmonic components in the frequency spectrum of a signal can be related to a specific fault type [14, 15, 16]. The following fault frequencies have been identified in the current spectra of the machine under study, and tracked over time. In the presence of rotor mechanical asymmetries, a set of new components at frequencies given by Eq. (1) appear:

$$f_{RFS} = f_s \pm \kappa s f_s \tag{1}$$

Rotor electrical unbalance can be identified by the components given by Eq. (2):

$$f_{FRU} = f_s \left| \frac{\kappa}{p}(1 - s) + l \right| \tag{2}$$

However, since a minimal degree of asymmetry always exists in an induction machine, the presence of the first pair of f_{RFS} and/or f_{FRU} is not sufficient to diagnose such fault(s). It is the amplitude increment of these faults, and the increment in the number of sub-harmonics ($\kappa > 1$) that will determine the presence of such fault(s).

There are other fault types, such as gearbox-related faults, whose frequency components will only be seen in the presence of a fault. Gearbox faults can originate from a gearbox shaft, bearing, gear, pinion, or a combination of these, generating characteristic rotor eccentricity components. Frequency components related to damaged teeth, scoring and debris can be calculated by Eq. (3):

$$f_{GBX} = f_s \left| 1 \pm \frac{\kappa}{G_r p} \right| \tag{3}$$

Once these fault-related frequencies have been identified, their behaviour is monitored throughout time with the purpose of identifying symptoms related to a developing failure. This work proposes the study of spectra waterfall with the goal of detecting any temporal variability. Spectra waterfall is a 3D representation of the signal spectra (current spectra in this case) at successive intervals of time, thus allowing to track changes in the current spectra over time. The chosen signals must be obtained at the same operating conditions in order to have consistency on the fundamental frequency and other derived frequencies of interest.

2.2. Wavelet Analysis

The second approach used in this work lies within the group of time-frequency techniques, as an alternative to overcome the loss of temporal information related to CSA. Wavelet analysis is the most widely applied time-frequency domain analysis technique. Over the past years, it has emerged as a powerful tool in many areas such as data analysis, image processing and fault detection, especially for electrical machines [17].

Like Fourier analysis, wavelet analysis is based on the principle of function expansion in terms of a set of basis functions. While in Fourier Analysis the basis functions are trigonometric, wavelet analysis decomposes functions in terms of *wavelets*, generated from translations and dilations of a prototype, called *mother wavelet* [18]. One of the main differences is the finite nature of this *mother wavelet* function versus the infinite character of sinus and cosinus. This makes time localisation possible in wavelet analysis. Unlike the FFT that provides a local representation of a signal, the wavelet transform represents changes in frequency throughout time facilitating the analysis of non-stationary signals and transient detection of abnormal events. This hybrid nature of time-frequency domain analysis techniques poses a challenge in time and frequency resolution, due to the Heisenberg uncertainty principle: higher precision in one quantity implies lower in the other. As a result, appropriate time resolution and poor frequency resolution are obtained at high frequencies and vice versa. The interested reader is referred to [18, 19] for more information about wavelet analysis and theory.

Since the faulty state in the analysed turbine is related to rotor mechanical asymmetries, the Morlet wavelet was selected, as its shape is similar to mechanical shock signals. This wavelet transform has already been used for WT condition monitoring, although applied to vibration signals [20]. In this work, the wavelet transform is computed for the current signals from the stator and then the temporal evolution of wavelet power levels at frequencies close to the frequencies of interest, previously identified (see section 2.1), is further analysed.

3. Case Study

A Spanish company who provide O&M services, specialising in power converters, generators, turbine controllers, CM systems and SCADA systems, has provided the data for the case study. The reported WT is equipped with a DFIG with the following characteristics, Table 1:

Table 1. Main characteristics of the DFIG under study.

Rated Power	1,500 kW
Pole Pairs, p	3
Synchronous Speed	1,000 rpm
Supply Frequency, f_s	50 Hz

The WT under study was diagnosed with rotor mechanical asymmetry [11], as previously mentioned. The database provided contained between 300 and 900 acquisitions per month, with each acquisition comprising six generator current measurements (three stator and three rotor currents) under different WT loading conditions. A sub-set of the database is given in Figure 1, showing the distribution of measurements for different values of speed and power.

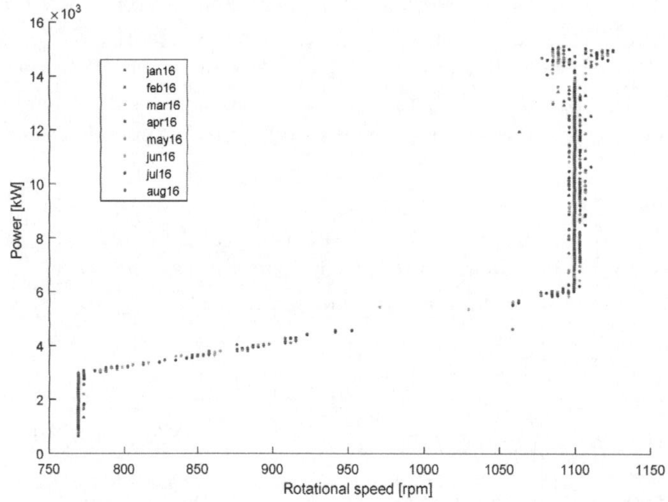

Figure 1. Selected WT measurements.

Of these, ten signals per month have been selected to illustrate the development of the fault over time, both at sub-synchronous and super-synchronous regimes. Such evolution is

IOP Conf. Series: Journal of Physics: Conf. Series **1222** (2019) 012042 doi:10.1088/1742-6596/1222/1/012042

explored by means of spectra waterfall and wavelet analysis, presented as follows in sections 3.1 and 3.2, respectively. Once the appearance of the fault is detected, advanced diagnosis is achieved through CSA (section 3.3).

3.1. Spectra Waterfall Results

Previous studies of vibration spectra waterfall have proven very useful in detecting changes in the spectra over time [11, 21]. The same methodology has been applied to the wind turbine under analysis, using the current spectra instead, as shown in Figure 2, for the eight-month period.

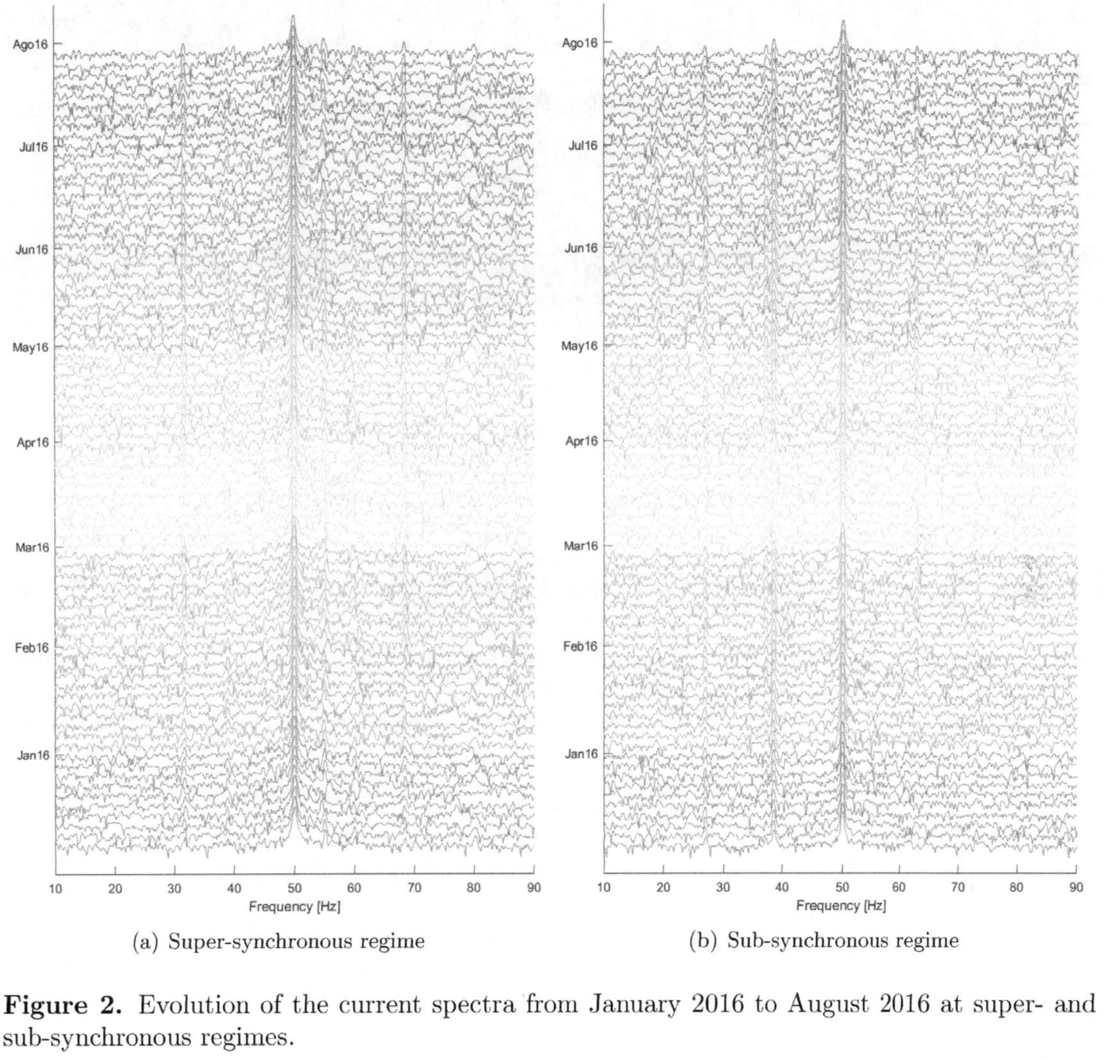

Figure 2. Evolution of the current spectra from January 2016 to August 2016 at super- and sub-synchronous regimes.

Since the faulty state in the turbine under analysis is related to rotor mechanical asymmetries, the objective is to track fault frequency components given by f_{RFS} (Eq. (1)) and f_{FRU} (Eq. (2)). Looking at the current spectra waterfall (Figure 2) the increment in amplitude or appearance of sub-harmonics related to f_{RFS} and/or f_{FRU} over time is not evident.

IOP Conf. Series: Journal of Physics: Conf. Series **1222** (2019) 012042 doi:10.1088/1742-6596/1222/1/012042

3.2. Wavelet Analysis Results

The wavelet transform was computed for the signals acquired separately, as well as concatenated. Figure 3 shows the resulting scalogram for one of the signals acquired in January 2016 at nominal power (super-synchronous regime), while Figure 4 illustrates the scalogram for the concatenated signals. As one can see in Figure 3, the scalogram shows the temporal evolution of the wavelet power levels for the covered range of frequencies. Additionally, the low temporal resolution at low frequencies is clearly seen at the top of the graph, and so is the high temporal resolution at high frequencies at the bottom. Apart from the first and second harmonics of the fundamental frequency, no significant wavelet power levels are observed.

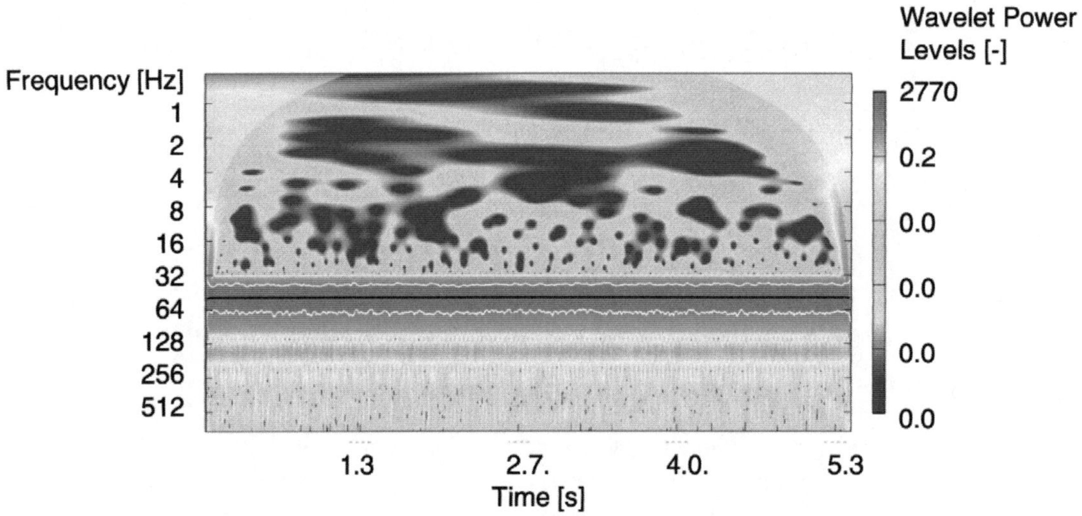

Figure 3. Wavelet scalogram of one of the current signals measured in January 2016.

Conversely, some peaks are observed in Figure 4, due to the discontinuities between the different samples analysed, and these are not meaningful for any deviation detection purposes. Ideally, a continuous signal should have been analysed for the purpose of tracking changes over time, which is currently not possible as can be seen in Figure 4.

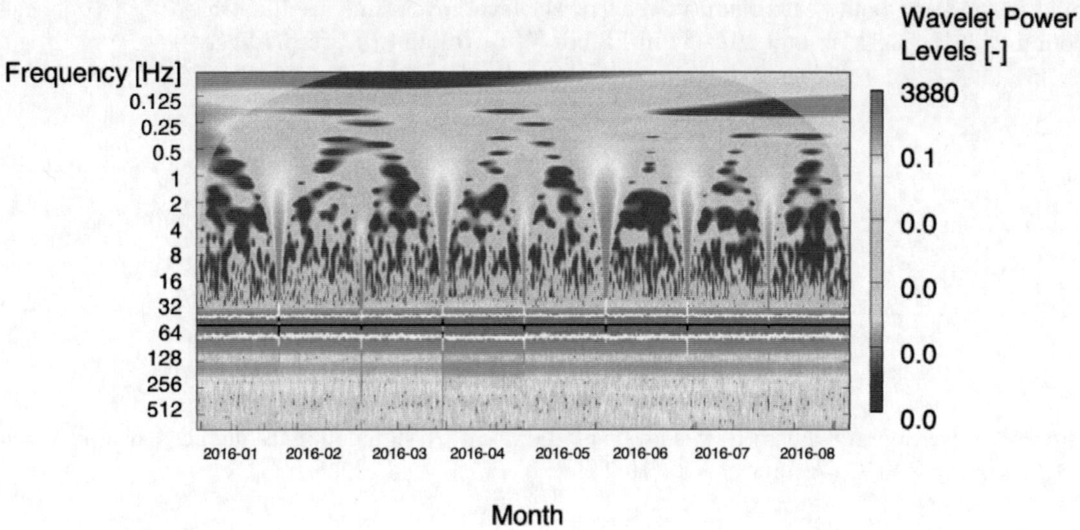

Figure 4. Scalogram of the signals measured over the eight months.

It should be noted that the limited resolution in frequency does not allow to track changes around the faulty frequencies identified with CSA. Also, the frequencies of interest are close to the supply frequency and this also tarnishes any significant deviation over time. To overcome this disadvantage, the wavelet power levels have been further analysed to detect any variability though time related to failure development. The average wavelet power was computed separately for the different monthly signals and for frequencies around 30 Hz (corresponding to the frequency of interest at nominal power). Results are illustrated in Figure 5 .

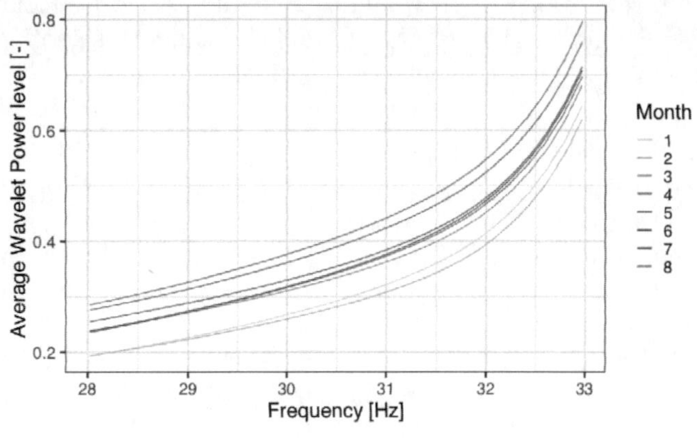

Figure 5. Average wavelet power levels of the monthly signals for frequencies around 30 Hz.

As one can see, as the failure develops the wavelet power level related to the faulty frequency increases. Notwithstanding the limited resolution in frequency, this supports the usefulness of the wavelet transform for tracking changes in electrical signals for monitoring purposes. Additionally, further harmonics were analysed also for signals at sub-synchronous regimes (lower loads), to identify regions where deviations could be more significant. Also, maximum wavelet power levels were retained, as this statistic is more likely to capture peaks than average, where deviations could be smoothed out. Maximum wavelet power levels over time are illustrated for frequencies around 26.9 Hz, 3.84 Hz and 19.2 Hz in Figure 6 (a), (b) and (c), respectively.

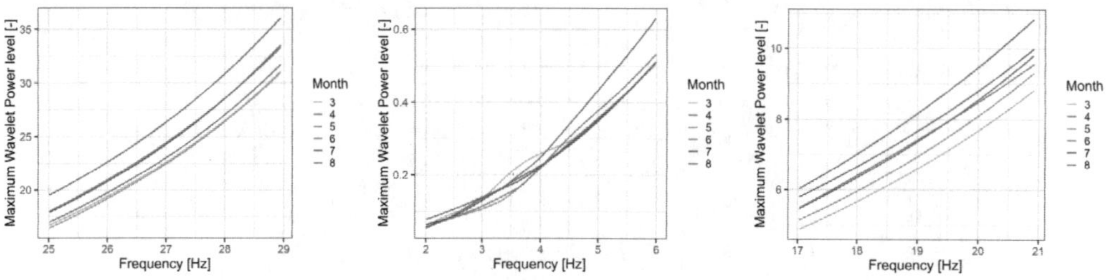

(a) $k = 1$ (b) $k = 2$ (c) $k = 3$

Figure 6. Maximum wavelet power levels of the monthly signals for frequencies around $k = 1$ (26.9 Hz), $k = 2$ (3.84 Hz) and $k = 3$ (19.2 Hz).

Although the increasing trend is doubtful for $k = 2$, it can however be clearly identified for the other cases in Figure 6. Furthermore, the magnitude of the observed increments are much higher than in Figure 5, where the average wavelet power levels were computed. Monitoring the maximum power levels at different frequencies over time opens the door to new faulty state detection possibilities for WT DFIGs. However, the limitation related to frequency resolution impedes a correct diagnosis, that could be overcome by punctual CSA.

Summing up, time-frequency domain analysis techniques, such as wavelets, allow the detection of changing behaviour of stature current signals over time but do not offer enough frequency resolution for an advanced diagnosis of the failure mode developing in the system. Once a deviation is detected through a wavelet analysis, diagnosis can be assessed through specific CSA that offers a very good precision in the frequency domain.

In practice, this solution could be implemented by performing the wavelet analysis for current data recorder at medium resolution (10 ms would be enough to cover frequencies around 30 Hz), that is readily available for modern WTs. CSA of higher frequency signals would only need to be performed in the event of a deviation detected by the first analysis.

3.3. Advanced Diagnosis through CSA

Once the deviation has been detected, an in-depth analysis is carried out through CSA. The wavelet analysis showed that fault-related frequency components given by f_{RFS} (Eq. (1)) increase in amplitude from January to August 2016, which was clear for the sub-harmonics $k = 1, 3$ and doubtful for $k = 2$. In order to validate these results, the current spectra of punctual measurements were analysed.

Figure 7 shows CSA for the sub-synchronous regime. As one can see, further peaks appear in August 2016 (Figure 7 (b)) compared to January 2016 (Figure 7 (a)). When analysing Figure 7 (b) one can observe one full pair for f_{RFS} for $k = 1$ and the left side for $k = 3$, plus two full pairs of f_{FRU} (for $k = 1, 3$). Whereas only the left side for f_{RFS}, $k = 3$ appears in Figure 7 (a). It must be noted that no pair corresponding to $k = 2$ is present, which explains why the wavelet analysis results were not able to confirm the increment for such component. It is very common that only the odd sub-harmonics appear in the faulty spectra.

The results at super-synchronous regime are shown in Figure 8. In this case, two full pairs of f_{RFS} (for $k = 1, 3$) and one full pair of f_{FRU} ($k = 1$) can be seen in August 2016 (Figure 8 (b)). Whereas in January 2016, Figure 8 (a), only the right-side components of f_{RFS} for $k = 1, 3$ and one full pair of f_{FRU}, ($k = 1$), are found. An increment both in amplitude and number of sub-

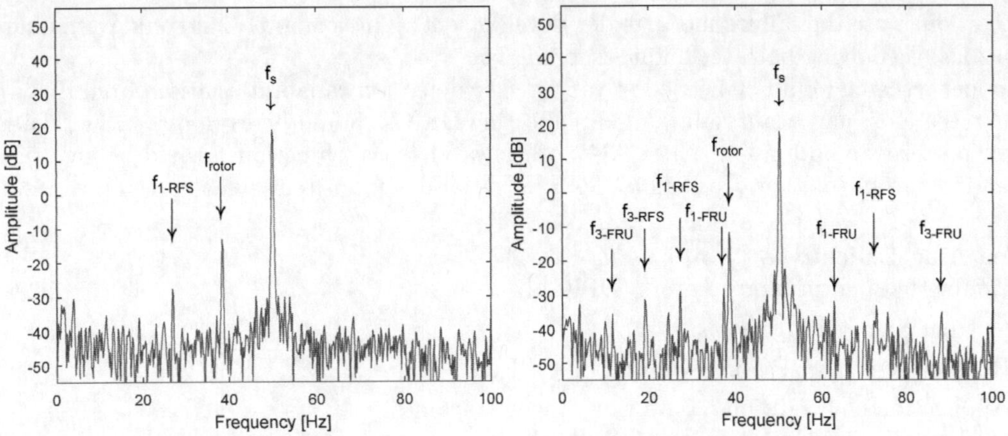

(a) January 2016 (b) August 2016

Figure 7. Advanced diagnosis via CSA at sub-synchronous regime.

harmonics is therefore appreciated in August 2016 compared to January 2016, which belongs to the sub-harmonics $k = 1, 3$; again, no sub-harmonics appear for $k = 2$.

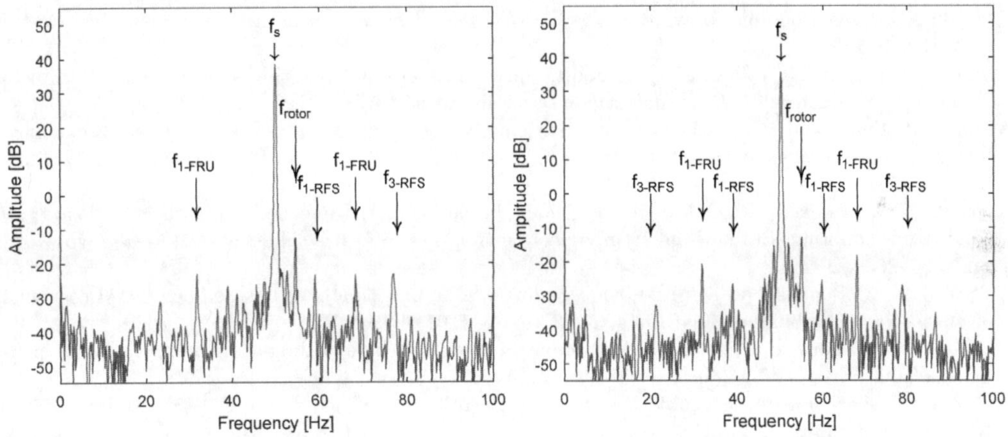

(a) January 2016 (b) August 2016

Figure 8. Advanced diagnosis via CSA at super-synchronous regime.

4. Conclusions

The current outlook of O&M in the wind energy sector has been briefly presented, introducing the motivation of the present work: to enhance O&M activities through CM in order to increase reliability and availability of WTs while reducing costs.

In this work, the basis of CSA and its application on WT DFIGs diagnosis have been presented. Two different approaches have been applied to an in-service WT equipped with a DFIG, for a period of eight months in order to assess the development of a failure. The variability of CSA through time by means of spectra waterfall and wavelet transform have been explored to address its usefulness to identify patterns related to failure development.

The wavelet transform, that is less computationally expensive for being run routinely, can contribute to the detection of abnormal behaviour and localisation in time. Then, the punctual

CSA can diagnose the failure mode under development. This combination could overcome the limitations of applying both techniques separately.

Further research would be needed to validate the detection capabilities arisen from this work. Namely, the most important failure modes affecting DFIGs should be explored, to be related to specific patterns in both approaches. Also, other wavelet basis functions should be investigated to identify suitable mother wavelets to describe variations in current signals.

5. Learning Objectives

- Doubly-fed induction generators (DFIG).
- Current Signature Analysis.
- Real analysis on operating wind turbine DFIG.
- Application on condition monitoring of wind turbines.

References

[1] L. Fried, "Global Wind Statistics 2017," tech. rep., Global Wind Energy Council, 2017.
[2] E. Zountouridou, G. Kiokes, S. Chakalis, P. Georgilakis, and N. Hatziargyriou, "Offshore floating wind parks in the deep waters of Mediterranean Sea," *Renewable and Sustainable Energy Reviews*, vol. 51, pp. 433–448, 2015.
[3] F. Spinato, P. J. Tavner, G. J. W. Bussel, and E. Koutoulakos, "Reliability of wind turbine subassemblies," *IET Renewable Power Generation*, vol. 3, no. 4, pp. 387–401, 2008.
[4] B. Hahn, M. Durstewitz, and K. Rohrig, "Reliability of wind turbines," in *Wind Energy*, pp. 329–332, 2007.
[5] M. I. Blanco, "The economics of wind energy," *Renewable and Sustainable Energy Reviews*, vol. 13, no. 6, pp. 1372–1382, 2009.
[6] F. P. G. Márquez, A. M. Tobias, J. M. P. Pérez, and M. Papaelias, "Condition monitoring of wind turbines: Techniques and methods," *Renewable Energy*, vol. 46, pp. 169–178, 2012.
[7] W. Qiao, P. Zhang, and M.-Y. Chow, "Condition monitoring, diagnosis, prognosis, and health management for wind energy conversion systems," *IEEE Transactions on Industrial Electronics*, vol. 62, no. 10, pp. 6533–6535, 2015.
[8] P. Tchakoua, R. Wamkeue, M. Ouhrouche, F. Slaoui-Hasnaoui, T. A. Tameghe, and G. Ekemb, "Wind turbine condition monitoring: State-of-the-art review, new trends, and future challenges," *Energies*, vol. 7, no. 4, pp. 2595–2630, 2014.
[9] J. Carroll, A. McDonald, and D. McMillan, "Failure rate, repair time and unscheduled O&M cost analysis of offshore wind turbines," *Wind Energy*, vol. 19, pp. 1107–1119, 2015.
[10] C. A. Rieg, C. J. Smith, and C. J. Crabtree, "Monitoring wind turbine loading using power converter signals," in *Journal of Physics Conference Series 749*, 2016.
[11] E. Artigao, S. Koukoura, A. Honrubia-Escribano, J. Carroll, A. McDonald, and E. Gómez-Lázaro, "Current signature and vibration analyses to diagnose an in-service wind turbine drive train," *Energies*, vol. 11, no. 4, 2018.
[12] M. E. H. Benbouzid, "A review of induction motors signature analysis as a medium for faults detection," *IEEE Transactions on Industrial Electronics*, vol. 47, no. 5, pp. 984–993, 2000.
[13] S. Nandi, H. A. Toliyat, and X. Li, "Condition monitoring and fault diagnosis of electrical motors - A review," *IEEE Transactions on Energy Conversion*, vol. 20, no. 4, pp. 719–729, 2005.
[14] C. Sheng, Z. Li, L. Qin, Z. Guo, and Y. Zhang, "Recent progress on mechanical condition monitoring and fault diagnosis," *Procedia Engineering*, vol. 15, pp. 142–146, 2011.
[15] Y. Amirat, M. E. H. Benbouzid, E. Al-Ahmar, B. Bensaker, and S. Turri, "A brief status on condition monitoring and fault diagnosis in wind energy conversion systems," *Renewable and Sustainable Energy Reviews*, vol. 13, no. 9, pp. 2629–2636, 2009.
[16] C. J. Crabtree, S. Djurovic, P. J. Tavner, and A. Smith, "Condition monitoring of a wind turbine DFIG by current or power analysis," in *Power Electronics, Machines and Drives (PEMD), 5th IET International Conference*, pp. 1–6, IET, 2010.
[17] L. Satish, "Short-time fourier and wavelet transforms for fault detection in power transformers during impulse tests," *IEE Proceedings-Science, Measurement and Technology*, vol. 145, no. 2, pp. 77–84, 1998.
[18] D. T. L. Lee and A. Yamamoto, "Wavelet Analysis : Theory and Applications," *Hewlett-Packard Journal*, no. December, pp. 44–54, 1994.
[19] A. Graps, "An Introduction to Wavelets," *IEEE Computational Science and Engineering*, vol. 2, no. 2, pp. 50–61, 1995.

[20] B. Tang, W. Liu, and T. Song, "Wind turbine fault diagnosis based on Morlet wavelet transformation and Wigner-Ville distribution," *Renewable Energy*, vol. 35, no. 12, pp. 2862–2866, 2010.
[21] S. Koukoura, J. Carroll, and A. McDonald, "Wind turbine intelligent gear fault identification," in *Annual Conference of the PHM Society*, 2017.

IOP Conf. Series: Journal of Physics: Conf. Series **1222** (2019) 012043 doi:10.1088/1742-6596/1222/1/012043

A Novel Probabilistic Long-Term Fault Prediction Framework Beyond SCADA Data - With Applications in Main Bearing Failure

Jürgen Herp

Applied Statistical Processing Group (πSeG), Embodied Systems for Robotics and Learning (ESRL), Maersk Mc-Kinney Moeller Institute, University of Southern Denmark, 55 Campusvej, 5230 Odense, Denmark

E-mail: herp@mmmi.sdu.dk

Niels L. Pedersen

Diagnostics, Siemens Gamesa Renewable Energy, Borupvej 16, 7330 Brande, Denmark

E-mail: niels.l.pedersen@siemensgamesa.com

Esmaeil S. Nadimi

Applied Statistical Processing Group (πSeG), Embodied Systems for Robotics and Learning (ESRL), Maersk Mc-Kinney Moeller Institute, University of Southern Denmark, 55 Campusvej, 5230 Odense, Denmark

E-mail: esi@mmmi.sdu.dk

Abstract. Prognostic and health monitoring addresses the issue of detecting faults and monitoring the current state of a wind turbine. Details about the fault's progression, and from there, the remaining useful lifetime, are key features in monitoring and industrial operation and maintenance planning. In order to avoid increase in operation and maintenance cost, as well as subjective human involvement, we present an online and automated monitoring framework for prediction of the remaining useful lifetime based on deep learning models. This framework includes training and re-training of predictive models with minimal oversight by the operators.

Further, we explore the dependency of various models' predictive abilities based on the input variables available, such as SCADA and secondary measurements. Especially deep learning approaches, such as neural networks, benefit greatly from the volume of data that can be extracted from modern-day turbines. This work utilizes upon the volume of data to present a case study on main bearing failures for 108 turbines. In the presented setting, predictions of the remaining useful lifetime of more than 90 days can be expected on average, outperforming the closest state-of-the-art estimate by almost a factor of two on average.

1. Introduction

Accurate asset health assessment, e.g. Remaining Useful Lifetime (RUL) estimations, is an essential part of industrial operation and maintenance strategies (O&M), be it for increased productivity and/or reduced O&M costs. As wind power constitutes an substantial part of renewable energies, availability, reliability and lifetimes are taken more and more into consideration by business economics to carefully handle O&M costs. Blanco [1], Bussel et al. [2], Wilkinson et al. [3], and Walford [4] describe the impact of the available O&M cost during operation of a fleet of wind turbines. Further, the independent identify the need to minimize such cost as a business case. Nielsen et al. [5] and Petersen et al. [6, 7] have identified areas within O&M tasks that can improove the maintenance of offshore wind turbines. In this context, Petersen et al. [7] has proposed a Lean approach to modularization of O&M tasks and resources. Facilitating the need of having an oversight of a physical system, condition monitoring have been employed in many fields, and are still subject to research the day to day, including bearing monitoring, as many publications over the resent years bare witness: Singleton et al. [8] propose both an experimental and computational approach for the remaining lifetime of bearing by taking the relationship between current discharge events and the vibration signal into consideration in their model. Herp et al. [9] describes a Bayesian approach to handling bearing model residuals as generated by a model proposed by Bach-Andersen et al. [10], this model includes temperature measurements. Another temperature based approach is presented by Kusiak et al. [11]. Other statistical approaches are given by Loutas et al. [12] and Li et al. [13]. Neural network based models and frequency spectral models are covered by Deutsch et al. [14], Ali et al. [15], Saruhan et al. [16], Elasha et al. [17], and Teng et al. [18]. Later on, the contribution of Teng et al. [18], and Herp et al. [9] are highlighted further in Section 2.

While in early years a human operate was needed to make sense of recorded signals, this practice has gradually made way to more and more automated systems. This is also the case for wind turbines; especially nowadays where the volume of data is shear unlimited and new machine learning algorithms outperform the human ability of identifying patterns.

In the present *Big Data* era, new statistical methods are needed to describe and learn from these large dimensional data sets. This work utilizes upon the volume of data by applying *Deep Learning* and statistical driven models to present a framework in which wind turbine can be monitored in on online manner, with minimum oversight by an operator. Main bearing failure are presented as case studies to illustrate the performance of the proposed framework. It will be elucidated how the proposed framework outperforms existing long-term fault prediction frameworks (Si et al. [19, 20], Herp et al. [9, 21], Teng et al. [18]), while archiving prediction horizons beyond 2 month.

Even though, much work has been done on bearing vibrations to determine bearing failure, less work is found [10], to the knowledge of the author, on the use or combination of other sensory inputs. This work will go beyond SCADA (Supervisory Control and Data Acquisition) data to include other relevant features in the predictive considerations, namely band energies in vibration spectra.

The aim of this work is twofold: (i) showing what potential lies within large volumes of data, and (ii) provide a flexible framework for wind farm operators to use in their condition monitoring and O&M efforts. It shows insight in the working of *Big Data* analytics for system monitoring and enables the interpret of the results in the proposed framework, i.e. understand the connection between a recurrent neural network, including their sequential training, and its connection to probabilistic distributions of the RUL of a turbine. Ultimately pointing out the proposed framework as a generalization of fault prediction and RUL estimation beyond the presented main baring case studies.

The paper is organized as follows: The methodology is presented in Sections 2 and 3, providing the necessary definitions on the predictive models and framework. Section 4 shows the implementation of the proposed model in a use-case of wind turbines, evaluating the model performance and comparing it to the state-of-the-art. Finally the work is concluded in Section 5.

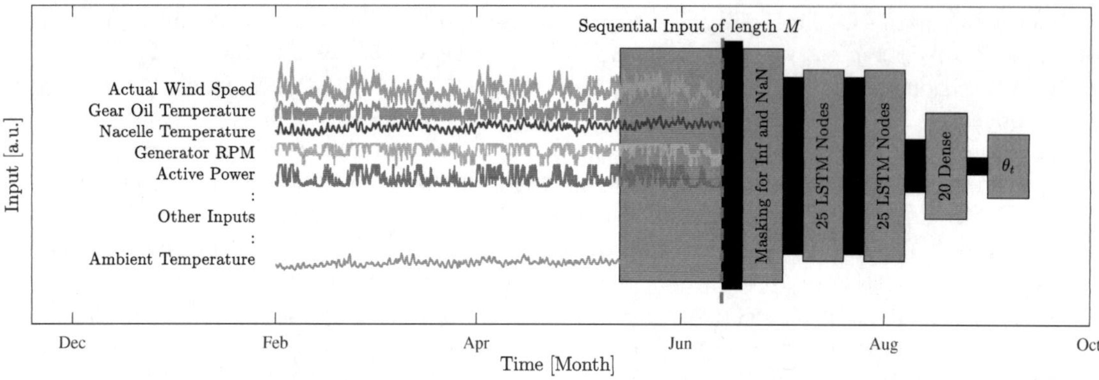

Figure 1: Illustration of the RNN used to predict the RUL. Selected timeseries shown as input reference.

2. Predictive Methodologies

Most wind turbine are equipped with SCADA systems (sampling 10 min averages), in addition thereto other means of measurements can also be available, one which this study is concerned with are the energy in selected bands of vibration Fourier spectra, other measures are preprocessed data, such as model temperature residuals as described by Bach-Andersen et al. [10]. These measurements make up a process $\{\mathbf{x}_t\}$, where t is a time instance, and $\mathbf{x}_t \in \mathbb{R}^m$ is a sample vector containing $m \leq 185$ features. This study comprises 108 turbines.

For the proposed framework models mapping from $\{\mathbf{x}_t\}$ to the RUL is denoted $M : \mathbf{X} \rightarrow \boldsymbol{\theta}$, where $\boldsymbol{\theta}$ is a measure for the RUL. These models are:

W Weibull Model for RUL [22]: As part of the proposed framework, this model was developed to cater the need of a highly flexible model with regards to the input. As the model is based on a Recurrent Neural Network (RNN) the input is limited to the number of features, $m \leq 185$. The model predicts the RUL as parametrization of the Weibull probability distribution:

$$\mathbb{P}(t) = \frac{\alpha}{\beta}\left(\frac{t}{\beta}\right)^{\alpha-1} e^{-\left(\frac{t}{\beta}\right)^{\alpha}}, \tag{1}$$

where the back-end of the model maximizes the likelihood of prediction by solving

$$\arg\max_{\boldsymbol{\omega}} \log \mathcal{L}(\boldsymbol{\omega}, \mathrm{RUL}, \Delta, \mathbf{x}_{[1,t]}) = \sum_{i=1}^{t} \left(\Delta_i \left[\alpha_i \log\left(\frac{\mathrm{RUL}}{\beta_i}\right) + \log(\alpha_i) - \log(\mathrm{RUL}) \right] - \left(\frac{\mathrm{RUL}}{\beta_i}\right)^{\alpha_i} \right) \tag{2}$$

for the topology of the RNN with weights $\boldsymbol{\omega}$. The underlying network topology is illustrated in Figure 1.

BM Brownian Motion Model for RUL [19, 20]: The model will assume that the bearing temperature is a nonlinear drift model driven by an underlying Brownian motion:

$$X(t) = x_0 + \lambda \int_0^t \mu(\tau; \mathcal{V}) d\tau + \sigma_{\mathrm{B}} B(t), \tag{3}$$

where $B(t)$ is the driving Brownian motion with nonlinear drift $\lambda\mu(t;\mathcal{V})$. Si et al. then compare Eq. (3) to a threshold, w, in order to estimate the remaining useful lifetime. The time for crossing a threshold w is be defined as:

$$T = \inf\{t : X(t) \geq w \mid X(0) < w\}, \tag{4}$$

As result, Si et al. derives a distribution for the remaining useful lifetime, referred to as Brownian Motion for Remaining Useful Lifetime (BM-RUL):

$$\mathbb{P}_{L_t|\mathbf{x}_{[0,t]}}(l_t \mid \mathbf{x}_{[0,t]}) \cong \frac{w_t\Lambda(l_t) - \alpha_t(l_t;\mathcal{V})\Delta(l_t)}{\sqrt{2\pi l_t^2 \Lambda^3(l_i)}} e^{-\frac{(w_t - \hat{\lambda}_t v(l_t))^2}{2\Lambda(l_t)}}, \tag{5}$$

where L_t is defined as the remaining useful lifetime, l_t is the residual measure corresponding to the remaining useful lifetime at t, $\omega_t = \omega - x_t$, $\hat{\lambda}_t = \mathbb{E}[\lambda_t \mid \mathbf{x}_{[0,t]}]$, $\Lambda(l_t)$ is associated with the variance of the process, $\alpha_t(l_t;\mathcal{V}) = v(l_t) - l_t\mu(l_t + t;\mathcal{V})$, $\Delta(l_t)$ is associated with the variance of the process and its drift, and $v(l_t) = \int_t^{l_t+t} \mu(\tau;\mathcal{V})d\tau$. See Si et al. [19, 20] for further details.

VS Vibration Spectral Model for RUL [18]: An artificial neural network to predict short-term tendencies of vibration energy bands. By combining the predicted and training features, a polynomial curve reflecting the long-term degradation process of bearings is fitted. Through solving the intersection between the fitted curve and a threshold, the RUL can be obtained.

GP Gaussian Process Model for RUL [9]: This is a statistical approach to abstract and predict turbine states in an online manner. The approach is based on the separability of the sufficient statistics and a hidden variable. By assuming that the feature space can be described by a multivariate Gaussian distribution, the prediction of the RUL is treated as a Gaussian process over the feature space. Any input can be provided to the model (but scales poorly with increasing input size), returning a non tractable probability distribution of RUL.

A list of the possible combination of inputs to each model is provided in Table 1, together with the model output. As seen in Table 1 the Weibull Model for RUL all necessary properties for all single input features and combinations thereof, as well as it provides full properties for the output space. In contrast, the Brownian Motion Model only provides properties for the single input feature space, but in practice is limited to selected SCADA input, mainly temperature, and requires careful initialization and threshold selection for vibration data. The output range is by construction limited to tractable probability distributions. For the Vibration Spectra Model the point-wise output is highly dependent on the chosen threshold. Due to poor scalability of the algorithm the Gaussian Process Model is limited to a small number of input features.

3. Online Monitoring and Training Framework

In this work an online monitoring framework is developed and illustrated by the flowchart in Figure 2 with $\mathbf{W}$ as an example of a model. In detail, the proposed approach starts with detecting a change in the turbine prior to its failure, this is referred to as opening a case. The turbine credential are then stored in a Library and mored as unprocessed. Following system will then acquire the relevant data for the monitoring based on $\mathbf{W}$. For each t, since the detection, the RUL is estimated. As long as the turbine still is operating, for each consecutive times-stamp, $t = t + 1$, the RUL can still be estimated as a probability distribution. This is illustrated in Figure 3a. The monitoring system provides the operator with a graphical representation of the RUL distribution at selected monitoring points. Further, for each t the first moment, median,

IOP Conf. Series: Journal of Physics: Conf. Series **1222** (2019) 012043 doi:10.1088/1742-6596/1222/1/012043

Table 1: Model comparison including model features.

Study	Input						Output			Avg. max(RUL) [days] over all turbines rounded to nearest days					
	SCADA	Vibration	Residuals	SCADA+Vibration	SCADA+Residuals	SCADA+Residuals+Vibration	Point-wise	Probability Distribution	Tractable	SCADA	Vibration	Residuals	SCADA+Vibration	SCADA+Residuals	SCADA+Residuals+Vibration
Weibull Model for RUL (**W**)*	●	◐	●	◐	●	◐	●	●	●	94	25	-	86	93	32
+Dense Layer*	●	◐	●	◐	●	◐	●	●	●	99	21	-	82	80	-
+LSTM Layer*	●	◐	●	◐	●	◐	●	●	●	95	19	-	94	98	-
+LSTM Layer and Dense Layer*	●	◐	●	◐	●	◐	●	●	●	64	22	-	82	70	-
Brownian Motion Model for RUL (**BM**)†	◐	◐	●	-	-	-	-	●	●	20	42	-	-	-	-
+Study State*	◐	◐	●	-	-	-	-	●	●	32	12	-	-	-	-
Vibration Spectra Model for RUL (**VS**)†	-	●	-	-	-	-	◐	-	-	-	54	-	-	-	-
Gaussian Process Model for RUL (**GP**)†	◐	◐	◐	◐	◐	◐	-	◐	-	15	8	33	-	-	-

● = provides property; ◐ = partially provides property; - = does not provide property;
†model has been published; *model has not been published yet

and maximum likelihood is provided for the operator to provide point-wise measures of the RUL, since case opening. When the turbines operation is terminated, either by the operator or a fault, Figure 3c is generated, comparing the RUL estimations against the real remaining lifetime.

If the error of the prediction is small compared to a threshold E_0, the case is closed. Else, the predictive model is updated, and the error re-checked. At any point the models residual error, over a distribution of the RUL given empirical evidence, $\mathbb{P}(\text{RUL} \mid \mathbf{x}_{[1,t]})$, is defined as:

$$E(t \mid \mathbf{x}_{[1,t]}) \propto \int_t^\infty (\text{RUL} - \hat{\text{RUL}})^2 \mathbb{P}(\text{RUL} \mid \mathbf{x}_{[1,t]}) d\text{RUL}. \tag{6}$$

The threshold E_0 is then defined as E such that the confidence of prediction is 90%, i.e. with respect to a normal distribution with mean, $\mu = \hat{\text{RUL}}$ and spread, $\sigma = 0.05\hat{\text{RUL}}$. If the error is lower than the threshold, the model is updated, else, user input is required to evaluate the case. The model update steps requires all unprocessed turbines and acquires their data. In the case of the Weibull model, the framework would would load the current RNN topology and weights, and starts the training process with the new data. At the end of the update a new model is formed and send forward to be used in the monitoring efforts.

4. Case Studies

For selected turbines the online predictions are shown in Figure 4a and 4c, including the real remaining lifetime for reference. These turbine's prediction likewise follow the real remaining lifetime.

In order to summarize the outcome, the proposed frame work will be evaluated in comparison to the above presented state-of-the-art RUL prediction approaches, **BM**, **VS**, and **GP**. For the

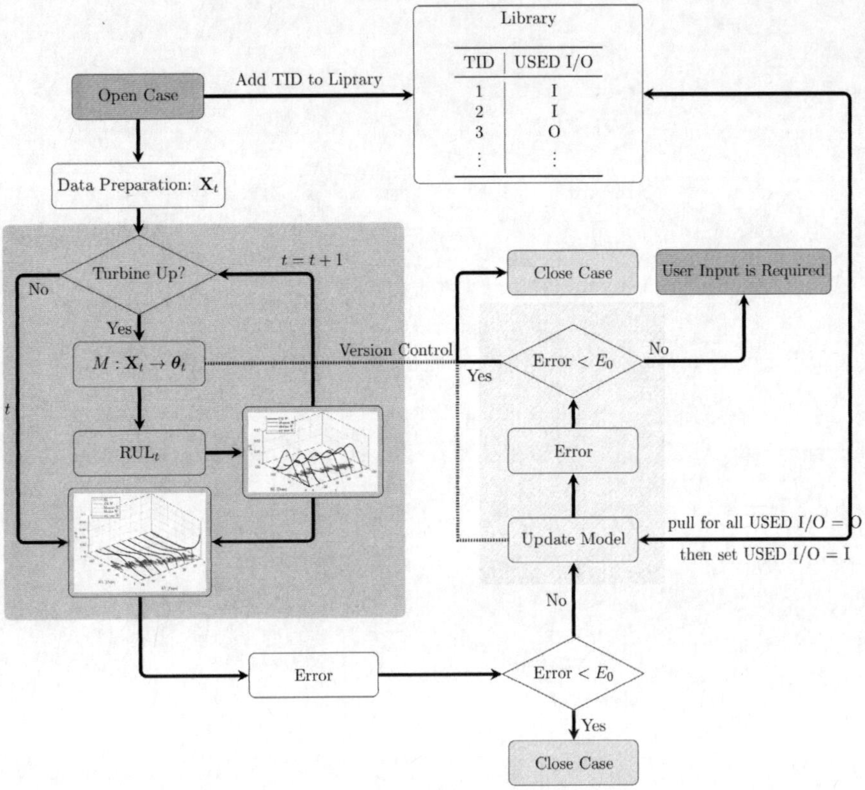

Figure 2: Flowchart of the proposed online monitoring framework.

study, the maximum RUL, rounded to the nearest day, is defined as the similarity with 90% confidence, between the true RUL with in a ±5% margin and the models prediction.

The studies are outlined in Table 1. Breaking down Table 1 by Input, Output, and max(RUL), it can be seen that **W**, and variation thereof, can be used on a wide set of input spaces, and is flexible in its output. For **BM** the input space is drastically reduced to $\mathcal{O}(m = 1)$, while maintaining a tractable probability distribution. **VS** is the least flexible model with respect to the input and output space, however, it performs reasonably well for the niche of vibration data with a prediction horizon of 54 days on average. The Gaussian process model (**GP**) is flexible in its input space, but the output space is not tractable and the algorithm scales poorly with increasing m. Thus, no results could be obtained for combinations SCADA, Vibration, and Residuals data.

In terms of the predictive capabilities of each model, ranking from best to worse, the **W** scores RUL above 90 days in most of the presented cases. One may wonder why for **W** and the combination of all three inputs leads to a dramatic drop in the predictive capability. At the current state of this work this must remain as a question to be answered in the future. **VS** achieves prediction up to 53 days, while **BM** and **GP** estimate the average RUL from 42 days to as low as 8 days.

5. Conclusion

This work has proposed a framework for predicting the RUL of wind turbines embedded in on online monitoring framework - combining concepts of statistics and the Turing properties of a neural network. The implementation of the W model and evaluation on wind turbine bearing

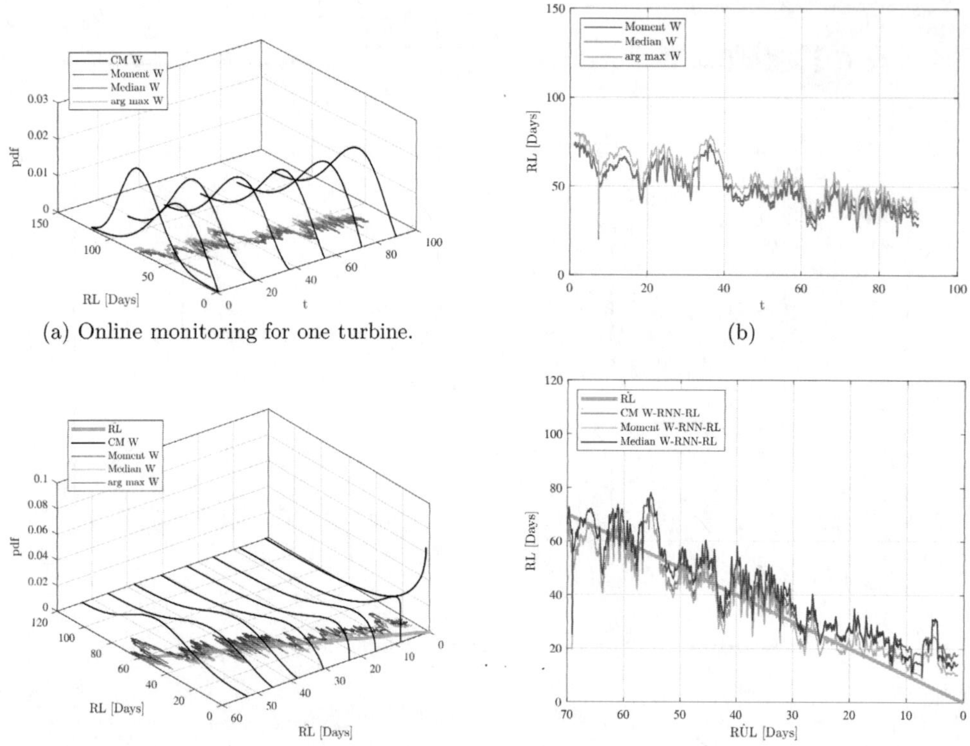

(a) Online monitoring for one turbine.

(b)

(c) RUL curve for the same turbine as in Figure 3a.

(d)

Figure 3: Prediction of RUL. Left column: three-dimensional representation of the predicted probability distribution at selected CM points. Right column: top-down view.

data has shown that the proposed model can predict the remaining useful life time beyond 60 days. Besides that, the proposed model shows better long-term prediction capabilities compared to the proposed models by Si et al. [19, 20] (**B**) and Herp et al. [9] (**GP**) and Teng et al. [18] (**VS**).

This study was limited to monitor wind turbine main bearings. In that context it is believed that the proposed model is generic enough to be trained and used on other wind turbine components, as well as in other applications than wind turbines.

References

[1] María Isabel Blanco. The economics of wind energy. *Renewable and Sustainable Energy Reviews*, 13(6–7):1372–1382, 2009.

[2] G. J. W. van Bussel and M. B. Zaaijer. Reliability, Availability and Maintenance Aspects of Large-Scale Offshore Wind Farms, a Concepts Study. In *MAREC 2001 Marine Renewable Energies Conference*, volume 113, pages 119–126, March 2001.

[3] M. Wilkinson, F. Spinato, and M. Knowles. Towards the zero maintenance wind turbine. In *Proceedings of the 41st International Universities Power Engineering Conference, 2006 (UPEC '06)*, volume 1, pages 74–78, 2006.

[4] Christopher A. Walford. Wind turbine reliability: Understanding and minimizing wind turbine operation and maintenance costs. Technical Report SAND2006-1100, Sandia National Laboratories, Sandia National Laboratories, Albuquerque, New Mexico 87185 and Livermore, California 94550, 2006.

[5] Jannie Jessen Nielsen and John Dalsgaard Sørensen. On risk-based operation and maintenance of offshore wind turbine components. *Reliability Engineering & System Safety*, 96(1):218–229, 2011.

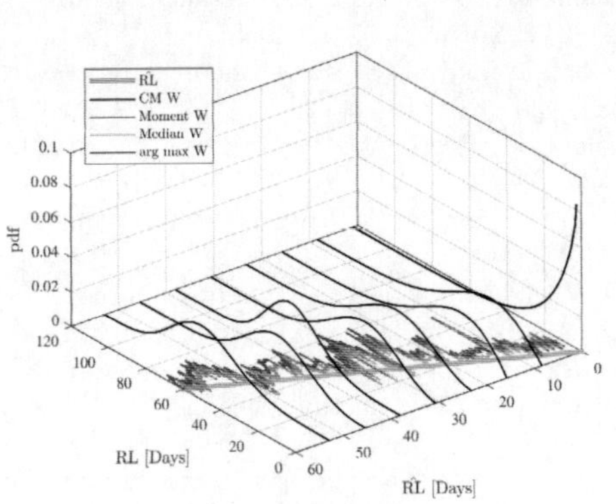

(a) Another turbine's RUL.

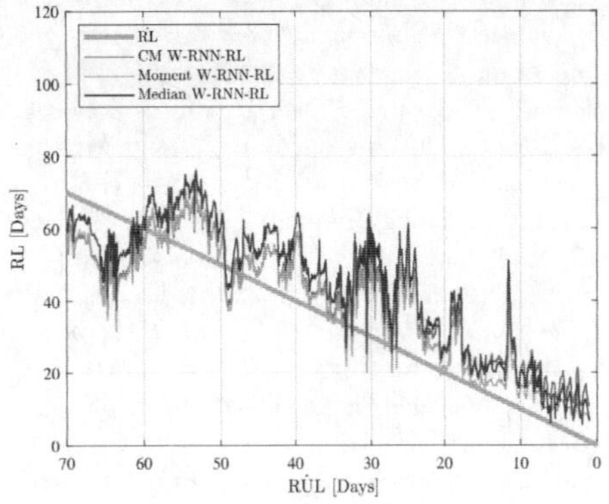

(b)

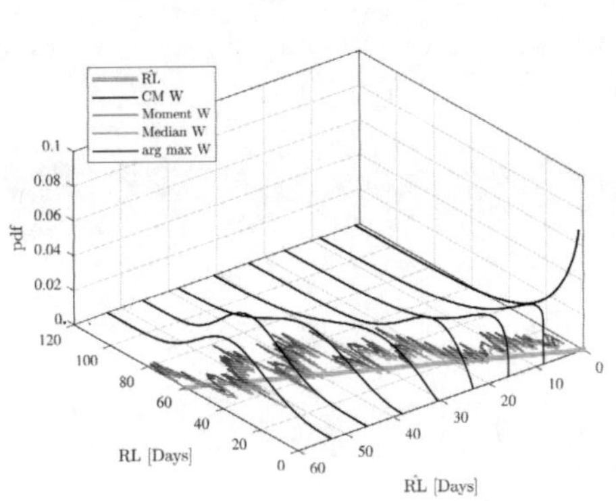

(c) A turbine with more variability in the prediction.

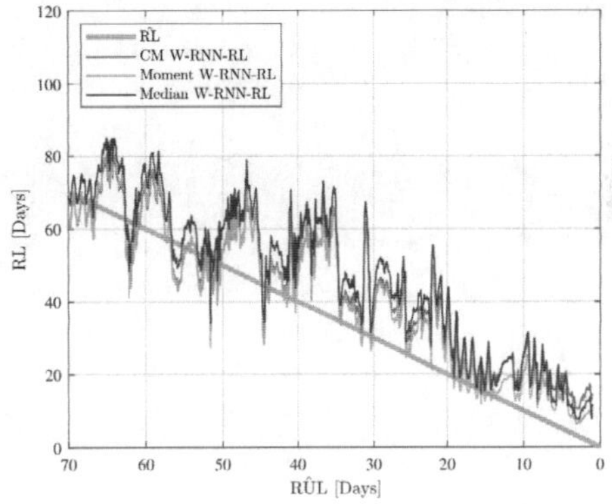

(d)

Figure 4: Prediction of RUL. Left column: three-dimensional representation of the predicted probability distribution at selected CM points. Right column: top-down view.

[6] Kristian Rasmus Petersen, Erik Skov Madsen, and Arne Bilberg. Offshore wind power at rough sea: The need for new maintenance models. In *20th EurOMA Conference*, 2013.

[7] Kristian R. Petersen, Erik Skov Madsen, and Arne Bilberg. First lean, then modularization: improving the maintenance of offshore wind turbines. *International Journal of Energy Sector Management*, 10(2):221–244, 2016.

[8] R. K. Singleton, E. G. Strangas, and S. Aviyente. The use of bearing currents and vibrations in lifetime estimation of bearings. *IEEE Transactions on Industrial Informatics*, 13(3):1301–1309, June 2017.

[9] Jürgen Herp, Mohammad H. Ramezani, Martin Bach-Andersen, Niels L. Pedersen, and Esmaeil S. Nadimi. Bayesian state prediction of wind turbine bearing failure. *Renewable Energy*, 2017.

[10] Martin Bach-Andersen, Bo Rømer-Odgaard, and Ole Winther. Flexible non-linear predictive models for large-scale wind turbine diagnostics. *Wind Energy*, 2016.

[11] Andrew Kusiak and Anoop Verma. Analyzing bearing faults in wind turbines: A data-mining approach. *Renewable Energy*, 48(0):110–116, 2012.

[12] T. H. Loutas, D. Roulias, and G. Georgoulas. Remaining useful life estimation in rolling bearings utilizing data-driven probabilistic e-support vectors regression. *IEEE Transactions on Reliability*, 62(4):821–832, Dec 2013.

[13] N. Li, Y. Lei, J. Lin, and S. X. Ding. An improved exponential model for predicting remaining useful life of rolling element bearings. *IEEE Transactions on Industrial Electronics*, 62(12):7762–7773, Dec 2015.

[14] J. Deutsch and D. He. Using deep learning-based approach to predict remaining useful life of rotating components. *IEEE Transactions on Systems, Man, and Cybernetics: Systems*, 48(1):11–20, Jan 2018.

[15] Jaouher Ben Ali, Brigitte Chebel-Morello, Lotfi Saidi, Simon Malinowski, and Farhat Fnaiech. Accurate

bearing remaining useful life prediction based on weibull distribution and artificial neural network. *Mechanical Systems and Signal Processing*, 56-57:150 – 172, 2015.

[16] H. Saruhan, S. Sandemir, A. Çiçek, and I. Uygur. Vibration analysis of rolling element bearings defects. *Journal of Applied Research and Technology*, 12(3):384 – 395, 2014.

[17] Faris Elasha, Matthew Greaves, David Mba, and Abdulmajid Addali. Application of acoustic emission in diagnostic of bearing faults within a helicopter gearbox. *Procedia CIRP*, 38:30 – 36, 2015.

[18] Wei Teng, Xiaolong Zhang, Yibing Liu, Andrew Kusiak, and Zhiyong Ma. Prognosis of the remaining useful life of bearings in a wind turbine gearbox. *Energies*, 10(1), 2017.

[19] X. S. Si. An adaptive prognostic approach via nonlinear degradation modeling: Application to battery data. *IEEE Transactions on Industrial Electronics*, 62(8):5082–5096, Aug 2015.

[20] X.S. Si, Z.X. Zhang, and C.H. Hu. *Data-Driven Remaining Useful Life Prognosis Techniques: Stochastic Models, Methods and Applications*. Springer Series in Reliability Engineering. Springer Berlin Heidelberg, 2017.

[21] Jürgen Herp and Esmaeil S. Nadimi. Dimensionality reduction by bayesian eigenvalue-analysis for state prediction in large sensor systems: with application in wind turbines. In *Proceedings of the 2018 Conference on Research in Adaptive and Convergent Systems, RACS 2018, Honolulu, HI, USA, October 09-12, 2018*, pages 1–5, 2018.

[22] Jürgen Herp, Niels L. Pedersen, and Esmaeil S. Nadimi. Using data driven learning for remaining probabilistic lifetime prognosis of wind turbine bearings. Unpublished manuscript, 2018.

IOP Conf. Series: Journal of Physics: Conf. Series **1222** (2019) 012044 doi:10.1088/1742-6596/1222/1/012044

Drivetrain reliability improvements from long-term field data processed in the cloud

Cédric Peeters[1,a], Nicoletta Gioia[1], Pieter-Jan Daems[1], Jonas Verbeke[1], Timothy Verstraeten[2], Ann Nowé[2], Jan Helsen[1]
[1]Vrije Universiteit Brussel, Acoustics and Vibrations Research Group (AVRG)/OWI-lab, Pleinlaan 2, B-1050, Brussel, Belgium
[2]Vrije Universiteit Brussel, Artificial Intelligence Lab (AI-lab), Pleinlaan 2, B-1050, Brussel, Belgium

E-mail: [a]Cedric.Peeters@vub.be

Abstract. This work describes an autonomous condition monitoring framework to process and analyze data measured on wind turbine gearboxes. Industry 4.0 and the Industrial Internet of Things open the door for much more elaborate measurement and data analysis campaigns thanks to the reduction in cost of sensors and of processing power. This increase in data acquisition and handling potential is especially useful considering that most current state-of-the-art methods in signal processing often lead to large quantities of health indicators due to the multiple processing steps. Such large numbers of indicators become unfeasible to inspect manually when the data volume and the number of monitored turbines increases. Therefore, this paper illustrates a hybrid analysis approach that combines advanced signal processing methods with machine learning and anomaly detection. This approach is validated on an experimental wind turbine gearbox vibration data set.

1. Introduction

The rise of Industry 4.0 and correspondingly the Industrial Internet of Things (IIoT), will facilitate an increase in the amount of instrumentation in wind turbines, either directly mounted by the Original equipment manufacturer (OEM) or installed during its lifetime by owner-operators [1, 2, 3]. The IIoT context in particular promotes sensors that are connected directly to the internet. A cloud data-center can thus continuously receive data coming from such sensors. This vast amount of sensor data can be employed for multiple goals. Continuous data streams have both longer signal length and a finer granularity compared to snapshot measurements which are often still common practice nowadays. These properties enable a more refined analysis and help to better understand the turbine characteristics and behavior changes a healthy turbine experiences.

While Industry 4.0 offers valuable opportunities, there are still some challenges that need to be overcome. Given that measuring continuously results in large data sets, automated processing approaches able to autonomously analyze these large amounts of data are necessary. Additionally, all these analyses result in a multitude of indicators that need to be fused into a joint conclusion about system health by applying sensor fusion approaches.

IOP Conf. Series: Journal of Physics: Conf. Series **1222** (2019) 012044 doi:10.1088/1742-6596/1222/1/012044

Gearbox condition monitoring of wind turbines (WT) is becoming increasingly more popular and advanced. Current practice in condition monitoring systems often revolves around tracking time-domain statistical indicators and more component-specific frequency-domain indicators [4, 5, 6]. The advantage of using simple scalar time-domain indicators is that no a-priori knowledge about the characteristic fault frequencies is required and thus the number and complexity of the components is not taken into account. While this does simplify the analysis procedure significantly, it also does not allow pinpointing which component is exhibiting anomalous behavior. Not having any knowledge about the exact location of a fault is a downside, but in an early analysis stage it is often easier and quicker to have a straightforward overview of which measurement sensor is picking up faulty behavior without needing to have an immediate answer regarding the exact component and location of failure. Afterwards, a more in-depth frequency-based analysis can follow up and try to determine the missing failure information.

This paper focuses on the challenges and opportunities related to the Industry 4.0 monitoring trend by demonstrating an automated multi-level approach designed for continuous measurements. The approach specified in Section 2 allows facilitating future steps towards prognosis by linking the specifics of the failure modes to the specifics of each analysis work-flow. The goal of this paper is to clarify the different techniques that are used to monitor the turbine and how to unify them into a single approach. The starting point for this monitoring approach are the failure modes and thus they form the base of what the monitoring scheme tries to track.

Section 3 shows how to use condition indicators optimally for the health monitoring of a WT gearbox. Thanks to the recent advances of machine learning (ML), it has now become more feasible to not only identify relevant condition indicators, but also relate them to the operating regime of the turbine for each measurement. It turns out that most of the commonly used statistical condition indicators are sensitive to changes in the operating parameters and thus can lead to erroneous conclusions if they are not compensated for these changes. This paper illustrates a combined use of statistical indicators with machine learning techniques to allow for normalization of the indicators making them independent of the operating regime and provide a streamlined methodology for efficient early fault analyses. Ideally, the provided indicators measure a certain property in the vibration signal that is linked to a specific failure type in order to provide the diagnostic ML models with physics-based and thus relevant features.

The performance of the methodology is illustrated on an experimental vibration data set measured on a WT gearbox. Short measurements were made for five years on the planetary stage of the gearbox providing enough data to extract a healthy behavior baseline. At the end of these measurements the diagnostic models trigger an alarm for anomalous behavior related to the planetary gear stages. This observation is then corroborated by a manual inspection of the gearbox which reveals severe indenting and micro pitting on the gears. This case study confirms that using a more advanced approach combining statistical condition indicators and machine learning can deliver on the promise of early detection of planetary gear damage in WT gearboxes. In turn, this will lead to a reduction in the turbine downtime and cost-of-energy. This paper serves as an overview of how a complex, hybrid monitoring approach can look like in the modern IIoT context, but it does not claim to cement the definite framework of a continuous monitoring scheme. Improvements and extensions to the framework will be implemented in the future to further increase the reliability and efficacy of the proposed approach.

2. Approach

The proposed monitoring approach consists out of multiple steps:

- Important failure modes of the turbine are extracted from a historical failure database.

IOP Conf. Series: Journal of Physics: Conf. Series **1222** (2019) 012044 doi:10.1088/1742-6596/1222/1/012044

The advantage of ranking the failure modes is that the monitoring system can target only the most relevant failure modes of the wind turbine. Furthermore, it allows adapting the monitoring methods for each failure mode specifically.

- Automated event detection is implemented in order to get a summary about all the impactful and dynamic events a wind turbine experiences. An example of such a dynamic event is the emergency braking of a turbine during grid loss. Automatically detecting these events and classifying their severity provides continuous insight into the loads that the drivetrain undergoes during its lifetime. Afterwards these events can be linked to specific failure modes.
- Features are calculated using vibration-based signal processing algorithms. Complex pipelines of signal processing algorithms are combined in order to extract characteristic health features.
- Information about the loads is used to normalize the extracted indicators from the vibration-based signal processing methods and the abnormal behavior health indicators. If there is no load information directly available, the load is extracted from operational parameters coming from SCADA data.
- Lastly, feature fusion approaches can be used now to integrate the health indicators and information from the previous steps into a joint assessment of turbine health. A single health indicator is defined using a fusion of the various calculated indicators by model ensemble. An overall health indicator for the turbine is defined by cascading the individual failure mode alarms.

3. Methodology

In order to be able to integrate the different monitoring approaches the data needs to be made available in a central data-storage location such that dedicated data processing pipelines can transform the raw data into health indicators. To facilitate this, we make use of an integrated big data platform. However, to keep the length of this paper concise, the interested reader is referred to [7] for more information about the practical implementation of this platform. This section provides an overview of the techniques used in the data processing and analysis scheme.

3.1. Vibration-based signal processing

Vibration signal processing allows for extracting complex health features from raw data [8]. Signals are processed by a sequence of data-cleaning and filtering steps. A typical flow is shown in Fig. 1. In case of wind turbine vibration signals, specific attention should be given to the compensation of speed variations, the variability of the wind, the operational speed will continuously change in the vicinity of the target value set by the controller [9]. Instantaneous speed estimation techniques are employed to estimate the speed signal without any need for an encoder signal.

The Multi-Order Probabilistic Approach (MOPA), originally developed by Leclère et al. [10, 11], is employed for this purpose. Figure 2a shows the spectrogram of a turbine gearbox signal [10]. Based on the spectrogram, a probability density function (pdf) map is constructed that combines contributions of all harmonics. This map for the considered signal is shown in Fig. 2b. Extracting the most probable estimate results in the speed signal illustrated in Fig. 2c. The MOPA method results in a more accurate speed estimation compared to a classical approach such as band-pass filtering. Figure 2d illustrates the effect of using the extracted speed to resample the signal in the angular domain, which leads to a substantial decrease in smearing.

Other processing methods are used to further clean the signal from unwanted components. Cepstrum-based techniques are employed for removing deterministic signal content [12, 13, 14]. Filtering the data is also performed to track the statistics in different frequency bands.

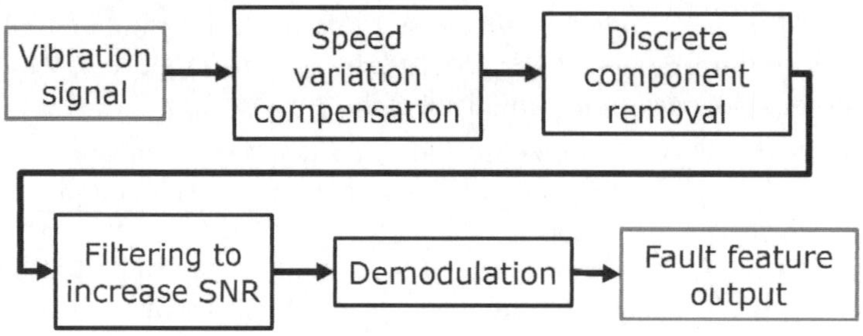

Figure 1. Schematic representation of the processing procedure for bearing failure detection.

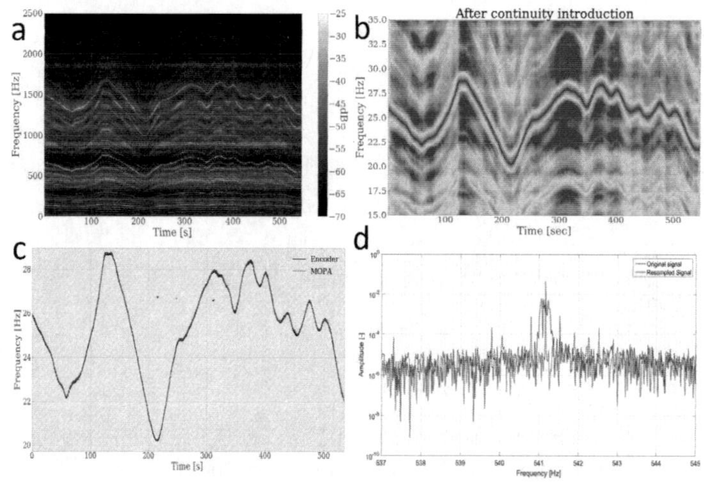

Figure 2. Speed estimation using MOPA, a) Spectrogram of WT gearbox vibration data, b) PDF map, c) Estimated speed profile, d) Spectra before and after angular resampling.

Demodulation of the signal is done after finding the optimal demodulation frequency band using cyclic spectral coherence [15]. Figure 3 shows an example coherence map detecting an outer race fault frequency in a vibration signal measured on a test turbine gearbox installed at the National Renewable Energy Laboratory (NREL). Lastly, condition indicators are calculated on the pre-processed signals to track the state of the WT gearbox. Both time- and frequency-domain indicators are extracted from the processed data. Examples of calculated indicators include the RMS, peak-to-peak value, crest factor, (Moors) kurtosis, and the peak energy index.

3.2. Event detection

The main goal of this step is to automatically detect events such as emergency braking, coast-downs and run-ups in time series data originating from SCADA 1-second from continuous encoder measurements. The developed approach is illustrated by an example where multiple months of SCADA of different turbines of an offshore wind farm have been automatically processed. Sequential pattern mining algorithms are used to extract relevant patterns from learned events. The main idea is to afterwards perform pattern recognition on speed profiles of the SCADA 1-second to detect all similar events automatically.

An example of an automatically annotated speed profile is shown in Fig. 4. The variability in the extracted run-up signals for the different turbines across the wind farm is shown in Fig. 5.

IOP Conf. Series: Journal of Physics: Conf. Series **1222** (2019) 012044 doi:10.1088/1742-6596/1222/1/012044

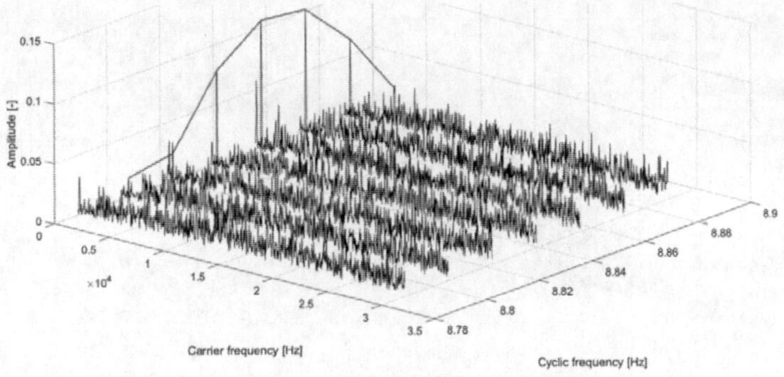

Figure 3. 2D-map of the cyclic spectral coherence. An outer race fault of the planet carrier bearing is detected at its characteristic frequency, of 8.84Hz around a carrier frequency of 4.6 kHz (red line) [12]

It can be seen that the variability of this event is quite large: there is almost 40% variation in the duration between the shortest and longest identified run-up. During the different run ups, the turbine decreases its acceleration around a normalized speed value of around 0.675. This corresponds to the first continuous operating point of the turbine. Afterwards, the acceleration increases once again until the turbine reaches its second continuous operating point. This information can be condensed into relevant summary statistics regarding typical event occurrence rates, maximum loads and event durations. This analysis step in the monitoring scheme provides us with the opportunity to detect events, classify them, and link them to failure modes of the drivetrain.

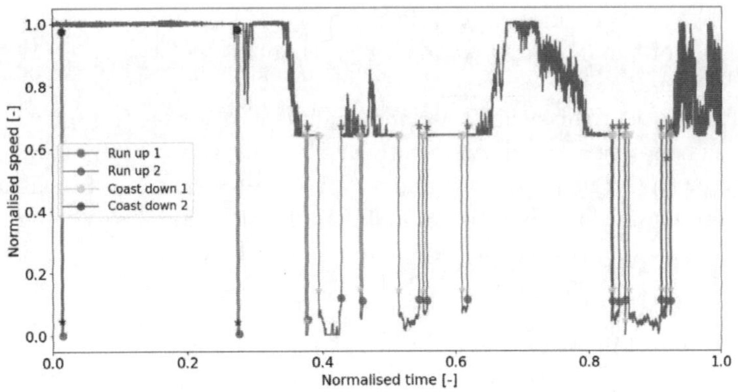

Figure 4. Example of an automatically annotated speed signal in which different events have been identified.

3.3. Health indicator fusion and anomaly detection

It is clear that automatic monitoring frameworks require all the indicators to be converted autonomously into a single health conclusion. Different fusion approaches are available to achieve this, e.g. principal component analysis and neural networks. We opt for an approach that is more physics-based and understandable: the model ensemble. Indicator processing pipelines are fed with different subsets of data. Each of the pipelines performs feature calculation and

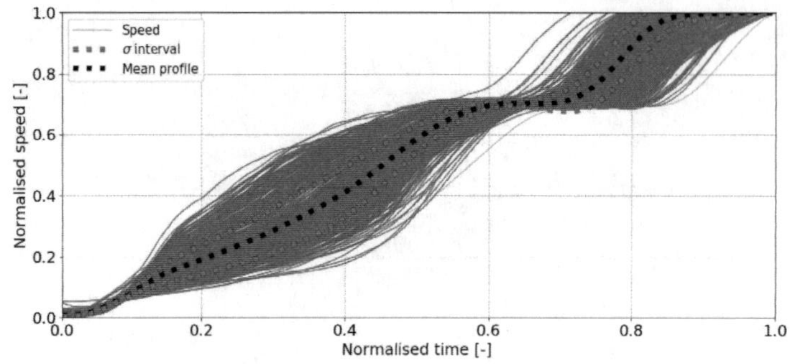

Figure 5. Variability in different run-up events.

anomaly assessment. Finally, a majority vote among the features decides on the anomalous state of the turbine. A health indicator is defined for each turbine subcomponent. These are propagated to the turbine level to indicate the overall health.

4. Experimental application

The proposed vibration-based anomaly detection method is validated on a case of damage initiation in a planet gear of a multi-megawatt WT gearbox. Due to confidentiality reasons, we cannot disclose any concrete details of the WT type or gearbox kinematics and the figures are displayed using anonymized axes. The measurements themselves span multiple years and consist of ten second vibration signals, captured every two to three days. Different measurement channels were mounted spread over the gearbox housing to increase the sensitivity for signatures coming from different gearbox components. Considering the fact that every single-channel, ten-second signal leads to a large number of indicators due to the pre-processing, it is impractical to manually assess all of the indicator trends for all channels.

The case study shown in this work describes incipient deterioration on the planetary gear stage that has a high potential to develop into much more severe damage. This damage was detected using the proposed methodology and validated afterwards. All measurement channels were employed in this analysis. For the analysis performed here, every signal leads to sixty (2x5x6) calculated indicators due to the pre-processing as follows:

- Raw data and cepstrum liftered data (x2).
- Unfiltered data and four frequency-filtered signals (x5).
- Six indicators per pre-processed signal (x6).

While for one turbine it is possible to manually go through all indicator trends of all measurement channels, it becomes much more cumbersome when this needs to be done for a whole wind farm. Therefore the sixty different indicator trends are used as input for the approach as described in Section 3.3.

The anomaly detection algorithm is validated through an in-depth analysis over all statistical features. In Figure 6, we present an example taken from this analysis, demonstrating the validity of the proposed method. It illustrates the anomaly detection and alarm fusion mechanism. The first level shows two peak-to-peak features A and B for the planetary stage channel and the same feature B for the high-speed stage channel with normalized values. Each data point in the time series is colored by the alarm level predicted by the ML model. At the bottom, the fused indicator over all peak-to-peak features and all channels is given. The y-value and color of the fused indicator shows the severity of the alarm. All time series span over multiple years. The results show a significant change in feature values for the planetary stage channel. This is successfully detected by the ML method. The high-speed stage channel is unrelated to the observed failure and therefore only shows stable behavior. The fused alarm is computed by taking the average alarm level over all features per statistical indicator type (e.g., peak-to-peak) and takes the maximum over all channels. This fusion is carried out over all 8 channels and 60 features, which results into 6 fused alarms (one per indicator type) and thus significantly reduces the workload of the analysis.

The effects of the failure on the behavior of the statistical features can be observed in the fusion, reaching an alarm severity of about 70%. A threshold can be set by the WT operator to automatically signify potential failures. The number of figures that need to be inspected also gets drastically reduced thanks to the feature fusion.

5. Conclusions

This paper illustrates an approach to extract event and turbine health information from large data measurement streams in a automated way. Advanced vibration-based signal processing methods are used to extract complex health indicators related to the gears and bearings in the turbine gearbox. Pattern recognition methods show the potential for automatically detecting transient events and thus gaining better insights into what a turbine experiences during its lifetime. Finally, these indicators are fused using an ensemble of models combined with majority voting to obtain an overall turbine health indication. The main goal of this paper is to provide a guideline for a scalable and effective automated monitoring approach of wind turbine drivetrains. Key monitoring aspects are illustrated on experimental wind turbine data.

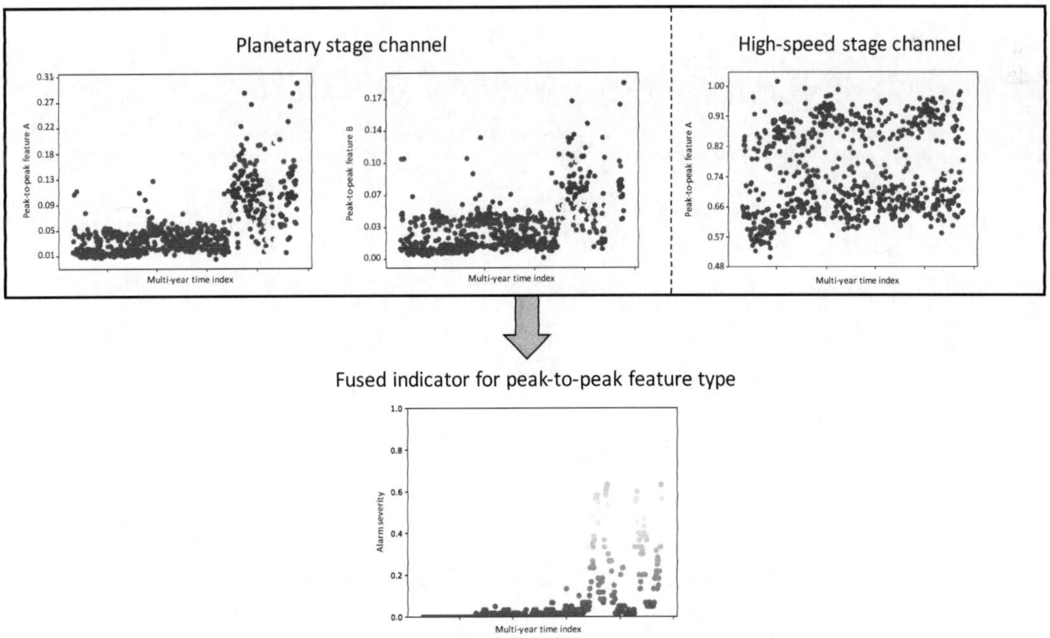

Figure 6. Feature fusion results for different channels and per feature type.

Acknowledgments
The authors of VUB would like to thank FWO for their support with an SB Ph.D. fellow grant and the agency for Innovation by Science and Technology in Belgium for supporting the SBO HYMOP project. They would also like to thank FWO for providing funding for a long stay abroad of Cédric Peeters at the university of INSA Lyon.

References

[1] Lee J, Kao H A and Yang S 2014 *Procedia Cirp* **16** 3–8
[2] Jeschke S, Brecher C, Meisen T, Özdemir D and Eschert T 2017 *Industrial Internet of Things* (Springer) pp 3–19
[3] Sadiku M N, Wang Y, Cui S and Musa S M 2017 *IJASRE* **3**
[4] de Azevedo H D M, Araújo A M and Bouchonneau N 2016 *Renewable and Sustainable Energy Reviews* **56** 368–379
[5] Lu B, Li Y, Wu X and Yang Z 2009 *Power Electronics and Machines in Wind Applications, 2009. PEMWA 2009. IEEE* (IEEE) pp 1–7
[6] Tchakoua P, Wamkeue R, Ouhrouche M, Slaoui-Hasnaoui F, Tameghe T A and Ekemb G 2014 *Energies* **7** 2595–2630
[7] Helsen J, Peeters C, Doro P, Ververs E and Jordaens P J 2017 *Big Data Computing Service and Applications (BigDataService), 2017 IEEE Third International Conference on* (IEEE) pp 179–184
[8] Carden E P and Fanning P 2004 *Structural health monitoring* **3** 355–377
[9] Yang W, Tavner P J, Crabtree C J, Feng Y and Qiu Y 2014 *Wind Energy* **17** 673–693
[10] Leclere Q, André H and Antoni J 2016 *Mechanical Systems and Signal Processing* **81** 375386
[11] Peeters C, Leclere Q, Antoni J, Guillaume P and Helsen J 2017 *Journal of Physics: Conference Series* vol 842 (IOP Publishing) p 012053
[12] Peeters C, Guillaume P and Helsen J 2018 *Renewable Energy* **116** 74–87
[13] Peeters C, Guillaume P and Helsen J 2017 *Mechanical Systems and Signal Processing* **91** 354–381
[14] Borghesani P, Pennacchi P, Randall R, Sawalhi N and Ricci R 2013 *Mechanical Systems and Signal Processing* **36** 370–384
[15] Antoni J 2007 *Journal of Sound and vibration* **304** 497–529

IOP Conf. Series: Journal of Physics: Conf. Series **1222** (2019) 012045 doi:10.1088/1742-6596/1222/1/012045

On the use of AI based vibration condition monitoring of wind turbine gearboxes

Sofia Koukoura, James Carroll, Alasdair McDonald

University of Strathclyde, Glasgow, UK

E-mail: sofia.koukoura@strath.ac.uk

Abstract. Condition monitoring (CM) systems are installed in wind turbines (WTs) in order to avoid component downtime and reduce maintenance costs. Vibration monitoring is widely used for the WT gearbox, which is a component with a significant downtime. Given that the installed wind capacity grows, the volume of CM data increases, making manual interpretation of vibration signals challenging. Therefore, there is a need for an efficient and automated maintenance decision support system. The aim to this paper is to propose an automated framework for gearbox incipient failure diagnosis. The framework utilises vibration signals and performs health estimation and fault isolation based on signal processing and artificial intelligence (AI) techniques. The methodology is demonstrated through a case study of vibration data from operating WTs with known gearbox failures. The study can be used to optimise wind turbine maintenance actions.

1. Introduction

Operation and maintenance (O&M) costs are a major contributor to the total cost of energy of large wind farms. Wind turbine (WT) failures and unscheduled maintenance actions increase the O&M costs and downtime, compromising therefore the annual energy production. In order to reduce those costs, condition monitoring (CM) systems are installed in wind turbines. The WT gearbox is one of the components with the highest downtime [1]. Vibration based CM has proven to be effective in this component. Accelerometers installed on the gearbox surface transmit high frequency signals that can give useful information about the health state of the gearbox.

A gearbox has various stages composed of gears, bearings and shafts, so the root cause of a gearbox failure can be attributed to many different failure modes. Vibration signals can give some understanding on what causes a failure. Vibration CM in wind turbine gearboxes has been widely researched in the literature. Various existing vibration analysis methods are evaluated and presented in [2]. However, as the installed wind capacity grows, the volume of CM data increases and manual interpretation of wind turbine vibration data becomes challenging. Therefore, in the era of industrial Artificial Intelligence (AI), it is possible to explore if machine learning techniques can aid the decision making process of wind turbine gearbox maintenance. AI has been applied in machine diagnostics [3] and there are a few studies in wind turbines [4], [5].

To this end, the objective of this paper is to present a framework that uses a stream of vibration data from wind turbine gearboxes and automatically isolates incipient faults. The novelty in the paper lies in the automatic wind turbine gearbox diagnostic framework

IOP Conf. Series: Journal of Physics: Conf. Series **1222** (2019) 012045 doi:10.1088/1742-6596/1222/1/012045

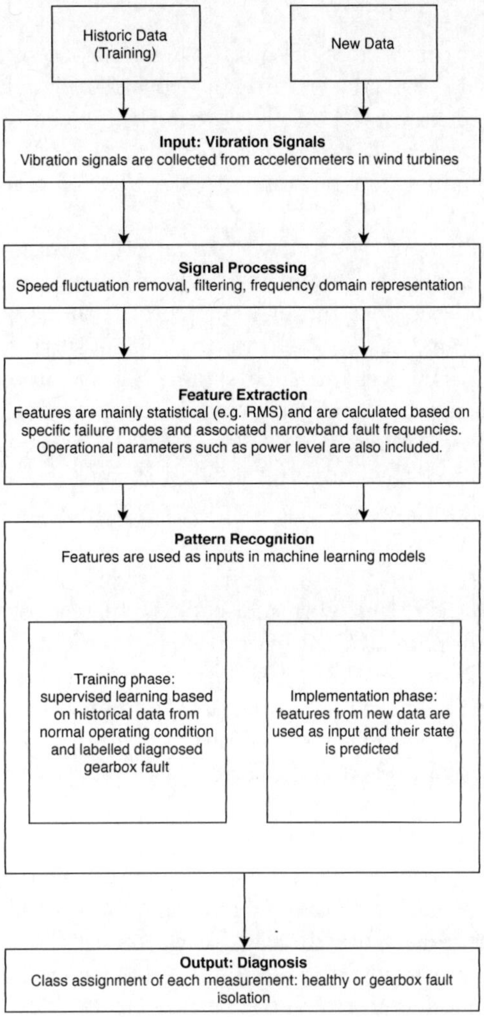

Figure 1. Methodology Flowchart

methodology using domain knowledge and its validation using vibration data from operating wind turbines with confirmed gearbox faults.

2. Methodology

The aim is to create an automated framework that is able to predict and diagnose incipient gearbox failures using AI. The inputs of the models are vibration signals and the outputs are health estimation of the gearbox and component fault isolation and diagnosis. First, signals are collected and processed in order to reveal potential fault signatures. Then, features are extracted from the processed signals around specific frequencies of interest related to gearbox component faults. Finally, these features are used as inputs in pattern recognition models that can determine if a fault is present and in which component and stage it occurs.

An overview of the methodology is given in Figure 1.

2.1. Signal Processing
Signal processing of the raw time series of vibration signals needs to be applied in order to remove speed fluctuations [6]. Diagnosis through vibration signals can be performed using time, frequency or time-frequency methods [7], with the two latter having shown the most promising results since they make it easier for specific frequency components of interest to be identified. Various techniques can be applied, like Fourier transform, Hilbert transform, cepstrum or envelope spectrum.

Gears faults affect the stiffness of the nearby teeth and this produces changes in the vibration signal. The changes are reflected as amplitude and frequency modulation [8] which can be detected as sidebands around the gear mesh frequencies in the spectrum. A simple sideband energy ratio [9] or cepstrum analysis is often used to diagnose these types of faults. Bearing faults excite high frequency resonances but their signatures are often weaker that the gear ones. A comprehensive tutorial on rolling element bearing diagnostics and advanced signal processing techniques that can be applied is given in [10].

Therefore, since the physics of the faults of the wind turbine gearbox is well understood, any appropriate signal processing method can be used to reveal potential fault signatures.

2.2. Feature Extraction
Features are extracted from historical vibration signals. Instead of using the whole vibration signal, the features provide a more straightforward health indication. A summary of vibration condition indicators can be found in [11]. These features are mainly statistical (e.g. RMS) and are calculated based on specific failure modes and associated narrowband fault frequencies. The operating conditions have a strong effect on vibration signals, so variables such as the power and operating speed are also considered as features.

2.3. Pattern Recognition
The aforementioned features are used as inputs in pattern recognition models. There exists a wide variety of supervised machine learning models that the interested reader can choose from [12]. These models are trained based on historical data, from both normal and abnormal operation, with the aim of predicting gearbox faults. The model is essentially a multi class classifier. The classes are the diagnostic state of the gearbox; either "healthy" or the failure mode of a gearbox component. The number of classes therefore depends on the number of failure mode examples used to train the machine learning model.

When new data samples become available, they are processed in a similar way as the historic signals used for training and the features extracted are used as inputs in the trained pattern recognition model. The output of the model is the classification of the sample that informs whether the gearbox is healthy or where a potential failure is located.

3. Case Study
Three wind turbines with three confirmed gearbox failures are used for the present case study. The wind turbines are offshore, same model and are rated between 2 and 4 MW. The gearbox examined is a double planetary stage with one parallel stage. The three confirmed gearbox faults examined are:

(i) High speed gear crack
(ii) Intermediate speed gear crack
(iii) High speed bearing outer race fault

A schematic is shown in Figure 2 with fault locations and the approximate accelerometer and tachometer locations.

IOP Conf. Series: Journal of Physics: Conf. Series **1222** (2019) 012045 doi:10.1088/1742-6596/1222/1/012045

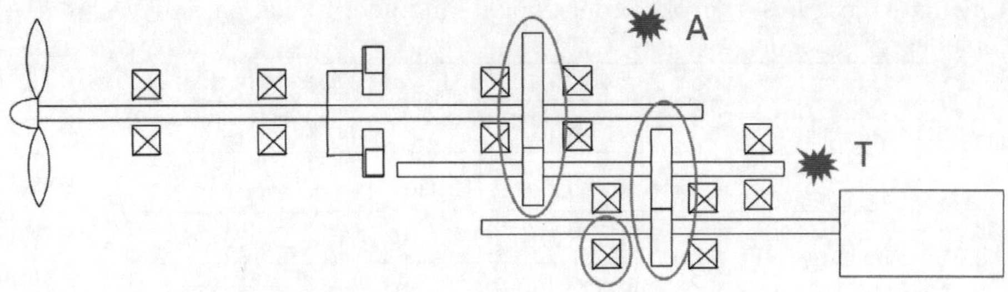

Figure 2. Gearbox diagram and fault locations.

Vibration signals are collected from an accelerometer positioned at the top of the gearbox close to the high speed stage. A tachometer pulse installed on the high speed shaft is used to resample the signals and they are transformed in the order domain. The results from processed signals progressively before failure are shown in Figure 3.

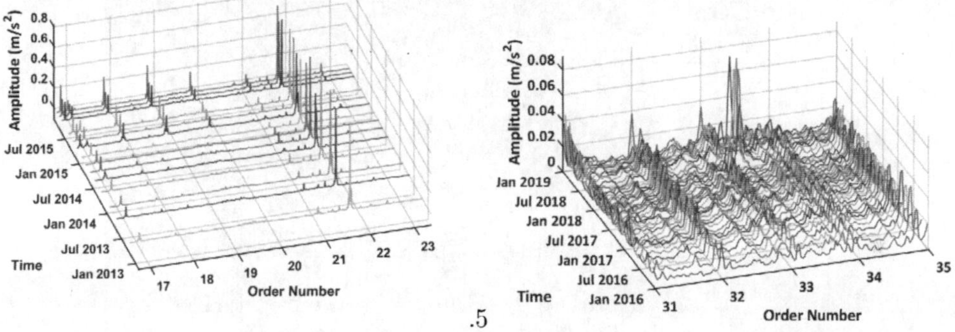

.5 .5

Figure 3. Order spectra leading up to a high speed pinion fault (right) and high speed bearing fault (left). Narrowband failure frequencies are shown, for gear mesh frequency and 5th harmonic of ball passing frequency

The features extracted are shown in Table 1. The gear meshing frequencies and the bearing fault frequencies are known parameters, as long as the design of the gearbox is known. The gear faults are located on the high speed and intermediate speed gears, so the gear frequencies considered are the high speed and intermediate speed gear meshing frequencies. Regarding the high speed bearing, the 5th harmonic of the ball passing frequency outer is used, because through observation of the signals it was concluded by the authors that it gives a good fault indication.

The sideband energy ratio sums the amplitudes of the first six sideband peaks on each side of the gear mesh frequency $\sum_{i=-6}^{6} A_{SB,i}$ and divides by the amplitude of the center mesh frequency A_F. For a healthy gear mesh the sidebands have a small amplitude compared to the center mesh frequency. As damage develops on a gear tooth, the sidebands rise in amplitude which results in a larger sideband energy ratio value. In a similar way, the energy of the bearing ball passing frequency is calculated, by considering a vector A which is centered at the bearing fault frequency and a window +-5% around it. The value is normalised with the length of the data segment. The RMS value can also provide a similar indicator and is calculated in a vector centered the gear mesh frequencies or bearing fault frequencies of interest and a window +-5% around it.

Table 1. Calculated features from the order domain and operational parameters used as inputs in the machine learning model.

	Feature	**Calculation**
Gear fault	RMS	$\sqrt{\frac{1}{N}\Sigma_{i=1}^{N} A(i)^2}$
	Sideband Energy Ratio	$\frac{\sum_{i=-6}^{6} A_{SB,i}}{A_F}$
Bearing fault	RMS	$\sqrt{\frac{1}{N}\Sigma_{i=1}^{N} A(i)^2}$
	Crest Factor	$\frac{max\lvert A(i)\rvert}{\sqrt{\frac{1}{N}\Sigma_{i=1}^{N} A(i)^2}}$
Operational	Power	Provided channel
	Speed	Provided channel

Table 2. Confusion matrix of gearbox fault classification.

Actual \ Predicted	**HS Bearing Fault**	**IS Gear Fault**	**HS Gear Fault**	**Healthy**
HS Bearing Fault	92%	<1%	<1%	7%
IS Gear Fault	1%	84%	4%	11%
HS Gear Fault	<1%	<1%	94%	5%
Healthy	2%	7%	4%	87%

Bootstrap-aggregated decision trees are chosen because they have proven to be a robust classifier [13]. They combine the results of many decision trees, thus reducing the effects of overfitting and improving generalization. The data is split into training and testing with a ratio 80/20%. The training set is used to build the classifier. The hyperparameters tuned are the maximum number of trees, the depth of the trees and the maximum number of features considered for splitting a node. These hyperparameters are tuned using random grid search and cross validation. Supervised learning is performed using labeled data. The labels are the health state of the component or fault isolation. Each sample is labeled as "healthy" if there was no incipient fault signature present in the vibration signals. If there are any frequencies present, indicating the developing of a fault, the signal is classified according to the type of failure mode, as enlisted above. The dataset was fairly balanced after downsampling the "healthy" data points.

The results of the multi class classifier for the testing dataset are shown in the confusion matrix Table 2. The vertical axis represents the actual values and the horizontal one represents the predicted ones. The aim is to have as many accurate predictions as possible (lots of points in the diagonal). Each row shows how the data of the testing set for each label are predicted using the machine learning model trained on the training set.

According to these results, there is a fairly accurate classification of the health state of the gearbox. It should be noted that the accuracy of the machine learning algorithm often depends on the quality of the features, and in this case the domain knowledge of the physics of failure of faults is used to extract features that are good condition indicators.

4. Conclusion

This paper presented an automated methodology for processing vibration signals, extracting features and using machine learning models to isolate incipient gearbox faults. The promising results suggest that AI can be used in order to make more informed maintenance decisions and avoid manual interpretation of a large amount of wind turbine CM data. The methodology is valid for vibration CM of WT gearboxes and could be utilised for optimising O&M actions. It should be noted however that sufficient number of training data and failure examples should be used to create pattern recognition models. Future research will focus on case studies of more failure modes and comparison of the suggested methodology against existing techniques used in industry.

5. Acknowledgements

The authors would like to acknowledge EPSRC funding number EP/L016680/1 for the funding of this project.

6. References

[1] James Carroll, Alasdair McDonald, and David McMillan. Failure rate, repair time and unscheduled o&m cost analysis of offshore wind turbines. *Wind Energy*, 19(6):1107–1119, 2016.

[2] Shuangwen Sheng. Wind turbine gearbox condition monitoring round robin study-vibration analysis. Technical report, National Renewable Energy Lab.(NREL), Golden, CO (United States), 2012.

[3] Arfat Siddique, GS Yadava, and Bhim Singh. Applications of artificial intelligence techniques for induction machine stator fault diagnostics. 2003.

[4] Pedro Santos, Luisa F Villa, Aníbal Reñones, Andres Bustillo, and Jesús Maudes. An svm-based solution for fault detection in wind turbines. *Sensors*, 15(3):5627–5648, 2015.

[5] Sofia Koukoura, James Carroll, and Alasdair McDonald. Wind turbine intelligent gear fault identification. In *Annual Conference of the PHM Society*, 2017.

[6] KR Fyfe and EDS Munck. Analysis of computed order tracking. *Mechanical Systems and Signal Processing*, 11(2):187–205, 1997.

[7] Andrew KS Jardine, Daming Lin, and Dragan Banjevic. A review on machinery diagnostics and prognostics implementing condition-based maintenance. *Mechanical systems and signal processing*, 20(7):1483–1510, 2006.

[8] PD McFadden. Detecting fatigue cracks in gears by amplitude and phase demodulation of the meshing vibration. *Journal of vibration, acoustics, stress, and reliability in design*, 108(2):165–170, 1986.

[9] J Hanna, C Hatch, M Kalb, A Weiss, and H Luo. Detection of wind turbine gear tooth defects using sideband energy ratio™. *China Wind Power 2011; October, 19-21, 2011, Beijing, China*, 2011.

[10] Robert B Randall and Jerome Antoni. Rolling element bearing diagnosticsâa tutorial. *Mechanical systems and signal processing*, 25(2):485–520, 2011.

[11] Junda Zhu, Tom Nostrand, Cody Spiegel, and Brogan Morton. Survey of condition indicators for condition monitoring systems. In *Annu. Conf. Progn. Heal. Manag. Soc*, volume 5, pages 1–13, 2014.

[12] Jerome Friedman, Trevor Hastie, and Robert Tibshirani. *The elements of statistical learning*, volume 1. Springer series in statistics New York, NY, USA:, 2001.

[13] Leo Breiman. Random forests. *Machine learning*, 45(1):5–32, 2001.

IOP Conf. Series: Journal of Physics: Conf. Series **1222** (2019) 012046 doi:10.1088/1742-6596/1222/1/012046

Framework for managing the operations and maintenance of wind farms

J. Izquierdo, A. Erguido, P.X. Zubizarreta, J. Uribetxebarria

Ikerlan Technology Research Center, Operations and Maintenance Technologies Area, 20500 Gipuzkoa, Spain

E-mail: jizquierdo@ikerlan.com

Abstract. The operations and maintenance of a wind farm constitutes a considerable source of expenditures therefore a holistic framework is herein presented aiming at optimising their management. The principles upon which the framework was developed belongs to the asset management discipline oriented towards the wind energy sector. Besides, it is understood from a cross functional perspective which also embraces information management and business aspects. The framework comprehends different modules conceived with the following purposes: (i) creation of an integral and comprehensive failure database, (ii) modelling of the wind turbines' reliability taking into account their operational context and (iii) defining of maintenance strategy within a life-cycle perspective. The practical aspects of the framework are demonstrated through a case study based on real-field data from the sector. Furthermore, the present research sets the foundation to consider all the proposed algorithms, processes and techniques within a Product-Service System approach, which entails important benefits for the OEMs an for the operators.

1. Introduction

The increasing importance and development of the energy sector in terms of installed capacity and technological solutions have arisen new challenges to face, in particular, related to asset's reliability and wind farm logistics [1]. In Europe, wind energy is the second largest form of power generation taking into account onshore and offshore wind farms. Operations and maintenance costs associated with onshore and offshore wind farms may rise to 32% and 12-30% of the total life-cycle cost of the wind farm respectively [2, 3]. It is reasonable to strive for reducing the operations and maintenance costs, but it is important to bear in mind that availability is a must to ensure the wind farms profitability. In such context, asset management stands as an advantageous discipline to maximize the profitability of the investment, a proper maintenance strategy helps to reduce maintenance costs whilst improving availability [4, 5].

Nowadays, the massive amount of data available may be transformed into useful information which may serve as a solid basis to optimize maintenance strategy. In order to translate the data into valuable information from the maintenance strategy perspective, reliability engineering provides useful techniques which allow characterizing the wind farm and wind turbines behaviours from operational data. Therefore the optimization of the maintenance strategy for the life-cycle of the wind farm is a heavily based data-driven decision-making process. This reality positions the Original Equipment Manufactures in a favourable position, their insights of the asset and its management enable an offer based on knowledge-intensive products and services which provide an important competitive advantage [6, 7]. This fact has given rise to Product-Service-Systems (PSS) where both, product and services, conform a single offer. PSS business model entails a closer relationship between the OEM and the customer bringing additional value to both.

In order to support the developments in the sector, the new emerging technologies and the PSS business models in the asset management field, the authors have conducted the research summarized in this paper. The main contribution here presented consists of a framework that comprehensively integrates data-structure and asset management innovations for maintenance strategy optimization within the scope of PSS business model for the wind energy sector.

1.1. Related works

In the literature, the reliability approach has proven useful when optimizing the definition of maintenance strategy [8]. In order to carry out a reliability study, the relevant information needed as an input to the decision should be at hand, in the right format and on time [9]. The creation of a common database for the wind energy industry is not a new problem in the literature, a description of the objectives of a Reliability-Availability-Maintainability (RAM) database in the sector can be seen in Hameed et al. [10] and in Zhao [11]. In the previous work, the challenges for creating such database are addressed by considering information sources such as operational data, equipment data, failure data, maintenance data and state information [8]. The need to integrate data coming from different sources with operations and maintenance purposes is also stated in [12], in order to exploit the value contained in that information it is critical a correct assessment of the failure process and therefore the selection of the proper time-to-failure model [13].

In the context of this paper, the reliability study of the wind turbines has to be performed under several conditions regarding data quality and quantity [8]. In order to achieve the required data standards, it is possible to combine data from similar assets [13, 14]. However, it is important to consider that some wind turbines may present different failure behaviours and frequencies caused by the operational conditions [15, 16]. In such context, it is important to take into account the reliability models which considers the influence of operational environment as well as dynamic conditions like the ones in [17, 18, 19]. The integration of operational context information, as well as dynamic considerations, has proven to provide more accurate reliability estimates and therefore better results of the maintenance strategy [18, 20]. In the literature, it can also be found research works which provide evidence of the improvements in the maintenance strategy by the integration of reliability models with clustering algorithms to deal with heterogeneity in failure behaviours [21, 22].

These reliability models serve as a valuable input in order to customize maintenance plans, which will be highly conditioned by the chosen maintenance policy [23]. The chosen maintenance policy has to be aligned with the organizational goals [24] and has to consider the specific aspects of the assets. One of the most studied maintenance policy in the literature for wind farms is the opportunistic maintenance [23]. The opportunistic maintenance policy makes the most of short-term situations to perform preventive maintenance on a system when corrective maintenance is needed in another one; the decision is grounded on a threshold regarding system's age, reliability or health. Besides, opportunistic maintenance is attractive in the context of wind energy sector because it considers wind turbines having economic, structural and stochastic dependencies [25, 26]. Some research of special interest can be seen in: the work of Erguido [26], where the maintenance plans are optimized for Lifecycle Cost (LCC) and availability according to weather conditions; in [27, 25] a multi-objective optimization is also propose to minimize maintenance cost and loss of production by grouping the WTs and their maintenance tasks; and the work of [28]explore the differences among opportunistic maintenance and two alternative maintenance strategies.

In the previous works, the optimization of the maintenance strategy is sought for the lifecycle of the assets. This life-cycle perspective is aligned with seeking recurrent income in the long-term whilst reducing costs, to this aim, it has attracted some attention the study of the conjoint offer of product and services known as Product-Service System (PSS) [29]. Moving towards a PSS business model ensures customer relationship as a competitive strategy [30] while the asset management discipline guarantees product performance and operational excellence [31], therefore aligning operational and strategic objectives. This comprehensive view of the maintenance strategy is supported by asset management discipline, accordingly, its importance its recognize by The International Standards Organization in [32] as well as by other asset management frameworks like2 [33, 35, 34].

1.2. Motivation, contribution and overview

It is the scope of this paper to provide a managing framework which helps OEMs and operators of wind farms dealing with operations and maintenance. The comprehensive approach aims at contributing with a cross-functional value proposition which starts by defining the data structure and database needs. Then, it provides guidelines on reliability engineering to describe the behaviour of the wind farm following a structured procedure and thus, enabling the optimization of the maintenance plans and a LCC analysis for the defined strategy. Besides, the approach is considered within the context of PSS scenarios, these scenarios entail services associated with the wind turbines which may condition the most favourable scenarios not only for the operator but for the manufacturer as well.

The contribution of the framework lies within the proposal of a management framework that integrates specific tools, mechanisms and algorithms in order to optimize the maintenance of a wind farm. Compared to other frameworks in the literature (see for instance [34, 33]), it provides a more pragmatical point of view facilitating its integration in the industry. Meanwhile, compared with more technical research regarding algorithms and specific methods, it contributes with a strategic view that provides the alignment among the mathematical methods and the organization goals enabling higher profitability.

In the remaining of the paper, the framework is first presented in section 2 and then every aspect of it is further explained in the corresponding subsections. The development of every aspect of the framework is set out from a managerial perspective but, in order to favour the applicability of the research, specific tools, algorithms and techniques are proposed as a guideline to serve the managerial purposes. As an example of the application of the framework, a case study is presented in section 3. In section 4, the results of the case study and the benefits derived from the implementation of the framework in the management of wind farms are discussed, they are mainly related to revenue, service level and maintenance optimization.

2. Proposed Framework

The presented framework is a holistic proposal whose departure point is the definition of the structure and information of the database needed for the reliability analysis of the wind turbines. Providing that the reliability of the wind turbines has been properly calculated and that it integrates the information regarding the operational environment of the wind turbines, the next stage consists of the definition of the optimal maintenance strategy. The maintenance strategy is determined by means of an optimization procedure embedded not only with the mathematical algorithms but with the logic proposed by dynamic opportunistic maintenance as well. Finally, the characterization of the wind turbines failure behaviour and the definition of certain maintenance strategy allow calculating the life-cycle costs (LCC) of the wind farm. The LCC calculation is performed by means of simulation tools, they have proven useful when determining the RAM statistics during the life-cycle as well as other important Key Performance Indicators (KPI) of the business.

It is important to take into consideration that this cross-functional approach is contemplated into a business model focused not only on selling products but the services associated with them as well. Therefore the framework is proposed aiming at enabling the organizational shift needed to maximize OEMs profitability and customer satisfaction. In Figure 1, a graphical representation of the framework can be seen and in the following subsections, every stage of the framework is further explained.

Figure 1. Proposed framework

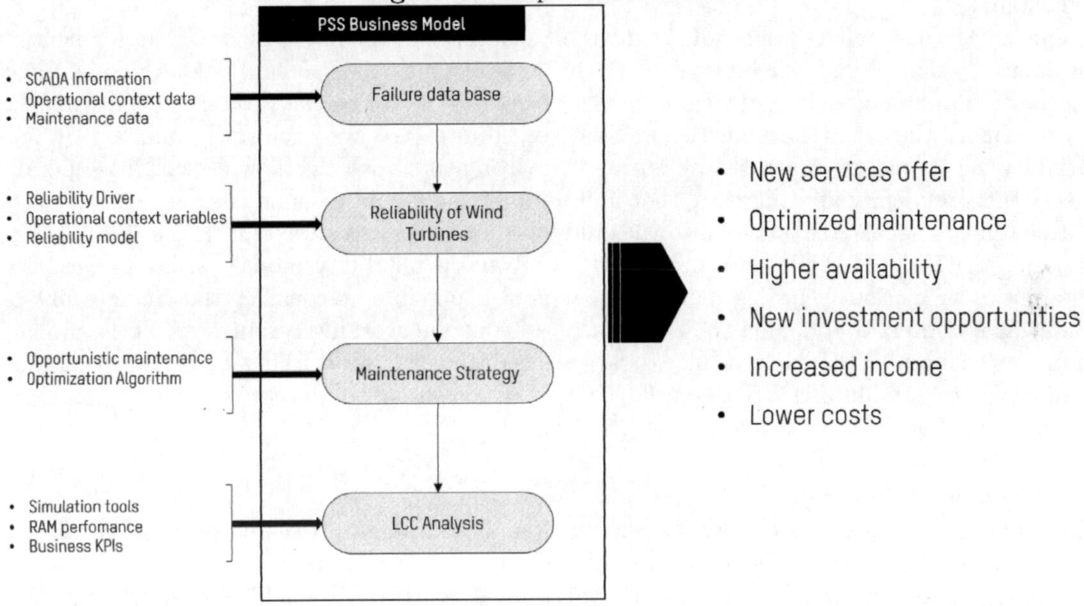

The framework representation no only includes the process to manage the maintenance and operations of wind farms, but it also gathers the inputs needed in every stage in terms of information, tools and techniques.

2.1. Failure database

Information and data quality are two of the essentials aspects to consider in every decision-making process, in the proposed framework the construction of a failure database is proposed as the starting stage.

As an initial step, it is important the definition of failure modes and their association with maintenance actions. The definition must ensure coherence among the maintenance actions indenture level and the failure modes of the components. Having defined them it is possible to deal with the historical records in a structured form.

In order to identify the failures modes in the records and information coming from the SCADA system, it is necessary to match the alarms with wind turbines states and functional failures whilst bearing in mind aspects concerning data scarcity, data censoring, data pooling and the effect of repair actions.

As the proposed reliability models integrate information regarding the operational context it is therefore vital that the failure database also includes the environmental conditions under which the wind turbines have been operating. Accordingly, in the framework, it is proposed to link the failures, alarms and states of the wind turbines with their specific operational condition which may influence the failure behaviour.

Once the database has been built and the data in it contained is consistent and represent the true behaviour of the wind farm, it is possible to select and fit a proper reliability model to exploit and understand the inherent value of the data.

2.2. Reliability of wind turbines

In an initial attempt to model the reliability of the components of the wind turbines, it is possible to rely on clustering approaches. In the proposed model, the k-medoids clustering algorithm is proposed as a mean for determining possible groups of turbines with different statistical behaviours.

One of the key pillars when adjusting the reliability model for each detected segment is the identification of the reliability driver, the variable that most conditions the component degradation and failure. It is of vital importance because it will be the variable to control since it will trigger the maintenance actions. However, failures are not exclusively affected by one variable; the characteristics of the operational environment in which the wind turbines operate affects their reliability and therefore their failure frequencies and behaviours.

Reliability models that take into account the operational context such as the Weibull Proportional Hazards Models are proposed. In advanced reliability models which harvest the capabilities of machine learning algorithms, it is also possible to consider the changes of the operational context overcoming the assumption of constant working conditions. These models enable to customize the maintenance strategy according to the working environment of the wind turbines establishing a differential competitive advantage, not only in the operations management but in the offer stage as well.

2.3. Maintenance strategy

In the paper, it is proposed to optimise the maintenance strategies based on previous reliability analysis since it allows assessing the risk of failure in every component. The identification of components' reliability in their specific operational context will enable to further optimise maintenance strategy.

The approach proposed for the maintenance strategy is dynamic opportunistic maintenance, it is especially interesting because it allows considering the economic, structural and stochastic dependencies among the components. In the proposed strategy different maintenance levels (imperfect/perfect) are defined which have associated certain reliability threshold. The different maintenance activities are triggered according to the reliability threshold, and the strategy considers the possibility of grouping several maintenance activities in order to optimize cost and performance. In order to integrate such considerations, dynamic opportunistic maintenance takes into account short term information regarding both internal and external factors, providing therefore, better decisions.

It is critical, given the nature of the dynamic opportunistic maintenance, to use simulation techniques to evaluate maintenance decisions according to their cost, availability or other key performance indicators. Consequently, it is necessary to develop simulation-based optimisation mechanisms, which will enable the combination of the simulation model developed for evaluating maintenance decisions and advanced meta-heuristic optimisation algorithms, such as particle swarm or genetic algorithms. Particularly, and considering the several objectives pursued in the context of maintenance and services, multi-objective optimisation algorithms should be implemented; they will allow decision-makers finding a trade-off among conflicting objectives and representing the objectives in their decisions.

Generally, in the case of Wind Farms, the conflicting interests regarding maintenance are due to the loss of production caused by the performance of maintenance actions, either corrective or preventive. The service-level disruptions caused by preventive maintenance actions are directly related to the service-level disruptions caused by the failure of components, i.e. the more preventive maintenance actions the fewer failures are likely to occur. Accordingly, the multi-objective techniques allow for balancing the service-level disruptions, which entail loss of production, with the life-cycle cost of the Wind Farm. The output of the multi-objective optimization will allow to select among several non-dominated solutions in which the service level cannot be improved without increasing the costs, thus a trade-off has to be chosen from the Pareto optimal solutions (all possible non-dominated solutions).

2.4. LCC Analysis

Regardless of the nature of the objectives that aforementioned maintenance strategies pursue, the decisions to be made should regard at the whole life cycle of the wind turbines, ranged between 20 and 30 years-long. In this context, it is critical to derive the simulation models so that the performance measures are evaluated along the assets' life cycle. Therefore, aided by the maintenance strategy derived in subsection 2.2, the simulation model should characterize the malfunction of the assets in a specific operational context and estimate the implications of operation and maintenance activities' planning in the corresponding KPIs.

A particularly interesting technique that lays the foundations for calculating such measures is life cycle costing, which attends both capital expenditures (CAPEX) and operational expenditures (OPEX). Capital expenditures correspond the ones related to the construction and installation of wind farms, including the cost of the turbine itself, its transportation, grid connection, civil work, R&D, etc. On the contrary, operational expenditures regard to the wind farm operation and maintenance, insurance, taxes, management, administration, etc. As a consequence, making decisions attending to both expenditure categories will significantly enhance not only operational decisions, but also strategic decisions, such as maintenance or investment planning respectively.

It should be noticed that this life-cycle cost analysis is critical to shift to service-oriented business models, such as wind farms performance-based contracting, where instead of selling wind turbines it is the energy they provide what is sold. In this context, asset management strategies enable the conjoint optimisation of LCC and RAM indicators, which is vital to find an accurate estimation of wind farm cost and service level along its life-cycle. Accordingly, previously derived dynamic opportunistic maintenance strategies have been complemented through the integration of the analytical modelling of wind farms CAPEX and OPEX in the simulation-based optimisation mechanism. As a result, optimal strategies that lead to highly efficient maintenance policies in terms of achieved energy-based availability and costs have been found, laying the foundations for defining an attractive service-oriented offer.

The approach, algorithms, techniques and tools proposed in the framework allows considering multiple PSS scenarios. By introducing changes in the simulation model and/or the cost structure it is possible to evaluate several offers which may entail different services. The proper characterisation of the wind turbines behaviour, as well as the definition of the maintenance strategy, allows to asses every scenario from a transparent and comprehensive perspective.

Nevertheless, the uncertainty sources and the stochastic nature of the algorithms and processes involved render non-deterministic solutions which may put at risk the profitability of chosen PSS. To overcome that difficulty, the framework considers uncertainty analysis that assesses the possible deviations from the different solutions and scenarios. By integrating the uncertainty analysis a risk-based decision approach is enabled, thus the results in terms of costs and services are enhanced since the decision making process regarding the operations and maintenance of wind farms is supported by a more solid knowledge.

3. Case Study

The proposed framework has been validated through a case study in the wind energy sector. From real field data, the life cycle of a wind farm has been simulated, i.e. 20 years of operation.

IOP Conf. Series: Journal of Physics: Conf. Series **1222** (2019) 012046 doi:10.1088/1742-6596/1222/1/012046

In order to optimize the maintenance strategy to be adopted in the operation of an onshore WF, a simulation based-optimization with the NSGA-II algorithm has been conducted. The Pareto optimal of maintenance strategies has been determined, in it four components, with three different failure modes each, are considered according to their criticality in terms of availability and costs: the gearbox, the blades, the pitch system and the yaw system. The multi-objective simulation seeks to minimize the OPEX of the WF whilst minimizing the loss of production for the life-cycle of the turbines, the Pareto optimal yielded by the optimization of the maintenance strategy can be seen in Figure 2. The maintenance strategy studied is the dynamic opportunistic maintenance, it determines whether to perform or not maintenance actions according to dynamic reliability thresholds that take into account the operational conditions of the WTs. As it can

Figure 2. Pareto Optimal for the dynamic opportunistic maintenance Strategy

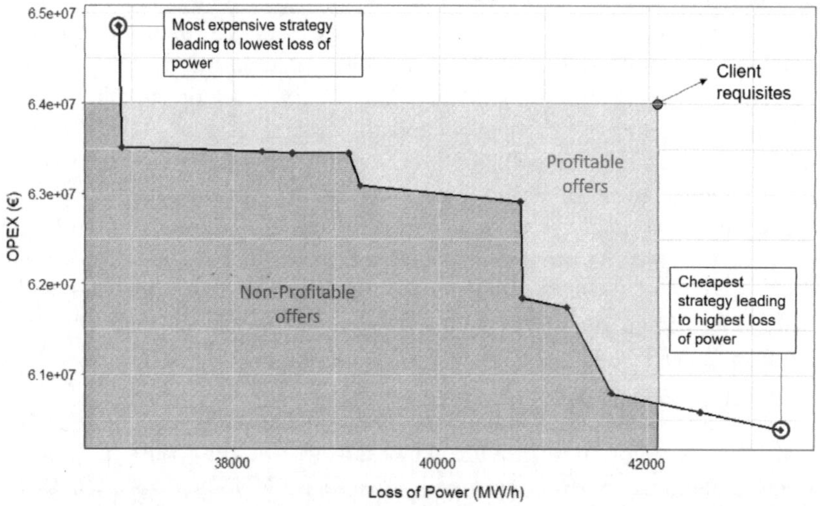

be seen in Figure 2, once the client has defined certain service level and certain OPEX they are willing to assume, it is possible to see the possible maintenance policies fulfilling such criteria. And it is also possible to determine those maintenance solutions which are unfeasible in terms of availability and price.

In order to provide deeper insights into the performance of the proposed approach, it is compared against static opportunistic maintenance and against a corrective strategy. To such aim one of the policies has been chosen a simulated through the life-cycle of the WF while comparing it with the other strategies. The results are represented for the objective functions of the multiobjective optimization, i.e. loss of production and OPEX. The representation of the results can be seen in Figure 3.

As it can be seen in the graphs, both of the opportunistic maintenance strategies yield better results in the OPEX objective than the corrective strategy. It is due to a correct assessment of the WT reliability taking into account the operational context, therefore the preventive maintenance actions are performed in the proper moment. Besides the dynamic opportunistic maintenance outperforms the static strategy when minimizing the loss of production because the actions are performed in those periods of time when it is more favourable for the business.

4. Concluding Remarks

The present paper proposes a comprehensive framework which aims at facilitating the management of operations and maintenance of wind farms. This holistic proposal covers several realities of interest regarding operations and maintenance; it deals with data structure and needs,

IOP Conf. Series: Journal of Physics: Conf. Series **1222** (2019) 012046 doi:10.1088/1742-6596/1222/1/012046

Figure 3. Results of the comparison

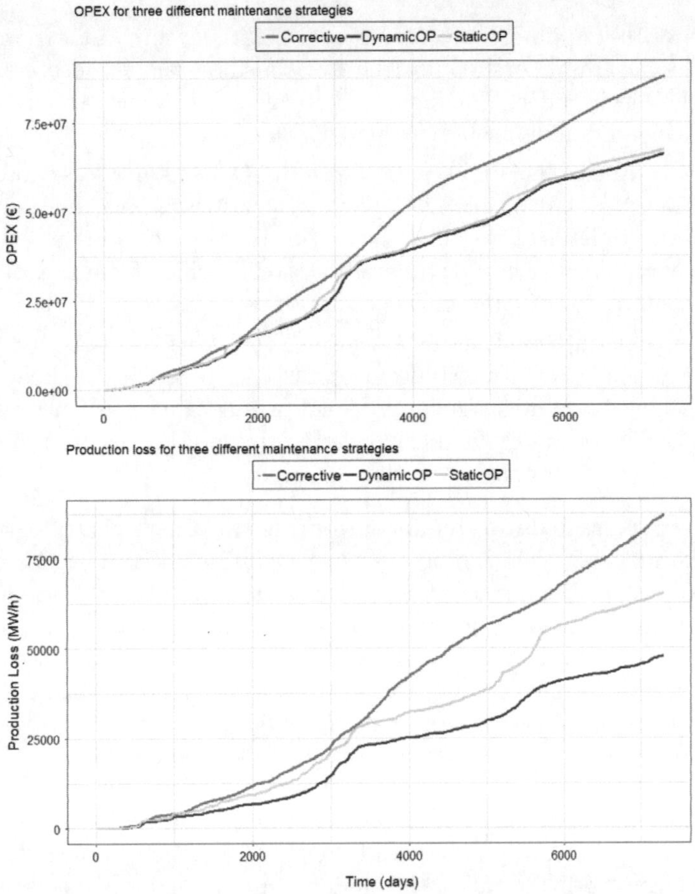

mathematical modelling of failure behaviours according to operational context, optimization of maintenance actions within a life-cycle perspective whilst maintaining desirable service levels and LCC calculation from a risk-based decision approach. Besides, the framework considers these realities within the context of servitization and therefore it provides a PSS business model perspective. The implementation of this value proposition in the management of wind farms will bring several benefits in terms of:

a. **Optimised maintenance-**Since the failure behaviour of the asset is characterized according to operational context, the maintenance actions are performed in the most appropriate time. The dynamic opportunistic maintenance allows grouping several maintenance actions in order to improve costs and availability. This is all considered not in the short term but for the whole life cycle of the wind turbines as it can be seen in the case study.
b. **Lower costs-**A better assessment of the WTs reliability leads to the optimisation of the maintenance actions entailing a cost reduction since the cost structure of the activities is considered and implemented in the simulation models. The improved characterisation of the failure events reduces also the costs derived from unavailability or security issues.
c. **Higher availability-**The optimisation is multi-objective, which implies that in spite of the cost reduction the service level provided by the wind farm will be maintained within the

desirable levels. Ensuring OEM's and operator's satisfaction during the life-cycle of the project.

d. **Increased income-**The improvements in availability ensures higher incomes. But the consideration of the framework within the PSS business model ensures recurrent income derived from the service offer. The income from the services ensures long term revenue for the OEM whilst maintaining customer satisfaction.

e. **New services offer-**By the simulations tools the OEMs can assess the offer of the new services they are interested in selling. They can provide the information rendered by the simulation in terms of KPIs to the customers. This information will present the new services offer in a transparent way such as the Pareto Optimal, which facilitates the sale of the new services and promotes loyalty and trust relationships.

f. **Investment opportunities-**The OEMs can use the provided tools in order to assess the profitability of possible investments from a risk-based decision-making approach. It is possible to see how investments in improving certain aspects of the wind turbines, e.g. component reliability or maintainability of failure modes, impact on the cost-performance metrics of the wind farm for the life-cycle.

The research herein presented is a step forward in the alignment of strategic and operational management by the proposal of a framework that integrates specific technologies across the different disciplines involved in the servitization process. The PSS business model considered in the framework facilitates the decision-making process of the OEMs and it provides multiple advantages, not only for them but for their clients as well. Highlighting the main advantages, it is remarkable how the maintenance strategies are aligned for the OEM, the client and the asset; the deployed technologies ensure that the operations provide the desired service level at an affordable cost; and the close cooperation between the OEM and the client results in a better feedback process with effects on product design, know-how, services offer and loyalty.

In fact, the principles in which PSS is based allow to identify and satisfy the clients' needs in terms of transparency, availability and product knowledge, whilst the innovative asset management technologies addressed allow the manufacturer to maintain his business profitability by optimizing costs. However it is important to bear in mind that further efforts are encourage to provide and develop more technological tools, algorithms and process to support the servitization process for OEMs, helping to overcome the barriers that hinder the adoption of PSS business models.

References

[1] Shafiee M 2015 *Renewable Energy* **77** 182–193 URL http://dx.doi.org/10.1016/j.renene.2014.11.045
[2] Byon E 2013 *IIE Transactions* **45** 1188–1201 URL http://dx.doi.org/10.1080/0740817X.2012.726819
[3] Kaldellis J and Kapsali M 2013 *Energy Policy* **53** 136–148 URL http://dx.doi.org/10.1016/j.enpol.2012.10.032
[4] Krokoszinski H 2003 *Renewable Energy* **28** 2165–2178 URL http://dx.doi.org/10.1016/S0960-1481(03)00100-9
[5] Sen Z 1997 *Renewable Energy* **10** 71–79 URL http://dx.doi.org/10.1016/0960-1481(96)00021-3
[6] Baines T, Lightfoot H, Evans S, Neely A, Greenough R, Peppard J, Roy R, Shehab E, Braganza A, Tiwari A, Alcock J, Angus J, Basti M, Cousens A, Irving P, Johnson M, Kingston J, Lockett H, Martinez V, Michele P, Tranfield D, Walton I and Wilson H 2007 *Proceedings of the Institution of Mechanical Engineers, Part B: Journal of Engineering Manufacture* **221** 1543–1552
[7] Banks J 2003 *Encyclopedia of Information Systems* (Elsevier BV) pp 663–671
[8] Leahy K, Gallagher C, OâĂŹDonovan P and OâĂŹSullivan D T J 2019 *Energies* **12** ISSN 1996-1073 URL http://www.mdpi.com/1996-1073/12/2/201
[9] Rausand M and HÃŭyland A 2004 *System reliability theory: models, statistical methods, and applications* vol 396 (John Wiley & Sons) URL https://www.wiley.com/en-us/System+Reliability+Theory%3A+Models%2C+Statistical+Methods%2C+and+Applications%2C+2nd+Edition-p-9780471471332

[10] Hameed Z, Vatn J and Heggset J 2011 *Renewable Energy* **36** 2154–2165 URL https://www.sciencedirect.com/science/article/pii/S0960148111000140
[11] Zhao Y, Li D, Dong A, Kang D, Lv Q and Shang L 2017 *Energies* **10** ISSN 1996-1073 URL http://www.mdpi.com/1996-1073/10/8/1210
[12] Nguyen T H, Prinz A, Friisø T, Nossum R and Tyapin I 2013 *Renewable energy* **60** 150–161 URL https://www.sciencedirect.com/science/article/pii/S0960148113002474
[13] Louit D, Pascual R and Jardine A 2009 *Reliability Engineering & System Safety* **94** 1618 – 1628 ISSN 0951-8320 URL http://www.sciencedirect.com/science/article/pii/S0951832009000891
[14] Stamatelatos M, Veseley W, Dugan J, Fragola J, Minarik J and Rialsback J 2002 - URL https://elibrary.gsfc.nasa.gov/
[15] Cao L, Qian Z, Zareipour H, Wood D, Mollasalehi E, Tian S and Pei Y 2018 *Energies* **11** ISSN 1996-1073 URL http://www.mdpi.com/1996-1073/11/12/3318
[16] Zimroz R, Bartelmus W, Barszcz T and Urbanek J 2014 *Mechanical Systems and Signal Processing* **46** 16 – 27 ISSN 0888-3270 URL http://www.sciencedirect.com/science/article/pii/S0888327013004974
[17] Okaro I A and Tao L 2016 *Reliability Engineering & System Safety* **156** 159 – 174 ISSN 0951-8320 URL http://www.sciencedirect.com/science/article/pii/S0951832016302903
[18] Peng W, Li Y F, Mi J, Yu L and Huang H Z 2016 *Reliability Engineering & System Safety* **153** 75 – 87 ISSN 0951-8320 URL http://www.sciencedirect.com/science/article/pii/S0951832016300199
[19] Izquierdo J, Crespo A, Uribetxebarria J and Erguido A 2017 *Safety and Reliability - Safe societies in a changing world* Proceedings of ESREL 2018 ed Press C (Trondheim, Norway: Taylor & Francis Group, London) pp 571–578 URL https://www.taylorfrancis.com/books/9781351174657
[20] Izquierdo J, Crespo A and Uribetxebarria J Dynamic artificial neural network-based reliability considering operational context of assets. unpublished
[21] Cannarile F, Compare M, Di Maio F and Zio E 2018 *Proceedings of the Institution of Mechanical Engineers, Part O: Journal of Risk and Reliability* **232** 140–150 URL https://journals.sagepub.com/doi/abs/10.1177/1748006X17716344
[22] Baraldi P, Di Maio F, Rigamonti M, Zio E and Redouane S 2015 *Journal of Intelligent & Fuzzy Systems* **28** 1723–1731 URL https://content.iospress.com/articles/journal-of-intelligent-and-fuzzy-systems/ifs1459
[23] Wang H and Pham H 2006 *Springer Series in Reliability Engineering* (Springer) pp 135–150 URL http://dx.doi.org/10.1007/1-84628-325-67
[24] Crespo A, Moreu P, GÃşmez J, Parra C and Lopez M 2016 *Quality and reliability engineering international* **32** 519–533 URL https://www.emeraldinsight.com/doi/abs/10.1108/13552510910961110
[25] Abdollahzadeh H, Atashgar K and Abbasi M 2016 *Renewable Energy* **88** 247–261 URL http://dx.doi.org/10.1016/j.renene.2015.11.022
[26] Erguido A, MÃąrquez A C, Castellano E and FernÃąndez J G 2017 *Renewable Energy* **114** 843 – 856 ISSN 0960-1481 URL http://www.sciencedirect.com/science/article/pii/S0960148117306304
[27] Atashgar K and Abdollahzadeh H 2016 *Energy Conversion and Management* **112** 445–458 URL http://dx.doi.org/10.1016/j.enconman.2016.01.027
[28] Zhu W, Fouladirad M and Bérenguer C 2016 *Reliability Engineering & System Safety* **153** 50–63 URL http://dx.doi.org/10.1016/j.ress.2016.03.020
[29] ÃŰner K, KiesmÃijller G and Van Houtum G 2010 *Quality control and applied statistics* **56** 397–399 URL https://www.sciencedirect.com/science/article/abs/pii/S037722171000038X
[30] Baines T and Lightfoot H 2013 *Made to serve: How manufacturers can compete through servitization and product service systems* (John Wiley & Sons) URL https://www.wiley.com/en-us/Made+to+Serve%3A+How+Manufacturers+can+Compete+Through+Servitization+and+Product+Service+Systems-p-9781118585313
[31] Hastings N A 2010 *Physical asset management* vol 2 (Springer) URL https://www.springer.com/us/book/9781447157472
[32] ISO55000 2014 *Asset Management - Overview, Principles and Terminology* second edition ed (International Organization for Standardization) URL https://www.iso.org/obp/ui/#iso:std:iso:55000:ed-1:v2:en
[33] Shahid M A, Mahamood M N and Das N 2016 *2016 Australasian Universities Power Engineering Conference (AUPEC)* pp 1–5 URL https://ieeexplore.ieee.org/abstract/document/7749310
[34] Brown K, Laue M, Tafur J, Mahmood M N, Scherrer P and Keast R 2014 *An Integrated Approach to Strategic Asset Management* (Cham: Springer International Publishing) pp 57–74 ISBN 978-3-319-02493-6 URL https://doi.org/10.1007/978-3-319-02493-6_5
[35] Council A M 2014 *Framework for asset management* (AMBoK Publications) URL http://www.amcouncil.com.au/news/60-general/637-free-download-framework-for-asset-management,-second-edition-pdf.html